3rd International Conference on Industrial Engineering and Applications (ICIEA 2016)

MATEC Web of Conferences Volume 68 (2016)

Hong Kong, China
28 - 30 April 2016

Editors:

Anand Jayakumar Arumugham
M. Mujiya Ulkhaq
Marek Kocisko
Ravi Kumar Goyal
Wan Ahmad Yusmawiza
Xual Qiu

ISBN: 978-1-5108-2746-2

Printed from e-media with permission by:

Curran Associates, Inc.
57 Morehouse Lane
Red Hook, NY 12571

Some format issues inherent in the e-media version may also appear in this print version.

Printed by Curran Associates, Inc. (2016)

For additional information, please contact EDP Sciences – Web of Conferences at the address below.

EDP Sciences – Web of Conferences
17, Avenue du Hoggar
Parc d'Activité de Courtabœuf
BP 112
F-91944 Les Ulis Cedex A
France

Phone: +33 (0) 1 69 18 75 75
Fax: +33 (0) 1 69 28 84 91

contact-edps@webofconferences.org

Additional copies of this publication are available from:

Curran Associates, Inc.
57 Morehouse Lane
Red Hook, NY 12571 USA
Phone: 845-758-0400
Fax: 845-758-2633
Email: curran@proceedings.com
Web: www.proceedings.com

TABLE OF CONTENTS

Aims & scope

MATEC Web of Conferences is an open-access journal devoted to the publishing of conference proceedings dealing with Materials science, Engineering and Chemistry.

The journal aims to make conference proceedings visible, citable and widely usable by the scientific community. It offers outstanding services for the publication and dissemination of contents in order to ensure long-term archiving and maximum exposure of scientific events.

Coverage includes all engineering disciplines: civil, naval, mechanical, chemical, and electrical engineering as well as nanotechnology and metrology. The journal also deals with the physical-chemical characterization of materials, their implementation, their resistance in their environment, etc. Others subdisciplines of chemistry, such as analytical chemistry, petrochemistry, organic chemistry, etc., and even pharmacology, are also welcome.

Why publish with MATEC Web of Conferences?

- **Open access:** conference papers available worldwide.
- **Fast publication:** less than 3 months.
- **Total flexibility:** no page restrictions, free unlimited use of colour in the online version, any kind of scientific document welcomed (traditional articles, posters, abstracts and slideshows).
- **High visibility:** all documents indexed in main databases and easily found online.
- **Fair author rights:** CC-BY Creative Commons license.
- **Range of additional publishing options:** printed proceedings, CD-ROMs, copy-editing, etc.

ISSN (electronic edition): 2261-236X

Contact: Isaline Augusto
matec@webofconferences.org

www.matec-conferences.org

edp open Open Access Publishing from EDP Sciences

ICIEA COMMITTEES

Conference Chair

Felix T. S. Chan, The Hong Kong Polytechnic University, Hong Kong

Organizing and Program Committees

Seoung Bum Kim, School of Industrial Mangement Engineering, Korea University
Henry Y.K. Lau, Industrial and Manufacturing Systems Engineering, The University of Hong Kong, Hong Kong

International Technical Committees

Marek Kocisko, Technical University of Kosice, Slovakia
Mohammad Arif Kamal, Dept. of Architecture, Aligarh Muslim University, India
Jowati Juhary, National Defence University of Malaysia, Malaysia
Zainal Abidin Akasah, Universiti Tun Hussein Malaysia, Malaysia
M. Mujiya Ulkhaq, Diponegoro University, Indonesia
Tamer Eren, Kirikkale University, Turkey
Byung Do Chung, Yonsei University, The Republic of Korea
Fernando Moreira da Silva, Faculty of Architecture of the Technical University of Lisbon, Portugal
S. H. Choi, University of Hong Kong, Hong Kong
Isti Surjandari, Industrial Engineering Department, University of Indonesia, Indonesia
Mehmet Savsar, Kuwait University, Kuwait
Yusri Yusof, Universiti Tun Hussein Onn Malaysia (UTHM), Malaysia
Xuan Qiu, The Hong Kong University of Science and Technology, Hong Kong
J.W. Wang, The University of Hong Kong, Hong Kong
Wan Ahmad Yusmawiza, Hail University, Saudi Arabia
Sureshkumar M.S, Sri Ramakrishna Engineering College, India
Mohamad Zairi Bin Baharom, Universiti Malaysia Pahang, Malaysia
Ravi Kumar Goyal, Jaipur Institute of Technology, India
Anand Jayakumar Arumugham, SVS College of Engineering, India
Sandeep Singhal, Department of Mechanical Engineering, National Institute of Technology, India
Muralindran Mariappan, Universiti Malaysia Sabah, Malaysia

edp sciences

MATEC Web of Conferences

Statement of Peer review

In submitting conference proceedings to MATEC Web of Conferences, I certify to the Publisher that I adhere to the **Policy on Publishing Integrity** of the journal in order to safeguard good scientific practice in publishing.

1. All articles have been subjected to peer review administered by the proceedings editors.
2. Reviews have been conducted by expert referees, who have been requested to provide unbiased and constructive comments aimed, whenever possible, at improving the work.
3. Proceedings editors have taken all reasonable steps to ensure the quality of the materials they publish and their decision to accept or reject a paper for publication has been based only on the merits of the work and the relevance to the journal.

Title, date and place of the conference

2016 The 3rd International Conference on Industrial Engineering and Applications (ICIEA 2016)

April28-30, 2016, Hong Kong

Proceedings editor(s):

Anand Jayakumar Arumugham,SVS College of Engineering, India, jay4upeople@gmail.com
M. Mujiya Ulkhaq, Diponegoro University,Indonesia, ulkhaq@live.undip.ac.id
Marek Kocisko,Technical University of Kosice, Faculty of manufacturing technologies,Slovakia, marek.kocisko@tuke.sk
Ravi Kumar Goyal,Jaipur Institute of Technology,India, ravikumargoyal@gmail.com
Wan Ahmad Yusmawiza,Hail University,Saudi Arabia, yusmawiza@iium.edu.my
Xuan Qiu,The Hong Kong University of Science and Technology, Hong Kong, xuanqiu@ust.hk

Date and editor's signature

23/6/2016

17, avenue du Hoggar - PA de Courtabœuf – BP 112 - 91944 Les Ulis Cedex A (France)
Tél. : 33 (0)1 69 18 75 75 - Fax : 33(0)1 69 07 45 17 – www.edpsciences.org

Transient Fuel-Film Compensation Strategy of PFI Gasoline Engine Based on Combined Film Model

Yongjun Huang, Guohua Chen

School of Energy and Power Engineering, Huazhong University of Scicence & Technology, 430070 Wuhan, China

Abstract. For port-injected gasoline engines, an innovative strategy of transient fuel-film compensation is proposed to eliminate the impact of fuel-film dynamic effects and improve the transient air fuel ratio control accuracy. A combined fuel-film model composes of a long-term fuel-film and a short-term fuel-film was established. A transient fuel-film compensator model was analyzed in detail. The respective term of the fuel-film model is calculated by a share factor. The decay coefficients control the fuel compensation quantity and the active time. The validity of above models were verified though engine bench tests. The results show that this new strategy can effectively suppress the λ offset during engine transient condition, the transient lambda deviation is less than 5% and stabilized within 1 second.

1 Introduction

PFI engine transient air-fuel ratio (AFR) accurate control is critical and difficult to achieve low emissions. Because during acceleration or deceleration, there are many factors that cause AFR deviation, as in [1], mainly including:

(1) The presence of dead time between actual load and measured load, as in [2]. The measured value can not truly reflect the actual value when the intake valve is closed.

(2) Intake lags. During accelerate process, since the actual intake air mass increases rapidly ,the calculated intake air mass is less than the actual intake air mass, causing a lean mixture; During decelerate process , since the actual intake air mass drops rapidly, the calculated intake air mass is greater than the actual intake air mass, resulting in a rich mixture, as in [3].

(3) Delay comes from sensor signal and processor computation, as in [4].

(4) Effect of fuel-film. General speaking, the evaporation and adhesion characteristics of fuel-film changes with pressure, as in [5]. In transition condition, changes of intake pressure causes fuel-film to change.

There are several existing solutions for fuel-film compensation: conventional PID controller, sliding mode controller, as in [6], neural network controller, as in [7], and model-based fuel-film compensator, as in [8]. Among them, the conventional PID controller is simple but too rough, its accuracy is limited. Neural networks and model-based controller usually involve complex mathematical theory using numerical integration for differential equations and a large number of matrix operations, as in [9].

This paper first proposes an application-oriented fuel-film model and a transient fuel-film compensation algorithm, and then verifies the correctness of the proposed strategy by engine bench tests.

2 Transient air fuel ratio control model

2.1 Non-dimensional pneumatic model

To simplify the calculations, the basic fuel injection quantity is calculated using a non-dimensional pneumatic model. Engine load is represented by relative intake charge which is defined as the ratio of the actual intake air quantity of a cylinder per cycle under current pressure and temperature to the ideal capacity of a cylinder under standard condition.

According to the ideal gas equation:

$$PV = MRT \tag{1}$$

The actual air charge of cylinder can be expressed as:

$$m_c = \frac{P_c V_c}{R T_c} \tag{2}$$

The ideal intake charge of a cylinder under standard condition (pressure P_0=101.3KPa, temperature T_0 = 0°C) is:

$$m_{std} = \frac{P_0 V_c}{R T_0} \tag{3}$$

According to the definition, the relative intake charge of cylinder is:

DOI: 10.1051/matecconf/20166801001

$$rl = \frac{m_c}{m_{std}} = \frac{P_c T_0}{P_0 T_c} \tag{4}$$

At the intake valve closing (IVC), the intake manifold pressure and the in-cylinder pressure can be treated as balance, namely:

$$P_c = P_s \tag{5}$$

The ideal gas equation is used again and it gets:

$$P_c = P_{in} + P_{int} \tag{6}$$

where, V_c is cylinder volume, m_c is in-cylinder air mass, T_c is in-cylinder air temperature, T_0 is cylinder air capacity under standard condition, P_C is in-cylinder air pressure , P_s is intake manifold pressure , P_{in} is fresh air pressure portion in-cylinder , P_{int} is residual gas pressure portion in-cylinder .

Therefore, the relative fresh air charge is described as:

$$rl_{in} = \frac{P_{in} T_0}{P_0 T_c} = \frac{(P_s - P_{int}) T_0}{P_0 T_c} = (P_s - P_{int}) f_{ps} \tag{7}$$

As shown in 0, the relative intake charge is linear to the intake manifold pressure.

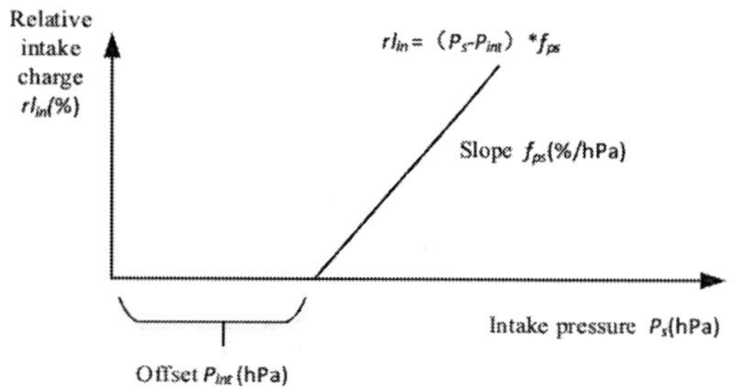

Figure 1. Relationship between relative the intake charge and intake pressure.

The slope of linear equation f_{ps} is calculated from intake air temperature, the intercept P_{int} is calibrated as function of engine speed and the valve overlap angle.

2.2 Calculation of steady-state film quantity and dynamic film delta

It has been pointed out that the steady fuel-film mass is influenced by various factors such as engine load, speed and temperature. In fact, during the film compensation calculation process, the delta between the fuel-film mass is more meaningful than the actual amount of fuel-film. When dynamically switching between two steady conditions, the same film delta value can be used in acceleration and deceleration.

Since the steady film amount can not be measured directly, it needs to be calibrated indirectly. After the engine heats up, to start with a speed of 1200 r/min and lambda = 1, switch between different loads, adjust the basic fuel-film amount according to the changes of air-fuel ratio, thereby change the film compensation amount correspondingly.

The calibration criteria is: During the process of changing the engine load, if the area of lean and rich part of lambda is severe asymmetry , it is considered as the film compensation is inadequate or excessive; If they are basically equal , the preset fuel-film is considered as the reasonable amount of steady fuel-film amount at the very operating point.

The adjustment idea is: If there is a large dilute peak at acceleration or a large peak at deceleration, it means that the fuel-film compensation is inadequate , so the amount of compensation should be increased; If there is a large rich peak at acceleration or a large lean peak at deceleration, it means that the film compensation is excessive, so the amount of compensation should be reduced; Keep tuning the fuel-film variables until the lambda peak value of acceleration and deceleration is basically symmetry. The characteristic curve of basic fuel-film quantity is shown in 0.

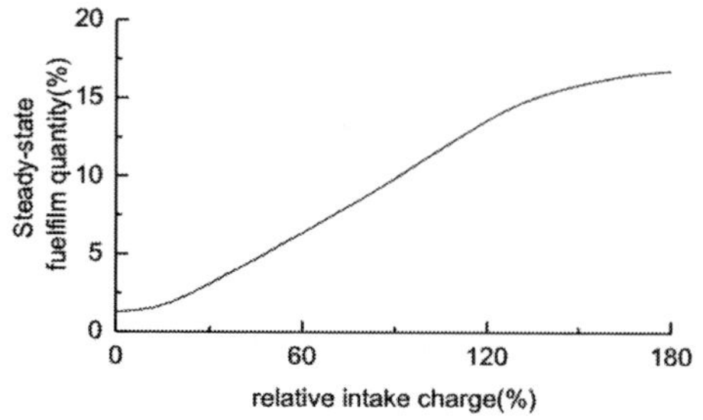

Figure 2. Characteristic curve of basic fuel-film quantity vs relative intake charge.

The basic amount of film is corrected according to the engine speed and coolant temperature. The correction coefficient is shown in 0.

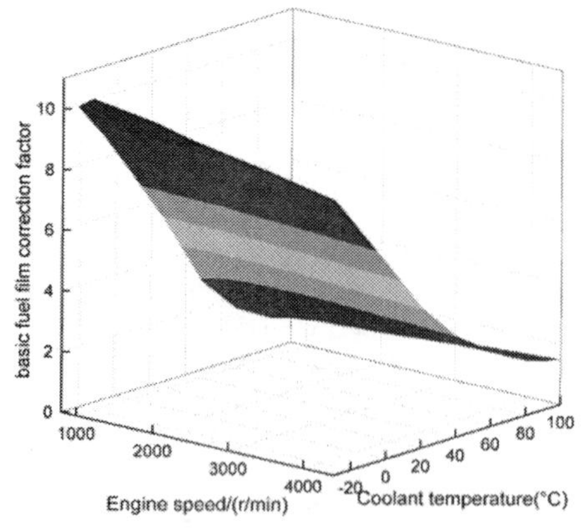

Figure 3. MAP of basic fuel-film correction factor.

2.3 Calculation of transient film compensation

Nowadays, the widely used film model x-τ model is derived based on physical model, as in [10]:

$$\begin{cases} \dot{m}_{ff} = x\dot{m}_{fi} - \dfrac{m_{ff}}{\tau} \\ \dot{m}_{f} = \left(1 - x\right)\dot{m}_{fi} + \dfrac{m_{ff}}{\tau} \end{cases} \tag{8}$$

The x-τ model is transformed to transfer function model using Laplace algorithm

$$G\left(s\right) = \frac{1 + \left(1 - x\right)\tau s}{1 + \tau s} \tag{9}$$

In order to achieve the desired film compensation effect, the transfer function of film compensator should be an inverse function of the transfer function of film

$$G\left(s\right) = G^{-1}\left(s\right) = \frac{1 + \tau s}{1 + \left(1 - x\right)\tau s} \tag{10}$$

where, $\dot{m}_{fi}$ is injection flow rate, $\dot{m}_{ff}$ is film variation rate, $\dot{m}_{f}$ is fuel flow rate into the cylinder, x is deposit factor, τ is evaporation time constant, s is Laplace operator.

According to the relationship between time-domain response and Laplace transform, making an inverse Laplace transform to the product of the film compensator transfer function and $1/s$ will yield a time-domain representation of the step response of the film compensator

$$G_c\left(t\right) = 1 + \frac{1 + \tau x}{\tau - \tau x} e^{-\frac{t}{\tau - \tau x}} \tag{11}$$

The above formula indicates that the ideal film compensation decays exponentially, the initial value and the decay rate are not only highly nonlinear but also time-varying. Since the exponential characteristic decays fast first then slow, the exponential decay film will be divided into two parts, a fast decay long-term film and a slow decay short-term film. They constitute a new combination film models.

The film model is shown in 0. According to the knowledge of geometry, the area enclosed by the solid line and the axis is approximated to the sum of areas enclosed by dashed and dotted lines and the axes. Both long-term film and short-term film have positive compensation and negative compensation.

Features of the film model: long-term film is less in quantity but decays more slowly and lasts for a longer duration, it makes the transition of injection fuel much smoother, so it is helpful in stabilizing the air fuel ratio; Short-term film is larger in quantity but decays more faster and lasts for a shorter duration, it functions for compensating the severe mutation film at rapid changes in throttle, avoiding large air-fuel ratio pulse. Short-term film is extremely important for suppressing the air-fuel ratio peak. During acceleration, short-term film increases rapidly to the maximum value to avoid producing large lean peak .Because the acceleration process is completed in a very short time, the short term film should quickly decay, to avoid producing obvious rich peak. During deceleration, short-term film quickly reaches the minimum value to avoid producing larger rich peak then decays quickly to avoid producing obvious lean peaks.

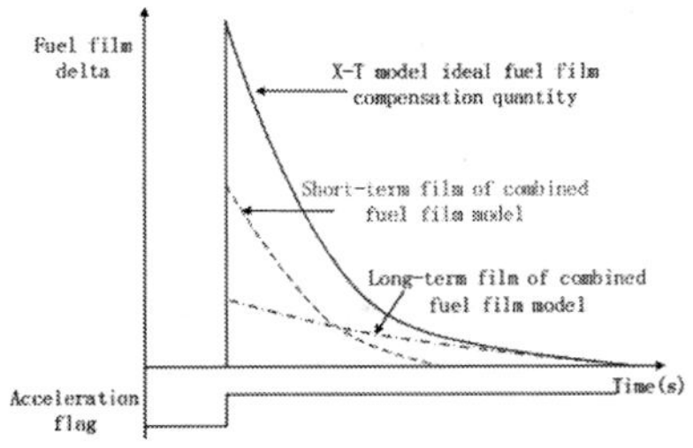

Figure 4. Combined fuel-film model.

The distribution of long-term film and short-term film is extremely important in derivation of film compensation. Upon the assignment of long-term film and short-term film of the film model, accurate calculations of compensation value for the film can be made. The long-term film and short-term film is decided based on different characteristics of engine (speed of film evaporation or adhesion). Because the film effects only affect the dynamic conditions, the steady conditions do not require compensation and the film compensation should be promptly ended after completion of dynamic behaviour. The overall scheme of transient film compensation is shown in 0.

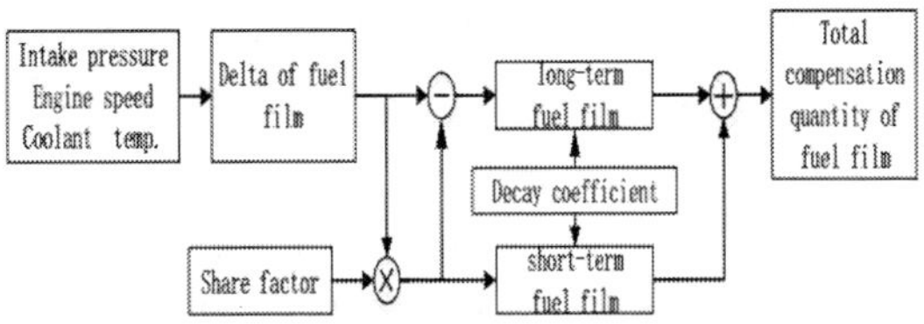

Figure 5. Overall transient compensation strategy.

2.4 Calculation of share factor of long-term/short-term film

The distribution ratio of long-term and short-term of the film is determined based on engine characteristics. The ratios of fuel-film distribution on acceleration and deceleration are different, and should be treated separately. The relative magnitude between long-term and short-term of the fuel-film varies with engine speed and temperature. Intake air pressure and air-flow mass are different at different engine speeds, resulting in different speeds of evaporation and adhesion of fuel-film. Wall temperature of Intake manifold is different at different engine coolant temperatures, resulting in different speeds of evaporation and adhesion of fuel-film too. The overall trend is that: as the speed increases, the film distribution

DOI: 10.1051/matecconf/20166801001

coefficient increases; as the temperature rises, the film distribution coefficient increases too. The MAP of film distribution coefficients is shown in 0 and 0.

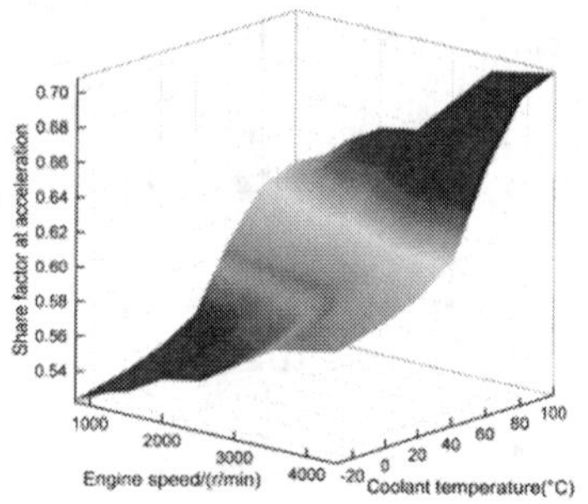

Figure 6. Share factor of fuel-film at acceleration.

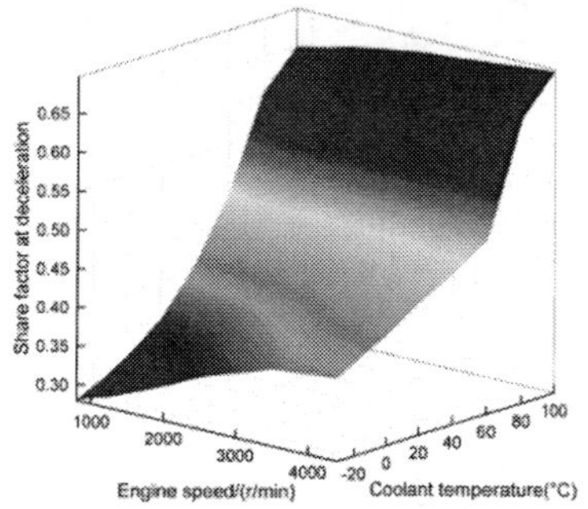

Figure 7. Share factor of fuel-film at deceleration.

2.5 Calculation of decay factor of long-term/short-term film

Upon finishing allocating the long-term film and short-term film, the decay coefficient is used to control the decay rate and the corresponding duration of film compensation for acceleration and deceleration condition respectively. The bigger the decay coefficient is, the shorter the film lasts for; on the contrary, the smaller the decay coefficient is, the longer the film lasts for. The film decay coefficient is shown in 0.

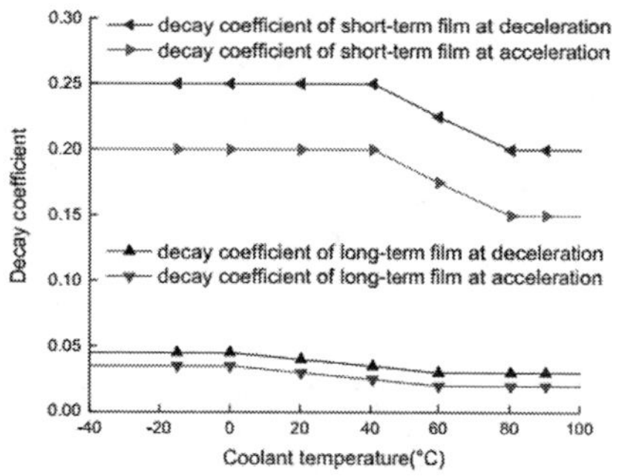

Figure 8. Decay coefficient of long-term/ short-term film.

On one hand, the decay coefficient mainly depends on the engine temperature. The respective decay coefficient of long-term film and short-term film varies in the same trend on both acceleration and deceleration working conditions. They decrease while temperature increases. Although the high temperature is conducive for the film to volatile, it seems that the decay coefficient should be increased. However, when the temperature is high, the basic film is usually small and the corresponding change of the magnitude of the film is small, hence the step size should be small.

On the other hand, the decay coefficient of both long-term film and short-term film is slightly bigger on deceleration than on acceleration. Since the intake manifold pressure decreases during deceleration, the film can evaporate faster than that on acceleration. The basic requirement is that the compensation value quickly decays when the film changes fast and the compensation value decays slowly when the film changes slowly. The film decay coefficient mainly affects the duration of the transient air-fuel ratio deviation. If the duration of deviation is too long, the film decay coefficient should be increased, to fasten the air-fuel ratio to recovery to normal value.

2.6 Calculation of total film compensation

The compensation quantities of long-term film and short-term film are calculated in the same way, and the way they are calculated are similar for acceleration and deceleration. If an injection occurs between the current calculation and the last calculation, it means that the compensation fuel previously calculated has entered into the cylinder, so the last compensation fuel should be subtracted off. If no injection occurs (e.g. the first cycle at recovery from fuel cut-off), it means that the fuel-film compensation of previously calculation is not performed, so the last compensation fuel can not be subtracted, but accumulated. Furthermore, during the calculation of compensation of the short-term film, a dead-zone function is added to filter the compensation value. When the compensation amount is less than the threshold, no compensation would be used, to prevent unnecessary compensation to result in frequency fluctuations in mixture when the change of engine load is too small. The calculation model of fuel-film compensation is shown in 0 and 0 respectively.

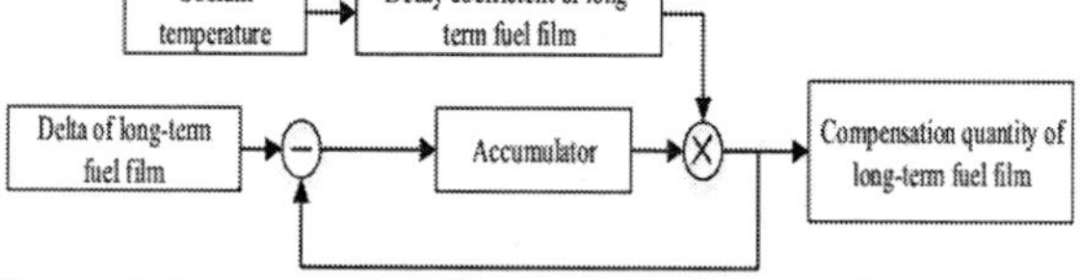

Figure 9. Long-term fuel-film compensation quantity computation module.

Both on acceleration and deceleration, the compensation value of long-term film $m_l(i)$ is calculated as:

$$m_l(i) = \begin{cases} m_{fi}(i) \times f_{lx}, & i = 1; \\ \Sigma\left[m_{fi}(i) - m_l(i-1)\right] \times f_{lx}, & i > 1 \end{cases} \quad (12)$$

where, $m_{fl}(i)$ is the delta of long-term film at current calculation time, $x = a, d$, f_{la} is decay factor of long-term film for acceleration, f_{ld} is decay factor of long-term film for deceleration.

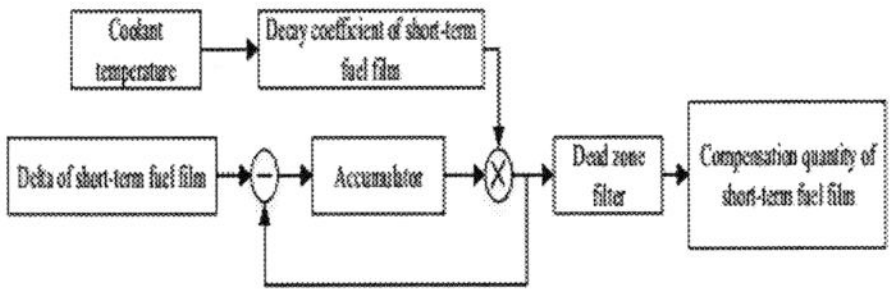

Figure 10. Short-term fuel-film compensation quantity computation module.

Both at acceleration and deceleration, the compensation value of short-term film $m_s(i)$ is calculated as:

$$m'_s(i) = \begin{cases} m_{fs}(i) \times f_{sx}, & i = 1; \\ \sum\left[m_{fs}(i) - m_s(i-1)\right] \times f_{sx}, & i > 1 \end{cases} \quad (13)$$

$$m_s(i) = \begin{cases} m'_s(i), & \left|m'_s(i)\right| \geq U; \\ 0, & \left|m'_s(i)\right| \leq U \end{cases} \quad (14)$$

where, $m_{fs}(i)$ is the delta of short-term film at current calculation time, $x = a, d$, f_{sa} is decay factor of short-term film for acceleration, f_{sd} is decay factor of short-term film for deceleration, U is filter threshold.

3 Experimental set-up

The tests were conducted on a 1.5L four-cylinder engine DFMA15 of Dengfeng Company. An overview of the engine bench configuration is shown in 0 and the engine specifications are listed in Table 1. Other devices included a bench control system F2000 of Xiangyi Power Instrument Company, an air-fuel ratio meter ES631, an exhaust analyzer LA4 and a CAN communication module ES581 of ETAS Company. The controller hardware is based on Freescale 32-bit processor MPC5634M. The advance software ASCET is used for graphically modelling and generating executable C code automatically. ES631 is installed upstream catalytic converter. Data are collected every 10ms using INCA7.0.

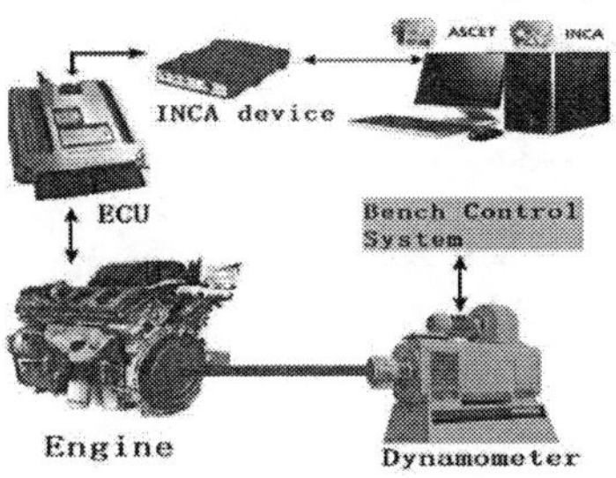

Figure 11. Sketch of engine bench configuration.

Table 1. Engine parameters.

Bore/Stroke	75 mm/ 84mm
Compression ratio	11
Displacement	1.5 L
Maximum torque	145 N m (at 4200 r/min)
Rated power	85 kW (at 6000 r/min)
Fuel injection system	Port Fuel Injection
Intake type	Normally Aspirated
Number of cylinder	4
Fire orders	1-3-4-2

4 Bench test results and analysis

Since throttle tips in and tips out are typical transient conditions of engine, a rapid step of the electronic throttle is applied. In order to verify the validity of proposed fuel-film model and the control model , the new compensation algorithm is compared to non-compensation algorithm and Delphy compensation algorithm. The test results are shown in 0, 0, 0 and 05.

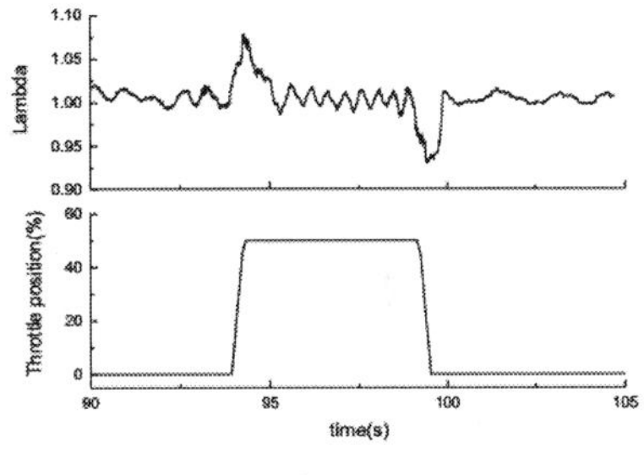

Figure 12. Throttle position varies 0-50%, non-compensation.

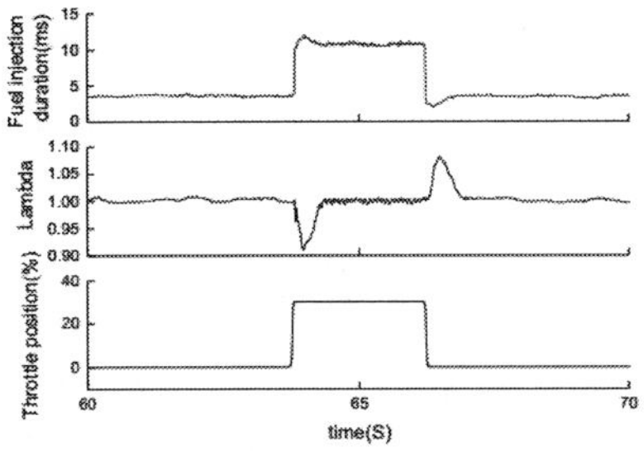

Figure 13. Throttle position varies 0-30%, film compensation of Delphy is applied.

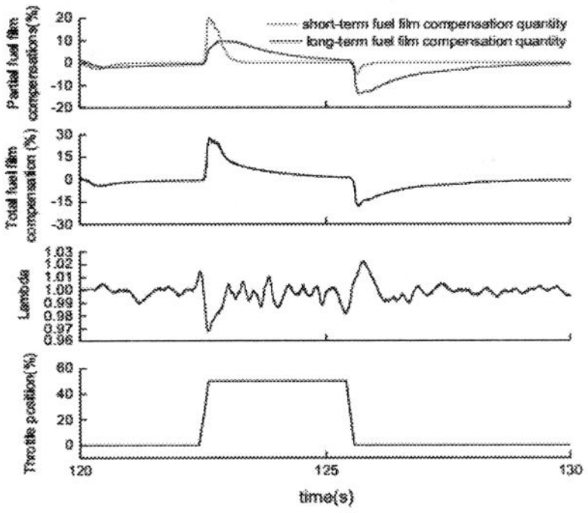

Figure 14. Throttle position varies 0-50%, combination film compensation is applied.

DOI: 10.1051/matecconf/20166801001

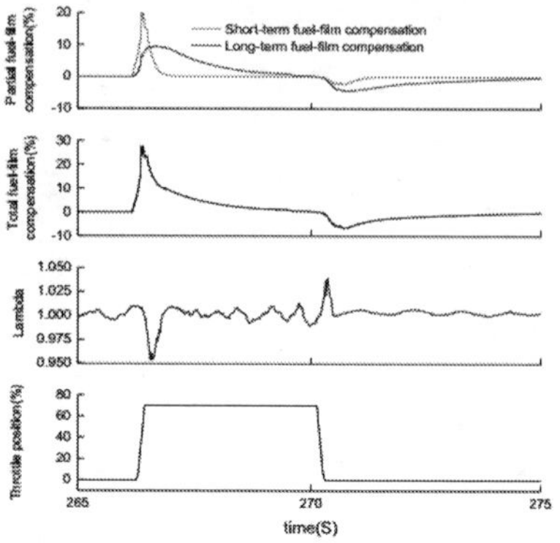

Figure 15. Throttle position varies 0-50%, combination film compensation is applied.

4.1 Comparison between combination film compensation algorithm and non-compensation algorithm

In 0, when no film compensation is used, the mixture is lean when acceleration and rich when deceleration, with a maximum deviation of about 8%. In 0 and 0, when the combination film compensation strategy is adopted, the mixture was enriched at acceleration and diluted at deceleration. The transient Lambda peak deviation is greatly reduced. The maximum deviation dropped to less than 5% and settling time is less than 1 second.

4.2 Analysis of control effect of the combination film compensation algorithm at tip-in/tip-out

As seen from the 0 and 0,when a tip-in occurs, both the long-term film and short-term film increases at the same time as the throttle opening increases until the throttle opening unchanged. The short-term film decays to zero rapidly, while the long-term film decays to zero slowly. The total compensation value first increases then decreases. When a tip-out occurs, the film compensation works in opposite direction. Both the long–term film and the short-term film decrease at the same time as the throttle turns smaller until the throttle opening unchanged. The short-term film decays to zero rapidly, while the long-term film decays to zero slowly. The total compensation value first increases then decreases. The total compensation value first reaches a minimum value then changes towards zero in an opposite direction.

The lambda is less than 1 and lasts for a short period of time at tip-in, it means that the acceleration has been enriched; The lambda is larger than 1 and lasts for a short period of time at tip-out, it means that the deceleration has been diluted; The lambda convergences to 1 rapidly after the throttle gets steady, it is controlled with the range from 0.99 to 1.01.

4.3 Comparison between combination film compensation algorithm vs other algorithm

According to the comparison of 0, 0 and 0, the new compensation algorithm on transient air fuel ratio control reached the same level as the existing mature system.

5 Conclusions

As can be seen from the dynamic experiment results, at the same operating conditions, lambda excursion decreased obviously when the combination film compensation algorithm is used. The change of fuel-film on transient working condition can be controlled correctly and precisely by the fuel-film compensation control strategy. The maximum deviation of transient lambda is less than 5%.

Transient On-line correction of the transition compensation in the event of coking on the valves, variations of individual engines, varying fuel quality etc will be investigated in future work. The transition compensation is corrected by means of an amplification factor which works on all parts of the transition compensation if considerable deviations in the air-fuel ratio are detected during load changes.

References

1. M. Takazawa, K. Komura, T. Kitamura, *SAE technical paper* (2015)
2. Y. Yildiz, A. Annaswamy, D. Yanakiev, *Control Engineering Practice*, **18**, 1369-1378 (2010)
3. C. Wang, J. Zhang, C. Yin, *Chinese Journal of Mechanical Engineering*, **27**, 2 348-357 (2014)
4. R. Maurya, A. Agarwal, *SAE technical paper* (2011)
5. T. Takiyama, *SAE Technical paper* (2014)
6. M. Ichiyanagi, T. Suzuki, *SAE technical paper* (2015)
7. H. Toshitani, T. Shen, and J. Zhang, *SAE technical paper* (2011)
8. A. Gaeta, S. Santini, L. Glielmo, *SAE technical paper* (2003)
9. L. Wu, J. J. Liu, *Applied Mechanics & Materials*, **643**, 66-71 (2014)
10. Y. Zhao, T. Shen, X. Jiao, *SAE technical paper* (2013)

Numerical Version of Self-Consistent Scheme for Heterogeneous Materials

Mengting Xu, Limei Xu and Xiaomei Xie

School of Astronautics and Aeronautics, University of Electronic Science and Technology, Chengdu China

Abstract. The self-consistent scheme (Budiansky, 1965) is one of the analytical methods to predict the effective properties of heterogeneous material. However, it has been noticed that this scheme cannot take the micro-structural information into the estimation. The full scale numerical simulation considering elaborately the micro-morphological characteristics, under certain conditions, needs too much computational resources (memory and time). A numerical version of self-consistent scheme based on finite element analysis is proposed to overcome the above drawbacks by combining the characteristics of numerical simulation and self-consistent scheme. As a demonstration, such scheme is applied to estimate the effective mechanical properties of bi-continuous composite in which all the constituent phases are on the equal status. The prediction is in good agreement with the full scale numerical simulation while the requirement on computational resource is reduced dramatically.

Keywords: bi-continues composite, effective mechanical properties, self-consistent scheme, numerical simulation

1 Introduction

The research on the effective elastic properties of heterogeneous materials has been attracting great attention in the last few decades. Various analytical methods, which can be classified into "bound method" and "approximated method" have been proposed. Voigt-Reuss bounds [1] and Hashin-Shtrikman bounds [2] are the two prominent pairs of lower and upper bounds. Approximated method includes self-consistent scheme (Budiansky [3], Hill [4]), three-phase model (Christensen [5]) and Mori-Tanaka method ([6]). For the self-consistent scheme, the essential point is to treat each constituent phase as an inhomogeneity embedded in a homogenized material with yet to be determined effective properties. In this context, each constituent phase is treated on the equal footing. And bi-continuous material is composed of multiple constituents which are on the equal status. It means the matrix and particulate cannot be clearly distinguished. Due to such micro-structural characteristics, the self-consistent scheme can be applicable for predicting the effective properties of the bi-continuous material. However, such scheme is not able to take the microstructural characteristics, such as the size, shape and distribution of its components, into the "over simplified" model.

With the progress of computational techniques and capabilities, numerical simulations, like finite element method, are widely employed to investigate the mechanical behavior and properties of heterogeneous material. A comprehensive discussion on the numerical simulation has been given by Cailletaud et al. [7] and Zohdi and Wriggers [8]. A full scale numerical simulation for the material's real microstructure has been considered as "exact" approach for the effective modulus prediction. However, it has been noticed that even with the most recent and ever improving development of computing and memory power, the real full scale numerical simulations are still limited if it is not impossible.

The aforementioned advantages and disadvantages of the pure analytical and the pure numerical approaches lead us to a possibility of mixing the analytical and the numerical schemes, namely, a numerical version of self-consistent scheme. The objective of the numerical version of self-consistent scheme is, on the one hand, to be able to deal with the various micro-morphological information, such as size, shape and their distributions, in the heterogeneous material which cannot be considered by the analytical scheme. On the other hand, this scheme requires much less computational resources, like time and memory space. In a finite element analysis, this means the proposed scheme needs much less number of elements in comparison with the full scale numerical analysis.

This article will be organized as followings. Section 2 introduces the idea of numerical self-consistent scheme when predicting the effective properties of heterogeneous materials. Based on the proposed scheme, finite element simulation for certain bi-continuous material are conducted using a commercial finite element package (ANSYS) in Section 3. Conclusion is given in the last section.

2 Numerical self-consistent scheme for simulation models

To illustrate the numerical self-consistent scheme, a two-phase bi-continuous material in two-dimension is considered. The specimen for modeling the macroscopically isotropic and homogenous composite material is defined as L×L, 'L' is the macroscopic dimension of the composite material. The specimen is modeled by $N \times N$ domains with equal size $d \times d$ and each constituent phase is isotropic. The domain size '*d*' is considered as a characteristic dimension with constant value of the microstructure. Following the finite element scheme described elaborately in our published paper [9], a realization of two-phase bi-continuous material used for the full scale numerical simulation is presented in Figure 1. The two phases are randomly distributed in the specimen according to their area fraction. It will be shown that reliable effective elastic modulus can only be achieved when domains reach a very large number.

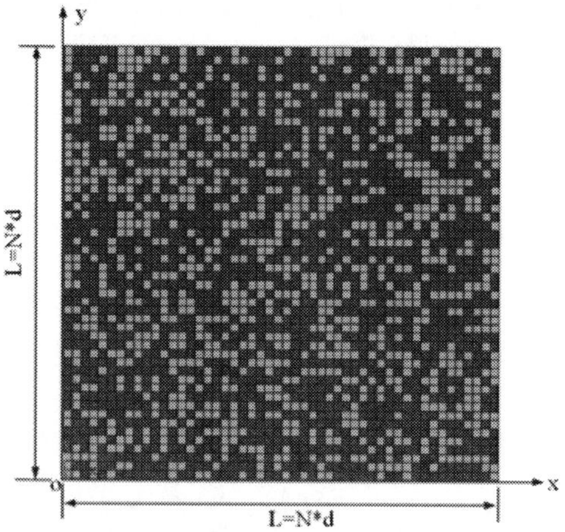

Figure 1. A realization of two-phase bi-continuous material. White color represents phase1 with area fraction is 40% while black color denotes phase 2 with 60% area fraction. 'N' is the number of domains of each edge. 'd' is the characteristic size for the domain.

Incorporating the essential point of analytical self-consistent scheme proposed by Budiansky, a finite element model, we name the numerical self-consistent model, is constructed. Figure 2 is a realization of the numerical self-consistent model corresponding to a two-phase bi-continuous material. The core of the model is made of the same structure as the full-scale numerical model, while the outer material is "yet to be determined" effective medium. Since the effective medium has uniform elastic constants, the number of elements used in the effective region is much less than the "composite region". Therefore, the self-consistent scheme saves a substantial amount of the computational resources. At the meantime, such model captures the details of the micro-structure of composite material. To evaluate the effective elastic constants, iterative procedure is employed following the framework of analytical self-consistent scheme. When the difference of the ith value and (i+1)th prediction is within the given error, the (i+1)th value is the effective Young's modulus.

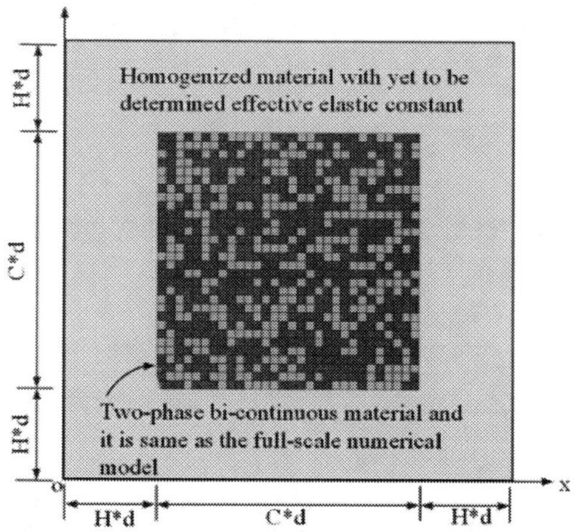

Figure 2. A realization of self-consistent model for the two-phase bi-continuous material. The inside body is simulated in micro-structural level and embedded in the outside body which is homogenized with yet to be determined effective elastic constants. The inside body is composed of domains with area fraction 40% for phase 1 (white phase) and 60% for phase 2 (black phase). 'C' is domains' number of the core structure's edge.

3 Finite element simulation and discussion

In order to avoid the numerical error due to the mismatch of the material properties between the adjacent domains, the process of mesh refinement is conducted during finite element simulation to relax the stress concentration and smooth the displacement variation between domains with different material properties. Besides, higher order element, namely twenty-node hexahedral element is adopted to improve the accuracy and accelerate the speed of convergence.

Furthermore, the concept of statistics is adopted due to the random distribution of two constituent phases over the established model. In the present work, the "population" refers to the specimen with a specific phase area fractions and material properties. For all the specimens in the population, the mean value of the effective modulus is denoted by μ and corresponding standard deviation is denoted by σ. A sample consisting of n (= 30) specimens is taken to infer the statistic data (μ and σ) of the population according to Central Limit Theorem (Mendenhall et al.[10]). Hence, a pair of statistical data, i.e., the sample mean value ($\bar{x}$) and the sample standard deviation (s), is determined. And, the mean value (μ) of population (the specimen including certain number of the grains) can be inferred within a 95%-confidence interval by

$$\bar{X} - 1.96\frac{s}{\sqrt{n}} \le \mu \le \bar{X} + 1.96\frac{s}{\sqrt{n}} \quad (1)$$

In evaluating the effective elastic Young's modulus of the selected bi-continuous material, the following loading and boundary conditions are imposed.
where a is prescribed displacement, u_x and u_y represent the displacement along the direction x and y in the global coordinate system.

$$\begin{cases} u_x = 0 & at\ x = 0 \\ u_x = a & at\ x = L \\ u_y = 0 & along\ x = 0, y = 0 \end{cases} \tag{2}$$

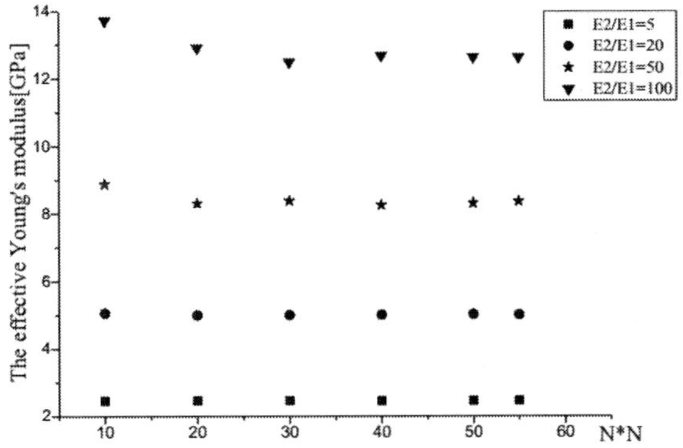

Figure 3. The effective Young's modulus in terms of the size of the specimen with different ratio of Young's modulus of two constituent materials.

Fig. 3 presents the effective Young's modulus obtained from different models with various Young's modulus ratio of two constituent materials, namely *E2/E1* = 5, 20, 50, 100, respectively, in terms of the size of the specimen. It is noted, with the increase of the specimen's size, the effective Young's modulus tends to a stable value, and then the size of Representative Volume Element (RVE) for each material can be determined. Table 1 lists the size of RVE for materials with different E2 / E1. It is seen the size of RVE is larger when the modulus ratio is bigger. As shown, to obtain the effective Young's modulus for material with *E2/E1* = 100, the corresponding RVE should have 50×50 domains.

Table 1. The RVE for materials with different E2 / E1

E2/E1	5	20	50	100
Size of RVE(N*N)	20*20	40*40	50*50	50*50

With reference to Fig. 2, the dimension of '*H*' is a critical size when establishing the corresponding self-consistent model. To determine the size '*H*', the two-phase bi-continuous material with *E2/E1* = 20 and 100 is taken into account, respectively. The corresponding specimen size is assumed to be identical with the predicted RVE size, namely 40×40 for *E2/E1* = 20, 50×50 for *E2/E1* = 100. In the process of estimation, the initial Young's modulus and Poisson's ratio are assumed to be the average value of the mechanical properties of the constituent phase. The obtained effective Young's modulus is depicted with the variation of '*H*' and iterative time in Fig. 4. The number of element and node in the finite element models and computational time for different models are summarized in Table 2.

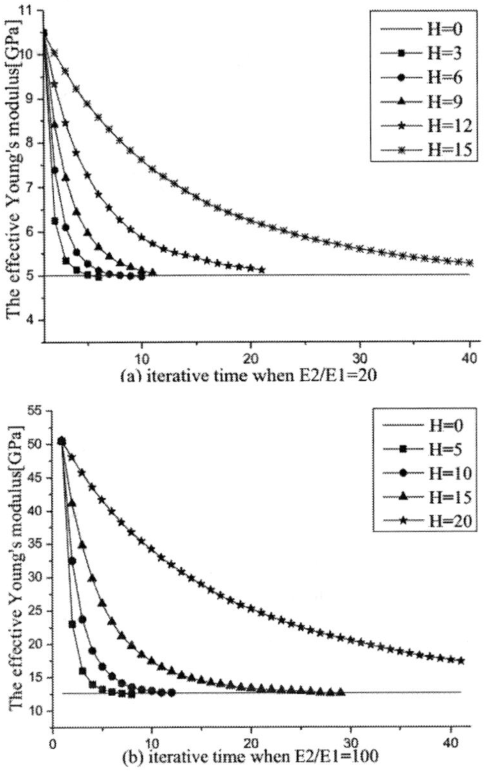

Figure 4. The effective Young's modulus of the composite material with changing '*H*', in terms of iteration number. (a) *E2 / E1* = 20, (b) *E2 /E1*=100

Table 2. The finite element models and computational time for different models while the specimen size is 50×50 and *E2/E1*=100.

The size of '*H*'	The effective Young's modulus [GPa]	Element number [Ten thousand]	Node number [Ten thousand]	CPU time for an iteration[s]	Number of iteration	Total time [s]
0(full-scale)	12.62	16	48.1601	62.853	1	62.853
5	12.46	15.84	47.6641	68.516	8	548.1
10	12.68	9.45	28.4581	47.315	11	520.5
15	12.60	4.68	14.1121	23.026	22	506.6
20	17.31	1.53	4.6261	7.675	46	353.1

As indicated in Figure 4, the effective Young's modulus will converge to the value obtained from the full scale model (H=0), no matter what the size of the H is, with the increase of the iteration. However, the speed of convergence becomes more and more slow as the size of '*H*' increases.

Reference to the data in Table 2, it is seen that the self-consistent scheme models need less number of elements and nodes in compared with the full-scale numerical analysis. When '*H*' is 15, the number of elements and nodes in full-scale model is about 4 times of those in the self-consistent scheme model.

As for the computational time, the full-scale model expends almost 3 times in one iteration more than the self-consistent models while the total time is much more for the self-consistent models. In this context, the proposed self-consistent model reduces the requirement of computational resource at the expense of computational time. Considering both calculating time and the convergence rate of the effective Young's modulus, the selected '*H*' and RVE size denoted by 'N' are summarized in Table 3, for different two-phase bi-continuous materials with various *E2/E1*.

Table 3. Suitable H and N for different ratio of Young's modulus of constituent phase in two-phase bi-continuous material

E2/E1	5	20	50	100
H/N	0/20	6/40	10/50	15/50

4 Conclusions

To predict the effective properties of heterogeneous material, a numerical self-consistent model is proposed based on finite element method. This model consider the real microstructural characteristics as the full-scale numerical model in the core, and the outer region is "yet to be determined" effective medium with uniform elastic constants. Such model overcome the shortcoming of the analytical version of self-consistent scheme by incorporating the interaction among the material's constituent phases. In comparison with the full-scale numerical model, the proposed numerical self-consistent method has less requirement on the computational resources, such as time and memory space. Thus, in the case of limited computational resource, the proposed numerical self-consistent scheme is effective in studying the effective properties of composites, especially for the materials which constituent phases have high contrast in their properties.

References

1. R. Hill, Proc. Phys. Soc. A65 (1952) 349-354.
2. Z. Hashin, S. Shtrikman, J. Mech. Phys. Solid **10** (1962) 335-342.
3. B. Budiansky, J. Mech. Phys. Solids. **13** (1965) 223-227.
4. R. Hill, J. Mech. Phys. Solids. **13** (1965) 213-222.
5. R. M. Christensen, K. H. Lo, J. Mech. Phys. Solids **27** (1979) 315–330.
6. T. Mori, K. Tanaka, Acta Metall. **21** (1973) 571-574.
7. G. Cailletaud, S. Forest, D. Jeulin, F. Feyel, I. Galliet, V. Mounoury, S. Quilici. Computational Materials Science. **27** (2003) 351-374
8. T. I. Zohdi, P. Wriggers, *An introduction to computational micromechanics*. Second edition. Springer-Verlag Berlin and Heidelberg GmbH Publishing (2008).
9. X. M. Xie, H. Fan, Materials Science and Engineering A. 527 (2010) 5452-5461.
10. W. Mendenhall, R. J. Beaver, B. M. Beaver, *Introduction to Probability and Statistics*, 12th edition, Thomson Learning Publishers, Singapore (2006).

Design of Accelerated Reliability Test for CNC Motorized Spindle Based on Vibration Signal

Chao Chen, Fei Chen, Yifeng Ye, Heng Zhang, Zhaojun Yang and Xuejiao Du

School of mechanical science and engineering, Jilin University, Changchun, 130025, China

Abstract. Motorized spindle is the key functional component of CNC machining centers which is a mechatronics system with long life and high reliability. The reliability test cycle of motorized spindle is too long and infeasible. This paper proposes a new accelerated test for reliability evaluation of motorized spindle. By field reliability test, authors collect and calculate the load data including rotational speed, cutting force and torque. Load spectrum distribution law is analyzed. And authors design a test platform to apply the load spectrum. A new method to define the fuzzy acceleration factor based on the vibration signal is proposed. Then the whole test plan of accelerated reliability test is done.

1 Introduction

Motorized spindle is the key functional component of CNC machining centers and is the typical device for high speed machining [1, 2]. The machining accuracy and machining efficiency of CNC machining centers are largely determined by motorized spindle. So it's necessary to test the reliability of motorized spindle. However, traditional reliability test must take long time and more samples to test products' performance [3]. As motorized spindle has long life, high precision and reliability, its reliability test may not only take a long time, but also increase the time and cost of development. Although it takes so long time, it rarely gets much failure data. So it's necessary to design accelerated life test for motorized spindle reliability. Accelerated life test is a stress testing methodology for accelerating product reliability by subjecting it to conditions (stress, vibration, load etc.) to uncover faults and potential failure modes in a short time [4-6]. Through the test data, MTBF of the normal field life and maintenance intervals of a product can be predicted. Accelerated life test is supposed to only change the failure rate without changing the failure mechanism of motorized spindle [7]. But as a complex electromechanical system, motorized spindle has several subsystems and multifarious failure mechanism [8, 9] which severely limits the design of accelerated reliability test for motorized spindle. Great effort will be taken to make an effective relationship model between failure rate in high load condition and normal load condition.

For the accelerated reliability test of motorized spindle, there are three main challenges. First, there is scarcity of real load spectrum data of motorized spindle in working condition. Secondly, it's difficult to build loading condition of motorized spindle in lab without changing the failure mechanism. Thirdly, it's extremely hare to determine the accelerating factor. For above 3 points, authors propose a practical and effective solution.

The rest of the article is organized as follows: a brief introduction to the motorized spindle is provided in Section 2. Section 3 introduces the load data acquisition from field reliability test in factory, such as workpiece material, cutting condition, cutting parameter and so on. Then load spectrum including rotating speed, cutting force and torque are calculated by cutting force formula in Section 4. In the following section, authors build simulation test platform of the motorized spindle for loading spectrum of accelerated reliability test. Next authors proposes a method to make sure the fuzzy acceleration factor and determine the accelerating load of accelerated reliability test for motorized spindle in Section 6. At this point, the design of accelerated reliability test for motorized spindle is finished. At last, a short conclusion is drawn.

2 Structure of motorized spindle

Motorized spindle combines the structures of frequency conversion motor and mechanical spindle in machining centers. The shell of motorized spindle is the stator of frequency conversion motor. The rotor of motor is the spindle and gripper [10]. It implements 'zero-drive chain' between motor and spindle. Motorized spindle has compact structure, light weight, small rotational inertia and outstanding dynamic characteristics. Motorized spindle consists of spindle shell, drive cooling, stator, rotor, bearing, lubrication module, power module, rotary feedthrough and detection module and so on. Its structure is shown in Figure 1.

DOI: 10.1051/matecconf/20166802001

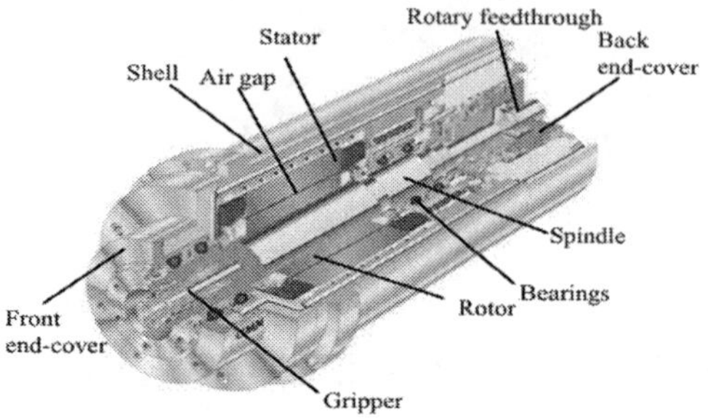

Figure 1. The structure of motorized spindle.

3 Load data acquisition

Load should be a generalized concept, such as force, torque, displacement, stress and strain, etc. Load spectrum of motorized spindle refers to the time-domain distribution of the load. Through statistics and analysis of the actual load data which is complex, chaotic in working conditions, load spectrum are shown in certain figures, tables or mathematical expressions [11].

To get the load data of motorized spindle in working condition, authors went to customers' enterprise using certain type CNC machining centers and carried out a large number of field reliability test for more than 3000h. The load data includes process data, cutting dosage, rotational speed of typical components. The load data acquisition form is shown in Table 1. As the blank materials and processing type of components are widely different, the load data have big differences. In order to facilitate the calculation, load data must be preprocessing. For instance, some processing steps rarely appear. And some processing steps, such as the chamfering process, need small cutting force, so the impact on motorized spindle is very tinny which has little significance for reliability research of spindle. These data cannot serve as the main basis for the reliability study of motorized spindle, only be used as auxiliary. Thus, in data preprocessing, these date should be removed. Then, by analyzing the processing, the load data after preprocessing can be used to calculate the cutting force and torque of motorized spindle in CNC machining centers.

Table 1. Load data acquisition form.

No.	1	2	3	4	…
Material of workpiece	Aluminum alloy	Aluminium alloy	Aluminium alloy	Aluminium alloy	
Process content	Rough milling convex	Precision milling convex	Precision milling gap	Countersink hole 68- ϕ16	…
Cutter	ϕ12 milling cutter	ϕ6 milling cutter	ϕ2.2 ball end mill	ϕ12 R0.9 milling cutter	…
Cutter material	Cemented carbide	Cemented carbide	Cemented carbide	Cemented carbide	…
Cutter type	End milling cutter	End milling cutter	End milling cutter	End milling cutter	…
Tool cutting edge angle K_r	90°	90°		90°	…
Rotational speed n(r/min)	800	900	3000	500	…
Feed speed V_c (mm/min)	36	40	10	48	…
Cutting depth a_p (mm)	3.5	0.5	1.59	3	…
Process redundancy h (mm)			1.6	3	…
Feed f(mm/r)	0.045	0.044444	0.003333	0.096	…
Feed per tooth f_z (mm/r)	0.01125	0.011111	0.001667	0.024	…
Cutting width a_e (mm)	12	6	2.2	16	…
Tool diameter d (mm)	12	6	2.2	12	…
Tool teeth z	4	4	2	4	…
Cutter life	60	60	66	60	…
Processing time t (min)	60	60	63	60	…

4 Load spectrum calculation and statistics

4.1 The calculation of cutting force

Motorized spindles in CNC machining centers include high-frequency/high-speed one and intermediate frequency one. When rough machining, the motorized spindle generally works at low or intermediate frequency. According to the previous data and experience, the force status of the motorized spindle is similar to turning. The decomposition and synthesis of cutting force is shown in Figure 2. And the key parameters of cutting fore are shown in Table 2.

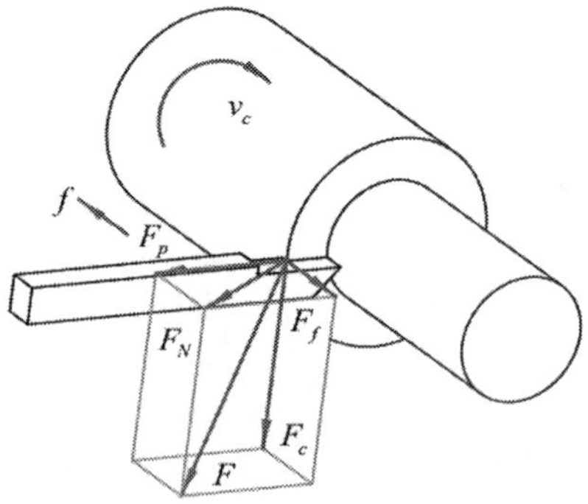

Figure 2. Cutting force of low/intermediate frequency motorized spindle.

Table 2. Key parameters of cutting force.

Symbol	Description	Unit	Direction
F_c	Main cutting force	N	Main cutting direction, tangential to transition surface, vertical to the base plane
F_p	Back force	N	Tangential to the base plane, vertical to the axis of workpiece
F_f	Feed force	N	Tangential to the base plane, opposite to the cutter forward direction

The decomposition of forces and power are expressed as follows [12]:

$$F_c = C_{F_c} a_P^{x_{F_c}} f^{y_{F_c}} v^{\eta_{F_c}} K_{F_c} \quad (1)$$

$$F_p = C_{F_p} a_P^{x_{F_p}} f^{y_{F_p}} v^{\eta_{F_p}} K_{F_p} \quad (2)$$

$$F_f = C_{F_f} a_P^{x_{F_f}} f^{y_{F_f}} v^{\eta_{F_f}} K_{F_f} \quad (3)$$

$$P_z = F_z v \times 10^{-3} \quad (4)$$

where P_z is the power of CNC machining center; K_{F_c} is the correction factor of cutting force, v is the milling speed, m/s; C_{F_c} , C_{F_p} and C_{F_f} are the coefficient determined by the workpiece material and cutting condition; x_F , y_F and η_F are the indexes of a_p , f and v .

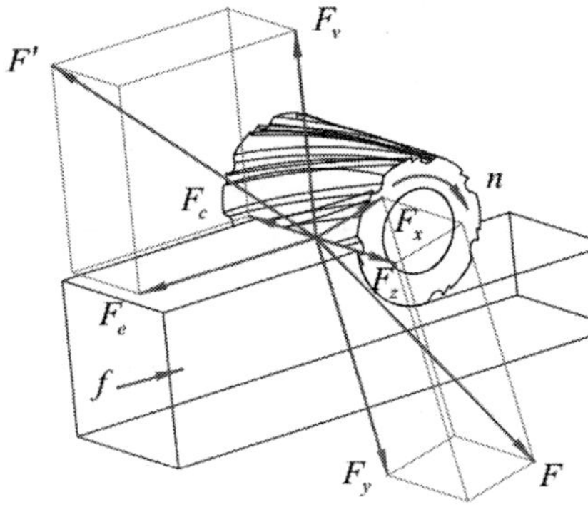

(a)

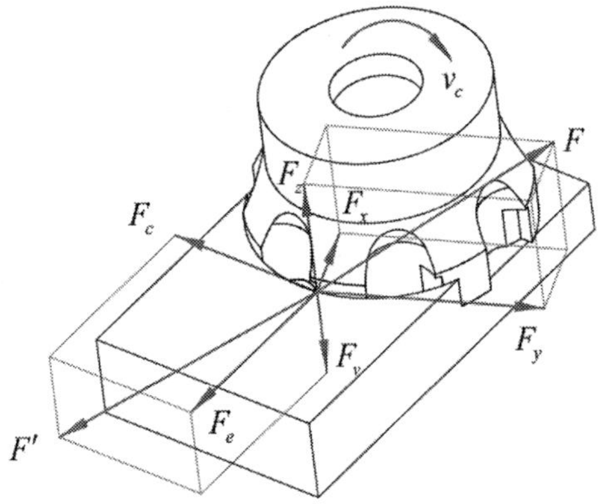

(b)

Figure 3. Cutting force of high frequency motorized spindle.

When precision machining, the motorized spindle generally works at high frequency/speed. According to the load data and experience, most of the processing technology is milling. The decomposition and synthesis of cutting force in milling is shown in Figure 3 (a) and (b) are 2 types of milling cutter. And the key parameters of cutting fore are as same as parameters in Table 2.

The milling forces, torque and power are expressed as follows:

$$F_c' = \frac{C_F a_p^{x_F} a_f^{y_F} a_w^{u_F}}{d_0^{q_F} n^{w_F}} k_{F_c} \quad (5)$$

$$M = \frac{F_c' d_0}{2 \times 10^3} \quad (6)$$

$$P_m = \frac{F_c' v}{6 \times 10^4} \quad (7)$$

where, F_c' is the peripheral milling force; C_F is the coefficient determined by the workpiece material and cutting condition; x_F , y_F , u_F , q_F and w_F are the indexes of a_p , a_f , a_w , d_0 and n ; k_{F_c} is the correction factor of cutting force; n is the revovling speed; M is the cutting torque; d_0 is the diameter of the milling cutter; P_m is the power of CNC machining center.

4.2 The load spectrum statistic of motorized spindle

According to formulas (1)-(7), based on the data from the field reliability test of certain type CNC machining center, authors calculated the sample data of motorized spindle in work condition. And the results are shown in Table 3. And Figure 4 illustrates the distribution law of rotational speed, torque, radial force and axial force in real work condition.

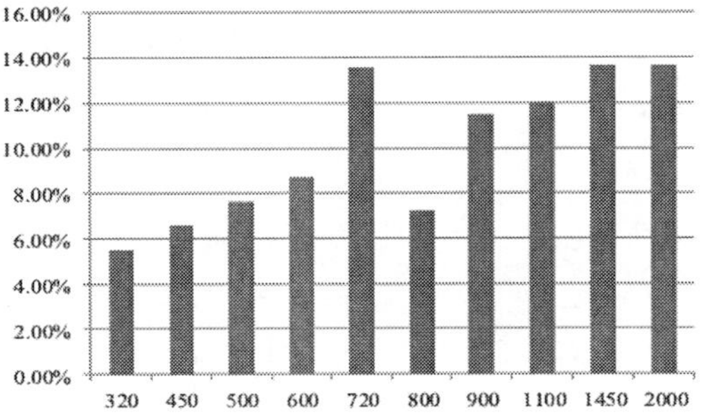

(a) Rotational speed /rpm

DOI: 10.1051/matecconf/20166802001

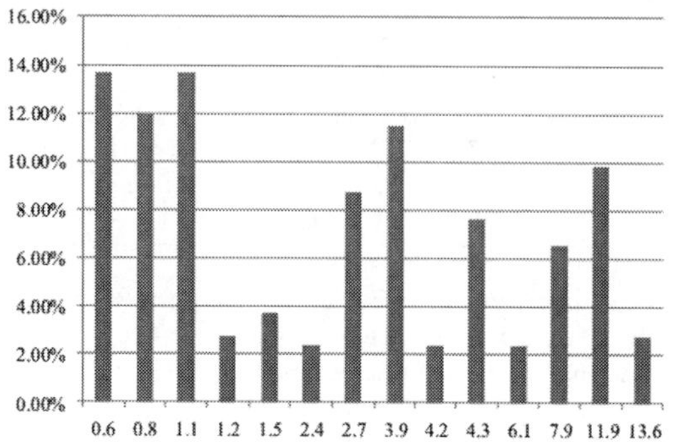

(b) Torque /N

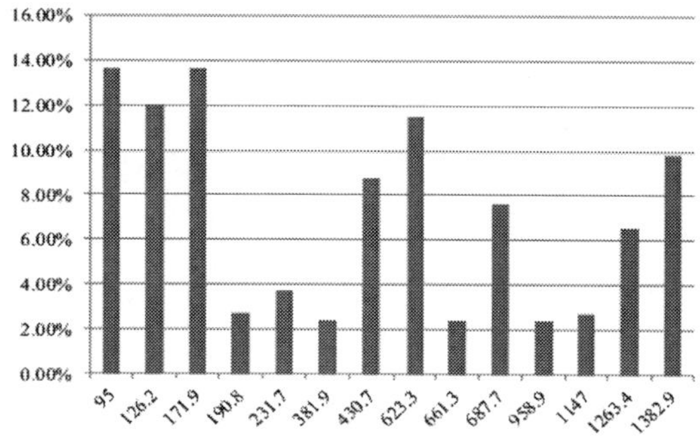

(c) Radial force / N

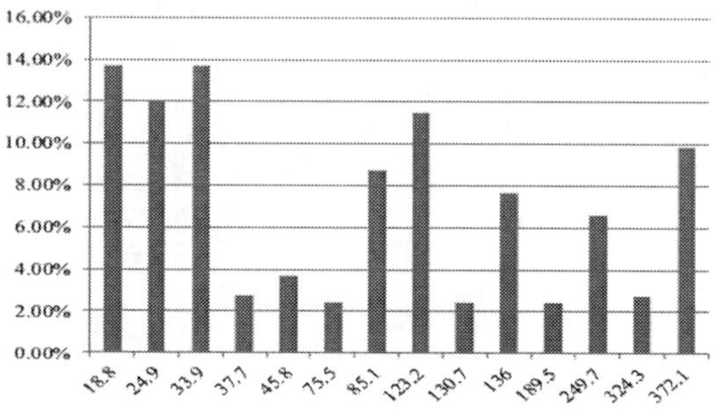

(d) Axial force /N

Figure 4. The distribution of load spectrum for certain type motorized spindle.

According to Figure 4, the processing time percentage of motorized spindle under light torque (3 Nm) is more than 56%, under light radial force (300N) is 45.79%, and under tiny axial fore (80N) is 48.20%. The percentage of motorized spindle in high speed (larger than 1000rpm) is nearly 40%. Combine with the result in Table 3, they reveals that motorized spindle general works at low-speed-high- force (26%) and high-speed-low-force (39%) conditions. To conduct accelerated reliability test of motorized spindle, the proportion of high-speed-high-force should be increased. Then the reliability test time of motorized spindle will be shorten. In the next section, the fuzzy acceleration factor between normal load condition and high load condition is deduced.

Table 3. Load spectrum of certain type motorized spindle.

Rotational speed n /rpm	Torque M /Nm	Radial force F_r /N	Axial force F_a /N	The proportion of cutting time
320	1.2	190.8	37.7	2.732240500%
	13.6	1147.0	324.3	2.732240500%
450	7.9	1263.4	249.7	6.557377000%
500	4.3	687.7	136.0	7.650273000%
600	2.7	430.7	85.1	8.743169000%
720	11.9	1382.9	372.1	9.836066000%
800	1.5	231.7	45.8	3.715846400%
	2.4	381.9	75.5	2.404371200%
	4.2	661.3	130.7	2.404371200%
	6.1	958.9	189.5	2.404371200%
900	3.9	623.3	123.2	11.475410000%
1100	0.8	126.2	24.9	12.021860000%
1450	0.6	95.0	18.8	13.661200000%
2000	1.1	171.9	33.9	13.661200000%

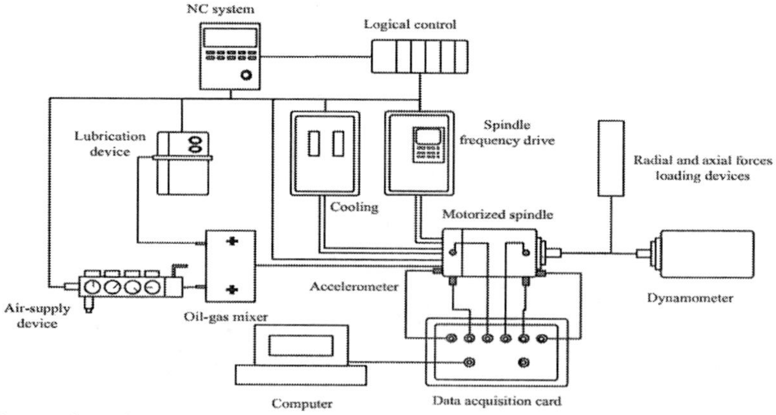

Figure 5. The simulation accelerated test platform of the motorized spindle.

5 Simulation accelerated test platform of the motorized spindle

The authors build simulation accelerated test platform of the motorized spindle which can load radicial force, axial force and torque. Figure 5 shows the skematic diagram of the test platform. The test platform can detect the vibration of bearing, frontend-cover and back end-cover. The radical and axial forces loading devices can apply these 2 forces to the spindle. And the torque are loaded and measured by the dynamometer. Figure 6 is the physical map of the loading devices.

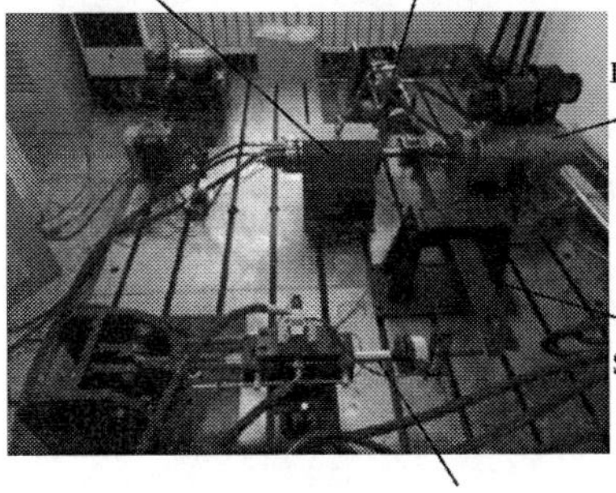

Figure 6. The physical map of the loading devices.

6 Fuzzy acceleration factor

Vibration is an index of the reliability test of motorized spindle [13].Torque, force and rotational speed all directly affect the vibration of motorized spindle. So the paper takes the vibration signal as the fuzzy standard of accelerated reliability test for motorized spindle. Assume the motorized spindle is a non-repairable system and it obeys cumulative damage (CD) law and inverse power law. Namely the life time of motorized spindle obeys the following formula:

$$\frac{T_1}{T_i} = \left[\frac{S_i}{S_1}\right]^N \tag{8}$$

where, T_1 is time to failure (TTF) of motorized spindle under average normal load S_1; T_i is TTF under high load S_i. The high load S_i is the variable of accelerated reliability test, namely the axial fore, radial force, torque and rotational speed on motorized spindle. N is the fuzzy acceleration factor.

However, the vibration and test time T satisfy the following formula:

$$\frac{T_p}{T_o} \propto \frac{V_o}{V_p} \tag{9}$$

where, V_p is the vibration under high load S_i; T_p is the test time when the vibration of motorized spindle is V_p under high load; T_o is the test time when the vibration of motorized spindle is V_p under normal load; V_o is the vibration of motorized spindle after test time T_o under average normal load S_1.

According to formula (8) and (9), the next formula can be obtained:

$$\frac{V_i}{V_1} \propto \left[\frac{S_i}{S_1}\right]^N \tag{10}$$

where, V_1 is the vibration under average normal load S_1; V_i is the vibration under high load S_i. The high load S_i is the variable of accelerated reliability test, namely the axial fore, radial force, torque and rotational speed on motorized spindle. N is the fuzzy acceleration factor.

According to formula (10), the relationship between vibration and load is set. And the fuzzy acceleration factor can be calculated after accelerated reliability test in simulation accelerated reliability test platform of motorized spindle. The series of high load S_i are determined by N and data in Table 3 and Figure 4. Finally, all parameters and test platform are ready and the design of accelerated reliability test for motorized spindle is finished. Next work is to conduct the accelerated reliability test to verify whether the fuzzy accelerated factor is correct and feasible.

7 Conclusions

The accelerated reliability test of the motorized spindle is of great significance to the research on motorized spindle reliability. The main conclusions about this study are as follows:

(1) Based on the field reliability test, authors collect load data in work condition, statistic and calculate the load spectrum;

(2) According to the load spectrum, the time distribution tables of load, such as the force, rotational speed and torque are obtained for CNC machine tools;

(3) Author proposed vibration signal as the index of accelerated reliability test for motorized spindle. The fuzzy accelerated factors on variables, namely force, rotational speed and torque is deduced.

(4) The accelerated test platform of motorized spindle is built which can apply the load spectrum and simulate the work condition of motorized spindle in CNC machining center. The accelerated reliability test also can be carried forward by changing the load which will shorten the reliability test cycle.

In the future, authors will carry out the accelerated test of motorized spindle to validate fuzzy accelerated factors.

Acknowledgements

This study is supported by 1, "Research on the reliability test system of key functional components of CNC machine tools" in special industry base project for Jilin

Province (20130302009GX); 2, "Reliability modeling of machining centers considering the cutting loads (3D513S292414)" in project "Science & Technology Development Plan for Jilin Province".

References

1. E. Abele, Y. Altintas, C. Brecher. CIRP Ann.-Man. Tec **59**,781 (2010)
2. H. Town. *The Design and Construction of Machine Tools*(Iliffe,1971)
3. D. Zhu, H. Liu, D. Yuan, L. Liu. J Mech. Str. **35**,493 (2013)
4. E. Elsayed. *Handbook of reliability engineering*(415-428) (2008)
5. W.Nelson. *Accelerated testing: statistical models, test plans, and data analysis* (Wiley Sons, New Jersey, 2009)
6. Q. William, J. Meeker. Technometrics **26**, 157 (1984)
7. C. Ebeling. *An introduction to reliability and maintainability engineering* (Waveland Press, Long Grove ,2009)
8. B. Lee, Y. Tang. Int. J Adv. Manu. Tech **15**, 238 (1999)
9. C. Lin, J. Tu. J Manu. Pro **9**,94-108(2007)
10. S. Auchet, P. Chevrier, M. Lacour, P.Lipinski. Int. J Mach. Tool. Manu **44**,1441 (2004)
11. Y. Wang, H. Liu, X. Jiang. Mech. Sci. Tech. Aero. Eng. **32**,279 (2013)
12. *MINFRE*(China machine press, Beijing, 2003)
13. X. Jiang, H. Liu, L. Liu, J. Zi. J Vib. Shoc. **32**,80 (2013).

Increasing of the Work Productivity of CMM Machine by Applying of Augmented Reality Technology

Jozef Torok, Marek Kocisko, Monika Teliskova[a] and Miroslav Janak

Technical University of Kosice, Faculty of Manufacturing Technologies with a Seat in Presov, Sturova 31, 080 01 Presov, Slovak Republic

Abstract. This article is focused on the area of metrology and possibility of connection between 3D coordinate measuring machine and augmented reality technology. The aim of this contribution was to bring close the possibilities of augmented reality for example by visualization of important zones during the measurement process. The use of this application is very useful and if the user needs to measure some object after the long time, he does not know, which points were defined for alignment or which position was set for measured object. In this case the use of augmented reality can be very useful.

1 Introduction

The main aim of all companies is to produce profit and for that every company tries to use the contained technology more effective to ensure minimization of production costs of their goods. This situation forces the companies to look for a solution to all steps of production process and also to output control by searching of defects on own products. One possibility for precise check is usage of coordinate measuring machines. They can relatively quickly and reliably detect the necessary dimensions and document compliance of production with manufacturing documentation. Nowadays, for example the automotive industry without such controls we cannot imagine. This group of modern coordinate measuring technique offers the universal usage in control of the various shaped parts. Universality of Coordinate Measuring Machine in short CMM is in accurate measurement on single machine be measuring length dimensions, angles, evaluation of form and location tolerances.

2 Coordinate measuring machine and measurement process

The Faculty of Manufacturing Technologies of Technical University of Kosice with a seat in Presov uses 3D measuring machine Rapid from company Thome Praezision. [10] This machine is equipped with a precise optimized conduction of granite and its thermal stability and high precision of conduction ensure the accurate measurement results without putting the software compensation. The conduct tracks are protected by flexible cover of accordion type. [10]

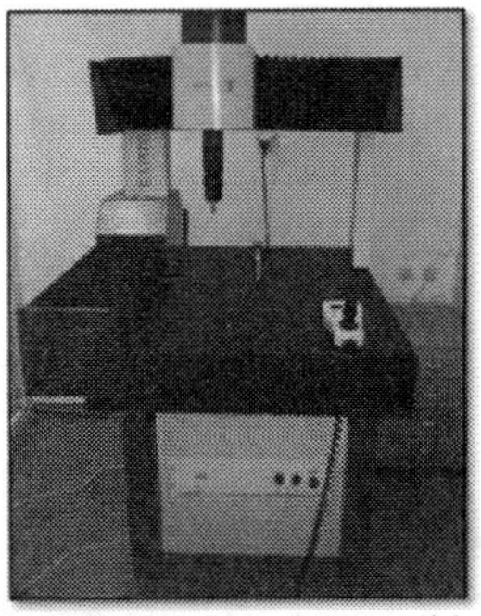

Figure 1. 3D CMM - Thome Praezision rapid.

Each measured object is composed of a number of elements, i.e. theoretical elements such as point, plane, cylinder or sphere in which it is possible to evaluate the several parameters - characteristics. These characteristics may describe one element, for example radius, length, flatness, etc., or may define the relationship between elements. The one element can detect several characteristics, e.g. the circle's diameter, coordinates of the center, roundness,... or the several elements can detect one common characteristic, e.g. angle between the planes or perpendicularity. CMM's can identify the geometry of the measured object by specifying the spatial coordinates of several points of the object. By usage of their information's can then be determined the replacement of object geometry [10].

The first necessary step in the measurement process is always drawing up of measurement process plan. It contains of documentation, clamping method of object, clamping products, etc.... The measurement plan should include a measurement strategy, from which must be

[a] Corresponding author: monika.teliskova@tuke.sk

clear the requirement on the accuracy of measurement and the optimal procedure for recognition of possible disparities 7.

Before each measurement user has to define a coordinate system which is usually somehow relevant to the measured object. By the initial definition of measurement process is always automatically unpacked menu with possibilities of what a user wants to measure, define or construct. Each measured object has a different dimensions and different shape. In the measurement process is necessary to determine what will be measured. During the measurement the measuring head moves towards the object and in the moment when it touches the object at the required point sensor sends an electrical signal with the information of coordinates of a point on object's surface. The measuring head is controlled manually by joystick or can be controlled by the computer (CNC mode).

It is necessary to determine the minimum number of points for definition of this element. For example to measure a circle it is needed to set at least three points but for each geometrical element it is required to set a different number of points depending on the required accuracy. For the most important and used geometrical features are available automatically categories of scanning (point, line, circle, radius, plane, sphere, cone, torus, flat points, cut, rectangle, groove,...).

From the measurement results can be calculated the angle, distance, cutting elements, symmetry,.... There are also shape and position tolerances as: straightness, flatness, roundness, cylindricity, surface shape, parallelism, perpendicularity, angularity, position, coaxiality, concentricity, symmetry, and runout.

2.1 Program

Creation of programs for inspection of components is currently very used mainly for control in mass production of components where individual measurement without some kid of automation would be very time-consuming. It is necessary to set the essential parameters for the correct measurement and then create a program. The basic steps to a good program are basically always the same. In the initial phase before the creation of the program is necessary to define the nature of probe - its diameter and all positions of rotation of measuring head and after that all these positions need to be calibrated.

At the beginning of the measurement it is necessary to define the coordinate system of the measured object. They can be used various methods of alignment: 1 point, geometric alignment, alignment on 6 points, with 3 centres, on the plane and 2 centres, Best-Fit and RPS,..., but the most common is 3-2-1 alignment.

Figure 2. The measured object with the created coordinate system

The first measured object is fixed on the granite table, user defines the object coordinate system - alignment and measures the necessary elements on the object. This first measurement is performed as a classical measure where the user manually selects the element and measures by measuring probe. In the measuring process is very important to define detour points to avoid later the collisions in the automatic measuring process. They are set during the measurement of elements, but also during the move of probe from one element to another. Measurement of this first object is then repeated using a CNC program.

The following objects for sensing are fixed on the granite table of CMM and measurement is done by CNC program (Fig. 3), but at the beginning of each new measurement cycle the needed points for alignment are defined manually by user. The method is advantageous because the user can measure whole series of identical objects.

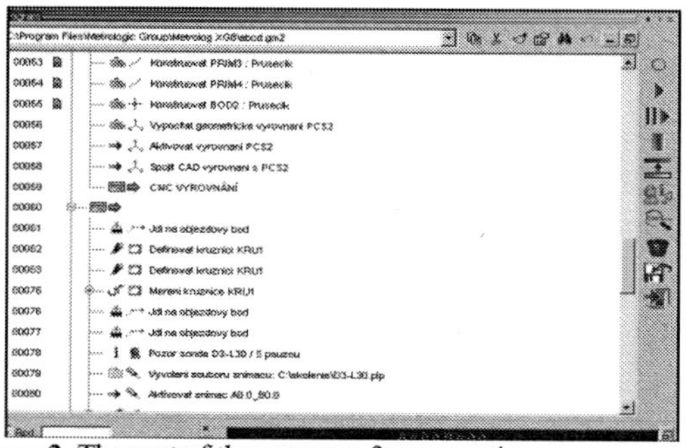

Figure 3. The part of the program for measuring

2.2 Augmented reality and its use in the metrology of CMM

Augmented reality (in short AR) is a live direct or indirect realistic view of the world with augmented elements generated by a computer. The presentation of reality can by enriched with digital elements like images text or nowadays more often by models of objects.

The camera, which is connected to a computing core, senses the fiducially marker (structured image) and special application detects the shooting scene, then places on the image supplementary information (objects, text, video,...) or can launch the pre-programmed actions in real time. For the detection of the object can be used also edge tracking technology which in some cases can be more precise.

The application is trying to identify a marker or object to determine its position and orientation in space of scene. Based on input information the application can add 3D model to the augmented scene at the right place and in the correct orientation. Some software can also "learn" to perform actions for example by tracking user´s hands or face (so called face tracking). These gestures can then be used to run the programmed actions and so to perform sort of application control (proper positioning of the scene).

Correct recognition of marker depends on the quality and resolution of the camera by the right lighting on scene and graphical look of marker. Therefore very often

DOI: 10.1051/matecconf/20166802002

are used simple black and white icons (symbols, QR codes, etc.).

Figure 4. Modeling of new production systems by using AR - solutions elements.

At our Faculty of Manufacturing Technologies with a seat in Presov we use the augmented reality as a support technology in assembling processes applications for example as in the paper "Open Source Tools in Assembling Process Enriched with elements of Augmented Reality" 2 The authors describe the current solutions in these field of AR and represent their own solutions to the position and orientation tracking and move of virtual objects by the equipment working not only on the marker principle. In this article are described Open Source software applications combinated with special rotating position table which connects the proper sensor data and virtual augmented environment elements. Contribution offers some examples of application of augmented reality by hand layering of composite fabrics (Fig. 5) and assembly process (Fig. 6) .

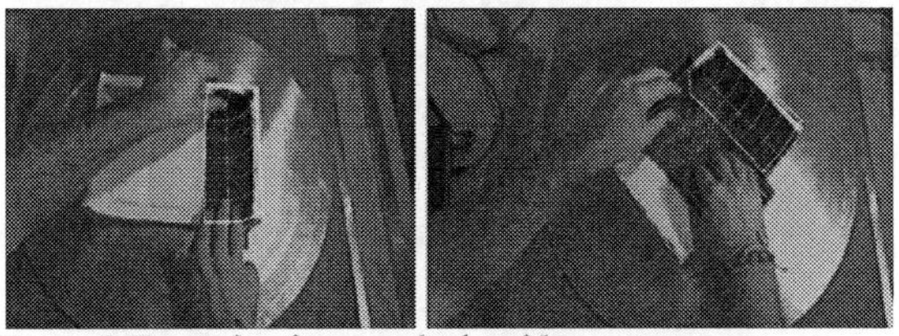

Figure 5. Example of composite hand lay - up process

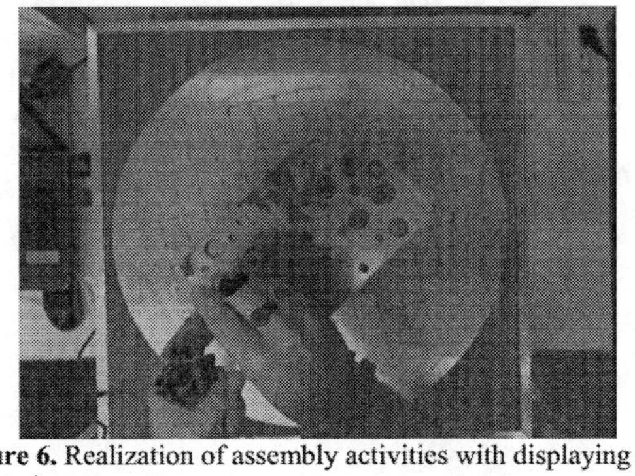

Figure 6. Realization of assembly activities with displaying of inserted parts

One of new trends in the field of metrology is the effort to involve a wider basis of variety specialized techniques suitable for the process of dimensional control of objects where the variety of included devices can simplify and facilitate the control process.

One example is hybrid contact-optical coordinate system, which was specially designed for measuring and analysis of an object geometry. This system combines the accuracy of contact measurement regarding the identification of measurement points and the speed of full-field structured light optical scanning methods. The measurement process starts from optical acquisition by structured light system positioned around the working space and the numerical analysis where is a set of surface points calculated which should be finally re-measured by the CMM. Both set-ups work in a common coordinate system created according to existing metrological strategies. Final metrological analysis is performed using classic certified metrological software. The mentioned article describes also the maximum permissible error of length measurement of the hybrid system 8.

Figure 7. Laboratory of hybrid system

The system can use more common CCD cameras for calibration so the paper 1 evaluates the accuracy of different calibrations and measurement methods used in 3D stereo vision with CCD cameras. This article describes several precision tests for different methods, determining their error limits under specified conditions of operation. Two practical applications are described as cost-effective system for the measurement of free-form surfaces able to generate CAD models and measuring programs for CMMs. The system aims to reduce some difficulties associated with stereo vision and to speed up the traditional digitizing process. The other application involves larger car frame measurement where new automatic measuring system has been developed which allows contactless car frame measurement through two rotating CCD cameras. The more detailed description is presented in 1.

Virtual reality technology in metrology is nowadays mainly used in the simulation of measurement process and for control of potential collision situations of measuring probe and sensed object caused by operator, simulation of hardly reachable access to the required elements of object,...

The most common measuring softwares for CMM offers the possibility of offline programming of CMM, some applications enable the simulation of CMM and for example a human operator or automated system for loading and unloading the measured object. These facts are described in article 6. The paper presents the features and advantages offered by Delmia V5 for simulating a complete measuring environment encompassing: the CMM, the human operator, a robot or automated system for manipulating the parts.

3 The possibility of connection augmented reality and 3D measurement machine

For demonstration of dimensional control and virtual reality technology connection was created real example, which uses 3D CMM Thome Rapid, metrology software Metrolog XG and software based on open source philosophy Blender, for augmented reality. For sensing of real world a classical web camera attached to the monitor was used.

For test measurement a proven component with few base shapes was chosen to provide necessary examples. By on-line programming the CNC program was created to measure the required dimensions. Without the use of the AR technology the program after start required six points for manual definition of object surface and then the automatic control process could start.

For concrete objects can be created applications which are prepared to show position of needed points not only for basic alignment. The real world is captured by the camera mounted on the back of the monitor which shows the augmented scene. So the user sees exactly how the real object is placed on the measuring table and the needed points are shown in order how the user need to define them on the object. The same effect can be provided with usage of HMD glasses.

On the start of the application camera senses the position of the marks, information about location of object placed near of the measured object important for the computing core and so in the next step (in real time) are generated virtual augmented objects for needed points. In this presented solution the measured object must be always placed in the relative same location respect to the marks placed on the table according to proper visualization of needed points. For the calibration process it would be good to use the preparation holes in the measuring table but there may cause confusion because they are same size and placed in regular intervals. Nowadays for faster and more proper placement recognition of measured product team of faculty works on advancer edge based tracking system.

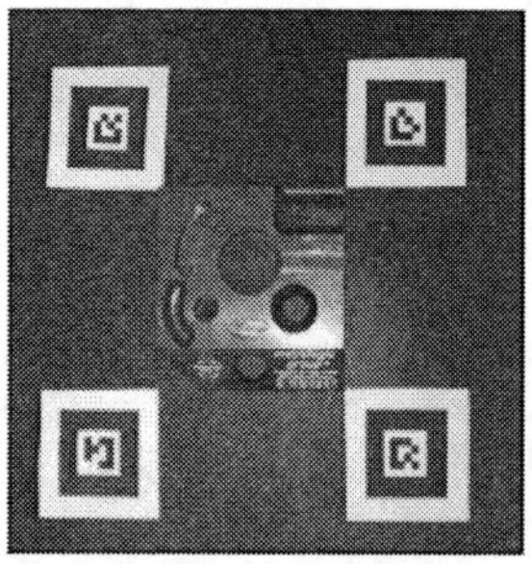

Figure 8. Location of object and calibration marks

Measured object should be placed in an exact fixture which ensures always the same position of the object. If the user measures without the fixtures the calibration marks is needed to place the object in exact location. An example of such a placement is shown in Fig. 10. The component is placed in a position where its three corners are touching the corners of calibration marks. It ensures always the same position of the measured object to the calibration marks so the user gets the correct delineation of the position of the required points needed for alignment of object and so for follow-up measurement.

After the detection of objects location the application shows needed surfaces for the measurement. The augmented surfaces are marked by red colour, three points up on the surface - to define the plane of which is defining the z-axis, 2 points on the side surface - to define a straight line, of which goes the y axis and the last point on the back surface which defines the last coordinate – x axis. This information together defines the coordinate system of the object. Application shows the position of all six points and then displays the animation of probe path to needed positions with defined points from first to the final point.

Fig. 9 shows running application and the calibration of object position by finished six needed points for alignment of the object.

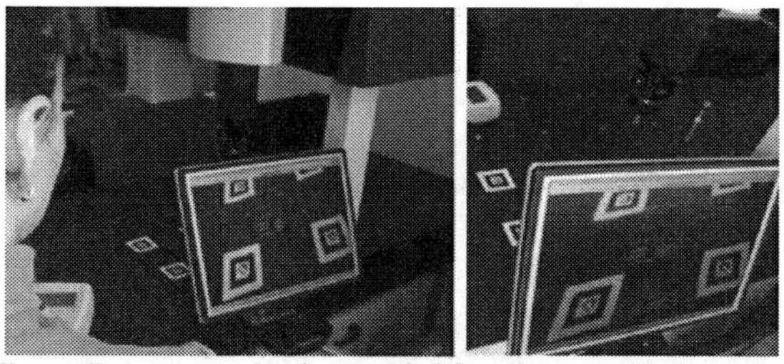

Figure 9. Augmented points needed for aligment of measured object

Fig. 10 illustrates the animation of required points for alignment. On the monitor user sees the real world with added AR application. This application shows him where he needs to define the alignment points. On the monitor screen user sees the real and augmented virtual measuring probe. If the user confirms some points software automatically will show the next points needed to measure.

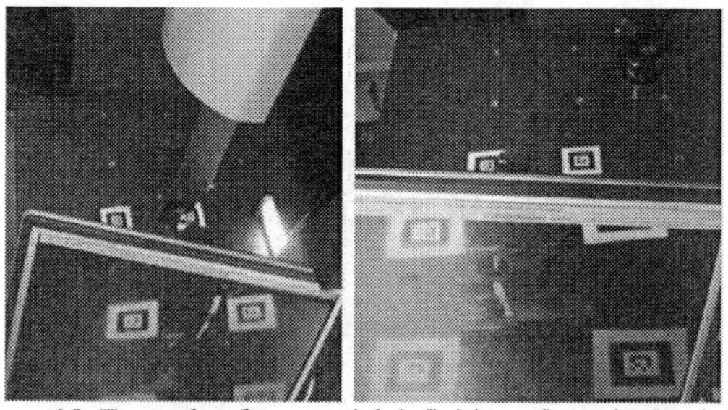

Figure 10. Example of sequential definition of required points

The operator according to the animation defines points and immediately after their definition he can start the automatic process of control which is already done without the operator assistance. The operator only is overseeing the measurement where he can change the speed of the measurement or to stop the process. If the program is correct designed and the object is correct clamped with well defined points during the measurement process shouldn't occur no collisions with the measuring object and measuring probe. Working operator and automatic measurement process on object by CNC program is shown in Fig. 11.

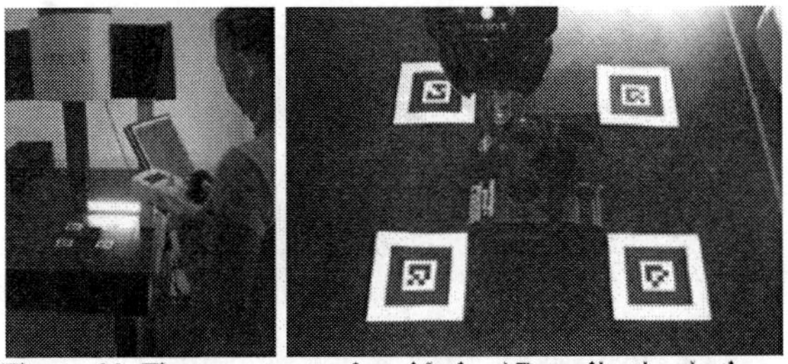

Figure 11. The operator works with the AR application in the measurement process and the automatic measuring of object

By the automatic measurement process we can use augmented reality to show the required values of actual measured dimensions. Fig. 12 shows the measurement of plane and AR application displays the estimated value of flatness for the measured plane according the drawing. Another picture shows the measurement of a circle. The monitor visualizes the real world with application which shows the value of diameter. This application can be used for animation of all important measured dimensions.

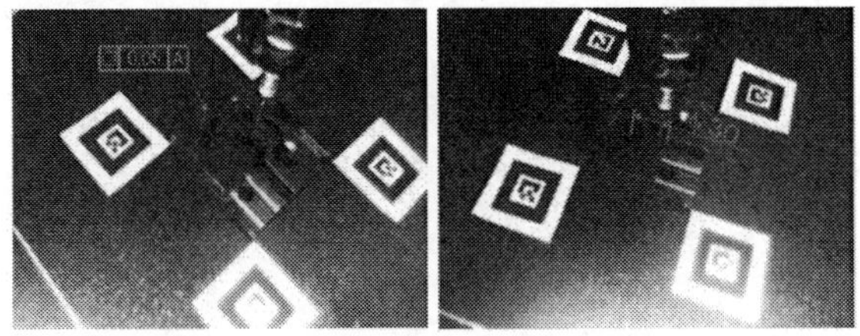

Figure 12. The measuring of flatness and diameter and their visualization with the AR application

4 Consolution

This article is focused on the possibility of improving the programming of CMM by involving augmented reality technology in the control process. Its main aim was to illustrate the points needed for alignment of measured object. If the user wants to measure old objects after a long time he does not know where to define the points, how he created the alignment and important surfaces. This application can be wary helpful not only for novices but also for advanced workers. It shows him where and in what order it is necessary to define these points for alignment and correct fast measurement.

The application during the control process shows the value of actually measured element according the drawing and can provide additional informations as room temperature or date about used probe. During the automatic measuring process the user does not need to follow the drawing but the application displays the values of the currently measured element step by step. This application for CMM was developed at the Faculty of Manufacturing Technologies with seat in Presov in Slovakia.

Acknowledgments

Ministry of Education, Science, Research and Sport of SR supported this work - ITMS project 26220220125.

References

1. J.J. Aguilar, F. Torres, M.A. Lope, Stereo vision for 3D measurement: Accuracy analysis, calibration and industrial applications, Measurement, **18**, 4, 193-200 (1996)
2. J. Barna, J. Novak-Marcincin, M. Janak, L. Novakova-Marcincinova, V. Fecova, J. Torok, Open source tools in assembling process enriched with lements of augmented reality, In: VRIC'12: Virtual Reality International Conference, Laval, France, New York: ACM, 2012, 1-8
3. W. Dangelmaier, M. Fischer, J. Gausemeier, M. Grafe, C. Matysczok, B. Mueck, Virtual and augmented reality support for discrete manufacturing system simulation, Computers in Industry, **56**, 371-383 (2005)
4. J. Dillinger, *et al.*, Moderní strojírenství pro školu i praxy, Praha: Europa –Sobotáles cz. s.r.o., 2007
5. A. Nafi, J.R.R. Mayer, A. Wozniak, Reduced configuration set for the multi-step method applied to machine and probe error separation on a CM, Measurement, **45**, 2321- 2329 (2012)
6. A. Nemtu, D. Hurgoiu, S. Poposcu, M. Dragomir, H. Osanna, Training in coordinate measurement using 3D virtual Instruments, Measurement, **45**, 2346-2358 (2012)
7. J. Novak-Marcincin, V. Fecova, J. Barna, Dimension control by CMM, In: Quality and Innovation in Engineering and Management, Cluj-Napoca: Technical University, 2011, 147-150
8. J. Sladek, P.M. Blaszczyk, M. Kupiec, R. Sitnik, The hybrid contact-optical coordinate measuring system, Measurement, **44**, 503-510 (2011)
9. K. Wendt, M. Franke, F. Härtig, Measuring large 3D structures using four portable tracking laser interferometers, Measurement, **45**, 2339-2345 (2012)
10. T. Präzision. http://www.thome-praezision.cz/1200_700_500.html

DOI: 10.1051/matecconf/20166802003

Band Saw Blade Crack before and after Comparison and Analysis of Experiments (2)

Jin-gui Gao, Zhao-fang Jiang, Jian Zhang, Hui Li and De-jun Shen and Hong-gang Zhao

Beihua University, Jilin, Jilin 132013, P.R.China

Abstract. Based on MJ3310 woodworking band saw machine as the research object, under the no-load and load of Vib system vibration signal acquisition, processing and analysis software of band saw blade transverse vibration test and the signal acquisition and analysis of the collected signals obtained: to determine the transverse vibration displacement 5.66μm ~ 7.86μm and the main vibration frequency between 624 Hz ~ 792 Hz, then saw blade crack at least 3 mm, need timely saw blade, cutting high hardness of wood band saw blade transverse vibration displacement and frequency will increase sharply. Can be generated according to the band saw blade crack before and after the changing rule of the horizontal vibration displacement and frequency of transverse vibration and scope, judgment and replacement time of saw blade saw blade defect types, which can fully rational utilization of saw blade work effectively.

1. Introduction

In order to eliminate or reduce safety accidents, improve labor productivity, improve the quality of band saw machine for processing, it is necessary for band saw blade falls from the teeth of the reasons and rules, forecast and prevention measures, work safety evaluation of multiple system research. The study of fallen tooth formation has the most overall and basic, it is the study of band saw blade is foundation of work safety, but also to build the core content of safety theory of band saw machine [1]-[3].

Therefore, this paper intends to band saw blade tooth loss were studied and summarized its cause, and changes to the rules, and deeply understanding out the intrinsic relationship between the teeth of each factor, and blade breakage mechanism, for improving the quality of sawn timber processing, the protection of the ecological environment, and prevent safety accident occurrence, reduce safety accident frequency and damage, ensure the safety of the lives and property of the people, have a very important practical significance [4]-[8].

2. Experimental

2.1 Laboratory instruments and equipment

2.1.1 Laboratory instruments

The following experimental instrument: Use of equipment and instruments are as follows: Data acquisition instrument WS-5942-2-50 Spectrum of Beijing century science and technology Development limited Woodworking band saw machine MJ3310 Foshan new woodworking machinery equipment company.

Instruments used in the experiment connection diagram is shown in Fig. 1.

2.1.2 Equipment

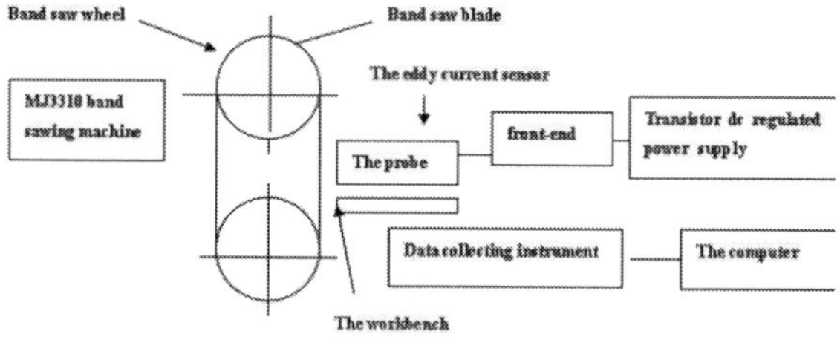

Figure 1. The main instruments and equipment connection diagram

2.2 Device and its parameter

2.2.1 Machine composition

The experimental research of forestry college at the BEIHUA University of woodworking machinery laboratory MJ3310 cars woodworking band saw. Its structure by saw wheel up and down, band saw blade, guide, tensioning system, power system, frame, security cover and other components. As shown in Fig. 2.

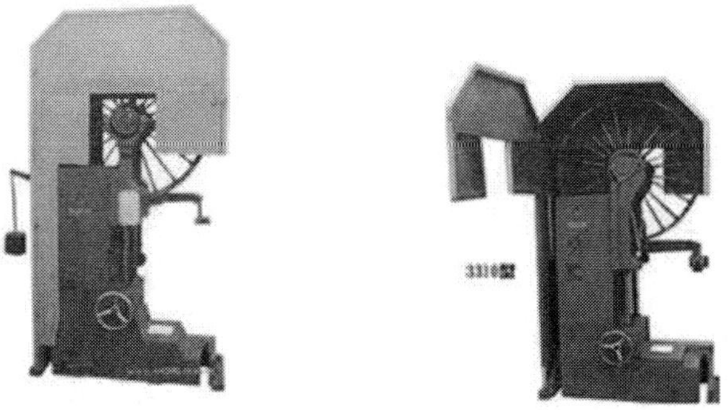

Figure 2. Wood working band saw machine

2.2.2 Machine parameters

Equipment: Fine woodworking band saw machine Type: MJ3310 Saw blade diameter: 1000mm Saw blade wheel speed: 650rpm Motor speed: 1400rpm Saw blade parameter: 7200mm x 120mm x 1.12mm (length x width x thickness GB/T 21690-2008 Fine woodworking band saw blade dimensions)

3. Results and discussion

3.1 Influence of wheel spindle speed of saw blade vibration falling tooth defect

In the band saw blade tension is 142 N, under the premise of belt tensioning force is 1435N, use of custom level 3 cone and the original belt pulley, adjusting the five speed, respectively on the full blade and two tooth defect of the saw blade for data acquisition, will now and with the increase of saw wheel spindle speed saw blade transverse vibration displacement collected data into curve as shown in Fig. 3. And will collect the data after the power spectrum analysis, get the frequency data after finishing drawing into curve as shown in Fig. 4.

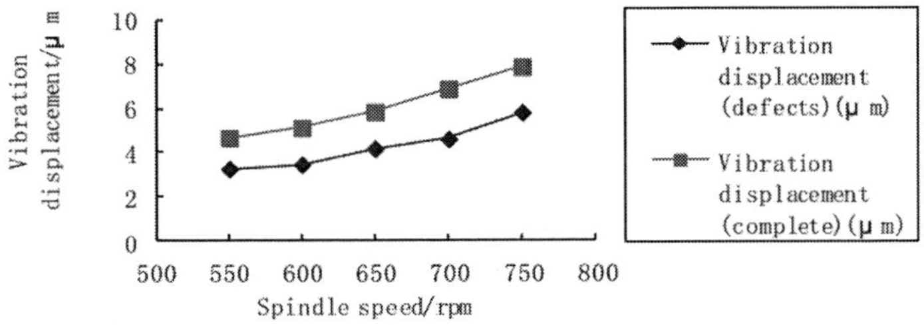

Figure 3. Different saw when complete blade wheel spindle speed and blade transverse vibration displacement of tooth defect

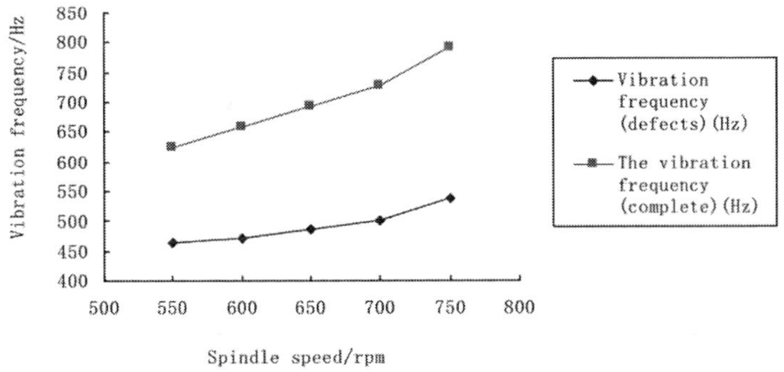

Figure 4. Sindle speed impact on the normal and defect of saw blade transverse vibration displacement

The Fig. 3 shows that,also saw, wheel spindle speed impact on the size of the band saw blade transverse vibration displacement is very obvious, which saw wheel spindle speed is 831 rpm the horizontal vibration displacement minimum, to achieve the best condition, and the other saw wheel spindle speed corresponds to the horizontal vibration displacement increases apparently. When band sawing machine saw wheel spindle speed is 550 rpm, the horizontal vibration displacement smaller, that speed (less than 550 rpm) than at high speed (greater than 650 rpm band saw blade transverse vibration displacement is small. Compared to two curves, the drop tooth defect of band saw blade transverse vibration displacement is slightly superior to complete band saw blade. Its drop tooth saw blade transverse vibration displacement between 5.66μm~7.86μm, through the

DOI: 10.1051/matecconf/20166802003

observation curve, shown in Fig. 4 will find in the band sawing machine high speed (greater than 700 rpm), drop tooth vibration main frequency band saw blade, almost no change in speed, both are the obvious contrast, produce tooth defect of band saw blade vibration main frequency will be lower than the whole band saw blade of about 50 Hz. The drop tooth saw blade vibration main frequency between 624 Hz and 792 Hz.

3.2 The influence of different tree species of band saw blade transverse vibration cutting

Blade in different kinds of wood with the case, we will collect the data of transverse vibration of band saw blade. When sawing, in determining the speed of the spindle 650 rpm, band saw blade for the integrity of the state and the band saw blade with 3 mm crack state, band saw blade tensioning force 1435N, speed and feed for 15 m / min, in belt tension tension on the premise of, respectively, the choice of the cutting of Larch and color wood, respectively, after whole band saw blade and crack of the band saw blade cutting data collection and complex analysis, under different conditions with saw blade lateral vibration displacement and vibration frequency of the data as shown in Table 1.

Table 1. Cutting blade transverse vibration displacement and main frequency contrast different tree species

Tree species	Larch (μm)	Color wood (μm)	Larch (Hz)	Color wood (Hz)
Have a crack defects of band saw blade transverse vibration displacement and main frequency	5.82	7.97	770	880
Complete blade vibration displacement main frequency	4.55	5.45	550	650

By Table 1 comparative analysis, we can know that, when sawing, the vibration displacement of the larch wood less than maple, which is due to saw wood hardness, no load is so. The increase of Larch bandsaw blade vibration displacement vibration displacement of the 1. The vibration displacement of transverse vibration displacement defects cutting band saw blade was significantly higher than that of no-load. From the frequency of the main value analysis and comparison shows that, the vibration frequency of the larch is higher than the color of wood cutting saw blade vibration frequency, which is due to an increase in the hardness of cutting wood, to increase the frequency of vibration; blade vibration frequency of the load is greater than that of empty vector of saw blade vibration frequency. In sawing process, when the vibration displacement of the rigid contact with the blade is larger.

4. Conclusions

Through the above type of MJ3310 sports car woodworking band saw band saw blade vibration analysis,

(1) using a 120 mm wide band saw blade, the impact of band saw blade vibration displacement and vibration frequency of the main factors are: saw wheel spindle speed, the blade tension, feed speed

(2) if the band saw blade vibration displacement between 5.66μm ~ 7.86μm and vibration frequency between 624 Hz ~ 792 Hz, band saw blade at least 3 mm crack defects, need to replace the band saw blade in time;

(3) cutting high hardness of wood can make the band saw blade transverse vibration displacement and frequency has increased dramatically, has the obvious difference compared with the perfect blade when idle, can according to the band saw blade transverse vibration displacement and frequency range of judging defect types and development degree, thus, can replace the saw blade scientifically, effectively use band saw blade.

Funding for the subject project

This topic by the Chinese national natural science fund project funding, its national fund name: based on the analysis of time series of woodworking band saw blade tooth crack and damage formation mechanism and prediction theory (31570556).

References

1. Guo-xi Zhu, Hua-bin Wang et al, Research on the production line of Chinese modern material production line. Harbin, Northeast Forestry University press, September, pp. 2-10, (1989)
2. Jin-gui Gao, Jian-guo Yu, Guo-xi Zhu, Xiao-min, Lai, Jian-hua Dou, The elastic dynamics analysis of band saw Tightening system. Journal of Forestry Research. **11**,pp. 27-32 (2000)
3. Jin-gui Gao, Band sawing machines for mechanical computer simulation and experimental study of tensioning system [dissertation], Northeast Forestry University. Library (2001)
4. Fuzhou Woodworking Machine Tool Research Institute, A foreign woodworking machine tool machinery industry basic situation. Beijing: Mechanical Industry Press, (1989)
5. [Su] Xie Lev, Effect of technical state of Des sawing machine for cutting process dynamics. (Translated) woodworking machine, **4**, pp. 8-42, (1989)

6. Zhang Tieling, The band saw blade tensioning force influence on the stability of the saw blade experimental research dissertation. Northeast Forestry University Library, (1988)
7. Wang Huabin, The band saw blade saw hanging position effect of stress distribution and stability of wood industry of blade width, **4**, pp. 5-11, (1989)
8. Pu Yongshou is in wood cutting. Fukui. At resistance in the analysis of the tool (first report), Journal of wood, **5**, pp. 46-51, (1984)

MATEC Web of Conferences **68**, 02004 (2016) DOI: 10.1051/matecconf/20166802004
ICIEA 2016

Meshing Stiffness Calculation and Analysis of the Cycloidal Pump's Rotor

Liu M. T. and Xue Y. N.

Tianjin University of Science & Technology, Tianjin, China

Abstract. In order to calculate the meshing stiffness of the cycloidal pump's rotor, a method of relationship that deduces two generating curtate epicycloid is proposed, the radius values of the inner rotor's actual tooth profile curvature was calculated and analyzed; Based on the formula of Hertz, the cycloidal pump rotor's meshing stiffness calculation formula is inferred, the changing curve of cycloidal pump rotor's meshing stiffness is made, some important parameters on its meshing stiffness are studied, the maximum value of meshing stiffness is when the outer rotor mesh with inner rotor's outer convex profile curve, and the minimum value is when the outer rotor mesh with the center of inner rotor's inner concave.

1 Introduction

Cycloidal pump is also called the cycloidal gear pump. Compared with other ordinary gear pumps, it has lots of advantages,such as compact structure, low noise, good self-priming performance, and suitable for high speed etc, and developed rapidly in recent years, widely used in food, textile, chemical, machinery and other industries. To accurately calculate the contact stress produced by the inner and outer rotor in cycloidal pump, the computation and analysis of the meshing stiffness is indispensable.

Reference [1] and [2] regard tooth inter-atomic forces and the displacement of mesh point in normal direction as direct ratio, which means the meshing stiffness is a fixed value. Due to the number of gear teeth of cycloidal pump rotor in the process of meshing is periodical change, its meshing force between the rotors is periodical. at the same time, the rotor tooth of the cycloidal pump's meshing stiffness is also cyclical change, solving the meshing stiffness of cycloidal pump rotors plays an important role in improving the transmission accuracy and analysis of the machine dynamic performance. Reference [3] formulates the mesh stiffness model of the RV reduction gears cycloidal-pin wheel's transmission pair based on Hertz formula and involute gear's transmission pair based on Shi Chuan formula, respectively. Reference [4] has an analysis and correction of single tooth meshing stiffness in reference [3], getting a new stiffness calculation formula; Reference [5] calculate ring-plate-type gear planetary drive's meshing stiffness by a tooth profile meshing analysis method taking clearance into account.

This paper researches the derivation relationship between two methods of generating short of epicycloid method, and then establishes the model of cycloidal pump rotor's meshing stiffness, getting each meshing tooth's meshing stiffness, and analyses its influencing factors.

2 The formation of curtate epicycloid

Cycloidal pump rotor pair is composed of a couple of internal meshing rotor as shown in Fig. 1. The outer rotor is internal gear and tooth profile curve is a certain number of equal distribution of circular arc curve; The inner rotor is external gear and its tooth profile curve is the equidistant curve of curtate epicycloid. There are two kinds of curtate epicycloid forming method: the forming method of two circle's inner gearing and the forming method of two circle's outer gearing [6], [7].

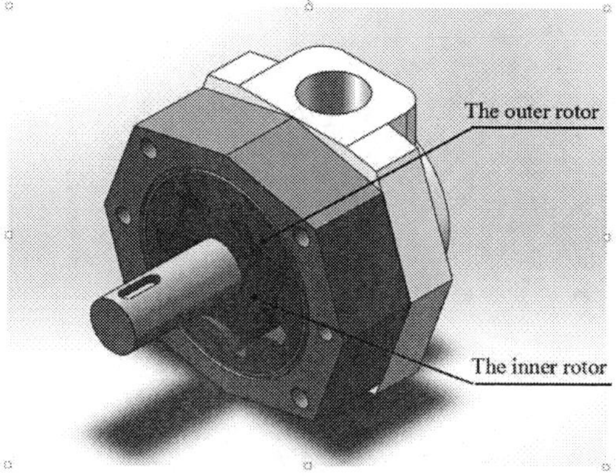

Figure 1. The cycloidal pump.

2.1 The relationship between two kinds of forming method of the cycloidal tooth profile.

Two ways of forming cycloidal tooth profile are different, but as long as reaching a certain conditions, it can form the same curtate epicycloid. This paper proposes a new proof method on the basis of reference [6] and [7] as shown in Fig. 2:

Point O_b, O_g is the center of circle 1 and circle 2, respectively. r_b, r_g is the radius of circle 1 and circle 2, respectively, what' s more, $r_b > r_g$, circle 1 and circle 2 inscribes. Point P_2 is the tangency point, draw a circle tangent to circle 1 (circle 2) go through point P_2; draw a circle using O_b as the center and $Rr = R + r$ as the radius, the above two lines intersect at point M. It can regard the track of point M as the track of point M connected to the rolling outside in fixed plane trajectory, when a radius of r_b rolling goes along the base circle with a radius of r_g for pure roll.

Connect the line O_bM and P_2M, respectively. Draw the parallel line of line O_bM through point O_g; Draw the parallel line of line O_bO_g through point M, the above two lines intersect at point O. It is easy to know that quadrilateral OO_gO_bM is a parallelogram. Draw a circle using O_g as the center and R as the radius; Draw a circle using O_g as the center and R as the radius; Therefore, it can also regard the track of point M as the track of point M connected to the rolling outside in fixed plane trajectory, when a radius of r rolling goes along the base circle with a radius of R for pure roll, the eccentricity $e = r_b - r_g$.

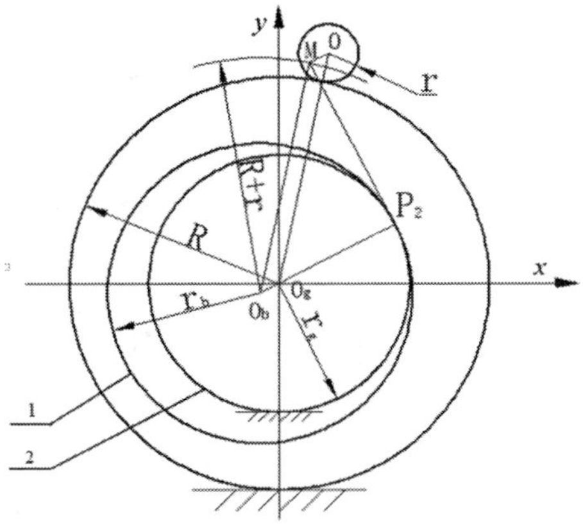

Figure 2. The Formation method of two kinds of curtate epicycloid.

Because of the quadrilateral OO_gO_bM is a parallelogram, two rolling radius r_b, r and two base circle radius r_g, R have the following relationships:

$$\triangle O_gP_1P_2 \backsim \triangle OP_1M \backsim \triangle O_bMP_2 \tag{1}$$

$$\frac{r_b}{O_bM} = \frac{r_b}{Rr} = \frac{e}{r} = \frac{r_g}{R} \tag{2}$$

Thus, the forming method of two circle's inner gearing and the forming method of two circle's outer gearing can be formed the same curtate epicycloid as long as meeting the above conditions.

2.2 The curve equation of inner rotor

The curve equation of the inner rotor's theoretic tooth profile can be expressed as:

$$\begin{cases} x = Rr\sin\theta - e\sin(Z_2\theta) \\ y = Rr\cos\theta - e\cos(Z_2\theta) \end{cases} \tag{3}$$

The curve equation of the inner rotor's actual tooth profile can be expressed as:

$$\begin{cases} X = Rr\sin\theta - e\cdot\sin(Z_2\theta) + a\cdot\dfrac{Z_2 e\sin(Z_2\theta) - Rr\sin\theta}{\sqrt{Z_2^2e^2 + Rr^2 - 2RreZ_2\cos(Z_1\theta)}} \\ Y = Rr\cos\theta - e\cdot\cos(Z_2\theta) + a\cdot\dfrac{Z_2 e\cos(Z_2\theta) - Rr\cos\theta}{\sqrt{Z_2^2e^2 + Rr^2 - 2RreZ_2\cos(Z_1\theta)}} \end{cases} \tag{4}$$

In the equation: Rr is the outer rotor tooth profile uniform circular radius, $Rr = R + r$, Z_2 is the tooth number of outer rotor, Z_1 is the tooth number of inner rotor, e is the eccentricity, a is the outer rotor tooth profile radius, θ is the parameter.

Fig. 3 shows that the inner rotor's practical and theoretical contour line with Rr =55mm, Z_2 =7, e =5mm, a =20mm:

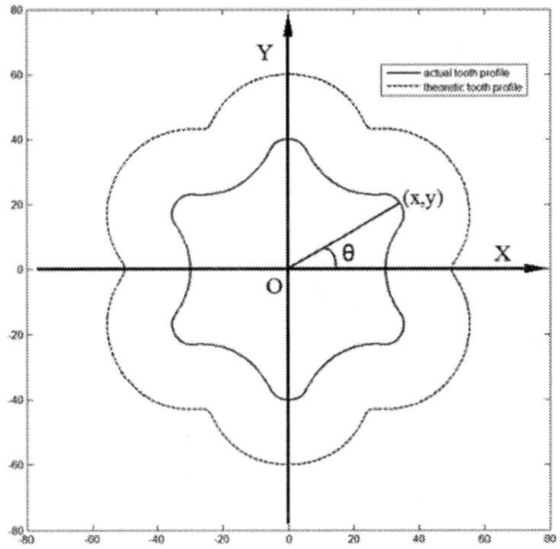

Figure 3. Practical and theoretical contour line of inner rotor

2.3 The curvature radius of the inner rotor's actual contour

The inner rotor's actual contour has outer convex and inner concave two cases as shown in Fig. 3. The curvature radius of the inner rotor's actual contour is equal to the curvature radius of the inner rotor's theoretic tooth profile plus the outer rotor tooth profile radius, then

$$\rho_b = \rho_a + a \tag{5}$$

Here:

$$\rho_a = Rr \cdot \frac{S^{3/2}}{T}$$

$$S = 1 + k^2 - 2k\cos(\theta - Z_2\theta)$$

$$T = k(1+Z_2)\cos(\theta - Z_2\theta) - (1 + Z_2 \cdot k^2)$$

In the equation: ρ_a is the curvature radius of the inner rotor's theoretic tooth profile, ρ_b is the curvature radius of the inner rotor's actual contour, k is the curtate ratio.

Fig. 4 shows that the curvature radius curve of the inner rotor's actual contour with k =0.64, Z_2 =7, a =20mm, Rr =55mm:

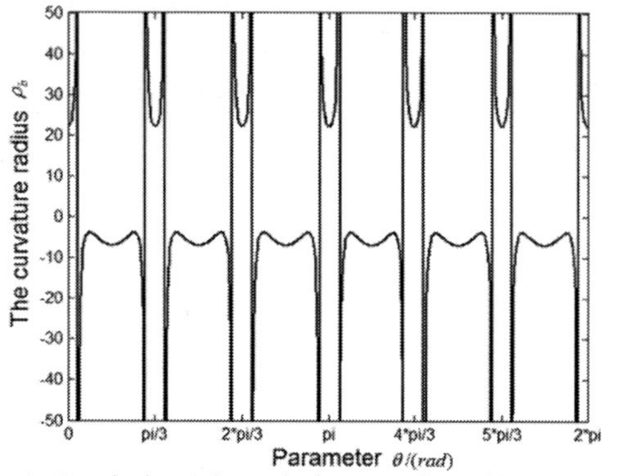

Figure 4. Practical and theoretical contour line of inner rotor.

Due to the curve equation of the inner rotor's actual tooth profile appears outer convex and inner concave two cases, alternately. The radius of curvature value also appears plus and minus two cases, alternately. The value is located at $k\frac{\pi}{3}+\frac{\pi}{270}, k=0,1,\ldots,5$ or $k\frac{\pi}{3}-\frac{\pi}{270}, k=1,2,\ldots,6$, in addition, this process requires very little time, it can approximate think that the curve has no sharp point and doesn't produce cross phenomenon. The radius of curvature value is positive on the inner concave, the minimum value is located at $k\frac{\pi}{3}, k=0,1,\ldots,5$, the radius of curvature value is negative on the outer convex, the minimum value is located at $k\frac{\pi}{3}+\frac{\pi}{40}, k=0,1,\ldots,5$ or $k\frac{\pi}{3}-\frac{\pi}{40}, k=1,2,\ldots,6$; Compared to the inner concave, the outer convex part's amplitude of variation is more less, the process is longer, and has two peaks of wave in a cycle.

3 The meshing stiffness between the cycloidal rotors

3.1 The analysis of the deformation length

The tooth profile curve what the outer rotor is meshing is the circular arc curve as shown in Fig. 1, and can be seen as cylinder. The tooth profile curve what the inner rotor is meshing has outer convex and inner concave two cases. The mesh transmission between the cycloidal rotors proceeds by line contact.In consideration of elastic deformation, the elastic deformation in the contact is a small area regional, the contact segment between inner rotor and outer rotor is arc. With regard to the mesh between inner rotor and outer rotor, it can assume that the deformation length is $2L$, the tooth width is B as shown in Fig. 5, so it can be calculated by Hertz formula[4].

$$L = \sqrt{\frac{4F}{\pi B} \cdot \frac{\frac{1-\mu_1^2}{E_1} + \frac{1-\mu_2^2}{E_2}}{\frac{1}{\rho}}} \tag{6}$$

Here:

$$\frac{1}{\rho} = \frac{1}{a} \pm \frac{1}{|\rho_b|}$$

When the inner rotor meshing profile is outer convex, the inner rotor and outer rotor is external contact, where the $\pm$ choose +.

When the inner rotor meshing profile is inner concave, the inner rotor and outer rotor is internal contact, where the $\pm$ choose -.

In the equation: a is the radius of curvature of outer rotor at contact point, that's the outer rotor tooth profile radius, F is the force at contact point, μ_1 , μ_2 is the cycloid inner rotor and outer rotor material's poisson ratio, E_1 , E_2 is the cycloid inner rotor and outer rotor material's elasticity modulus, B is the rotor tooth width.

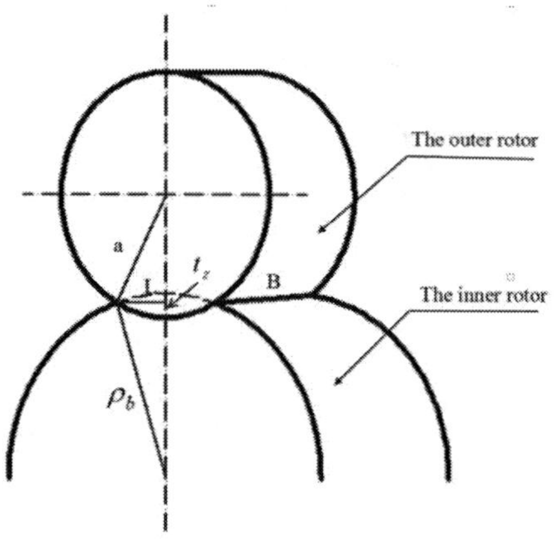

Figure 5. The extrusion deformation of the inner and outer rotor.

Due to the cycloid inner rotor and outer rotor material is the same, it can define $\mu = \mu_1 = \mu_2$, $E = E_1 = E_2$, then the (6), can be simplified as follows:

$$L = \sqrt{\frac{8F\rho\,(1-\mu^2)}{\pi BE}} \tag{7}$$

3.2 The calculation of meshing stiffness

3.2.1 The mesh between the outer rotor and inner rotor's outer convex profile curve

The mesh between the outer rotor and inner rotor's outer convex profile curve. From the analysis of single outer rotor meshing stiffness in Fig. 5:

$$(a-t_Z)^2+L^2=a^2 \tag{8}$$

In the equation: t_Z is the extrusion deformation degree of the outer rotor in the radial direction.

$$t_Z=a\cdot[1-\sqrt{1-(\frac{L}{a})^2}] \tag{9}$$

Use Taylor formula at the point $(\frac{L}{a})^2=0$, ignore the higher order infinitesimal, and put the (7), into it:

$$t_Z=\frac{4F\rho(1-\mu^2)}{\pi BEa} \tag{10}$$

Therefore, single outer rotor meshing stiffness is obtained:

$$C_Z=\frac{\pi BE}{4(1-\mu^2)}\cdot\frac{RrS^{3/2}+2Ta}{RrS^{3/2}+Ta} \tag{11}$$

Similarly, use Taylor formula at the point $(\frac{L}{\rho_b})^2=0$, ignore the higher order infinitesimal, and single inner rotor meshing stiffness is obtained:

$$C_b=\frac{\pi BE}{4(1-\mu^2)}\cdot\frac{RrS^{3/2}+2Ta}{Ta} \tag{12}$$

In conclusion, single inner rotor and outer rotor meshing stiffness can be written as

$$K_{outer}=\frac{C_bC_Z}{C_b+C_Z}=\frac{\pi BE}{4(1-\mu^2)} \tag{13}$$

3.2.2 The mesh between the outer rotor and inner rotor's inner concave profile curve

From the analysis of single outer rotor meshing stiffness in Fig. 5:

$$(a-t_Z)^2+L^2=a^2 \tag{14}$$

In the equation: t_Z is the extrusion deformation degree of the outer rotor in the radial direction.

$$t_Z=a\cdot[1-\sqrt{1-(\frac{L}{a})^2}] \tag{15}$$

Use Taylor formula at the point $(\frac{L}{a})^2=0$, ignore the higher order infinitesimal, and put the (7), into it:

$$t_Z=\frac{4F\rho\ (1-\mu^2)}{\pi BEa} \tag{16}$$

Therefore, single outer rotor meshing stiffness is obtained:

$$C_Z=\frac{\pi BE}{4(1-\mu^2)}\cdot\frac{RrS^{3/2}}{RrS^{3/2}+Ta} \tag{17}$$

Similarly, use Taylor formula at the point $(\frac{L}{\rho_b})^2=0$, ignore the higher order infinitesimal, and single inner rotor meshing stiffness is obtained:

$$C_b=\frac{\pi BE}{4(1-\mu^2)}\cdot\frac{RrS^{3/2}}{Ta} \tag{18}$$

In conclusion, single inner rotor and outer rotor meshing stiffness can be written as

$$K_{inner}=\frac{C_bC_Z}{C_b+C_Z}=\frac{\pi BE}{4(1-\mu^2)}\cdot\frac{RrS^{3/2}}{RrS^{3/2}+2Ta} \tag{19}$$

3.2.3 The cycloid rotors meshing stiffness

From the above, the cycloid rotors meshing stiffness K_s is given as

$$K=\begin{cases}\frac{\pi BE}{4(1-\mu^2)}, & mesh\ with\ the\ innerrotor's\ outer\ convex\\ \frac{\pi BE}{4(1-\mu^2)}\cdot\frac{RrS^{3/2}}{RrS^{3/2}+2Ta}, & mesh\ with\ the\ innerrotor's\ inner\ concave\end{cases} \tag{20}$$

4 Analysis Example

According to (20), the rotor tooth width B, the cycloid rotor material's poisson ratio and elasticity modulus determines the maximum mesh stiffness between cycloid rotors. The rotor tooth width B and meshing stiffness are proportional relationship, poisson ratio μ and elasticity modulus E ratio is associated with the material of cycloid rotors. The majority of cycloid pumps' inner rotor teeth is 6 in practical application, this paper uses this kind of cycloidal pump to analyze. According to reference [8], the better value range of curtate ratio is from 0.50 to 0.75. In this range, cycloidal pump has higher bearing capacity, less wear and tear. The selection of system parameters is shown in Table 1.

Table. 1 The design parameters

Name	Symbol	Number
inner rotor's tooth number	Z_1	6
outer rotor's tooth number	Z_2	7
the eccentricity (mm)	e	5
the rotor tooth width (mm)	B	44
poisson ratio	μ	0.25
elasticity modulus (Pa)	E	115x109
the outer rotor tooth profile uniform circular radius (mm)	Rr	Z2e/k

DOI: 10.1051/matecconf/20166802004

As shown in Fig. 6, use the curtate ratio k=0.55, k=0.64, k=0.73 as variables to make a , b , c three curve. The theoretical value of the meshing stiffness cycle is $\pi/3$, the meshing stiffness cycle in Fig. 6 is also $\pi/3$. When the outer rotor mesh with inner rotor's outer convex profile curve, the meshing stiffness is constant, as well as the maximum value, the minimum value is when the outer rotor mesh with the center of inner rotor's inner concave, it's $\theta = k\frac{\pi}{3}+\frac{\pi}{6}, k=0,1,\ldots,5$. In a cycle, the meshing stiffness remains the same in most of the time, its value varies considerably in other time, the greater the curtate ratio, the smaller the minimum value of meshing stiffness, and smaller variation as well.

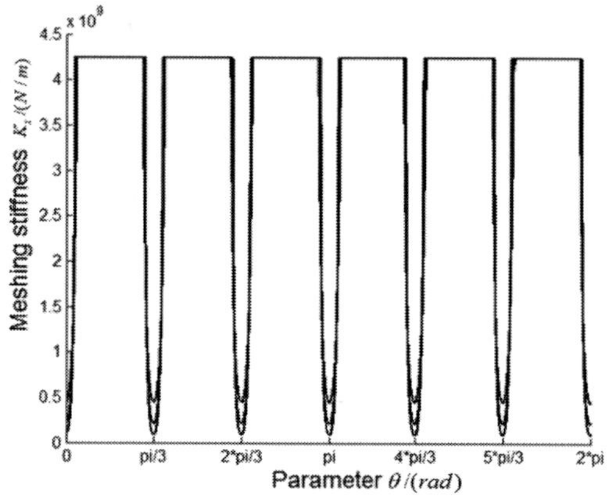

Figure 6. The meshing stiffness changing curve of different curtate ratio.

5 Conclusions

This paper researches the cycloidal pump, and gets the following conclusions:

Proposes a new method of relationship that deduces two generating curtate epicycloid.

Deduce the method of cycloid pump rotor's meshing stiffness, and get the rotor tooth width and meshing stiffness are proportional relationship.

The maximum value of meshing stiffness is when the outer rotor mesh with inner rotor's outer convex profile curve, and the minimum value is when the outer rotor mesh with the center of inner rotor's inner concave,it's $\theta = k\frac{\pi}{3}+\frac{\pi}{6}, k=0,1,\ldots,5$. The greater the curtate ratio,the smaller the minimum value of meshing stiffness, and smaller variation as well.

References

1. Cao R. , Wuhan University of Technology, (2006).
2. Wu D. , Zhang Y. C. , Coal Mine Machinery. **29**, 60-62, (2008).
3. Zhang D. W. ,Wang G. ,Huang T. , Journal of Mechanic Engineer. **37**, 69-74, (2001).
4. Zhang Y. H. , Xiao J. J. ,He W. D. , Journal of Dalian Jiaotong University. **31**, 20-23, (2010).
5. Shan L. J. , Tian L. Y. , Guan T. M. , He W. D. , Journal of Dalian Jiaotong University. **29**,14-16, (2010).
6. Rao ZH. G. , *The Design of Planetary Transmission Mechanism*, (National Defend Industry Press. 1994).
7. An Z. J. , Yanshan University. (2000).
8. Mao H. Y. , Li G. X. , Liu H. T. , Hu Y. P. , Wang W. , Transactions of the Chinese Society of Agricultural Machinery. **37**, 45-47, (2006).

Research for Billet Limited Weight Cutting Based on Behavior Stateflow

Wang Fubin , Xu Ao, Zeng Kai, Chen Zhikun, Zhou Yaluo

Electrical Engineering College, North China University of Science and Technology. 063009 Tangshan, China

Abstract. It is critical to improve the yield of rolled products and reduce the energy consumption based on billet limited weight cutting. The effect on billet limited weight by dimensional accuracy of billet image and billet casting speed is analyzed, and the behaviour control architecture of billet limited weight cutting is established. The goals of billet limited weight cutting is decomposed into cutting tasks and cutting actions based on finite-state machine theory. The theoretical weight, the integral weight and the predicted weight as the state switching conditions of behaviour control model, so that the billet limited weight cutting is realized. The model of construction is beneficial for the autonomous behaviour control of billet limited weight cutting.

1 Introduction

It is critical to improve the utilization ratio of steel billet based on billet limited weight cutting. The decrease of billet limited weight cutting error not only can improve the yield of downstream steel rolling process, but also can reduce the energy consumption and the cost of per ton steel, and enhance the economic benefits of enterprise. There are many factors that affect the billet limited weight cutting in the process of steel billet casting, such as the length measurement precision of billet, the speed of billet casting, the wear degree of slab-casting mold, the temperature of casting molten steel, the response sensitivity of cutter clamping and so on. It has high engineering value to improve the accuracy of billet cutting to find the main factors of effecting the billet weight cutting and consider the interactional result of many factors comprehensively, also to promote the intelligent of billet limited weight cutting [1].

This paper analyzed the effective factors of billet limited weight cutting and established the architecture of billet limited weight cutting control system. A finite-state machine behavior logic control strategy of fitting for billet limited weight cutting, which to cut the target decompose step by step for cutting behavior and actions, is proposed. And it explored the autonomous behavior of billet limited weight cutting target.

2 Parameter Analysis of Stateflow Model for Billet Limited Weight Cutting

Finite-state machine (FSM) is a discrete system mathematical model, through the realization of state transition in event-driven way. It consists of many state of specific objectives will be completed, trigger events between states and transferred condition. Using the FSM can realize complex logic control function conveniently. Analysis the several main factors of influencing the billet limited weight cutting as follows:

2.1 Parameter Analysis of Billet Casting Speed

Many factors that effects the billet limited weight cutting in the process of steel billet casting, such as the speed of billet casting, the temperature of casting molten steel, the wear degree of slab-casting mold and so on. The function relation between the weight of billet and the speed of billet casting can be described as:

$$G=\sum_{i=1}^{n} S\rho v_i \frac{1000}{60} \tag{1}$$

where, S is the area of billet, ρ is the density of billet, v_i is the real-time speed of billet casting.

Through the analysis of a large number of experimental data, it shows that, due to changes in the cross section of billet, the weight of billet and the length of billet appear a phenomenon of dispersion corresponding distribution. Fig.1 illustrates the dispersion relation between weight and length of billet. From Fig.1, the same weight of billet can be corresponding to different range of billet length, the biggest difference between 15 mm. And corresponding to the billet that cross section is 165×165 mm, the error of weight will reach 3.185 kg. If only consider the length of billet as basis in the process of billet cutting, the billet weight will produce a large error. So in the process of billet visual measurement, it can use method of piecewise integral to calculate the weight of billet. This method not only to consider the length of billet, and also consider the cross

Fund project: Science and technology plan projects of Hebei province. No:15212112, No: 15271611

DOI: 10.1051/matecconf/20166802005

section [2].

Acquisition the speed of billet casting every second, Fig.2 illustrates the statistical distribution between speed of billet casting and the length of billet. From Fig.2, in a certain speed range, the weight range of billet is concentrated. So the speed sum of billet casting is an important factor that affects the weight of billet.

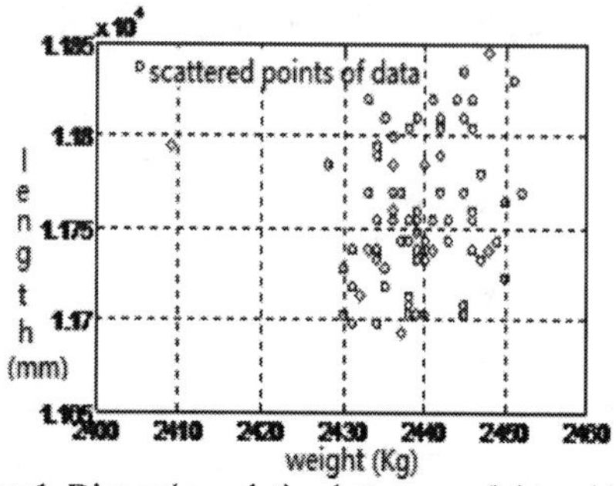

Figure 1. Dispersion relation between weight and length of billet

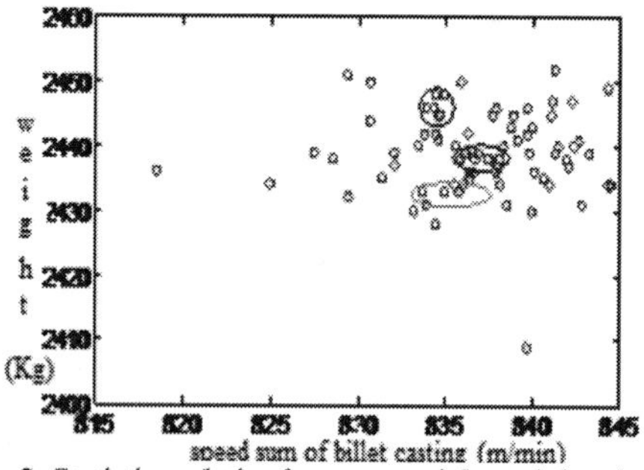

Figure 2. Statistics relation between weight and drawing speed of billet

2.2 Fourier Transform of Billet Image

Using the visual measurement to obtain the length of billet, if under the ideal conditions, such as consistent area, stable temperature, and uniform of composition and so on, the length of billet and the weight of billet exists a good linear relationship. However, many factors, such as the speed of billet casting, the length measurement precision of billet and so on, can lead to changes in cross section of billet in the actual process of billet casting. Then these factors can results in a nonlinear corresponding relation between the length of billet and weight of billet. The length measurement precision of billet has a great influence on billet limited weight cutting. When use the visual to measure the length of billet, due to exist the factors such as dust and high temperature turbulence, the images of billet happen degradation, then the factors can effects the segmentation precision and measurement error of images [3].

Using the method of Fourier Transform to filter the original image in the frequency domain. This treatment can remove high the frequency interference and realize image recovery. Then using Fourier inverse transformation to obtain the processing result of image [4,5]. Image $f(x,y)$ corresponding to the Fourier spectrum is $F(u,v)$, the two-dimensional discrete Fourier transform of this image can be expressed as:

$$F(m,n)=\frac{1}{N}\sum_{i=0}^{N-1}\sum_{k=0}^{N-1} f(i,k)e^{-j2\pi(m\frac{i}{N}*n\frac{k}{N})} \tag{2}$$

The inverse transformation can be expressed as:

$$f(i,k)=\frac{1}{N}\sum_{i=0}^{N-1}\sum_{k=0}^{N-1} F(m,n)e^{j2\pi(m\frac{i}{N}*n\frac{k}{N})} \tag{3}$$

Fig.3 shows the Fourier transform processing of image, Fig.3(a) is the original image of billet, Fig.3(b) is the discrete Fourier spectrum, Fig.3(c) is the image after discrete Fourier inverse transformation.

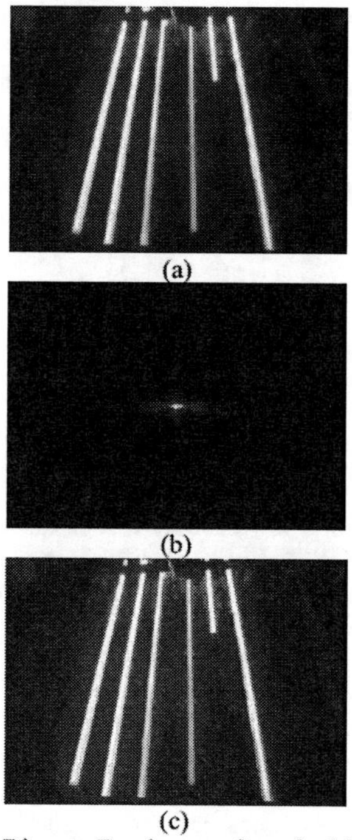

Figure 3. Discrete Fourier transform for billet image

After Fourier transform, the image can effectively remove noise and improve the reliability of image segmentation, then the visual measurement precision can be improved. The cross section of billet can change because of the wear of slab-casting mold, for small billet. The higher measurement accuracy of visual, the less billet cutting error. So in the process of vision measurement, not only to measure the length of small billet, and also measure the cross sectional area of small billet. Thus the changes of billet weight can be real-time tracking.

3 Behaviour Stateflow Model of Billet Limited Weight Cutting

The structure of small billet limited weight cutting is defined a structure model for achieving specific cutting goals by perceiving and processing the information of environment and the control of cutting machine. Behavior

DOI: 10.1051/matecconf/20166802005

control have three basic control architectures: deliberating control structure, reaction control structure and mixed control structure. In deliberating control structure, every module is performed sequentially, and each operation should be repeated through planning, decision-making, implementation, so the response of environment is poor; in reaction control structure, task and behavior are separated, specific target task is decomposed into the simple behavior and actions that they can parallel response, so it has a good real-time; mixed control structure is a kind of combination with planning of global goal, decision-making and Reactive task execution.

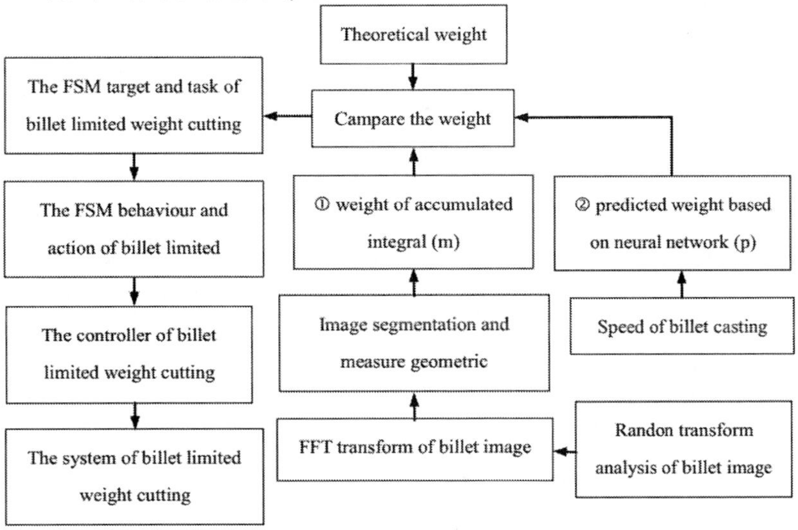

Figure 4. Behavior control architecture for billet weight cutting

Reaction control structure suitable for behavior control of billet limited weight cutting, Fig.4 shows the behavior control architecture of billet limited weight cutting system. It combines perception, planning, decision-making, execution and other links, and it formed a certain intelligence architecture [6,7].

The control system includes modules with measuring perception, planning, decision-making and control. The task plan of limited weight cutting is obtained by FSM, the task of cutting is decomposed into cutter holder control action and flame cutting control action. According to the measured geometric parameter of billet image for real-time accumulated integral the weight of billet, then the weight of billet P_1 is obtained; at the same time, the speed of billet casting as the input vector, it can forecast the weight of billet, then the weight of billet P_2 is obtained. The difference between the two weights as the instruction of cutting machine control action, billet limited weight cutting can be realized.

4 Behavior Stateflow of Weight Cutting Object

Based on the principle of FSM, make the flame cutting moment of cutting machine as the stable control target, and translate to the control behavior of flame cutting machine. According to the difference value and direction between the forecast weight and the weight of accumulated integral, the action moment of cutting machine is determined, and the control model based on the behavior of FSM is established, also make the different operation state of operation parameter as the event-driven transfer conditions of FSM model.

The state-flow visualization dynamic logical model and simulation toolbox based on FSM are very suitable for establishing and simulating the model of event-driven system and complex logic system [8]. Firstly, creating the model based on the flow diagram, state diagram and graphics functions; then, the theory of finite-state can be realized easily in combination with the event, action of states, action of conditions, transferred action, event broadcast, etc. The generated logic of condition monitoring by state-flow can be connected with the Simulink model by the I/O interface of Chart diagram. So that the behavior control be achieved completely.

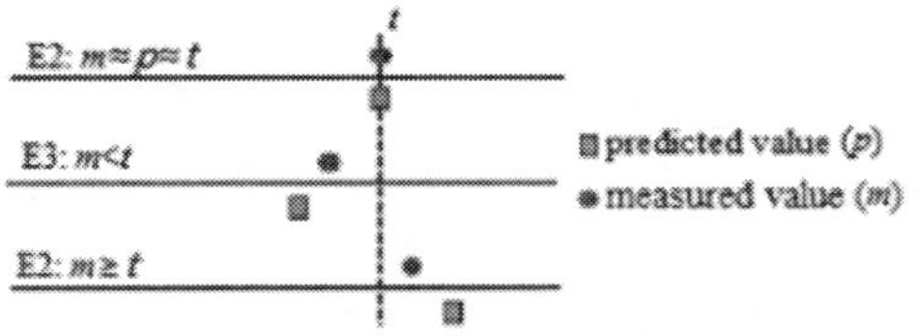

Figure 5. Variable variation of billet weight

Fig.5 illustrates the change process of the theoretical values and the predicted values and the measured values of billet limited weight. Where, t is the theoretical weight value that the ideal length of billet corresponding to billet weight; m is the weight of accumulated integral; p is the predicted weight of billet. For a given billet, the value of t and p are relatively consistent, but m *is* a real-time variation, so that it can be used as the state transition condition of FSM model.

Table 1. Finite state and it's transition condition

state name	State description	State transfer condition
S0	initial state	E1： start condition
S1	state of normal cutting	E2： the condition of maintaining the normal
S2	predicted value < theoretical value	E3： $p<t$， $m<t$， when $m\approx t$
S3	predicted value > theoretical value	E4： $p>t$， when $m\geq t$

Table 1 illustrates the various states and the corresponding state transition conditions of FSM model. E1 is the start condition of system normal operation; E2 is the condition of maintaining the normal cutting that he theoretical value, the forecast value and the measured value are close; E3 is the state of system that he measured values and the predicted value are less than the theoretical value, when the measured value and the theoretical value is closing, trigger the state transition; E4 is state that the predicted value is more than the theoretical value, when the measured value is more than the theoretical value or equal to the theoretical value, trigger the state transition.

When the model of FSM is established, open the figure of Chart into the interface of state diagram editor. According to the theory of FSM, the goal of limited weight cutting can be broken into four states (S0-S3), the name of states are listed in table 1. Fig.6 shows the state flow of limited weight cutting.

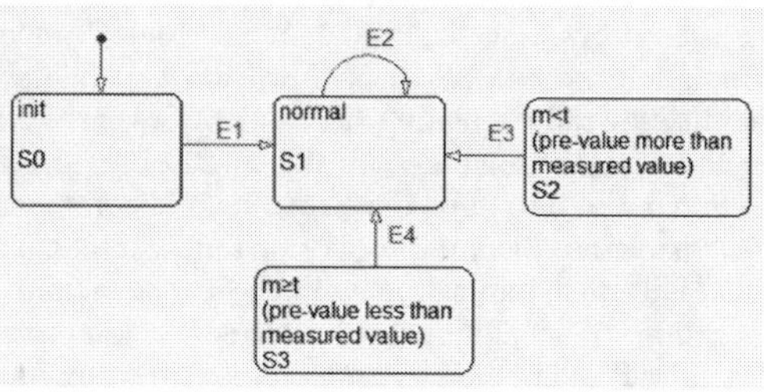

Figure 6. The stateflow for billet weight cutting

The system is in the initial state S0. When the trigger condition E1 of system is meted, it shows that the theoretical value, the forecast value and the measured value are close, so it can perform the task of cutting, in state S1. If the current theoretical value, the measured values and the prediction value is close, also the condition E2 is meted, State S1 is maintained. When the system is in the state S2, it shows that the measured values and the prediction value are less than the theoretical value, the condition E3 is meted, the state happen transition. When the system is in the state S3, it shows that the predicted value is more than the theoretical value, it should wait for the measured values more than or equal to the theoretical value, when the condition E4 is meted, the state happen transition.

5 Conclusion

The ability of autonomous behavior control for billet limited weight cutting is the important embodiment of billet cutting system automation level. This paper established behavior control architecture for billet limited weight cutting on the basis of analysis of the steel billet images and analysis of the effect of billet casting speed on billet limited weight cutting. Based on the theory of finite-state machine, achieve behavior state model of billet cutting independently by the stateflow technology. By constructing the transferred condition of different states, achieve transformation between different states in billet cutting. This is a beneficial e'exploration for the billet limited weight cutting control method based on behavior.

References

1. Ge Yadong, Dong Chao. Measures to raise products ratio of bar. Hebei Metallurgy, **12**:27-29 (2011)
2. Wang Fubin, Chen Zhikun, Wu Baolin, et al . Forecast for Billet Weight Based on SVM Optimized by Niching Particle Swarm. Control Engineering of China, **22(3)**:554-558 (2015)
3. Jin Guoping, Xu Fuhao, Wang Yong. Application of the machine vision technology in the billet cut-to-length auto measuring system. Bao-Steel Technology, **3**:76-80 (2011)
4. Zhang Chao, Hu Shaohai, Xiao Yang. 2-D DFT-DWT hybrid transform and its application in removing speckle noise for medical ultrasound images . Signal Processing, **25(3)**: 464-468 (2015)
5. Xiao T., Hu S.H., Xiao Y., 2-D DVF-DWT application to multidimensional signal processing, Proc. ICSP 2006, PP.1306-1309.
6. Jian Liu, Zhong Liu1,Ying Xiong. Method of Parameters Optimization in SVM based on PSO. Transactions on Computer Science and Technology, **2(3)**:9-16 (2013)
7. Su Mang, Li Fenghua, Shi Guozheng. Action-Based Multi-level Access Control Model. Journal of Computer Research and Development, **51(7)**:1604-1613 (2014)
8. Mei Yang, Yi Zichen, Wang Lipeng, et al. An improved modulation strategy for the five-leg inverter based on finite state machine

MATEC Web of Conferences **68**, 02006 (2016)
ICIEA 2016
DOI: 10.1051/matecconf/20166802006

Serious Gaming for Improvised Explosive Device Neutralization Training

Christopher C.K. Chan[1], Alexander Ferworn[1] and Const. Matthew Young[2]

[1]*Dept. of Computer Science, Ryerson University, 350 Victoria St. Toronto, Ontario, Canada*
[2]*Ontario Provincial Police, USAR/CBRNE Response Team, Bolton, Ontario, Canada*

Abstract. An improvised explosive device (IED) is a "homemade" bomb intended to cause great harm when it explodes. The public safety task of identifying and neutralizing IEDs falls to military and police services often called explosive disposal units (EDU) who act to neutralize the threat associated with the IED either rendering it inoperable or destroying it safely. EDUs train in various aspects of explosive handling and investigation but are limited in the tools available for safely analyzing real world bombs. This paper describes a game based approach to IED training that employs an interactive 3D simulation to spatially identify key IED components of interest. We give an example of how this approach might be used and provide a preliminary evaluation of its potential effectiveness. We employ images formed from a Digital Imaging and Communications in Medicine (DICOM) system captured using Magnetic Resonance Imaging (MRI) technology to a virtual IED in a game. Empirical evaluation and EDU testimony suggest accurate representation of the IED and the potential efficacy of the proposed approach for successfully identifying components in the bomb for the purposes of EDU training.

1 Introduction and Background

An improvised explosive device (IED) is a "homemade" destructive device designed to destroy, incapacitate, harass, or distract [1]. Many commonly available materials such as fertilizer and common chemicals can be used as explosive materials in IEDs. They can be constructed in many forms, ranging in size and power from small and simple pipe bombs causing severe injuries to a few people to large and sophisticated devices capable of causing massive damage and loss of life.

In crisis and emergency management, IEDs have a long history, however, the Industrial Revolution created a surplus of preprocessed materials that enabled the development and diversity of IEDs [2]. Chemicals for explosives, and materials and products that can be modified to function as casings, shrapnel, triggering mechanisms and power sources have become readily available on a large scale. The potential number, variety and complexity of IED designs can make it difficult for EDU personnel to safely investigate a bomb's characteristics and effectively neutralize its threat without endangering themselves.

Regardless of design peculiarities, all IEDs are composed of a main explosive charge, a method of detonation, a triggering device, a power source and may include shrapnel [2]. There have been many examples of explosive charges constructed from everything from gun power [3] to fertilizer [4]. Detonators (the part of the bomb that sets off the main explosive charge) can be as small as a ½ in. in diameter. Triggering mechanisms (the timer or communication mechanism for the bomb) can be readily improvised through the use of a clock or cell phone and depend on the intended use of the IED, available resources and the knowledge of the builder. The power source (supplying electricity to the other components) can be made from common batteries [5].

Our work focuses on visually determining regions of an IED that may resemble one of the key elements listed, for the purposes of bomb deactivation (neutralization). We propose an approach that involves the sensing of physical IEDs and rendering their virtual equivalents in software. For the purposes of our initial work, we consider IEDs hidden within a simple package (a cardboard banker box). Our intent is to provide a game mechanism for EDU members to practice, identifying key components of an IED and interacting with them to neutralize their threat in a safe environment. Figure 1 depicts our simulated IED that takes the form of a pipe bomb.

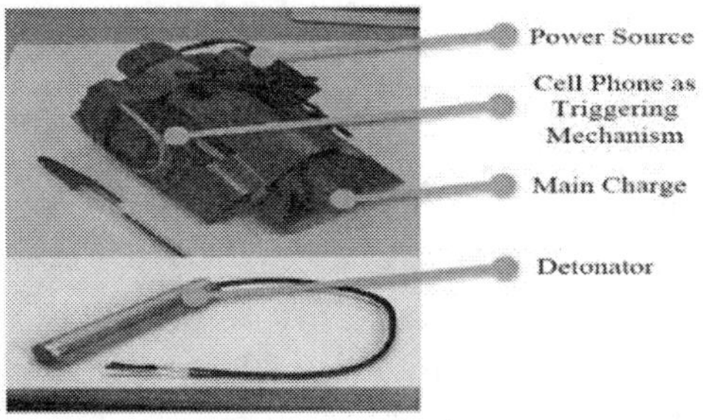

Figure 1. Simulated pipe bomb with a battery power source, cell phone trigger, commercial explosive main charge simulated by wooden sticks and a detonator.

DOI: 10.1051/matecconf/20166802006

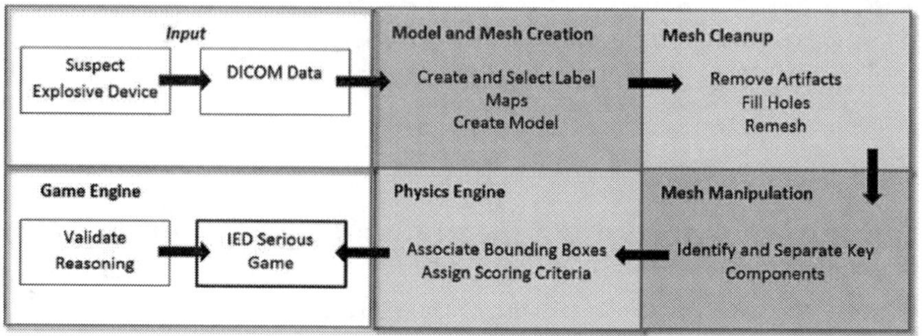

Figure 2. An overview of the proposed pipeline to create a virtual bomb from a real suspect explosive device real bomb.

Our goal is to use our virtual IED and concealing container as input into a serious game for purposes of EDU training, providing objective scoring concerning the neutralization task. In order to achieve our goal we must provide a physically accurate representation of the IED that maintains all spatial relationships between all of its components and the container it is in.

We utilize an interaction model browser [6] for exploring and visualizing relationships between structures that are often layered and visually occlude one another and we separate these models into independent parts for the 3D simulation within a game engine.

Figure 2 depicts the generalized overview of our pipeline. This work stems from our previous research in which we use data collected from an Unmanned Aerial Vehicle (UAV) fitted with an RGB-D sensor to create a high fidelity 3D point cloud model of scanned rubble terrain [7] and inject it into another serious game. Our previous work, in this broad area, focused on time-sensitive operations with technology that can be applied to support USAR operations [7–9]. Stemming from this work, we present an approach to separate parts of an IED within a model and provide a game-based simulation that can potentially assist in the training of EDU personnel with the benefits of lowering exposure to actual IEDs, and providing visually and spatially accurate 3D model representations.

2 EDU Operations and related work

EDU operations involving a suspect explosive device (SED) normally begin by ensuring the three important safety factors are addressed; minimizing the amount of time spent in proximity to the device (since the device could be activated at any time by an integral timer, a remote trigger, or some other cause). Additionally, the response team and everyone else must comply with appropriate standoff distances calculated according to the apparent size of the IED. Finally, sufficient shielding must be provided to protect from a potential blast wave [10].

Essentially, the steps involved in dealing with a suspected SED-according to EDU professionals and the Federal Explosives Law and Regulations [11] involve many steps. One of these steps is to x-ray the SED (if possible) and determine where the power source/trigger is. It is important to make this determination to facilitate a means to neutralize the device by taking aim at one of the non-explosive components of the SED with a "disruptor" to disarm the SED. The process will be repeated if necessary.

The technology and software available to most EDUs are x-ray scanners, x-ray viewers [12], EOD disruptors [13] and various explosives to safely control and manage the SED through controlled detonation.

Figure 3. A 20mm Neutrex Waterjet Disruptor by Proparms [13].

Having taken an x-ray image, EDU personnel attempt to quickly determine if a SED is an IED, and if it is an IED, identify the primary components. They use this information to determine the best approach to neutralizing the device.

The closest related work to our proposed approach is represented by the Scanview [12] software package--an application developed by Scanna MSC in the UK. Scanview provides a number of image manipulation and enhancement tools that can help EDU specialists identify the components in an IED, using a 2D image. It provides a "pseudo-3D" feature that can provide some depth representation of the scanned image.

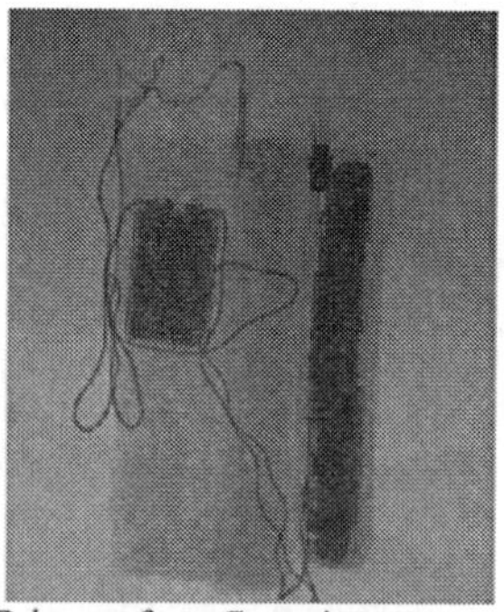

Figure 4. The 2D image from Scanview.

Serious games have been developed in a number of arenas such as defense, education, scientific exploration, health care and emergency management [14]. The main purposes of a serious game can be to train, educate or promote learning for a particular topic. A serious game requires a player to solve a problem while being evaluated (rewarded) through a scoring mechanism. Common games range from air defence [15] to forms of military training [16] to rehabilitation [17]. Serious games provide realism and presence in a virtual environment sometimes employing physics [18-19] to facilitate reality simulation. A physics engine computes how physical objects should move and interact with each other within the game.

Previous work related to IEDs focuses on locating IED threats [20-21], determining the blast effects of IEDs [22], IED threat assessment [23], and determining where IEDs might be placed [24]. The closest related work to interactive 3D simulation is USARSim [25]. Few software systems have been created to aid in neutralizing an IED.

Our work revolves around the use of non-traditional, noncontact sensing and recording of physical IED devices, their modeling and interaction with trained operators who act as players in serious game settings in which a player is asked to neutralize the threat and is scored on their ability to do so using devices within the game.

3 Methodology

Our proposed algorithm is comprised of three steps. The first step involves scanning the IED in a process we call the model creation pipeline (MCP). In contrast to the high fidelity point cloud framework described in our previous work [7-8], this process utilizes open source projects, Slicer 3D [26], Blender [27] and Aperio [6] to convert Digital Imaging and Communications in Medicine (DICOM) data into separate mesh models as object files compatible with game engines such as Unity [28] or Unreal [29] for interactive 3D simulation. This MCP allows for the ability to inject arbitrary real models into a game engine and reason about them.

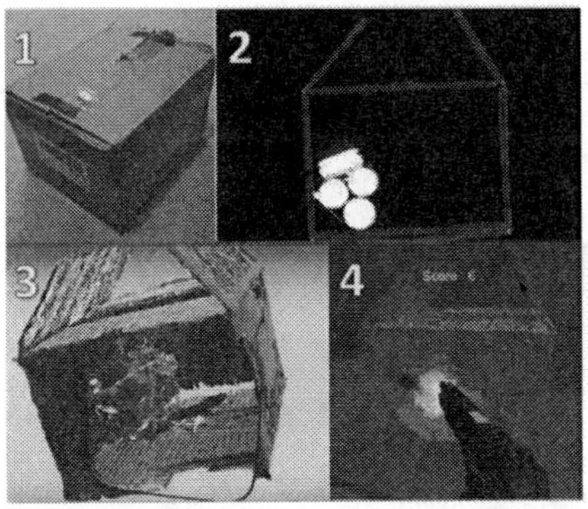

Figure 5. 1). A simulated IED within a banker box; 2). A DICOM cross sectional image of the bomb; 3). The bomb in Aperio for the purposes of revealing occlusions and separating the bomb into components; 4). The model inside the game engine for interactive training.

The second step is to create an injectable model of IED components in order to make a virtual representation of the entire IED within a game engine. In addition, other models of necessary tools and artifacts can be created to interact with the virtual IED. One such tool — known as a disruptor — is used to remove key components of an IED by accurately firing a bolt of water at them. In our simulation, we model the 20mm Neutrex disruptor sold by Proparms [13]. In addition, we create a model of what we call "the rig" which is an assemblage of the objects around the IED. As all the models are physically accurate and the game engine supports physics-based interactions, we are able to reason about them inside the game. The results of reasoning take the form of data such as the distance from the Neutrex disruptor to the container and the angle to the target with respect to the normal plane of the disruptor. The reasoning occurs by the player interacting with the artifacts inside the game simulation and can be validated against a real-world IED and rig.

The third part of the system is a game scoring algorithm which measures the effectiveness of the disruptor's target location with a projectile blast in the game. We take advantage of known heuristics. For example, if the disruptor fails to hit the desired target the score will be low while hitting the target leads to a high score. Of course, hitting the main charge, causes the game to end abruptly.

The simulation provides feedback to a player through scoring points as to the effectiveness of their targeting. We suggest that this interactive part of the simulation can be used to safely train response personnel by providing visual cues and feedback for important and specific regions of interest of simulated IEDs.

Our initial system is informed by input from subject matter experts from the Ontario Provincial Police (OPP) UCRT from their Chemical, Biological, Radiological, Nuclear explosive (CBRNe) team and the Deltic Group [30] who supply a range of specialty equipment such as x-ray technology for CBRNe applications.

Based on initial feedback, we feel that future versions of our training simulation may be able to augment the training for responders to identify bomb neutralization targets — supplementing the identification process that ordinarily is performed in the real world. The advantage in using our proposed methodology is that the simulated IEDs can be easily shared between EDUs throughout the world. Once scans are obtained and injected into a game, EDU reasoning about the virtual model as a precise physical representation of the real model is facilitated in a safe environment.

Our initial work is described below.

3.1 Create Representative IED

The real world inert IED was constructed using a 9V battery as a power source, an old cell phone as a trigger, small wooden pieces as the main charge (see figure 1) with the assembly wrapped in electrical tape. These components were chosen from commonly available material with varying densities which are characteristic of pipe bombs according to CBRNe team members [30]. On x-ray technologies [12] (see figure 4) components are differentiated by their densities, the darker the component is, the denser the material. In our algorithm, the components are segmented based on signal intensity.

3.2 Place the IED within a representative container

The inert pipe bomb was placed within a small banker box with dimensions of 21 cm × 25 cm × 16cm. Packing air bags were placed around the pipe bomb as supports to avoid movement in of the IED in transit.

3.3 Scan the IED within the container

Magnetic Resonance Imaging (MRI) technology was used to capture data of the container and the bomb. MRI is one of many modern diagnostic imaging techniques that produces cross-sectional images of an entity using energy from an oscillating magnetic field at the appropriate resonance frequency [31]. This process excites hydrogen atoms and causes them to emit a radio frequency signal which is measured by a receiving coil. The MRI machine decodes densities of different components by the rate at which excited hydrogen atoms return to the equilibrium state by varying the main magnetic field using gradient coils.

This method of scanning has many advantages for EDU personnel over the current x-ray methodology as there is no risk of radiation exposure. In addition, actual scans take less than 30 seconds and an operator can change the imaging plane without moving the object being scanned. However, the technology cannot readily be used for EDU operations as no magnetic components can exist within the IED being scanned and MRI machines are exceedingly expensive, relatively large and problematic as to how they could be employed operationally.

For the moment, medical imaging is beyond the capabilities of most EDUs who have to make due with x-ray equipment. However, the technology can be used to create the virtual IEDs for a game.

3.4 Create representative IED

To create a model from the DICOM data, we use 3D Slicer [26] to create and select label maps using threshold ranging. The ranges we chose were from -976.00 to -490.00. DICOM data captured from the MRI machine shows three hypothetical planes used to transect the entity; axial, sagittal and coronal.

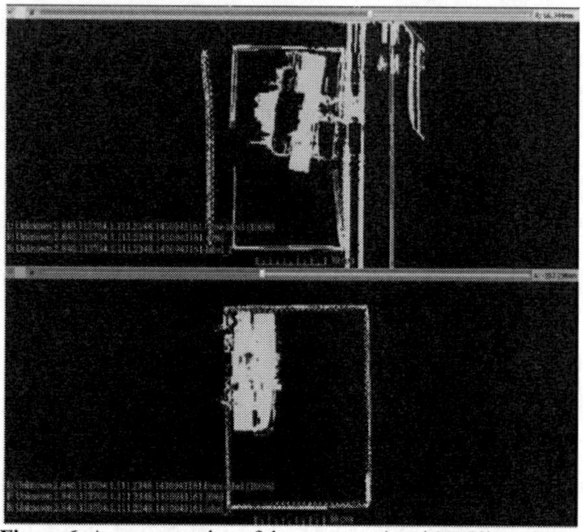

Figure 6. A representation of the cross sectional 3D Slicer.

The threshold feature in the editor allows for a user to choose what entities should be merged into a model.

However, the model created in 3D Slicer often includes artifacts which are caused by a variety of factors that may be related to scanning during motion, metallic components or foreign bodies [32]. Artifacts, as shown in figure 7, can be identified as zippers, strips, fringes or overflow marks which can be caused by Finite sampling, k-space encoding, and Fourier transformation may cause aliasing and Gibbs artifacts [33].

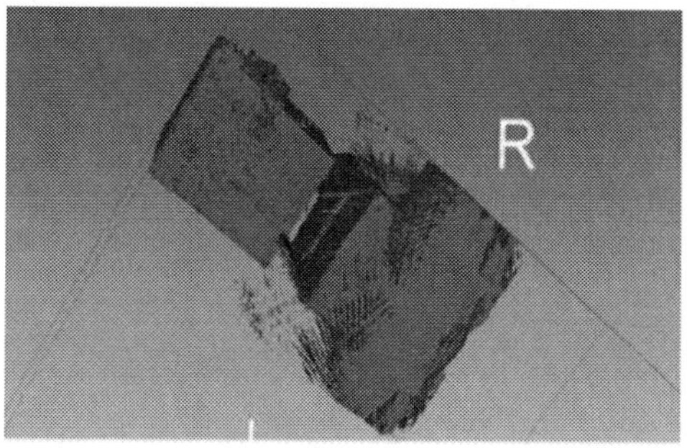

Figure 7. The 3D model produced by combining cross sections in 3D Slicer.

3.5 Mesh cleanup

To cleanup artifacts and extraneous entities such as the support surface, we use Blender [27] to select and delete parts of the mesh. We then use a hole filling algorithm [34] to generate a new surface mesh that follows the same shape as the input, but with more regular topology. The output of this algorithm is blocky re-meshes that are not too small or irregular--useful for separating into components in Aperio. The most important part of this process is to remove nonmanifolds-- geometries that cannot exist in the real world such as 3 or more planes intersecting on an edge. By using this algorithm, it also removes small disconnected pieces and fills holes in the mesh, the resulting mesh appears much like figure 8.

DOI: 10.1051/matecconf/20166802006

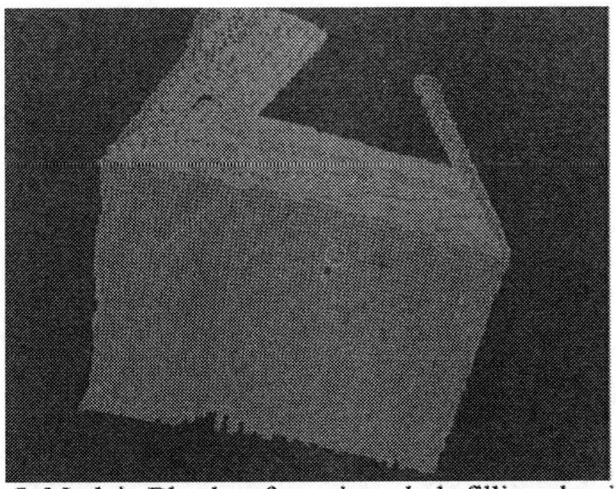

Figure 8. Mesh in Blender after using a hole filling algorithm.

3.6 Mesh manipulation

We identify key components by using Aperio [6], a mesh editor that manages 3D scene occlusion using tools similar to mechanical tools like a cutter, a ring and a rod to manipulate meshes. We identify which parts of the mesh visually represent IED components and separate them using these tools. The output is an object and material file compatible for importing into a game engine.

Figure 9. Mesh modified by tools in Aperio.

3.7 Physics engine

After importing models into the Unity game engine [28], we associate bounding boxes to certain components and assign unique scoring criteria to each box to indicate which are key regions of interest. The algorithm for the bounding box has three components-a damage receiver script, a bullet penetration script and a crosshair script.

Figure 10. Visually associating a bounding box to the component object.

3.8 Interaction with the game

Inside the game, the player has a choice of picking up different models of disruptors. The objective of the game is to aim and shoot a projectile to neutralize the IED. As the projectile hits the object and its bounding box, an effectiveness score appears indicating whether it has successfully hit a key component or not. The player must look for visual cues shown within the bomb model. As the player hovers the crosshair near key components with the disruptor, the crosshair changes color, and as the projectile hits the bounding box, the distance and the angle to the target with respect to the normal plane of the disruptor are shown at the bottom of the display.

Figure 11. The 3D interactive game with the effectiveness score, distance and angle to the target with respect to the normal plane of the disruptor.

4 Experiments and results

We validated the simulation by comparing dimensions of the models in the simulation to dimensions of the model in the real-world. To determine the dimensions of the models in the game engine, we use a distance measuring tool [35] to measure game objects in the scene. This tool displays the ratio between object scale in pixels based on a user-defined pixel-per-unit setting.

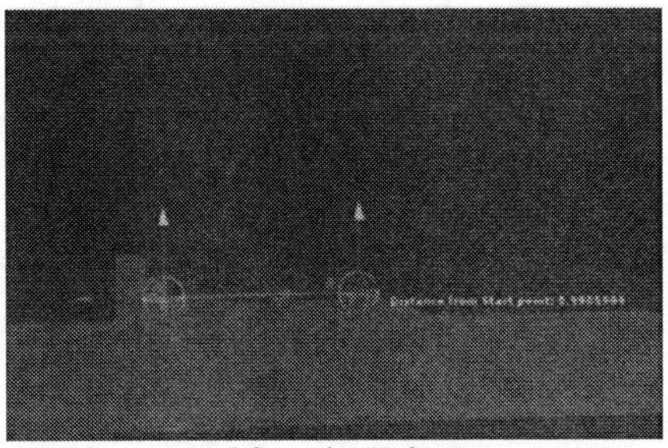

Figure 12. Distance tool from the Unity asset store to measure game objects in the scene.

The pixel-per-unit was set to 9.5714 px = 1 cm, which means a pixel measurement of 100px would equal 10.4477 cm. The results of our measurements are:

Table 1. Comparing Dimensions of the Container

Sim vs. Reality	Container Dimensions	
	Model in Sim	Model in Reality
Height	153.5224 px = 16.0397 cm	16 cm
Width	201.6814 px = 21.0712 cm	21 cm
Length	239.5571px = 25.0284 cm	25 cm

Table 2. Comparing Dimensions of the Bomb

Sim vs. Reality	Bomb Dimensions	
	Model in Sim	**Model in Reality**
Height	59.6987 px = 6.2372 cm	6 cm
Width	67.1586 px = 7.0166 cm	7 cm
Length	143.5744 px = 15.0003 cm	15 cm

In Table 1, the comparison in terms of width suggest a 1.0033% difference when comparing the result of 21.0712 cm to 21 cm. In Table II, the result for height shows a 1.0395% difference when comparing 6.2372 cm to 6 cm. According to EDU professionals, they direct and aim disruptors (to the best of their ability) at the centre of their target. We suggest that we can provide a virtual model to accurately represent a real IED model within 1-2% degree of difference.

5 Conclusion and future work

The motivation for our methodology has been to set the stage for safe investigation of IEDs for EDU professionals. In the field, EDU professionals use the Neutrex disruptor to fire high velocity water to separate the components of an IED before it explodes in an uncontrolled manner.

With our system, we simulate this by converting real world arbitrary models into virtual models and allow reasoning to occur about them, leading to game interaction and a score being produced—an inherently safe approach. We create and show an accurate representation of the real-world IED, and enable EDU professionals an additional framework that can provide them with information such as distances and angles from a safe virtual environment.

Our future work will include real-world validation by creating a situation where our game simulation is one of a real-world experimental rig involving an inert IED within a container and aiming and firing a real disruptor at one of the key components of the IED using only data available from our game simulation. In essence, we will use the game to inform our real-world decisions. If we hit the target, we win the game and our approach will be valid for at least one instance. If we miss the target, our score will be low or the game will end abruptly.

References

1. Department of Homeland Security (2015, Jun . 15). *IED Attack Improvised Explosive Devices* [Online]. Available: https://www.dhs.gov/xlibrary/assets/prep_ied_fact_sheet.pdf.
2. A. Farazmand, *Handbook of crisis and emergency management.* CRC Press, 2001.
3. V. Smirnov, "Possibilities of Creating Improvised Explosives and Explosive Devices," in Detection and disposal of improvised explosives, *Springer*, 2006, pp. 217–221.
4. S. Mallonee, S. Shariat, G. Stennies, R. Waxweiler, D. Hogan, and F. Jordan, "Physical injuries and fatalities resulting from the Oklahoma City bombing," *Jama*, **vol. 276,** no. 5, pp. 382–387, 1996.
5. F. Craig (2015 Jun. 15) *Anatomy of an IED.* [Online]. Available http://science.howstuffworks.com/ied1.htm
6. D. Tran (2015 May. 23) *Aperio by eternallite, 3D Interactive Tool for Revealing Occlusions in Models* [Online] Available http://eternallite.github.io/Aperio/
7. A. Ferworn, S. Herman, C. Kong, A. Ufkes, and J. Tran, "Interacting with a virtual Destroyed Environment constructed from real disaster data," in *Safety, Security, and Rescue Robotics (SSRR), 2014 IEEE International Symposium* on, 2014, pp. 1–6.
8. A. Ferworn, S. Herman, J. Tran, A. Ufkes, and R. Mcdonald, "Disaster scene reconstruction: Modeling and simulating urban building collapse rubble within a game engine," in Proceedings of the *2013 Summer Computer Simulation Conference*, 2013, p. 18.
9. M. Coatsworth, J. Tran, and A. Ferworn, "A hybrid lossless and lossy compression scheme for streaming RGB-D data in real time," in *Safety, Security, and Rescue Robotics (SSRR), 2014 IEEE International Symposium* on, 2014, pp. 1–6.
10. Bureau of Alcohol, Tobacco, Firearms and Explosives (2015 Jun 3) *ATF Explosives Handling Handbook* [Online] Available https://www.atf.gov/file/58741/download
11. Arson. Enforcement (2015 Jun 15) *Arson Explosives Enforcement* [Online] Available https://www.atf.gov/content/Arson/arsonexplosives-enforcement
12. Scanna (2015 Jun 15) *Scanna portable x-ray and postal x-ray systems.* [Online] Available http://www.scanna-msc.com/uk/aboutscanna/
13. Proparms (2015 Jun 16) EOD Tools, *Waterjet Disrupters and Forced Entry Tools - Proparms Ltd.* [Online] Available http://www.proparms.com/site/product_10.html
14. R. Ratan and U. Ritterfeld, "Classifying serious games," *Serious games: Mechanisms and effects*, pp. 10–24, 2009.
15. C. C. Abt, *Serious games.* University Press of America, 1987.
16. M. Zyda, "From visual simulation to virtual reality to games," *Computer*, **vol. 38**, no. 9, pp. 25–32, 2005.
17. P. Rego, P. M. Moreira, and L. P. Reis, "Serious games for rehabilitation: A survey and a classification towards a taxonomy," *in Information Systems and Technologies (CISTI), 2010 5th Iberian Conference* on, 2010, pp. 1–6.
18. A. Seugling and M. Rölin, "Evaluation of physics engines and implementation of a physics module in a 3d-authoring tool," *Umea University*, **2**, 2006.
19. A. Boeing and T. Bräunl, "Evaluation of real-time physics simulation systems," in *Proceedings of the 5th international conference on Computer graphics and interactive techniques in Australia and Southeast Asia*, 2007, pp. 281–288.
20. W. B. Stafford, "Sequential pattern detection and time series models for predicting IED attacks," Monterey, California. Naval Postgraduate School, 2009.

21. D. E. Brown and M. Smith, "Terrorist Threat Prediction in Urban Environments."
22. L. Nilsen-Nygaard and C. W. Johnson, "Using mathematical models to guide the simulation of improvised explosive devices in public spaces," in *System Safety, 2008 3rd IET International Conference* on, 2008, pp. 1–6.
23. M. C. Association (2015 Jun 16) *Counter-IED.* [Online] Available https://www.mca-marines.org/gazette/2014/05/counter-ied
24. A. Shankar, "Spatial and Temporal Modeling of IED Emplacements against Dismounted Patrols," George Mason University, 2014.
25. S. Carpin, M. Lewis, J. Wang, S. Balakirsky, and C. Scrapper, "USARSim: a robot simulator for research and education," *in Robotics and Automation, 2007 IEEE International Conference* on, 2007, pp. 1400–1405.
26. A. Fedorov, R. Beichel, J. Kalpathy-Cramer, J. Finet, J.-C. Fillion- Robin, S. Pujol, C. Bauer, D. Jennings, F. Fennessy, M. Sonka, and others, "3D Slicer as an image computing platform for the Quantitative Imaging Network," *Magnetic resonance imaging*, **30**, no. 9, pp. 1323–1341, 2012.
27. B. O. Community, Blender - *a 3D modelling and rendering package*. Blender Institute, Amsterdam: Blender Foundation.
28. Unity Technologies. (2014, Jul. 18). Unity Game Engine-Official Site [Online]. Available: http://unity3d. com
29. Epic Games. (2014, Jul. 18). Unreal Development Kit (UDK) [Online]. Available: https://www.unrealengine.com/products/udk/
30. D. Group, "Defense and Security Speciality Equipement." .
31. R. R. Edelman and S. Warach, "Magnetic resonance imaging," *New England Journal of Medicine*, **328**, no. 10, pp. 708–716, 1993.
32. "MRI artifacts | Radiology Reference Article | Radiopaedia.org." .
33. E. M. Bellon, E. Haacke, P. Coleman, D. Sacco, D. Steiger, and R. Gangarosa, "MR artifacts: a review," *American Journal of Roentgenology*, **147**, no. 6, pp. 1271–1281, 1986.
34. T. Ju, F. Losasso, S. Schaefer, and J. Warren, "Dual contouring of hermite data," *ACM Transactions on Graphics (TOG)*, **21**, no. 3, pp. 339–346, 2002.
35. W. McDermott (2015 Jun 28) Distance Tool. [Online] Available https://www.assetstore.unity3d.com/en/#!/content/307

MATEC Web of Conferences **68**, 02007 (2016)
ICIEA 2016
DOI: 10.1051/matecconf/20166802007

Design Method of Hull Structure Model for Ultimate Strength Test

Huanxiang Sun[1,2], Jin Gan[2]

[1]*Key Laboratory of High Performance Ship Technology of Ministry of Education, Wuhan University of Technology, China*
[2]*Departments of Naval Architecture, Ocean and Structural Engineering, School of Transportation, Wuhan University of Technology, China*

Abstract. Based on the several practical research on real ship hull ultimate strength model test, a design method of hull structure model for ultimate strength test, which combine the numerical simulation analysis with a similar model design are proposed.

1 Introduction

Ships are large complex engineering structures. The ultimate strength of hull structure is considered as an important factor for evaluation of hull structure safety margin, and it must be checked in structure design and strength evaluation [1], [2]. There are two main methods, include numerical calculation and structural test, which are used to evaluate the ultimate strength of hull structure. Represented by Smith method and nonlinear finite element method, the numerical calculation has made abundant achievements [3], [4]. However, due to the complexity of hull structure and its load bearing, the numerical approach is hard to predict the ultimate strength of hull girder accurately. Consequently, the structure model experimental study is particularly important for ship engineering. The failure process is very complex, because hull girder is a complex space thin-walled structure. Through the model test of hull structure, the collapse process could be directly observed and its failure mechanism could be analysed [5], [6]. Additionally, the available experimental results could be used to verify and improve the accuracy and reliability of other numerical methods.

The ship structure laboratory of Wuhan University of Technology has been working on the hull structure strength tests for long time. Combined with model tests and numerical calculation methods, many experimental studies on the ultimate strength of various ship types, such as mono-hull ship, SWATH ship [7], Trimaran ship [8] and river-sea ship [9], have been completed in recent years. The test photos are shown in Fig. 1 to Fig. 4. Meanwhile, abundant research achievements have been obtained both in design and manufacture of similarity models, and the ultimate strength of several ships with special types have been predicted. In this paper, the design method of hull ultimate strength text model should be introduced in detail.

Figure 1. Experimental study on mono-hull ship.

Figure 2. Experimental study on SWATH ship.

Figure 3. Experimental study on trimaran ship.

Figure 4. Experimental study on river-sea ship

2 Similarity analysis of structural model

The reliability and accuracy have been confirmed by a lot of experiments that setting the length L and thickness δ as the basic parameters in thin-walled structure modelling design. Usually, this kind of model is called distortion model. When design the ship hull ultimate strength experiment model, the length L, thickness δ and force F are chose to be the basic physical quantities of dimensional analysis, therefore the basic dimensional system could be expressed as [F, L, δ]. According to the characteristics of the ultimate strength of hull structure experiment, the similarity criterion relation in model design could be derived via dimensional analysis.

Due to the length dimensions [F] of the dimension matrix in the index of various physical quantities are zero. And modulus of elasticity E is the intrinsic parameter of material and has no relationship with the external force. In this case, the value of the external force should not essential influence on the inner mechanism of the experiment. In the test condition, it is hard to make the load as same as the actual one, because the ships are subjected to the shear force and the bending moment under the real condition with the static water and the wave. Although the amplitude and the distribution of shear force and bending moment in the full range of ship could be got in the design phase, it is usually limited by test conditions to achieve this parabolic form distribution of the shear force and bending moment in the model. For load similarity, the load distribution is simplified to apply to the model in practice. For the boundary conditions similarity, the actual movement of the ship in water could not be simulated in the laboratory. Considering the factors above, in the model test, the way of model design is that guaranteeing the boundary as far as possible from the core area of the model, imposing the boundary conditions at both ends of the extended range, and t minimizing the influence of boundary conditions on the experimental results according to the Saint-Venant principle.

3 Structure model design

3.1. Scale cabin model design

If the cabin model was designed based on several sections, the stiffness in different parts is variable, and a small scale will decrease the reliability and accuracy of model structure. So if we apply the bending load to the model, the buckling must be occurred in the part where it's structural strength weakest. But in actual situation, the buckling of real ship depends on specific load conditions of each section. So if we want to locate the dangerous section exactly, we should adopt multi-node loading to simulate the real load situation. Because of the complexity and uncertainties of the ultimate strength research, the selection of risky section of hull girder is one of the keys of the model design. Because of the complexity of the ship lines and structure, the section structures in different part vary greatly. Under different working conditions the load distribution along the ship will also be different, so we should calculate the ultimate strength of typical sections to identify the location of dangerous ones for the further design. Before determining the experimental cabin model, we could use numerical calculation method (Smith method, nonlinear finite element method) or standard method to identify dangerous sections of whole ship under different working conditions. The experimental cabin model could be obtained by extending the section, then four-point bend loading experiment could be conducted and the ultimate strength of the model could be observed. Furthermore, the real ship's capacity could be predicted. According to this method, we avoid experiment result error leading by the complexity of model. Also, numerical calculation method could be optimized according to the experiment results.

3.2 Model similarity scale

After determining the risky section, the structure design of the model section should be carried out under the condition of similarity criterion, which is the basic requirement of the general static strength model in structure design. Considering the characteristics of the ultimate strength, the size and number of bars of the stiffened plate should be rationally allocated in the scale model, which could ensure the stiffened plate grids have the approximate stability between the model and ship. Finally the model of the real ship section and the test section model could meet the similar conditions in the essence of the similar phenomenon. Considering the several factors such as the manufacture process of steel model, the actual purchase of plate models, the laboratorial loading capacity and the restrictions of field testing, the value of the similar thickness constant Cδ is suggested between 3:1 and 5:1, and the value of the similarity scale constant CL is suggested between 6:1 and 10:1 by the experience of hull structure model tests in our research team for years.

3.3 Model structure design

The most dangerous section of real ship is chosen to be the research object in test, and the test results could be used to predict the ultimate bending moment of the target section. In view of the above approach, the whole test model consists of one test section and two load sections,

DOI: 10.1051/matecconf/20166802007

and the test load is four-point pure bending. The test section of model structure is designed as the dangerous section of real ship by the similarity theory. The load sections added at both ends of the test section could provide suitable loading and boundary conditions, and could ensure the accuracy of the test results.

By the conditions of similarity criterion, the section of real ship and model should meet the following similarity ratio:

$$C_I = C_L^3 C_\delta \tag{1}$$

where CI is the ratio of section moment of inertia, CL is the similarity ratio of the section geometry and Cδ is the similarity ratio of the section thickness of real ship to model.

If the above similarity criterion is met, the section bending rigidity of real ship and model will be similar, which makes the same section deformation and curvature of real ship and model under similar bending moment. According to the assumption of plane section, the strain of profiles in different location is shown as follows:

$$\varepsilon_i = \frac{y_i}{\rho} \tag{2}$$

where ε i is the strains in different section area, yi is the distance between component and neutral axis in different section area and ρ is the bending radius of curvature for hull girder.

According to the formula Eq. 1 and Eq. 2, the height of natural axis and the stress distribution of real ship and model must have the similarity ratio. The plate thickness in different locations has different thickness ratio, and the geometry sizes of section also meet the geometric similarity ratio. The preliminary determination of stiffeners sizes should satisfy the requirements as the following equation:

$$C_I = C_L^3 C_\delta = \frac{I_{gs}}{I_{gm}} \tag{3}$$

where Igs is the section moment of inertia for stiffeners with ribband in real ship, and Igm is the section moment of inertia for stiffeners with ribband in test model.

In accordance with the model section design criteria above, the model section structure could be designed according to the type and size of the dangerous section of real ship structure. On this basis, the test section, transition section and load section of ship hull girder test can be designed.

4 Model processing technology

Since the hull structure model is a complex thin-walled structure, the model processing precision affects the result of the test directly, especially has a greater influence to the testing results. Therefore, the hull model should be conducted on high precision processing platform. In the process of machining, undeserved technological process and processing method could affect the precision of the model. For guaranteeing the precision of hull model, several measures are advised as follows:

All components of the test model are trying to use shear or frilled edge process, and less to use cutter knife and welding set.

Prefabricate and correct the shell parts and internal frames, and choose the proper assembly procedures which need to control the welding procedure strictly on the assembly platform.

Lines lofting and frame assembly should be located strictly in the process. Minimize the weld length under the premise of reliable connection and effective force transfer, and select the effective electrode and the best welding current at the same time.

Conduct dimensional checks in every working procedure during the components assembling, and correct the detected problems timely.

5 Conclusions

Structure model design and manufacture are vital to the ultimate strength test of hull structures. The ship hull structure models are designed by the method, which combined numerical calculation and structural similarity analysis, have been applied to ultimate strength model tests and the similarities prediction of real ships successfully. The method has been used effectively to evaluate the ultimate strength of ship structures. Due to the complexity of ultimate strength problem for ship structures, the study is not completed yet and more work need to be done.

The similarity model is designed based on the linear elastic theory, but the structure model of ultimate strength test involves the geometrical nonlinearity and material nonlinearity, so, the conversion from the experimental result to real ship is the main problem need to overcome in further study.

With the development of ship type and the application of new material, the ultimate strength problem of ship structure should become more and more complex. Considering the coupled effects longitudinal bending, transverse bending and torsion, the ultimate strength of hull structure still need our further study.

References

1. Vasta J: Lessons learnt from full-scale ship structural test, Trans. SNAME, 1958, **66**:165-243.
2. Sugimura T, NoZaki M, SuZuki T: Destructive experiment on ship hull model under longitudinal bending, Journal of the Society of Naval Architects of Japan, 1966, **119**: 209-220.
3. Caldwell J B: Ultimate longitudinal strength, Trans RINA, 1965, 107: 411-430.
4. Chen Y K, Kutt L M, Piaszezyk C M, et al: Ultimate strength of ship structures, Trans.SNAME, 1983: 149-168.
5. Hughes O F: Ship structural design: a rationally-based, computer-aided, optimization approach, New York: Wiley, 1983.

6. Bai Y, Bendiksen E: Collapse analysis of ship hulls, Marine Structures, 1993, **6**: 485-507.
7. Liu Bin, Wu Weiguo and Huang Yanling: Ultimate strength test study of SWATH ships, Proceedings of the Nineteenth International Offshore and Polar Engineering Conference (2009).
8. Wu Weiguo, Xu Shuangxi and Liu Weiqin: Study on the ultimate strength test of high speed trimaran, Proceedings of the 3rd International Conference on Marine Structures (2011).
9. Wu, Haochen， Wu, Weiguo, Gan, Jin and Sun Huanxiang: Ultimate strength analysis of a river-sea ship under combined action of torsion and bending, 32nd International Conference on Ocean, Offshore and Arctic Engineering (2013).

MATEC Web of Conferences **68**, 03001 (2016)
ICIEA 2016
DOI: 10.1051/matecconf/20166803001

Objects Classification for Mobile Robots Using Hierarchic Selective Search Method

Cheng Xu[1], Ting Cao[1], Zhiqi Li[2] and Xiongren Xiao[1]

[1]*Information Science and Engineering, Hunan University, 410082 Changsha, China*
[2]*Civil Engineering, Hunan University, 410082 Changsha, China*

Abstract. Aiming at determining the category of an image captured from mobile robots for intelligent application, classification with the bag-of-words model is proved effectively in near-duplicate/planar images. When it comes to images from mobile robots with complex background, does it still work well? In this paper, based on the merging criterion improvement, a method named hierarchical selective search is proposed hierarchically extracting complementary features to form a combined and environment-adaptable similarity measurement for segmentation resulting a small and high-quality regions set. Simultaneously those regions rather than a whole image are used for classification. As a result, it well improved the classification accuracy and make the bog-of-word model still work well on classification for mobile robots. The experiments on hierarchical selective search show its better performance than selective search on two task datasets for mobile robots. The experiments on classification shows the samples from regions are better than those original whole images. The advantage of less quantity and higher quality object regions from hierarchical selective search is more prominent when it comes to those special tasks for mobile robots with scarce data.

1 Introduction

With the rapid development of mobile internet, mobile robots has played increasingly important roles in our daily life. Object classification as important parts for the mobile robots brings us much convenience and high efficiency, such as the traffic monitoring, agricultural products or fruit picking, mobile E-Commerce and so on. Regard a smart phone as a mobile robot according to the definition of robot in [1]. You can imagine, when you're walking on the street, your sight is deeply attracted by a cloth, you just need pick up your smart phone and click the screen, then the similar cloth jump into your eyes.

General methods on objects classification are based on the Bag of Words model (BOW) [2-3] to construct a visual vocabulary. The BOW is a popular form of object representation well applied in the near-duplicate/planar images. However the images captured from mobile robots are easily influenced by the light, viewpoint, complex background and other factors, the same object from different images in various viewpoint or light condition often change great, so how does BOW work well for object classification for mobile robots? In most cases, the construction of a good classifier relies on many samples. When it comes to some specific tasks for mobile robots in which capturing samples becomes hard and the labeling becomes time-consuming, how can an efficient classifier be constructed using those small samples?

Among mobile robot environment, the key to improve the performance of object classification is accurately and efficiently to obtain the object in the image. Under the premise of reducing the computational cost, the accuracy and real-time performance of classification should be guaranteed. Since the object can be anywhere in an image, the computation complexity of a traditional exhaustive search method in the whole image is heavily dependent on the size of the image.

Take these into account, in this paper, a method based on Hierarchic Selective Search (HSS) comes up aiming at achieving better classification for the mobile robots. The method uses the small and high-quality object regions set segmented by HSS instead of the whole image to build visual vocabulary and construct classifier. HSS is a newly improved image segmentation method based on region merging. Considering the optimizing of merging criterion, the HSS uses multiple and complementary similarity measurements to get a combined similarity measurement with adjustable weights adapted to the environment of mobile robots during its once running. As a result, HSS method generates a smaller, higher-quality object regions set through an image and reduce the later computation cost for classification. Figure 1 shows the structure of objects classification for mobile robots using HSS.

2 Theoretical background

2.1 Graph-based image segmentation model

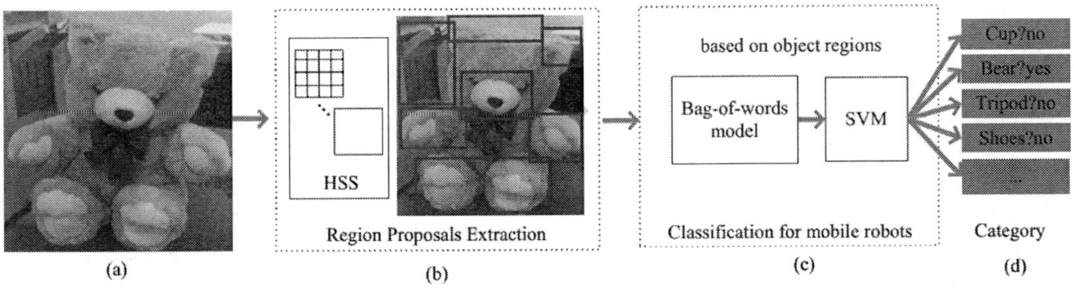

Figure 1. Classification for mobile robots using HSS .

Many scholars use the graph theory to describe the image segmentation problem. For an undirected graph $G=(V,E)$, each node $v_i \in V$ corresponds to a pixel of the image, the edge $(v_i, v_j) \in E$ connects a pair of adjacent pixels v_i and v_j, the weight of edge $w((v_i, v_j))$ measures dissimilarity (e.g., intensity, color, texture or other attribute) between the neighboring elements v_i and v_j with a non-negative value. A segmentation S is a dividing of V into different regions. The general methods on evaluating the quality of image segmentation require the pixels in a region to be similar, that is, the weight of two adjacent pixels in the same region is low and the weight of two neighboring pixels in different regions is high.

Define the *internal difference* $Int(C)$ of one region $C \subseteq V$ as the largest weight in minimum spanning tree [4] of the region, $MST(C, E)$. that is,

$$Int(C) = \max_{e \in MST(C,E)} w(e) \tag{1}$$

Define the *difference* between two regions $C_1, C_2 \subseteq V$ to be the minimum weight edge connecting the two regions. That is,

$$Dif(C_1, C_2) = \min_{v_i \in C_1, v_j \in C_2, (v_i, v_j) \in E} w((v_i, v_j)) \tag{2}$$

By comparing the minimum of the two regions *internal difference* $MInt(C_1,C_2)$ and $Dif(C_1,C_2)$ to check if there exists real boundary $D(C_1,C_2)$. 'true' denotes there exists boundary between the two regions while 'false' denotes not exists. The pairwise comparison predicate is defined as follow:

$$D(C_1, C_2) = \begin{cases} true & if\ Dif(C_1, C_2) > MInt(C_1, C_2) \\ false & otherwise \end{cases} \tag{3}$$

The threshold function τ controls the degree to which the difference between two regions must be greater than their *internal differences*. k is a constant number used to control the scale of region.

$$MInt(C_1, C_2) = min(Int(C_1) + \tau(C_1),\ Int(C_2) + \tau(C_2)) \qquad \tau(C_i) = k/|C_i| \tag{4}$$

The segmentation criteria has a great impact on the complexity of the image segmentation. In order to make it more robust to those outliers, finding an effective segmentation method has proved to be a NP-hard problem [5]. Based on the pairwise comparison predicate shown in (3) and greedy decision, a segmentation method is proposed in [5] which is proved neither too coarse nor too fine, and its running time is nearly linear in the number of image pixels, so it was often used to produce the initialized segmentation regions. The algorithm is well described in Figure 2.

Input: Graph $G=(V,E)$, n vertices and m edges
Output: A segmentation of V into regions $S=(C_1,C_2,...,Cr)$

1. Sort E into $\pi=(o_1,...,o_m)$ by non-decreasing weight;
2. Start with a segmentation S^0, where each v_i represents its region;
3. Initialize $q=1$, while $(q<m+1)$ repeatedly construct S^q given S^{q-1} with step 4;
4. Let $o_q=(v_i,v_j)$ where v_i in $C_i^{q-1} \in S^{q-1}$ and v_j in $C_j^{q-1} \in S^{q-1}$, if $w(o_q)<MInt(C_i^{q-1}, C_j^{q-1})$ and $C_i^{q-1} \neq C_j^{q-1}$ then merge C_i^{q-1} and C_j^{q-1} in S^{q-1} to form S^q otherwise $S^q=S^{q-1}$; update $q=q+1$;
5. Output $S=S^m$.

Figure 2. Segmentation algorithm.

2.2 Bag of words model

The Bag of words model was initially proposed in [2]. It can be described as four steps: (1) extract visual words from images by local feature descriptors like SURF [6], SIFT[7-8], PCA-SIFT [9], (2) construct the visual vocabulary by the clustering algorithms like k-means [10-11] and the random clustering forests [12], (3) quantify the images by the histogram of the extracted words in the vocabulary, and (4) use the sample images to train and test the classifier. The direct using of BOW model has been proved to be well applied to the classification of the repeated/near-duplicate images. However the irrelevance of space information to vocabulary makes the accuracy vulnerable to the impact of complex background from mobile robots.

As an extension of the BOW, in this paper, the hierarchic object regions with spatial information are used to be the inputs of the classification method based on BOW and SVM instead of the whole image. For the less background information interferes the construction of the vocabulary, it is exactly well applied to deal with the various and complex background of images captured from mobile robots.

2.3 Classification of object regions

The classifiers appeared in the recent papers on the objects or scene recognition focusing on the K-nearest neighbor algorithm, Support Vector Machine (SVM) [13-14] and so on. The SVM dexterously applies the kernel function to avoid the search of mapping algorithm and the calculation of high-dimensional space. By using different kernel functions and parameters, the SVM get the curved interface to the input space.

In this paper, the LIBSVM [15] is used for the selected object regions classification. For the minimizing of SVM shrinking and caching, the L1-norm is used when dealing with the regions and an optimized sampling strategy is used to remove the redundant regions before the BOW model.

Input: Color image
Output: Object location boxes $\boldsymbol{L}$

1. Initialize the thresholds K and the similarity measurements set *similarityset*;
2. Concurrently extract object location boxes for each threshold $k_l \in \boldsymbol{K}$ through step 3-6;
3. Get start regions $\boldsymbol{R}=\{r_1, ..., r_n\}$ with k_l using Figure 2 and initialize similarity set $\boldsymbol{S}=\emptyset$, $\boldsymbol{S}_{new}=\emptyset$ and $S_{struct}(r_i, r_j)=\emptyset$;
4. For each neighboring region pair (r_i, r_j), calculate the similarity structure set $S_{struct}(r_i, r_j)$ and update $\boldsymbol{S}=\boldsymbol{S} \cup S_{struct}(r_i, r_j)$, each structure contains five similarities $S_{simi}(r_i, r_j)$ with $simi \in similarityset$ and a combined similarity $S_{combined}(r_i, r_j)$;
5. While $\boldsymbol{S} \neq \emptyset$ do
   a. Get the structure with the highest combined similarity $S_{struct}(r_i,r_j)=\max(\boldsymbol{S}, combined)$;
   b. Merge r_i and r_j to form $r_{new}=r_i \cup r_j$;
   c. Dropout the similarities related to r_i: $\boldsymbol{S}=\boldsymbol{S} \setminus \boldsymbol{S}_{struct}(r_i,r_*)$;
   d. Dropout the similarities related to r_j: $\boldsymbol{S}=\boldsymbol{S} \setminus \boldsymbol{S}_{struct}(r_j,r_*)$;
   e. Calculate similarity structures $\boldsymbol{S}_{new}$ between r_{new} and its neighbors;
   f. Update $\boldsymbol{S}=\boldsymbol{S} \cup \boldsymbol{S}_{new}$;
   g. Update region set $\boldsymbol{R}=\boldsymbol{R} \cup r_{new}$;
6. Extract object region boxes $\boldsymbol{L}_l$ from all regions in $\boldsymbol{R}$;
7. Combine all the extracted boxes $\boldsymbol{L}_l$ from step6 to form and output $\boldsymbol{L}$.

Figure 3. Hierarchic selective search algorithm.

3 Hierarchic Selective Search

There are many methods to extract the object regions from various images. The popular methods contain Edge Box [16], Selective Search [17-18], Randomized Prim [19], Objectness [20] and so on. They are based on methods of threshold, edge, region, graph or energy functional. Each method has its own special advantages. Considering the computation cost and the various background of mobile robots images, in this paper, the graph-based image segmentation in Figure 2 is used to initialize the start regions, then the HSS based on region merging is used to form a higher quality, multiple scales and class-independent regions set.

The HSS currently deals with two start regions set for higher quality, the grouping procedure of a start regions set works as follows: firstly calculate the combined similarities between all neighboring regions; then the two most similar regions are merged together, and new similarities are calculated between the resulting region and its neighbors. The process of grouping the most similar regions is repeated until the whole image becomes a single region. The start regions with Figure 2 is well satisfying the requirement of a single object per box. For the regions are hierarchically merged into larger regions, it naturally generates regions at all scales. The detailed description of HSS is present in Figure 3.

3.1 Combined similarity measurement

While the images captured from the vision sensors on the mobile robot are device-dependent and vulnerable to the influence of illumination, occlusion and so on. Using multiple complementary similarity measurements has obvious advantages compared with a single measurement. In Figure 3, the combined similarity measurement is made of color and texture similarity measurements. A structure is used to keep all five single similarity in *similarityset* and a combined similarity *combined.*

3.1.1 Color space and texture feature

As for the color models, the RGB, Lab and HSV are used in this paper. The RGB simplifies the architecture and design but relies on the devices heavily. The RG channels of the normalized RGB plus intensity denoted as RGI. The Lab well describes the human visual perception. For its device-independent, it exactly appeals to the various mobile robots and covers the shortage of RGB. The HSV and RGI well describes the hue, saturation, light and shadows for invariance properties.

Texture feature is the intrinsic characteristic of all the surface of the object, it contains important information about the structure of object surface and the relationship with the surrounding environment. Also it reflects visual characteristics of the homogeneous phenomenon in the image without depending on the color and brightness information. In this paper, the fast SIFT-like method [21] is used to extract texture features.

3.1.2 Feature similarity measurement

From the similarity measurements such as Euclidean distance, histogram intersection and so on. In this paper, the distances of histograms on two regions' color and texture are calculated to measure similarities. The formula for computing the components of the combined similarity measurement in Figure 3 is shown below.

$$S_{simi}\left(r_i, r_j\right) = \sum_{l=1}^{n} min\left(simi_i^l, simi_j^l\right) \tag{5}$$

Subscript *simi* refers to the color or texture features in *similarityset={HSV, Lab, RGI, Intensity, Texture}*. For the color similarity measurement, a one-dimensional color histogram is obtained by using 25 bins for each color channel in each region, so the resulting color histogram for a three channel image is $C_i = \{c_i^1, \cdots, c_i^n\}$ where n=75. As for the texture similarity measurement, the method is based on fast SIFT-like and takes Gaussian derivatives in eight orientations using σ=1 for each color channel. Each color channel extracts a histogram using a bin size of 10, so the texture histogram for a three channel image is $T_i = \{t_i^1, \cdots, t_i^n\}$ where n=240. Both the color histogram and the texture histogram are normalised using the L1-norm. The color and texture histograms are effectively passed through the hierarchy. The formula is shown in (6) where CT_{result} refers to the color or texture histogram of the merged region.

$$CT_{result} = \frac{size(r_i) \times CT_i + size(r_j) \times CT_j}{size(r_i) + size(r_j)} \qquad (6)$$

3.2 Adjustable weights for various environments

While the images are captured in various scenes and conditions, the emphasis of a similarity measurement is not the same. In this paper, a flexible weight adapted to the environment of the mobile robots is given to the corresponding similarity measurement. For example, when the mobile robot is in an environment with similar color in foreground and background, a higher weight is given to the texture similarity measurement and lower to color similarity measurements. The combined similarity measurement is detailed described in (7) where num=5 is the number of similarity measurements, $simi_k$ refers to one similarity measurement in *similarityset*.

$$S_{combined}(r_i, r_j) = \sum_{k=1}^{num} a_k * S_{simi_k}(r_i, r_j) \qquad (7)$$

For the distribution of weights, it depends on the real environment of mobile robot, the paper does not give a standard distribution strategy, mobile robots can check environment to dynamically adjust weights in real scene.

In a word, HSS has changed the merging criterion by using a combined similarity measurement adapting to mobile robots various environment compared to Selective Search (SS) in [17]. While the SS contains multiple color similarity measurements and starting regions for many image cases. But for the color similarity measure in a special starting regions set, they just take a single color space into account during the whole grouping algorithm. HSS uses various color spaces during its once running. It seems that HSS get a intersection among the regions repeatedly running the SS with single color space. So the HSS get a smaller regions set with nearly the same quality compared with SS.

4 Experiments Evaluation

The SS performs well in the general object classification and image segmentation on the dataset Pascal VOC 2007 [22] with its fast and multiple strategies. So, in this paper, the HSS is compared to the SS on both Pascal VOC 2007 and OCS [23]. The OCS is a mobile product images dataset used for retrieving which contains 854 training samples (total 20 categories) and 60 testing samples (total 6 categories). While the Pascal VOC 2007 was an official dataset for image classification and segmentation in the ImageNet challenge [24], many efficient algorithms are evaluated on it, besides the same type of images have various status and most of images contain multiple objects. To some certain extent, these samples can be regarded as the images captured by the mobile robot in a better environment. Some testing images are shown in Figure 4. All experiments are done on the computer with Pentium Dual-Core E6700 3.2GHz, with 4G RAM using the Matlab, C and C++ development languages.

Figure 4. Some testing images on the dataset.

4.1 Experimental method

In the paper, the segmentation results are evaluated firstly. The HSS algorithm is tested just with thresholds 50 and 100. The color spaces include HSV, RGI, I (Intensity) and LAB. The weights corresponding to each similarity measurement are set manually according images global properties. For the OCS dataset only provides the sample images, in order to reuse the code, the OCS dataset is organized using the form of Pascal VOC 2007 dataset and firstly get the OCS ground truth annotations, and then calculate overlaps between the resulting region and the true object annotation as the assessment of the quality of segmentation algorithm. In this paper, the mean average best overlap (MABO) and mean number of object regions (MNOR) for all samples are used to comprehensively evaluate HSS, and then compared with SS. To calculate the Average Best Overlap for a specific class c, firstly the best overlap between each ground truth annotation $g_i^c \in G^c$ and the object regions L generated for the corresponding image is calculated.

$$Overlap(g_i^c, l_j) = \frac{area(g_i^c) \cap area(l_j)}{area(g_i^c) \cup area(l_j)} \qquad (8)$$

$$ABO_c = \frac{1}{|G^c|} \sum_{g_i^c \in G^c} \max_{l_j \in L} Overlap(g_i^c, l_j) \qquad (9)$$

$$MABO = mean\left(\sum_{c \in C} ABO_c\right) \qquad (10)$$

Later the results of object classification for mobile robots are evaluated based on the produced regions. Considering the real-time requirement of mobile robots, all regions are sampled to get the new training and testing regions set. Then the BOW model the represents object features from regions using the SIFT [8] descriptor and k-means method [10], finally the SVM classifier [13] is trained and the accuracy of prediction test regions is calculated. The same procedure is used for regions segmented from SS.

While scholars often use accuracy to evaluate method object classification. Suppose there are P positive samples and N negative samples, TP refers to the positive samples correctly predicted to positive sample, TN refers to the negative samples correctly predicted to negative sample, the accuracy is described as below.

$$Accuracy = (TP + TN) / (P + N) \tag{11}$$

4.2 Evaluation and analysis

4.2.1 Performance of HSS segmentation algorithm

Table 1 and 2 respectively shows the MNOR and MABO for all samples on Pascal VOC 2007 (VOC) dataset and OCS dataset, in which HSS50 and HSS100 separately refer to the condition of using HSS algorithm in threshold k_i= 50 and k_i= 100, SS100 refers to the condition of using SS in threshold k_i=100. The combined similarity is based on a color space and texture, Table 1 and 2 don't list the texture similarity measurement for SS just used a single color similarity measurement. Other tables are the same.

Table 1. Comparison under single color space on VOC dataset.

Similarity measurement	MNOR HSS50	MNOR HSS100	MNOR SS100	MABO HSS50	MABO HSS100	MABO SS100
HSV	471	288	463	0.691	0.686	0.693
Lab	328	189	318	0.698	0.690	0.690
RGI	370	214	362	0.693	0.684	0.693
Intensity	408	241	399	0.688	0.675	0.670

Table 2. Comparison under single color space on OCS dataset.

Similarity measurement	MNOR HSS50	MNOR HSS100	MNOR SS100	MABO HSS50	MABO HSS100	MABO SS100
HSV	489	303	429	0.583	0.572	0.580
LAB	277	162	214	0563	0.551	0.560
RGI	334	186	254	0.572	0.562	0.558
Intensity	378	219	304	0.560	0.552	0.566

Table 3 shows the hierarchical combination strategy result on VOC and OCS dataset. Considering the real object annotations from OCS dataset are based on the boxes rather than the shapes of objects, so the MABO is lower in OCS compared to its value in VOC dataset.

Table 3. Comparison under hierarchical combination strategy.

<table>
<tr><th>Dataset</th><th>Strategy(k_θ=50|k_I=100)</th><th>MNOR</th><th>MABO</th></tr>
<tr><td rowspan="3">Pascal VOC 2007</td><td>HSV+LAB|HSV+LAB SS</td><td>2147</td><td>0.799</td></tr>
<tr><td>LAB+RGI | HSV+I HSS</td><td>1976</td><td>0.815</td></tr>
<tr><td>HSV+RGI | I+LAB HSS</td><td>1988</td><td>0.818</td></tr>
<tr><td rowspan="3">OCS</td><td>HSV+LAB|HSV+LAB SS</td><td>1789</td><td>0.611</td></tr>
<tr><td>LAB+RGI | HSV+I HSS</td><td>1206</td><td>0.616</td></tr>
<tr><td>HSV+RGI | I+LAB HSS</td><td>1120</td><td>0.621</td></tr>
</table>

Observing from all table, HSS proposed in this paper is better than SS in the number of object regions. From the comparison under single color space when k_i= 100, HSS has the MNOR value less than SS nearly 1/3, while the change of MABO value is less than 0.01. Some MABO values from HSS are even better than the MABO from SS. From the comparison under the hierarchical combination strategy, HSS shows its superiority in both MNOR and MABO.

While SS repeatedly runs single color space strategy leading to the object regions redundancy, HSS gets the higher quality regions with less quantity using combined and environment-adjustable similarity measurement. The subsequent classification relies much on the segmented regions, so the quantity and quality of object regions have an important influence on the computation and accuracy of classification. Under the special mobile robots task and Pascal VOC 2007, HSS is relatively better than SS .

4.2.2 Performance of classification for mobile robots

Table 4 shows the results of the classification for mobile robots based on a special task dataset OCS. 'HSS-ORS' refers to the regions from HSS; 'SS-ORS' refers to the regions from SS; 'OI' refers to the original images. By changing the samples for classification among the three sources, the accuracy along with the quantity of regions used in training and testing are compared.

Table 4. Classification for mobile robots based on OCS.

Samples source	Training samples	Testing samples	Accuracy (%)
SS-ORS	5881	1101	48.97
HSS-ORS	3977	660	51.36
OI	186	60	45.00

Observing the data from above tables, HSS uses less regions than SS. Therefore, during the whole process of classification for mobile robots based on regions, the computational cost of HSS is lower than that of SS. From the accuracy, HSS is also better than SS. Compared with the classifier constructed using original images, the classifier based on object regions has higher accuracy. This is due to the improvement in the quality and quantity of samples based on object regions. The OCS dataset is relatively small and the background of training and testing samples belonging to the same category varies much. Less background interference information and more high-quality object region samples make the accuracy improved but at the expense of time cost. At present, while the computing abilities to mobile robots are quickly upgraded, most of the classification tasks tend to pursue the improvement in accuracy, so the advantages of HSS will be reflected more obviously.

5 Conclusion

Image segmentation plays an important role in improving the performance of object classification for mobile robots. With respect to the feature extracted in the whole image, the method based on object regions effectively reduces the interference of the background and computation cost; accurate object regions can significantly improve the classifier training and testing process and polish up the accuracy of classification in a large extent. As a method of regions capture, HSS used a combined similarity measurement to get a smaller and higher-quality regions set. At the same time, it was applied to the classification for mobile robots based on regions which relieved the background interference in mobile environment. When using the learning methods for object classification, the accuracy is much related to the quantity and quality of samples. In a certain extent, HSS can also solve the problem of data scarcity on some special task for mobile robot cooperating with the method based on regions. The later of this work will be extended to more other datasets for mobile robots tasks.

Acknowledgments

This work is supported by the National Science and Technology major project (No. 2011ZX05027-005). Thanks to all of the relevant persons.

References

1. S.B. Niku, Introduction to Robotics: Analysis, Control, Applications. *Prentice Hall/Pearson*, (2004).
2. J. Sivic, A. Zisserman, Video Google: A Text Retrieval Approach to Object Matching in Videos. *IEEE International Conference on Computer Vision*, 1470, (2003).
3. A. Sampath, A. Sivaramakrishnan, K. Narayan, A Study of Household Object Recognition Using SIFT-Based Bag-of-Words Dictionary and SVMs. *Proceedings of the International Conference on Soft Computing Systems. Springer India*, (2016).
4. C.T. Zahn, Graph-theoretic methods for detecting and describing gestalt clusters. *IEEE Transactions on Computing*, **20**:68-86, (1971).
5. P.F. Felzenszwalb, D.P. Huttenlocher, Efficient Graph-Based Image Segmentation. *International Journal of Computer Vision*, **59**(2):167-181, (2004).
6. B. Herbert, T. Tinne, V. G. Luc, SURF: Speeded Up Robust Features. *Computer Vision & Image Understanding*, **110**(3):404-417, (2006).
7. D. Lowe, Object recognition from local scale-invariant features. *IEEE International Conference on Computer Vision*, **2**:1150-1157, (1999).
8. Q. Zhu, C. Liu, C. Cai, A Novel Robot Visual Homing Method Based on SIFT Features. *Sensors*, **15**(10):26063-26084, (2015).
9. Y. Ke, R. Sukthankar, PCA-SIFT: a more distinctive representation for local image descriptors. *IEEE Computer Society Conference on Computer Vision and Pattern Recognition*, **2**:506-513, (2004).
10. N. A. M. Isa, S. A. Salamah, U. K. Ngah. Adaptive fuzzy moving K-means clustering algorithm for image segmentation. *IEEE Transactions on Consumer Electronic*, **4**(55): 2145-2153, (2009).
11. C.H. Lin, C.C. Chen, H.L. Lee, Fast K-means algorithm based on a level histogram for image retrieval. *Expert Systems with Applications*, **41**(7): 3276-3283, (2014).
12. G. Fanelli, M. Dantone, J. Gall, Random Forests for Real Time 3D Face Analysis. *International Journal of Computer Vision*, **101**(3): 437-458, (2013).
13. C. Cortes, V. Vapnik, Support-vector networks. *Machine Learning*, **3**(20): 273-297, (1995).
14. B.C. Kuo, H.H. Ho, C.H. Li, A Kernel-Based Featur-e Selection Method for SVM With RBF Kernel for Hyperspectral Image Classification. *IEEE Journal of Selected Topics in Applied Earth Observations & Remote Sensing*, **7**(7): 317-326, (2014).
15. C.C. Chang, C.J. Lin, LIBSVM:A library for support vector machines. *Acm Transactions on Intelligent Systems & Technology*, **2**(3): 389-396, (2011).
16. C. We, C. X. xin, and Z. Li, Box-counting dimension analysis of edge images detected with classical edge detector. *International Conference on measuring technology and mechatronics automation*, Changsha , China, 756-758, (2010).
17. J. Uijlings, K. Sande, T. Gevers, A. Smeulders, Selective search for object recognition. *International Journal of Computer Vision*, **104**(2): 154-171, (2013).
18. X. Wan , X. Zheng , X. Pang, Spatial Pyramid Pooling in Deep Convolutional Networks for Visual Recognition. *IEEE Transactions on Pattern Analysis & Machine Intelligence*, **37**(9): 1904-1916, (2014).
19. S. Manen, M. Guillaumin, L.V. Gool, Prime object proposals with randomized prim's algorithm. *International Conference on Computer Vision,* 2536-2543, (2013).
20. B. Alexe, T. Deselaers, V. Ferrari, Measuring the objectness of image windows. *IEEE Transactions on Pattern Analysis & Machine Intelligence*, **34**(11): 2189-2202, (2012).
21. J. Rabin, J. Delon, Y. Gousseau, A contrario matching of SIFT-like descriptors. Pattern Recognition, *IEEE International Conference on. Pattern Recognition*, 1-4, (2009).
22. M. Everingham, L.V. Gool, C.K.I. Williams, J. Winn, A. Zisserman, The PASCAL Visual Object Classes Challenge 2007 (VOC2007) Results. http://www.pascal-network.org/challenges/VOC/voc2007/works-hop/index.html, (2007).
23. S.X. Hui, Z. Lin, B. Jonathan, W. Ying, Mobile product image search by automatic query object extraction. *European Conference on Computer Vision*, 114-127, (2012).
24. J. Deng, A. Berg, S. Satheesh, H. Su, A. Khosla, L. FeiFei, ImageNet Large Scale Visual Recognition Competition 2014. http://www.imagenet.org/challen-ges / LSVRC/ 2014/, (2014)

DOI: 10.1051/matecconf/20166803002

Industrial Robotics Platform for Simulation Design, Planning and Optimization based on Off-line CAD Programming

Khelifa Baizid[1, 2], Amal Meddahi[3], Ali Yousnadj[4], Saša Ćuković[5] and Ryad Chellali[6]

[1]*Mines Douai, IA, F-59508, Douai, France*
[2]*Univ. Lille, F-59000 Lille, France*
[3]*DIEI, University of Cassino and Southern Lazio, Cassino, Italy*
[4]*DGRSDT, Advanced Technology Development Center, Algeria*
[5]*FE, University of Kragujevac, Kragujevac, Serbia*
[6]*CEECS, Nanjing Robotics Institute, China*

Abstract This paper presents IRoSim: Industrial Robotics Simulation Design Planning and Optimization platform which we developed based on SolidWorks API. The main objective is to integrate features from mechanical and robotics CAD software into the same platform in order to facilitate the development process through a friendly interaction interface. The platform provides important steps to develop a given robotized task such as: defining a given task, CAD learning of the end-effectors' trajectory, checking the manipulator's reach-ability to perform a task, simulating the motion and preventing the trajectory from possible collisions. To assess the usability of the proposed platform, a car's doors painting task using a 6 Degree Of Freedom industrial manipulator has been developed.

1 Introduction

Nowadays, programming of industrial robots is of importance to many manufacturing industries, because changing and reinventing their production systems continuously appear within the manufacturing environment [1, 2]. However, industrial robots programming generally is a tedious and time-consuming task that demands significant technical expertise, and require a tremendous amount of programming to make them useful. Also, in order to accomplish such task skilled and experienced programmers are often a scarce resource. Nevertheless, new and effective programming approaches that are easier, faster and advanced with low cost are constantly sought such as conventional teaching pendant [3] and Off-Line Programming (OLP) environments. Usually, such environments are based on graphical simulation platforms, where programmers need only to learn the simulation language and not any of the robot programming languages. Moreover, OLP environments include various libraries such as tools for the simulation scenarios and some predefined applications of robotics tasks like spot-welding and painting. And the most important is that it can allow the kinematics control to enable users to plan collision-free trajectories. The simulation may also be used to determine the cycle time of execution of the robotics task.

In the last decades, OLP coupled with Computer Aided Design (CAD) leaded to learning trajectory through teach pendant [3, 4]. As a result of combining these technologies make it possible to reduce both the time and the cost of development process, and improve the quality of the product. Therefore, CAD systems have been in use for several industry process including, automobile, aircraft manufacture and shipbuilding. These fields, employ in their systems CAD and embedded knowledge to design the robot mechanism and to simulate its motion [5]. However, many of these systems do not rely on standard CAD packages such as SolidWorks and CATIA, which presents supplementary difficulties for novice users to model and to export their 3D models to simulate a given task. Whereas, standard CAD systems that are used for design do not provide non-free packages that can be used for robotics task simulation. Thus, our objective in this approach is to provide an easier process, to define a robotized task, integrated into a 3D CAD system.

In this paper, we present a novel Industrial Robotics Simulation (IRoSim) CAD robotics system that has been realized to assist users in designing and developing industrial robotized tasks. IRoSim is accessible to anyone with basic knowledge of CAD and robotics. As shown in Figure 1, the system offers users to simulate several steps to fulfill the desired objectives.

IRoSim is integrated into SolidWorks and can easily be upgraded to future versions. Unlike programming packages [3, 6], IRoSim provides several features such as 3D design of the robotic manipulator, definition of the task to be accomplished, accessibility verification, optimization of the task time execution [7], graphical simulation, detection of the eventual collisions and generation of a collision-free trajectory to be directly

 DOI: 10.1051/matecconf/20166803002

mapped to the real cell. In order to show the effectiveness of the proposed software we implemented a robotized task which is the painting (by dipping) using a 6-DOF robot manipulator.

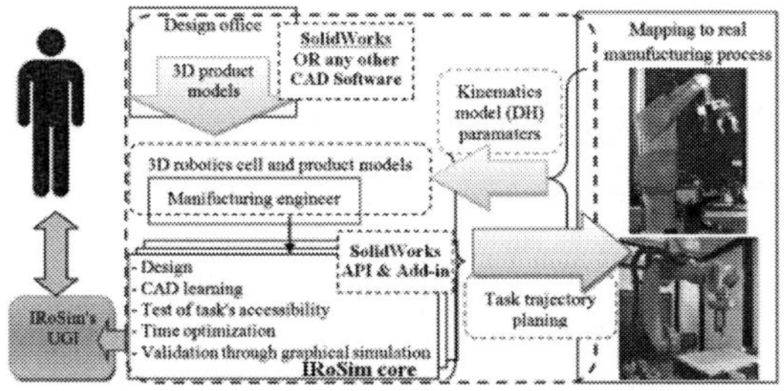

Figure 1. Offline-CAD programming schema of IRoSim.

2 CAD based approach

Nowadays, the CAD model databases are widely used in the industry to design, model and programming robotics products [3, 6] such us AutoCAD, Ideas, Cimatron, CATIA and SolidWorks API (Application Programming Interface). We used this last to offer the feature to calculate the robotics task parameters based on a virtual 3D model of the real robotics cell. Moreover, the modification and the redefinition of these parameters are easily possible with the developed user-friendly integration.

2.1 Data Acquisition

In order to evaluate the robot control through an OLP, data acquisition systems must be built into the control program. In our approach, it consists of the 3D virtual representation of the physical model of the robotized cell, followed by the application of several algorithms based on CAD learning to extract useful information from the virtual environments. This gives definition of the task and delimits some kinematics constraints between the robotic manipulator and the task. Figure 2 summarizes the process of data acquisition.

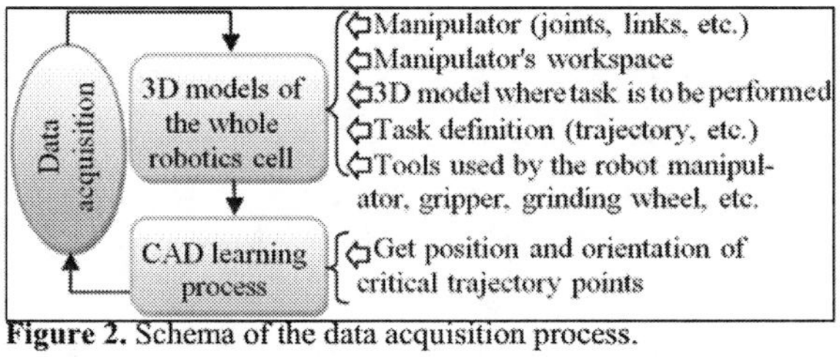

Figure 2. Schema of the data acquisition process.

2.2 3D Modelling of Robot and Environment

2.2.1 3D modeling

3D modeling of the robot cell [8] can be contacted based on a volume and/or surface. This can include the robots' libraries, End-EFfectors (EEF), the 3D model where the task is to be performed and the remaining additional parts. For instance, a robot's WorkSpace (WS) can be modeled only by volume. This allows having more geometrical information that can be used during the simulation process. However, other 3D objects can be modeled by surface if no need to a volume representation. Figure 3 shows some parts of a robotics cell which contains a robot manipulator, gripper, two conveyors and drying machine. The manipulator was designed in a way that allows possible kinematic joints representation of the corresponding links subject to certain numerical constraints (e.g., DH: Denavit–Hartenberg parameters).

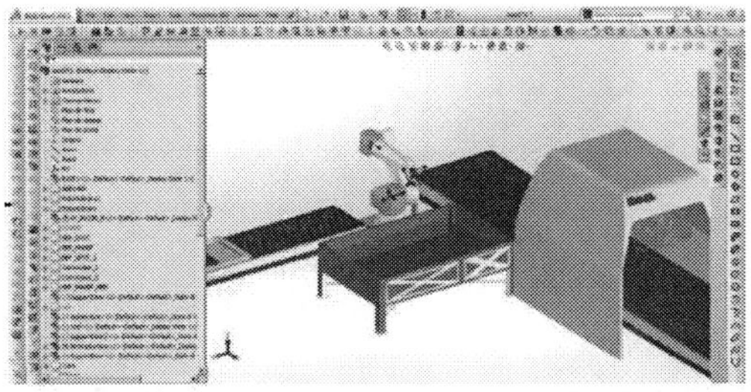
Figure 3. IRoSim development environment with 3D models of robot manipulator, conveyors, etc.

2.2.2 Task definition

Robotic task can be represented as a 2D/3D set of points to be visited by the manipulator [9]. We distinguish two kinds of trajectories: Point-to-Points (P2P) and Continuous. P2P trajectories may include spot welding and drilling tasks while arc-welding, laser cutting, painting, polishing [10] etc. are modeled using continuous trajectories where we used a 3D spline. Figure 4 presents critical points of grasping, painting and releasing car's doors.

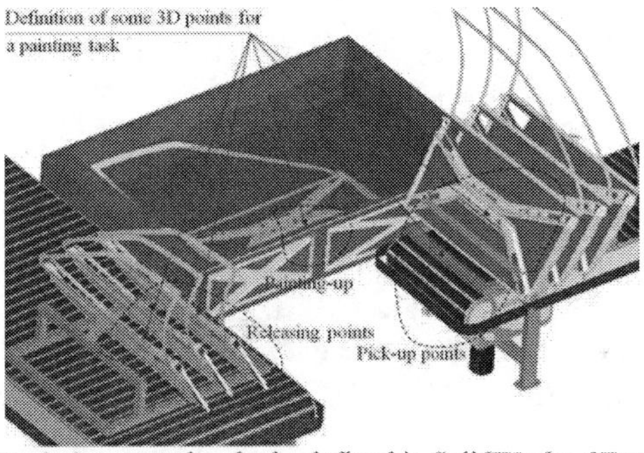

Figure 4. An example of asks defined in SolidWorks, 3D points are the picking-up, painting and releasing locations.

2.2.3 Robotics library

IRoSim platform includes a set of robot manipulators such as Staubli and KUKA KR6-2. 3D models of these manipulators can be integrated and assembled easily into SolidWorks; and the platform is open to integrate new manipulator models. Moreover, the library contains several other 3D tools that are needed for the simulation.

3 CAD programming stages

IRoSim platform is integrated as a plug-in with SolidWorks software based on Microsoft Visual Basic

Application (VBA). As shown in Figure 5, we present it as Product Lifecycle Management (PLM) software that develop a set of stages (Figure 5) in order to allow the user to define, simulate and validate a given industrial robotics task, without exporting all models from the design environment to the simulation software.

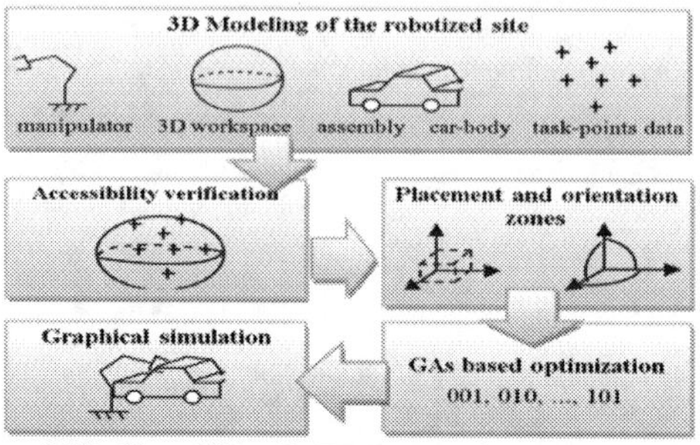

Figure 5. IRoSim offline-CAD programming steps.

3.1 CAD learning

The proposed learning process is provided by GUI of IRoSim platform. We select first: a 3D CAD model of the task (e.g., Figure 3 and 4) and a realistic 3D model of the EEF whose dimensions are specified in the main GUI of the IRoSim. User can move the 3D model of the EEF in the graphical area by translations and rotations around X, Y and Z axes using IRoSim's GUI. Then, we need to get the critical information that is helpful for learning more complex information from the graphical area. Figure 6 presents steps and pseudo-code to involve this process. Considering a P2P trajectory, we first get coordinates of the task points. From these point's coordinates we can extract the orientation at each point as the normal of the surface holding the point (i.e., a coincidence constraint between the surface and the point). So, we have 4-DOF (1 orientation and 3 positions) which will help in the next stage of learning.

The second step (Figure 7) involves tuning of the EEF 3D model in the virtual environment. The designed GUI of IRoSim platform offers features to displace and rotate the EEF around all possible axes. In case of a task involving a P2P trajectory e.g. grasping, this process is executed for two locations; the pre-grasping and the grasping itself.

3.2 Task accessibility checking

In order to accomplish the robotics task accurately and precisely, the robot should be able to reach all task points defined by the user. The task's accessibility check, in our platform, is based on: generating the 3D model of WS, representation of the task points and finally checking the interference between them.

The shape of 3D WS of a robotic manipulator is a function of its mechanism that includes DH parameters, link lengths, joints' range of motion etc. Since several industrial manipulators have the first three joints used for positioning, while the last three are mainly for orientation [11], we considered, in our approach, only the first three joints to generate the 3D model of the WS from DH parameters. Generally, we used the second and the third joint to generate a sketch in a vertical plane (parallel to the z axis of the first joint). So this joint is used for creating the volume function (e.g., revolution or extrusion according to the joint type revolute or prismatic, respectively). The algorithm that depicts the main sequence involved in this step is given in [5]. A set of different mechanisms for a manipulator and their corresponding 2D sketches and 3D WS are also given in [5] based on this algorithm.

Figure 6. CAD learning of the critical task points with some pseudo-codes.

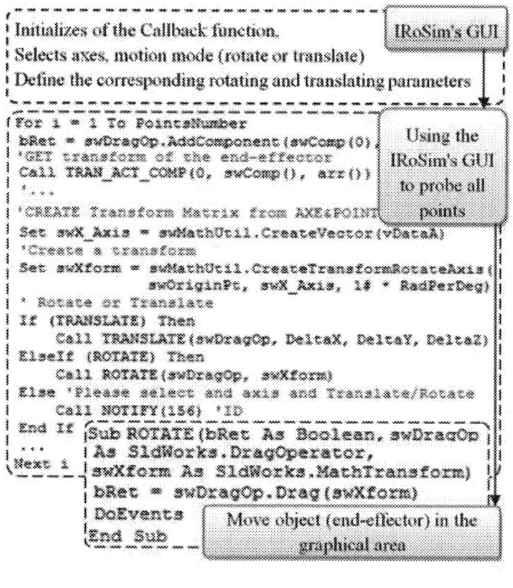

Figure 7. CAD learning of the critical transformation at each task point with some pseudo-codes, the objective here is to define the homogenous transformation matrix at each point.

In the proposed platform, tasks accessibility verification process is based on fully graphical. Using Oriented Programming Objects (OPO) of SolidWorks's API, we use geometrical features of the task points and the 3D WS model of the manipulator to check the interference between them. The interference is checked between parallelogram-shaped temporary 3D bodies that we create around each task points and the WS body of the

 DOI: 10.1051/matecconf/20166803002

manipulator. In case the task points are not reachable, the IRoSim GUI offers possibility to redefine the task or to load another manipulator with different DH parameters if necessary to accomplish the task.

3.3 Placement and orientation zones

We developed platform that offers the possibility to optimize the best time of the task execution. In order to achieve this we need to find the shortest distance travelled by the manipulator in the coordinate space. It is reported previously in [12] that the placement and the orientation of the robotic manipulator are of the most importance in time optimization. In other words, the manipulator can be displaced on three axis coordinate system and can also rotate around these axes to obtain the suitable placement and orientation relative to the task by ensuring the task reachability.

3.4 Time optimization

The main motivation to have an industrial framework providing all necessary stages needed by a robotics task lead us to include time optimization algorithm in our IRoSim platform. This optimization depends, in the first case, on the distance traversed by the manipulator's joints, which is related to the sequence of the task-points visited by the EEF. Also, the manipulator executes a specific task in the operational space and performs the motion in the coordinate space, which makes the travelled distance a function of manipulator's IKM. In that sense, the optimization problem is concerned with finding the minimum distance between each two consecutive trajectory-points. Moreover, any solution of a IKM joint is strongly affected by the placement and the orientation of manipulator. Thus, in order to have a complete optimization method which performs the strategies presented by authors in [12, 7] we proposed an optimization method based on Genetic Algorithms (GAs) that incorporates the best cycle time of task execution as objective function, while the chromosome is composed of: order of achievement of the task-points, IKM at each task-point, relative placement and orientation between the manipulator and the task-points. The cycle time has been computed based on joints' displacement between each successive pair of points and the average velocity of the corresponding joint.

4 Graphical simulation

All graphical objects, modeled as 3D model can be displaced or moved by acting on its transform. During the motion of the 3D objects we solicit API futures to detect the collision with the surrounding. Figure 8, bellow, shows a simplified algorithm of motion of the robot manipulator's joint from one configuration to another. It start by adding the corresponding joints that are considered to be moved, gets the current joints axes (Z), creates an axis object, transforms and finally moves the joints by updating the 3D graphical area, in the main time obstacle avoidance algorithm is running on the fly.

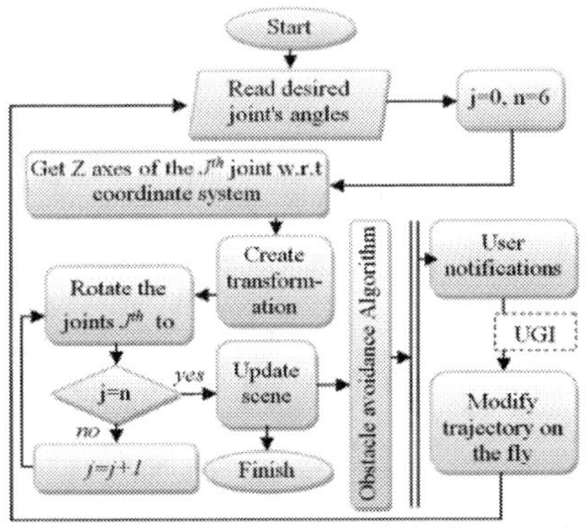

Figure 8. Algorithm of 6-DOF robot manipulator motion.

5 Case of study

Based on IRoSim, we present a simulation case study that involves an industrial task which is painting of a car's doors (by dipping).

5.1 Task definition, accessibility and CAD trajectory learning

As shown in Figure 4 each point of the task is constrained on the given surface of the corresponding car's door. Figure 9 shows snapshot of CAD trajectory learning GUI.

In details Figure 9a shows the IRoSim main GUI and the corresponding CAD learning trajectory. It is worth to note that Euler angles between the initial EEF configuration *Conf_0* and the first configuration *Conf_1* are given in [11]. As shown in Figure 9c user must rotate the EEF using the GUI to reach the grasping configuration *Conf_g*,. When this is done, user translates backward following Z axis to reach the pre-grasping configuration *Conf_p*. At the mean time he/she mast save the two matrices of *Conf_g* and *Conf_p* (Figure 9d) and finally he/she checks the task accessibility (Figure 9e) by moving the workspace manually around the task and checking if all the points are reachable, or automatically by running the corresponding algorithm.

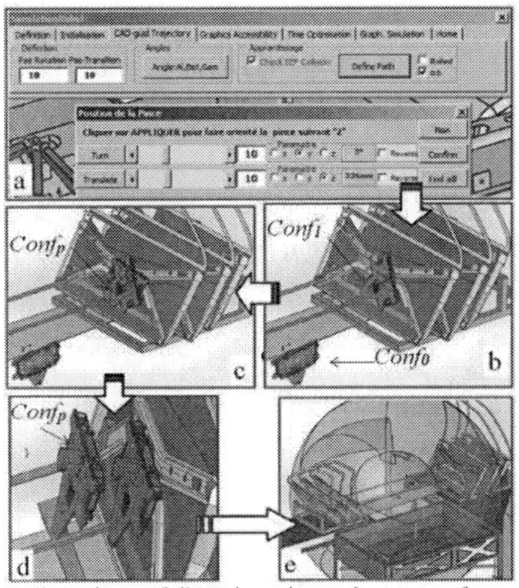

Figure 9. Snapshots of CAD learning trajectory and accessibility verification process.

5.2 Simulation

After optimizing the task's time, the last step is the graphical simulation and validation of the task trajectory.

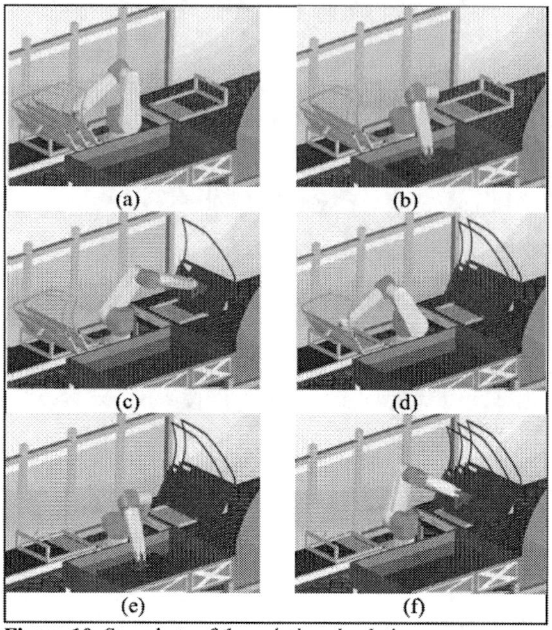

Figure 10. Snapshots of the painting simulation process.

It is worth to mention that the order of visiting points is fixed because of the highly constrained task, where we have only one sequence. For instance, the manipulator must grasps the support from the first conveyor and puts it on the second one, then starts transporting the first door (Figure 10a) and paints it in the basin (Figure 10b) and then dropping it on the second support posed, earlier, on the second (big) conveyor (green color model). Such operation is repeated till the last door (Figure 10c to Figure 10f). Finally, and after doors arrive to the drying machine the manipulator transports the last support to prepare for the next a round of the task. It is worth to note that collision avoidance is resolved by adding intermediate points in the trajectory of the robotic manipulator. A video can be found in this link: https://www.youtube.com/watch?v=im5qGO7Ky5Y

6 Conclusion

A novel CAD-based OLP platform IRoSim for industrial robots has been presented. A platform that is able to define, optimize, simulate and validate an industrial task and then map it into a real site. The proposed GUI of this platform is fully integrated with SolidWorks API and it provides many utilities that can cope with several industrial tasks including but not limited to; drilling, painting, spot welding, pick and place and other object manipulation tasks. Since the creation of the CAD models and robot programming task are performed in the same software the entire robot programming process becomes faster, easier and cheaper. In addition, the case of study of the painting task showed that the proposed platform is intuitive to novice use in industrial environments. In near future, we are looking forward to integrate some feedback control via cameras, sensors and other embedded systems.

References

1. P. Zengxi, Recent Progress on Programming Methods for Industrial Robots, *Robotics and Computer Integrated Manufacturing*, **28**(2), 87-94 (2012)
2. B. Hein, M. Hensel, H. Wrn, Intuitive and model-based on-line programming of industrial robots: a modular on-line programming environment, *IEEE International Conference on Robotics and Automation*, pp. 3952-3957 (2008)
3. P. Neto, N. Mendes, Direct off-line robot programming via a common CAD package, *Robotics and Autonomous Systems*, **61**, 896-910 (2013)
4. S. Mitsi, K. D. Bouzakis, G. Mansour, D. Sagris, G. Maliaris, Off-line programming of an industrial robot for manufacturing, *The International Journal of Advanced Manufacturing Technology*, **26**(3), 262-267 (2004)
5. A. Meddahi, K. Baizid, A. Yousnadj, J. Iqbal, API based graphical simulation of robotized sites, *14th IASTED Robotics and Applications Conference*, 458-492 (2009
6. H. Chen, W. Sheng, N. Xi, M. Song, Y. Chen, CAD-based automated robot trajectory planning for spray painting of free-form surfaces, *Industrial Robot: An International Journal*, **29**(5), 426-433 (2002)
7. K. Baizid, R. Chellali, A. Yousnadj, A. Meddahi, B. Toufik, Genetic Algorithms Based Method for Time Optimization in Robotized Site, *IEEE/RSJ International Conference on Intelligent Robots and System (IROS)*, 1359-1364 (2010)
8. Zeghloul, B. Blanchard, M. Ayrault, SMAR: A Robot Modeling and Simulation System, *Cambridge journal*, **15**(1), 63-73 (1997)
9. P. Th. Zacharia, N. A. Aspragathos, Optimal Robot Task Scheduling based on Genetic Algorithms, *Robotics and Computer-Integrated Manufacturing journal*, **21**(1), 67-79 (2005)
10. P. Neto, N. Mendes, R. Arajo, J.N. Pires, A.P. Moreira, High-level robot programming based on CAD: Dealing with unpredictable environments, *Industrial Robot*, **39**(3), 294-303(2012)
11. W. Khalil, E. Dombre, Modlisation identification et commande des robots, *HERMES Science Publications 75004*, (1999)
12. K. Baizid, A. Yousnadj, A. Meddahi, R. Chellali, J. Iqbal, Time scheduling and optimization of industrial robotized tasks based on genetic algorithms, *Robotics and Computer-Integrated Manufacturing Journal*, **34**, 140-150 (2015)

DOI: 10.1051/matecconf/20166804001

Effects of Door Width and Human Body Size on Walking Speed

Siwalee Jetthumrong and Phairoat Ladavichitkul

Industrial Engineering Department, Faculty of Engineering, Chulalongkorn University, Bangkok, Thailand

Abstract. Door width is one of the important factors to concern in layout or facilities design because it affects directly to traffic speed and overall traffic time simultaneously. Nowadays, common assessment method is computer simulation which is still not realistic due to the unchanged speed of model while walking through a door. This research aims to study an effect of door width to individual walking speed. Sixty subjects participated in the experiment and performed task by walking through the door that is set the width as 40, 50, 60, 70, 80, 90 and 100 centimetres. The optical motion capture system was used to determine walking speed. The results showed that Fitts' law was applied to the participants with high weight. Door width below 70 centimetres significantly affected to changing speed at 0-0.5 m. before the door. Additionally, human size also affected changing speed. The factors include shoulder breadth, weight and interaction between shoulder breadth and weight were found to be significant. These factors explained 54.2% of changing speed.

1 Introduction

In places and buildings usability, passageway width is one of the important factors to be considered because it affects to overall traffic time. As we know that time is a valuable resource. In the city, people usually travel hurriedly but the problem of overcrowded people in public area is very common. Especially at the rush hours in some public places such as subway stations or even airports, amount of passengers decrease extremely but the traffic area is restrict. This may cause retardation in pedestrian traffic. As a result, it causes late in transportation that may affect to various issues such as society or economics. Moreover, passageway width becomes a critical factor in case of emergency because traffic time represents the evacuation time. If the traffic area especially the exit is not wide enough, it cannot support evacuation of all people lead to injuries and losses. Thus, an assessment of passageway width in public place is required when the situations change. If the impropriety between amount of people and traffic area is indicated, this will lead to further solutions.

There are two common assessment methods including hand calculation and computer simulation. Hand calculation usually follows the equations defined by National Fire Protection Association Handbook [1]. The equations treat the occupants as fluid particles which their individual speed depends on the density of the area.

$$S = K - 0.266KD \qquad (1)$$

When S is the walking speed throughout the line, K is the velocity factor which depends on egress components such as corridor or stairway. D is the density of the area in person/m^2.

Computer simulation program, which is widely used to support layout and facilities design, will simulate occupant's behaviour and finally calculate the overall egress time. However, the movement of human model is still not realistic. The walking speed of the model will decrease only when the density of the area becomes higher [2].

In fact, human have a complex cognitive process called perception. When human perceive that they are approaching to the narrower area such as a door, they will automatically decrease their walking speed in order to make sure they are able to control themselves. In other words, door width has an influence on human walking speed. This follows the concept of the speed and accuracy trade-off theory which describes human's performance in both physical and cognitive task. According to Fitts' law [3], the movement time from target to target increases by the difficulty of the task which depends on target width and target amplitude. Similarly, narrower door should cause more speed reduction. The accuracy is needed to maintain safety so walking speed should decrease by target width which refers to door width.

Generally, human walking speed depends on both physical factors such as age, height [4], weight or body proportions and psychological factors include emotion or even personality type. Moreover, walking through the door is also related to motor control ability which is affected from various factors such as age, fatigue, or gender. Men are likely to response faster [5], combine with the risk-taking behaviour of women that lower than men [6].

If the effect of door width on walking speed can be identified, the actual walking speed can be predicted. This will lead to more accuracy in pedestrian simulation. Human characteristics which can affect walking speed include

DOI: 10.1051/matecconf/20166804001

gender, weight, height and body width are also considered in this experiment.

2 Method

2.1 Participants

Sixty college students (33 males and 27 females) participated in this experiment. Age range was 20-29 (21.98±2.03 years). All participants were healthy and did not have any medical records in hip and knees. Body strength was normal, not athlete. Normal sights or wore glasses/contact lens through the experiment.

2.2 Apparatus

Martin-type anthropometer was used for measuring body proportions of all participants. In order to determine walking speed, OptiTrack motion capture system was used for collecting data at sampling rate of 50 frames per second (fps). The system consisted of twelve OptiTrack FLEX 13 cameras and tracking software, Motive:Body version 1.7.5 (NaturalPoint, Inc., Corvallis, OR). The capture area was 4.5 metres x 3.9 metres, completely covered by non-slip mat.

2.3 Procedure

2.3.1 Experimental setup

First, the system was set in the Human Factors and Ergonomics laboratory, Chulalongkorn University. Before the experiment each day, system calibration was performed. The calibration procedure consisted of wanding the entire capture volume and setting ground plane. The mean error of calibration was needed to be around one millimetre, otherwise the calibration process should be repeated.

The door was formed within the capture area by two rigid rectangle Polypropylene-corrugated boards with 22 centimetres width and 1.9 metres height. Each board had reflective markers (diameter = 14 mm.) on it as the reference. Seven different door widths were considered in this experiment: 40, 50, 60, 70, 80, 90 and 100 centimetres. For each reconfiguration, the width was measured by measuring tape and would be accepted if the difference from desired width was not exceed five millimetres.

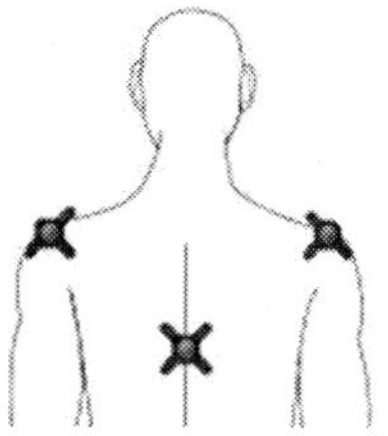

Figure 1. Position of markers.

The following anthropometric data of all participants were measured: weight, height, shoulder breadth and hip breadth. Height was a stature from top of the head to floor. Shoulder breadth was the width between left and right acromion. Hip breadth was the width between left and right pelvis. After the measurement, each participant wore t-shirt or tight-fitting clothes. Three reflective markers were attached on participants' body using velcro tape as shown in Figure 1.

2.3.2 Data collecting

The task was walking through the door. All participants were briefed to walk as fast as they can but did not think of emergency situation. Each participant performed the task one by one in random sequence of door width. At the beginning of each task, the participant stood at start point which was three metres away from capture area. All tasks were captured by motion capture system since participants appeared in the capture area until they walked through the door. The total distance was 6.9 meters: 5.5 meters before the door and 1.4 meter after the door (see Figure 2).

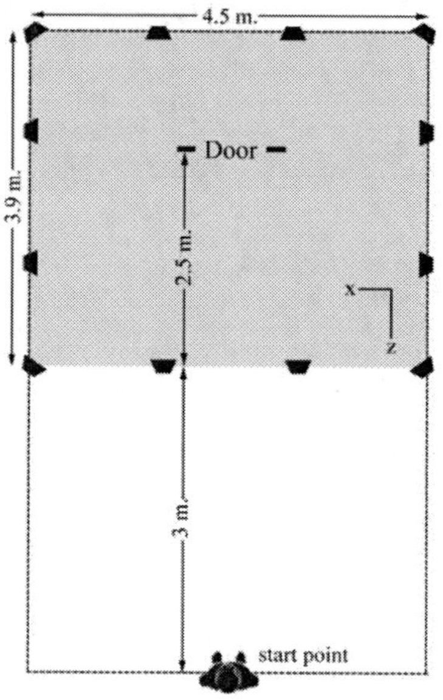

Figure 2. Experimental layout.

Data verifying was carried out right after each trial in order to prove that participant's body did not hit the door. If the markers on the boards moved from their positions more than five millimetres, that trial would be rejected. The repetition would not exceed three times in order to avoid learning-by-doing behavior.

The results consisted of :

60 Participants ×

7 Conditions of door width ϵ {40 cm., 50cm., 60 cm., 70 cm., 80 cm., 90 cm., 100 cm.}

= 420 trials

2.3.3 Data processing

After the experiment, all data obtained would be processed and manually calculated to walking speed in the walking direction. Since the data is 3D positions of markers, walking speed in z-axis was;

$$v_i = (z_{i+1} - z_{i-1})/(t_{i+1} - t_{i-1}) \quad (2)$$

When v_i is an instantaneous speed at frame i, z_{i+1} and z_{i-1} are positions in z-axis at frame i+1 and i-1, respectively. t_{i+1} and t_{i-1} are time at frame i+1 and i-1, respectively.

Changing speed at specific distance was provided by the difference of initial speed and terminal speed. In order to normalize value, changing speed was then divided by initial speed.

$$\Delta v = (v_1 - v_0)/v_0 \quad (3)$$

When v_1 is the terminal speed and v_0 is the initial speed.

2.4 Variables

In this study, the dependent variable is changing speed (Δv). The independent variables considered in this experiment were divided into two groups. First group is the factor of passageway, which is door width. Second group is human characteristics which consists of gender, weight, height and body width that includes shoulder breadth and hip breadth.

3 Results and Discussion

The results are presented as walking speed versus distance before the door. It is found that walking speed of all participants is not constant along the distance. The trend is likely to decrease when participants approach to the door. It clearly shows that the door has an effect on participants' response. In addition, the distance before the door is divided to 4 intervals as 1.5-2 m., 1-1.5 m., 0.5-1 m. and 0-0.5 m. It is found that the speed change or slope at 0-0.5 m. is the lowest and also has the highest variance. Hence, most participants reduced their walking speed at 0.5 m. before the door.

After data cleaning and excluding the outliers, all participants are divided to three groups as follows.

I. Walking speed declines with door width at 0.5 m. before the door.
II. Walking speed slightly declines with door width since 2 m. before the door.
III. Walking speed does not decline with door width.

Table 1. Participants characteristics, according to groups.

	I	II	III
Shoulder breadth (cm.)	41.35±2.79	40.42±3.31	39.43±3.70
Hip breadth (cm.)	32.41±3.44	29.71±3.60	32.17±5.03
Height (cm.)	167.70±6.57	168.39±8.76	164.75±10.06
*Weight (kg.)	68.33±9.83	60.45±10.91	57.70±11.75

Results are reported in mean ± SD.
*Significantly different among groups (p<0.05)

General characteristics of participants are shown in Table 1. It is found that weight is significantly different among groups. Participants in group I have higher weight compared with group II and III. In this paper, group I will be considered afterwards since the walking speed is mainly influenced by door width. Twelve participants (20% of all participants) are classified into group I (eight males and four females).

3.1 Walking behaviour

As Figure 3 illustrates, walking speed of participants fluctuate at 0.5-2 m. before the door then hardly reduce at 0.5 m. This obviously occurs at the narrow door, around 40-60 cm. For the door width of 70 cm. upwards, the graphs appear to be flat along the distance. This implies the constant walking speed when the door is wide enough. This corresponds with minimum standard door width of 81.3 cm [7].

3.2 Factors affect walking speed

Factors include door width, gender and body size are analysed subsequently.

3.2.1 Door width

One-way ANOVA was performed in order to test the difference of changing speed among seven door widths. It is found that changing speed at 0-2 m. is not different but significantly different at 0.5-2 m. (p=0.001) and 0-0.5 m. (p=0.006).

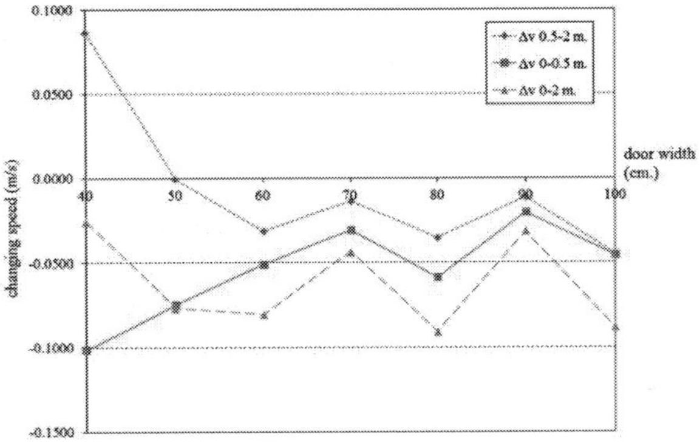

Figure 4. Main effects of door width.

As can be observed in Figure 4, changing speed at 0.5-2 m. seems to be different only at the door width of 40 cm., consistent with the results obtained from post hoc test (Tukey's HSD test). Conversely, there is the direct variation between changing speed at 0-0.5 m. and door width. This is comparable to Fitts' law. Additionally, linear relationship is found approximately at the door width of 40-70 cm.

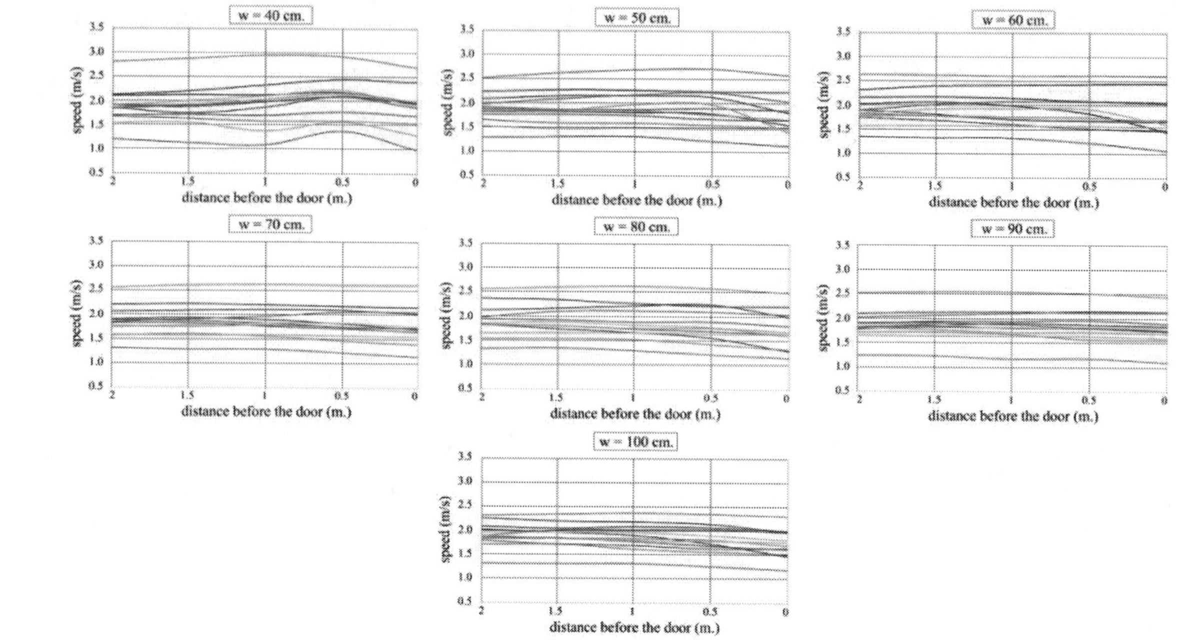

Figure 3. Walking speed along the distance of participants in group I, according to door width.

3.2.2 Gender

The independent sample t-test proved that there is no significant difference of changing speed between male and female. This probably due to the task was less risky. As a result, risk perception between male and female are not different. Other than that, the stimulus in this experiment did not appear unexpectedly so participant was able to notice and early beware. Ability of response between male and female are therefore similar.

3.2.3 Body size

Body size considered in this study consists of weight, height and body width that includes shoulder breadth and hip breadth. Changing speed at 0-0.5 m. shows strong correlations with weight ($r=-0.528$, $p<0.001$) and moderate correlations with shoulder breadth ($r=-0.404$, $p<0.001$). No significant correlations between changing speed and hip breadth or height are observed. Although both shoulder breadth and hip breadth are considered in the interest of body width, hip breadth is always shorter and proximal among all participants. The effect of height is insignificant in this study due to the fixed height of door.

3.3 Multiple regression analysis.

After multiple regression analysis, shoulder breadth and weight are selected as the attributes of changing speed at 0-0.5 m before the door. Moreover, it is found that shoulder breadth highly correlates with weight ($r=0.742$, $p=0.006$) Thus, interaction between shoulder breadth and weight is also significant. These factors explain 54.2% of this variable (See Table 2).

Table 2. Results of multiple linear regression analysis to predict changing speed at the door width of 40-70 cm.

	Coefficient (β)	Standard error (S.E.)	95% Confidence interval		p-value
			Lower	Upper	
Constant	-2.597	1.150	-4.916	-0.277	0.029
door width	0.002	0.001	0.001	0.004	<0.001
shoulder breadth	0.062	0.027	0.008	0.117	0.026
weight	0.041	0.019	0.003	0.078	0.035
shoulder breadth x weight	-0.001	0.000	-0.002	-0.000	0.021

Changing speed in m/s = -2.597 + (0.002 x door width) + (0.062 x shoulder breadth) + (0.041 x weight) – (0.001 x shoulder breadth x weight); $R^2 = 0.542$.

4 Conclusions

From this study, it has found following conclusions.

In walking through the door, the speed reduction is found mostly at 0.5 m. before the door. Walking behavior rather depends on weight. Participants with high weight reduce their walking speed when they are approaching the door that narrower than 70 cm. Thus, the current standard door width around 80 cm. is assigned properly.

Other than door width, human body size also affects to speed reduction. The changing speed presents weight-related decline as well as shoulder breadth. Since shoulder breadth associates with weight, interaction between shoulder breadth and weight is also significant.

For the future study, the investigation of other factors that can affect visual perception such as age would be conducted. Older or younger people may produce the difference in results. More complex situations such as increasing amount of people walking altogether or panic are also interesting.

Acknowledgment

The authors would like to thank all participants that willing participated in this experiment.

References

1. CFPA-Europe. *Fire Safety Engineering Concerning Evacuation from Buildings* (Milano, 2009)
2. E. D. Kuligowski, R. D. Peacock, B. L. Hoskins, *A Review of Building Evacuation Models (2nd ed.)* (NIST, Washington DC, 2010)
3. S. MacKenzie, Hum-Comput. Interact. **7**, 91-139 (1992)
4. M. M. Samson, A. Crowe, P. L. de Vreede, J. A. G. Dessens, S. A. Duursma, H. J. J. Verhaar, Aging-Clin. Exp. Res. **13**, 16-21 (2001)
5. D. K. Spierer, R. A. Petersen, K. Duffy, B. M. Corcoran, T. Rawls-Martin, J. Strength. Cond. Res. **24**, 957-963 (2010)
6. C. R. Harris, M. Jenkins, Judg. Decis. Making. **1**, 48-63 (2006)
7. International Code Council, Inc. *International Building Code* (Illinois, 2011)

Vibration Analysis of 5-DOF Vehicle Model under Stochastic Road Surface Excitation

Yanlong Zhang[1], Lixia Chen[1], Li Wang[2] and Wenxue Zhao[1]

[1]*School of Mechatronic Engineering, Lanzhou Jiaotong University, Lanzhou 730070, China*
[2]*Department of Mathematics, Lanzhou City University, Lanzhou 730070, China*

Abstract. Considering human body vertical motion, vehicle body vertical motion, pitch movement and vertical jump of front and rear wheels, a five-degree-of-freedom vehicle model is established to study basic driving characteristics of the vehicle. Using Fourier transform method, acceleration power spectral density of the seat and the mean square value curves of seat vertical weighted acceleration are obtained by numerical simulation. Combined with comfort provision standards, the influence of vehicle model parameters and speed on seat acceleration power spectral density and vertical root-mean-square value of seat weighted acceleration are analyzed. Results show that the stiffness and damping of the seat have no significant effect on seat acceleration power spectral density, and seat acceleration PSD increases with increasing front or rear suspension stiffness, but it decreases with increasing front or rear suspension damping. It should also be concluded that the model stiffness and the mean square value of seat vertical weighted acceleration present positive correlation in general, but seat vertical weighted acceleration decrease first and then increase when model damping increase. Such analysis results can provide reference for the parameter optimization design of the automobile.

1 Introduction

Vehicle ride comfort is a major aspect of evaluating vehicle comfort properties. Commonly, 1/4 three degrees of freedom vehicle model does not take into account body pitching motion, vehicle sports performance does not reflect fully [1]. Therefore, it must to establish a multi-degree-of-freedom vehicle model, which can better reflect human, vehicle body and wheels vibration.

Many literatures have studied on a five-degree-of-freedom (5-DOF) vehicle model. Xiaoliang Li analyzed 5-DOF vibration of automobile ride comfort using pseudo-excitation method [2]. Lijun Zhang studied the non-stationary dynamics and control of 5-DOF vehicle [3]. Zhigang Huang analyzed 5-DOF mini-car vibration using calculation method [4]. Lixin Guo analyzed vehicle vibration in Changeable Speeds [5]. Jie Li found a kind of new method used for analyzing random vibration of vehicles [6]. But those studies did not analyze the effect of model parameter values on seat acceleration PSD and seat vertical root-mean-square value of weighted acceleration. Therefore, a 5-DOF vehicle vibration model which stochastic road surface excitation is conducted as the main source of vehicle vibration is established in this paper, and the ride comfort of the passengers is analyzed by frequency-domain simulation.

2 The 5-DOF vehicle model

As the vehicle suspension parts such as springs, shock absorbers and tires are symmetrical, a 5-DOF vehicle models used to describe automotive car vibration, which is shown in Figure 1, is established considering the man-chair plane.

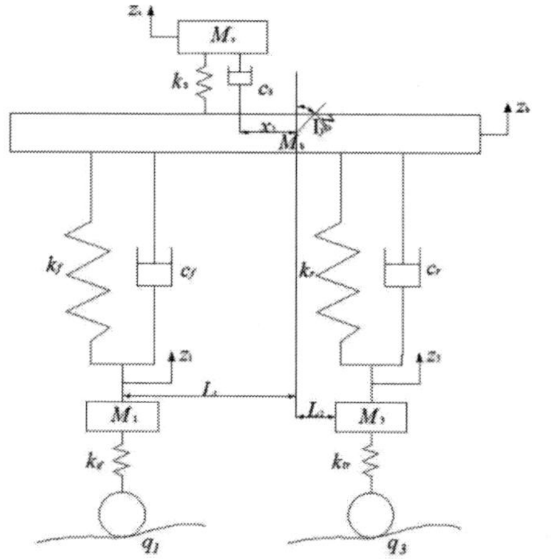

Figure 1. The 5-DOF vehicle model

In Figure 1, z_s, Z_b, z_1 and Z_3 are the vertical displacement of seat, vehicle body, front, rear wheels. Z_p is body pitch angle displacement. $q_1(t)$, $q_3(t)$ are the front and rear wheels of road surface roughness. M_s, M_b,

Support by National Natural Science Foundation of China (11302092).

M_1, M_3 are human-seat mass, vehicle body mass, front suspension mass, rear suspension mass, and their values are respectively 70 kg, 1100 kg, 50 kg, 65kg. I_y, K_s, K_f, K_r, K_{tf}, K_{tr} are body moment inertia, seat stiffness, front suspension stiffness, rear suspension stiffness, front tyre stiffness and rear tyre stiffness, and their values are respectively 1840kg.m^2, 12000N/m, 17600N/m, 22300 N/m, 180000N/m and 180000N/m. C_s, C_f, C_r are damping coefficients of the seat, the front suspension, rear suspension, and their values are respectively 1250 Ns/m, 1500Ns/m, 1700Ns/m. L_1, L_2, x_1 are horizontal distance from body center mass to the front axle, the horizontal distance from body mass to the rear axle, horizontal distance from body center mass to the seat, and their values are respectively 1.30m, 1.15m, 0.779m.

According to the Lagrangian method, the differential equations of the system can be given by

$$[M]\{\ddot{Z}\}+[C]\{\dot{Z}\}+[K]\{Z\}=[F]q \quad (1)$$

where, $M=\mathrm{diag}\begin{bmatrix}M_s & M_b & I_y & M_1 & M_3\end{bmatrix}$, $q(t)=\begin{bmatrix}q_1(t)\\ q_3(t)\end{bmatrix}$,

$$C=\begin{bmatrix} C_s & -C_s & C_s x_1 & 0 & 0\\ -C_s & C_f+C_r+C_s & -C_s x_1-C_f L_1+C_r L_2 & -C_f & -C_r\\ C_s x_1 & -C_s x_1-C_f L_1+C_r L_2 & C_s x_1^2+C_f L_1^2+C_r L_2^2 & C_f L_1 & -C_r L_2\\ 0 & -C_f & C_f L_1 & C_f & 0\\ 0 & -C_r & -C_r L_2 & 0 & C_r \end{bmatrix},$$

$$K=\begin{bmatrix} K_s & -K_s & K_s x_1 & 0 & 0\\ -K_s & K_f+K_r+K_s & -K_s x_1-K_f L_1+K_r L_2 & -K_f & -K_r\\ K_s x_1 & -K_s x_1-L_1 K_f+L_2 K_r & K_s x_1^2+L_1^2 K_f+L_2^2 K_r & K_f L_1 & -K_r L_2\\ 0 & -K_f & K_f L_1 & K_f+K_{tf} & 0\\ 0 & -K_r & -K_r L_2 & 0 & K_r+K_{tr} \end{bmatrix},$$

$$Z=\begin{bmatrix}Z_s & Z_b & Z_p & Z_1 & Z_3\end{bmatrix}^T,\ F=\begin{bmatrix}0 & 0 & 0 & K_{tf} & 0\\ 0 & 0 & 0 & 0 & K_{tr}\end{bmatrix}^T.$$

According to the literatures [7, 8], we can get $\dot{q}_1(t)+\beta V q_1(t)=\omega(t)$, $\dot{q}_3(t)=-\frac{2}{\Delta_2}q_3(t)+\left|\frac{2}{\Delta_2}+\beta V\right|q_1(t)-\omega(t)$, $\Delta_2=L/V$, where $q_1(t)$, $q_3(t)$ is front and rear wheel road displacement; β is road roughness coefficient; V is vehicle speed; $\omega(t)$ is white noise; Δ_2 is time delay; L is wheelbase.

3 Seat acceleration PSD

Using Fourier transform method to derive system response power spectral density, we can get

$$\left(-\omega^2[M]+j\omega[C]+[K]\right)\{Z(\omega)\}=[F]\{q(\omega)\},$$

$$[H(\omega)]=\frac{\{Z(\omega)\}}{\{q(\omega)\}}=\left(-\omega^2[M]+j\omega[C]+[K]\right)^{-1}[F], \quad (2)$$

where, ω is circular frequency, $Z(\omega)$ is Fourier trasform of displacement vector $\{Z\}$, $q(\omega)$ is Fourier transform of road vectors $\{q\}$, $j=\sqrt{-1}$.

According to random theory, power spectral density of the vibration response can be obtained

$$G_x(f)=|H(f)|^2_{x-q}G_q(f) \quad (3)$$

where, $H(f)$ is amplitude-frequency characteristics of vibration responses when $\omega=2\pi f$, $G_q(f)$ is power spectral density of the road. Seat vertical mean square value of weighted acceleration a_{wk} can be calculated by

$$a_{wk}=\left[\int_{0.5}^{80}W_k^2(f)G_x(f)df\right]^{\frac{1}{2}}. \quad (4)$$

where, $W_k(f)=\begin{cases}0.5 & (0.5<f<2)\\ f/4 & (2<f<4)\\ 1 & (4<f<12.5)\\ 12.5/f & (12.5<f<80)\end{cases}$.

The relationship between the total weighted acceleration RMS value and human subjective feelings is shown in Table 1[7, 8].

Table 1. Relationship between a_{wk} and human feelings.

Weighted acceleration RMS a_{wk} m/s^2	**Human subjective feelings**
<0.315	Comfortable
0.315 ~0.63	A little uncomfortable
0.5 ~1.0	Quite uncomfortable
0.8 ~1.6	uncomfortable
1.25~2.5	Very uncomfortable
>2.0	Extremely uncomfortable

We choose the C level pavement, and the vehicle speed is $v=20$m/s. Using 4-order Runge-Kutta method, we can get the frequency domain response comparison graphs as shown in Figure 2. In Figure 2(a-b), the seat acceleration PSD has a tendency increase with the increase of front and rear suspension stiffness, the frequency of peak also increase, and the influence of front suspension stiffness take a larger proportion than back suspension stiffness. According to the calculation formula of seat natural frequency, we can get the natural frequency is 2.085Hz. In order to avoid the resonance of seats, front and back suspension, the stiffness can't be increased without any limit. But if the suspension stiffness of front and back is too small, it will lead to increase the suspension dynamic trip so that the passengers will feel uncomfortable. In Figure 2(c-d), the seat acceleration PSD will reduce with the increase of front and rear suspension damping. But the larger front and back suspension damping maybe leads the passengers to sustain more impact vibration, so suspension damping should be not too big. In Figure 2(ef), the increasing of seat stiffness and damping have a little effect on the seat acceleration PSD, except for the resonance frequency area which should avoid.

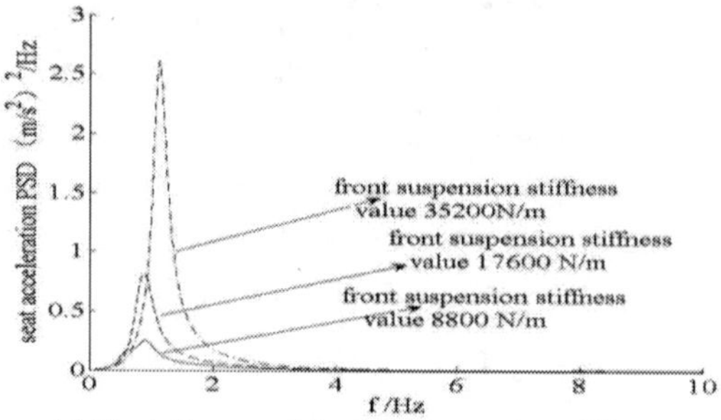

(a) The influence of the front suspension stiffness

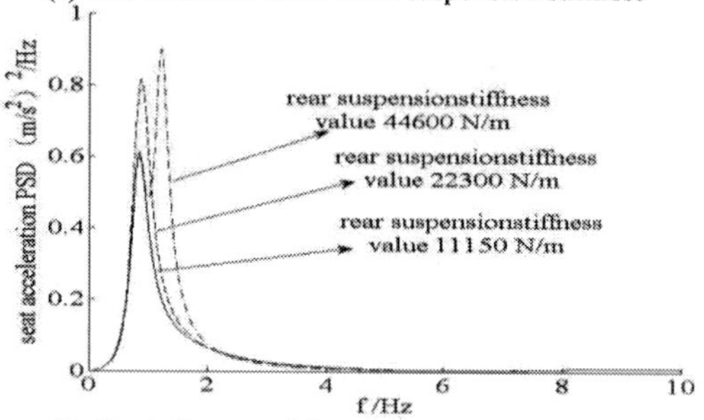

(b) The influence of the rear suspension stiffness

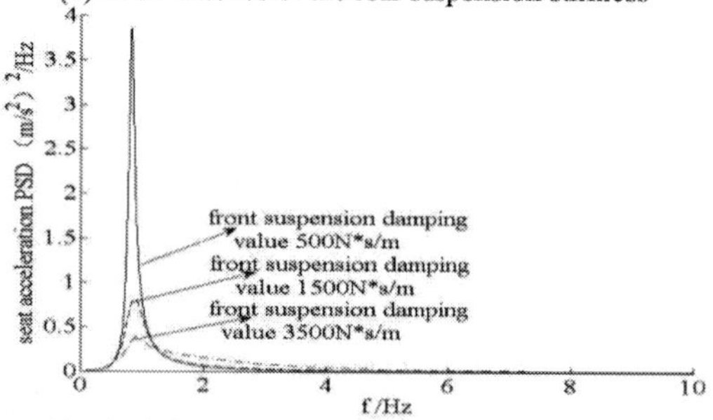

(c) The influence of the front suspension damping

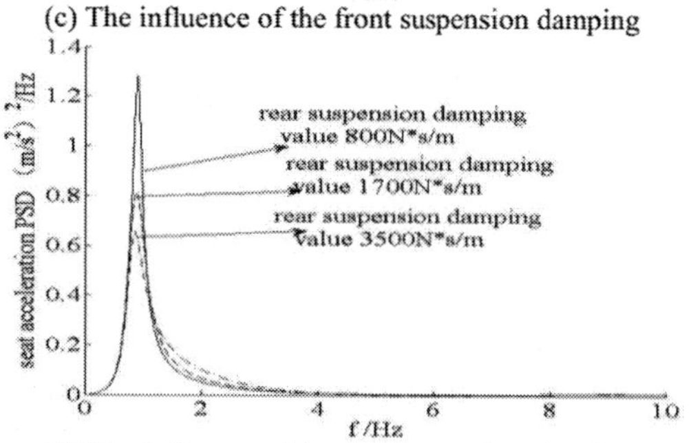

(d) The influence of the rear suspension damping

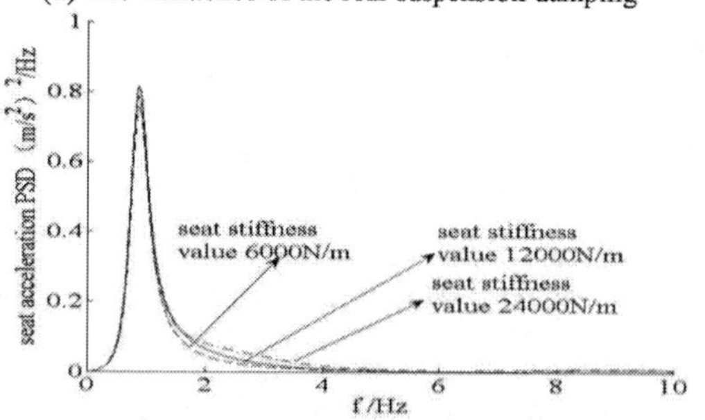

(e) The influence of the seat stiffnes

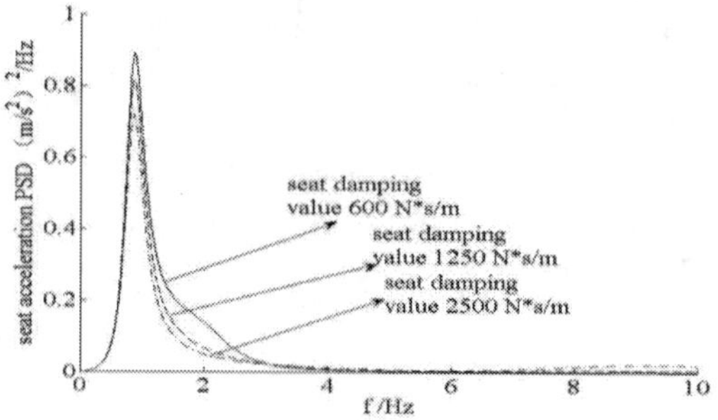

(f) The influence of the seat damping

Figure 2. Parameter's influence on seat acceleration PSD

4 Seat weighted acceleration RMS

The influence of vehicle parameters and speed on the root mean square value of weighted acceleration of the seat is analyzed in Figure 3.

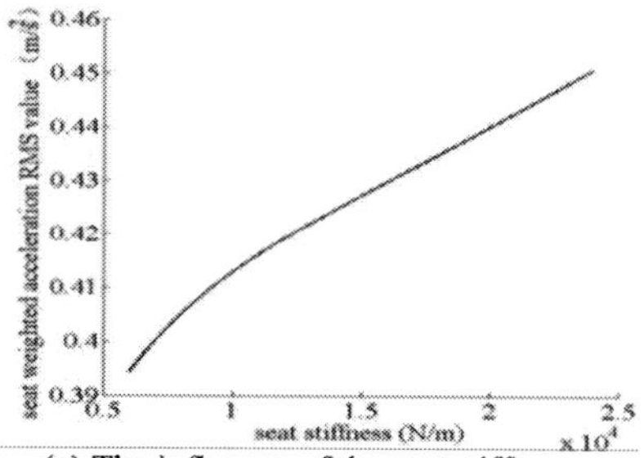

(a) The influence of the seat stiffness

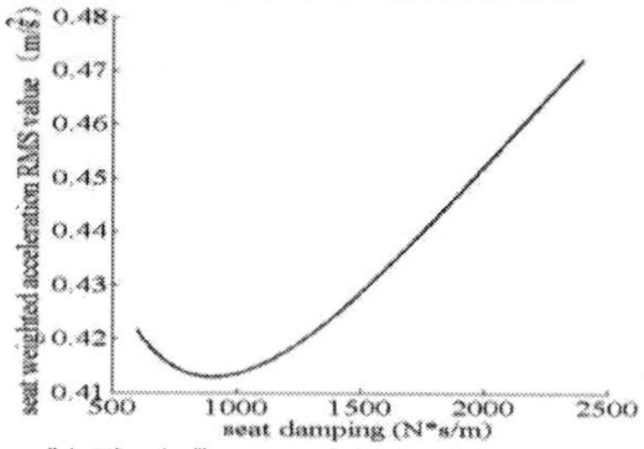

(b) The influence of the seat damping

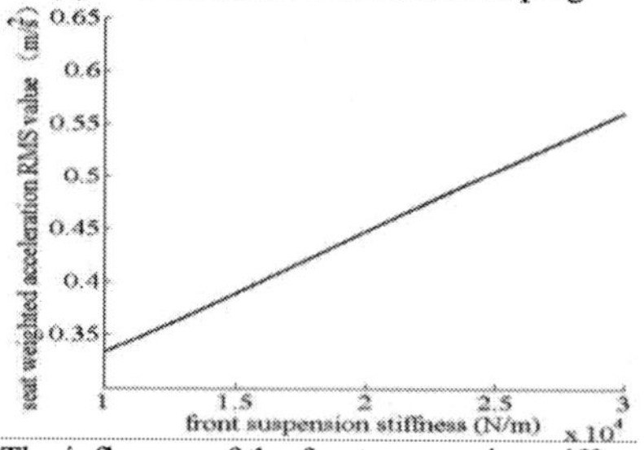

(c) The influence of the front suspension stiffness

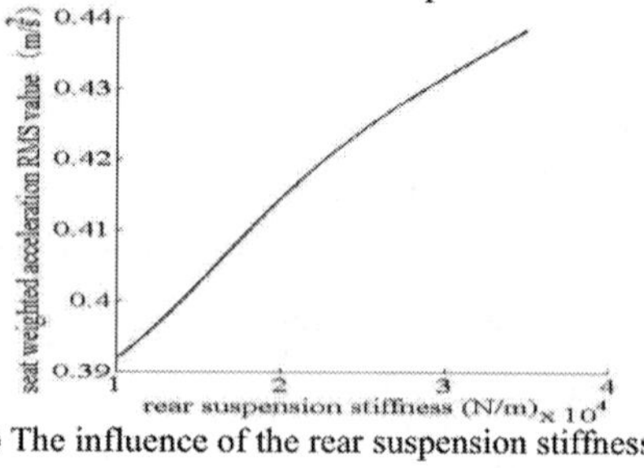

(d) The influence of the rear suspension stiffness

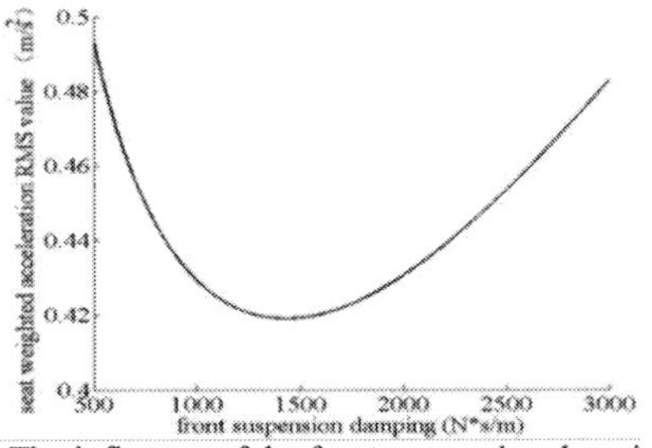

(e) The influence of the front suspension damping

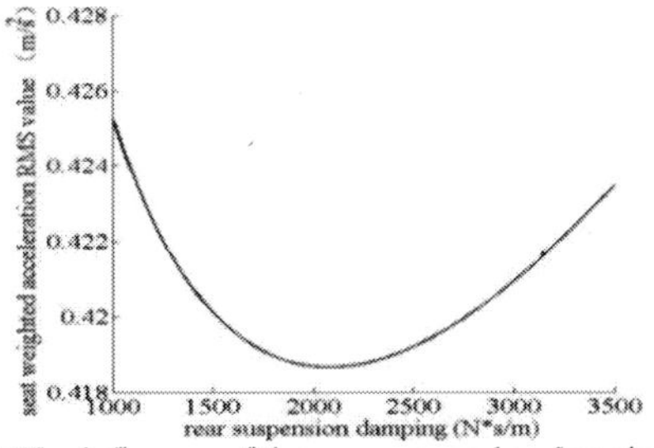

(f) The influence of the rear suspension damping

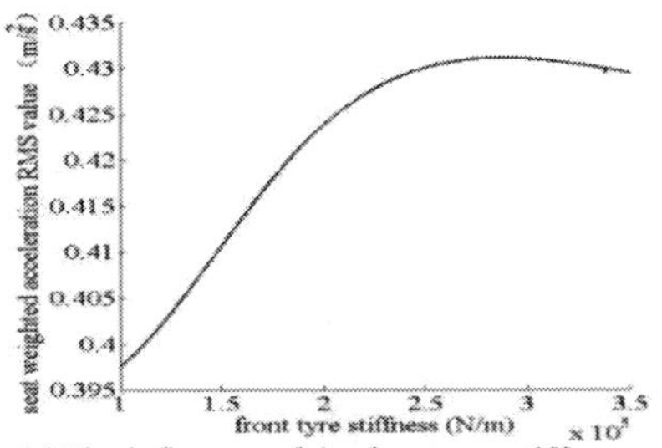

(g)The influence of the front tyre stiffness

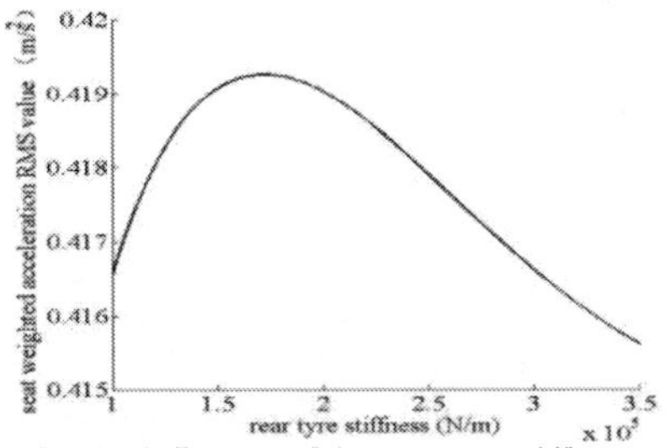

(h)The influence of the rear tyre stiffness

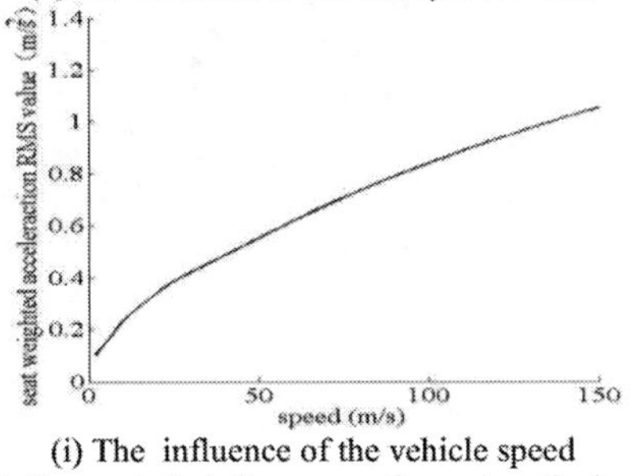

(i) The influence of the vehicle speed

Figure 3. Parameter's influence on the seat vertical weighted acceleration RMS

With increasing the stiffness of seat, front suspension, rear suspension and the vehicle speed, the root-mean-square value of weighted acceleration a_{wk} increases as shown in Figure 3(a), (c), (d) and (i). With increasing the damping of seat, front and rear suspension, a_{wk} decreases first and then increases as shown in Figure 3(b), (e) and (f). With increasing the stiffness of front and rear tyre, a_{wk} increases first and then decreases as shown in Figure 3(g) and (h). Referred to table 1, the suitable vehicle parameter ranges and the suitable vehicle speed range can be obtained according to the comfort requirements of the passengers. The above analysis results can provide the reference for the vehicle's parameter optimization design.

5 Conclusions

Considered the stochastic road surface excitation, a 5-DOF vehicle model is established to study basic driving characteristics of the vehicle. The acceleration power spectral density of the seat and the mean square value curves of seat vertical weighted acceleration are obtained at different system parameters by numerical simulation. Combined with the comfort provision standards ISO2631-1:1997(E), the influence of vehicle model parameters and speed on seat acceleration power spectral density and vertical root-mean-square value of seat weighted acceleration are analyzed. Results show that the stiffness and damping of the seat have no significant effect on seat acceleration power spectral density, and seat acceleration PSD increases with increasing front or rear suspension stiffness, but it decreases with increasing front or rear suspension damping. It should also be concluded that the model stiffness and the mean square value of seat vertical weighted acceleration present positive correlation in general, but seat vertical weighted acceleration decrease first and then increase when model damping increase. The suitable vehicle parameter ranges and the suitable vehicle speed range can be obtained according to the comfort requirements of the passengers. Such analysis results can provide the reference for the vehicle's parameter optimization design.

References

1. Hongliang Zhang, Binggang Wang, Wanqiao Yang, J. of Chongqing Jiaotong Univ. (Natural Science), **28:6**, 1041 (2009).
2. Xiaoliang Li, *Vibration simulation and analysis for automobile ride comfort based on pseudo excitation method*(Wuhan univ. of technology, 2011).
3. Lijun Zhang, *Study on non-stationary ride dynamics and control of vehicle*(Northeastern univ., 2006).
4. Zhigang Huang, Muqiao Zheng, J. of Beijing Institute of Light Industry, **13:1**, 68 (1995).
5. Lixin Guo, Liping Zhang: Mathematical Problems in Engineering, **2010**, 1 (2010).
6. Jie Li, Yuyin Qin, Qi Zhao, Wei Zhang, J. of Machine Design, **26:4**, 14(2009)
7. P Michelberger, L Palkovics, J Bokor, Int. J. of Vehicle Design, **14:22**, 145(1993).
8. F Oueslati, S Sankar, SAE, 534(1994).

MATEC Web of Conferences **68**, 04003 (2016)
ICIEA 2016
DOI: 10.1051/matecconf/20166804003

Gasoline New Timing and Flux Adjustable Rotary Valve Design (Hereinafter: Rotary Valve)

DU huiqi, DU dong

Tianjin Sino-German Vocational Technical College, Tianjin 300200

Abstract. Conventional gasoline engine with an umbrella valve control cylinder intake and exhaust, in order to achieve sealing effect, the valve is driven by the spring force; at the same time, when the cam opens the valve to overcome the spring force acting. Sealing the better, the more power consumed in the engine mechanical losses, the valve mechanism consumes about 30%, which is not a small loss! This article describes a new type of rotary valve is to significantly reduce mechanical losses, so as to achieve energy saving purposes.

1 Introduction

At present, the gasoline engine is mainly used umbrella ventilation valve (see Fig. 1) controlled cylinder internal deflagration of the moment, the pressure is very high. If Mifengbuyan umbrella valve will leak, the piston will be reduced thrust, the engine efficiency is low. To allow the umbrella valve closing tightly, it requires a large spring force (see Fig. 2); Of course, in order to allow the valve to open the umbrella, with a cam drive link also large force, the more unwanted power consumption.

At present, the mechanical efficiency of the gasoline engine is about 35%, meaning that only 35% of the energy is converted to useful work we need to rest for about 65% of energy are wasted as mechanical loss. The main manifestations of waste heat and mechanical friction do attack like. Mechanical friction loss accounts for 62-75% loss; loss of driving various accessories accounted for 10-20% of mechanical losses; driven mechanical supercharger loss accounts for 6-10% loss; pumping loss accounts for 10-20% mechanical loss.

Umbrella valve mechanism and drive 30% mechanical loss. When mainly at low speeds, lubrication is in a critical state, the friction loss will significantly increase the proportion; and the load acting on the main valve spring force caused, integrated into the output shaft of the engine drag torque is most obvious at this time, accounting about 14% of the total energy (65% * 75% * 30% = 14%)! Five-valve four-cylinder gasoline engine, 20 gas valve opening and closing mechanical energy to be consumed can be imagined. To this end, we carried out a bold attempt to break through the traditional design concepts, the concept of the rotary valve.

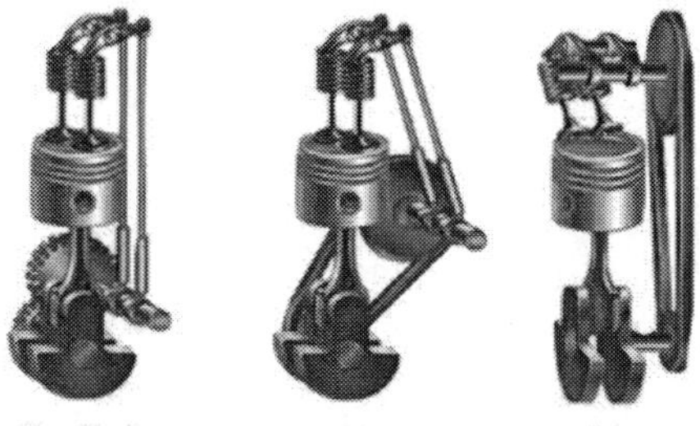

Figure 1. shaped valve.

Figure 2. shaped valve.

2 A historical causes gasoline umbrella valve

Before the introduction of the new rotary valve concept, let's look at why the conventional gasoline engine is an umbrella organization of ventilation? We believe that the umbrella organization was based on the ventilation machine control level. Machinery and equipment from simple to complex, from single to comprehensively carry out development, with more and more complex machinery, the relevance of their actions, coordination is becoming a problem. At that time people can only continue to explore more specific and more complex mechanical mechanism along the mechanical structure design to meet the relevant mechanical and coordination of action. Such machinery is designed for mechanical drive mechanism and a mix of mechanical control mechanism, constitute the so-called complex machinery. Gasoline engine design also continue to use this method, but also a mix of mechanical drive mechanism and mechanical control mechanism. E.g., piston, connecting rod and crank mechanism is mechanically driven, controlling the opening and closing of the umbrella valve cams, connecting rods, levers, springs and other mechanical control means [1] (see Fig. 3,4).

Figure 3. cam, connecting rod, lever.

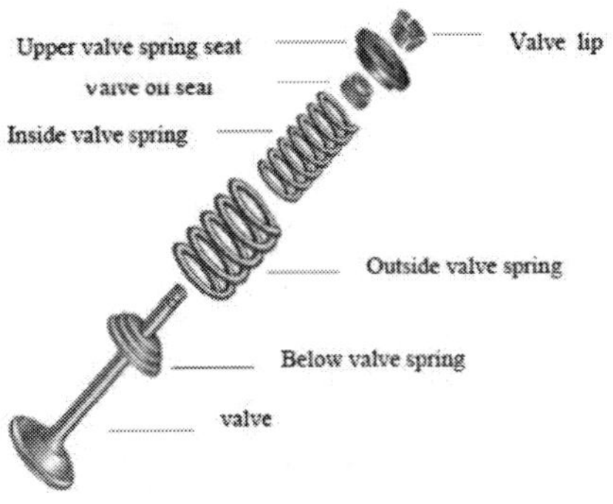

Figure 3. cam, connecting rod, lever.

At that time, electric drive and electronic control logic level limitations, most of the mechanical drive and control are provided by a driving source. Starting from a drive source, according to the control logic or timing requirements, mechanical drive and control mechanism to achieve the main driver of redistribution. This design concept can be seen everywhere in the traditional machinery, such as latitude typewriter pendulum control, the loom control, control, Shaper reciprocating mechanism inside the watch like a ratchet. Currently, these mechanical control mechanism and has been developed to a near-perfect situation. However, with increased levels of social development, the traditional mechanical design can not satisfy people more advanced automated machinery and equipment requirements, then the emergence of automated machinery and equipment mechatronics design. The following detailed description of this design concept used in the engine valve control.

3 Gasoline rotary valve design ideas

Faced with an increasingly complex machinery and equipment, the complex mechanical control mechanism becomes more and more bloated, but encountered a bottleneck for further development, then there is urgent need of a new control mechanism to replace the traditional mechanical control. Due to the rapid development of modern electronic control technology, we can now use the rotary encoder to replace the traditional mechanical cam mechanism, accurate and complete perspective of the angular velocity control. Decentralization drive element to the development of relations between the respective drive element action by an electronic control device such as a PLC and other coordination. Conventional gasoline engine intake and exhaust relationship with the spindle rotation is to rely on mechanical control mechanism, which does not meet modern mechatronics design. To make the gasoline engine become mechatronic devices, you must use the drive element decentralization principle, each intake and exhaust operation with a motor drive control, for eight-valve four-cylinder engine, eight little need to control the spindle motor coordinated action, complete with the best four-cylinder burst when suction pressure discharge valve.

Because the demand for gasoline engine performance improvement, such as combustion efficiency, energy saving, etc. the mechanical efficiency. Specifically, and puts forward some requirements of more and more precise control of angle and angular velocity, into the cylinder air intake time t, intake air quantity Q, closing valve time t, ignition time t fire, exhaust time t row, row exhaust quantity Q, these quantities are with the volume of three related, that is the shaft spindle position angle phi, angular velocity Omega shaft and spindle torque was axis. There is a definite link between them, in order to achieve the best control according to the need, to obtain the best output power, speed, acceleration, need to establish their function [2]. For example, P power (kw) = torque T (nm) x speed omega (RPM) * 2 pi / 60, the simplified calculation become: P power (kw) = torque T (nm) x speed (RPM) /9549. Which torque T (nm) = f (t burning, into the T, Q, t); speed = f (phi axis, Omega axis, as axis, t burn, into the T, Q, t,). And according to the relevant variables of the function requirements of real-time adjustment. In order to adjust the real-time adjustment device, the traditional air intake and exhaust valve opening and closing time and the amount is fixed or there is a very adjusted, it is difficult to meet the requirements of the above functions for extreme operation requirements. Therefore, it is necessary to design a can meet the real-time adjustment of the device and the valve action time and action quantity can be stepless adjustment, realize the spindle angular position diameter shaft, spindle angular velocity Omega axis and spindle resistance torque was axis function.

4 Gasoline rotary valve structure and how it works

Based on the above analysis, design a device to complete this function is running relationship, which is the main problem to be solved herein. Conventional internal combustion engine intake and exhaust valve is completed by an umbrella, which the valve is closed by spring force to achieve, in order to prevent the umbrella valve leakage, we must use a larger spring force mushroom valve and air inlet in close contact , the valve opening by the cam against the spring force of the inlet valve and the umbrella separately. Especially in the five-valve four-cylinder gasoline engine, 20 gas valve opening and closing mechanical energy to be consumed is considerable, accounting for about ten percent of the mechanical energy output [3].

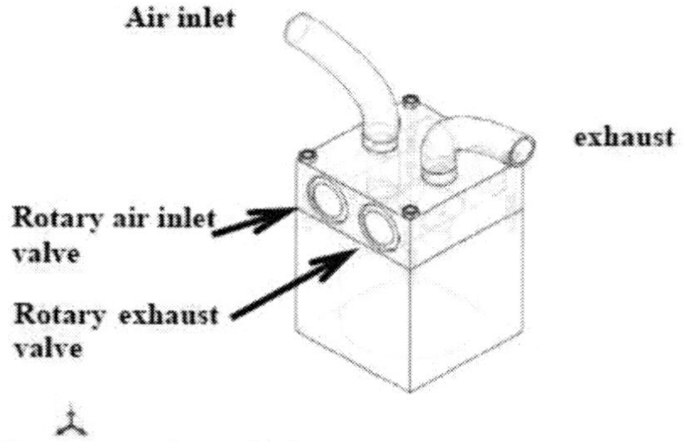

Figure 4. new shape of valve.

To overcome the umbrella valve mechanical energy consumption, we have designed a new type of valve, its shape shown in Figure 3, which changed the traditional mushroom-valve structure, replaced by a rotary valve,the structure is open on a cylindrical spool a cylindrical spool perpendicular to the axis of the through-holes. In the body also has a corresponding diameter of the airway, the valve spool is rotated 90 degrees, you can complete the valve opening and closing operation, namely the valve opening and closing is done by rotating the spool. Since the rotary valve is completed by sealing the structure itself, and can resist the instantaneous deflagration tremendous pressure required to overcome the additional spring, so reducing the valve opening and closing operation of the energy consumption.

Internal combustion engine is a sophisticated machine, various mechanical motion mechanism of an organic whole, used to be on mechanical contact the various agencies, between each driven mechanical institutions now have no fixed mechanical mechanism, is to rely on a kind of invisible things connect each drive, they are single chip microcomputer, PLC and other logical control unit, to coordinate the relationship between various actions between the drive, which is driving the decentralized design concept. Rotary valve implementation method is described below.

Rotary valve valve core rotation is by small step motor or servo motor drive, because the motor is vector control motor, control motor pulse output terminal Q0.0 issued by PLC, the pulse frequency is up to 20 KHZ, if choose motor subdivision control pulse number is 10000 / r, Angle precision can reach 360 ° / 10000 p = 0.036 ° / 1 p, namely each pulse can make the motor around 0.036 °. If the opening of 90 ° and 0.036 ° is open for this 0.036 ° 90 ° / = 0.0004 times. So, in terms of internal combustion engine valve control, the accuracy is high, the Angle and speed can be precisely controlled. In fact, open unnecessarily high precision requirement, can choose motor subdivision control pulse number is 400 / r, under the condition of the pulse frequency of 20 KHZ, speed = 20000/400 = 50 r/s = 3000 r/min. Due to the rotation gas every turn 180 ° open gate valve core, so the rotation of the 3000 r/min 6000 r/min of internal combustion engine valve support. Rotary valve when opening or closing, is with the spindle angular position and angular velocity and torque at that time. Angular position information acquisition is to rely on and principal axis coaxial rotary encoder, it can rotate in a week, to launch 36000 pulse, namely each pulse represents 0.1 °. Angular velocity is calculated, by the number of pulses through calculating unit of time, such as computer read 100 pulses within 0.1 seconds, rather read 1 seconds 1000 pulses angular velocity omega = 1000 ° / s. According to the same method can also get angular acceleration, and then according to the angular acceleration can get the output torque and power. Of course, also can be used to linear motor rotating valve adjusted by the amount of the valve core to open, adjusted according to the mathematical model, the internal combustion engine achieve ideal running status.

In the hardware set up is completed, according to the theory of fuel combustion and the actual operating experience, in the PLC program can be easily write control program. For example, the main shaft turn to any Angle, make a revolving valve motor start or stop. In the actual debugging engine, can be easily in the program by rewriting the pulse number value, to the best control state. Of course also have control of the fuel delivery, air door and oxygen, etc. In a word, through the above control method, can complete accurate control, convenient for the internal combustion engine at full combustion, and reduce the fuel consumption and environmental protection.

5 Technical innovations is mainly manifested in two aspects of hardware and software of the five points, below shows its advantages respectively

Using rotating valve control of internal combustion engine intake and exhaust Mechanical structure of the rotating valve itself ensures that can easily turn, open breath consumed less mechanical energy, and can be multi-layer type O sealing ring to prevent the leakage of high pressure gas.

rotating valve with step motor or servo motor drive Rotary valve of the valve core with stepper motor or servo motor drive directly, belong to this kind of motor vector control model, can accurately control the motor shaft Angle, especially suitable for rotary valve drive.

Rotating valve opening area size can adjust Rotary valve open area size can be adjusted with step motor or servo motor, so that you can under the condition of different speed setting different cross-sectional area, and as the output control functions, through the PLC control of the motor, the adjustment of the complete circulation.

Flux valve opening and closing time and open closed digital; Valve open and close and cross-section of the valve is controlled by step motor or servo motor, the motor is belong to the vector control of motor, the motor stator coil magnetic field is produced digital vector synthesis of a rotating magnetic field, magnetic field and the Angle of the rotating magnetic field can accurately adjust speed, it's digital provides the basis for internal combustion engine rotating valve control, intelligent control created the conditions for internal combustion engine [4].

Control valve opening because of the stepper motor or servo motor motor belongs to the vector control motor, the controller is based on PC control from the PLC pulse sequence. So from the pulse number and pulse frequency can be operating conditions such as omega axis, Phi axis, as axis, burning T, t, Q, t function. The mathematical model is established in the combustion laboratory experimental data and experience for people to grasp, with PLC programming language, and constantly improve and optimize.

6 Concluding remarks

This paper presents a new type of internal combustion engine valve, which can accurately control the time and the flow of air intake and exhaust. The valve and the traditional umbrella valve to reduce driving the power consumed by the valve, to realize the digital and intelligent control, improve the combustion quality, reduce pollution emissions, there is a very high popularization value.

References

1. Chen Jiarui, *Automobile structure*,(Beijing: Mechanical Industry Press ,2009)
2. Wang Wangyu,*Automobile design*(Beijing: Mechanical Industry Press, 2011)
3. Zhang Jilong,Development status and trend of optimization technology of gas distribution mechanism. Mechanical engineer **11**, 83-84 (2015)
4. Chen Xinquan,The design principle and development of variable valve system from VVT-i to VVTL-i and cam control technology: equipment manufacturing technology **10**, 104-106 (2008)

MATEC Web of Conferences **68**, 05001 (2016) DOI: 10.1051/matecconf/20166805001
ICIEA 2016

A Contradiction-Based Approach for Innovative Product Design

Yao-Tsung Ko, Chia-Chen Lu and Li-Hui Lee

Department of Industrial Design, Tunghai University, Taiwan
No.1727, Sec. 4., Taiwan Boulevard, Xitun District, Taichung City, 40704

Abstract. Without creativity in design there is no potential for innovation. This paper investigates the role of contradictions in enhancing creativity in product design. Based on the inventive principles of TRIZ, this paper presents a novel design method by integrating technical and physical contradiction analysis methods into the conceptual design activities of new product development (NPD). Despite the recognized importance of innovative design, there is a lack of the systematic and effective design-thinking process that can covers all conceptual design activities. To address this gap, a sharper and fundamental model of the problem-solving is created for innovative product design based on the contradiction-oriented concept. Eventually, one case study is employed to illustrate the method and the result validates that it can help designers produce more creative outcomes in product design.

1 Introduction

When studying the complexity and numbers of constraints and contradictions in product design projects, to the observer it sometimes seems miraculous how designers can work in such environments–let alone produce outstanding creative outcomes. The importance of conflicts and constraints in design is a well-studied subject [1], but the relationship between contradictions and creativity seems to be fairly unexplored. It not only within design but also in the creativity literature in general. Sternberg and Kaufman [2] point out the contradictions exist on creativity in their recent research. The engineering design literature seems to pay less attention to it, despite engineering design being recognised as a constraint-based domain [3], [4] where creativity plays an important role. The term "constraints" often implies the restrictions of free and creativity [5], but numerous researchers pointed out that without conflicts and constraints there can be no creativity [6], [7]. Not only can contradictions enhance creativity, they are key elements for all creative activities [1], [2]. Some types of contradictions can be highly limiting for creativity [8], but without contradictions and constraints there is no problem to resolve and no potential for creativity. Regarding contradictions as important drivers for creative design, it becomes important and interesting to study how the contradictions in design process affect creativity. Seeking to explore this relationship in product design, this paper proposed a creative design-thinking approach based on technical and physical contradictions concept. This study focused on three main areas: (1) how can contradictions be created and managed to give the best possible room for innovation, (2) how do designers think with contradictions as part of their creative process and (3) the creative role of contradictions and their abstraction level. Creativity and contradictions are inevitably related. To understand this complex relationship is not only important for realizing creativity in product design, it is also essential to support real-world design activities. Finally, work within this topic can contribute explicitly to understanding and supporting creativity in product design.

2 Methodology

2.1 Technical contradiction

The core concept of TRIZ is engineering contradiction. An engineering contradiction is a situation, in which an attempt to improve one parameter of an engineering system leads to the worsening of another parameter. Engineering contradiction also known as "Technical Contradiction". Altshuller's Matrix is a problem-solving tool that recommends inventive principles for solving engineering contradictions. What are Altshuller's findings? He studied the engineering problems and their resolution by analyzing thousands of patent documents in Russia. The followings are his key findings. He identified that 39 typical parameters are involved in most engineering contradictions. For example, length of a stationary object, weight of a stationary object etc. Typical solutions to typical contradictions seem to work successfully in most situations. He generalized these solutions and called them "Inventive Principles". There are 40 Inventive Principles in TRIZ [9].

2.1.1 Altshuller's matrix

Only the engineering contradiction cannot help designers create novel ideas. The users must employ the Altshuller's Matrix to identify the improve and worsen parameters then produce the advised inventive principles. The Altshuller's Matrix also known as "TRIZ Matrix". How to use the TRIZ Matrix? The first step is to transform a specific technical contradiction to a typical one. It is critical to formulate a typical contradiction using the 39 universal technical parameters from TRIZ Matrix. Next, we identify and apply recommended principles from the matrix. The result is a set of general recommendations for resolving the identified contradiction. Finally, those general recommendations must be translated into specific technical ideas that can solve the initial technical contradiction.

2.1.2 Process of resolving technical contradictions

Inventive problems are written in the form of "If-then-but". The procedure of resolving technical contradictions are listed as below.
(1) Identifying specific parameters.
(2) Identify the parameters in the engineering contradiction.
(3) Identifying typical parameters.
(4) Identify from Altshuller's list those Typical Parameters that are similar in meaning to the specific parameters or are derivatives of specific parameters.

2.2 Physical contradiction

A physical contradiction is two opposite requirements placed upon a single physical parameter of an object. These requirements are caused by the conflicting requirements of an engineering or technical contradiction. There is Moe power and uniqueness in physical contradiction analysis. The engineering contradictions are formulated for technical parameters of the engineering system. Physical contradictions represent the next step of abstraction. Physical contradictions are formulated for physical parameters which can be resolved in Time, Space, Affiliation, and System Hierarchy. This allows TRIZ Inventive Principles to be used more effectively. Physical contradictions present a sharper and fundamental model of the problem-solving than the traditional TRIZ engineering contradictions, thereby enabling more powerful solutions.

The steps of physical contradiction analysis are presented as follows:

(1) Formulate two justified contradictory (opposite) requirements for one of the parameters of the engineering system or its components.

(2) Determine what typical approaches are applicable to the physical contradiction at hand.

(3) Identify the inventive principle needed relevant to the chosen approach. Applying inventive principles leads to ideas that solve the initial physical contradiction.

A physical contradiction is a situation in which an engineering system demands contradictory values from the same parameter. For example, to improve the quality of nail penetration, the contradiction is that the hammer should be made heavy, but to improve its handling it should be made light. The design-thinking process will be shown as follows:

To (achieve goal 1) (the subject parameter) should be (+A); But

To (achieve goal 2) (the subject parameter) should be (-A);

In order to formulate the requirement of the system we must indicate the definitions including:

(1) The demanded parameter

(2) Goal 1 and Requirement 1; Goal 2 and Requirement 2

There are three categories of ways to solve physical contradictions:

(1) Separating contradictory demands

(2) Satisfying contradictory demands

(3) Bypassing contradictory demands (System transition)

2.2.1 Separating contradictory demands

Five methods of separation in the separating contradictory demands are proposed as below.

(1) Separation in Space

(2) Separation in Time

(3) Separation in Relation

(4) Separation in System Levels

(5) Separation between Parameters

Separation removes the contradiction and enables each demand to be met. The designers can get the initial solutions for design problems using these five separation methods.

2.2.2 Separation in space

If two contradictory demands are required at different locations within an engineering system, separate the contradictory demands. The keyword of design-thinking is "where". Ask where do the system want +A ? and where do the system want –A ? If the answer can be different, we can separate the opposite demands by space. Resolving the physical contradiction using "Separation in Space", there are six Inventive Principles of TRIZ recommended for separation concepts in space:

#1-Segmentation
#2-Separation (taking out)
#3-Local quality
#7-Nested doll
#4-Symmetry change (asymmetry)
#17-Dimensionality change (another dimension)

2.2.3 Separation in time

If the contradictory demands are required at different times, separate them using "Separation in Time" method. For example, how to resolve the design problem of the eye of the needle. The contradiction is that the eye of the needle shoule be large for directing the thread easily into it, but the eye of the needle should be small to avoid damage to the clothes. The keyword of design-thinking is "when". Ask when do I want +A? and when do I want –A? The separation method in time can be identified as follows:

(1) Separation in Space cannot be applied

Location of contradictory demands is same i.e., the hole of the needle

(2) Separation in Time can be applied

"Hole should be large" is required before sewing

"Hole should be small" is required while sewing

Since the contradiction demands are for different times, this physical contradiction can be separated using "Separation in Time". Five Inventive Principles are recommended for resolving the physical contradiction using "Separation in Time":

#15-Dynamization

#34-Discarding and recovering

#10-Preliminary action

#9-Preliminary counteraction

#11-Prior compensation

2.2.4 Separation in relation

If the contradictory demands are required for different components of the supersystem, separate them using 'Separation in Relation' method. For instance, how to design a "Window" with good function. The contradiction is that a window should be open to allow the air to come in but the window should be closed to not allow the sun rays to come in. The window is a system. Sun and air belong to supersystem components. The method of Separation in Relation can be proposed as follows:

(1) Separation in Space cannot be applied

The contradictory demands are required for the same window

(2) Separation in Time cannot be applied

"Allowing air" and "not allowing sunlight" must happen at the same time.

(3) Separation in Relation can be applied

"Window should be open" is required for air

"Window should be closed" is required for sunlight

Sunlight and air are two different components of the supersystem. There are five Inventive Principles proposed for "Separation in Relation".

#40-Composite material

#31-Porous materials

#32-Optical properties changes

#3-Local quality

#19-Periodic action

#17-Dimensionality change

2.2.5 Separation in system level

If one of the contradictory demands is required at the Subsystem or Supersystem level, separate them using "Separation in System Level" method. Separate the contradictory demands in separate system levels. For example, how to design a "Strong Rope". The engineering problem is to produce a rope which can meet the following requirements:

(1) As strong as cast iron

(2) Flexible enough to be folded and stored in a small place

There are four Inventive Principles are recommended for resolving the physical contradiction using "Separation in System Level" method:

#1-Segmentation

#5-Merging

#33-Homogeneity

The designer can apply the Inventive Principle #1-Segmentation to resolve the problem of Strong Rope. The Inventive Principle #1-Segmentation means that:

(1) Divide an object into independent parts

(2) Make an object easy to disassemble

(3) Increase the degree of fragmentation or segmentation

The final design solution is based on the "chain" concept. The components of the chain are made of cast iron and are rigid but the whole chain is flexible and can be folded for storage purpose.

2.2.6 Separation between parameters

When +A and -A are required for one parameter, check if the parameter is dependent on multiple independent parameters. Let +A requirement apply on one composing parameter and let –A requirement apply on another composing parameter. The formulation of separation between parameters is shown as follows:

Y ➔ big; Y ➔ small

Actually Y ➔ big is X1 to be big or Y ➔ small is X2 to be small.

Y = function (X1, X2, …)

Let X1 and X2 handle the opposing requirements separately.

2.3 Converting technical contradictions to physical contradictions

When two parameters are functions of same parameter, a technical contradiction can be converted into a physical contradiction. That means we located a core fundamental parameter. In general, converting from a technical contradiction to a physical contradiction makes the case more inventive as we are solving a more controlling and core problem. In that case, the solution are usually "stronger". The relationships between technical contradiction and physical contradiction is presented as Fig. 1.

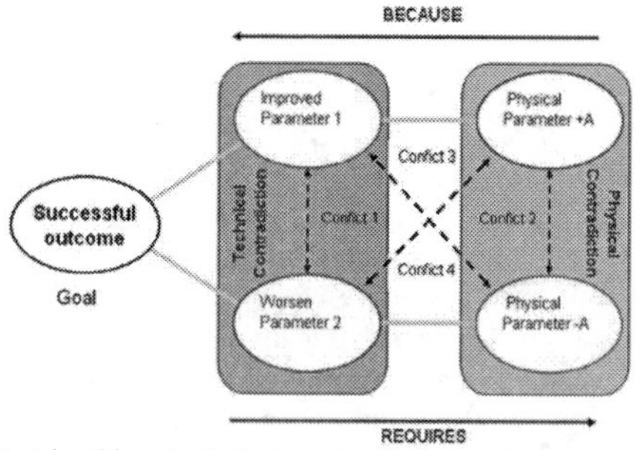

Figure 1. Identify a technical contradiction and corresponding physical contradiction.

2.3.1 Four paths of resolving design problems

In Fig. 1, the design problems can be resolved by the two ways of considering technical contradiction and physical contradiction. There are four paths of resolving design conflicts between technical and physical contradiction as shown follows.

(1) Resolve technical contradiction (conflict 1) between improved and worsen parameters (parameter 1 and 2).

(2) Resolve physical contradiction (conflict 2) between physical parameter +A and –A.

(3) Resolve technical contradiction and physical contradiction (conflict 3) between improved parameter and physical parameter +A.

(4) Resolve technical contradiction and physical contradiction (conflict 4) between worsen parameter and physical parameter -A.

The contradiction matrix and inventive principles of TRIZ can be used to resolve conflict 1, 3 and 4. The related algorithm for resolving physical contradiction mentioned above can be used to resolve conflict 2.

3 A case study

The creating process of a innovative screwdriver is presented as an example in this case study. The technical and physical contradictions are adopted to resolving the screwdriver design problems for creatiing product innovation at the same time. The main function of the screwdriver is to turn or release screws. It will raise insufficient torque when driving bigger screws. Meanwhile the screwdriver is not good in the driving efficiency when driving longer screws. The innovative goal of designing screwdriver is to improve the torque and efficiency in operation.

3.1 Identify the contradiction of the system

According to the goal of design screwdriver, the contradiction of the system can be obtained. The contradiction is when we drive a screw with high rotation speed, it will raise a low operation toeque. These two conflict parameters are "operation efficiency" and "operation torque". The improving technical parameter is "#9-Speed" and the worsen technical parameter is "#10-Force".

3.2 Analyze the design problem and conflict

After identifying the contradictions the conflict matrix is used to analyze the design problem and solve the technical contradictions. The related inventive principles is obtained from the contradiction matrix. Based on the 39 technical parameters and 40 inventive principles of TRIZ, the design solution trigger and specific solution can be obtained as follows:

#13-The other way around
#15-Dynamization
#19-Periodic
#28-Mechanics substitution

3.3 Convert technical contradiction to physical contradiction

In order to produce more creative outcomes we must convert technical contradiction to physical contradiction for more shaper and advanced design solutions. There are two conflict technical parameters in the design problem. When the screwdriver drives the longer screw, it need higher rotation speed to add operation efficiency. When the screwdriver drives the bigger screw, it need lower rotation speed to add operation torque. The physical contradiction is a situation in which the screwdriver design demands the contradictory values from the same parameter "Rotation Speed". Therefore, the technical contradiction can be converted to the physical contradiction as shown as following Fig. 2.

3.3.1 Use separation principles

The physical contradiction parameter "rotation speed" can be produced from the analysis of the above conevrting process. The condition meets the following design-thinking semantics.

To improve the operation efficiency Requires higher rotation speed; BUT

To improve the operation torque Requires lower rotation_speed.

3.3.2 Transition to a different system

The transtion principles can be also adopted to resolve the design problems unless the separation principles. The design thinking semantics can be shown as follows.

+A: high rotation speed; -A: low rotation speed

Can high rotation speed (+A) and low rotation speed (-A) be transferred to another system? The answer is positive. To establish a new gear module system which can offer different rotation speed can resolve this design problem.

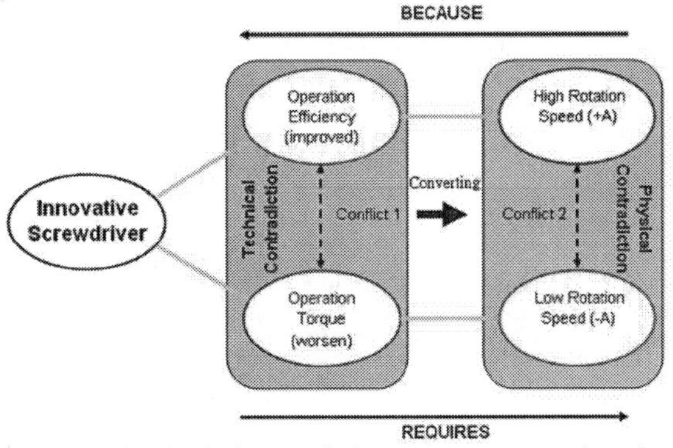

Figure 2. The physical contradiction parameter converting of innovative screwdriver.

3.4 Visualize the design concept

Based on the analysis of technical and physical contradiction, the concrete design concept for innovative driver can be presented as Fig. 3.

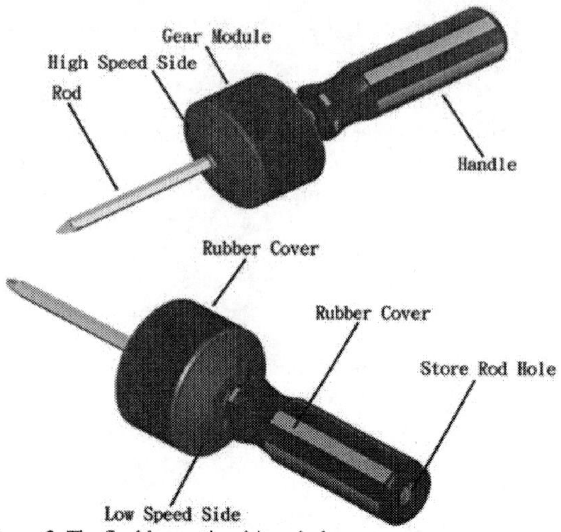

Figure 3. The final innovative driver design.

4 Conclusions

Various methods focus on increasing creativity in conceptual design. Which methods are capable of producing more creative outcomes is an important practical problem. Creativity is an integral and essential part in the product design process. Without creativity in design there is no potential for innovation, which is where creative ideas are actually implemented and transformed into commercial value. Also, without contradiction in design there is no novel ideas for creativity. As the contradictions were used both to assess and to explain creativity, it shows that in complex designs, creativity is closely linked to the contradictions used to reach the result. Considering this crucial creative role of contradictions, the reviewpoint of this paper is not considering contradictions as fixed single entities that can each be understood and researched in isolation, but as complex, dynamic and interdependent sets of contradictions. Based on the TRIZ concept, this paper presents a novel design-thinking approach by integrating technical and physical contradiction analysis methods into the conceptual design activities of New Product Development (NPD). The proposed design problem-solving method really can help designers create novel ideas for new product design by the verification of the case study. Meanwhile, by employing a multi-dimensional understanding through exploring contradictions beyond binary typologies and singular dimensions, creativity research can offer a more realistic and relevant starting point for both researchers and practitioners seeking to stimulate creativity through contradictions.

References

1. M. Darlington, Cognition and the engineering design requirement, Thesis (PhD), University of Bath, (2002)
2. R.J. Sternberg and J.C. Kaufman, Constraints on creativity, Cambridge handbook of creativity. New York: Cambridge University Press, 467-482, (2010).
3. S. Ajit, Constraint capture and maintenance in engineering design, AI EDAM, **22** (4): 325-343 (2008)
4. M. Stacey and C. Eckert, Reshaping the box: Creative designing as constraint management, International Journal of Product Development, **11** (3-4): 241-255 (2010).
5. P.D. Stokes, Creativity from constraints: What can we learn from Motherwell? From Modrian? From Klee? The Journal of Creative Behavior, **42** (4): 223-236 (2008).
6. P.D. Stokes, Creativity from constraints, New York: Springer, (2005)
7. J.H. Dyer, H.B. Gregersen and C.M. Christensen, The innovator's DNA. Harvard Business Review, **87** (12): 60-67 (2009).
8. A. Salter and D. Gann, Sources of ideas for innovation in engineering design, Research Policy, **32** (8): 1309-1324 (2003).
9. G. Altshuller, And Suddenly the Inventor Appeared TRIZ, the Theory of Inventive Problem Solving, 2nd ed., (1996)

Product Design Improvement Using Human-Computer Interaction Method Based on Modal Analysis

Lei Liu , Hongying Liu and Aijun Ma

China Astronaut Research and Training Center, Beijing, China

Abstract. A new kind of product design improvement method was proposed, which was mainly based on modal analysis using human-computer interactive way. And it was used in design improvement of an aerospace assembly to verify the effectiveness. The method made full use of the advantages of human and computer in product design and the work was mainly focused on modal analysis by computer and structure modification by human-computer interaction method. The final design got a better solution than the initial design as the results showed that the base frequency increased by 34.7% from 43.2Hz to 58.2Hz after design improvement and the vibration feature improved a lot. Also the experimental results were close to the simulation results with the relative error was 2.5%. The human-computer method can save cost and time and can provide a reference for other design of new products at the same time.

1 Introduction

With the development of society and industry, more and more products will be designed and produced. In the design phase of products, a lot of factors should be considered. And in the aerospace field, one of most important factors is vibration environment [1]. Although people wish the structure of the product can be the most suitable in the initial process of the development, the products either cannot satisfy the constraints or have large useless material allowance constantly at the beginning [2, 3, 4].

In the original ways, people produce the test specimen product and do the experiment based on the test part under the given test environment [5]. If the test pieces cannot satisfy the environment, people will improve the design and reproduce the test specimen product and test on it in a series of cycles. Obviously, the original product design means will expenditure a big amount cost and waste too much time at the same time. But with the development of CAD/CAE/CAM software, people can modify the structure in the product design software such as Pro/E, CATIA, SolidWorks and so on. Besides, by analysis of finite element software, people can find the weak point of the structure and improve the design in the software [6, 7, 8]. By cycles of design improvement, people will get a better solution than the initial design, and only need to do verified experiment on the final product. Based on the development of CAD/CAE/CAM software, we proposed a human-computer interaction method [9, 10] of product design improvement.

2 Product design improvement using human-computer interaction method

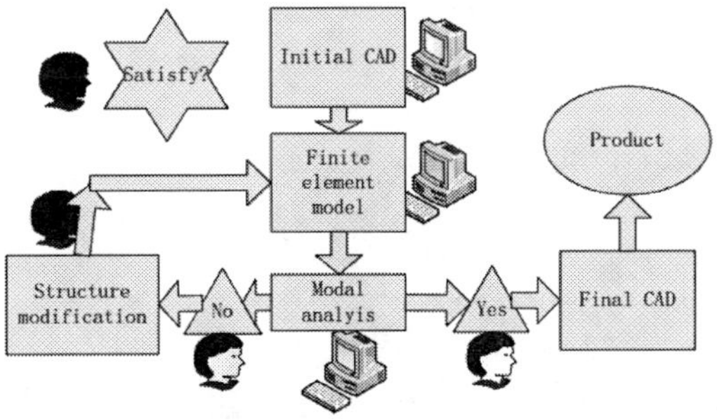

Figure 1. Flow chart of human-computer interaction method.

In this paper, a new kind of product design improvement using human-computer interaction method based on modal analysis was proposed and could be described as Fig. 1. As people design the product, they usually have initial CAD in the preliminary stage. Then they can import the initial CAD to finite element software such as HyperMesh, PATRAN, ANSYS and so on to get the finite element model. By modal analysis, the vibration feature and vibration frequency of the structure will be achieved. Then the weak point of the structure will be got by analysis of the vibration feature and people can improve the design by modification of the structure [11]. The biggest characteristic of the method is human-computer interaction; human and computer play a different part in the design improvement progress, but have sufficient interaction to finish the final job. Human not only work in the structure modification phase, but also help in the decisions whether the design is satisfied or not with the optimum scheme.

3 Improvement design of an aerospace assembly

DOI: 10.1051/matecconf/20166805002

3.1 Descriptions of the aerospace assembly

The aerospace assembly was consist of external framework and internal function products. As the internal function products were hard to modify, then the improvement work was mainly focused on the external framework which was consist of thin-walled beams. The objective was increasing the base vibration frequency and improving the vibration feature. To get a better look at the improvement point of the framework and in considering of privacy, the following figures just displayed the framework of the aerospace assembly and hided part of internal function products. But in the design and computing phase, all parts had been considered. The initial CAD could be described as Fig. 2. This phase was mainly finished based on computer using CAD software.

Figure 2. Initial CAD of the aerospace assembly.

3.2 Creating finite element model

Figure 3. Initial finite element model of the aerospace assembly (model a).

Then the initial CAD was imported to the finite element software. And the finite element model of the frame was divided by two dimension grids while the internal function products were divided by three dimension grids. Also this phase was mainly finished based on computer and the finite element model can be described as Fig. 3.

3.3 Modal analysis

As the finite element model was achieved, then modal analysis could be conducted and the first vibration feature could be got as fig.4. And the base frequency of the assembly was 43.2Hz which was less than the desired value. Worse still, vibration of the cared internal function products displayed as black circles in fig .4 was too large to satisfy the demand. Also this phase was mainly finished based on computer by finite element software.

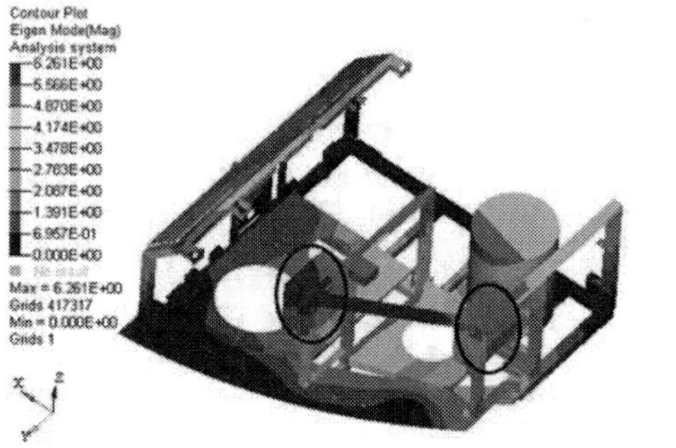

Figure 4. First vibration feature of the aerospace assembly (model a).

3.4 Improvement of the finite element model based on modal analysis

Then the design could be improved in the finite element software by human-computer interaction, which meant that modal analysis and structure modification was conducted in an interactive style and the assembly would make a better performance by cycles of structure modification again and again. In this phase, the work was finished by human-computer interaction. The improvement design of the aerospace assembly in the finite element software can be described as Fig. 5.

The design model of the aerospace assembly was improved from model 'a' to model 'd' after a cycle of structure modification. And Table 1 listed the base vibration frequency of four structures and the change of cared parameters.

Table 1. Parameters change chart of four structures.

Parameters	Topo a	Topo b	Topo c	Topo d	Rate change
Base frequency/Hz	43.2	44.2	45.8	58.2	+34.7%

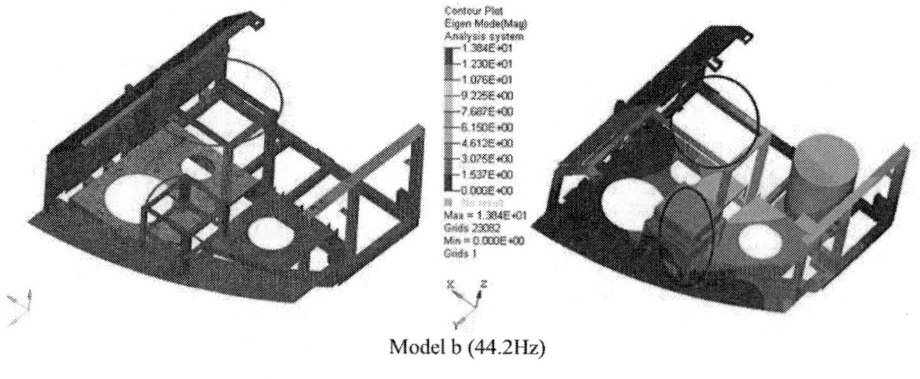

Model b (44.2Hz)

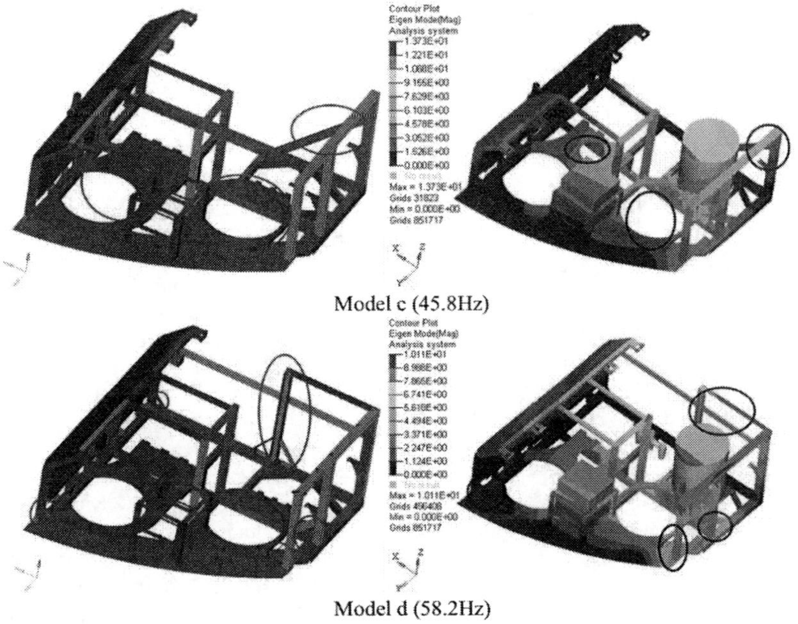

Figure 5. Improvement design of the aerospace assembly in finite element software.

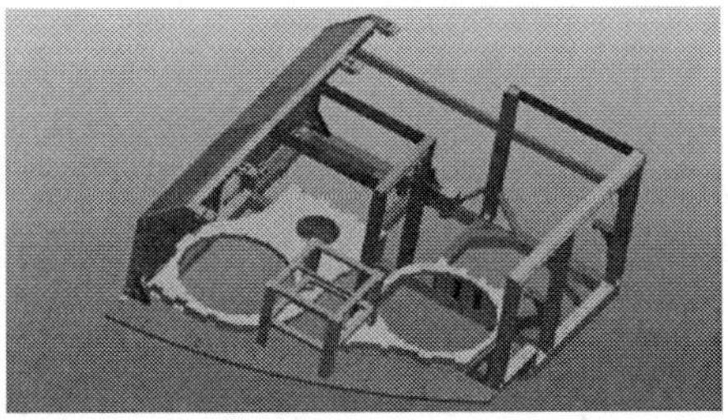

Figure 6. Final CAD of the aerospace assembly.

3.5 The final CAD and product.

By cycles of improvement by human-computer interaction method, the final finite element model could satisfy the requirement by modal analysis results. Besides, the base frequency of the final model improved from 43.2Hz to 58.2Hz and the weak points had been modified, so model 'd' could be regarded as the final finite element model [12]. Then we used the Pro/E software to generate the final CAD described as Fig. 6 and produced the product.

3.6 Experimental results.

The final product was produced based on the final CAD of Fig. 6. And vibration environment experiment was established on the 10t platform vibrator. The experimental results of the base vibration frequency was 59.7Hz. Table 2 listed the comparison of the simulation results with the experimental results.

4 Results & discussion

The simulation data from table 1 showed that the base frequency of aerospace assembly had improved from 42.3Hz to 58.2Hz after human-computer interaction product design improvement based on modal analysis, and the rate change between final model 'd' and initial model 'a' was 34.7%, which showed that the improvement design made a good performance and the effect was obvious.

And the experimental results of the base frequency of the real product from table 2 was 59.7Hz comparing to the simulation results which was 58.2Hz, and the relative error was about 2.5%, which proved the effectiveness of the simulation analysis by finite element software. From this case, the following conclusion could be drew that product design improvement using human-computer interaction method could make a great contribution in design of similar product, and was not limited to the aerospace product.

Table 2 Comparison of the simulation results and the experimental results.

Analysis situation	Simulation results	Experimental results	Relative error
Value	58.2Hz	59.7Hz	2.5%

5 Conclusions

In this paper, a new kind of product design improvement using human-computer interaction method was proposed and was used to an aerospace assembly to verify the effectiveness. The results showed that the design improvement increased the base frequency by 34.7% from 43.2Hz to 58.2Hz and the vibration feature improved a lot. Also the experimental results were close to the simulation results with the error was 2.5%. The method was mainly based on modal analysis on the finite element model by computer and structure modification by human-computer interaction. The method made full use of the advantages of human and computer in product design and could get a better solution than the initial design. In a word, the human-computer method can save cost and time and can provide a reference for other design of new products at the same time. This human-computer ideology might also be used for other fields such as key engineering material and so on.

Acknowledgement

Funded by the manned space flight project of China was greatly appreciated.

References

1. Tibert, Gunnar. Deployable tensegrity structures for space applications. Royal Institute of Technology (2002)
2. Nguyen, Trung Thanh, Shengxiang Yang, and Juergen Branke. "Evolutionary dynamic optimization: A survey of the state of the art." Swarm and Evolutionary Computation **6** (2012): 1-24.
3. Schlegel, Martin, and Wolfgang Marquardt. "Detection and exploitation of the control switching structure in the solution of dynamic optimization problems." Journal of Process Control **16**, 3 (2006): 275-290.
4. Åkesson, Johan, K-E. Årzén, Magnus Gäfvert, Tove Bergdahl, and Hubertus Tummescheit. "Modeling and optimization with Optimica and JModelica. org—Languages and tools for solving large-scale dynamic optimization problems." Computers & Chemical Engineering **34**, 11 (2010): 1737-1749.
5. Wegener, Joachim, A. Baresel, and H. Sthamer. "Evolutionary test environment for automatic structural testing." Information & Software Technology **43**, 14 (2001): 841-854.
6. Peeters, Bart, and Guido De Roeck. "Reference-based stochastic subspace identification for output-only modal analysis." Mechanical systems and signal processing **13**, 6 (1999): 855-878.
7. Ewins, David J. Modal testing: theory and practice. Letchworth: Research studies press, **6** (1995)
8. Peeters, Bart, and Guido De Roeck. "Stochastic system identification for operational modal analysis: a review." Journal of Dynamic Systems, Measurement, and Control **123**, 4 (2001): 659-667.
9. Preece, Jenny, Yvonne Rogers, Helen Sharp, David Benyon, Simon Holland, and Tom Carey. Human-computer interaction. Addison-Wesley Longman Ltd., (1994)
10. Cohen, Ira, Fabio G. Cozman, Nicu Sebe, Marcelo C. Cirelo, and Thomas S. Huang. "Semisupervised learning of classifiers: Theory, algorithms, and their application to human-computer interaction." Pattern Analysis and Machine Intelligence, IEEE Transactions on **26**, 12 (2004): 1553-1566.
11. Kitazaki, Satoshi, and Michael J. Griffin. "A modal analysis of whole-body vertical vibration, using a finite element model of the human body." Journal of Sound and Vibration **200**, 1 (1997): 83-103.
12. Nefske, D. J., J. A. Wolf, and L. J. Howell. "Structural-acoustic finite element analysis of the automobile passenger compartment: a review of current practice." Journal of Sound and Vibration **80**, 2 (1982): 247-266.

DOI: 10.1051/matecconf/20166806001

A Hybrid Algorithm for Strip Packing Problem with Rotation Constraint

Huan Chen[1], Furong Ye[1] and Yain-Whar Si[2]

[1]*Department of Computer Science, Xiamen University, Xiamen, China*
[2]*Department of Computer and Information Science, University of Macau, Macau*

Abstract. Strip packing is a well-known NP-hard problem and it was widely applied in engineering fields. This paper considers a two-dimensional orthogonal strip packing problem. Until now some exact algorithm and mainly heuristics were proposed for two-dimensional orthogonal strip packing problem. While this paper proposes a two-stage hybrid algorithm for it. In the first stage, a heuristic algorithm based on layering idea is developed to construct a solution. In the second stage, a great deluge algorithm is used to further search a better solution. Computational results on several classes of benchmark problems have revealed that the hybrid algorithm improves the results of layer-heuristic, and can compete with other heuristics from the literature.

1 Introduction

Strip packing is a well-known NP-hard problem and has many practical applications in the engineering fields. For example, strip packing can be applied for tasks such as placing goods on shelves in the warehouses, arranging articles and advertisements during the type-setting of the newspapers, cutting rectangular pieces from large sheets of material in the wood or glass industries, and laying very-large-scale integration (VLSI) in VLSI floor planning industry. These applications can be formalized as a strip packing problem according to different constraints and objectives [1]. The interested reader is referred to the literature [1]-[3] for more information on packing problems.

This paper considers a two-dimensional orthogonal strip packing problem: Given a rectangular sheet of given width and unlimited height, and a set of rectangles with arbitrary length and width, the orthogonal strip packing problem is to place each rectangle on the sheet such that no two rectangles overlap and the used height h of sheet is minimized. Let W be the width of the rectangular sheet, and n is the number of rectangles, let hi and wi be the length and width of rectangle i (i=1, 2,..., n) respectively. Where, we assume that the edges of each rectangle are parallel to the edges of the sheet. In addition, all rectangles except for the rectangular sheet are permitted to be rotated when they are placed. A formal definition for this problem can be found in [4].

The two-dimensional orthogonal strip packing problem mentioned above belongs to a subset of classical cutting and packing problems and has been shown to be NP-hard [5]. Some exact algorithms for orthogonal two-dimension cutting and packing problem were proposed in [6] and [7]. However, they might be impractical for large problems because large amount of computational time is needed to obtain an optimal solution. Therefore, heuristic algorithms, which can produce a good approximation solution within an acceptable time, are preferred to solving this class of problems. Examples of heuristic algorithms include the well-known bottom-left (BL), bottom-left-fill (BLF) and other heuristic methods [8]. In addition, some new heuristic algorithms such as constructive approach [9], new placement heuristics [10], and heuristic recursion algorithm [4] were developed to solve this class of packing problems. These heuristic algorithms can obtain a good solution in a short time. In order to obtain a better solution, heuristic algorithms are often combined with meta-heuristic algorithms such as genetic algorithms [11]-[13], neural network and GA [14], simulated annealing algorithms [15]-[17] and other meta-heuristics [18]-[23]. An empirical investigation of meta-heuristic and heuristic algorithms for the orthogonal packing problem of rectangles is given by Hopper and Turton [24]. Recently, some excellent hybrid algorithms were developed [25]-[27]. In particular, a class of deterministic heuristics is developed by a number of researchers and they are successful in obtaining the better results [20], [28], [29]. Leung and Zhang [30] proposed a fast layer-heuristic algorithm, which have many applications in routing problem with loading constraints [31], [32]. Several efficient heuristic algorithms for the variants of packing problem were also developed [33]-[35] recently. In this paper, a hybrid algorithm for solving the orthogonal strip packing problem with rotation constraint is developed by combining layer-based heuristics with a great deluge algorithm. Computational results on several classes of benchmark problem instances have shown that the hybrid algorithm can compete with other heuristics.

2 Imprroved layer-based heuristic

Layer-based heuristic was proposed by Leung and Zhang [30]. It is a fast heuristic algorithm for strip packing problem. The idea of layer-based heuristic is as follows:

(1) Select a reference item r from unpacked items.

(2) Stack some unpacked items above the item r to determine a reference line.

(3) Pack the available space under the reference line as follows:

(3.1) Determine the lowest available space

(3.2) Select an unpacked item i with maximal fitness value from unpacked items.

(3.3) Pack the item i and update the available space s;

(3.4) If there is an unpacked item can be packed into the available space s, go to (3.1);

(4) if there exists an unpacked item, go to (1).

The detailed process of layer-based heuristic based on the idea of fitness value is given in Leung and Zhang [30]. There are four cases when the fitness value is computed in the process of layer-based heuristic [30]. In fact, there are two special cases that should be given a bigger fitness value when $h1 \geqslant h2$ and $h1<h2$, respectively. The case for $h1 \geqslant h2$ is shown in Figure 1(a). And for the case given in Figure 1(b), the item R should be given higher fitness value than the item R from Figure 1(a). Also, the item R in Figure 1(d) should be given higher fitness value than the item R from Figure 1(c). Meanwhile, there also are two special cases as $h1<h2$, and they are shown in Fig. 2.

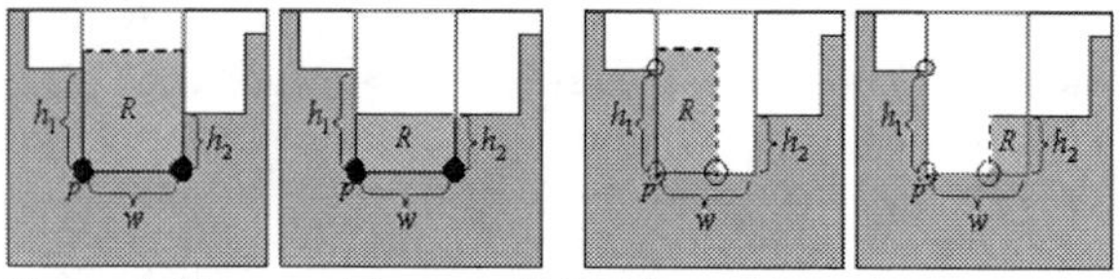

Figure 1. Two special cases of h1 ⩾h2.

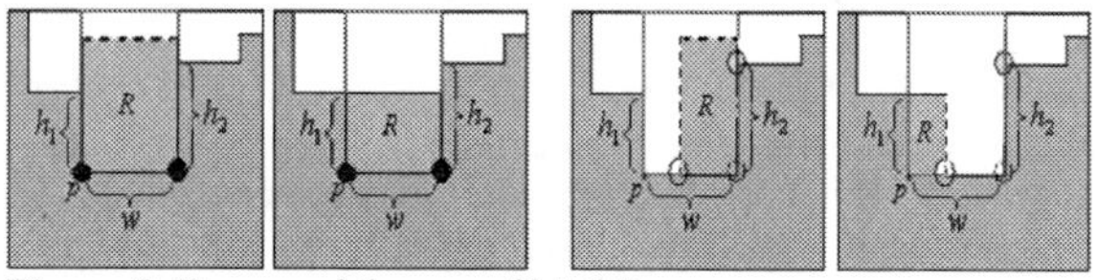

Figure 2. Two special cases of h1<h2.

3 Hybrid algorithm

The great deluge algorithm (GDA) is a general algorithm which is applied to solving optimization problems. It is similar to the hill-climbing and simulated annealing algorithms in many ways. GDA comes from the analogy that in a great deluge a person climbing a hill will try to move in any direction that does not get his/her feet wet in the hope of finding a way up as the water level rises. In a typical implementation of the GDA, the algorithm starts with an initial solution which is a poor approximation S of the optimum solution. A numerical value called "badness" is computed based on S and it measures how undesirable the initial approximation is. The higher the value of badness, the more undesirable the approximate solution is. Another numerical value called "tolerance" is calculated based on a number of factors, often including the initial badness.

For a given solution *x*, we can construct a neighborhood *N*(*x*). A new approximate solution *x*' which is called a neighbor of *x*, is selected from *N*(*x*). The badness b' of *x*' is computed and compared with the tolerance. If b' is better than tolerance, then the algorithm is restarted by setting *x*:=*x*', and tolerance:=decay(tolerance), where decay is a function that lowers the tolerance (representing a rise in water levels). If b' is worse than tolerance, a different neighbor *x** of *x* is chosen and the process is repeated. If all the neighbors of *x* produce approximate solutions beyond the tolerance value, then the algorithm is terminated and *x* is returned as the best approximate solution obtained.

```
GDA()
best=besth,level=3;
while(1)
 for(i=0;i<2*typen;i++)
     k1=rand()%(typen-1);
     do k2=rand()%typen;
     while(k2==k1);
     swap(x[k1],x[k2]);
     flag1=Heuristicplacing(x);
     up=flag1-best;
     if(up<=level)  best=flag1;
     if(best<besth)
         besth=best;
         level=level-up;
     else
         swap(x[k1],x[k2]);
     end = clock();
     double runtime=(float)(end - start) / CLK_TCK;
     if(runtime>60.0)
         return besth;
 rrr=rand()%3;
 if(rrr==0) sort1(x);
 else if(rrr==1)sort2(x);
 else if(rrr==2)sort3(x);
```

The detailed GDA for strip packing problem is as follows:

where *x* is an array and records the orderings of rectangular pieces. Function Sort1(*x*) sorts a sequence of rectangle pieces according to the non-increasing ordering of area size. Sort2(*x*) sorts a sequence of rectangle pieces according to the non-increasing ordering of length size. Sort3(*x*) sorts a sequence of rectangle pieces according to the non-increasing ordering of perimeter size. The parameter "besth" denotes the current minimal height.

Function Heuristicplacing(x) is the same heuristic algorithm used in Leung and Zhang [30]. Therefore, the proposed algorithm based on GDA (FH+GDA) is as follows:

```
FH+GDA()
  Fastheuristic();
  GDA();
```

where Fastheuristic() is the function proposed in Leung and Zhang [30]. Note that the performance of heuristic algorithm significantly depends on the packing ordering X of the rectangles. Some results have shown that the packing orderings affect the performance of the presented algorithm [31]. In this paper, we consider three orderings to avoid the search process from being trapped in a local minimum

4 Computational results

In order to test the performance of the proposed algorithm (FH+GDA), a large amount of benchmark are selected and the results of the proposed algorithm are compared with results of some published heuristics. The benchmark instances (C11~C73) in [24] include 21 problem instances ranging from 16 to 197 items. And the instances 50cx~15000cx are taken from [36], whose sized ranges from 25 to 5000. Considering more constraints in strip packing problem, instances zdf1~zdf16 with fix-orientation and instances beng1~beng10 with small-scale are also selected. All problem classes can be downloaded from
http://algorithm.xmu.edu.cn:10000/Download.aspx#p4.
In the section, FH+GDA is with compared BF+TS, BF+SA, BF+GA, HRP, FH and DHA. BF+TS, BF + SA, BF + GA were tested on a Pentium IV machine with 2.0 GHz. The algorithms are executed for 10 times with a time limit of 60s per run. HRP is implemented in C#.net programming language and it was tested on a 2 GHz Pentium 4 notebook computer with 2048 MB of RAM. FH is a deterministic algorithm and it was tested on a 2 GHz Pentium 4 notebook computer with 2048 MB of RAM with the time limit of 60s per run. DHA was implemented in C programming language and executed as a single process/thread, and all tests were performed on a personal computer with 3.0 GHz CPU and 2.0 GB memory with a time limit of 1000s. FH+GDA was coded in C++ and executed on a TECRA M2V computer with a 1.6GHz CPU and 256 MB of RAM. The proposed algorithm is executed five times and each execution last for maximum of 60 seconds. The comparison of the performance of different algorithms is based on h and *time*, where, h and *time* denote the height and running time (in second) respectively.

4.1 The computational results on the test problems permitting rotation

In the part, the computational results are from the instances permitting rotation. Meanwhile, BF+GA, BF+SA and BF+TS in [16], HRP [20], FH [30] and DHA [29] are introduced for comparison. The results of BF+GA, BF+SA and BF+TS are from [16], the results of HRP and FH are from [30], and the results of DHA are from [29]. In tables, "op" represents the optimal height, OPTMA# denotes the number of the optimal height obtained by different algorithms. BEST# denotes the number of the best height obtained by different algorithms.

Since the previous paper did not supply the time of experiment. The comparison among BF+TS, BF+SA, BF+GA and FH+GDA is only concentrated on the h, and the results is shown in Table1. Obviously, FH+GDA achieves the best result in every instances among four strategies. The comparison among FH+GDA and other three algorithms on *time* and h is presented in Table 2. Also, FH+GDA shows efficiency on all instances, FH+GDA gets best results in all instances and do not cost much time. Table 1 and Table 2 clearly indicate that FH+GDA performs better than other algorithms on instances permitting rotation.

Table 1. The results of BF+TS, BF+SA, BF+GA and FH+GDA on instances permitting rotation.

in	op	BF +TS(h)	BF +SA(h)	BF +GA(h)	FH+GDA
C11	20	**20**	**20**	**20**	**20**
C12	20	21	**20**	21	**20**
C13	20	**20**	**20**	**20**	**20**
C21	15	16	16	16	**15**
C22	15	16	16	16	**15**
C23	15	16	16	16	**15**
C31	30	31	31	31	**30**
C32	30	32	31	32	**30**
C33	30	31	31	31	**30**
C41	60	62	61	62	**60**
C42	60	62	61	62	**60**
C43	60	61	61	62	**60**
C51	90	92	91	92	**90**
C52	90	92	91	92	**90**
C53	90	92	92	92	**90**
C61	120	122	122	122	**120**
C62	120	121	121	121	**120**
C63	120	122	122	122	**120**
C71	240	245	244	245	**241**
C72	240	244	244	244	**240.6**
C73	240	245	245	245	**241**
OPTMA#		2	3	2	18
BEST#		2	3	2	21

Table 2. The results of HRP, FH, DHA and FH+GDA on instances permitting rotation.

in	op	HRP		FH		DHA		FH+GDA	
		h	*time*	h	*time*	h	*time*	h	*time*
C11	20	**20**	0.03	**20**	0	**20**	0	**20**	0.01
C12	20	**20**	0.19	**20**	0.01	21	0.3	**20**	0

DOI: 10.1051/matecconf/20166806001

C13	20	**20**	0.03	21	0	**20**	0.02	**20**	0
C21	15	**15**	0.27	16	0.02	**15**	0.59	**15**	0.07
C22	15	**15**	0.06	**15**	0	**15**	0.23	**15**	0
C23	15	**15**	0.05	**15**	0	**15**	0.06	**15**	0
C31	30	31	0.61	31	0.02	31	3.27	**30**	0.01
C32	30	31	0.61	31	0.02	32	5.14	**30**	0.97
C33	30	31	0.63	32	0.02	**30**	0.25	**30**	0.08
C41	60	61	1.84	61	0.16	61	20.86	**60**	0.20
C42	60	**60**	0.23	61	0.11	61	16.95	**60**	1.20
C43	60	61	2.02	61	0.08	**60**	14.62	**60**	0.05
C51	90	91	4.3	91	0.28	**90**	2.02	**90**	0.18
C52	90	**90**	1.34	**90**	0	**90**	15.62	**90**	0.07
C53	90	91	4.3	91	0.3	**90**	39.18	**90**	1.38
C61	120	121	9.84	121	0.83	121	169.53	**120**	0.04
C62	120	121	8.38	121	0.89	121	139.25	**120**	2.07
C63	120	121	9.94	121	0.76	**120**	2.7	**120**	4.21
C71	240	241	61.48	241	13.91	241	1000	**241**	60.00
C72	240	241	58.91	241	11.64	241	1000	**240.6**	1.73
C73	240	241	62.9	241	15	241	1000	**241**	60.00
OPTMA#		8		5		11		18	
BEST#		10		7		13		21	

In order to extensively test the performance of FH+GDA, A large-scale problem is considered. Table 3 presents the comparison of computational results of four algorithms with instances 50cx~15000cx. According to the table, FH+GDA outperforms other three algorithms. FH+GDA find smaller *h* than FH in 50xc and 100cx, and solutions of other instances obtained by FH+GDA are the same as FH. It reveals that FH+GDA can improve the solutions obtained by FH. We can observe that BEST# and OPTIMA# of FH+GDA is highest, which means FH+GDA shows efficiency with large-scale problems as well. Meanwhile, the *time* of FH+GDA is smaller than the *time* of HRP and DHA for most instances.

Table 3. The results of large-scale instances.

in	op	HRP		FH		DHA		FH+GDA	
		h	*time*	*h*	*time*	*h*	*time*	*h*	*time*
50cx	600	615	10.09	624	0.33	644	26.91	**606.8**	60.00
100cx	600	615	47.88	619	3.56	637	266.64	**609**	60.00
500cx	600	611	87.31	**600**	2.00	601	1000	**600**	4.62
1000cx	600	607	86.08	**600**	0.02	**600**	628.48	**600**	1.38
5000cx	600	607	9256.37	**600**	0.05	**600**	3.92	**600**	0.01
10000cx	600	—	—	**600**	0.05	**600**	6.5	**600**	0.02
15000cx	600	—	—	**600**	0.06	**600**	9.88	**600**	0.02
OPTMA#		0		5		4		5	
BEST#		0		5		4		7	

4.2 The computational results on test problems with fixed orientation

In this paper, similar to [30], we test the performance of FH+GDA with several large-scale instances with fixed orientation. The well-known GRASP, SVC, and FH algorithms are selected for comparison in the Table 4. The computation results of other three algorithms can be accessed from [30] directly. Likewise, FH+GDA gets best results among four algorithm in most instances. Due to the large scale of problems, FH+GDA only achieve better solution in two instances and improve FH slightly, since FH+GDA is executed in limited 60 seconds.

Table 4. The results of large-scale instances with fixed orientation.

in	op	GRASP	SVC	FH	FH+GDA
zdf1	330	333	331	**330**	**330**
zdf2	357	360	358	**357**	**357**
zdf3	384	387	385	**384**	**384**
zdf4	407	410	408	**407**	**407**
zdf5	434	437	**434**	**434**	**434**
zdf6	4872	5300	**5085**	5311	5102
zdf7	4852	5163	**5083**	5311	5102
zdf8	5172	5544	5386	**5360**	**5360**
zdf9	5172	5476	5468	**5402**	**5402**
zdf10	5172	5570	5462	**5178**	**5178**
zdf11	5172	5608	5516	**5174**	**5174**
zdf12	5172	—	5651	5175	**5173**
zdf13	5172	—	5600	**5172**	**5172**
zdf14	5172	—	5468	5257	**5244**
zdf15	5172	—	5960	**5402**	**5402**
zdf16	5172	—	5931	**5402**	**5402**
OPTMA#		0	1	**6**	**6**
BEST#			3	**12**	**14**

4.3 The computational results on problems with non-zero waste

Similar to [29], we apply the benchmark Beng [37] which includes ten small-scale instances ($20 \leqslant n \leqslant 200$) with fixed orientation constraint. The lower bounds of these problems are known but the optimal solutions are still not yet known.

Table 5 reports the results achieved by the exact staircase algorithms [38], DHA and FH+GDA. The staircase algorithms were executed on a 3.0 GHz Pentium 4 computer with 1.0 GB memory with the time limit of 3600s. W denotes the width of the rectangular board in the table. From Table 5, we could clearly observe that three algorithms obtain the same height as the lower bounds, namely the optimal height for all the instances. However, FH+GDA is much faster than DHA, for all instances, the *time* of FH+GDA do not beyond 0.01s.

Table 5. The results on problems with non-zero waste.

in	n	W	op	Staircase		DHA		FH+GDA	
				h	time	h	time	h	time
beng1	20	25	30	**30**	0.08	**30**	0.78	**30**	0.003
beng2	40	25	57	**57**	0.12	**57**	6.28	**57**	0.001
beng3	60	25	84	**84**	0.1	**84**	17.06	**84**	0.001
beng4	80	25	107	**107**	0.08	**107**	34.74	**107**	0.002
beng5	100	25	134	**134**	0.08	**134**	52.34	**134**	0
beng6	40	40	36	**36**	0.1	**36**	7.08	**36**	0
beng7	80	40	67	**67**	0.11	**67**	36.17	**67**	0.001
beng8	120	40	101	**101**	0.17	**101**	85.78	**101**	0.001
beng9	160	40	126	**126**	0.23	**126**	145.11	**126**	0
beng10	200	40	156	**156**	0.81	**156**	231.27	**156**	0.001
OPTMA#				10		10		10	

5 Conclusions

A hybrid algorithm based on a layer-based heuristic and a great deluge algorithm for the strip packing problem is presented. The proposed algorithm is able to solve the strip packing problem efficiently and further improve the packing results of FH. Extensive computational results show that FH+GDA can compete with meta-heuristics in terms of both solution quality and execution time. FH+GDA is superior especially for large test problems. Therefore, FH+GDA may be of great practical value to the rational layout of the rectangular objects from the engineering fields, such as the wood-, glass- and paper industry, and the textile and leather industry. Future work is to further improve the performance of this algorithm and extend it to solve other packing problems.

Acknowledgment

The work was partially supported by the National Nature Science Foundation of China (61272003).

References

1. A. Lodi, S. Martello, M. Monaci, Eur. J. Oper. Res. **141**, 241–252 (2002)
2. D.Pisinger, Eur. J. Oper. Res. **141**, 382–392 (2002)
3. E. Silva, J.F. Oliveira, G. Wäscher, Eur. J. Oper. Res. **237**, 846-856 (2014)
4. D. Zhang, Y. Kang, A. Deng, Comput. Oper. Res. **33**, 2209–2217 (2006)
5. J.E. Beasley, Oper. Res. **33**, 49–64 (1985)
6. S. Martello, M. Monaci, D. Vigo, INFORMS J. Comput. **15**, 310–319 (2003).
7. Y. Cui, L. Huang, Comput. Optim. Appl. **33**, 287-301 (2006)
8. B. Chazelle, IEEE Trans. Comput, **32**, 697–707 (1983)
9. M. Hifi, R.M Hallah, Int. Trans. Oper. Res. **10**, 1–22 (2013)
10. E.K. Burke, G. Kendall, G. Whitwell, Oper. Res. **52**, 655–671 (2014)
11. S. Jakobs, Eur. J. Oper. Res. **88**, 165–181(1996)
12. D. Liu, H. Teng, Eur. J. Oper. Res. **112**, 413–419 (1999)
13. D. Zhang, S. Chen, Y. Liu, Acat Automatica Sin. **33**, 911-916 (2007)
14. C.H. Dagli, P. Poshyanonda, J. Intell. Manuf. **8**, 177–190 (1997)
15. L. Wei, D. Zhang, Q. Chen, Comput. Oper. Res. **36**, 1608-1614 (2009)
16. E.K. Burke, G. Kendall, G. Whitwell, INFORMS J. Comput. **21**, 505–516 (2009)
17. S.C.H. Leung, D. Zhang, K. M. Sim, Eur. J. Oper. Res. **215**, 57-69 (2011)
18. D. Zhang, A. Deng, Y. Kang, *Lecture Notes in Computer Science 3514*, 783–791 (2005)
19. D. Zhang, Y. Liu, S. Chen, X. Xie, *Lecture Notes in Computer Science 3612*, 1235–1241 (2005)
20. W. Huang, D. Chen, R. Xu, Comput. Oper. Res. **34**, 3270–3280 (2007)
21. D. Zhang, S. Han, W. Ye, Chin. J. Comput, **23**, 509–515 (2008)
22. R. Alvarez-Valdes, F. Parreño, J.M. Tamarit, Comput. Oper. Res. **35**, 1065–1083 (2008)
23. G. Belov, G. Scheithauer, E.A. Mukhacheva, J. Oper. Res. Soc. **59**, 823– 832 (2008)
24. H. Hopper, B.C.H. Turton, Eur. J. Oper. Res. **128**, 34–57 (2001)
25. D. Zhang, L. Wei, S.C.H. Leung, Q. Chen, INFORMS J. Comput. **25**, 332-345 (2013)
26. Y. Wang, L. Chen, Expert Syst. Appl. **42**, 3297-3305 (2015)
27. L. Wei, W.C. Oon, W. Zhu, A. Lim, Eur. J. Oper. Res. **215**, 337-346 (2011)
28. K. He, W. Huang, Y. Jin, Comput. Oper. Res. **39**, 1355-1363 (2012)
29. K. He, Y. Jin, W. Huang, Expert Syst. Appl. **40**, 5542-5550 (2013)
30. S.C.H. Leung, D. Zhang, Expert Syst. Appl. **38**, 13032-13042 (2011)
31. S.C.H. Leung, X. Zhou, D. Zhang, J. Zheng, Comput. Oper. Res. **38**, 205-215 (2010)
32. S.C.H Leung, J. Zheng, D. Zhang, X. Zhou, Flex. Serv. Manuf. J. **22**, 61-82 (2010)
33. S. Hong, D. Zhang, H.C. Lau, X. Zeng, Y. Si, Eur. J. Oper. Res. **238**, 95-103 (2014)
34. L. Wei, T. Tian, W. Zhu, A. Lim, Eur. J. Oper. Res. **239**, 58-69 (2014)
35. K. He, P. Ji, C. Li, Eur. J. Oper. Res. **241**, 674-685 (2015)
36. E. Pinto, J.F. Oliveira, *Second ESICUP Meeting*, **29**, 531– 540 (2015)
37. B.E. Bengtsson, Comput. J. **25**, 353–357 (1982)
38. M. Kenmochi, T. Imamichi, K. Nonobe, M. Yagiura, H. Nagamochi, Eur. J. Oper. Res. **198**, 73–83 (2009)

Supplier selection by Using AHP-TOPSIS and Goal Programming - A case Study in Casumina Rubber Company - Vietnam

Ha Thi Xuan Chi[a], Do Hoang Ngoc Trinh

International University – Vietnam National University-HCMC, Vietnam

Abstract. Suppliers, one of the important factors, directly affect to trade strategy of a company which frequently changed. Evaluating the supplier's abilities is a required step to select the appropriate supplier who ensure meet the demands of the company flexibly. This paper integrates both MADM and MODM approaches considering quantitative and qualitative techniques to support decision makers select the best solution under criteria and goals suggested by the company in different period. Analytic Hierarchy Process (AHP) and Technique for Order Preference by Similarity to Ideal Solution method (TOPSIS) are used to weigh criteria and obtain total score for each alternative, respectively. In addition, Goal Programming (GP) allows defining priorities of goals and conflicted goals as the company expects to achieve. This study combines three methods solving the supplier section issue for Casumina Rubber Company in Vietnam.

1 Introduction

Purchasing management is an essential part that constitutes to a successful business. According the study done by Ghodsypour and O'brien [1], the cost of raw materials and resources can account up to 70% of final cost price. Therefore, the company can manage to select the supplier based on the lowest possible cost; it can result in a significant reduction in final products price and increase profit gained. However, selecting the best suppliers not only based on the lowest price that offers from suppliers, but regard to other factors such as quality, durability, delivery, cost. Thomas Saaty [2] firstly introduced a well-known method named Analytic Hierarchy Process (AHP) to help solving complex problems and conflicting criteria in decision making. AHP allows building the hierarchy tree including level of goals, criteria, sub-criteria and alternatives. In practice, many companies have successfully selected the best supplier by applying AHP. In general, AHP, a great support tool, is widely applied in many different fields; for instance, in manufacturing system, in layout design, in supplier selection. However, AHP suffers from number of items in each level no more than nine. Thus, AHP is unable to deal with complicated situations. In order to overcome this shortcoming, TOPSIS, as an effective tool for making multi-criteria and introduced first by Hwang and Yoonc[3]. TOPIS method has been successfully applied by Omid Jadidi, Fatemeh Firouzu, Enzo Bagliery for supplier selection [4]; Vijay Manikrao Athawale, Shakar Chakraborty [5] for machine tool selection ; Amir Naderi, Mohanmad Saeed Zaeri [6] for evaluation suppliers in supply chain management; Birsen Karpak, Rammohan E. Kasuganti [7] for an application of visual interactive goal programming. Moreover, the combination of AHP-TOPSIS brings useful outcome for business in case of Min Wu [8] proposed simulation model and its application to supply chain management; Jayaram C Sasi, Dr, Abhijeet K Digalwar [9] offered solution for supplier selection between India & China in Textile Industry.

Multi-criteria decision-making (MCDM) is divided into two subtypes named Multi-Objective Decision Making (MODM) and Multi-Attribute Decision Making (MADM). MODM can determine the optimal solution for both single objective and multiple objectives under satisfying the constraints and preferences priorities. Methods can be considered as MODM are Linear Programming (LP), Goal Programming (GP), Mix Integer Programming (MIG), Data Envelopment Analysis (DEA). Unlike MODM, MADM limits a number of predetermined alternatives; numerous methods for solving decision-making problems are AHP, TOPSIS, ANP, MAUT, Outranking and others. Therefore, a perfectly combination between MADM and MODM for evaluating performance of suppliers are necessary. Indeed, there are many researches highlight integration of several methods that their advantages can be utilized. Ghodsypour, S.H., O'Brien [1] used an integrated analytical hierarchy process and linear programming for supplier selection. Sanayei, A., Mousavi, S.F Abdi, M.R and Mohaghar, A.,[10] used an integrated group decision making process for supplier selection and order allocation using muli –attribute utility theory and linear programming ; Razmi, L., Rafiei, H. [11], proposed "An intergrated analytic network process with mixed- integer

[a] Corresponding author: htxchi@hcmiu.edu.vn

Non-linear Programming to supplier selction and order allocation. Weber, C.A., Ellram, L.M. [10] used multi-objective programming: a decision support system approach for supplier selection. Therefore, this paper employs AHP, TOPSIS and GP to solve selection supplier for the Casumina rubber company.

This paper is organized as follows. Section 1 presents introduction. Section 2 shows reviewing existing methods. A case study are described in Section 3. In Section 4, the results of the case study are displayed. The rest is conclusion.

2 Reviewing existing methods

In this section, three main relevant methods have been considered. Firstly, AHP is used to simply obtain the weight of each criterion. Secondly, total scores of alternatives are gained by using TOPSIS. Finally, GP model shows the priorities, conflicting goals of the company that allow evaluating suppliers appropriately.

2.1 AHP method

As being released the first time by Thomas Saaty in 1980, The Analytic Hierarchy Process (AHP) is widely known for its power for being used as a helpful tool to solve the matter of complex decision making. It possesses the advantages in evaluating and selecting the best solution among alternative solutions, based on list of criteria. AHP procedure has been presented as the following steps:

Step 1: Constructing the decision hierarchy

Step 2: Constructing the pair wise comparison matrix

$$A_{n\,x\,n} = \begin{matrix} A_1 \\ A_2 \\ \vdots \\ A_n \end{matrix} \begin{bmatrix} 1 & a_{12} & \dots & a_{1n} \\ 1/a_{12} & 1 & \dots & a_{2n} \\ \vdots & \vdots & \vdots & \vdots \\ 1/a_{1n} & 1/a_{2n} & \dots & 1 \end{bmatrix}$$

Step 3: Constructing normalized decision matrix

$$c_{ij} = {a_{ij}}\Big/{\sum_{j=1}^{n} a_{ij}} \quad (1)$$

With i=1, 2, 3…n, j= 1, 2, 3…n

Step 4: Constructing the weighted, normalized decision matrix

$$w_i = \sum_{j=1}^{n} {c_{ij}}/{n} \quad (2)$$

Step 5: Calculating Eigenvector and Row matrix

$$E = {N^{th}\, roothvalue}\Big/{\sum N^{th}\, roothvalue} \quad (3)$$

$$Row\ matrix = \sum_{j=1}^{n} a_{ij} * e_{j1} \quad (4)$$

Step 6: Calculating the maximum Eigenvalue, λ_{max}

$$\lambda_{max} = Row\ matrix\,/\,E \quad (5)$$

Step 7: Calculating the consistency index and consistency ratio

Consistency index

$$CI = (\lambda_{max} - n)/\,(n - 1) \quad (6)$$

Consistency Ratio

$$CR = CI/\,RI \quad (7)$$

2.2 TOPSIS method

Yoon and Hwang firstly invented TOPSIS (Technique for Order Preference by Similarity to Ideal Solution) method. It is well-known as useful tool to aid in decision making. The basic idea of this technique is choosing the alternative should have a shortest distance from the ideal solution and farthest distance from the negative distance. TOPSIS is presented as follows:

Step 1: The structure of matrix

$$D = \begin{bmatrix} & X_1 & X_2 & \dots & Xj \\ A_1 & X_{11} & X_{12} & \dots & X_{1j} \\ A_2 & X_{21} & X_{22} & \dots & X_{2j} \\ . & . & . & \dots & . \\ A_i & X_{i1} & X_{i2} & \dots & X_{ij} \end{bmatrix}$$

Step 2: Calculate the Normalized the matrix D by using the following formula

$$r_{ij} = {x_{ij}}\Big/{\sqrt{\sum_{j=1}^{J} xij^2}} \quad (8)$$

Step 3: Construct the weighted normalized decision matrix by multiplying the normalized decision matrix by its associated weights

$$v_{ij} = w_{ij}\, r_{ij} \quad (9)$$

Step 4: Determine the positive ideal solution and negative ideal solution

$$A^* = \{(\max v_{ij} \mid j \in J), (\min v_{ij} \mid j \in J')\} \quad (10)$$

$$A^- = \{(\min v_{ij} \mid j \in J), (\max v_{ij} \mid j \in J')\} \quad (11)$$

Step 5: Calculate the separation measure

$$S_i^* = \sqrt{\sum_{j=1}^{n} (v_{ij} - v_j^*)^2} \quad (12)$$

$$S_i^- = \sqrt{\sum_{j=1}^{n} (v_{ij} - v_j^-)^2} \quad (13)$$

Step 6: Calculate the relative closeness to the ideal solution

$$C_i^* = {S_i^-}\Big/{S_i^* + S_i^-}, \; 0 \le C_i^* \le 1 \quad (14)$$

Step 7: Calculate the total score

2.3 Goal programming

After getting total scores of the alternatives, a mathematical model for allocating the right quantities to be purchased from each supplier is built. In goal programming these goal are set as a target in GP, the lower priority goals are considered only after higher priority goals have been satisfied. In addition, GP also shows the flexibility in changing the priorities result in reaching the company's targets.

3. A case study

In this part, the Casumina rubber company, a big manufacturer, produces many types of tire for vehicles in Vietnam. They need huge quantity raw material that ensures to meet their demands, selecting the right suppliers in order to production happen smoothly is essential. In recent, the company considers 15 suppliers (S1, S2, S3... S15) supplying two type of raw materials (SVR10 and SVR20). To evaluate these suppliers, the company bases on 6 criteria such as quality (X1), delivery (X2), profile (X3), service (X4), distance (X5), price (X6) collected from a group of 5 experts in the company shown in Table 1.

Table 1. Supplier selection criteria.

Main criteria	Sub-criteria
Quality	
Delivery	1. On-time
	2. Right quantity
Profile	3. Certification ISO 9000
	4. Long term relationship
	5. Reputation and position in industry
Service	6. Flexible
	7. After sale service
Distance	
Price	

The average weight of criteria is obtained from the group decision maker for item 1: *X1*= 0.44, *X2*= 0.12, *X3*= 0.09, *X4*=0.05, *X5*= 0.03, *X6*= 0.27

The average weight of criteria is obtained from the group decision maker for item 2: *X1*= 0.45, *X2*= 0.12, *X3*= 0.08, *X4*=0.05, *X5*= 0.03, *X6*= 0.26.

Then inputting these weighted into TOPSIS, result for item 1 and item 2:

$C_1^* = 0.22$, $C_2^* = 0.21$, $C_3^* = 0.23$, $C_4^* = 0.33$, $C_5^* = 0.32$, $C_6^* = 0.53$, $C_7^* = 0.21$, $C_8^* = 0.39$, $C_9^* = 0.25$, $C_{10}^* = 0.14$, $C_{11}^* = 0.22$, $C_{12}^* = 0.50$, $C_{13}^* = 0.27$, $C_{14}^* = 0.27$, $C_{15}^* = 0.33$.

$C_1^* = 0.19$, $C_2^* = 0.16$, $C_3^* = 0.22$, $C_4^* = 0.37$, $C_5^* = 0.38$, $C_6^* = 0.49$, $C_7^* = 0.29$, $C_8^* = 0.28$, $C_9^* = 0.23$, $C_{10}^* = 0.24$, $C_{11}^* = 0.21$, $C_{12}^* = 0.41$, $C_{13}^* = 0.28$, $C_{14}^* = 0.24$, $C_{15}^* = 0.33$.

Then, total weighted score from TOPSIS are used as one of the goals in GP model.

3.1 Indices

i: index of item *i=1,2,...i;*
j: index of supplier *j= 1,2,...,j;*

3.2 Parameters

q_{ij}: Defects ratio of item i from supplier j in period t (%)
d_{ij}: Delivery delay of item i from supplier j (days)
D_{it}: Total demand for item i in the period t (ton)
P_{ijt}: Unit price of ordered item i from supplier j in period t ($/ton)
TS_{ij}: Total score of supplier j for item i from TOPSIS
Δ_{ij}: Flexible level of supplier j supplying i item
N: Minimum number of selected suppliers supplying item i
B_{it}: Total budget for purchasing item i in period t
$Q_{ijt\,min}$: Minimum order quantity of item i from supplier j in the period t
$Q_{ijt\,max}$: Maximize order quantity of item i from supplier j in the period t
PC: purchasing cost ($)
DT: delivery time (days)
DE: defective amount (%)
CA: level of capacity (%)
I: ideal total weighted score
d_k^+: Amount of deviation above the goal
d_k^-: Amount of deviation below the goal
3.4 Decision variables:
X_{ijt}: Order quantity item i from supplier j in period t
Y_{ij}: Binary variable if the jth supplier supplying item i is selected or not

$$Y_{ij} = \begin{cases} 1: if\ Xij > 0 \\ 0: otherwise \end{cases}$$

$d_k^+ \geq 0$, k= 1, 2, 3, 4, 5
$d_k^- \geq 0$, k= 1, 2, 3, 4, 5

3.3 Mathematical model

Objective functions:

$$Z = min\,\{d_1^- + d_2^+ + d_3^+ + d_4^- + d_5^-\}$$

<u>Goal constraints:</u>
Maximize the total weighted value of purchase:

$$\textstyle\sum_t \sum_i \sum_j TS_{ij} * X_{ijt} + d_1^- - d_1^+ = I \quad (15)$$

In this constraint we use the scores obtained from TOPSIS technique. Thus, the purpose of this constraint is to try to allocate more unit order to the supplier with the higher total scores in the period of time.

Minimize total cost of purchasing:

$$\textstyle\sum_t \sum_i \sum_j P_{ijt} * X_{ijt} + d_2^- - d_2^+ = PC \quad (16)$$

Purchasing cost account for 70% of complete product value so one of the most important duty for purchasing department is try to minimize the cost as much as possible. This constraint is minimizing the expense in the period t by the total cost offered by each supplier and the orders that company placed.

Minimize the average delivery time delay:

$$\textstyle\sum_i \sum_j d_{ij} * Y_{ij} + d_3^- - d_3^+ = DT \quad (17)$$

Delivery delay may cause the bad influence to company such as manufacturing slackness as well as the supplier's prestige. Thus, this constraint is expected to minimize the delivery time delay for each supplier.
Minimize the average defect ratio:

$$\textstyle\sum_t \sum_i \sum_j q_{ij} * X_{ijt} + d_4^- - d_4^+ = DE \quad (18)$$

This constraint is tried to minimize the defective ratio by collecting the defective percentage rate of each supplier for item i and orders that company placed

Maximize the supplier's capacity flexibility:

$$Y_{ij} + d_5^- - d_5^+ = CA \quad (19)$$

Each supplier has their own capacity. However, the company expects that whenever they need, supplier can satisfy the demand. Thus, this constraint is maximized the supplier's capacity flexibility by total the percentage level flexible and order that company placed.

Hard constraints:

Minimum order quantity:

$$X_{ijt} \geq Q_{ijt\,min} * Y_{ij}, \forall\, i,j \quad (20)$$

This constraint is the orders that company placed in period *t* should larger than the minimum order quantity of item *i* from supplier *j* in the period time *t*.

Maximum order quantity:

$$X_{ijt} \leq Q_{ijt\,max} * Y_{ij}, \forall\, i,j \quad (21)$$

This constraint is the orders that company placed in period *t* should less than the maximum order quantity of item *i* from supplier *j* in the period *t*.

Production demand:

$$\sum_j X_{ijt} * (1 - q_{ij}) \geq D_{it}, \forall i \quad (22)$$

The amount of items no defective after production in the period *t* should larger than the demand for order item in period *t*.

Minimum number of suppliers:

$$\sum_j Y_{ij} \geq N, \forall i \quad (23)$$

The minimum number of selected suppliers supplying item *i* should less than number of supplier selected.

Purchasing budget:

$$\sum_j P_{ijt} * X_{ijt} \leq B_{it} \quad (24)$$

The cost for buying material in the period time *t* should be less than the budget that company have for purchasing item *i* in period t

$$X_{ijt} >= 0 \text{ and integer } \forall\, i,j \quad (25)$$

Binary constraints:

$$Y_{ij} = 0 \text{ or } 1\ \forall\, i,j \quad (26)$$

3.4 Data collection

The strategy of company is shown as follows:

- There are four quarter period per year.
- Two raw materials SVR10 and SVR20 are considered.
- Purchasing cost that company spends for material in year is $998000.
- Delivery time days that company is allow for each supplier to be late is 3 days.
- Defective amount is to be expected in 8%
- Flexible level capacity that company expect supplier is 10%.
- Total weighted scored is 95.

4 Results

4.1 Goal achievement

In this section, this study shows goal achievement based on goal priority of the company. As shown in Table 2, the both objectives total weighted value and purchasing cost are achieved with $d_1^-=0$ and $d_2^+=0$, the rest of objectives cannot reach the goals with $d_3^+=3$, $d_4^+=1.067$, and $d_5^-=55$. In other words, the third, fourth, fifth goals are sacrificed in order to satisfy the first and the second goal of the company's priorities.

With goal achievement result, Table 3 displays the detailed results of five goals corresponding with each supplier that supports decision makers make a right decision and control their solution. The total weighted scores of all suppliers are greater than 95 satisfying the first priority,

Table 2. Goal achievement.

Priorities	Goals	Objectives	Results
1	Total weighted value for all suppliers	Min d_1^-	d_1^-=0
2	Purchasing cost ($)	Min d_2^+	d_2^+ =0
3	Delivery time delay for all suppliers (days)	Min d_3^+	d_3^+ =3
4	Defect ratio for all suppliers (%)	Min d_4^+	d_4^+ =1.067
5	Capacity flexibility for all suppliers (%)	Min d_5^-	d_5^- =55

Table 3. Results of total weighted value, capacity level, defective, purchasing cost.

Suppliers	I	PC($)	DE (%)	DT(days)	CA (%)
1	113.1	83190.3	6.4	3	13
2	110.3	85765	7.2	2	11
3	127.6	84656.2	7.3	3	5
4	97.0	43568	2.9	3	4
5	163.3	69258	5.8	3	6
6	124.6	34901	2.4	1	3
7	106.4	64869	8.0	4	6
8	96.8	37199	3.3	1	6
9	122.6	75547	5.6	5	10
10	95.1	72904	9.1	3	8
11	120.4	83557	6.1	3	3
12	117.5	35047	3.3	3	6
13	95.1	51923	4.1	3	5
14	118.6	69678	6.2	3	11
15	169.9	77472	3	5.7	3
Total		969531			

The company spent $998000 to purchase two items but the results obtained from this study shows that the company only needs pay $969531, results in saving $28469.

5 Conclusions

The main purpose of the paper focuses on supplier selection problem based on three main techniques, namely Analysis Hierarchy Process (AHP), Technique for Order of Preference by Similarity to Ideal Solution (TOPSIS) and Goal Programming. Firstly, AHP is applied to determine weight for each criterion that could be quantitative or qualitative proposed by experts, and then the total weighted score for each supplier is obtained by the TOPSIS. Finally, the Goal Programming shows flexible ability by allowing decision makers establish their business strategies in different periods with certain priorities of goals.

Applying these methods, the selecting supplier problem in Casumina Company is completely solved. The results show number of goals of company can be achieved and suggest number of suppliers should be considered for purchasing. This thesis also analyze for different scenarios by changing priorities that brings decision makers more options support for their final decision.

In summary, the solution should been chosen based on company condition and the manager would make final decision to solve the present issues.

References

1. Ghodsypour, S.H. and C. O'Brien, *A decision support system for supplier selection using an integrated analytic hierarchy process and linear programming*. International Journal of Production Economics, (1998). **56–57**: p. 199-212
2. Saaty, T.L., *The Analytic Hierarchy Process: Planning, Priority Setting, Resource Allocation*. (1980): McGraw-Hill
3. Hwang, C.L.Y., K. , *Multiple Attribute Decision Making: Methods and Applications*. (1981): New York: Springer-Verlag
4. Omid Jadidi, F.F., and Enzo Bagliery *TOPSIS Method for Supplier Selection Problem* World Academy of Science, Engineering and Technology, International Journal of Social, Behavioral, Educational, Economic, Business and Industrial Engineering, **4**, 11 (2010).
5. Athawale, V.M., *A TOPSIS method-based Approach to machine tool selection*. The 2010 international coference on industriaengineering and operations management, (2010)
6. Amir Naderi, M.S.Z., *Application of multi criteria decision making technique to evaluation suppliers in supply chain management*. **4** (2011)
7. Birsen Karpak, R.E.K., *Multi-objective decision: An application of visual interactive goal programming*. The journal of applied business research. **15**
8. Wu, M., *Topsis-AHP simulation model and its application to supply chainmanagement*. World Journal of Modelling and Simulation, (2007). **3**(3): p. 196-201.
9. Jayaram C Sasi1, D.A.K.D., *Application of AHP and TOPSIS Method for Supplier Selection between India & China in Textile Industry*. International Research Journal of Engineering and Technology, **2**, 4 (2015)
10. Sanayei, A., et al., *An integrated group decision-making process for supplier selection and order allocation using multi-attribute utility theory and linear programming*. Journal of the Franklin Institute, (2008). **345**(7): p. 731-747
11. Jafar Razmi, H.R., *An integrated analytic* network *process with mixed-integer non-linear programming to supplier selection and order allocation*. The International Journal of Advanced Manufacturing Technology, (2010). **49**(9): p. 1195-1208

Assessing Service Quality: A Combination of SERVPERF and Importance-Performance Analysis

Dyah R. Rasyida, M. Mujiya Ulkhaq, Priska R. Setiowati and Nadia A. Setyorini

Department of Industrial Engineering, Diponegoro University, Semarang 50275, Indonesia

Abstract. Service providers in this highly competitive market are demanded to find ways to attract new customers, retain existing customers, and pursue customers' loyalty. The quality of the service which has to deliver to the customers has been considered as a critical factor for the success of the service providers by a reason of its close connection to customer satisfaction. This paper aimed to assess the service quality of the service firms and identify what dimensions they have to prioritize to attain the customer satisfaction. SERVPERF model was employed to achieve the first objective while importance-performance analysis was utilized for the second objective. A case study was conducted in Ahmad Yani International Airport, Indonesia, to demonstrate the applicability of the method.

1 Introduction

Service quality currently has been considered as a strategic tools for positioning and means of attaining operational efficiency, improving business performance [1], as well as a key factor for the success of service providers. The improvement of service quality will result in the satisfaction improvement of service customers [2]–[4]. Furthermore, tremendous service precedes customers' retention and leads to repeat customers purchase behaviour [5] which can increase the service provider's market share and generate high incomes [6].

Although assessing service quality is challenging due to service's unique characteristics, such as intangible, heterogeneous, inseparable, and perishable, there are numerous researchers in the field of service quality and the number is still growing. Among them, perhaps SERVQUAL by [7] and SERVPERF model by [8] are the most popular ones.

In SERVQUAL model, service quality is linked to the concepts of disconfirmation or gap between customers' perceptions and expectations. Even though it is intuitively appealing and conceptually sensible, the ability of these scores to provide additional information beyond that already contained in the perception component is under doubt. While the perception is definable and measurable in a straightforward manner as the customers' belief about service is experienced, expectation is subject to multiple interpretations and as such has been operationalized differently by different researchers [9]–[11]. It is argued that the conceptual basis of the SERVQUAL scale is confusing with the service satisfaction and suggested to leave the perception alone [8]; hence the SERVPERF model plays its role.

This research attempted to combine the SERVPERF model with importance-performance analysis (IPA) technique by [12]. IPA technique can be used to prioritize the service attributes based on the importance and the performance, which are the results of SERVPERF model.

To exhibit the applicability of the method proposed, a case study was conducted in Ahmad Yani International Airport (AYIA) which is located in Semarang, Indonesia. Even if there are several areas of service that this method can be applied, however, investigation into airport service have remained rather limited. Perhaps this is due to conventionally perceived notion that airports are natural monopolies, and have often been seen as a take-it-or leave-it proposition [13]. The absence of competition among airports makes the quality of the services become less noticed. Customers often do not have a choice among airports, regardless of the price and the quality levels of the service. Even if one found the parking arrangements poor, the terminal facilities somewhat puzzling, the restaurants and retail outlets high-priced, or the ground transportation facilities congested, the customers are forced to accept the situation offered by airport authorities [14]; or at least this somewhat happened in the developed countries.

The objectives of this research are twofold. The first is to show how to assess the service quality using the SERVPERF model. It may provide some insights on how patrons rate the service quality; hence enabling the managers to position their service quality in accordance to their competitors and to discover dimensions of service that they need to improve. The second is to identify the service firms' selection attributes perceived as important to offer strategic implications that should be pursued to boost their competitiveness and attract more customers.

DOI: 10.1051/matecconf/20166806003

2 Research method

SERVPERF model consists of five service dimensions: tangibles, reliability, responsiveness, assurance, and empathy, with two sets of 22 item statements for the importance and perception sections of the questionnaire. Tangibles is about physically visible aspects. It could be viewed from the cleanliness of washroom, car parking facilities, waiting lounge comfort, and the appearance of the employees. Reliability is the ability to provide reliable service immediately and accurately. It relates to the adequacy of flight information, such as information display regarding baggage location and the availability of dependable officers. Responsiveness is willingness to respond to the wishes or needs of the customers' support and fast services. It includes the rapid airport access, as well as effective and efficient security clearance. Assurance includes the knowledge, courtesy, skills, and trust-worthiness of the employees, as well as free from danger, risk, or doubt. The last dimension, empathy, means the ease of relationships, good communication, personal attention, and understand the need of customers. It can be observed from the availability of waiting time for check-in and immigration inspection and reclaim service for every single customer. The item statements to assess the performance of the service quality can be seen in Table 1.

Service quality is measured by multiplying the weights with the perception scores:

$$SQ_i = \sum_{j=1}^{k} W_{ij} \cdot P_{ij} , \quad (1)$$

where SQ_j is the service quality of item statement j, W_{ij} is the weighting factor of item statement j to an individual i, and P_{ij} is perception of individual i with respect to the performance on item statement j. The weighting factors is the normalized importance score:

$$W_{ij} = \frac{I_{ij} - Min}{Max - Min}, \quad (2)$$

where I_{ij} is the importance score obtained from the questionnaire, *Min* is the minimum score and *Max* is the maximum score of the importance score.

The 22 item statements then were used to analyse the differences between the importance and performance in AYIA selection attributes: the IPA technique [12]. It is the two-dimensional state space where the vertical axis described the importance of the selection attributes, while the horizontal axis described how well the service firm is performing the service.

The IPA's two-dimensional state space were categorized into four quadrants: concentrate here, keep up with the good work, low priority, and possible overkill. The first quadrant, i.e. concentrate here, which is located in the north-west corner, had the attributes that become the priority of the management due to having high importance but indicates low performance ratings. The second quadrant: keep up with the good work, identified that both importance and performance of the customers already high rated and should be maintained well by the management. Attributes that were rated low in both importance and performance were put in the third quadrant: low priority, which is located in the south-west corner. The last is possibly overkill, where there were unnecessary attributes that need to be maintained by the management due to having low importance but high performance rating.

Table 1. Dimensions and item statements of SERVPERF

Dimensions	Item Statements
Tangible (4 items)	Q1: Service firm has up-to-date equipment Q2: The physical facilities are visually appealing Q3: The employees are well dressed and appear neat Q4: The appearance of the physical facilities is in keeping with the type of services provided
Reliability (5 items)	Q5: When the employees promise to do something by a certain time, it does so Q6: The employees are sympathetic and reassuring when customers have problems Q7: The employees are dependable Q8: The employees provide their services at the time they promise to do so Q9: The employees keep their records accurately
Responsiveness (4 items)	Q10: The employees are expected to tell customers exactly when services will be performed Q11: It is realistic for customers to expect prompt service from the employees Q12: The employees are expected to always help the customers Q13: It is a problem if the employees are too busy to respond to customers' requests promptly
Assurance (4 items)	Q14: The customers can trust the employees Q15: The customers are able to feel safe in their transactions with the employees Q16: The employees are polite Q17: The employees should get adequate support from the service firm's management to do their jobs well
Empathy (5 items)	Q18: The employees are expected to give customers individual attention Q19: The employees are expected to give customers personal attention Q20: It is expected that the employees know what the needs of their customers are Q21: It is expected that the employees have their customers' best interests at heart Q22: The employees are expected to have operating hours convenient to all their customers

3 Case Study

The first aim of this research is to investigate the performance of the service quality using the SERVPERF model. A case study was conducted in AYIA which is located in Semarang, Indonesia. The airport was opened for commercial flights on 31 August 1966 and gained its international status in August 2004 with its inaugural flight from Semarang to Singapore. A hundred participants were randomly selected from various sources to guarantee a diversity of opinions on the service quality of the airport. The participants include students, lecturers, business people, housewives, employees, and civil servants, indicates plenty diversity for the purpose of the research. In addition, participants were required to be over 18 years of age and have been experienced in getting benefit and perceiving the service from AYIA within the previous six months. Potential participants were first approached and asked if they wished to participate in the survey. All item statements were measured on a 7-point Likert-type scale, ranging from 1 which is indicated strongly disagree to 7 for strongly agree.

The reliability test was conducted to check whether or not the respondents' scores on any item statement tends to be related to their scores on the others. The Cronbach's alpha for each dimension are computed using PASW Statistics 17.0; the results are shown in Table 2. Note that all of the dimensions have the value of Cronbach's alpha [15] more than 0.6, indicated that the questionnaire being utilized is reliable [16].

The average values are then computed for each item statement throughout all respondents. The results are shown in Table 3, respect to each item statement and each section: importance and perception. The weighting factors are calculated using Equation (2) and the performance of service quality is computed using Equation (1).

The attributes with the highest score of importance for each dimension are: Q1 of tangibles, Q5 of reliability, Q12 of responsiveness, Q16 of assurance, and Q20 of empathy; whereas reliability has the largest average score, i.e. 6.616. It seems that customers expected more on the reliable and accurate service. On the other hand, the attributes with the lowest score of importance for each dimension are: Q2 of tangibles, Q8 of reliability, Q10 of responsiveness, Q14 of assurance, and Q19 of empathy; while empathy has the lowest average score, i.e. 4.470. The customers did not seek so much for personal attention and good communication from the employees. It is evidenced that the Q18, i.e. the employee should not be expected to give customers individual attention, has the lowest score of importance.

In the perception section, assurance has the highest average score which is 4.492. The attributes with the highest score are: Q3, Q9, Q11, Q16, and Q20 of tangibles, reliability, responsiveness, assurance, and empathy respectively. It looks the airport has the knowledgeable, skilled, and trustworthy employees as being the best among the other dimension. Q16 is considered as the highest score which implies that the airport has polite employees. Contrarily, the dimension which has the lowest average score is tangibles, i.e. 3.924. The attributes with the lowest score for each dimension are: Q2 of tangibles, Q6 of reliable, Q12 for responsiveness, Q17 of assurance, and 22 of empathy. The attribute which is considered as the worst performance of the airport is the appearance of the physical facilities, Q2, i.e. airport's physical facilities should be visually appealing. It is a signal that the airport must improve its performance using the information shown in the SERVPERF analysis.

Table 2. Cronbach's alpha for each dimension of SERVPERF

Dimensions	Cronbach's Alpha
Tangible	0.886
Reliability	0.853
Responsiveness	0.638
Assurance	0.794
Empathy	0.649

Table 3. Questionnaire result

Dimensions		I_i	W_i	P_i	SQ_i
Tangible	Q1	6.505	0.876	3.707	3.247
	Q2	6.091	0.773	3.505	2.709
	Q3	6.354	0.785	4.444	3.489
	Q4	6.162	0.721	4.040	2.913
Reliability	Q5	6.758	0.939	4.303	4.041
	Q6	6.586	0.862	4.263	3.675
	Q7	6.606	0.869	4.535	3.941
	Q8	6.546	0.848	4.566	3.872
	Q9	6.586	0.862	4.687	4.040
Responsi-veness	Q10	5.000	0.667	3.700	2.468
	Q11	5.687	0.737	3.889	2.866
	Q12	6.010	0.835	3.424	2.859
	Q13	5.374	0.729	3.748	2.732
Assurance	Q14	4.808	0.635	4.343	2.758
	Q15	5.748	0.791	4.626	3.659
	Q16	6.700	0.899	4.707	4.232
	Q17	6.141	0.857	4.293	3.679
Empathy	Q18	3.869	0.478	3.950	1.888
	Q19	3.828	0.471	4.010	1.889
	Q20	5.293	0.659	4.505	2.969
	Q21	4.616	0.603	3.929	2.369
	Q22	4.748	0.549	3.545	1.946
Average		**5.727**	**0.748**	**4.124**	**3.088**

The average score of 3.088 indicated that the AYIA is slightly lack of capability for providing the "best" service for the customers and must do something in order to achieve customer satisfaction. The IPA technique can be employed to build strategies to attain it based on the importance and the performance from the customers' point of view. The average score for each item statement were then plotted in the two-dimensional state space. The horizontal axis refers to the perception or how well the airport is performing its activity, i.e. providing the service to the customers; while the vertical axis refers to the importance of the activity. The IPA of service quality of AYIA is depicted in Figure 1.

The item statements belong to the first quadrant are the ones with low performance but importantly perceived by the customers. Therefore, the attributes should receive the most investment to boost the customers' satisfaction. It is suspected to bring the maximum effect with the minimum investment. The item statements belong to this quadrant are Q1, Q2, and Q4, all belong to the tangibles dimension. It is recommended to the management to improve the physical appearance with the latest equipment and remove the distraction that could interfere the customers' attention toward the physical facilities.

Several item statements are embraced in the second quadrant: Q3, Q5, Q6, Q7, Q8, Q9, Q12, Q15, Q16, and Q17, means that the attributes are considered important and customers are fond with the performance of the management. It indicates that: the employees of AYIA have well-dressed and appear neat and are willing to help customers; the customers are feel safe in their transaction in the airport. All item statements belong to reliability dimension are in this quadrant, designated that the airport can provide reliable service immediately and accurately. In summary, the management must maintain these aspects to keep and maintain the customers.

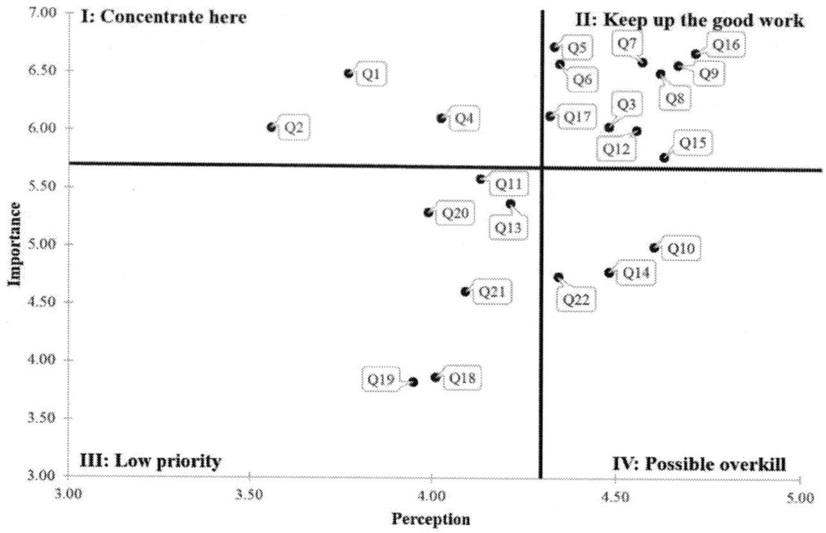

Figure 1. Importance-performance analysis of service quality of AYIA.

The low priority quadrant identifies that the attributes belongs here are performing adequately but customers perceive them as less important when compared with other airport attributes. This quadrant consists of the attributes from responsiveness and empathy dimensions: Q11, Q13, Q18, Q19, Q20, and Q21. Although the result showed that both responsiveness and empathy dimensions did not perceived importantly, this does not mean that the management should reduce their efforts to improve the service. The management could give a reward to the employees who are able to maintain their responsiveness and empathy skills to the customers. If the customers are satisfied with the quality of the attributes, such satisfaction will lead them to spread favourable word-of-mouth publicity.

The attributes in the possible overkill quadrant are considered less important by the customers and felt too excessive, so that need to be reduced due to the excessive investment. If these attributes are applied to other areas, it is suspected to bring better results.

4 Discussion

Parasuraman, Zeithaml, and Berry were among the earliest researchers to put interest on the service quality assessment. One main contribution of their works was to provide a concise definition of service quality as a global judgment, relating to the superiority of the service, and explained it as involving evaluations of the outcome and process of service act. It is a difference between customers' expectations of what they want and their perceptions of what they get. According to these conceptualization and operationalization, they proposed a measurement scale called SERVQUAL. It establishes an important milestone in the service quality literature and has been extensively applied in several service settings.

However, SERVQUAL has been criticized on both theoretical and operational grounds. Some major issues regards with the use of disconfirmation model or the gap scores, length of the questionnaire, predictive power of the instrument, and the validity of the five-dimensions structures [8]–[11], [17], [18].

Cronin and Taylor were among the researchers who levelled maximum strike on the SERVQUAL scale. They though that expectation can be discarded and the performance can be used alone. They proposed what is referred to as the SERVPERF scale and provided an empirical evidence across four industries: banks, pest control, dry cleaning, and fast food to corroborate the superiority of their performance-only instrument over disconfirmation-based SERVQUAL scale.

The SERVPERF can be considered suppress over the SERVQUAL scale. When it is applied in conjunction with the SERVQUAL scale, the SERVPERF has outperformed the SERVQUAL scale [8], [11], [18], [19]. It is also able to explain greater variance in the overall service quality measured through the use of single-item scale [20]. Although the application is still lagging behind the SERVQUAL, researchers have increasingly started utili-zing the performance-only measure of service quality.

Due to the superiority of the SERVPERF scale to the SERVQUAL, this research applied the SERVPERF scale to assess the service quality and then combined it with the

DOI: 10.1051/matecconf/20166806003

IPA technique. Because the IPA consists of two aspects: importance and performance, consequently, to combine with the SERVPERF, we used weighted version. SERVPERF which stands alone does not comprise the importance aspect thus it cannot be merged with the IPA technique.

The result came from the SERVPERF model can assist the managers to identify where the performance improvement can best be targeted. The performance improvement in some attributes would be addressed for the lowest score. Rather, if some attributes have high scores, the managers could review whether they may be "over-supply" the particular features.

The IPA technique used the scores form SERVPERF questionnaire and then plotted it into the two-dimensional state space. IPA is considered simple; it is used to understand customer satisfaction and prioritise areas for improvement. It has been widely used in service industries, see for example: [21], [22]. In short, this IPA evaluation tool is used to prescribe the prioritisation of attributes for improvement and it can also provide guidance for strategic development [23].

In previous case study, several attributes have been plotted in the certain quadrant of IPA. This could help the managers to find out attributes that are doing well and attributes that essentially to be improved. It has to be pointed out that IPA is not a tool for final decisions; it is a tool for problem conceptualization and initial analysis. It does not, and will not, yield quantitative information that should be the basis for final decisions. Rather, it assists to limit options thereby identifying which studies of scale, scope, benefits, etc. are needed to determine if resources should be expended on particular actions.

5 Conclusion

The research has demonstrated the assessment of service quality using weighted SERVPERF and identified the attributes of the service providers that perceived important by the customers and its performance. A case study was managed in the airport service, a capital intensive establishment. The SERVPERF model which consists of importance and perception aspects has been found to provide a relative simple and inexpensive means of doing service quality assessment. The results indicated that the assessment of service quality has many potential benefits for airport managers. Identifying customers' perceptions of service quality for a particular establishment allows management to better tailor its marketing efforts to ensure patrons' expectations are met. This includes identifying, prioritizing and improving areas of service weakness and ensuring that valuable resources are allocated in the most effective areas. In addition, promotional messages can be refined so that customers have realistic expectations of the service offered.

Based on the overall performance scores, it was found that the service quality is low. Customers do not perceived high performance service delivered by the service provider. Combining the performance score and the result from the IPA techniques, we suggest the management to immediately improve the service quality. All aspects of service quality, including service efficiency, politeness and friendliness, as well as assurance should be maintained and consistently reviewed to see whether any improvements are required. The physical facilities should be improved in order to reach customer satisfaction. For instance, in-house training programs could be arranged to improve employee courtesy, helpfulness, understand ability, language skills, appearance, and service skills. Furthermore, the management should ensure that all employees are required to become involved in setting quality standards, and should realize that maintaining service quality is part of their jobs.

References

1. S. C. Mehta, A. K. Lalwani, S. L. Han, Int. J. Retail Distrib. Manag. **28**, 62–72 (2000)
2. A. Parasuraman, V. A. Zeithaml L. L. Berry, J. Marketing **49**, 41–50 (1985)
3. G. R. Gilbert, C. Veloutsou, J. Serv. Marketing **20**, 298–308 (2006)
4. I. H. Chow, V. P. Lau, T. W. Lo, Z. Sha, H. Yun, Hosp. Manag. **26**, 698–710 (2007)
5. R. Ladhari, I. Brun, M. Morales, Int. J. Hosp. Manag. **27**, 563–573 (2008)
6. X. Luo, C. Homburg, J. Marketing **71**, 133–149 (2007)
7. A. Parasuraman, V. A. Zeithaml, L. L. Berry, J. Retailing **64**, 12–40 (1988)
8. J. Cronin, S. A. Taylor, J. Marketing, **56**, 55–67 (1992)
9. K. R. Teas, J. Marketing **57**, 18–34 (1993)
10. K.R. Teas, J. Marketing **58**, 132–139 (1994)
11. P. A. Dabholkar, D. C. Shepherd, D. I. Thorpe, J. Retailing **76**, 139–173 (2000)
12. J. A. Martilla, J. C. James, J. Marketing **41**, 77–79 (1977)
13. M. S. Sohail, A. S. Al-Gahtani, Int. J. Serv. Stan. **1**, 482–493 (2005)
14. D. Rhoades, B. Waguespack, S. Young, Managing Serv. Qual. **10**, 257–262 (2000)
15. L. J. Cronbach, Psychometrika **16**, 297–334 (1951)
16. J. Nagpal, A. Kumar, S. Kakar, A. Bhartia, J. Assoc Phys. India **58**, 295–304 (2010)
17. P. Asubonteng, K. J. McCleary J. E. Swan, J. Serv. Marketing **10**, 62–81 (1996)
18. E. Babakus G. W. Boller, J. Bus. Res. **24**, 253–268 (1992)
19. S. K. Jain, G. Gupta, Vikalpa **29**, 25–37 (2004)
20. M. K. Brady, J. Cronin, R. R. Brand, J. Bus. Res. **55,** 17–31 (2002)
21. M. J. Enright, J. Newton, J. of Travel Res. **43**, pp. 339–350 (2005)
22. J. E. Blose, W. B. Tankersley, L. R. Flynn, Qual. Manag. J. **12**, 7–24 (2005)
23. M. S. Wong, N. Hideki, P. George, J. Theor.. Elec. Commer. Res. **6**, 17–30 (2011)

MATEC Web of Conferences **68**, 06004 (2016) DOI: 10.1051/matecconf/20166806004
ICIEA 2016

Measuring Thermal Stress of Dairy Cattle Based on Temperature Humidity Index (THI) in Tropical Climate

Sugiono[1], Hardiningtyas Dewi[1] and Soenoko Rudy[2]

[1]*Industrial Engineering Department, Brawijaya University, Malang, East Java, Indonesia 65145*
[2]*Mechanical Engineering Department, Brawijaya University, Malang, East Java, Indonesia 65145*

Abstract. Thermal comfort for workers is very important factor to increase their performance, as well as the comfort level of dairy cattle will influence in milk productivity. The purposes of the paper is to measure the level of heat stress and then use the information to design the dairy cattle house for increasing thermal comfort. The research is started with literature review of heat stress and early survey of environment condition e.g. temperature, wind speed and relative humidity. The next step is using the information to determine the temperature humidity index (THI) level for dairy cattle with maximum THI = 86 and 84 (moderate stress). The 3D CAD model and Computational Fluid Dynamic (CFD) simulation are employed to looking for solution for reducing the discomfort thermal of dairy cattle. A scenario (fan) air conditioning to get better condition of thermal comfort have been successfully presented with final THI index = 76 and 78 (mild stress). Finally, the paper shows how to reduce heat stress of cattle house by installation 3 exhaust fans in tropical climate.

1 Introduction

Dairy cattle in Indonesia is generally has less productivity of milk comparing to animal husbandry in sub–tropical countries like Australia, USA and Europe. It is not only about the quality of food, but also the level of cow comfort (environment condition) and design facility. The characteristic of tropical weather with higher relative humidity and warm temperature creates trend to increase heat stress. Moreover, currently dairy house design did not consider the needs of physic and psychology of dairy cattle. All the factors will affect the productivity of dairy cattle. Michel J. Brouk and friends on their research suggested that due to increasing the milk production, facility and climate effects should be a priority program in better dairy cow management [1].

There are a lot of researches about heat stress as one of comfort indicator for dairy cattle correlated with milk productivity. Animal science instructor Adrew P. Fidler and Karl Van Devender (2011) in his article stated that during the summer in Arkansas raises swordfish comfortable environmental conditions that can cause milk production of dairy cows has fallen dramatically with the average productivity is only 50% [2]. Even these conditions can disrupt the mating process a cow with a success rate only in the range of 10-20% only. J.W. West (2003) in his research entitled "Effect of Heat Stress on Production in Dairy Cattle" states that the temperature and high relative humidity will lead to high levels of heat stress in dairy cows [3]. High relative humidity values that cause cows difficult to release heat to the environment with regard to body temperature is much hotter than the environment, it will cause a disturbance in appetite/drinking, which in turn lowers milk production. Moreover, West (2003) noted that milk production declines as air temperature exceeds 24^{0}C or falls below -12^{0}C. Further, Craig Thomas (2012) in his research on heat stress in cattle states that heat stress is a valuable strategy in the management of critical empowerment, especially dairy cows against temperature and relative humidity factor [4].

To cover the heat stress problem of dairy cattle, the physiological ergonomics knowledge is needed. Tropical climate with identical to hot and humid condition is important factor in analysing the heat stress level in order to increase cow comfort. Computer Aided Design (CAD) and Computational Fluid Dynamic (CFD) are beneficially help to looking for the best solution of thermal comfort for dairy cattle.

2 Research theory

Dairy cattle generally has a physical characteristic of mammary glands is larger than cow's beef cattle. Dairy cows is only benefitted for producing milk as maximum as possible. The best of dairy cattle is categorized as higher productivity, long-term productive and producing quality of milk. There are several factors that influence the total of milk production such us environment, species (breeds), individual factors, genetic, common factors, lactation, pregnancy, hormonal, lactation level, the persistence of production and body size of dairy cattle.

The increasing of body weight of dairy cattle will add the production of milk proportionality.

The heat index was developed in 1978 by George Winterling and adopted by the USA National Weather Service a year later. It is derived from the work done by Robert G. Steadman [5]. As the wind chill index, heat index contains assumptions about the human body mass and height, clothing, amount of physical activity, blood type, exposure to sunlight and ultraviolet radiation, and wind speed [6]. Significant deviations from this will produce heat index values that do not accurately reflect the perceived temperature [7]. The thermal humidity index is an index that combines air temperature and relative humidity in an attempt to determine the level of heat that is felt by the animal body or generally known as the feeling temperature.

Table 1 shows temperature humidity index (THI) for dairy cattle. The THI with yellow explain that dairy cattle will feel start to stress. Medically, it characterized by respiratory exceed 60 beeps per minute (bpm). Based on previous research, the amount of milk in this area will be decreased at 1:13 Kg per day for each cow. The reproductive system is not functioning properly. But, the yellow region will not give negative impact brown colour is categorized as the region of mild to moderate stress. Medically characterized by respiratory exceed to 75 beeps per minute (bpm). The next is in the red region, it is moderate to severe stress situation of dairy cattle. Medically, it is characterized by respiratory exceed to 85 beeps per minute (bpm). The purple is in phase of heavy stress. Medically, it is indicated by respiratory exceed to 120 beeps per minute (bpm). Finally, the white areas explains that the dairy cattle cannot life in this area as temperature and relative humidity too high for animals

Table 1. Temperature humidity Index (THI) evaluation for dairy cattle in tropical situation [8].

Temp. °F	Humidity (%)							
	55	60	65	70	75	80	85	90
72	69	69	69	70	70	70	71	71
74	70	70	71	71	72	72	73	73
76	72	72	73	73	74	74	75	75
78	73	73	74	74	75	75	76	76
80	75	76	76	77	78	78	79	79
82	76	77	77	78	79	79	80	80
84	78	78	79	80	80	81	82	83
86	79	80	81	81	82	83	84	84
88	81	81	82	83	84	85	86	86
90	82	83	84	85	86	86	87	88
92	84	85	85	86	87	88	89	90
94	86	86	87	88	89	90	91	92
96	87	88	89	90	91	92	93	94
98	88	89	90	91	92	93	94	95
100	90	91	92	93	94	95	96	98
102	91	92	94	95	96	97	98	100
104	93	94	95	96	98	99	100	101
106	94	95	97	98	99	101	102	103
108	96	97	98	100	101	103	104	105
110	97	99	100	101	103	104	106	107

Temperature humidity index for human is explained in Table 2. Similar with dairy cow, the human heat stress is divided into 6 regions for comfort area (green), caution area (yellow), extreme caution (brown), danger (red), extreme danger (purple) and deadly area (white). Purple area will impact heat syndrome on the human body, red area will impact stroke and light brown will impact heat cramps.

A study of the heat stress charts for cows and humans show that cows are much more sensitive to heat stress than people. People with environment temperature 26.67°C and relative humidity 40% are just start to feel heat stress, but the cow have been suffering for some time with THI = 73 (light brown) or the dairy cattle get impact of moderate heat stress.

Computational Fluid Dynamics (CFD) is a calculation method with a control of dimensions, area and volume in special computer program as a simulation process [9]. The principle of the CFD is a chamber that contains the element of fluid which is presented into several parts, it is often called the cell and the process to build it is called meshing [10]. CFD is a calculation that specializes in fluid ranging from fluid flow, heat transfer and chemical reactions that occur in the fluid. According to the basic principles of fluid mechanics, the conservation of energy, momentum, mass, and species can be counted by the CFD simulation. As example, when a model will be analyzed by involving temperatures, it means that the model contents of the energy equation or the conservation of energy. Initialize the beginning of the equation is the boundary condition. Boundary condition is a condition in which the controls are defined as a calculation of the initial definition to be included into the calculation. In general, the CFD calculation process consists of three main parts: preprocessor, processor, post processor.

Table 2. Temperature humidity Index (THI) evaluation for human [8].

Temp °F	Humidity (%)							
	55	60	65	70	75	80	85	90
72								
74								
76								
78								
80	81	82	82	83	84	84	85	86
82	84	84	85	86	88	89	90	91
84	86	88	89	90	92	94	96	98
86	89	91	93	95	97	100	102	105
88	93	95	98	100	103	106	110	113
90	97	100	103	105	109	113	117	122
92	101	105	108	112	116	121	126	131
94	106	110	114	119	124	129	135	
96	112	116	121	126	132			
98	117	123	128	134				
100	124	129	136					
102	130	137						
104	137							
106								
108								
110								

The preprocessor is defined as activities to entry the input any information that are needed in simulation. The preprocessor data such as the information of environment condition (temperature, relative humidity, wind speed, wind direction), quality of meshing, defined axis, units,

DOI: 10.1051/matecconf/20166806004

fluid, and goals of the simulation. The processor process is explained as the simulation process which is indicated by iteration process. The duration of simulation is depending on the complexity of the 3D model and the level quality of 3D meshing. The post processor process is used to show the simulation results as contour, 3D animation flow, and output of the goals.

3 Research methodology

According to the aim of the research as mentioned in introduction, the research will produce some recommendations of reducing THI for dairy cattle by modified air conditioning. The research is started with references of cow thermal comfort and correlated with Computational Fluid Dynamics (CFD). The research object location of dairy farmer is in Blitar district, East Java province, Indonesia. The collecting data of temperature, relative humidity, wind speed have been done in two periods for raining season (5 days) and dry season (5 days).To do collecting data some equipments are prepared such us: aerometer, thermometer, camera, stopwatch, etc.

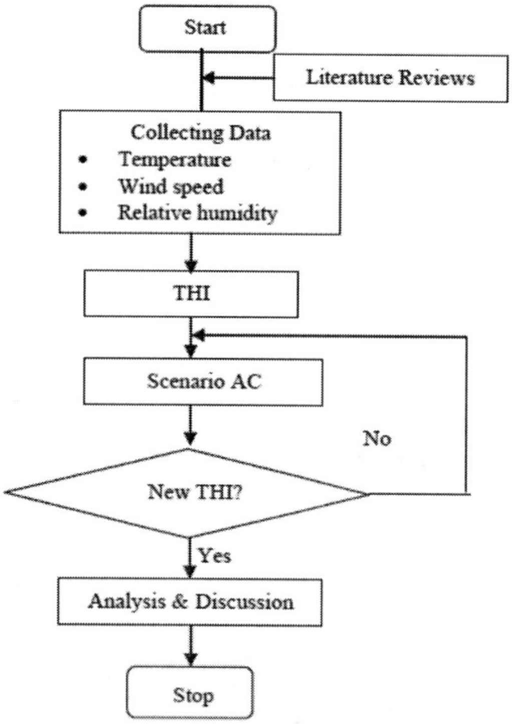

Figure 1. Flowchart of measuring and modifying THI for dairy cattle.

To control the steps of the research are logically right, flowchart of measuring THI and developing air conditioning (AC) for dairy cattle house are presented as can be seen in figure 1 below. According to the graph, the next step after collecting environment data is measuring the THI existing. The THI is used as basic modification of design cow house for controlling the inner air quality (temperature and relative humidity).Analysis and discussion of reducing heat stress are presented in the last part of the research.

4 Results and discussion

Data of tropical environment condition (temperature, relative humidity, wind speed, wind direction) were collected during two seasons, the dry season and the rainy season. Generally, dry season in Indonesia occurs in May to October and wet season in November to June. During the dry season, Indonesia does not have rain. But sometimes, it rains with little rain-fall. The land is dry and it is hot during the season. Generally, there is no difference air quality (temperature and relative humidity) between inside and outside the dairy cattle house as low air circulation. The low air circulation is caused by bad layout and bad design dairy cattle house.

Table 3 shows some temperature averages data measurements which were collected in 3 difference time for fresh air (morning), hot temperature (day) and warm temperature (afternoon) in 5 days randomly during May to October. According to the table, the overall temperature average for dry season is = 29.6 °C. The data of temperature is higher 1.3 °C than overall average in the Blitar district.

Table 3. Temperature data in dray season.

Day	Temperature (^{0}C) –Dray Season		
	Morning (07.00 – 10.00)	**Day (10.00 – 15.00)**	**Afternoon (15.00 – 17.00)**
1	27.3	30.9	27.6
2	26.6	32.6	27.2
3	27.8	31.2	27.5
4	27.4	29.1	27.9
5	26.4	32.7	27.7

At the same time, the relative humidity (RH) data were measured in 15 times with results as shown in table 4. The data of relative humidity is the amount of water vapour contained in the air around the dairy cattle house. The relative humidity in the research area tends to be higher in comparison with other areas in the surrounding districts. From the table, it can be shown that HR maximum is 89.7% and average HR is 87%. The data of relative humidity is higher 6% than overall average in the Blitar district.

Table 4. Relative humidity data in dry season

Day	Relative Humidity (%) –Dray Season		
	Morning (07.00 – 10.00)	**Day (10.00 – 15.00)**	**Afternoon (15.00 – 17.00)**
1	89.7	85.6	86.1
2	87.5	87.3	85.4
3	89.4	86.2	85.4
4	87.9	87.4	85.6
5	88.4	85.4	85.8

Based on the explanation before, the temperature and relative humidity average are 29.6 °C and 87%. By using Table 1, the THI value for dry season is ≈ 84 (red colour) or located in moderate – severe stress area (level 4 out of

6 levels). It is characterized by respiratory exceed to 85 beeps per minute (bpm), feed intake will decrease and water consumption will increase, core body temperature will increase and milk production will decreased. Burgos Zimbelman and Collier (2011) [11] reported that the THI in red colour (80 – 89) will reduce milk production around = 3.9 Kg/cow/day. The air quality should be maintained immediately to increase milk production and to eliminate the other impact of heat stress.

On the other hand, the air quality (temperature and relative humidity) in wet season are presented in Table 5 for temperature and Table 6 for relative humidity. According to the Table 5, the temperature data collected during the rainy season is at average of = 27°C with maximum data = 29.8°C.The overall temperature average for dry season is = 29.6 °C. The data of temperature is higher 2.1 °C than overall average in the Blitar district. Table 6 explains the HR maximum is 93.4% and average HR is 91%. The data of relative humidity is higher 4 % than overall average in the Blitar district.

Table 5. Temperature data in rainy season.

Day	Temperature (°C) –Raining Season		
	Morning (07.00 – 10.00)	Day (10.00 – 15.00)	Afternoon (15.00 – 17.00)
1	25.9	29.8	25.3
2	26.8	29.2	24.9
3	25.5	28.1	24.6
4	25.7	29.1	24.9
5	26.7	27.4	24.7

Table 6. Relative humidity data in rainy season.

Day	Relative Humidity (%) –Raining Season		
	Morning (07.00 – 10.00)	Day (10.00 – 15.00)	Afternoon (15.00 – 17.00)
1	92.2	89.8	91.0
2	91.5	89.2	90.6
3	91.3	89.1	90.7
4	93,4	89.7	90.6
5	92.1	89.9	90.4

Based on the average temperature = 27 °C (80.60°F) and average relative humidity = 91% has the THI value ≈ 86 (red colour) or located in areas of moderate – severe stress area (level 4 out of 6 levels) with reducing of milk production around = 3.9 Kg/cow/day. In sort, the THI in wet season is higher than THI in dry season as the impact of RH in wet season is higher than dray season. It is clearly that THI in existing condition for both of dray season and wet season give higher alarm for famer to fix the problem sooner.

Workers get less heat stress than cow in both of wet and dry season. The table 2 can be used to examine the heat index (HI) level for human. The wet season with air temperature quality = 27 °C (80.60°F) and average relative humidity= 91% has the heat index (HI) value ≈ 86 (yellow colour) or located in areas of light stress area (level 2 out of 6 levels). The dry season with air temperature quality = 29.6 °C (85.28°F) and average relative humidity = 87 % has the heat index (HI) value ≈ 102 (yellow colour) or located in areas of light stress area (level 2 out of 6 levels). All the seasons show there is no significant impact for human thermal comfort to disturb the human performance.

There are some strategies to increase milk production of dairy cattle as the impact of heat stress. Some recommendations to eliminate the effect of heat stress are:

- Plant some trees surrounding the dairy cattle house to make natural shade. Trees are excellent source of shade and effective to block direct solar radiation. Moreover, trees photosynthesis can reduce heat energy and produce cold moisture. As consequence, cow thermal comfort will better than without tress near the cow house.
- The second recommendation of economically reducing heat stress is fan installation. Fan works to circulate fresh air (lower relative humidity and temperature) inside the cow house. In addition, fan can remove moisture which produced by cow body and others source get out the cow house. The fan specification and fan position determine the level of success for reducing heat stress. Figure 2a shows 3D CAD model for existing dairy cattle house design. The model is used to define the new scenario to reduce THI risk by installation low volume high speed (LVHS) – axial fan φ50cm. The fan works to reduce temperature and relative humidity inside the cow house. It exploited the positive difference of air quality between inner and outer cattle house with average difference temperature ΔT =1.3°C and relative humidity (RH) = 6 % in dry season and ΔT = 2°C and relative humidity (RH) = 4 % in wet season. Figure 2b explains the result of CFD simulation test for dairy cattle house with 3 fan in wet season. According to the test, the THI near cow is ≈ 76 in light brown colour (mild stress) with average T = 25.74°C and relative humidity (RH) = 90.37%. The new THI will reduce milk production to ≈ 2.7 Kg/cow/day. The same way, the CFD simulation test for dry conditions gives new THI index ≈ 78 in light brown colour (mild stress) with average T = 27.4°C (81.32 °F)and relative humidity (RH) = 82. 56%.

Figure 2. a. 3D design of existing dairy cattle house, b. CFD simulation for dairy cattle house using 3 fan.

Beside installation fan and planting some trees, there are some methods to avoid heat stress for dairy cattle e.g. provide plenty of water for drink, sparkling water into cow body, increasing the roof venting, etc. The sparkling water is efficient method to reduce heat stress (increase milk production) but still need more investigation as requirement more electrical energy.

4 Conclusions

The research is successfully presented the measurement of temperature humidity index (THI) for dairy cattle in tropical climate, Indonesia. According to the THI calculation, the existing condition has high risk of thermal stress with class area in red colour (level 4). It will significantly reduce milk productivity around 3.9 kg/cow/day or decreasing 23% of standard milk production. One scenario to reduce heat stress is presented by benefited from the difference of T and RH between inner and outer cow house. It will reduce THI from 86 to 76 in wet season and from 84 to 78 in dry season. To reduce the heat stress, further action should be done in the future such as water sparkling treatment to reduce temperature and increasing the roof venting. Converting methane gas from cow waste to be electrical energy to generate fan or water sparkling is a big opportunity in concerning with green energy.

Acknowledgment

Thanks to the Ministry of National Education of the Republic of Indonesia for supporting this paper. The authors are also grateful to the laboratory of work design and ergonomics, Industrial Engineering Department, the University of Brawijaya (UB), Malang Indonesia for their extraordinary courage.

References

1. B. J. Michael, S. F. Jhon, and H. P. Joseph, FCEDMI, *Kansas State University Press*, (2001).
2. Adrew P. Fidler, Karl VanDevender, *Agriculture and Natural Resources Dept.*, (2016)
3. J.W. West, EHSPDC, *Journal of dairy Science*, **86**, 2131–2144, (2003).
4. Craig Thomas, HSDC, *Michigan State University Extension*, (2012).
5. R. G. Steadman, TAS, *Journal of Applied Meteorology*, **18**, 861–873, (1978).
6. Mikell P. Groover, WSM, *Pearson Education, Inc.*, (2007).
7. Khalid Setaih at all, AOTC, *13th CIBPSA*, (2013).
8. Reynolds Jim, DFCC, *Veterinary Medicine Teaching and Research Center*, (2016)
9. Chung T.J., CFD, *Cambridge University Press,* (2011).
10. Doyle B. Joshua, GA, *Auburn University Press,* (2007).
11. Laurent Dussert, Aurelien Piron, HS, *Lallemand Animal Nutrition,* (2012)
12. V. Gantner et al, THI, *Daily production of dairy cattle*, **61**, 56-63, (2011).
13. Ghavi Hossein-Zadeh, N., Mohit, A.and Azad, N., THI, *Iranian Journal of Veterinary Research, Shiraz University*, 14, 106-112, (2013).
14. Krawczel P., Grant Rick, CC, *NMC Annual Meeting Proceedings*, (2009).
15. Holter, J.B., J. W. West, M. I. McGilliard and A.N. Pell..DCI. *J. Dairy Sci.*, **79**, 912-921, (1996).

MATEC Web of Conferences **68**, 06005 (2016)
ICIEA 2016
DOI: 10.1051/matecconf/20166806005

A Buffer Analysis in a Transfer Production Line

Yonlanan Chomnawung, Suksan Prombanpong and Chanakarn Klavohm

Dept. of Production Engineering, King Mongkut's University of Technology Thonburi (KMUTT), Thailand

Abstract. The objective of this paper is to demonstrate a determination of buffer to increase line efficiency in a transfer line where several workstations are linked together by a conveyer. One of the common problems of a transfer line is minor stoppages i.e. part short, machine adjustment, and so on are the typical problems which result in low uptime efficiency and are detrimental to productivity. Thus, buffer stock is designed to mitigate the problem; however, knowledge in determining optimal number of buffers is not prevalent. The buffer analysis using constant downtime distribution is employed in this paper.

1 Introduction

A transfer production line namely flow production line is one of the manufacturing configurations where two or more workstations consecutively link together and a series of parts smoothly flows through the workstations. It is suitable for mass production with a high volume. A task must be able to divide in small tasks and assigned to each workstation in a balance manner. The task in each workstation must be finished within the designed cycle time and normally the working part is designed to move simultaneously. Therefore, once any workstation interrupts or halts, all of the workstations must also be stopped. As one can imagine, during a production there will be various minor stoppages due to part short, tool or machine adjustment, quality inspection, equipment malfunction and so on. These minor stoppages drastically reduce the productivity. Thus, a buffer between the workstations is designed to mitigate the abovementioned problem with an aim to allow the upstream and downstream workstations to continue their tasks until the buffer is running out or the recovery of the ceased workstation is completed. A number of literatures are investigated. Buzacott formulated equations for a determination of the optimal sizes of intermediate buffers in a continuous flow transfer line with three machines and two buffers [1]. Prombanpong *et al.* presented a buffer design in an automated transfer line for automobile part manufacturer using geometric downtime distribution to calculate the appropriate buffer size [2]. Hillier et al. investigated a production line performance during machine breakdown and then studied an effect of buffer size to the line performance [3]. Wijngaard compared line efficiency between constant down time distribution and geometric down time distribution of a two-stage transfer line with finite buffer [4]. Enginarlar *et al.* demonstrated a calculation of minimum buffer sizes to maintain the production rate [5]. It can be seen that a number of research has been conducted dealing with effect of buffer to production line efficiency during machine break down. The objective of this paper also attempts to demonstrate a calculation of appropriate buffer size using constant downtime distribution assumption to generate a solution for various buffer sizes. Then, the slope of the graph between obtained efficiency and buffer size is determined and the steepest slope will be used as a criterion which will be described next.

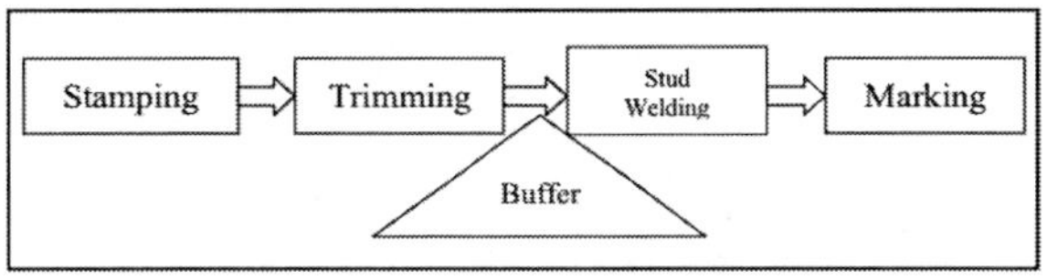

Figure 1. Separation of the line by a buffer.

Table 1. Frequency and average down time.

Machine Adjustment	Frequency (stop /cycle)	Average down time (min)
Stamping machine and Trimming machine (F_1)	0.000280	27.97
Automatic welding machine and marking machine (F_2)	0.000140	31.59

2 Problem scenario

The interested transfer line is consisted of four work stations stamping, trimming, stud-welding, and marking which is linked by a conveyor as shown in Fig. 1. The machine adjustment task can be divided into two groups, group 1 (stamping and trimming) and group 2 (stud-welding and marking) based on the machine adjustment plan. Therefore, a buffer stock between these two groups is proposed to reduce an effect of downtime losses due to machine adjustment. Therefore, the whole transfer line is divided into two stages as stage 1 and stage 2. Thus, the research question is that how large the buffer should be so that it will effectively optimize the overall line efficiency [6, 7]. Therefore, a collection of data such as frequency of line stop per cycle and average down time per line stop was conducted and can be tabulated in Table 1.

3 Methodology

In the automated production line, if there is no break down or line stop, the line will ideally operate at 100% uptime efficiency. Line efficiency can be referred to the proportion of uptime on the line and can be used as performance measurement. The line efficiency can be calculated from equations as the following [8]. Equation (1) is to calculate the line efficiency when it has no buffer stock.

$$E_0 = \frac{T_c}{T_c + (F_1 + F_2)T_d} \tag{1}$$

where E_0 is line efficiency without buffer stock, T_c is cycle time (sec), F_1 and F_2 is frequency of downtime stop at stage 1 and stage 2 respectively (stop/cycle), T_d is average downtime per line stop (sec/stop).

It can be explained that the second term of the denominator is an additional time to the normal cycle time whenever there is any line stop, as a consequence, the efficiency will be less than 100 percent.

The efficiency at stage 2 can be calculated by:

$$E2 = \frac{T_c}{T_c + (F_2)T_d} \tag{2}$$

Thus, when buffer stock is implemented, the line efficiency will be improved. In addition, it is assumed that the frequency of breakdown of each stage is equal. Thus

$$E_b = E_0 + D_1' h(b) E2 \tag{3}$$

Equation (3) shows the efficiency improvement gained from buffer b additional to the efficiency with no buffer stock E_0. The second term on the right hand side indicates additional efficiency improvement. D_1' is the proportion of expected downtime of stage 1 to the total production time and can be formulated in equation (4). The term h(b) is the rate of improvement obtained from buffer b and can be formulated as in equation (5) and (6).

$$D_1' = \frac{F_1 T_d}{T_c + (F_1 + F_2)T_d} \tag{4}$$

In the calculation for h(b), it is assumed that downtime is constant because the mentioned downtime is machine adjustment and it is less variation. According to the formula given by Buzacot, there are two cases depending upon the value of F_1/F_2 [8]. If F_1/F_2 is equal to 1, case I will be applied; otherwise, case II will be used. The formula of both cases are as follows:

Case I

$$h(b) = \frac{\mathrm{B}}{\mathrm{B}+1} + \left(\mathrm{L}\frac{\mathrm{T_c}}{\mathrm{T_d}}\right)\frac{1}{(\mathrm{B}+1)(\mathrm{B}+2)} \tag{5}$$

Case II

$$h(b) = r\left(\frac{1-r^B}{1-r^{B+1}}\right) + \left(L\frac{T_c}{T_d}\right)\frac{(r^{B+1})(1-r)^2}{(1-r^{B+1})(1-r^{B+2})};$$

$$r = \frac{F_1}{F_2},\ b = B\frac{T_d}{T_c} + L,\ B \le b\frac{T_c}{T_d} \tag{6}$$

where r is the proportion of frequency per line stop of stage 1 and stage 2. B is largest integer and L is left over stock.

Equation (5) and (6) indicate an ability of buffer b in the improvement during down time occurrence. The term h(b) is multiplied with E2, as a result, the rate of improvement obtained from buffer is mainly due to the value of h(b).

The equation (1) to (6) can be used to calculate efficiency for any b buffer. For this case study, the relevant information is as following, cycle time (T_c) = 0.17 min., the frequency of line stop in stage 1 and 2 F_1= 0.000280, F_2=0.000140. (stop per cycle), the downtime (T_d) = 27.79 min. Thus, the line efficiency E_0 and E_2 can be calculated from equation (1) and (2) by substitute T_c, F_1, F_2, and T_d in the equation respectively.

$$E_0 = \frac{0.17}{0.17 + (0.00028 + 0.00014)27.79}$$

$$= 0.9357$$

Thus, the line efficiency without buffer is 93.57%. Next, the efficiency of stage 2 namely E2 can be calculated as:

$$E2 = \frac{0.17}{0.17 + (0.000140)\,27.79}$$

$$= 0.9776$$

Next, h(b) will be calculated by substitute T_c, T_d, F_1, and F_2 in (4) obtain :

$$D_1' = \frac{0.00028(27.79)}{0.17 + (0.00028 + 0.00014)27.79}$$

$$= 0.042$$

In order to determine appropriate buffer size, efficiency of different buffer sizes must be investigated. Thus, buffer sizes ranged from 50 up to 250 pieces are designed to calculate line efficiency for comparison.

Let assume b =50 pieces, calculate r and B as:

$$r = \frac{0.000280}{0.000140}$$

$$= 2$$

$$B \leq 50\frac{0.17}{27.79}$$

$$\leq 0.306$$

Thus, the largest integer B which less than 0.306 is 0. Then, L can be computed as:

$$L = 50 - 0\,(\frac{27.79}{0.17}) = 50$$

Since F_1/F_2 or r is equal to 2, the equation (6) in case II will be used and, therefore, substitute r, B and L in equation (6), obtain:

$$h(50) = 2\left(\frac{1-2^0}{1-2^{0+1}}\right) + \left(50\frac{0.17}{27.79}\right)\frac{(2^{0+1})(1-2)^2}{(1-2^{0+1})(1-2^{0+2})}$$

$$= 0.203$$

Substitute $E_0 = 0.936$, h(b) = 0.203, $D_1' = 0.042$, and $E_2 = 0.977$ in the equation (3) obtain;

$$E_{50} = 0.936 + (0.042x0.203x0.977)$$

$$= 0.9443$$

Thus, the line efficiency when placing 50 pieces of buffer in the line will be 94.43 percent.

Likewise, h(b) value of b at 100, 150, 200 and 250 can be calculated as follows:

$$h(100) = 2\left(\frac{1-2^0}{1-2^{0+1}}\right) + \left(100\frac{0.17}{27.79}\right)\frac{(2^{0+1})(1-2)^2}{(1-2^{0+1})(1-2^{0+2})}$$

$$= 0.407$$

$$E_{100} = 0.936 + (0.042x0.407x0.977)$$

$$= 0.9527$$

$$h(150) = 2\left(\frac{1-2^0}{1-2^{0+1}}\right) + \left(150\frac{0.17}{27.79}\right)\frac{(2^{0+1})(1-2)^2}{(1-2^{0+1})(1-2^{0+2})}$$

$$= 0.611$$

$$E_{150} = 0.936 + (0.042x0.611x0.977)$$

$$= 0.9611$$

$$h(200) = 2\left(\frac{1-2^1}{1-2^{1+1}}\right) + \left(36.53\frac{0.17}{27.79}\right)\frac{(2^{1+1})(1-2)^2}{(1-2^{1+1})(1-2^{1+2})}$$

$$= 0.702$$

$$E_{200} = 0.936 + (0.042x0.528x0.977)$$

$$= 0.9648$$

$$h(250) = 2\left(\frac{1-2^1}{1-2^{1+1}}\right) + \left(86.53\frac{0.17}{27.79}\right)\frac{(2^{1+1})(1-2)^2}{(1-2^{1+1})(1-2^{1+2})}$$

$$= 0.767$$

$$E_{250} = 0.936 + (0.042x0.767x0.977)$$

$$= 0.9675$$

Similarly, the efficiency due to a buffer of 50, 100, 150, and 200 can be computed and they are tabulated in Table 2.

Table 2. Summary of efficiency of each buffer size.

Buffer Size (pc)	$H_{(b)}$	Efficiency (%)
0	-	93.57
50	0.203	94.43
100	0.407	95.27
150	0.611	96.11
200	0.702	96.48
250	0.767	96.75

From Table 2, the line efficiency at zero buffer stock is 93.57 percent. When the 50 pieces buffer size is placed, the line efficiency can reach upto 94,43 percent and so on. Thus, it is obvious that the more buffer, the higher line efficiency. In order to determine the appropriate buffer size, a graph of buffer size against line efficiency is plotted as shown in Figure 2.

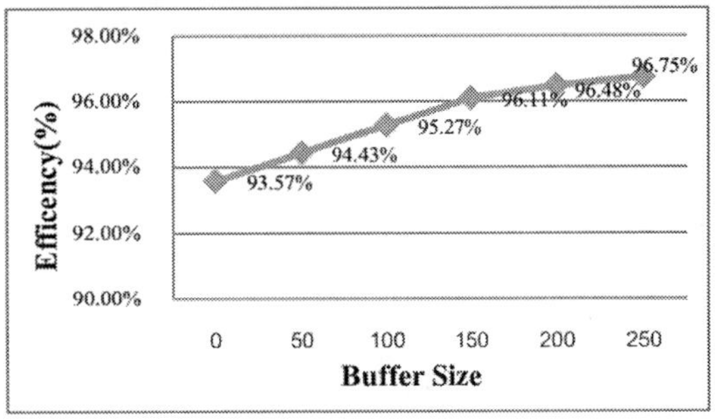

Figure 2. Productivity of the production line

A slope can be used to determine appropriate buffer size and summarized in Table 3. It is obvious that a buffer upto 50 piece-buffer size provides the steepest slope; therefore, the 50 piece-buffer size is the most appropriate size since the slope after this point becomes flat. Consequently, it is not worth to increase more buffer while the increment rate of efficiency is lower.

Table 3. Slope of each buffer

Buffer Size (pc)	Slope
0-50	0.0176
50-100	0.0164
100-150	0.0168
150-200	0.0074
200-250	0.0052

4 Conclusion

As mentioned earlier, different buffer sizes are considered to calculate the line efficient at various buffer capacities. The result of line efficiency at buffer size of 50 to 250 is summarized in Table 2 and plotted in Figure 2. It can be seen from the graph that at 50 piece-buffer size is the most lucrative point in this case study.

5 Discussion

This paper demonstrates the numerical approach to determine an appropriate number of buffer stocks to mitigate the adverse effect of down time in an automated transfer line. When machine adjustment occurs during the production, an efficiency will be decreased if no buffer is provided. The machine adjustment oss can be considered as constant downtime since this adjustment is a routine task and normally becomes a standard time. Therefore, a constant downtime equation is used.

Typically in any factory, a buffer is used to placed between work stations without calculating an appropriate number of buffer. Thus, a maximum efficiency may not be obtained if buffer size is not enough. On the contrary, too large buffer sizs does not proportionally increase the line efficiency.

References

1. J. A. Buzacott, "Automatic Transfer Lines with Buffer Stocks," International Journal of Production Research, Vol. 5, No.3, pp. 183-200. (1967).
2. S. Prombanpong, J. Kaewyu, N. Thanadulthaveedech, and M. Matwangsang. " A Buffer Design for Mitigation Downtime Effect in an Automated Transfer Line", International Journal of Innovation, Management and Technology, vol.4, pp. 155-158. (2013).
3. F. S. Hillier and K. C So, "The effect of machine breakdown and internal storage on the performance of production line systems", International Journal of Production Research, vol. 29, pp. 2043-2055, (1991).
4. J. Wijngaard," The effect of inter-stage buffer storage on the output of two unreliable production units in series, with different production rates," AIIE Transaction,vol. 11, no.1,(1979).
5. E. Enginarlar, J. L, Semyon, M. Meerkov, and R. Q., Zhang, " Buffer capacity for accommodating machine downtime in serial production lines, International Journal of Production Research, vol. 40, no. 3, pp 601-624, (2002).
6. Y. Dallery, R. David and X. Xie, "An efficient algorithm for analysis of transfer lines with unreliable machines and finite buffers," IIE, Transactions, vol. 20, no. 3, pp. 280-283,(1988).
7. Kubat, P. and Sumita, U, Buffers and backup machines in automatic transfer lines, International Journal of production Research, vol.23, pp.1259-1270,(1985)
8. M. P. Groover, Automation, Production Systems and Computer-Integrated Manufacturing, 2nd ed. Prentice Hall International Inc. NJ, ch.18, pp. 565-595,(1987).

MATEC Web of Conferences **68**, 06006 (2016) DOI: 10.1051/matecconf/20166806006
ICIEA 2016

Stochastic Dynamic Programming for Three-Echelon Inventory System of Limited Shelf Life Products

Noha M. Galal

Arab Academy for Science, Technology, and Maritime Transport, Industrial and Management Engineering Department, 1029 Alexandria, Egypt

Abstract. Coordination of inventory decisions within the supply chain is one of the major determinants of its competitiveness in the global market. Products with limited shelf life impose additional challenges in managing the inventory across the supply chain because of the additional wastage costs incurred in case of being stored beyond product's useful life. This paper presents a stochastic dynamic programming model for inventory replenishment in a serial multi-echelon distribution supply chain. The model considers uncertain stationary discrete demand at the retailer and zero lead time. The objective is to minimize expected total costs across the supply chain echelons, while maintaining a preset service level. The results illustrate that a cost saving of around 17% is achievable due to coordinating inventory decisions across the supply chain.

1 Introduction

Coordinating and integrating inventory decisions across the supply chain is the key difference between managing inventory in a single facility and in a supply chain environment [1, 2]. Inventory models studying the effect of coordinated replenishment decisions are referred to in the literature by joint economic lot size models [1]. Extending the classical inventory models is necessary to address the inventory decisions in a supply chain environment; in this respect, special attention should be given to the multi-product and multi-echelon inventory models [3].

Products with limited life time, as food, pharmaceuticals, and blood supply in healthcare are highly affected by the ordering decisions. The major problem faced in managing such products is wastage when holding the product for more than its life time. Ordering an adequate quantity to avoid both overstocking and understocking of products is the main target of decision makers striving for achieving higher service levels at minimum total cost. In the case of ordering excess amounts, additional costs are incurred to dispose of excess items or to keep them in inventory. On the other hand understocking causes a decrease of service level.

Various research efforts have been dedicated to the perishable inventory; applications include food products [4], blood supply chains [5], and fashion supply chain [6] among others. Making inventory decisions in a multi-echelon supply chain network is challenging, especially in the presence of uncertainties such as uncertain customer demand and/ or replenishment lead time.

In order to incorporate the variability in a multi-period time horizon decision making process, a model has been formulated and solved via stochastic dynamic programming. The dynamic programming approach is suitable to represent the multi-period decision environment. Incorporating the variability of customer demand facilitates studying the uncertainty inherent to the inventory problem.

The aim of this work is to apply stochastic dynamic programming to a multi-echelon supply chain distribution network of products with a limited shelf life under stochastic demand and a service level constraint. The current research contributes to the literature by investigating the effect of joint replenishment in a supply chain environment on total supply chain inventory cost.

The remainder of this paper is structured as follows. In Section 2 a literature review is presented covering the multi-echelon inventory management problem with an emphasis laid on the application of stochastic dynamic programming to inventory items with limited life time. A stochastic dynamic programming model for a three echelon supply chain with stationary demand is developed in Section 3. In order to illustrate the stochastic dynamic programming model, a numerical instance is presented and solved in Section 4. Results are analyzed and discussed in Section 5. Finally the paper is concluded and directions of future research are indicated.

2 Literature review

The purpose of this review is to trace published research on inventory management in the context of supply chain of items with limited life time. The main interest here is in mathematical modelling approaches so far applied, and more specifically the dynamic programming approach. Thus the following review addresses three areas: inventory policies of perishable and deteriorating products, joint economic lot sizing, and stochastic dynamic programming. An emphasis is laid on the joint consideration of all three research areas.

Perishable inventory management has been extensively reviewed in [7, 8]. A review on inventory modeling techniques for multi-echelon supply chains with uncertainties in demand and lead time is presented in [9]. The authors in [9] made the finding that papers adopting the mathematical modeling technique considered stochasticity in demand only. Lead time was neglected or assumed to be fixed, zero, or constant. In their review, only the work of Rau [10] has been identified to address the perishable item inventory. A review of joint economic lot sizing has monitored the application to deteriorating items [1]. The authors classified the work done in this respect into two categories; items constantly deteriorating during storage and items having a constant life time after which they have to be disposed of. The current work concentrates on the latter category.

When dealing with inventory management in supply chain context, the majority of research modelled a two echelon supply network; the maximum number of echelons considered in the multi-echelon network is three [8].

In [10] a model was developed to determine the optimal order quantity across the supply chain for deteriorating products. Single supplier-single producer-single buyer situation was modeled using deterministic demand, negligible lead time, and constant deterioration rate of items. Shastri *et al.* [11] developed a multi-echelon inventory model for deteriorating items with partial backlogging and considering inflation rate.

The inventory replenishment policy problem for products with limited life time has been modeled via the dynamic programming approach in the literature [12–15]. All the models targeted the minimization of costs taking into consideration purchasing, holding and disposal costs to achieve a specific service level [14, 15]. The developed models were multi-period models to benefit from the capability of dynamic programming in modeling multiple stages. Both deterministic and stochastic demands have been considered [14]. As to the lead time aspect, authors either assumed it fixed, positive, zero or very long. Thus the effect of lead time and its variability have not been explicitly addressed in the reviewed models. This is due to the additional complexity that would result when considering an additional random parameter. All of the reviewed models considered the inventory policy for a single stage, thus the consideration of the joint replenishment for perishable products seems to need more research effort. The current work investigates the effect of joint replenishment on total supply chain inventory cost for a multi-echelon supply chain of limited life time products.

3 Model development

3.1. Problem definition

The problem addressed in this work is that of identifying an inventory replenishment policy for a product with a limited shelf life in a distribution network with serial configuration; the three echelons addressed are a single supplier, single distributor and single retailer as depicted in Figure 1. The retailer has a random stationary demand d_t given by a discrete probability distribution. Based on retailer's random demand, the distributor d orders a quantity Q_t^d in period t from the supplier s, who in turn orders a quantity Q_t^s in period t from his supplier. Distributor and supplier purchase products at a unit cost of c^d and c^s , respectively. The lead times for replenishing the supplier and distributor are assumed to be zero. Quantities are stored at both the supplier and distributor, whose capacity is assumed to be infinite. The products have a shelf life of M periods. Quantities reaching their shelf life are disposed of at a unit disposal cost w. Holding costs are incurred at both supplier and distributor and amount to h^s and h^d per unit per period, respectively. Furthermore, there is a fixed ordering cost k associated with each order. The inventory level at distributor and supplier in any period t is given by I_t^d and I_t^s, respectively. Any unmet demand is lost, and the initial inventory levels at both supplier and distributor ($I_{1,0}^s$ and $I_{1,0}^d$) are zero. A predetermined service level α is targeted. The service level is calculated as the probability of no stock out during a replenishment cycle.

The objective is to identify replenishment quantities for supplier Q_t^s and distributor Q_t^d per period t so as to minimize the total expected cost across the echelons and satisfy a service level constraint.

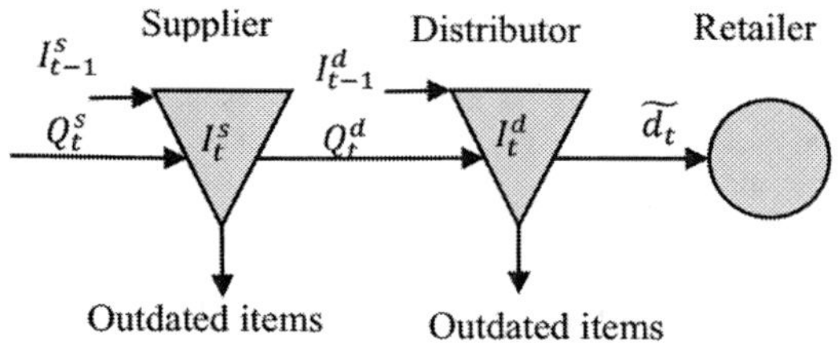

Figure 1. Model description for period *t*.

3.2 Dynamic programming formulation of the problem

Since the problem is characterized by making decision at each period based on the state of the system in previous stages and the decisions made, it lends itself to the dynamic programming approach. Having uncertain demand, the problem is modeled via stochastic dynamic programming. To this end the following elements have to be identified: the states, the decision variables, the contribution function, and the objective function. States are the inventory levels I_t at each echelon. The decisions made are the quantities ordered in each period t. The state transition function describes the inventory level in period t based on the inventory level in previous period (t-1) and the decision made in the current period t. The expected

contribution to the objective function is comprised of fixed ordering cost, purchasing cost of quantities Q_t^s and Q_t^d, the expected holding and wastage costs. Backorders are not allowed in this model. The elements defining the stochastic programming model are described as follows:
Stage (t): each stage presents a time period t, t=1,...,n, where n is the planning horizon.

State (s_t): Inventory level at the beginning of period t. To meet a predefined service level, quantities on hand should satisfy the following equation:

$$P(I^d_t \geq 0) \geq \alpha \quad (1)$$

Decision variables: the decision in stage t is the amount to order at the beginning of period t. Two decision variables are present Q_t^s and Q_t^d.
Contribution function: the function providing the cost at stage t, given that the decision Q_t^s and Q_t^d is made.

Optimal value function$(f_t(s_t))$: minimum expected cost from the beginning of period t to the end of the planning horizon, period n, given that the inventory on and is s_t at the beginning of period t.

Optimal policy $(p_t(s_t) = Q_t^{s*})$: optimal replenishment policy for period t, given that the on hand inventory is s_t.

Transformation function: the change of inventory level (state) for the next stage based on the current state, stage, and decision.

$$I_{t+1}^d = (I_t^d + Q_t^d - d_t)^+ \quad (2)$$

$$I_{t+1}^s = (I_t^s + Q_t^s - Q_t^d)^+ \quad (3)$$

Recurrence relation: the following equation indicates the optimal policy at stage t, given that the optimal policy at stage t+1 is known.

$$f_t(s_t) = \min_{Q_t^d} \begin{bmatrix} E[k^d Y^d + p^d Q_t^d + h^d I_t^d \\ +k^s Y^s + c^s Q_t^s + h^s I_t^s + f_{t+1} \end{bmatrix} \quad (4)$$

$$t = 1, \cdots, n$$

where Y^d and Y^s are binary variables which are equal to 1 if the order quantity is greater than zero, and E[.] denotes the expected value.
Boundary conditions: $(f_{n+1}) = 0$.

4 Numerical illustration

4.1. Problem description

The following example illustrates the application of the model to a serial multi-echelon distribution system. It is required to identify the order quantities of the supplier and distributor so as to satisfy the retailer's demand with a service level (α) of 90% at minimum expected total cost. Demand is assumed to be a stationary discrete random variable with a known probability distribution. Demand values are 0, 1, 2, and 3 units with probabilities of 0.25, 0.4, 0.25, and 0.1, respectively. The planning horizon is 2 periods and the product life time is 2 periods, after which the product is disposed of at a wastage cost of \$500 per unit. Ordering costs for supplier and distributor are \$3,000. Unsold quantities are kept in inventory at a holding cost of \$100 per unit per period. Initial inventory at distributor and supplier is assumed to be zero. From the cumulative probability it becomes evident, that it is necessary to have 2 or more units in stock to achieve a service level of 90%. Thus, orders are placed only, if at the beginning of the period less than 2 units are in stock.

For this example quantities may be ordered in each period or aggregated and placed once at the beginning of the first period. The limited shelf life is considered by allowing to aggregate demand over only two periods, which is the product shelf life.

4.2. Model solution

The model is solved via backward induction. Starting at stage 2, the possible states (i.e. inventory levels) are 0, 1, 2, and 3. These represent the closing inventory, and are thus disposed of at the wastage cost. Thus the contribution function valuation for states 0 through 3 will be \$0, \$500, \$1,000, and \$1,500, respectively.

For stage one, the possible states are 0, 1, 2, and 3, since there is only one stage to go and the maximum anticipated demand is 3 units. Based on the quantity on hand, a decision is made regarding how much to order. To achieve the required service level, orders are placed only if the quantity on hand is below 2 units. The order quantity is selected so as to increase the inventory level to 2 units. The consideration of the service level constraint helps decreasing the number of feasible scenarios to consider. For example, when the state is zero, the minimum amount to order should be 2, to satisfy the service level constraint. Similarly, if only one unit is on hand at the beginning of period t, at least one unit is to be ordered. To facilitate model solution, it is further assumed that the inventory levels at beginning of period t at supplier and distributors are identical. The associated costs of ordering, purchasing, inventory and valuation of previous stage are calculated for each scenario. These include cost data for both supplier and distributor. The expected cost of each decision made based on the current inventory level is the sum of the weighted total cost by the respective probability. A sample of the detailed calculations of stage 1 is displayed in Table 1. The calculations shown in the table display all the possible decisions if the state at the beginning of the period is 2. In such a case, two decisions are possible either to order nothing or order 1 unit. The order quantities are chosen so as to preserve the service level. After making the decision, the demand is revealed. And either 0, 1, or 2 units are sold. Note that since no backorders are allowed, any demand exceeding the available inventory level is lost. Here it is also assumed that the demand is equal to the amount sold. Accordingly, the state, i.e. inventory level at the end of the period may be calculated. The probability in the 4th column indicates the probability of satisfying the demand, i.e. probability of sales. It is derived from the given demand probabilities. All the cost elements of the distributor and supplier are calculated and

DOI: 10.1051/matecconf/20166806006

summed along with the previous stage valuation function f_0. The expected contribution is the sum of the weighted total cost by the respective probability. Hence, for a given state of 2 units and a decision of no orders to place, the expected cost is $740. From the previous calculations, the resulting optimal value function for stage 1 is attained as shown in Table 2.

Table 1. Sample of calculations made for stage 1.

state I_t^d	$Q_t^d = Q_t^s$	d_t	Prob.	I_t^d	k^dY^d	$p^dQ_t^d$	$h^dI_t^d$	k^sY^s	$p^sQ_t^s$	$h^sI_t^s$	f_0	total cost	expected contribution
2	0	0	0.25	2	0	0	200	0	0	200	1000	1400	**740***
2	0	1	0.4	1	0	0	100	0	0	200	500	800	
2	0	2	0.35	0	0	0	0	0	0	200	0	200	
2	1	0	0.25	3	3000	1000	300	3000	900	100	1500	9800	9080
2	1	1	0.4	2	3000	1000	200	3000	900	100	1000	9200	
2	1	2	0.25	1	3000	1000	100	3000	900	100	500	8600	
2	1	3	0.1	0	3000	1000	0	3000	900	100	0	8000	

For stage zero, only one state is possible, since we assumed an initial inventory level of zero. Decisions made are to order 0 through 3 units. The resulting optimal valuation function and the corresponding decision are $15,665 and 3 units, respectively.

Table 2. Optimal value function and policy for stage 1.

States	**0**	**1**	**2**
f_1 ($)	10,340	8,440	740
Q_t^{d*} (units)	2	1	0

4.3. Experimentation

In order to illustrate the effect of coordination between the different supply chain echelons, the problem has been solved via two approaches. The first approach is the one described in Section 4.2, where the replenishment policy is determined based on integrating the decision for both the distributor and supplier (Experiment 1). By adopting this approach it is assumed that both echelons order the same quantity. The second conducted experiment (Experiment 2) is solving the model sequentially for one echelon at a time disregarding the multi-echelon structure. More specifically, the order quantity of the distributor is determined based on the uncertain demand of the retailer; these quantities are used as demand for the supplier. Since the decision made regarding the order quantities at the distributor are based on expected values, it becomes necessary to determine the probability of the demand of the distributor. This is achieved by the following approximation. For each stage, all feasible decision alternatives for all the possible states are identified, and the relative frequency of each decision is determined. For example in stage 2, there is only one state, which is to have no quantities on hand. In such a case, either 2 or 3 units are to be ordered to achieve the required service level. Quantities sold will turn out to be 0, 1, or 2, if two units are ordered; and 0, 1, 2, or 3, if three units are ordered. Thus there are seven feasible decision alternatives. The decision of ordering 3 units occurs in four out of these seven alternatives. Hence, the probability of having the distributor ordering 3 units from the supplier in t=2 is 4/7=0.57. In this example the demand of the distributor in stage one is 0, 1, 2, and 3 with a probability of 0.51, 0.19, 0.19, and 0.11, respectively. Similarly, in stage 2, the probability of the distributor ordering 2 and 3 units is 0.43 and 0.57, respectively. Based on the probability and cost values, the corresponding expected total cost is calculated. The approach of backward induction is applied twice; one time for supplier and the second for the distributor. Total supply chain inventory cost is the sum cost of the optimal policy for each of the supplier and distributor. The results of the two approaches are compared in the following section.

5 Results and discussion

The results obtained from the dynamic programming model are presented in this section. Table 3 summarizes the results for the two experiments. Experiment 1 considers the joint economic replenishment, while Experiment 2 neglects the coordination between the supply chain echelons.

Table 3. Summary of results.

	Expected Cost ($)		
	Distributor	Supplier	Total
Experiment 1 (with coordination)	9,275	6,390	15,665
Experiment 2 (without coordination)	8,355	10,729	19,084
% improvement via coordination	-11.0%	40.4%	17.92%

The joint replenishment has a benefit of decreasing the total supply chain cost by 17.92%. Although an increase in cost at the distributor is noticeable, a larger decrease of cost at the supplier causes an overall improvement in the supply chain. This is in line with the findings in [1]. The current study has showed that the same results are attainable with the presence of uncertainty in demand. The main cost reduction for the supplier are due to decreasing the holding cost. In fact coordinating the replenishment policy shifts the inventory from the supplier to the distributor.

6 Conclusions and future work

This work has presented a stochastic dynamic programming approach to manage inventory in a simple serial three-echelon supply chain for a product with limited shelf life. Demand is assumed to be a stationary discrete random variable with known probability distribution. The model has been solved via backward induction for a small size problem of two time periods and one retailer, one distributor and one supplier. Two experiments have been conducted to illustrate the impact of coordination on expected cost at each echelon and across the supply chain. Coordination has proved to decrease expected total cost across the supply chain. A cost savings of around 17% was attained due to coordination between the supplier and distributor.

The current approach may be extended to study the effect of coordination in more complex network structures with multi-products. The further consideration of non-stationary and continuous demand will allow to approach the real life case.

References

1. C. H. Glock, Int. J. Prod. Econ., **135**, 2, 671–686, (2012)
2. I. Z. Karaesmen, A. Scheller–Wolf, and B. C.-K. Deniz, Planning Production and Inventories in the Extended Enterprise, **151** (2011)
3. M. A. B. and A. Guiffrida, M. Y. Jaber, and M. Khan, Manag. Res. Rev., **38**, 3, 283–298 (2015)
4. R. Haijema, Int. J. Prod. Econ., **157**, 158–169 (2014)
5. U. Abdulwahab and M. I. M. Wahab, Comput. Ind. Eng., **78**, 259–270 (2014)
6. I. Konstantaras and K. Skouri, Math. Probl. Eng., (2009)
7. S. K. Goyal and B. C. Giri, Eur. J. Oper. Res., **134**, 1, 1–16 (2001)
8. M. Bakker, J. Riezebos, and R. H. Teunter, Eur. J. Oper. Res., **221**, 275–284 (2012)
9. A T. Gümüs and A F. Güneri, Proc. Inst. Mech. Eng. Part B J. Eng. Manuf., **221**, 1553–1570 (2007)
10. H. Rau, M. Y. Wu, and H. M. Wee, Int. J. Prod. Econ., **86**, 155–168 (2003)
11. A. Shastri, S. R. Singh, D. Yadav, and S. Gupta, Procedia Technol., **10**, 320–329 (2013)
12. K. G. J. Pauls-Worm and E. M. T. Hendrix, Lect. Notes Comput. Sci. **9157**, 397–412 (2015)
13. C. L. Williams and B. E. Patuwo, Eur. J. Oper. Res., **116**, 352–373 (1999)
14. E. M. T. Hendrix, R. Haijema, R. Rossi, and K. G. J. Pauls-Worm, Comput. Sci. Its Appl. - ICCSA 2012, Pt III, **7335**, 45–56 (2012)
15. R. Rossi, IFAC Proc., 2021–2026 (2013)

MATEC Web of Conferences **68**, 06007 (2016) DOI: 10.1051/matecconf/20166806007
ICIEA 2016

A Review of Granulation Process for Blast Furnace Slag

Pengfei Yu, Shuzhong Wang, Yanhui Li and Guixi Xu

Key Laboratory of Thermo-Fluid Science and Engineering, Ministry of Education, School of Energy and Power Engineering, Xi'an Jiaotong University, Xi'an, Shaanxi ,710049, China

Abstract. Molten slags of blast furnace is a second resources with great value of 1600~1 800 MJ sensible heat per ton. At present, water-quenching process plays a leading role in recovering waste heat of the molten slags. However, this method not only cost lots of water, but also recover little sensible heat and can pollute the surrounding environment. Dry granulation process, as an environmentally friendly method with high-efficiency heat recovery, have attracted widespread attentions. In this paper, the water quenching and dry granulation processes were discussed in detail. After a thorough comparative analysis of various treatment technologies, it can be concluded that centrifugal granulation affiliated with dry granulation is the optimum process, with smaller slag particle size (about 2mm), more glassy phase and higher recovery rate.

1 Introduction

The steel industry is one of the most energy-intensive industries in China. Based on statistics of World Steel Association, China's crude steel output reaches 823 million tons in 2014, accounting for 49.4% of the global proportion. In the first eight months of 2015, China's produced 543.02 Mt of crude steel [1]. Although the energy consumption get a slight decrease this year, the production condition and structure are different among different enterprises as well as the energy conservation, and the inefficient capacities account for a comparatively great proportion, which result in a certain gap of the average energy consumption in China compared with the developed countries level. Therefore there is a great energy-saving potential in China. The Blast Furnace (BF) energy consumption account for the largest percentage, and responsibly, it should bear more energy conservation and emissions reduction such as BF slag sensible heat recovery.

2 Blast furnace slag and its utilizations

Blast furnace slag is a by-product of blast furnace (BF) ironmaking produce. Every pig iron production is accompany by 0.3 ~ 0.6 tons of blast furnace slag produced with the temperature of 1350 ℃ ~1450 ℃ (1600~1 800 MJ sensible heat, equivalent to 55~61kg standard coal combustion after the heat generated). On this basis, the BF slag is up to 247 million tons in 2014, corresponding to 11 million tons standard coal, approximating 14.0% of consumption in the BF process.

Generally, according to the cooling rate, blast furnace slag processing can be divided into quenching, half-quenching and slow cold. Because of their different properties, the products, which mainly consist of pelletized blast-furnace slag, expanded slag ball and dry slag, have different function.

3 Water-Quenching process

So far, water treatment method is widely used in the cooling method of blast furnace slag. There are many kinds of water treatment method for BF SLAG, such as Advection Sedimentation Tank, RASA, INBA, MTC, TYNA, and RASA. The method of RASA is adopted extensively in the steel plant in China as well as INBA.

3.1 RASA

Dehydration facilities of the RASA (Fig. 1) is composed of several filtering basins, which is made up of layers of different particle size pebble bed at the bottom. In blunt box, high pressure water from porous nozzle jet will quench the molten BF slag. In the next case, water quenching slag pours through graining tank and flows into the settling pond. After filtration, the slag is grabbed by grabbing crane and stacked in the SLAG field to further dewater. The water and suspended solids in the pond flow into the filter tank, filtered by the pebble bed. Filtered water collect and pump into the cooling tower, recycling. Water loss added by new water.

3.2 INBA

In the early 1980s, the method of INBA was first designed by P&W of Luxembourg and SIDMAR of Belgium. Baosteel group first introduced the technology

DOI: 10.1051/matecconf/20166806007

and many steel plants still uses this process which is shown in Fig. 2. It can be divided into hot INBA, cold INBA and environmentally friendly INBA. The greatest advantage is low sulfur emissions, which was most transported to the water-cycle system.

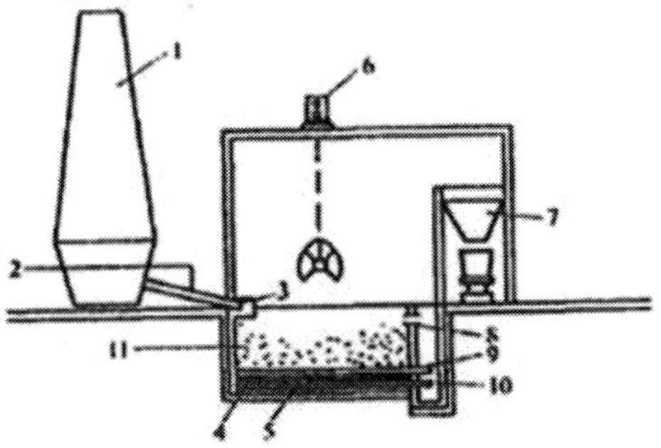

1—clinker ejector, 2—water slag ditch, 3—water slag tank, 4—distributor, 5—drum filter, 6—buffer tank, 7—collecting tank, 8—thermal basin, 9—cooling tower,10—water tank, 11—belt conveyor, 12—product bin
Figure 1. RASA process.

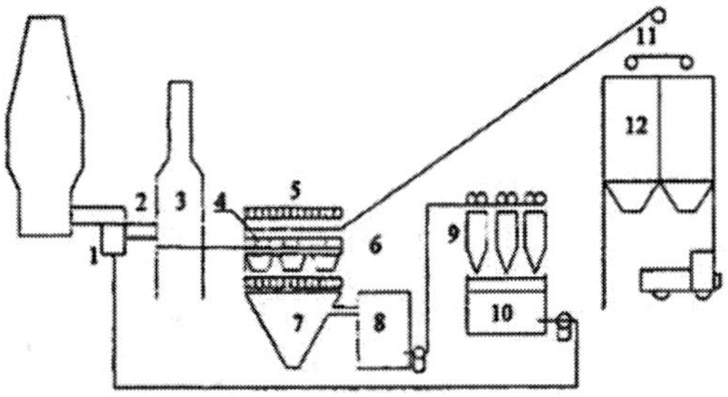

1—blast furnace, 2—clinker ejector, 3—granulator, 4—guard rail, 5—sewerage of RASA, 6—grabbing crane, 7—storage hopper, 8—overflow, 9—inlet of flushing air, 10—water outlet, 11—granulated slag
Figure 2. INBA process.

4 Dry granulation process

On the blast furnace slag treatment technology, for a long time, water quenching process has token the leading position. However, there are many defects of water-quenching slag method, which can be roughly concluded as follow: a mass of sensible heat unrecovered, high water consumption, air pollution, huge system maintenance workload and high power consumption. These problems greatly reduce the utilization value of BF slag as, which go against the energy recycling and sustainable development. As a result, it is the inevitable trend to develop an environment friendly and control reliable dry granulation process with high value-added products and high rate of waste-heat recovery.

Dry granulation is a technology of no new water consumption, direct or indirect contact with blast furnace slag to recover sensible heat. And at the same time it will use mechanical power or high pressure air to make slag granulating, without harmful gas discharge, a new type of environmentally friendly technology for slag treatment. According to the slag crushed process, many methods have been proposed. Those which have great influence are rotating drum process, air blast method and centrifugal granulate process.

4.1 Rotating drum process

There are two main forms of this technology: single drum and double drums.

NKK [2] built a pilot blast furnace to research the double rotating drum technology (Fig. 3) at Fukuyama. It simulates the working principle of dry quenching. Liquid slag is poured into the space between the two opposite rotating drum, and cooled by the heat transfer medium in the drum, along with the slag crushed. Recycling heat is collect to the heat recovery boiler and used to generate electricity. Although the glassy phase we can get reach about 80%, the slag particle size is much bigger and must be grinded if used for cement. What's more, the slag often adheres to the drum, hard to control the thickness of slag film. As efficient organic solvents is difficult to choose, the exergy recovery is low (about 40%). So the process has not been industrial applications.

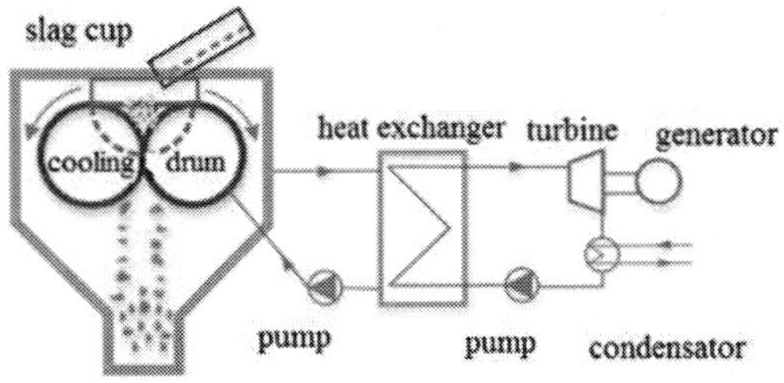

Figure 3. NKK cooling drum process.

The single rotating drum process was first put forward by Ishikawajima-Harima Heavy Industries and Sumitomo Metal [3] and the process is shown as Fig. 4. The drum with non-wetting coating on the surface rotates at a certain speed, while melted slags are poured onto the drum and thrown out, broken to particle. Meanwhile, the broken particles are gathered to fluidized bed by the falling water-quenched slag and then heat exchange with air occurred. It is reported that this process can achieve a temperature of 500℃ heated air and slag diameter less than 10 mm (about 95%). The heat recovery rate can be as high as 60%. However, this method belongs to half-quenching treatment and the product mainly used for the concrete aggregate, low added-value. And the processing capacity is insufficient, hard to industrialize.

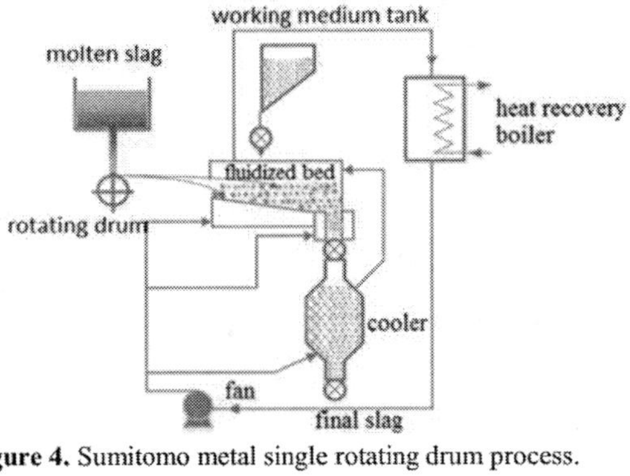

Figure 4. Sumitomo metal single rotating drum process.

DOI: 10.1051/matecconf/20166806007

4.2 Air blast method

High speed and high pressure air plays a pivotal role to discharge the molten slag with a rapid cooling in this process, along with absorbing the sensible heat [4]. And then recycle heat are sent to heat recovery boiler to produce steam for power generation. Japan, Germany, Russia, etc. have studied on this technology, among which Japan have outstanding performance with industrial application. Since 1982, six major Japanese steel companies develop this process and successively carried on the experiment, half of industrialization and mass production with slag handing capacity of 40t/h and heat recovery rate of 60%. The technological process is shown in Fig. 5. Liquid slag from tilting slag ditch poured into the wind hole, granulates under the action of the air from the bottom pelletizer fan. Particles solidify during its flight, the temperature depressing from 1500 ℃ to 1100℃, and then in the heat exchanger further cooled to about 300℃. The finished slag particles less than 5mm account for 95%, whose glassy phase is more than 95, similar to that of water treatment slag. And the heat recovery rate is approximate 60%. However, there is no follow-up report due to its drawbacks, mainly summary as follow:

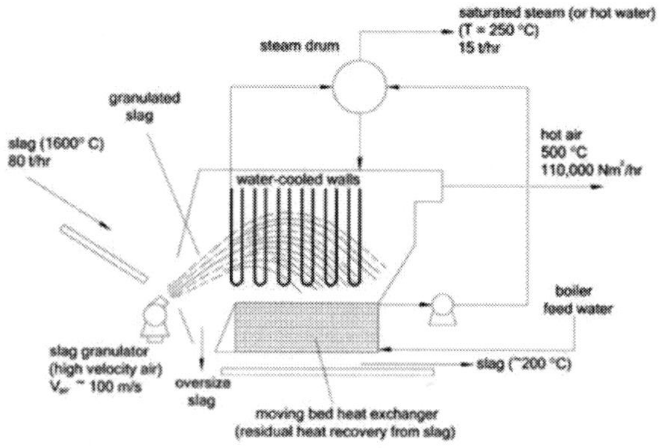

Figure 5. Schematic of molten slag heat recovery by air.

• As a result of the lower heat capacity far less than that of water, large quantities of air are required, thus result in high energy consumption of the fan and high operation cost.
• As a reason of the lower cooling speed than that of the water quenching, it is necessary to increase the size of the equipment in case the slag adhere to the surface of device. Thus the whole system covers a large area and need long payback time.

4.3 Centrifugal granulation method

In early 1980s, Pickering [5] and his colleagues first invented the centrifugal granulating device at Redear blast furnace of British Steel Corporation Based on the variety of the granulator, the critical component of this process, it can be divide into several different forms: Rotating Cup Atomizer (RCA) [6], Spinning Disk Atomizer (SDA), Rotating Cylinder Atomizer (RCLA) [7]. The mechanism of RCA is similar to SDA.Depend on the way to recover heat, it can also separate into physical and chemical method. From the moment it emerges, the centrifugal granulation method has attracted the eyes of the world. The process invented by Pickering is shown as follow (Fig. 6). BF slag pours into the high speed rotary cup in the heat recovery vessel from the slag supple launder. And then molten slag are thrown out to grain under the centrifugal force at the edge of the cup. Meanwhile, annular air are set at the surround of the cup for the purpose of instability wave of slag membrane so as to break up the slag. And the particle slags flow out along radial direction, slightly upward. As the slag particle flying in the air, their surface are rapidly cooled with the temperature depress (100K~200K). There are cooling water pipe in the wall. After heat transfer, the slag particle rebounded to the primary fluidized bed, with a heat recovery rate of 43%. Grain slag overflow to the second fluidized bed due to the swelling of primary fluidized bed and further recycling residual heat. The process obtains a heat recovery rate of 59% and average slag particles less than 2mm.The finally slag particles are of high glassy rate (more than 95%) and cooled down to 250℃.

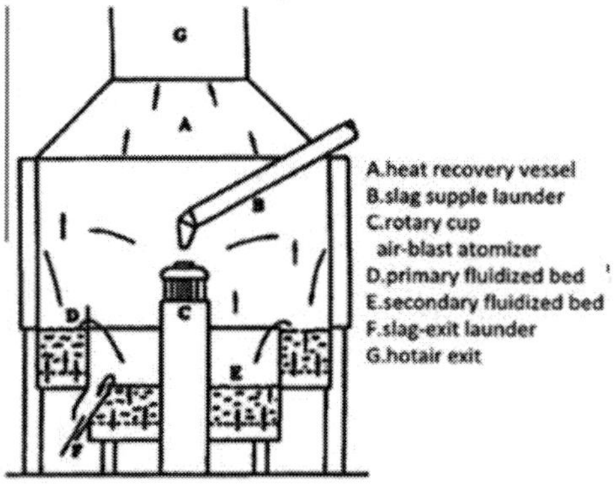

Figure 6. RCA process developed by picking.

Figure 7. The semi-industrial scale (3 m in diameter) integrated heat recovery and dry granulation pilot plant at CSIRO with design throughput of 6 tonnes/hour of slag [8].

A growing number of people have focus on this method. In 2001, Mizuochi [2] focused on the Centrifugal granulation method and studied on its feasibility. In 2004, Purwanto [3] had a further research on the characteristics of glass beads frSom molten slag obtained in the RCA process, using high speed camera to observe the experimental process. From 2007 to 2008, Xie et al. at CSIRO forced on the researches on utilization of slag waste heat in steel plants. Furthermore, a semi-industrial scale (3m in diameter with spinning disc of up to 250 mm in diameter and slag tapping rates up to 100kg/min) integrated dry granulation and heat recovery pilot plant at CSIRO has been built (Fig. 7). The granulated products were highly glassy (~98%) and were found to have good cementitious properties.

5 Conclusions

Although there is no dry blast furnace slag processing technology to realize industrial application, the existing research has accumulated quite a lot of relevant theoretical knowledge and practical experience. From the analysis above, including rotating drum process, air blast method and centrifugal granulation method, can meet the requirements of cement material. Among them, particle size of the rotating drum method is larger, slag piece often bonding on the drum which need to clean using scratch residue board. Thus the processing capacity is limited. As to air blast method, the air as the most common heat transfer medium has a low heat transfer coefficient, causing the bigger air compressor equipment and uneconomical heat recovery process. Centrifugal granulation method is the optimum process, with smaller slag particle size (about 2mm), higher glassy phase and recovery rate. The process capacity and the cooling speed can match the blast furnace production well. So the centrifugal graining attract the most considerable attention in the world. With the promotion of Supply-side reforms in steel industry, the advance of the centrifugal granulation method in energy saving and environmental protection will be more evident. No doubt that centrifugal granulation method is the main trend and more industrial trial will be conducted both in China and abroad, once succeed, it will be widely applied.

References

1. September 2015 crude steel production. http://apps.webofknowledge.com
2. B. Du, G. L.Ruo, R. Q. Jiang, The development of molten slag dry granulation and waste heat recovery technology, The 14th national meeting of drying technology, 3-13 (2013)
3. M. Yoshinaga, K. Fujii, T. Shigematsu, T. Nakata, Dry granulation and solidification of molten blast furnace slag, Transactions of the Iron and Steel Institute of Japan, 823-829 (1982)
4. M. Barati, S. Esfahani, T. A. Utigard: Energy recovery from high temperature slags, Energy, 5440-5449 (2011)
5. S. J. Pickering, N. Hay, T. F. Roylance, G. H. Thomas, New process for dry granulation and heat-recovery from molten blast-furnace slag. Ironmaking & Steelmaking, 14-21 (1985)
6. H. Purwanto, T. Mizuochi, H. Tobo, M. Takai, T. Akiyama, Characteristics of glass beads from molten slag produced by Rotary Cup Atomizer, Materials Transactions, 3286-3290 (2004)
7. Y. Qin, X. Lv, C. Bai, P. Chen, G. Qiu, Dry granulation of molten slag using a rotating multi-nozzle cup Atomizer and Characterization of Slag Particles, Steel Research International, 852-862 (2013)
8. S. Jahanshahi, D. Xie, Y. Pan, et al., Dry slag granulation with integrated heat recovery; proceedings of the 1st Int Conf on Energy Efficiency and CO_2 Reduction in the Steel Industry (EECR Steel 2011), F, (2011)

 DOI: 10.1051/matecconf/20166806008

Neural Networks and Genetic Algorithms Applied for Implementing the Management Model "Triple A" in a Supply Chain. Case: Collection Centers of Raw Milk in the Azuay Province

Bermeo M. Juan Pablo[1], Castillo C. Hilda Piedad [2] and Serrano V. Santiago[3]

[1] *UPS Universidad Politécnica Salesiana, Electronic Engineering Career, 1230 Calle Vieja, Cuenca, Ecuador*
[2] *UPS Universidad Politécnica Salesiana, Accounting and Auditing Career, 1230 Calle Vieja, Cuenca, Ecuador*
[3] *UPS Universidad Politécnica Salesiana, Business Administration Career, 1230 Calle Vieja, Cuenca, Ecuador*

Abstract. To get successful the companies need a combination of several factors, the most important one is the management of Supply Chain. This paper proposes the use of intelligent systems such as Artificial Neural Networks (ANN) and Genetic Algorithms as support systems together with monitoring indicators and monitoring, in order to implement the management model Triple A, which is focused on Agility-Adaptability-Alignment, where the "Agility" is the speed of response to changes in demand, "Adaptability" is the ability to tailor the supply chain front market fluctuations and "Alignment" is to align the chain between consumers and suppliers. The Neural Network was trained to work as a predictor of demand and will improve the "agility" of the supply chain, the genetic algorithm is used to obtain optimal routes of pickup from providers, this support to the "alignment" the product of suppliers in the supply chain to final customers; the Neural Network with the Genetic Algorithm together serve as support to "adapt" the supply chain to variations of demand and the suppliers, however, for successful of the model are need other factors such as the use of indicators and training of staff on the administration of management model triple A in the supply chain.

1 Introduction

Many companies even though they provide an excellent product and have an excellent human talent they fail for mistakes in making logistical decisions that is where tools of Artificial Intelligence such as Artificial Neural Networks (ANN) [1] and genetic algorithms [2] they can serve support in making decisions, in order to reduce losses and increase profits considering the line of business, this paper shows a practical application of such tools for the implementation of the Management Model Triple A [3] in the supply chain of the collection center for raw milk.

In Ecuador, the profit margin of the company is around 20%, so a loss of revenue for errors in logistics management can be the difference between success and failure. Logistics management includes management of physical flows, administrative and information, so that all projects of strategic planning should involve decisions directly in the supply chain. One proposal of logistic management was propounded by Hau Lee in 2004, it called the Triple-A supply chain management model that guides management in three main edges that are Agility, Adaptability and Alignment.

2 The "Triple-A" management model

The management Model Triple A for supply chain, covers several concepts that are not relatively new, for example in 1994 Goldman [4], proposed the "agile" concept refers to the speed of response of the company to changes in demand and supply for calibrating the production process so you can respond quickly to changing market information. The Toyota production model is a practical application of management models "agile" [5]. Another important concept is "lean" used for creating new companies called "lean star-up's" [6] that among its main ideas are:
- Testing ideas and finding customers in a short time and with few resources.
- Apply to the development of a market and a constant focus on the customer.
-The term "minimum viable product" to analyse the market and adapt the product.
-The principle is start small, rapid product launch and get customers.

In short, the "lean" is a philosophy and "agile" is a management methodology they can be supplemented appropriately, so that appears the model "leagile"[7], that integrates both concepts, later in 2003 Stratton [8] proposes an improvement and creates the "mobility" concept which aims to properly plan small inventory levels throughout the supply chain with limited levels of production and distribution so that they are not a constraint for get a quick supply chain and agile with zero products expired or obsolete.

Although they represent great strides in managing the supply chain, however, it is born in 2004 Management Model Triple A [9] in the Harvard Business magazine, proposed by Hau L. Lee, who raises with the concept of "Agility", two new concepts: "adaptability" and "Alignment", where adaptability is related to the design of a supply chain, and the "Alignment" raises the integration with suppliers and customers through the use of technology are needed, in short the "mobility" would be charged in a short time to respond to changes in demand, "Adaptability" makes the relevant design and planning of the supply chain in accordance with market changes and the "Alignment" manages the relationship with suppliers and customers in order to improve the efficiency of the supply chain.

In the following graph is outlined the main edges to be considered in the implementation of a management model "Triple A", even though they are raised as independent principles in reality there is a common area between them as seen in the figure below.

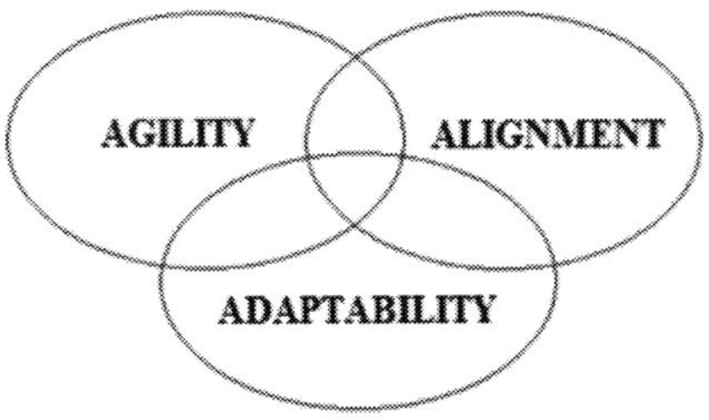

Figure 1. Management model "Triple A" proposed by HAU L. LEE.

3 Intelligent systems and its application to implement the model "Triple A"

Intelligent systems for the implementation of the management model "Triple A" in the supply chain to be used are the RNA's and Genetic Algorithms, where the RNA's are used as a predictor of demand in order to use the prediction to respond in short time for potential changes in demand ("Agility") and make appropriate modifications and adjustments in the supply chain in accordance with variations of the market ("Adaptability"); the genetic algorithm is used to route optimization in order to "align" the milk collection processes in order to get the optimal route for collection of the product from suppliers.

Although intelligent systems are powerful management tools, however, the real success in logistics management is in the skill and knowledge of the processes of all staff involved in the flow of the supply chain, combined with an adequate level of empowerment in making critical decisions in order to achieve economic, financial and business objectives.

3.1 Neural network

Among the various applications of neural networks that serve as support in logistics operations, has been very successful prediction of future demand based on historical behavior, this document uses an Artificial Neural Networks in forecast the milk sales.

A neural network is an infinite connection from neurons, in Figure 2 the relationship between a biological neuron and artificial neuron are shown.

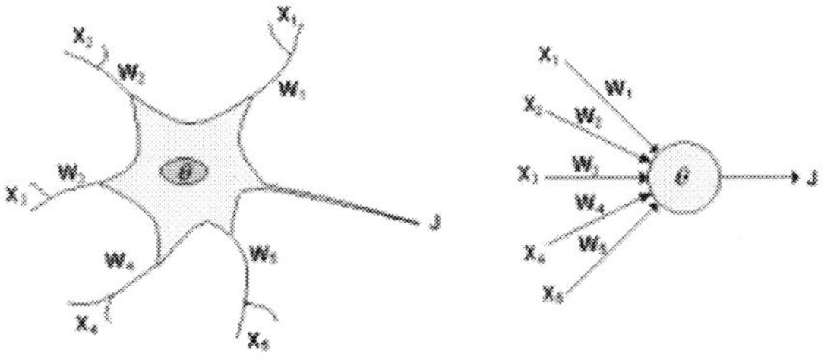

Figure 2. Biological Neuron and its mathematical representation [10].

Inside the human brain there are about a hundred billion (10^{11}) of neurons [11], in the case of artificial neural networks, are handled in the order of hundreds[a] and thousands, grouped by layers with an input layer, an output layer and one or more hidden layers, the connections are made between neurons of adjacent layers, in Figure 3, an example of a ANN shown as artificial neurons are connected.

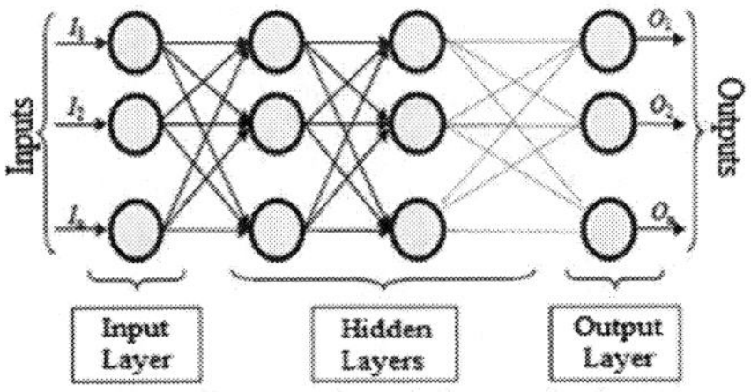

Figure 3. Artificial Neural Network (ANN) Example.

In the case of the implementation of the Model "Triple A" was done a prediction of liters of milk sold, with real information of collection center for a period of four years, from January 2008 until December 2012. Figure 4 shown movement of liters diaries in the collection center.

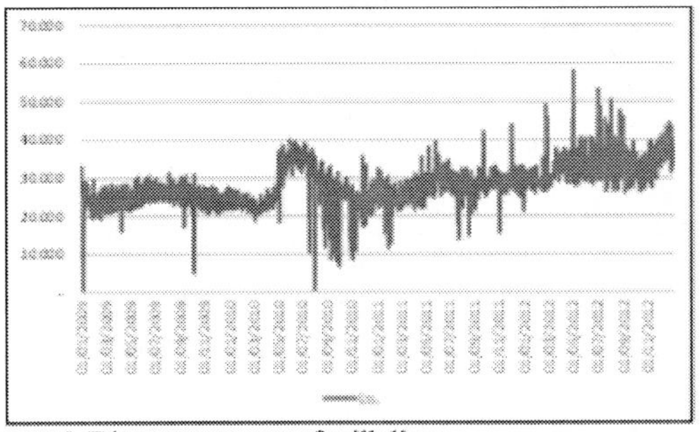

Figure 4. Diary movement of milk liters.

[a] IBM in 2015 conducted a simulation with Compass which was 65 billion neurons, but they are isolated cases and not used commonly in applications with Artificial Neural Networks, for more detail can review the document in http://www.modha.org/blog/SC12/SC2012_Compass.pdf

In order to improve the prediction of liters of milk, it must perform a pre-processing to remove noise, for which was used the moving average of thirty days before, with the formula.

$$\overline{X_k} = \frac{1}{30}\sum_{j=1}^{30} X_{k-j+1} \quad (1)$$

The following graph shows a comparison between the original values in blue and the average value of liters in red where you can see the elimination of fluctuations by the use of moving average.

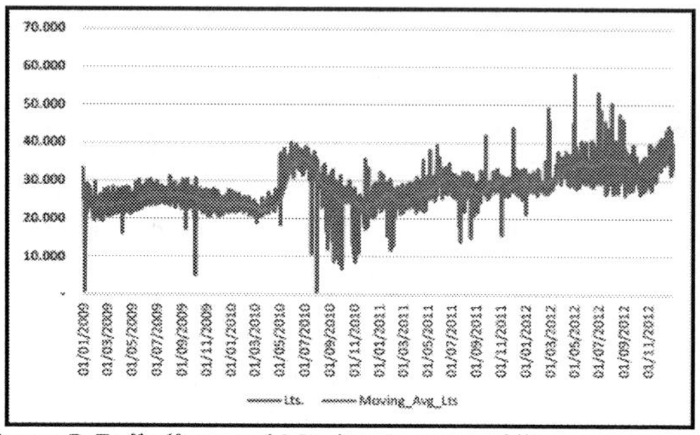

Figure 5. Daily liters and Moving Average of liters.

For the collection center may have an "agile" response, the time prediction required is a month as output, for this reason was used ninety-four inputs that are described in the Table 1.

According to what indicated in the previous paragraph, outputs would be the prediction of thirty days from the day zero to twenty-nine.

For training, 1371 pairs of input-output samples were used, of which 90% were used in the training and 10% for validation.

Table 1. Inputs for the Artificial Neural Network.

Inputs	Description
Year	year of the zero day
Month	Number of month of the zero day (1-January, 2-February …)
Num_week	Week number of the zero day
Weekday	Week day of the zero day (1 is Monday,. 7 is Sunday)
Lt_i	Liters of milk in the i-th day prior to the day zero, where i varies from one to ninety (three months).

3.2 A Genetic algorithms for optimizing routes

Genetic algorithms [12] perform the following steps:

1. Encode information in genes and chromosomes.
2. Define the adaptation function
3. Generate a base population
4. Evaluation and Selection
5. Reproduction
6. Mutation
7. Termination condition

In the seven steps, the genetic algorithm performs an iterative process until meet the termination condition in order to get the best possible solution for complex systems and nonlinear.

There are several models for the detection of optimal routes such as the proposed by Eslava and Quiroz [13]. In the case of route optimization analysis, the coding it is performed according to the distances matrix between collection points. For modelling were taken five points, which generated the matrix shown in Table 2.

Table 2. Distances in kilometers between the collection center (Point A) and collection points.

Km.	A	B	C	D	E
A		9.4	15.2	10.4	14.3
B	9.4		20	11.4	7.2
C	15.2	20		7.8	10.7
D	10.4	11.4	7.8		6.5
E	14.3	7.2	10.7	6.5	

Distances by road were taken from the site: http://maps.google.com as can be seen in the Figure 6 below.

According the trajectory which cover all points in Table 2, the information is encoded such as shown in Table 3, in order to separate it in genes and chromosomes, each section between two points was represented by one bit where if the section is used in the trajectory is "one" and if it not be used is "zero".

Figure 6. Measured distance between two points by road in Maps.Google.com.

Table 3. Coding the trajectory of collection center

AB	BC	CD	DE	EA	AC	BD	AD	BE	CE
B10	B9	B8	B7	B6	B5	B4	B3	B2	B1

Then it is defined the adaptation function to minimize by:

$$f(W) = g(W) - 0.1 \cdot h(w) \quad (2)$$

where:

$$g(W) = 1 - \frac{|R - \sum_{i=1}^{L} B_i|}{R} \quad (3)$$

DOI: 10.1051/matecconf/20166806008

$$h(W) = \frac{\sum_{i=1}^{L} B_i \cdot D_i}{\sum_{i=1}^{L} D_i} \quad (4)$$

R: is the number of waypoints
L: Number of bits used (L=10)
B_i: i-th bit of the byte "W"
D_i: Cost/distance of ith section, see table 2

With encoding and adaptive function, an script was executed in Matlab with Genetic Algorithms for the optimal route, a hundred runs were performed with the case information on the best analysis results are given in Table 4, where the journey begin and ends at E, following the sequence A-C-B-D-A-E. For more collection points only must recalculate the distances matrix and encoding of vector binary and thus it can be run in Matlab. To get the optimal route, other possible encodings are proposed by Bildhardt [14], but in the case of analysis the best result is raised in this document.

Table 4. Results of calculated optimal route by Genetic Algorithms.

AB	BC	CD	DE	EA	AC	BD	AD	BE	CE
0	1	0	0	1	1	1	1	0	0

4 Key indicators of management "Triple A"

Although the intelligent systems can serve for support to improving the supply chain management, however for the success of Management Model "Triple A" is necessary to monitor, measure and interpret information through indicators to manage the entire supply chain [15], in Table 5 the processes most common of average company are shown, for which are defined the subprocesses which exist in the collection center.

Table 5. Processes of Supply Chain.

Processes	Sub processes
Supplying	• Collection - Shopping • Storage • Commercial
Production	Does not apply to the collection center.
Dispatch	• Distribution to milk processors
Support Processes	• Preventive Maintenance • Human Talent Management • Accounting • Strategic management

Then, an Internal Factor Analysis Summary (IFAS) and External Factor Analysis Summary (EFAS) were done in order to make a SWOT analysis where the strength was used to overcome weaknesses and the opportunities to overcome threats, after that, the sub processes of the supply chain the indicators from Tables 6, 7 and 8 were proposed, which will be used to monitor and manage the supply chain in order to obtain "mobility", "Alignment" and "Adaptability" that is a management model "Triple A".

Table 6. Indicators for sub processes of Suppliers.

Sub-processes	Indicator
Collection-Shopping	• % Savings for optimal route selection by Genetic Algorithms • Earnings per collection route • Minimum number of liters collected per route • % non-compliance Delivery by supplier • Delay of deliver by supplier • Milk quality by supplier • Strategic alliances
Storage	• Using Warehouse • % Using cooling equipment
Commercial	• Level of deliveries on time • Increase in new customers

Table 7. Indicators for sub processes of Dispatch.

Sub processes	Indicators
Distribution to milk processors	• Earnings per trip • Accuracy of the Forecast of the Artificial

Table 8. Indicators for sub processes of Support.

Sub processes	Indicators
Preventive Maintenance	• Level of implementation of the maintenance plan • Stops by mechanical failure
Human Talent Management	• Level of knowledge of the processes of the supply chain • Penetration of the use of intelligent systems.
Accounting	• Increase Sales • Costs by downtime of machinery • Costs for non-compliant milk. • % of logistic costs • Percentage of saving in provisioning
Strategic management	• Degree of implementation of the strategy.

In the tables above, the core indicators for the implementation of the Management Model "Triple A" for the Collection Center are proposed. However, the proposed intelligent systems such as an Artificial Neural Network and Genetic Algorithms only would impact on two indicators, that is, although they are excellent systems the real impact depends on many other factors including both staff and the management of all the processes involved in the chain supply.

5 Conclusions

The supply chain covers the entire management chain from suppliers to the end customer. So an efficient supply chain management is vital importance for the success of a company, since it allows an overview of end to end across the enterprise, in the case of the Collection Center the chain starts from the milk collection to delivery to large processors so the implementation of Model management "Triple A" should cover the whole entire chain.

In the management model of triple A, the "agility" is responsible for respond in short time to the changes in demand. Instead the "alignment" manages the relationship between suppliers and end customers for this reason the most indicators are proposed in order to align the processes of supply chain. Finally the "adaptability" has adjustments in the design of the supply chain as to adapt to changes in the market. Indicators, Artificial Neural Network and Genetic Algorithms were designed to implement the Model Management "Triple A", however, they are useless tools if the company fails to implement in processes, or in to promote its application, nor has clearly defined its goals short and long term.

All strategic planning in their fundamentals must have an analysis of the supply chain in order to identify and define the competitive advantages and disadvantages of the chain to define the strategies appropriate for sustainable success of the business.

The success of intelligent systems is based on it can be implement in these systems, the criteria of all knowledge and experience of experts for optimal solution to solve problems in a company.

In summary, the decision-making in the supply chain management is a key factor in the success of businesses, incorrect decisions can generate losses that could lead to bankruptcy, the question arises: is there any way can provide support in logistics decisions and minimize losses ?, the proposal herein is management model "Triple A" with the aid of the use of intelligent systems such as tools to support decision-making, that enable the company to "Agile", "Align" and "Adapt" its supply chain to market needs, the company and customers and get increase their economic returns, however they are not enough at all, the contribution of strategic planning accompanied by the training of all personnel involved in the supply chain is always necessary.

References

1. M. T. Hagan, H. B. Demuth, M. H. Beale and O. De Jesús, *Neural Network Design* (M. Hagan, Oklahoma, 2014)
2. M. Mitchel, *An Introduction to Genetic Algorithms Complex Adaptive System* (Bradford, 1998)
3. H. L. Lee and Y. Lee, *Building Supply Chain Excellence in Emerging Economies – International Series in Operations Research & Management Science* (Springer, Washington, 2006)
4. S. L. Goldman, R. N. Nagel and K. Preiss, *Agile Competitors and Virtual Organizations: Strategies for enriching the Customer* (John Wiley & Sons Inc, 1994)
5. Y. Monden, *Toyota Production System: an integrated approach to Just-In-Time* (Productivity Press, Oregon, 2011)
6. E. Rodriguez, I. Carreras and M. Sureda, *Innovar para el cambio Social de la idea a la acción* (El Tinter, Madrid, 2012)
7. J. Naylor, M. Naim and D. Berry, Leagility: integrating the lean and agile manufacturing paradigms in the total, *International Journal of Production Economics*, **62**, 107 (1999)
8. R. Stratton and R. D. H. Warburton, The strategic integration of agile and lean supply, *International Journal of Production Economics*, **85**, 183 (2003)
9. H.L. Lee, The triple A supply chain, *Harvard Business Review Online*, Available: https://hbr.org/2004/10/the-triple-a-supply-chain (Last access: December 12,2015)
10. A. González, Modelo para la predicción de la radiación solar a partir de redes neuronales artificiales, *Escuela de Ingeniería en Antioquia* Online, Available: http://repository.eia.edu.co/bitstream/11190/326/1/MECA0108.pdf (Last access: December 12, 2015)
11. M. F. Bear, B. Connors and M. Paradiso, *Neurociencia: explorando el cerebro* (Masson, Barcelona, 2002)
12. P. Ponce Cruz, *Inteligencia Artificial con aplicaciones en ingeniería* (Alfaomega, México, 2010)
13. F. Eslava and M. L. Quiroz, *Computación emergente en la detección de rutas de distribución óptimas* (EAE, Caracas, 2012)
14. R. H. Ballou, *Inteligencia Artificial – Ejercicios Resueltos* (Universitaria Ramón Areces, Madrid, 2015)
15. R. H. Ballou, *Logística, Administración de la cadena de suministro* (Pearson, México, 2004)

Determining the Optimal Parameter of Coordinate Measuring Machine with Design of Experiment

Kusalin Sangnuan and Wimalin S. Laosiritaworn

Department of Industrial Engineering, Faculty of Engineering, Chiang Mai University, Thailand

Abstract. Coordinate measuring machine (CMM) is a device used to measure physical geometry of object. It has been used widely in industry for quality control and monitoring of production process due to its high accuracy and precision. Measurement accuracy is very crucial in quality inspection as measurement result is used to decide if the product should become defect. CMM accuracy depends on several factors such as probe size, number of touching point and measurement speed. This research applied design of experiment technique to optimize setting of those parameters to obtain high measurement accuracy. A case study from swage hole measurement of hard disk drive actuator arm was use to demonstrate the proposed technique. The result showed that CMM accuracy of the case study company was significantly increased after CMM has been set with parameters optimized with design of experiment.

1 Introduction

Measurement systems is very crucial in controlling product quality. Coordinate measuring machine (CMM) is a device used to measure geometry characteristic of an object by moving a measuring probe to contact points on object surface. It is widely used in most manufacturing factory as it can be used to measure almost any kind of object 1. CMM has been an important inspection tool in quality control due to its high accuracy and precision 2. It is the main measurement instrument used to measure object geometry in the case study company who is a manufacturer of actuator arm for hard disk drive component (Fig. 1a). Actuator arm is a part of hard disk that read/write arm is attached. It is used to move the head arm to different track on the surface of the platter. One of the most important part in actuator arms is swage hole (Fig. 1b). Swage hole attach head gimbals assembly (HGA) with carriage. In assembly process, swage ball is pressed into swage hole with swage pressing machine. If the swage hole is too big, HGA will not able to hold on to carriage. On the other hand, if the swage hole is too small, it will not be possible to insert HGA into swage hole.

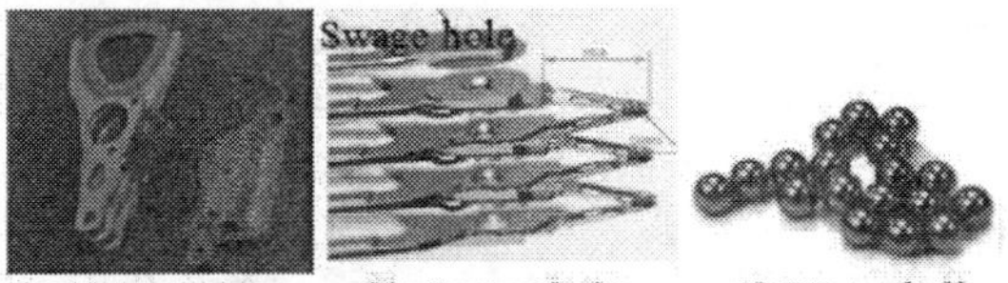

a) Actuator arm b) Swage hole c) Swage ball

Figure 1. Component of hard disk drive actuator

Design of experiment (DoE) is a "process of planning the experiment so that appropriate data that can be analyzed by statistical methods will be collected, resulting in valid and objective conclusions" 3. It can be used to statistically investigate the effect of independent factors on dependent factors. DoE has been applied successfully to optimize actuator productions for example design of a small linear actuator 4, actuator coil baking 5.

This paper investigated the effect of CMM measurement setting on the swage hole measurement accuracy. The objective of this research is to find the optimum setting that provide swage hole geometry as close as possible to reamer tool used to manufacture that hole.

2 Research methodology

This research starts with the review of literature about the factors affecting CMM measurement accuracy. After preliminary analysis three factors namely probe size, number of measurement point, measurement speed were selected. Length of probe was fixed at the case study company therefore was not included in the experiment.

The level setting of these parameters. For factor A, the probe size, as there are only two size of probe , Probe size level was set at 0.5 millimeter for low level and 1 millimeter for high level. Factor B is number of measurement point. It was set at 8 points for low level and 16 points for high level. Finally factor C, the measurement speed. This factor was set to the lowest and highest possible setting of the machine, which was low level at 1 millimeter/sec and high level at 8 millimeter/sec.

2^3 factorial design was used to generate experimental setting. Minitab software was used for design and analysis of experiments in this research. Sample of actuator arm was prepared with the same reamer to block the effect of tooling variation. Diameter and roundness of this reamer was measured with smart scope which has accuracy of 0.0005 millimeter. This standard reamer has 2.668 millimeter in diameter and 0.001 millimeter roundness. This reamer was installed into CNC machine and used to drilled numbers of swage hole for each experimental setting. We set three sample were taken from the group of actuation that has flatness, roll and pitch at medium, higher and lower than specification. Measurement by CMM for each sample was repeated 5 times.

Two responses were consider in this experiment. The first response was diameter deviation which was calculated from the difference between actual reamer diameter (2.668 millimeter) and the diameter measurement from CMM. Second response was roundness deviation which was the difference between actual reamer roundness (0.001 millimeter) and roundness measured from CMM.

3 Results and discussions

2^3factorial design requires 8 experimental run. For each experimental setting, 3 actuator arms were sampled. Measurement of diameter and roundness were repeated five times.

3.1 Residual analysis

ANOVA analysis has basic assumption that the observations are adequately described by the model, the errors are normally and independently distributed with mean zero and constant but unknown variance 3. To test those assumption, residual analysis was performed. Normal probability plot for both diameter deviation and roundness deviation form straight lines which suggested that the residuals are normally distributed. The residuals versus fitted values for both diameter deviation and roundness deviation do not has particular trend. This is the same as residual versus the order of the data graph, which suggested that residuals were independent from observation order. After all assumptions were checked, ANOVA can be used for data analysis.

3.2 Analysis of variance (ANOVA)

Table 1 shows analysis of variance result of diameter deviation. Blocks, main effects, and 2-way interactions have significant effect on diameter deviation at 95% confidence as they have P value less than 0.05.

Table 2 shows estimated effect of coefficients for diameter deviation. Factor A, C and AC interaction has significant effect on diameter deviation at 95% confident as they have P value less than 0.05.

Regression equation was fitted with significant terms from Table 2 (factor A, C and AC interaction). The equation to predict Y_1 (diameter deviation) is as follow;

$$Y_1 = 0.00415 + 0.002033A + 0.001567C + 0.000217AC$$

Table 1. Analysis of variance for diameter deviation

Source	DF	Seq SS	Adj SS	Adj MS	F	P
Blocks	2	2.32E-05	2.32E-05	1.16E-05	71.6	0.000*
Main Effects	3	0.000791	0.000791	0.000264	1630.24	0.000*
2-Way Interactions	3	5.7E-06	5.7E-06	1.9E-06	11.75	0.000*
3-Way Interactions	1	0.000	0.000	0.000	*	*
Residual Error	110	1.78E-05	1.78E-05	1.6E-07		
Lack of Fit	14	4.98E-06	4.98E-06	3.6E-07	2.67	0.002
Pure Error	96	1.28E-05	1.28E-05	1.3E-07		
Total	119	0.000837				

Table 2. Estimated effects of coefficients for diameter deviation.

Term	Effect	Coef	SE Coef	T	P
Constant		0.00415	0.000037	113.07	0.000
Block 1		-0.000025	0.000052	-0.48	0.631
Block 2		0.00055	0.000052	10.60	0.000*
A	0.004067	0.002033	0.000037	55.40	0.000*
B	0.000000	0.000000	0.000037	0.00	1.000
C	0.003133	0.001567	0.000037	42.68	0.000*
A*B	0.000033	0.000017	0.000037	0.45	0.651
A*C	0.000433	0.000217	0.000037	5.90	0.000
B*C	0.000033	0.000017	0.000037	0.45	0.651

DOI: 10.1051/matecconf/20166806009

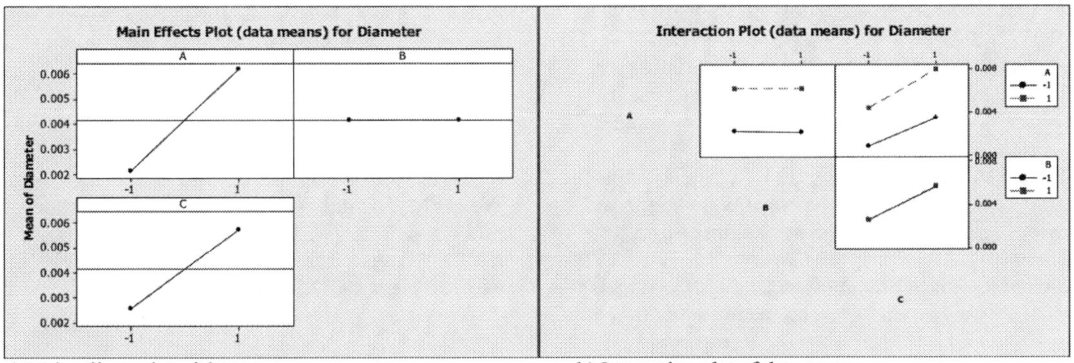

a) Main effects plot of data mean b) Interaction plot of data mean

Figure 2. Main effects plot and interaction plot of data means for diameter deviation response

Fig. 2a) is the main effects plot that show the different between means of each level for every factors. As factor B is not significant, it shows almost no different in mean between low and high level (the plot is horizontal). On the other hand, factor A and C are significant as the line is not horizontal. Smaller mean deviation is more desirable, so factor A and C should be set at low level.

Fig. 2b) is the interaction plot of data mean represents the interaction between a pair of factors. For example, interaction between factor A and B is calculated from the average over the remaining parameter. AB and BC interaction do not show significant interaction as both line are in parallel. AC interaction, on the other hand, is significant and the optimum setting of AC is at low level for both factor A and factor C.

The same procedures were repeated for the second response (roundness deviation). Residual analysis showed that the residual are normally distributed with no particular trend in both residual versus fitted values and residual versus the order of the data, therefore, passed ANOVA assumption. ANOVA results suggested that AC interaction is significant to roundness deviation at 95% confidence. The equation to predict Y_2 (roundness deviation) is as follows,

$$Y_2 = 0.001292 + 0.000075A + 0.000025C - 0.00013AC$$

Response optimizer in Minitab was used to find the optimum setting. It was found that the best setting is at diameter of 0.5 millimeter, 16 measurement point with measurement speed of 1 millimeter/sec. 15 confirmation experiments were conducted using the optimum setting. The average response for both diameter deviation and roundness were 0.001 millimeter which is considered very accurate.

4 Literature References

There has been a number of reports on improving CMM efficiency. For example, Miller et al. 6 studied the effect of stylus diameter used for probing in CMM in order to obtain the effective stylus diameter. Mian *et al.* 2 reduced measurement cycle time by generating the most suitable measurement path. A number of factors that could have impact on CMM measurement accuracy can be summarized as follows;

Probe size. Rajamohan *et al.* [7]. studied effect of probe size on measuring free form profile and conclude that smaller probe size provide more accuracy. This result is consistent with Ayub *et al.* 8 where roundness tolerance was investigated.

Number of measurement point. Measurement error is high if small number of measurement point were taken. This is due to the lack of stability in the approximation area of the interpolation methods 8.

Measurement speed. Measurement speed is the time taken for measurement probe to touch the object. With fast measurement speed, the probe might touch the object harder which could leads to higher measurement error [8].

Length of probe. Probe errors usually increase with the length of the stylus 9.

5 Summary

This paper present an improvement of CMM accuracy with design of experiment technique. Factors that could have impact on CMM measurement was surveyed in literature. Three factors adopted were probe size, number of measurement point and measurement speed. It was found that number of measurement point has no significant effect on accuracy while probe size and measurement speed are significant. Smaller probe size (0.5 millimeter) provide significant improve in measurement accuracy. This result is in consistent with Rajamohan *et al.* [7]. and Ayub et al. 8. Lower measurement speed (1 millimeter/sec) provides better accuracy. These new setting results in both diameter and roundness deviation of 0.001 millimeter which is considered significant improvement in this case.

References

1. J. Ni and F. Waldele, Coordinate measuring machines, in: J. A. Bosch (Ed), Coordinate measuring machines and system, Marcel Dekker Inc., New York (1995)
2. Mian, S.H. and A. Al-Ahmari, Enhance performance of inspection process on Coordinate Measuring Machine. Measurement, **4**,78-917, (2014)
3. D.C. Montgomery, Design and Analysis of Experiments. John Wiley & Sons, New York, (2005)
4. D.J. Lee, K.S. Wo, N.C. Park, and Y.P. Park. Design and optimization of a linear actuator for subminiature

optical storage device. IEEE Trans of Magnatics. **41**, 1055-1057, (2005)
5. Punnarungsri, P. and W. Laosiritaworn. Optimization of coil baking parameters using design of experiments. in Lecture Notes in Engineering and Computer Science. 933-937, (2013)
6. J. Miller, S. Dutta, E. Morse, and J. Yague-Fabra, Effective stylus diameter determination using near zero-width reference, Precision Engineering, **35**, 500-504, (2011)
7. Rajamohan, G., M.S. Shunmugam, and G.L. Samuel, Effect of probe size and measurement strategies on assessment of freeform profile deviations using coordinate measuring machine. Measurement, **4**, 4832-841, (2011)
8. Ayub, M.A., A.B. Mohamed, and A.H. Esa, Experimental Investigation on the Measurement of Roundness Tolerance. European Journal of Scientific Research, **65**, 304-311, (2011)
9. Wozniak, A. and M. Dobosz, Factors influencing probing accuracy of a coordinate measuring machine., IEEE Transactions on Instrumentation and Measurement, **54**: 2540-2548, (2005)

MATEC Web of Conferences **68**, 07001 (2016) DOI: 10.1051/matecconf/20166807001
ICIEA 2016

A Numerical Method Based on the Range-Discrete Grid for One-Dimensional Buckley-Leverett Equation

Zhiwei Cao, Zhifeng Liu and Xiaohong Wang

University of Science and Technology of China, Department of Thermal Science and Energy Engineering, Hefei, Anhui, P. R. China

Abstract. A numerical method for solving the one-dimensional Buckley-Leverett equation arising in the process of displacement of oil by water is presented. Instead of using the traditional spatial discrete grids, the numerical algorithm is built on a "range discrete" grid, which is obtained by direct subdivisions in the saturation domain. The range discrete grid describes the discontinuities explicitly and completely, and has an adaptive nature in smooth regions. Grid points are divided into two classes: continuous points and discontinuous points. Numerical solution of the Buckley-Leverett equation is achieved by moving continuous points by tracing characteristics and moving discontinuous points by tracking discontinuities. Numerical examples are presented, and the solutions obtained by the proposed method are found of high precision. Especially, shocks are solved with no dissipation, and the sharpness is maintained.

1 Introduction

The displacement of oil by water in porous media can be described by the Buckley-Leverett (B-L) theory [1], [2], which is an important theory for underground water follow and oil recovery. When capillary effect and gravity is ignored, the B-L equation becomes

$$\frac{\partial S}{\partial t}+\frac{\partial f(S)}{\partial x}=0 \tag{1}$$

where S is water saturation and $f(S)$ is the fractional-flow function, which is typically S-shaped. Eq. (1) is a nonlinear hyperbolic equation and discontinuities may appear in finite time even the initial condition is smooth enough [3], [4]. When discontinuities appear, numerical methods based on traditional spatial discrete grid, i.e. Finite Difference Method (FDM), suffer either massive numerical dissipation or unphysical oscillations [5]-[10]. This is because that the spatial discrete grid describes a discontinuity within a finite-small cell, which, with no doubt, causes numerical problems.

Here, we proposed a numerical method that maintains sharp discontinuities for the one-dimensional B-L equation by introducing a uniformly distributed "range-discrete" grid. The range-discrete grid shows an adaptive nature in smooth regions. Moreover, discontinuities can be solved with no diffusion. The proposed method combines the method of characteristic and shock tracking method. The numerical method is substantially a moving mesh method and thus abandons the use of traditional spatial discrete grid.

2 The range-discrete grid

Considering that $S\in[0,1]$ in the B-L equation, a series of discrete saturation values can be obtained by a direct subdivision in the saturation domain, that is $S_i\,(i=0,1,2,\cdots,m)$, where $S_0=0, S_m=1, S_{i+1}=S_i+\Delta S_i$. Next, as shown in Fig. 1, intersections $A_j\,(j=0,1,2,\cdots,n)$ of the line sets $S=S_i$ and the curve $S=S(x)$ form a discrete grid by employing each intersection as a grid point.

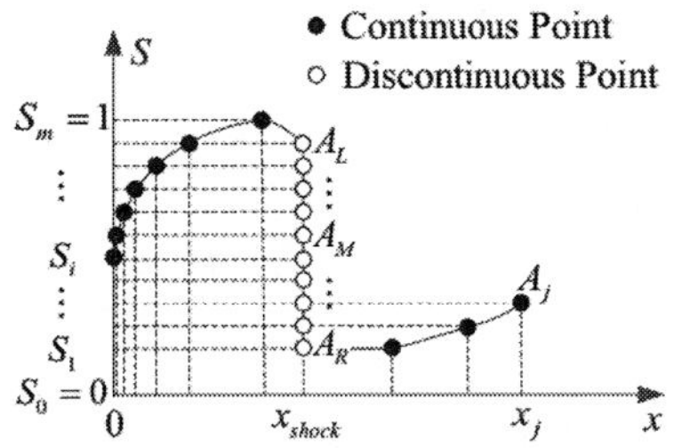

Figure 1. Description of the range-discrete grid. The solid dots represent continuous points and hollow dots are discontinuous points and the same below.

As shown in Fig. 1, a set of intersection points $A_L, A_M, \cdots, A_R$ exist at a discontinuity. These points are treated as different grid points while sharing the same spatial coordinate value, which is $x_L=x_M=\cdots=x_R=x_{shock}$.

Using the rules explained above, a new discrete grid for a general saturation field is constructed. The saturation field is numerically approximated when x_j, the spatial coordinate of the grid points $A_j\left(j=0,1,2,\cdots,n\right)$, with each assigned to a discrete saturation value S_i, are known. To differ from the spatially discretized Eulerian grid, we call this grid system a "range-discrete" grid.

Grid points in the range-discrete grid go into two classes intuitively: points in smooth regions are continuous points; while points with a same spatial coordinate are marked as discontinuous points to represent a discontinuity. In this way, the traditional way that describes a discontinuity within limited small space region is abandoned, so the sharpness will be maintained. Furthermore, grid points will gather where saturation various quickly in space and vice versa, which is analogous to the adaptive mesh refinement strategy [11], [12]. While complex refinement strategies have to be processed in these methods, the range-discrete grid is born with an adaptive nature.

3 Numerical method based on the range-discrete grid

Using the technique illustrated in section 2, the saturation field at $t=0$ can be approximated by a range-discrete grid. The movements of the grid points A_j can be tracked for a small time step Δt. Once each $x_j^{t+\Delta t}$ calculated, approximation of the saturation field at $t+\Delta t$ is obtained. The B-L equation can be solved by repeating this. One should notice that, grid points move with their saturation values unchanged, which is the main difference from the numerical methods based on spatial discrete grids.

In each time step, movements of the grid points can be established explicitly and with little computation cost. However, this may result in multi-valued curves in space due to the nonlinear hyperbolic nature of the B-L equation, so that a correction procedure has to be introduced. Consequently, there are two steps involved in each time step. The first is a prediction step to move these points explicitly and the second is a correction step to solve the multi-value problem.

3.1. The prediction step

If marked as a continuous point at t, point A_j will remain to be a continuous point and can be traced along its characteristic line, giving $v_j=f'\left(S_j\right)$.

As a set of discontinuous points represents a discontinuity, their movements are determined by the evolution of this discontinuity, which can be established by entropy condition [4]. That is, the points in the part that will change into a shock wave remain to be discontinuous points, moving at the velocity of the shock v_{shock} achieved from Rankine-Hugoniot condition; the points in the part that will change into a rarefaction wave will become continuous points, moving along its characteristic line. Altogether, the spatial coordinates of points A_j at $t+\Delta t$ are

$$x_j^{t+\Delta t}=x_j^t+v_j\Delta t \tag{2}$$

with $v_j=\begin{cases} v_{shock} & \text{if } A_j \text{ is a discontinuous point after } t \\ f'\left(S_j\right) & \text{if } A_j \text{ is a continuous point after } t \end{cases}$.

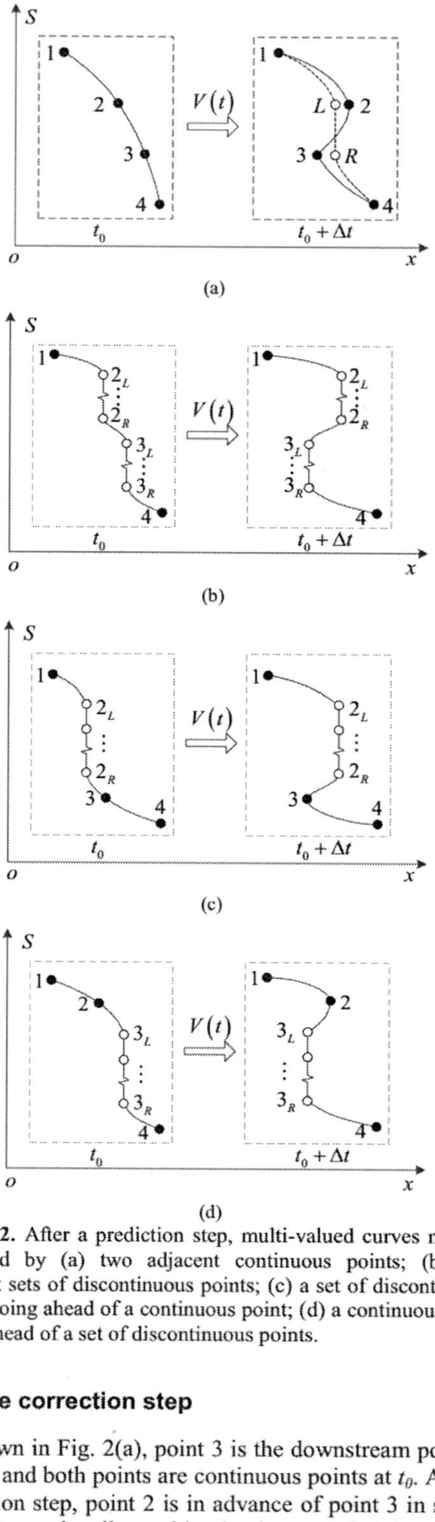

Figure 2. After a prediction step, multi-valued curves may be produced by (a) two adjacent continuous points; (b) two adjacent sets of discontinuous points; (c) a set of discontinuous points going ahead of a continuous point; (d) a continuous point going ahead of a set of discontinuous points.

3.2 The correction step

As shown in Fig. 2(a), point 3 is the downstream point to point 3 and both points are continuous points at t_0. After a prediction step, point 2 is in advance of point 3 in space, producing a locally multi-valued curve that has to be corrected. This is the result of the nonlinear hyperbolic nature of the B-L equation, which is physically impossible. The real saturation curve is spatially discontinuous [1], [13]. Considering that $f(S)$ typically have inflection points, we combine the material balance method suggested by Buckley and Leverett with entropy condition to solve the multi-value problem after the prediction step.

3.2.1 Correction procedure of two continuous points

As a numerical method, it is quite straightforward to mark the two points in Fig. 2(a) as discontinuous points with their spatial coordinates assigned to a same value x_s, representing the discontinuity that will be introduced to correct the multi-valued portion to a locally single-valued one, as shown in dash in Fig. 2(a).

We should mention that only discontinuities that satisfy entropy condition can be produced in this correction step, so that an entropy checking is needed. It is easy to prove that if one continuous point goes ahead another one, that is point 2 advances faster than point 3 in this situtation, the entropy condition is always checked. So, these two points are admitted to form a new discontinuity here.

After this, x_s can be calculated using the material balance principle:

$$\int_{S_1}^{S_L} x(S)dS + x_s(S_R - S_L) + \int_{S_R}^{S_4} x(S)dS = \sum_{i=1}^{4}\int_{S_i}^{S_{i+1}} x^*(S)dS \quad (3)$$

where x^* and x represent the spatial coordinates before and after the correction procedure. Eq. (4) maintains the conservation of mass in the correction step. The integral in Eq. (4) can be calculated by linear interpolation or a higher resolution can be expected with the interpolation introduced by Forjoun etc. [14].

$$x(S) = x_a + \frac{f'(S) - f'(S_a)}{f'(S_b) - f'(S_a)}(x_b - x_a) \quad (4)$$

where the subscript a and b represent two adjacent points. Thus, with the position of the new discontinuity calculated, the unphysical multi-value problem is solved.

3.2.2 Correction procedures when discontinuous points involved

After the prediction step, the multi-valued problem happens not only between continuous points(see Fig. 2(a)), but also exits when discontinuous points occurred (see Fig. 2(b) , 2(c) and 2(d)). Actually, as a set of discontinuous points represents a single discontinuity, it is reasonable to treat the set of discontinuous points as a whole here. If we do so, things in Fig. 2(b), 2(c) and 2(d) become exactly the same as the situation in Fig. 2(a). Consequently, the correction procedures when discontinuous points involved are actually the same as the procedure illustrated in 3.2.1, with only one more thing to notice.

That is, if we mark all the points in the multi-valued portion of the curve as discontinuous points, it may produce a discontinuity that does not satisfy the entropy condition. So, from the most upstream point in the multi-valued portion to the most downstream point, an entropy checking is processing after marking them as discontinuous points: take the discrete saturation values of each point and the most downstream point in the multi-valued portion as the upstream and downstream value of a discontinuity respectively. If this discontinuity does not satisfy the entropy condition, this point will be marked as a continuous point until an entropy satisfied discontinuity occurs. In this way, the resulting discontinuity is enforced to satisfy the entropy condition.

If the time step Δt is small enough, the multi-value problem only exists between adjacent points so that all possible situations are shown in Fig. 2. So, with the technique illustrated above, the multi-valued curve after the prediction step can be corrected to a single-valued saturation curve, giving the approximation of the saturation curve at $t_0 + \Delta t$. With mass balance principle and entropy condition applied explicitly in the correction step, the solution achieved is thought to be mass conserved and satisfy the entropy condition.

4 Numerical Examples

4.1. The appearance of a shock

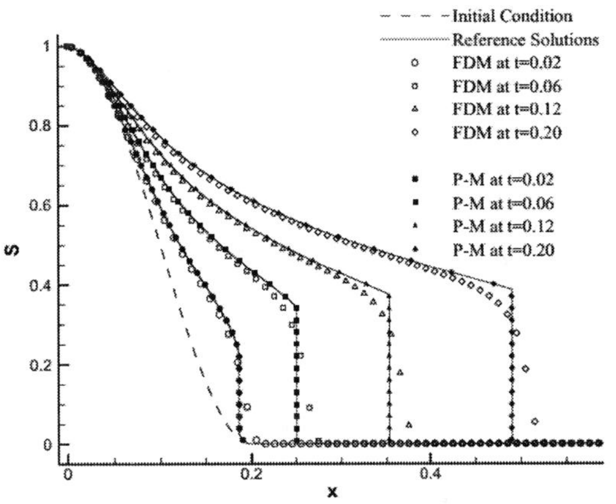

Figure 3. Solutions at different time from a continuous initial condition. (P-M is short for the proposed method.)

Here, the fractional-flow function is set as $f(S) = (S^2)/(S^2 + 5(1-S)^2)$. The initial condition is set to with no discontinuities as

$$S(x,t=0) = \begin{cases} 1 & x \le 0 \\ 250x^3 - 75x^2 + 1 & 0 \le x \le 0.2 \\ 0 & x \ge 1 \end{cases}.$$

The solutions obtained by FDM and the proposed method at different time are shown in Fig. 3, where the reference solution is obtained by the FDM with a refined grid. As we can see, the solution develops a moving shock. The shock is solved by the proposed method with no diffusion. Using equal number of grid points, first order FDM gives almost a smooth solution. We should mention that, as no linear equations has to be solved, the proposed method cost even much less computation effort than the first order FDM.

4.2. Two initial discontinuities

In this example, $f(S)$ is designed to be a third order polynomial as $f(S)=3S^2-2S^3$. The initial saturation field consists of two discontinuities given by

$$S(x,t=0)=\begin{cases}1 & 0\le x\le 0.2\\ 0 & \text{elsewhere}\end{cases}$$

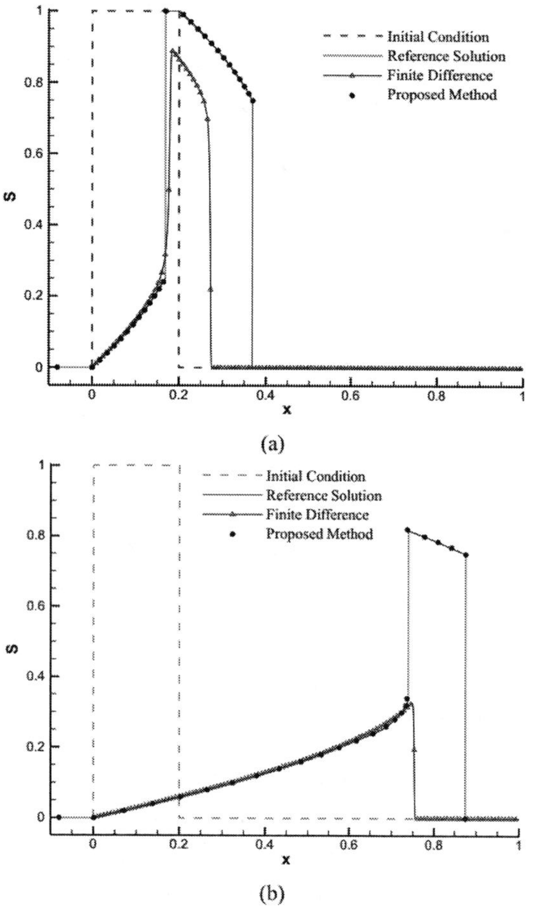

Figure 4. Solutions at different time from the two initial discontinuities. (a) t=0.15 (b) t=0.6. Grid points of the proposed method at a discontinuity is represented by two grid points at both ends for simplicity.

At $t>0$, each of these two discontinuities will change into a shock followed by a rarefaction and will interact with each other. The solutions at t=0.15 and t=0.6 are shown in Fig. 4. While the FDM gives rather wrong answers, the proposed method solves the position and strength of the shocks accurately with even fewer discrete points, with sharpness of the shocks kept and grid points located adaptively.

5 Conclusions

A range-discrete grid system and a numerical method based on the grid for the one-dimensional nonlinear hyperbolic B-L equation is presented. The numerical method solves the discontinuities sharply, and has an adaptive nature in smooth regions. With mass balance principle and the entropy condition employed explicitly, the solution is mass conserved and satisfies the entropy condition.

Our future work is to expand this method to higher dimensions, as it is straightforward that, in two dimensional problems, the range-discrete grid becomes contour lines in the saturation map.

References

1. S.E. Buckley, M.C. Leverett, Trans. AIME. **146**, 107–116 (1942).
2. J.W. Sheldon, B. Zondek, W.T. Cardwell Jr., Pet. Trans. AIME. **216**, 290–296 (1959).
3. P.D. LAX, *Hyperbolic Systems of Conservation Laws and the Mathematical Theory of Shock Waves*, SIAM, Philadelphia, (1973).
4. A. Bressan, Rev. Mat. Complut. **12**, 135–198 (1999).
5. R.J. LeVeque, *Finite volume methods for hyperbolic problems*, Cambridge University Press, (2002).
6. M. Allen, G. Pinder, Soc. Pet. Eng. J. **25**, 363-370 (1985).
7. Z. Xu, C.W. Shu, J. Comput. Phys. **205**, 458–485 (2005).
8. X. Zhang, C.W. Shu, J. Comput. Phys. **229**, 3091–3120 (2010).
9. X.D. Liu, S. Osher, T. Chan, J. Comput. Phys. **115**, 200–212 (1994).
10. C.W. Shu, SIAM Rev. **51**, 82–126 (2009).
11. X. H. Wang, M. Quintard, G. Darche, Numer. Heat Transf. Part B Fundam. **50**, 231–268 (2006).
12. W. Huang, R.D. Russel, *Adaptive Moving Mesh Methods*, Springer Science & Business Media, (2010).
13. Y. Ha, Y.-J. Kim, Q. Appl. Math. **73**, 661–678 (2015).
14. Y. Farjoun, B. Seibold, J. Eng. Math. **66**, 237-251 (2010).

MATEC Web of Conferences **68**, 07002 (2016) DOI: 10.1051/matecconf/20166807002
ICIEA 2016

Simulation on Heavy Oil Production from Steam-Flooding

Suthon Srochviksit and Kreangkrai Maneeintr

Carbon Capture, Storage and Utilization Research Laboratory, Department of Mining and Petroleum Engineering, Faculty of Engineering, Chulalongkorn University, 10330 Bangkok, Thailand

Abstract. Heavy-oil is one of the main energy sources in the future. However, with the high viscosity of heavy oil, steam-flooding (SF) is one of the major techniques to produce this oil. To determine a favorable operating condition, many recent studies have evaluated the different methods like varying the well spacing and steam injection rates. However, there is no single optimum value for all reservoirs or all modes of operation. In this study, it's aimed to investigate the effects of well spacing, injection rate, and perforation on heavy oil production with low permeability in multi-layered heterogeneous reservoir. The practical field data was applied to simulate the SF process by using STARS, a CMG program. Production parameters such as injector – producer (inj. – prod.) distance from 141 m to 282 m and steam injection rates from 30 m3/d to 180 m3/d were evaluated to obtain the optimum conditions with 80 % steam quality. The entire project was considered for 20 years of production. From the results, it showed that the higher oil recovery can be achieved by increasing injection rate and/or shorten well spacing. In order to lower the steam consumption, the selective perforation in bottom layers was compared with full perforation strategy. With the balance between oil production and steam injection, about 60 % of steam can be saved in terms of water barrels (bbls). Meanwhile, it's possible to achieve a longer project life due to the slow depletion rate.

1 Introduction

Steam injection is one of the most successful enhanced oil recovery techniques for heavy oil production. However, the drawback of heavy oil is the high viscosity thus making it difficult and expensive to produce. To overcome this problem, steam-flooding (SF) is applied to produce this oil [1]. Injected steam heats the formation around the well-bore to reduce the oil viscosity and recover towards the producing wells.

In terms of the production effectiveness, steam must be injected at sufficient rates [2]. Moreover, a proper flood pattern can provide good sweep efficiency and more producers to achieve high rates. Chu [8] compared the 5-spot pattern with inverted 7-, 9-, and 13-spot patterns. He found that the oil recovery in a 5-spot (normal or inverted) is the best based on the assumptions that the drainage areas of all producers are the same and that the steam rate is proportional to the pattern size. Therefore, inverted 5-spot pattern was also applied in this study (Fig. 1).

Steam-Oil Ratio (SOR) is a parameter used to monitor the efficiency of oil production with respect to steam injection. The lower SOR, the more efficient steam is utilized. However, maximum oil recovery or high recovery factor (RF) is also favorable to be considered as one of the decision criteria. Although the heat balance model demonstrated that oil/steam ratios are improved with increased steam quality [3], the steam injection rate is a significant parameter to affect the performance of team injection. In general, higher oil recovery can be achieved by higher injection rate, but it is required more energy consumption for steam generation. Also, well spacing is another critical consideration affecting the efficiency and the number of well drills [4]. The more wells drilled, the more fuel energy consumption and hence the higher project cost could be accounted. Also, the heat transfer would be poor due to heat loss if the distance between injectors and producers is far away [5]. Apart from that, the strategy of selective perforation was tested and compared in the modeling studies.

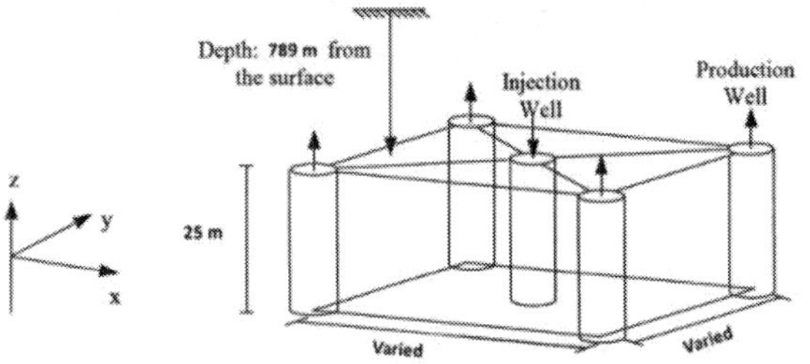

Figure 1. Schematic diagram of inverted 5-spot pattern.

Due to the reservoir heterogeneity, Hong [6, 7] concluded that no single value of steam injection rate or quality can be optimum for all reservoirs or all modes of operation. As such, the same set of values should not be used in all reservoirs. Therefore, in this nameless heterogeneous heavy oil reservoir, a hypothetical study has been performed in order to evaluate the favorable SF conditions such as well spacing, injection rate, and number of perforations based on this particular field data by using simulation model.

2 Simulation

2.1 Reservoir model

All simulation studies were performed by CMG simulator. Multi-layered heterogeneous reservoir was constructed from the actual field data as shown in Table 1. By calculating the average permeability and porosity from the low permeability zone of the field, totally 5 interested layers are selected for this study. Depending on the well spacing, the grid dimension in 5 x 5 x 5, 7 x 7 x 5, and 9 x 9 x 5 were created for 3 different injector-producer distance, 141.4 m, 212 m, and 282.8 m, respectively. Inverted 5-spot pattern with 80 % steam quality was used in this study. These three different well spacings were evaluated by considering the production efficiency and RF with various injection rates. The simulation was run by four different injection rates at 30 m^3/d, 60 m^3/d, 120 m^3/d, and 180 m^3/d with those three well spacing cases separately. Initially, full perforation of 5 layers was run in these twelve cases. In order to compare the results consistently and dynamically with the strategy of selective perforation, the same sets of cases (another twelve cases) were simulated by opening the layer 4 and 5 only, where the permeability and porosity are relatively higher than other layers. The reservoir model is illustrated in Fig. 2.

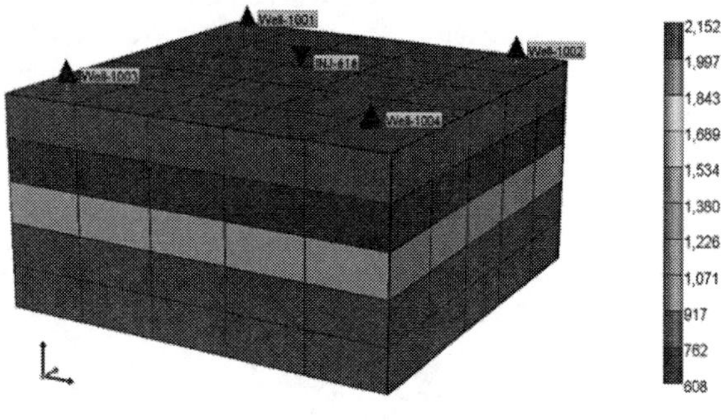

Figure 2. Reservoir model simulation: Permeability, model configuration, and wells location.

Table 1. Operating parameters for simulation

Operating Conditions	Values
Grid dimension (block)	(5 × 5 × 5); (7 x 7 x 5); (9 x 9 x 5)
Grid size (m)	50 × 50 × 5
Top of reservoir (m)	789
Depth of well (m)	25
Effective porosity (%)	26 - 31
Horizontal permeability (md)	608 – 2152
Vertical permeability (md)	608 – 2152
Initial oil saturation (%)	80
Initial water saturation (%)	20
Reference pressure at datum depth (kPa)	8000
Fracturing Pressure (kPa)	11066 – 12713.9
Reservoir Pressure (kPa)	7825.8
Reservoir temperature (°C)	62.78
Oil gravity (API)	23.9
Oil Viscosity (cP) at 25 °C	2075

Since one of the tasks is to evaluate the perforation strategies, oil RF and cumulative SOR would be the factors for comparison. All in all, the various steam-injection rates from 30 m^3/d to 180 m^3/d and well spacings were evaluated for different scenarios as proposed in the following:

1. Scenario 1 – Full perforation in all layers. The inverted 5-spot steam-flooding is injected and produced from layer 1 to layer 5 starting from the first day to the end of 20 years.

2. Scenario 2 – Selective perforation in bottom layers. The inverted 5-spot steam-flooding is injected and produced from layer 4 to layer 5 starting from the first day to the end of 20 years.

3 Result and discussion

3.1 Effect of well spacing

From the result of 3 different injector-producer distances, the shorter the distance provides higher oil recovery than the longer distance. However, the cumulative SOR is decreasing with the increasing distance. In general, both perforation scenarios showe the same trend of this result.

3.2 Effect of steam injection rates and perforation strategy

Regarding the full perforation scenario, Fig. 3 and Table 2 exhibit the effects of total oil recovery and the cumulative SOR for various steam injection rates. It's noted that the higher the steam injection rate can increase both the total oil recovery and SOR. Considering together with the well spacing, the injection rate at 180 m^3/d yields the highest RF at 46.6 % if injector – producer distance is 141.4 m. However, the cumulative SOR is also the highest which means the cost of steam would be higher as well.

Regarding the selective perforation scenario, Fig. 4 and Table 3 exhibit the effects of total oil recovery and the cumulative SOR for various steam injection rates. It's observed that the oil recovery is increasing with the injection rate increased from 30 m3/d to 60 m3/d for injector – producer distance at 212 m and 282.8 m. When the injection rate rises to 120 m3/d and 180 m3/d, the RF does not increase and the value is the same as the case of 60 m3/d. Similarly, the oil recovery is increasing with the injection rate increases up to 120 m3/d for injector – producer distance at 141.4 m. Its RF keeps at the same value as the case of 120 m3/d.

Based on such phenomenon, it can be interpreted that the maximum injection rate cannot be reached during the 20 years period for selective perforation scenario, especially in longer well spacing. That means the bottom-hole pressure is still high during the period because higher steam injection rate provides more volume of fluid to the reservoir and hence the pressure can be maintained.

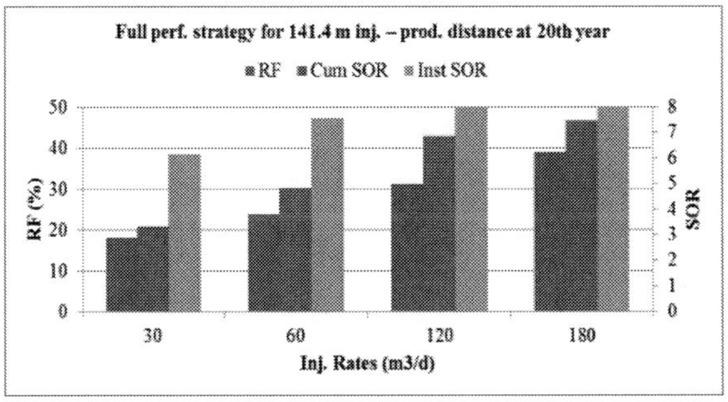

Figure 3. Result of full perforation for 141.4 m inj. – prod. distance at 20th year.

Table 2. Numerical results of full perforation for 212 & 282.8 m inj. – prod. distance at 20th year

	Inj. – Prod. Distance: 212 m (7x7_full perf.)			Inj. – Prod. Distance: 282.8 m (9x9_full perf.)		
Inj. Rate (m^3/d)	*RF (%)*	*Cum. SOR*	*Inst. SOR*	*RF (%)*	*Cum. SOR*	*Inst. SOR*
30	14.30	2.11	4.48	10.71	1.66	3.29
60	19.80	2.87	5.78	14.47	2.28	4.64
120	26.28	3.66	6.74	17.45	2.78	4.98
180	28.03	3.90	6.87	17.47	2.79	5.22

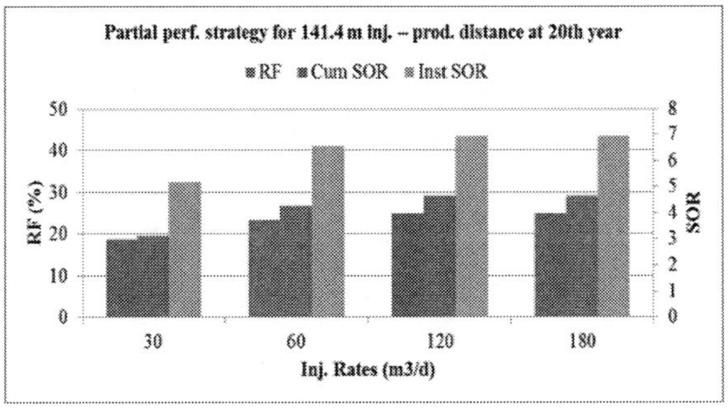

Figure 4. Result of partial perforation for 141.4 m inj. – prod. distance at 20th year.

In other words, the oil recovery should increase with the increase of injection rate if the end time of project is longer than 20 years. In this study, project time was considered in 20 years only as the objective is to compare the scenarios between full perforation and selective perforation based on the particular field data. Furthermore, the effects of varying different operating parameters can be fully observed in the full perforation strategy.

Table 3. Numerical results of partial perforation for 212 & 282.8 m inj. – prod. distance at 20th year

	Inj. – Prod. Distance: 212 m (7x7_partial perf.)			Inj. – Prod. Distance: 282.8 m (9x9_partial perf.)		
Inj. Rate (m^3/d)	*RF (%)*	*Cum. SOR*	*Inst. SOR*	*RF (%)*	*Cum. SOR*	*Inst. SOR*
30	12.76	2.22	4.09	9.14	1.74	3.11
60	15.37	2.62	4.76	9.76	1.93	3.49
120	15.37	2.62	4.83	9.76	1.93	3.49
180	15.37	2.62	4.83	9.76	1.93	3.49

4 Conclusion

Steam-flooding provides efficiently ultimate oil recovery for heavy oil. However, the huge energy consumption can impact to the production efficiency as well as project economics. Although the higher oil recovery could be obtained by increasing the steam injection rate and/or shorten the well spacing, the higher SOR would impact to the project cost at the same time. In terms of oil recovery, injection rate at 180 m^3/d and injector – producer distance in 141.4 m gave the best RF in this study. In contrast, injection rate at 30 m^3/d and injector – producer distance in 282.8 m gave the best result in terms of SOR. In comparison with the full perforation for injector – producer distance in 141.4 m, the selective perforation scenario can save about 60 % of steam with the balance between oil production and steam injection. Besides, the project life can be lengthened by slow reservoir depletion. It's because the high injection rate and less perforation zone can yield a stronger pressure to the field. This can be expressed by the aforesaid numerical results that both RF and Cum SOR remained in the same value with the increasing of injection rate at the 20th year. As such, it may be a better option if high cash flow is not critical in the early stage of the project.

Acknowledgment

The authors would like to gratefully acknowledge Chevron Thailand Exploration and Production Ltd., for financial support of this project.

References

1. Chierici, Gian Luigi. Enhanced Oil Recovery Processes: A State-of-the-Art Review. Azienda generale italiana petroli, (1980)
2. Ali, S. M., and R. F. Meldau. Current Steamflood Technology. Journal of Petroleum Technology **31**, 10 1-332 (1979)

3. G.L. Messner, A Comparison of Mass Rate and Steam Quality Reductions to Optimize Steamflood Performance. SPE Annual Technical Conference and Exhibition. Society of Petroleum Engineers, (1990)
4. Roberts, Tom. Economics of Well Spacing, (1961)
5. Doscher, Todd M., and Iraj Ershaghi. Current Economic Appraisal of Steam and Combustion Drives. SPE Symposium on Improved Methods of Oil Recovery. Society of Petroleum Engineers, (1978)
6. K.C. Hong, Effects of Steam Quality and Injection Rate on Steamflood Performance. SPE Reservoir Engineering, **9**, 290-296 (1994)
7. K.C. Hong, Steamflood Reservoir Management: Thermal Enhanced Oil Recovery. PennWell Books, (1994)
8. Chu, Chieh. "Pattern configuration effect on steamflood performance." Journal of Petroleum Technology **31**, 09 1-101 (1979)

MATEC Web of Conferences **68**, 07003 (2016) DOI: 10.1051/matecconf/20166807003
ICIEA 2016

Study on Productivity Numerical Simulation of Highly Deviated and Fractured Wells in Deep Oil and Gas Reservoirs

Liangchuan Li, Jianyi Huang, Cheng Luo, Fanglan Du, Yi Liu

Drilling and Production Technology Research Institute, PetroChina Jidong Oilfield Company, Tangshan, China

Abstract. This paper establishes the model of sandstone, porosity and permeability on single well in allusion to 10 highly deviated and fractured wells in deep oil and gas reservoirs of Jidong Oilfield, which forms a numerical simulation method of highly deviated and fractured wells in deep oil and gas reservoirs of Jidong Oilfield. The numerical simulation results of highly deviated and fractured wells productivity in deep oil and gas reservoirs are given out under different layers (layer ES1, layer ES3, layer ED2,and layer ED3), different deviation angles(60°and 75°), different fracture parameters and producing pressure drops. Through the comparison with testing data getting from exploration wells, we know that the calculation results of numerical simulation are consistent with practical testing results.

1 Introduction

Large-scale hydraulic fracturing of highly deviated and horizontal wells is the most effective method to develop low permeability sandstone and carbonate reservoir, tight oil and gas reservoir, unconventional oil and gas reservoir [1-3]. The productivity simulation and influencing factor analysis of highly deviated and fractured wells are the scientific basis of fracturing design and productivity prediction [4-5]. The computational difficulties of numerical simulation of highly deviated and fractured wells are mainly reflected in deviation angle, long-term flow conductivity, the processing of natural fracture, etc [6-8]. At present, the researches on highly deviated fracturing issues mainly focus on the mechanism of fracture initiation and propagation, deviated well fracturing, indoor simulation experiments, field tests and so on [9-11]. There are few researches on the productivity numerical simulation of highly deviated and fractured wells [12-14]. This paper forms a deviated and fractured wells productivity numerical simulation method of natural fracture development formation in deep oil and gas reservoirs, which provides a basis for the optimization design of the fracturing construction parameters and the evaluation of fracturing effect on highly deviated wells, so as to guide the actual field work for oilfield.

2 Establishment of representative well geological models

For 10 representative highly deviated wells of different layers in deep oil and gas reservoirs of Jidong Oilfield, on the basis of porosity and permeability deriving from single well logging interpretation, establish the geological models based on single well sandstonesection distribution, which provides the basis for numerical simulation.

For 10 representative highly deviated wells of different layers in deep oil and gas reservoirs of Jidong Oilfield, on the basis of porosity and permeability deriving from single well logging interpretation, establish the geological models based on single well sandstone section distribution, which provides the basis for numerical simulation.

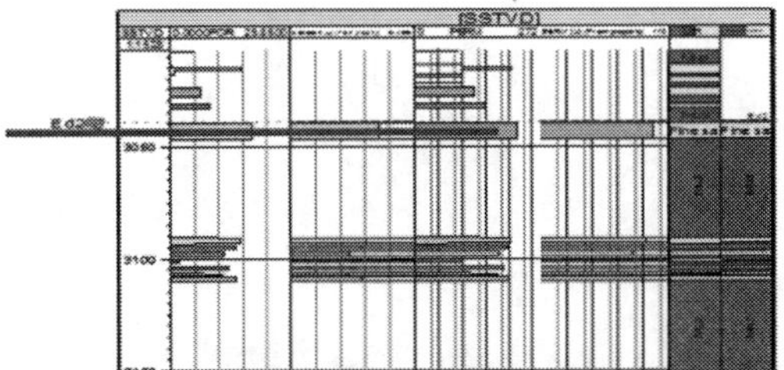

Figure 1. Log interpretation profile.

DOI: 10.1051/matecconf/20166807003

The sample well is located in layer ED_3 of Jidong Oilfield deep reservoirs, and the log interpretation profile is shown in Figure 1. The plane grid number of the sample well is set to 54 *44, with a step length of 30m, and the number of vertical grid is set to 59. There is no interference between the sample well and the surrounding wells, so a rectangle boundary can approximatively be used as the boundary of a single well model. Figure 2 shows the plane grid graph of the sample well, and Figure 3-Figure 6 respectively represent for the structure model, porosity model, permeability model and sandstone model of the sample well.

The geological model shows that the single well model is faithful to the single well data, and the goodness of fit is more than 95%.

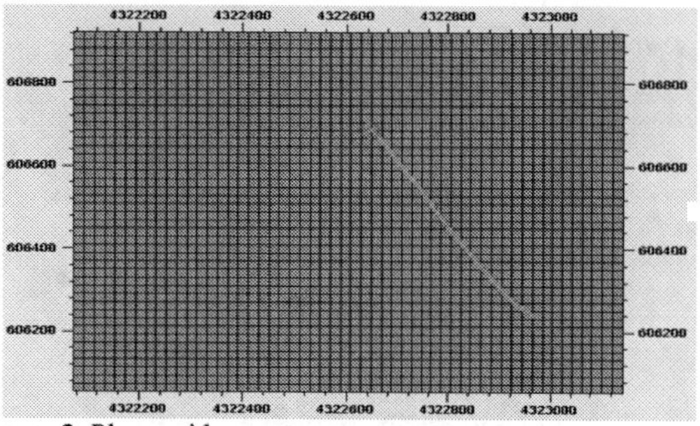

Figure 2. Plane grid.

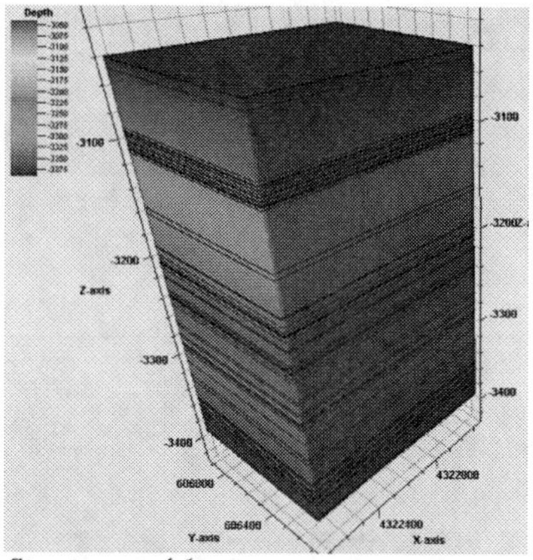

Figure 3. Structure model.

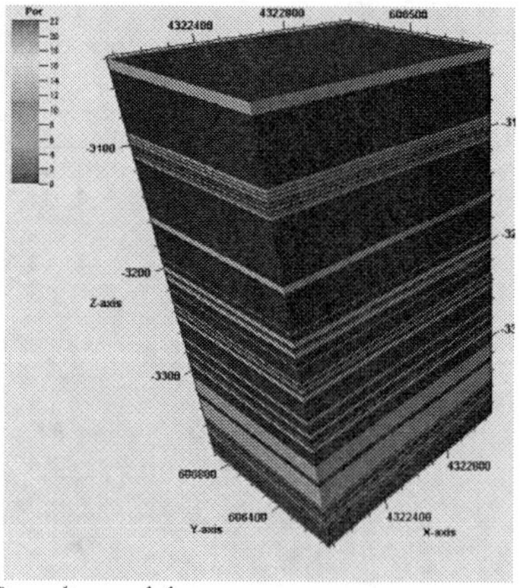

Figure 4. Porosity model.

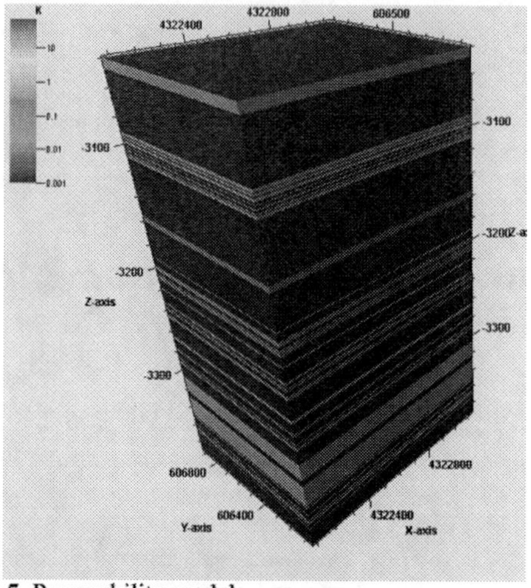

Figure 5. Permeability model.

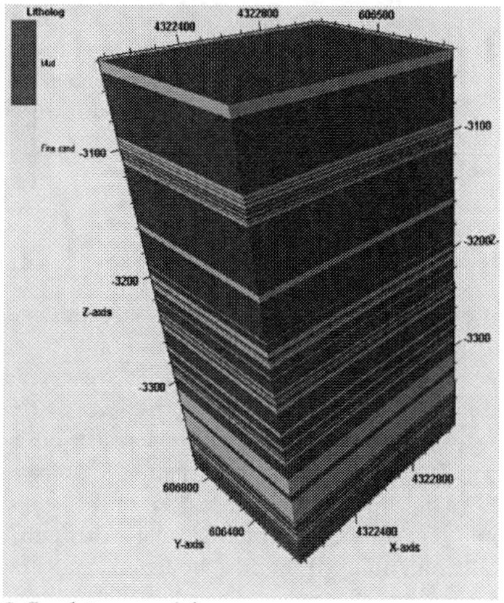

Figure 6. Sandstone model.

3 Numerical Simulation of Highly Deviated and Fractured Wells

3.1 Simulated horizons and the treatment of natural fracture development

Table 1 shows the testing and log interpretation results, natural fracture development and some others of the target layer section of 10 representative wells, and we can get the following conclusions from Table 1: layer ES_3 of deep oil and gas reservoirs in Jidong Oilfield is natural fracture development, while the fractures of layer ES_1, layer ED_2 and layer ED_3 do not develop. Through practical experience and the analysis of calculation results, for the intervals of natural fracture development, the goodness of fit is higher when the effective permeability is dealt with logging permeability, and for the intervals that do not develop, when the effective permeability is calculated according to 1/8 of the logging permeability, the result is more reliable.

Table 1. Simulation horizons and natural fracture development situation.

Serial Number	Horizon	Interval (m)	Testing/Logging	Natural Fracture	Treatment of Effective Permeability
1	ES_3	4340.8-4379.4	oil layer	development	effective permeability = logging permeability
	ES_3	4676-4691.8	oil layer	development	effective permeability = logging permeability
	ES_3	4834.4-4852.2	oil layer	development	effective permeability = logging permeability
	ES_3	4890.8-4902	hydrocarbon zone	development	effective permeability = logging permeability
2	ES_1	4078-4124	second type reservoir	undevelopment	effective permeability = 1/8 of logging permeability
	ES_1	4337-4393	second type reservoir	undevelopment	effective permeability = 1/8 of logging permeability
3	ES_{2+3}	4744.6-4801.6	second type reservoir	development	effective permeability = logging permeability
4	ES_{2+3}	4077.8-4138	gas layer	development	effective permeability = logging permeability
	ES_{2+3}	4284-4305.4	poor gas layer	development	effective permeability = logging permeability
	ES_{2+3}	4317-4373.8	poor gas layer	development	effective permeability = logging permeability
5	ES_3	4309.8-4318.8	oil layer	development	effective permeability = logging permeability
	ES_3	4663.8-4741	oil layer	development	effective permeability = logging permeability
6	ED_2	3405.8-3415.6	poor oil layer	undevelopment	effective permeability = 1/8 of logging permeability
7	ED_2	4661.6-4673.6	oil layer	undevelopment	effective permeability = 1/8 of logging permeability
Serial Number	Horizon	Interval (m)	Testing/Logging	Natural Fracture	Treatment of Effective Permeability
7	ED_2	4700.8-4718.4	hydrocarbon zone	undevelopment	effective permeability = 1/8 of logging permeability
8	ES_1	4272.2-4280.8	poor oil layer	undevelopment	effective permeability = 1/8 of logging permeability
9	ED_3	4533.6-4544.6	oil layer	undevelopment	effective permeability = 1/8 of logging permeability
10	ED_3	3404.6-3447.4	oil layer	undevelopment	effective permeability = 1/8 of logging permeability

3.2 Experiment of long-term flow conductivity

The experiment of long-term flow conductivity can reflect the true flow conductivity of the fracture under the condition of oil and gas reservoir, and provide reliable reference for fracturing design and construction. The Fracturing Center of PetroChina Research Institute of Petroleum Exploration gives the experimental result of long-term flow conductivity (20-40 mesh ceramic), which is shown in Figure 7.

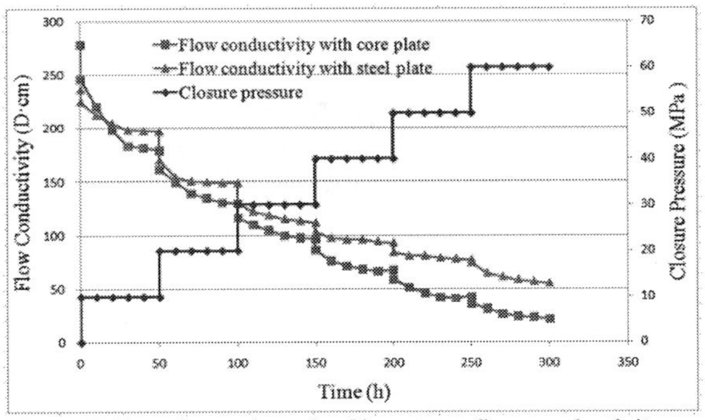

Figure 7. Experimental result of long-term flow conductivity.

The long-term flow conductivity fitting model of core plate is shown as following according to the experimental results:

$$w_f \cdot k_f = 258.69 / e^{0.0078t} \quad (1)$$

where w_f is fracture closure width (cm), k_f is fracture permeability (D), t is time (d).

According to the model, the flow conductivity is about 5.0 D.cm at 22 - 20 days (Table 2).

Table 2. Predicting results of long-term flow conductivity.

Time (d)	Flow Conductivity(D.cm)
2	177.90
4	122.34
6	84.14
8	57.86
10	39.79
12	27.36
14	18.82
16	12.94
18	8.90
20	6.12
22	4.21
24	2.89
26	1.99
28	1.37
30	0.94

3.3 Skin factor caused by well deviation

The highly deviated well is different from vertical and

horizontal wells. In the process of production, the well deviation will cause formation fluid disturbance and generate additional pressure drop, which can be represented by the skin factor. Figure 8 is a schematic diagram of partially perforated deviation well. Where θ is deviation angle (°), r_w is well diameter (m), z_w is the central coordinate of deviation well (m), h_w is the perforation length (m), h is reservoir thickness (m).

Define the dimensionless quantity:

$$z_{wD} = z_w / r_w,\ h_{wD} = h_w / r_w,\ h_D = h / r_w \quad (2)$$

When all the intervals are perforated:

$$z_{wD} / h_D = 0.5,\ h_{wD} \cos\theta_w / h_D = 1.0 \quad (3)$$

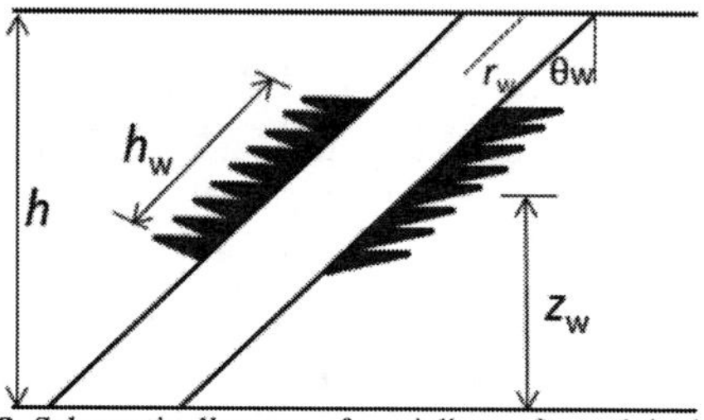

Figure 8. Schematic diagram of partially perforated deviation well.

At this point, the skin factors of different deviation angles are shown in Table 3. Where $s_{\theta+p}$ is the skin factor caused by well deviation and perforation, s_p is the skin factor caused by perforation, and s_θ is the skin factor caused by well deviation.

Table 3 shows that when all the intervals are perforated, the skin factor caused by perforation is 0, that is to say, the skin factor is only related with the deviation angle. In this paper, the skin factor caused by well deviation is integrated into the grid computing of numerical simulation.

Table 3. Skin factors of different deviation angles (Cinco-Ley)

θ_w	$S_{\theta+P}$	S_P	S_θ
15	-0.128	0	-0.128
30	-0.517	0	-0.517
45	-1.178	0	-1.178
60	-2.149	0	-2.149
75	-3.577	0	-3.577

4 Analysis of Calculation Results

The numerical simulation results of highly deviated and fractured wells productivity in deep oil and gas reservoirs is given out under different layers, different deviation angles, different fracture parameters and producing pressure drops. When the producing pressure drop is 24MPa, the production estimation ranges of different layers are shown in Table 4.

Table 4. Production estimation ranges of different simulation schemes

Simulation Scheme	Permeability (mD)	Reservoir Thickness	Porosity	Initial Production (m^3/d)	Production after1 Year (m^3/d)
Oil well of layer ES_1	2.6	7	9.7%	3-4.5	0.8-1.4
Gas well of layer ES_1	0.26	56	5.46%	1.5×10^4-3×10^4	0.5×10^4-1×10^4
	4.8	46	12%	15×10^4-25×10^4	6×10^4-10×10^4
Oil well of layer ES_3	0.5	15-39	6.8%	5-25	0.5-9
	7	78	12%	150-225	40-70
Gas well of layer ES_3	0.5	11	6.8%	5×10^4-10×10^4	2×10^4-4×10^4
	7	9	12%	25×10^4-45×10^4	10×10^4-18×10^4
Oil well of layer ED_2	0.68-1.5	12-17.6	13%-15%	1-4	0.25-1.25
	18	9.8	16%	7.5-11	2-3.5
Oil well of layer ED_3	6.2-13.2	11-42	12.7%-15.1%	8-12	2-4

5 Summary

The numerical simulation results of highly deviated and fractured wells productivity in deep oil and gas reservoirs are given out under different layers, different deviation angles, different fracture parameters and producing pressure drops.

DOI: 10.1051/matecconf/20166807003

When the producing pressure drop is 24MPa, the production estimation ranges of different layers and production parameters are obtained through the statistical results of numerical simulation: for the oil wells in layer ES_1, when the permeability is about 2.6 mD, the initial production is between 3 and 4.5 m^3/d, and for the gas wells in layer ES_1, when the permeability is about 0.26 mD, the initial production is between 1.5×10^4 and 3×10^4 m^3/d. For the oil wells in layer ES_3, when the permeability is about 0.5 mD, the initial production is between 5 and 25 m^3/d, and for the gas wells, the initial production is between 5×10^4 and 10^5 m^3/d. For the oil wells in layer ED_2, when the permeability is about 0.68 to 1.5 mD, the initial production is between 1 and 4 m^3/d. For the oil wells in layer ED_3, when the permeability is about 6.2 to 13.2 mD, the initial production is between 8 and 12 m^3/d. The results of this research have an important significance for the optimization design of the fracturing construction parameters and the evaluation of fracturing effect on highly deviated wells.

References

1. Lang X.J., Li F.X., Liu H., Oilfield Chemistry, **21** (2004).
2. Lv C.B., Yu D.H., Zhao J.Z., Well Testing, **15** (2006).
3. Zhu X.G.,. Northeast Petroleum University, (2012).
4. Rogers E J, Economides M J., SPE 37068, (1996).
5. Sun E.H., Li X.P., Zheng L.N., Journal of CQ University of Science and Technology, **14**, 6 (2012).
6. Daneshy A A. SPE 4062, (1973).
7. Niu Z.Q., Sui X.Y., Zhang P., Oil Drilling & Production, **27** (2005).
8. Dinh A V, Tiab D., SPE120540, (2009).
9. Jia C.G., Li M.Z., Journal of Southwest Petroleum University, **29**, 2, (2007).
10. Hu J.Q., Petroleum Geology and Engineering, **24** (2010).
11. Liu W., Li L., Petroleum Geology and Engineering, **24**, 1, (2010).
12. Roemershauser A E, Hawkins M F., Journal of Petroleum Technology, **7**, 2, (1995).
13. Cinco L H., Stanford University, (1974).
14. Van der vlis A C, Haafkens R, Schipper B A. SPE56937, (1975).

MATEC Web of Conferences **68**, 07004 (2016) DOI: 10.1051/matecconf/20166807004
ICIEA 2016

A Study of Coal Miners' Safety Psychological Elements Based on Questionnaire Survey

Zeng-bo Li [1,2], Long-kang Wang [1,2], Dong Zhang [1,2], Bin Peng [1,2], Jia-bin Zhang [1,2], Ting Hong [1,2], Han Shen [1,2] and Liang Dong [1,2]

[1]*China University of Mining & Technology (Beijing), School of Resources & Safety Engineering, 100083 Beijing, China*
[2]*China University of Mining & Technology (Beijing), State Key Laboratory of Coal Resources and Safe Mining, 100083 Beijing, China*

Abstract. During the production process, the instability of safety psychological elements of coal miners contributes to unsafe behaviors that may result in fatal accident. To search the dominant psychological elements, three types of unsafe psychology, comprising 18 unsafe psychological elements, are obtained by theoretical analysis and site survey. Then, an assessment model is established, and a matched questionnaire is applied to two large modern coal mines in Inner Mongolia Autonomous Region. Based on questionnaire survey results and analytic hierarchy process (AHP), assessment judgment matrixes are constructed, effect weight is calculated, and consistency check is conducted. The analysis results show that safety psychology elements of defective type are the dominant elements that trigger unsafe behaviors of coal miners.

1 Introduction

In recent years, with the increasing investment in production safety, the improving adjustment in energy structure, and the improving macroeconomic situation, Death rate per million tons of coal mines is declining year by year in China. The rate in 2012 was 0.374, a rate lower than 0.5 for the first time in history [1]. The situation in China's coal mines keeps turning better, however, in comparison with the advanced international level, the gap is still evident. Death rate per million tons is roughly 0.02~0.03, and even zero in some years in the developed nations. This indicates that safety situation of coal mines is still severe. Researches show that up to 80% of the coal mine accidents is the result of unsafe behaviours that are the outward manifestation of unsafe psychology. Therefore, to study the safety psychology of coal mine workers is of important theoretical and realistic significance to improve safety situation and protect their life.

2 Establishment of Assessment Model on Safety Psychology for Coal Mine

2.1 Theoretical Analysis of Safety Psychology

Safety psychology is a science that aims to minimize and eliminate accidents from the source [2]-[4] by interpreting, regulating and predicting workers' behaviours based on the law of the deeply rooted psychological activity behind behaviours. Safety psychology is an independent interdisciplinary subject on the basis of psychology and safety science. Its essence is the new application of psychology in the field of safety science [5]-[7]. Safety psychology of coalmine is a special and important research and application field of safety psychology. It aims to prevent coalmine accidents and protect life and health of coalminers [8], [9]. It is an emerging application science that investigates the occurrence of coalmine accidents and prevents the psychological elements of mankind and their exerting law by combining with the production activity law and specific practices of work safety in coal mines [10].

Investigation of the influence of the safety psychology of coalminers on their safety behaviours concentrates on their safety psychological process, which covers cognition and safety, emotion and safety, wills and safety, attention and safety. The four aspects are correlated, mutually promoted, and unified in safety psychological process [11]. At first, let us analyse cognition and safety. Safety cognition process, coalminer-oriented, is a very complicated processing course of cognition message. As the processor of message, a coalminer will, through his physiological organs like eye, nose, ear, hand and foot, receive the relevant information of the object in production process, and conduct effect information process, and thus make wise and correct information judgment and guarantee the implementation of safety behaviours [12], [13]. In China, because of the low-threshold of coalminers' work, they have poor education background and technique competence, and lack cognition on the underground working environment, different marks and geographic environment, as a result,

DOI: 10.1051/matecconf/20166807004

they lack cognition on the objective loopholes in work safety such as hidden dangers, have no means to do qualified jobs and cannot ensure work safety [14].

Second, let us analyse emotion and safety. Safety emotion is a kind of emotion produced in the body of coalminers in a given timeframe under the influence of a given event or scenario. Typical emotional states include mental state, passion and stress [15]. Practices show that coalminers are affected by kinds of underground severe environment such as dim light, black coal and rock, large dust concentration in workface, threat from gas explosions and coal-gas outburst. On the one hand, this circumstance make coalminers form a team unit similar to a small family where they care for and influence each other, which is good for formation of the safety psychology of coalminers. On the other hand, this circumstance creates their bad-tempered character, which is adverse to work safety, and sometimes is the direct trigger of accidents [16], [17].

Third, let us analyse wills and safety. Wills is the preparatory state of behavior. Coalminers have endurance and persistence in special underground environment. For example, a water-inrush accident, occurred in Jingyang No.2 mine of Yiyang County of Henan Province, because of strong wills, coalminers had persisted for 21 days until they were saved by the rescuers.

At last, let us analyse attention and safety. Attention means conscious psychological activity's orientation and concentration on a given matter. Safety attention directly influences the safety behaviours of coalminers [18], [19]. Research indicates that coalminers' inattention leads to a large portion of coal mine accidents among quantities of coalmine safety accidents.

2.2 Assessment model of safety psychology for coal mine

The assessment model of safety psychology is designed on the basis of the theoretical analysis of coal mine safety psychology. In this model, unsafe psychological elements are divided into types of situational, defective and endogenous. The first is motive-typed where different levels of need cannot be satisfied, the second is a mental process-typed where safety cognitive ability is inadequate, emotion and feeling are confused, and attention is not concentrated, The third is a personality difference-typed based on safety character and temperament[20,21]. By analysing the unsafe psychological elements of three types, it can be found that both their generation and evolution mechanism are different (shown in Table 1), so that further analysis on countermeasures should be conducted.

In the assessment model of safety psychology, the three unsafe psychological types include numerous safe psychological elements. The elements mean the psychological factors of coalmines which change under external environments such as the nature, society, coal mine enterprises, family and operating environment, and their own physiological and psychological fluctuations in the production process. Through theoretical research, site survey and statistical analysis, it is concluded that there are 18 safety psychological elements that influence the safety behaviours of coalminers. They include carelessness, emotion flaw, inertial violation, inadequate cognitive level, conformity, poor safety awareness, safety character, stress state, risky mind, safety temperament, reversal mind, lack of experience, over fatigue, safety need, boredom, poor environment, distraction, biorhythm, etc. The assessment model of safety psychology is as shown in Fig. 1.

Table 1. Types of unsafe psychological elements and their features.

Type	Generation mechanism	Stability	Actor
Situational	Triggered by external cause	Apt to change	Third party
Defective	Internal and external influence	Change after efforts	Object affects subject to act
Endogenous	Determined by internal cause	Hard to change	To guide

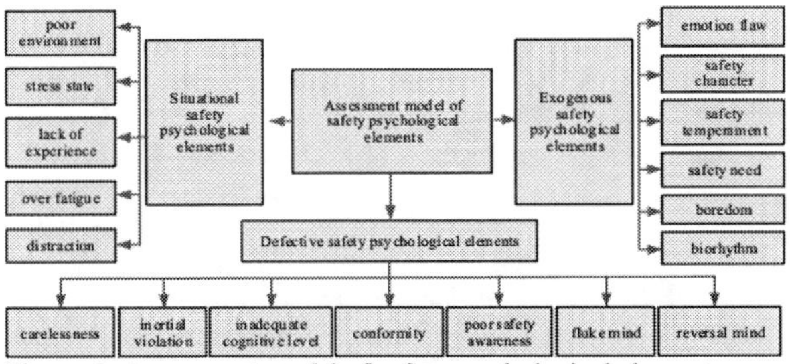

Figure 1. Assessment model of safety psychological elements.

3 Design of survey questionnaire and survey analysis

3.1 Design and implementation of survey questionnaire

Design of survey questionnaire is unfolded from the theoretical analysis of coal mine safety psychology and the assessment model of safety psychological elements. Based on this, experts are interviewed in coal mine industry. A preliminary survey questionnaire on coalminers' safety psychology is designed. Then, small scale test is made, the insufficiencies are found and improved. 32 questions are defined for survey questionnaire that cover 18 safety psychological elements, where 11 questions on situational type, 11 questions on defective type, and 10 questions on endogenous type.

After the survey questionnaire is well-designed, the staffers of Inner Mongolia Zhongbei Coal Chemicals Company (Zhongbei) and Inner Mongolia Dongdaneng Mining Co. Ltd. (Dongdaneng) serve as the assessed subjects. The controlling party of Zhongbei is a state-owned enterprise, and its management system can representatively embody the current state of the central government-owned enterprises and other state-owned enterprises in Inner Mongolia, while Dongdaneng is one of privately-run enterprises in the region which are the largest in scale and best in economic benefits, and its management system can fully embody local and private enterprises.

The implementation of questionnaire survey takes into account sampling method and survey mode. Probability sampling method is selected for the survey. Probability sampling comprises stratified sampling, cluster sampling, random sampling, etc. Errors can be calculated. This can better avoid errors in sampling and coverage. In the implementation process, a large sample database is established at first, then a survey is conducted by probability sampling. Prior to implementation of the survey questionnaire, the purpose was explained to the coalminers, and the coal mine officials were requested to lead in the survey. Then, coalminers were asked to complete the questionnaire. The timeframe of completing the questionnaire was regulated. Finally, the contact information of the respondents was left on each questionnaire for later interviews and improvement of survey information.

3.2 Analysis of questionnaire survey

Through arrangement and statistics of the survey questionnaires, it is concluded that the scores of safety psychological elements of Zhongbei's coal mines are higher than Dongdaneng's. Moreover, the higher the academic degree, the more stable safety psychological quality, so is the position. However, among situational, defective and exogenous safety psychological elements, it is hard to determine the major element that triggers unsafe behaviours of coalminers. Therefore, analytical hierarchy process (AHP) is adopted to determine the weight of safety psychological elements. The purpose is to decide what safety psychological element exerts the most influence on the coalminers. AHP consists of construction of judgment matrix, calculation of weight and consistency check.

3.2.1 Construction of judgment matrix

As illustrated in Figure 1, the assessment model of safety psychological elements is a hierarchical-structured model. The target layer is the assessment result of safety psychological elements (A). The criterion layer is the safety psychological elements of Situational, Defective and Endogenous, set as B_1, B_2 and B_3, respectively. The grade-II influential factors of Situational safety psychological elements (B_1) include poor environment, stress state, lack of experience, over fatigue (B_{14}) and distraction (B_{15}). The grade-II influential factors of Defective safety psychological elements (B_2) include carelessness (B_{21}), inertial violation (B_{22}), inadequate cognitive ability (B_{23}), conformity (B_{24}), poor safety awareness (B_{25}), fluke mind (B_{26}) and reversal mind (B_{27}). Grade-II influential factors of Endogenous safety psychological elements (B_3) include emotion flaw (B_{31}), safety character (B_{32}), safety temperament (B_{33}), safety need (B_{34}), boredom (B_{35}) and biorhythm (B_{36}).

By use of paired comparison method, the safety psychological elements in the three types are scored in pairwise comparison to construct judgment matrix A. 9-level scoring method is used for assessment criteria. Scoring criteria are shown in Table 2. In the judgment matrix A, if element i is compared with element j, its importance is a_{ij}, the ratio of representation between element j and element i is $a_{ij}=1/a_{ij}$.

Table 2. Meanings of AHP scores.

Score	Meaning
1	Two elements are equally important.
3	Former element is slightly important than latter one.
5	Former element is more important than the latter one.
7	Former element is significantly important than latter one.
9	Former element is absolutely important than latter one.
2,4,6,8	The mean value of the above adjacent judgments

The data of Zhongbei was processed. In order to overcome the influence of subjective factors on AHP results to the largest extent, the mean scores of 100 samples in situational, defective and endogenous were analyzed, and pairwise comparison was done. By giving quotients and differences and amplifying appropriate multiples, they were made to accord with the requirements of 9-level scoring method. The results are as shown in Table 3. In like manner, the data of 100 samples of Dongdaneng were calculated. The results are as shown in Table 4.

Table 3. Matrix of pairwise judgment for three safety psychological types in Zhongbei.

Assessment result (A)	Situational (B_1)	Defective (B_2)	Endogenous (B_3)
Situational (B_1)	1	1/4	1/3
Defective (B_2)	4	1	2
Endogenous (B_3)	3	1/2	1

Table 4. Matrix of pairwise judgment for three safety psychological types in Dongdaneng.

Assessment result (A)	Situational (B_1)	Defective (B_2)	Endogenous (B_3)
Situational (B_1)	1	1/2	3
Defective (B_2)	2	1	5
Endogenous (B_3)	1/3	1/5	1

3.2.2 Calculation of weight

Matrix A obtained from Table 3 is just the judgment of AHP. Generally, the eigenvector of the largest characteristic root is considered as weight vector. The weight vector is just the weight index of the elements of this layer in respect to a factor of its upper layer. At first, each column of the judgment matrix A is normalized, namely,

$$B_{ij} = \left\{ B_{ij} / \sum_{i=1,j=1}^{3} B_{ij} \right\}, i = 1,2,3\ ,\ j = 1,2,3 \tag{1}$$

Normalized judgment matrix B is obtained. Matrix B is added by row, namely

$$W_i = \sum_{j=1}^{4} B_{ij}\ ,\quad i = 1,2,3 \qquad (2)$$

Hence weight $\beta = [\beta_1, \beta_2, \beta_3]$ is figured out, where $\beta_i = W_i / \sum W_i$, i=1,2,3.

3.2.3 Consistency check

In order to assess the reasonable degree of the judgment matrix, it is needed to make consistency check for the matrix. After the weight is obtained, the largest eigenvector of the matrix A is attained.

$$\lambda_{\max} = \frac{1}{n}\left(\sum_{i=1}^{3} (A \times W)_i / W_i\right) \qquad (3)$$

where, $\lambda_{\max}$ is the largest characteristic root of matrix A; n is the number of factors; W_i is the weight of factor i. The values of $\lambda_{\max}$ and W_i are shown in Table 5.

Table 5. Values of $\lambda_{\max}$ and W_i.

	$\lambda_{\max}$	W_1	W_2	W_3
Zhongbei	3.0183	0.1220	0.5584	0.3196
Dongdaneng	3.0037	0.3090	0.5819	0.1095

Then consistency index is

$$CI = (\lambda_{\max} - n)/(n-1) \qquad (4)$$

According to the value of the average random consistency index *RI* of same order matrix as shown in Table 6, 3rd order matrix's =0.52. Then consistency ratio *CR* is calculated with formula 5

$$CR = CI / RI \qquad (5)$$

Table 6. Values of average random consistency index *RI*.

Order	2	3	4	5	6	7
RI	0	0.52	0.89	1.12	1.26	1.36
Order	8	9	10	11	12	
RI	1.41	1.46	1.49	1.52	1.54	

If CR<0.1, the consistency of the judgment matrix is acceptable. It can be known from the above calculations, *CR* of Zhongbei's judgment matrix is 0.0176, and *CR* of Dongdaneng's judgment matrix is 0.0036, both less than 0.1. Therefore, the coefficients of the judgment matrixes shown in Table 3 and Table 4 are reasonable and acceptable.

According to Table 3 and Table 5, In Zhongbei's judgment matrix, weight coefficient of defective psychological elements is the largest, endogenous elements come second, and situational elements are in the last place. This demonstrates that, although the state-owned coal mine has higher quality coalminers and pays them well, and has better coal mining mechanization, it does not adequately train the staff on safety, and the coalminers generally have defective mind such as fluke/risk-taking mind.

According to Table 4 and Table 5, in Dongdaneng's judgment matrix, the weight coefficient of defective psychological elements is the largest, situational elements come second, and endogenous elements are in the last place. It can be seen that, whether for the state-owned coal mine or for the private-owned one, defective psychological elements are the most important factors that influence the coalminers' safety psychology. Therefore, strengthening safety training, improving safety management level and formation of sound corporate safety culture in coalmine, enterprises can effectively reduce the unsafe behaviours of coalminers, and effectively prevent safety accidents. Meanwhile, it can be noticed that, because the private-owned Dongdaneng's miners are of lower academic degree and poorer quality, and their daily work intensity is higher, situational psychological elements also significantly influence their safety psychological state, and cannot be ignored.

4 Conclusions

Theoretical analysis of coal mine safety psychology and thorough analysis of coalminer's safety psychological process lay a solid foundation for determining the factors of safety psychological elements.

According to the working environment of coalminers and their own condition, 18 coal mine safety psychological elements are put forward, and the assessment model and survey questionnaire for coalminer's safety psychological elements are designed.

By statistical analysis of the survey questionnaires, it is concluded that the safety psychological level of workers in state-owned coalmine is better than that of workers in private-owned coalmine; the higher the academic degree of coalminers, the longer their service life, the better their safety psychological elements. AHP analysis reveals that safety psychological elements of defective type are major factors that influence the safety behaviours of coalminers.

The questionnaire and assessment model maybe has limitation to some extent, we need explore more psychological elements based on site experience and development of safety science and psychology.

References

1. Jingchao C. Mortality rate per million tons of coal dropped to below 0.4% in 2012 [EB/OL]. http: //news.xinhuanet.com/politics/2013-02/07/c_124335688.htm.2013-2-7.
2. Hui S., Kaiquan W. *Safety Psychology*[M]. Beijing: Chemical Industry Press, 2004.
3. Edelhauser V. Psychological examination and reexamination necessity for underground worker safety[J]. Mine, petrol size, **35,** 6 (1984): 284-300.
4. Whitaker Leslie A., Shoptaugh Carol F. Mine safety, productivity and job design[C]. *Proceedings of the Human Factors Society 28th Annual Meeting: New Frontiers for Science and Technology*, TX, USA, 1984.
5. Shichang Z. The study and exploration on the relationship between miners' psychology safety in production[C]. *Proceedings of the 1998*

International Symposium on Safety Science and Technology, Beijing, China, (1998)

6. Camm T., Girard-Dwyer J. Economic consequences of mining injuries[C]. *2004 SME Annual Meeting Preprints*, Denver, United states, (2004)
7. Seth Ayim G. Causal attributions of Ghanaian industrial workers for accident occurrence: Miners and non-miners perspective [J]. Journal of Safety Research, **34**, 5(2003):533-538.
8. Ghosh A. K., Bhattacherjee A. Predictors of occupational injuries among coal miners: causal analysis [J]. Transactions of the Institutions of Mining and Metallurgy, **116**, 1(2007):16-24.
9. Jizu L., Shaohong Z. Research on psychological evaluation system of safety personnel in coal mine[C]. *1st International Conference on Information Science and Engineering*, Nanjing, China, (2009)
10. Jizu L., *Safety Psychology* [M]. Beijing: China Social Guarantee of Labor Press, (2007).
11. Yajing M. Thinking about the safe psychological problems of the miners [J].Coal Mine Safety, 7 (2007): 87-89.
12. Yiqin Y. Miner's psychological health and coal mine safety production [J].China Coal, 4 (2000):49-51.
13. Hong C, Jing L, Ting W, et al. Regulation strategy design of coalmine management staff's psychological contract violation [J]. Proceeded Earth and Planetary Science, **1**, 1 (2009):1730-1736.
14. Theron W. J., Van Heerden G. M. J. Fatigue knowledge-a new lever in safety management [J]. Journal of the Southern African Institute of Mining and Metallurgy, **111**, 1 (2011):1-10.
15. Lujia G. Expanding explorations of psychological methods [J]. Journal of Nanjing Normal University (Social Science), **1** (2005): 84-91.
16. Xiaojuan H., Chao W. Review on the studying methodologies of human safety psychological characteristics[J].China Safety Science Journal, **19**, 7 (2009): 5-13.
17. Jizu L., Lixun K., Zhili Z., etc. Study on psychological evaluation system of safety personnel in coal mine[J]. China Safety Science Journal, 3 (2004): 95-97.
18. Margolis K. A, Westerman C. Y. K., Kowalski-Trakofler K. M. Underground mine refuge chamber expectations training: program development and evaluation[J]. Safety Science, **49**, 3 (2011): 522-530.
19. Baoyue H., Fei C. Safety psychology application on coal mine safety management based on information system[C]. *International Conference on Advanced Research on Electronic Commerce, Web Application, and Communication,* Guangzhou, China, 2011.
20. Weimin C., Gang Z., Gang W., etc. Psychological measurement and analysis on human unsafe behavior[J]. China Safety Science Journal, 6 (2009): 29-34.
21. Guanglei Z. The building of psychological evaluation system and its application [J]. International Journal of Digital Content Technology and its Applications, **7**, 3 (2013): 151-157.

Research on Flow Field Characteristics Affected by Obstacles of Pull-Push Ventilation System of Plating Tanks through Numerical Simulation

Hu Hong[1,2] and Yi Can-nan[2]

[1] *Human Factors Institute, University of South China, Hengyang, China*
[2] *Ergonomics and safety management Institute, HuNan Institute of Technology, Hengyang, China*

Abstract. Pull-push ventilation system, with high capture efficiency, is universally used to capture contaminants from containment generation where can't be enclosed and where the working area is large. When there are obstacles between the pull hood and the push hood, the air jet will disperse, leaving contaminants out of control, causing atmosphere polluted. This paper aims to discuss flow filed of pull-push ventilation system, located in some tank of some plating workshop, with obstacles between pull hood and push hood. Mathematical model, involving physical model, assumptions, governing equations, boundary conditions and grid separation is established. By CFD code FLUENT, flow field, containing velocity field and concentration field, is achieved. To verify the simulation models, velocity field for simulation results is compared with the theory results, while concentration field for simulation results with experimental results. The comparison results show that the results matches well and that mathematical model proposed in this paper is reasonable, helpful and realistic significant in designing this kind exhaust ventilation systems where obstacles presence.

1 Introduction

Push-pull ventilation system, involving a jet of air that is blown from one side of the tank and collected by an exhaust hood on the opposite side, is such empowered local exhaust hoods with significant advantage over local ventilation systems in capturing contaminants and resisting disturbing airflow that they are widely used to control contaminants derived from containment generation where can't be enclosed owing to the necessary to produce the products, nor could other hoods be chosen due to the large working areas. When controlling pollutant, push-pull ventilation system can save up to 50% of flow rate than lateral exhaust [1], and it is prior to any other hoods when the hood is far away from pollutant generation (eg, more than 2m). Nowadays, push-pull ventilation system is widely chosen to capture harmful gas and vapor [2] and fumes [3]. Since it controls the pollutants through the associative action of pull hood and push hood, we should determine the best effective and matched velocities of them. It is suggested use numerical methods to analyze flow field instead of experimental work which can't easily be extended to different operating conditions [4]. Michael Robinson present a semi - analytical approach which exploits the self - similar nature of much of the fluid flow [5], R. Rota discussed the design guidelines for push-pull ventilation systems through computational fluid dynamics modeling [6], Hunt [7], HE Su-yan [8] and CHEN Jiang-ping [9] deal with the mathematical model, flow-field test, and tuburlence character. Existing simulation methods only focus on the flow character of the hoods or big obstacles between hoods (desk, computer, chair redistribute [10], ceiling mounted obstacles [11] or woker [12]), nor do they consider that there are some small obstacles existed between pull hood and push hood located in plating tanks. When there are obstacles, the matched velocities of the two hoods become crucial for us to determine: if they are slower than the demand, the containments won't be captured entirely; while they are faster, the air jet will be dispersed to outer space when it shot the obstacles.

This paper aims to analyze the flow field of push-pull ventilation system over plating tanks with obstacles between them: setting the pollutant gas emitted on plated pieces, with volatile liquid adhesives from the tanks, of sulfate electroplating tinning product line of some factory as the physical model, combining with some theories, such as hydrodynamic, aerodynamics and so on, mathematical model is built based on some assumptions; by CFD commercial code FLUENT, flow field, including velocity field and concentration field, of the push-pull ventilation system with various matched velocities is simulated, helping design this kind of ventilation system.

2 Mathematical models

2.1 Physical model

Harmful gases generated in the plating process due to the high concentrations of liquid adhered over the plated pieces, moving among the tanks with cranes, will pose great potential hazard to workers and environment. The main gas in plating line of some factory is NO_2 given off by pickling tanks, with high concentrations of nitracidium, a mixture of nitrate solution and water with the proportion of 1:3. We find that the acid mist produced on moving plated pieces is 800～3 000 g/(m^2·h), so the maximum value3 000 g/(m^2·h), seeing as a constant value and not varying with time, is chosen as the target contaminant we should capture. Related parameters are set as follows: mass flow (mass source term)is 0.000 084 kg/(m^3·s), while momentum source term 0.000 055 8 N/m^3(momentum equals to velocity times mass flow), summer, room temperature T=310K, tank liquid temperature T=313 K, simulating the near-source zone only, physical model, as shown in Fig. 1, is enlarged up to 3.5×3×3 m^3 to neglect the outer space's influence.

2.2 Assumptions

1)Steady flow field.

2)Neglecting the density variations of the contaminants and air when velocity is quite slow, namely, seeing them as incompressible fluid.

3)Exception of gravity effects.

4)Assuming that the selected control volume is large enough when adopt control volume method to calculate.

5)Treating the hoods as wall and considering it is radiationless and thermal isolated.

6)Pollutant diffuses evenly.

7)The plated pieces is simplified as cuboids shown in Fig. 2.

8)Neglecting the influence of the plating crane to the flow field.

9)When plating, the crane moves slowly, with the velocity of 0.2 m/s. Push-pull hood located in carne moves with plated pieces simultaneously, remaining relatively static.

10)Setting the obstacles as wall at ordinary temperature , emitting contaminants with temperature of 313K and velocity of 0.000 055 8 N/m^3.

11)Ignoring the disturbing air currents.

2.3 Governing equations

There are many equations over simulating contaminants emissions [13]: standard $k-\varepsilon$ turbulence model realizable $k-\varepsilon$ turbulence model, and large-eddy simulation(LES). HE Su-yan found that the standard $k-\varepsilon$ turbulence model works well when simulating contaminants. So we choose standard $k-\varepsilon$ turbulence mode. General governing equation is [14], [15]:

$$div(\rho \vec{u}\phi) = div(\Gamma grad\phi) + S$$

where ϕ is general variable, Γ diffusion coefficient, S source term. Every symbol and its corresponding form is as Table 1 shows.

2.4 Boundary conditions

Inlet boundary: velocity-inlet

Outlet boundary (face of the open space where push hood exists): pressure-outlet

Outlet boundary (push hood): outlet-vent

Outlet boundary(faces of open space): pressure-inlet

Interior surface boundary: auxiliary surface, established by separating grid, is defined as interior surface where flow can pass through freely

Source term (plated pieces): According to the results of in-situ measurements, mass flux of NO_2 is 8.4×10^{-5} kg/(m^3·s), with the momentum flux of 5.58×10^{-5} N/m^3.

Table 1. Detail forms for the symbol of general governing equations

Equations	ϕ	Γ	S
Mass conservation equations	1	0	0
Momentum equations	u_i	$\mu_{eff} = \mu + \mu_t$	$-\frac{\partial P}{\partial x_i} + S_i$
Species mass-conservation equations	c_s	$D_s \rho$	S_s
Turbulent kinetic energy	k	$\mu + \frac{\mu_t}{\delta_k}$	$G_k + \rho\varepsilon$
Turbulent dissipation rate	ε	$\mu + \frac{\mu_t}{\delta_\varepsilon}$	$\frac{\varepsilon}{k}(C_{1\varepsilon}G_k - C_{2\varepsilon}\rho\varepsilon)$
Where : $\mu_t = \rho C_\mu \frac{k^2}{\varepsilon}$, $G_b = \beta g_i \frac{\mu_t}{Pr_t}\frac{\partial T}{\partial x_i}$, $G_k = \mu_t\left(\frac{\partial u_i}{\partial x_j} + \frac{\partial u_j}{\partial x_i}\right)\frac{\partial u_i}{\partial x_j}$, $Y_M = 2\rho\varepsilon M_t^2$,			
Constant numbers [16], [17] : Pr_t=0.85, $C_{1\varepsilon}$=1.44, $C_{2\varepsilon}$=1.92, C_μ=0.09, σ_k=1.0, σ_ε=1.3			

2.5 Grid Separation

The computational domain is 3.5×3×4.5m^3, hood 0.4×0.6 m^2, columned plated pieces with 0.05m in diameter and 0.4m in height. To simplify the calculation, little columned plated pieces is treated as a little cuboids, with 0.055×1.8×0.4 m^3, defined as wall. The space expect hoods and cuboids is divided into 2 fluid zones (Fig. 2): fluid zone 1 is a very thin shield of flow, circling the cuboids, which is assumed to be contaminants generation, with mesh interval of 0.025m; fluid zone 2 is the residue space. The mesh interval of the hoods is 0.1m. The face is separated by pave triangle grid and the volume Tgrid. Partial grid near hoods and plated pieces are separated with shorter intervals to enhance the precision and convergence speed of calculation, and the whole computational domain is separated into 300,000 computation cells as Fig. 3 shows.

3 Simulation Results and Flow Field Profile

FLUENT code is used to analyze push-pull ventilation system's flow field. The matched velocities of push unit and pull unit calculated by flow ratio method are: v_1=2 m/s, v_3=10.42 m/s; v_1=2.5 m/s, v_3=11.67 m/s; v_1=3 m/s, v_3=12.74 m/s; v_1=3.5 m/s, v_3=14.0 m/s, respectively. Velocity contours and mass fraction contours are as shown from fig. 4 to fig.7.

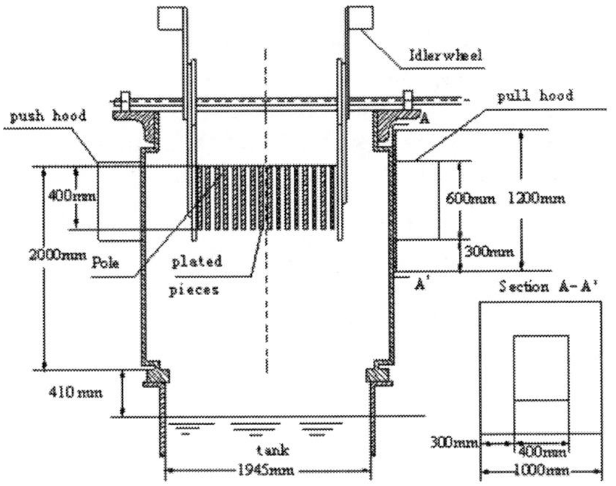

Figure 1. Pull-push ventilation system fixed on crane

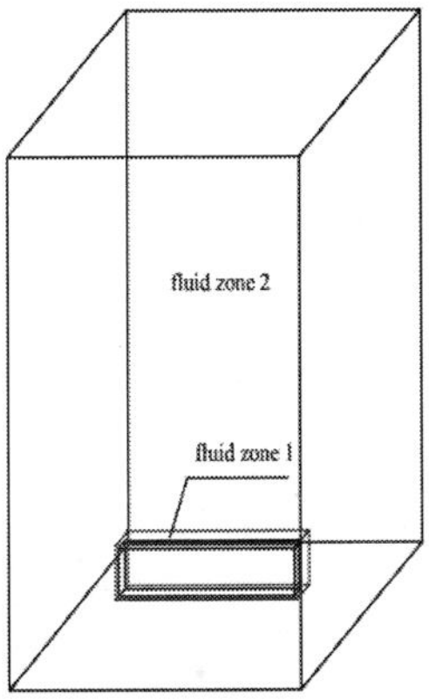

Figure 2. Fluid zones

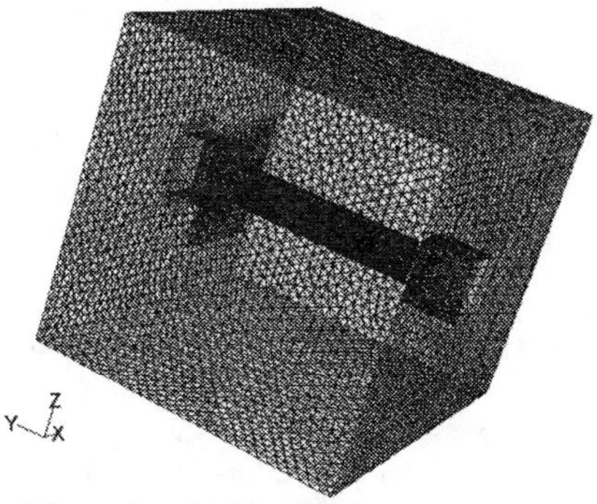

Figure 3. Schematics of grid separation

1)Velocity contours

(1)The air jet disperses due to the obstacles, the greater the velocity is, the more obvious the dispersion is.

(2)The best matched velocities are v_1=2.5 m/s and v_3=11.67 m/s, forming a gas curtain covering the plated pieces completely with the smallest air flow.

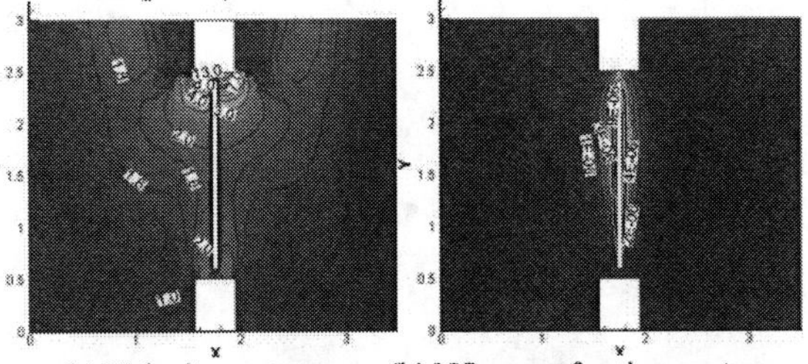

(a) Velocity contours (b) NO_2 mass fraction contours

Figure 4. Velocity contours and NO_2 mass fraction contours at section z=1.5 m (v_1=2 m/s)

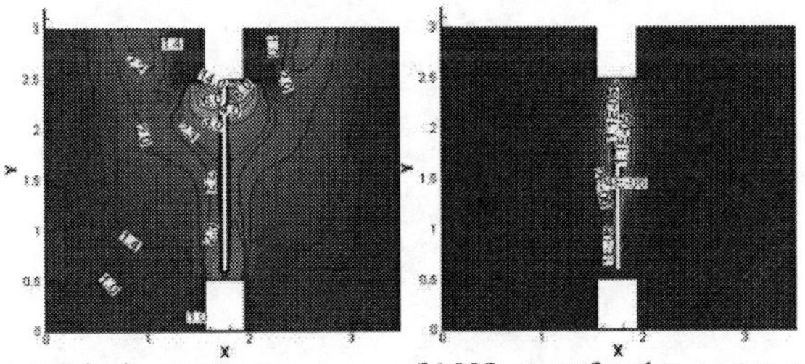

(a) Velocity contours (b) NO_2 mass fraction contours

Figure 5. Velocity contours and NO_2 mass fraction contours at section z=1.5 m(v_1=2.5 m/s)

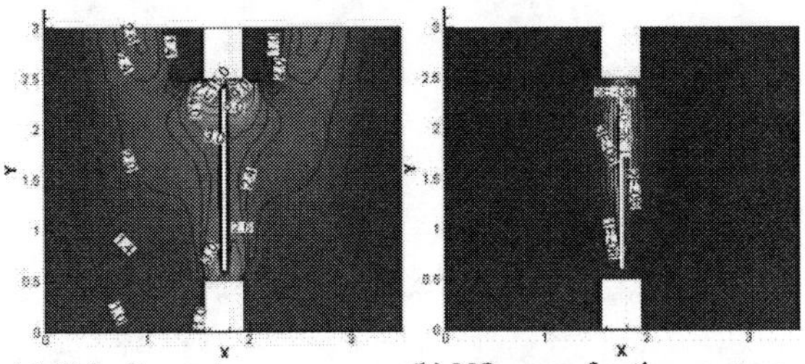

(a) Velocity contours (b) NO_2 mass fraction contours

Figure 6. Velocity contours and NO_2 mass fraction contours at section z=1.5 m(v_1=3 m/s)

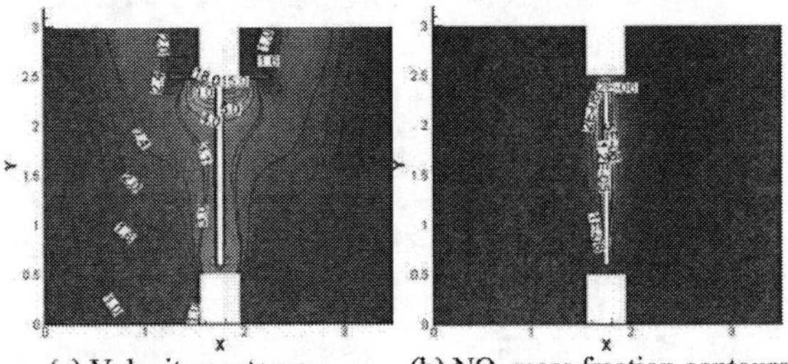

(a) Velocity contours (b) NO_2 mass fraction contours

Figure 7. Velocity contours and NO2 mass fraction contours at section z=1.5 m(v1=3.5 m/s)

2)NO_2 mass fraction contours

From Fig. 4 to Fig. 7, we can find that the greater the velocity is, the closer the contours presses to the plated pieces, and the smaller the contaminants emitting radius is. Obviously, the larger the air flow rate is, the higher the contaminants capture efficiency is. But large air flow

DOI: 10.1051/matecconf/20166808001

means higher costs, including equipments outlays, operation costs and maintenance costs. What we pursued is the smallest air flow rate with contaminants captured entirely. From this point of view, when contaminant emits at 0.5 m/s, the velocity of push hood should be greater than or equal to 2.5 m/s.

4 Comparative Analysis for Results

4.1 Velocity field comparison

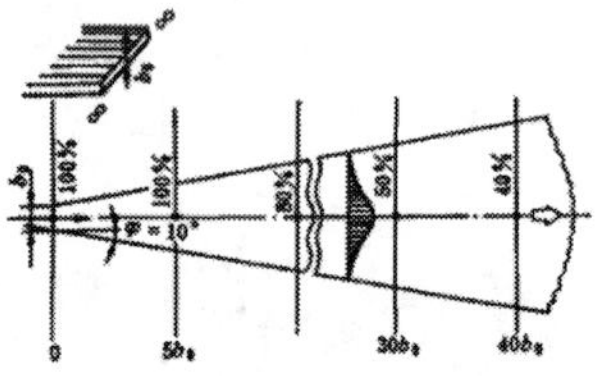

(a) 2D velocity profile of blowing air flow

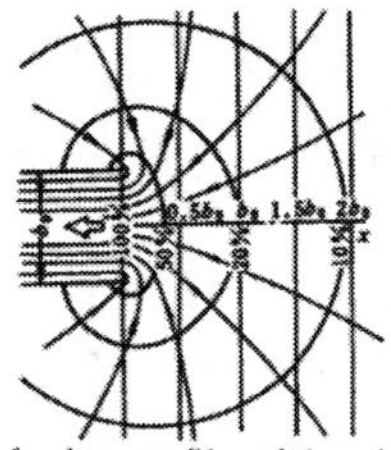

(b) 2D velocity profile of drawing air flow

Figure 8. Velocity profile correlation for pulling and pushing air flow

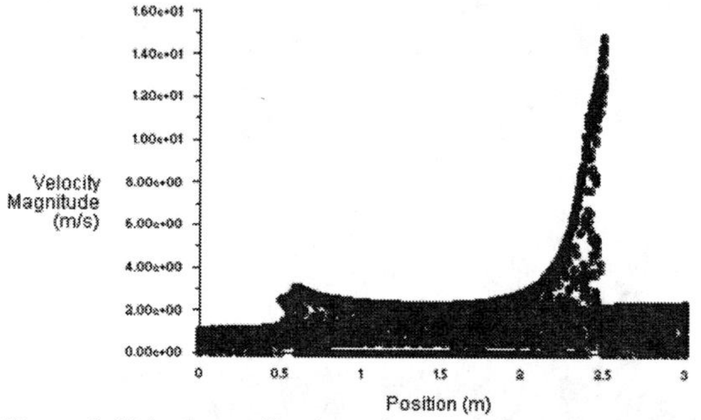

Figure 9. Velocity profile along the x coordinate axis at section z=1.5m

Fig. 8 [18] is the universally accepted velocity profile schematics for push jet and pull air flow which attenuate with the increase of distance. As we can see from the figure that velocity of push jet remains unchanged within $5b_0$ away from push hood; while the velocity of pull air flow attenuates to be 20% of face velocity at b_0, that is, the velocity at pull hood would be 12m/s if the velocity at b_0 was 2.4m/s. Fig. 9 is the schematics for velocity profile along the x coordinate axis at section z=1.5m, the push hood lies in the location of x=0.4, while the pull hood x=2.6. It is easily to be found that the push jet velocity attenuates slowly, the velocity at x=1.6, that is $3b_0$ away from push hood, keeps in 2.5m/s, which matches closely with the results from Fig. 8; while the induced air velocity of pull hood attenuates rapidly, the velocity at x=2.2m, that is b_0 away from drawing cover, attenuates to be 2.5m/s, which basically matches with the results from Fig. 8. Therefore, the simulation results appear to be reasonable.

4.2 Concentration field comparison

To verify the simulation models, combining with BW–GAXT NO_2 tester, NO_2 concentration at face z=1.5 m with location A(1.6,0.8), B(1.6,1.6), C(1.6,2.4), D(2,0.8), E(2,1.6), F(2,2.4), as shown in Fig. 10, and velocities of I(v_1=2 m/s, v_3=10.42 m/s), II(v_1=2.5 m/s, v_3=11.67 m/s), III(v_1=3 m/s,v_3=12.74 m/s), IV(v_1=3.5 m/s, v_3=14.0 m/s), are tested (Table 2). From this table, we can find that simulation values of NO_2 concentration nearly keeps with experimental data, with errors less than 9%.

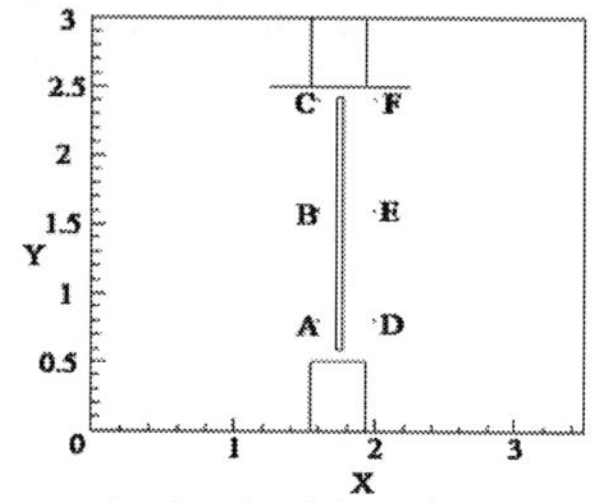

Figure 10. Diagrammatic sketch of detecting position of NO_2

Table 2. simulation values and experimental data of NO_2 concentration at z=1.5m×10^{-6}

	I		II		III		IV	
	SV	ED	SV	ED	SV	ED	SV	ED
A	1	1.3	1	1.2	1	1	1	1
B	9	10	4	5	3.3	3	2	1.8
C	2	2.2	2	2.1	2	2.1	1.6	1.7
D	0	1	0	1	0	0	0	0
E	2	3	1	2	0	0	0	0
F	0	1	0	1	0	0	0	0

Where : SV :simulation values , ED :experimental data

5 Conclusions

(1) Taking the volatile contaminants given off in plating process of some enterprise as subject, combining with some assumptions, mathematical model is established. By CFD code FLUENT, flow field, under v_1=2 m/s, v_3=10.42 m/s; v_1=2.5 m/s, v_3=11.67 m/s; v_1=3 m/s, v_3=12.74 m/s; v_1=3.5 m/s, v_3=14.0 m/s, of pull-push ventilation system, over plating tank, with obstacles between them is obtained. When obstacles presenting, air jet from push hood will disperse. Our research find that in this circumstance the best matched velocities are v_1=2.5 m/s and v_3=11.67 m/s.

(2) Computer simulation is useful to simulate pull-push ventilation system's performance under various conditions, and benefit to determine the best matched velocities of push air flow and pull air flow to avoid equipment costs due to the change of the size, leaving less experimental cost and shortened experimental period.

(3) Results show that simulated velocity field matches accepted theory well, while simulated concentration field keeps with experimental data nearly. So mathematical model presented in this paper is helpful for design.

Acknowledgements

The research presented in this paper is supported by Hengyang scie-tech Bureau (2014KG65), leading academic discipline project of Hunan Institute of technology (HXK13002).

References

1. Malin B S. Practical pointers on industrial exhaust system. Heat & Vent, **42**: 75–82(1945).
2. Enrique Gonza¡ä lez, Francisco Marzal, Agustı¡än Min˜ ana,etc. Influence of Exhaust Hood Geometry on the Capture Efficiency of Lateral Exhaust and Push–Pull Ventilation Systems in Surface Treatment Tanks. Environmental Progress, **27**(3):405-411, (2008).
3. Yi Zheng, Magesh Thiruvengadam, Hai Lan. Design of push–pull system to control diesel particular matter.inside a dead-end entry. Int J Coal Sci Technol, **2**(3):237–244 (2015).
4. Guo Peishan, Fu Haiming, Zhao Youjun, et al. Development and present status of push-pull ventilation technology. Journal of HV & AC, **38**(4): 57–61 (2008).
5. Michael Robinson, Derek B. Ingham. Design and operating parameters for push - pull ventilation of open surface tanks. International Journal of Energy Research, **27** (8):757-770(2003).
6. R. Rota, G. Nano, L. Canossa. Design Guidelines for Push-Pull Ventilation Systems through Computational Fluid Dynamics Modeling. AIHAJ: A Journal for the Science of Occupational & Environmental Health&Safety. **62** (2):141-148.(2001)
7. Hunt G R, Ingham D B. Long range exhaustion--a mathematical model for the axisymmetric air flow of a local exhaust ventilation hood assisted by a turbulent radial jet. The Annals of Occupational Hygiene, **40**(2): 171–196(1996).
8. He Suyan, Wang Yingmin, Li Yunfei. Flow-field test and simulation of a three-dimensional Aaberg exhaust hood. Journal of Shanghai Jiaotong University, **36**(10): 1492–1495(2002).
9. Chen Jiangping, Feng Xin, Mu Jingyang. Numerical simulation of turbulence character the 2-layer non-isothermal blow-draw air-curtain. Journal of Refrigeration, **4**: 16–20 (2001).
10. Nielsen PV. Computational fluid dynamics and room air movement. Indoor Air, **14**:134–143 (2004).
11. K Svidt. Numerical prediction of airflow in a room with ceiling-mounted obstacles.Denmark, Aalborg Universitets center, (1992).
12. Welling I, Andersson M, Rosen G, Ra¨isa¨nen J, Mielo T, Marttinen K, Niemela¨ R Contaminant dispersion in the vicinity of a worker in a uniform velocity field. Ann Occup Hygiene, **44**(3):219–225 (2000).
13. Xian Xiangli, Chun Holiu, Dennis Y.C.Leung, K.M.Lam. Recent progress in CFD modeling of wind field and pollutant transport in street canyons. Atmospheric Environment, **40**:5640-5658(2006).
14. J.D. Anderson, McGrawHill.Computational Fluid Dynamics: The Basics with Application. Tinghua University Press (1995).
15. Kambiz Nazridoust, Goodzarz Ahmadi. Airflow and pollutant transport in street canyons. Journal of wind Engineering and Industrial Aerodynamics, **94**:491-522 (2006).
16. Tao Wenquan.Numerical Heat Transfer (the 2nd edition).Xi'an: Xi'an Jiaotong University Press (2002).
17. Wang Fujun. Computational Fluid Dynamics. Beijing: Tsinghua University Press (2004).
18. Sun Yijian. Industry Ventilation (the 3rd edition). Beijing: China Building Industry Press (2000).

Hydraulic Analysis of Biochannels for Sustainable Urban Drainage Systems

Muhamad Nurfasya Alias[1], Muhammad Mujahid Muhammad[2], Khamaruzaman Wan Yusof[2], Muhammad Raza Ul Mustafa[2], and Aminuddin Ab. Ghani[1]

[1]River Engineering and Urban Drainage Research Centre (REDAC), Universiti Sains Malaysia (USM), Malaysia
[2]Civil & Environmental Engineering Department, Universiti Teknologi PETRONAS (UTP), Malaysia

Abstract. This study focuses on employing channel lining materials that includes Turf Reinforcements Mat (TRM) and Erosion Control Blanket (ECB) as bioengineering techniques for erosion control. Hydraulic characteristics of flow through these materials were determined in terms of flow velocities and shear stresses. The experimental results indicated that the average velocity for TRM ranges between 0.6 to 3.5 m/s, while that of ECB varies from 0.7 to 4.1 m/s. Also, TRM has shear stress values of 1.950 N/m^3 for bed and 1.575 N/m^3 for bank respectively of test channel when the flow depth was 15cm. While ECB has shear stresses of 2.138 N/m3 for bed and 1.721 N/m^3 for bank for same flow depth of 15cm. However, when the flow depth is 5 cm, TRM has values of 0.972 N/m^3 for bed and 0.769 N/m^3 for bank, and ECB gives 0.957 N/m^3 for bed and 0.766 N/m^3 for bank of test channel. Hence, the greater the flow depth, the higher the velocities and shear stresses respectively. Conversely, lower flow depths lead to velocities and shear stresses that are within the allowable standards for designing a stable channel. Generally, TRM performed the best for having higher shear stresses compared to ECB.

1 Introduction

Bioengineering is a construction method of using living plant or plants combined with dead or organic materials. The practice results in the combined use of biological, ecological and engineering concepts to produce a living, functioning system to prevent erosion and provide particulate contaminant removal through sedimentation and soil infiltration, and improves the aesthetics of a site [1, 2]. In addition, natural vegetation can be applied on hilly land surfaces to increase slope stability against failure due to erosion by rainfall. Erosion occurs when the hydraulic forces in the flow exceed the resisting forces of the channel boundary. The amount of erosion is a function of the relative magnitude of these forces and the time over which they are applied. The interaction of flow with the boundary of open channels is only imperfectly understood [3].

The use of plants for river bank protection and erosion control has a long tradition in Europe. Recently, existing bank erosion and stabilization techniques have been rediscovered and improved [4]. These old techniques have been rediscovered and improved in the first half of nineteenth century, with new technical and biological knowledge. Today, there is a variety of bioengineering techniques available to suit different situations and requirements. Biotechnical engineering techniques have become part of geotechnical and hydraulic engineering and have helped bridge the gap between classical engineering disciplines, land use management, landscape architecture and biological sciences [5].

These biotechnical engineering (bioengineering) techniques are developed by a system of trial and error. Most of these techniques are based on long-term practical experience. These "soft" engineering practices can provide possibilities to complement and improve or in some cases even replace traditional "hard" river-training constructions, such as placement of gabions or rock. These also offer a more ecologically acceptable way of bank stabilization that still compiles the land use and safety requirements.

Many governmental agencies favored stone or concrete riprap because over time, a high degree of precision and confidence in construction has developed from research and analysis. In engineering viewpoints, these techniques have been successful or their immediate protections of properties or infrastructure adjacent to the stream after projects were completed. What was thought successful in the past is being reevaluated in context of impacts resulting from excessive and rapid urbanization, and public awareness of these new environmental issues. Increasing failures of traditional channelization and armoring techniques are generating questions as to whether traditional practices are appropriate in every setting [6]. The interest in natural techniques called biotechnical engineering was raised, and the benefits and advantages of biotechnical engineering were gradually re-examined [7].

The purpose of this study was to investigate the effectiveness and applicability of bioengineering techniques for bank stability and protection against erosion as a viable alternative to traditional channelization and hard armoring techniques. Also, the suitability of the technique was assessed which can be used in urban environments like Malaysia.

2 Theoretical concepts

2.1 Vertical velocity profiles

Velocity varies approximately as a parabola from zero at the channel bottom to a maximum near the surface. A typical vertical velocity profile is shown in Figure 1. It has been determined empirically that for most channels the velocity at six-tenths of the total depth below the surface is a close approximation to the mean velocity at that vertical line. However, the average of the velocities at two-tenths and eight-tenths depth below the surface on the same vertical line provides a more accurate value of mean velocity at that vertical line.

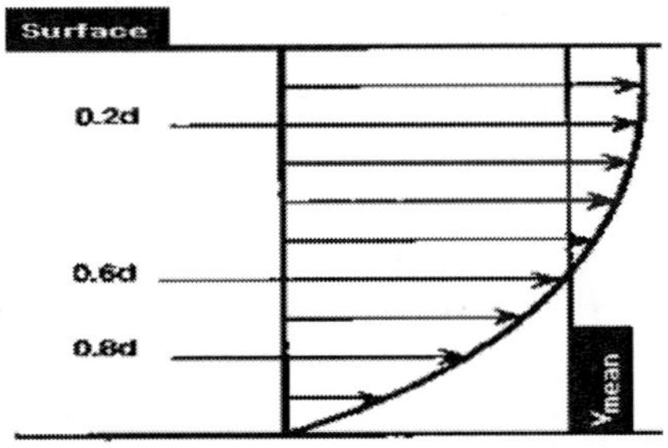

Figure 1. Typical river velocity profile in the vertical plane.

2.2 Channel stability

Two basic types of lining classes are rigid linings such as concrete, and flexible linings such as rock rip raps, rock mattresses, and all RECPs [8, 9]. Rigid linings are non-erodible, permanent and long life, but they are susceptible to failure from foundation instability and unreliable hydraulic pressure release. Construction of rigid linings requires specialised equipment and costly materials. As a result, the cost of rigid channel linings is typically higher than an equivalent flexible channel lining[8].

The stability of a channel or the suitability of various channel linings can be determined by first calculating both the mean velocity and tractive stress. Additionally, product literatures from manufacturers can provide information on allowable tractive stresses or velocities for various types of erosion control products. A general procedure for designing a stable channel has been presented by Fischenich [3]. Thus, after estimating the mean hydraulic conditions, the shear stress may be computed, for example, the local maximum shear stress, τ_{max}, for straight channels can be calculated from the following simple equation (1):

$$\tau_{max} = 1.5\ \tau \quad (1)$$

For sinuous channels, the maximum shear stress, τ^*_{max}, should be determined as a function of the platform characteristics using Equation (2):

$$\tau^*_{max} = 2.65\ \tau\,(R_c/W)^{-0.5} \quad (2)$$

where, R_c = Radius of curvature (m) and W = Top width of the channel (m).

Moreover, existing stability should be assessed by comparing estimates of local and instantaneous shear and velocity to values. If the existing conditions are deemed stable and are in consonance with other project objectives, then no further action is required. Otherwise recomputed flow values, Table 1 presents velocity limits for various channel boundaries conditions, while in order to account for flow duration Figures 2 and 3 may be used as a guide.

Table 1. Stability of channel linings for given velocity range by fischenich [3].

	Velocity range (m/s)				
Lining	**0 – 0.61**	**0.61 – 1.22**	**1.22 – 1.83**	**1.83 – 2.44**	**> 2.44**
Sandy Soils					
Firm Loam					
Mixed Gravel and Cobbles					
Average Turf					
Degradable Rolled Erosion Control Products (RECPs) Stabilizing Bioengineering					
Good Turf					
Permanent RECPs Armoring Bioengineering CCMs & Gabions					
Riprap					
Concrete					
Key:					
	Appropriate				
	Use Caution				
	Not Appropriate				

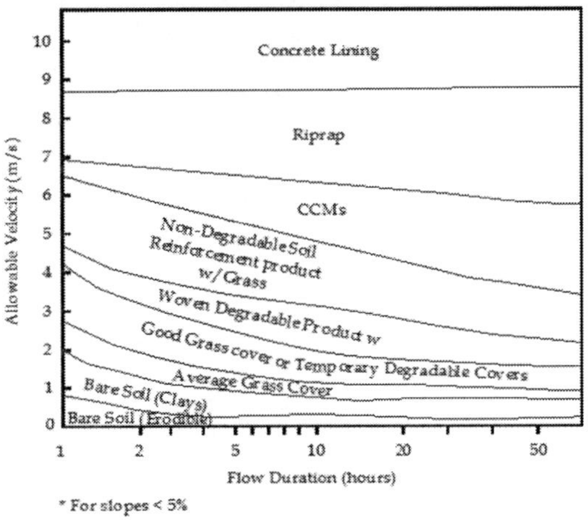

Figure 2. Erosion limits as a function of flow duration by Fischenich and Allen [10].

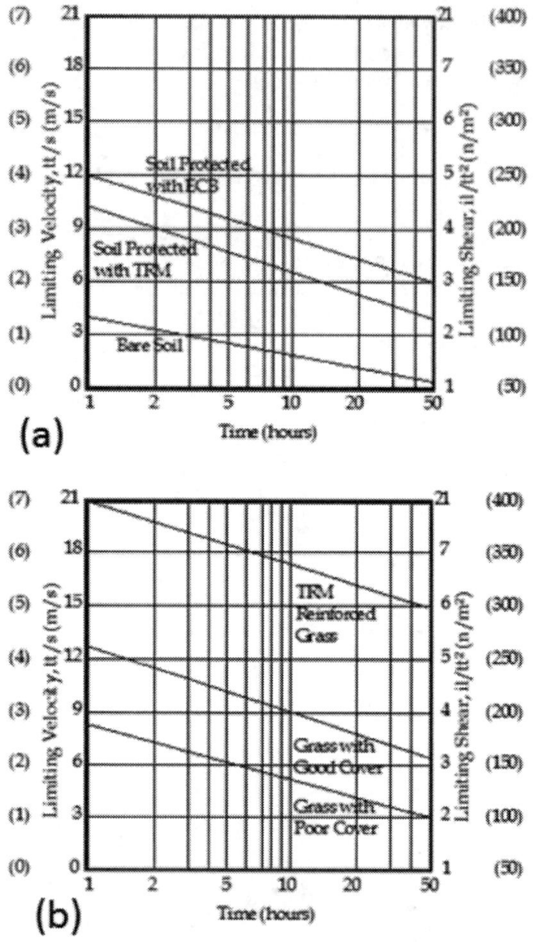

Figure 3. Limiting value for velocity and shear for (a) Bare soil, TRM and ECB ; (b) Grass and/or TRM [11].

2.3 Shear stress on streambed and banks

On any given bank, the material and vegetation types required to resist erosion may vary with location. Lane's diagram, Figure 4, shows theoretical distribution of shear stress on streambed and banks on a straight section of trapezoidal channel. Based on Lane's diagram, materials and plants of greater shear resistance are required lower on the bank, while a lighter-duty treatment may be sufficient near the top of the bank.

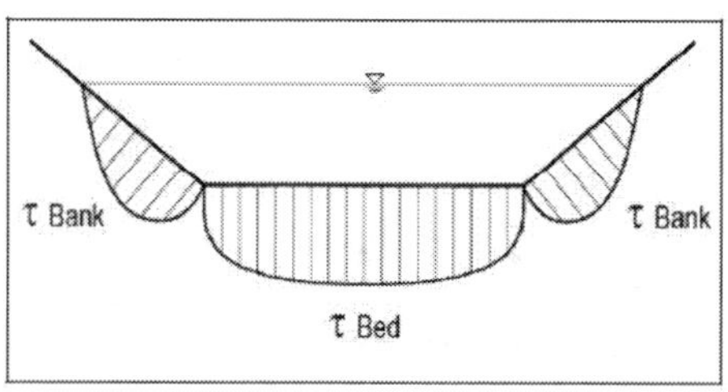

Figure 4. Distribution of shear stress on streambed and banks.

Shear stress is an important parameter in habitat rehabilitation design, because all materials, whether manufactured or natural, used for habitat rehabilitation must be able to withstand the expected shear stress at the design discharge [12].

A common application of the equation for maximum shear stress in a straight reach will be apply for streambed and bank of the test channel. For maximum shear stress on the bed is:

$$\tau_{bed} = \gamma RS = \gamma (A/P)S \quad (3)$$

$$\tau_{bed} = 9806\, RS \quad (4)$$

where, τ_{bed} = maximum bed shear stress (N/m^2), γ = the specific weight of water (N/m^3), R = hydraulic radius in m, A = flow cross-sectional area (m^2), P = wetted perimeter (m) and S =energy slope.

This calculation gives a quantitative measure of the erosive force acting on the bed of the channel.

By approximating the channel cross-section as a trapezoid or rectangular, the maximum bed shear stress can be used to estimate the maximum bank shear stress. This stress acts approximately one-third of the distance up the bank (from the bed) and can be approximated by multiplying the maximum bed shear stress by a factor (see Lane's Diagram, Figure 5). This factor, K_1, varies based on channel side slope and the ratio of bottom width to depth as shown in Figure 5. This approximation applies only to a relatively straight channel reach.

For maximum shear stress, τ_{bank}, on the bank is:

$$\tau_{bank} = K_1 \tau_{bed} \quad (5)$$

where, τ_{bed} = maximum bed shear stress (N/m^2) and K_1 = ratio obtained from Figure 5. It should be noted that τ_{bed} is estimated by knowing the aspect ratio width (B) divide by flow depth (d) = B/d and the side slope z using Figure 5 as well.

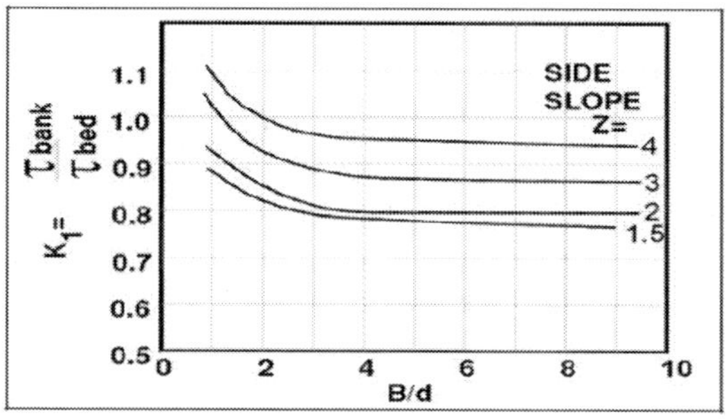

Figure 5. Coefficient K_1 vs. side slope z, and width/depth B/d ratio.

3 Methodology

3.1 Selection of the bioengineering materials for channel bank stabilization

Two (2) types of channel lining materials of bioengineering technique such as Turf Reinforcement Mat (TRM) and Erosion Control Blanket (ECB) were selected in this study. Both products were designed for erosion protection and the establishment of vegetation up to 12 months. Others technical specification for both selected material are presented in Table 2.

Table 2. Technical specification of tested material (fibromat).

Bioengineering Techniques Material	Technical Specification			
Turf Reinforcement Mats (TRM C350) 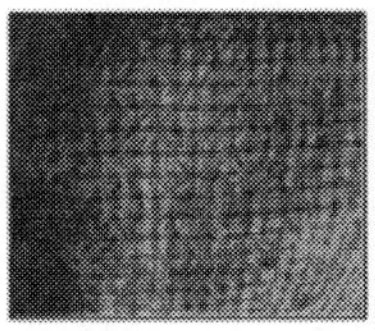	Matrix	: Coir Fibre (C350) / Coir Fibre + Paddy Straw (SC350)		
	Content	: 100% (C350) / 70% + 30% (SC350)		
	Roll Weight	: 2.4m		
	Roll Length	: 21m		
	Area	: 50m²/roll		
	Functional Longevity	: Approx. 36 months		
	Applications	: 3:1 to 1:1 Slopes		
	Permissible Shear Resistance	: 15.6kg/m² or 152.98N/m²		
	Flow Velocity	: 3.8m/s (estimated)		
	"n" Value Roughness Coefficient	: 0.018 (estimated)		
	Water Absorption	: 25.0% (estimated)		
	Swell	: 233.0% (estimated)		
	Netting:			
		Top	Center	Bottom
	Material	Black PP	Balck PP Corrugated	Black PP
	Weight	3.9kg/100m²	11.7kg/100m²	3.9kg/100m²
Erosion Control Blanket (ECB DS250) 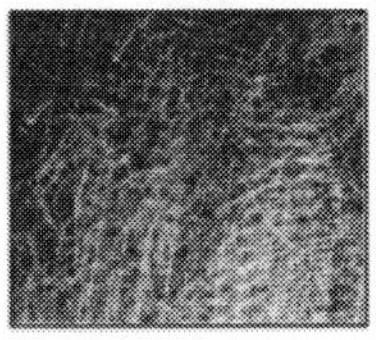	Matrix	: Agriculture Straw		
	Content	: 100%		
	Roll Weight	: 2.4m		
	Roll Length	: 42m		
	Area	: 100m²/roll		
	Functional Longevity	: Approx. 12 months		
	Applications	: 3:1 to 2:1 Slopes		
	Permissible Shear Resistance	: 10.25kg/m² or 100.52N/m²		
	Netting	: PP UV Stabilized		

3.2 Design considerations of the test channel

The channel for this research project was constructed by using the design criteria as follows:

Bed slope, So = 1 / 500 or 0.002

Channel bank slopes = 1:2 (left bank of the channel) and 1:1 (right bank of the channel)

Channel length = 10 m

Channel cross section = Trapezoidal

Bottom width = 0.6 m

Top width = 2.1 m

Depth = 0.5 m

After preparing the platform and backfilling the channel cross section with soil, Figure 6 (a) – (f) shows the laying works of lining material on the soil surface in the proposed test channel.

3.3 Cross section measurements

Bank full width method was used in order to get the cross section profile of the test channel. A tape measure been stretched from one bank to the other at 90° to the course of the channel. Figure 7 shows the measuring works of test channel cross section, which represents the measurement locations of the cross section profile at 1m intervals from the channel inlet, with CH 0.0, CH 1.0, CH 2.0, CH 3.0, CH 4.0, CH 5.0 and CH 6.0 denoting 0 m, 1 m, 2 m, 3 m, 4 m, 5 m and 6 m respectively.

Figure 6. Laying and stapling works of lining material in the proposed test channel.

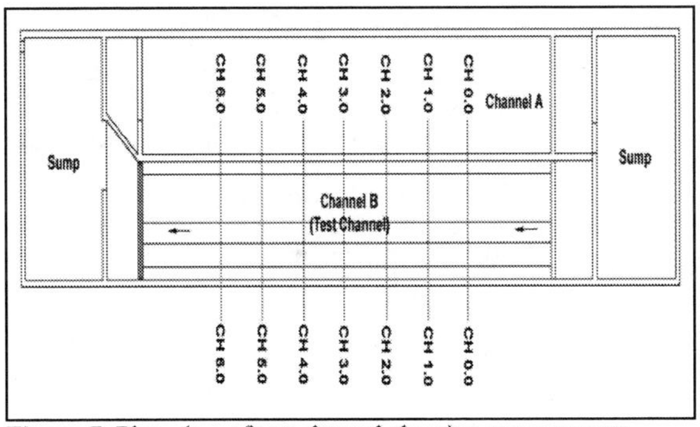

Figure 7. Plan view of test channel showing measurement locations.

3.4 Experimental procedure for testing the bioengineering materials

The water was pumped into the test channel from the existing overhead water tank. An underground sump was to collect and store the water from channel before it is being circulated to the water tank. Figure 8 (a) – (b) shows the flow of water through the test channel.

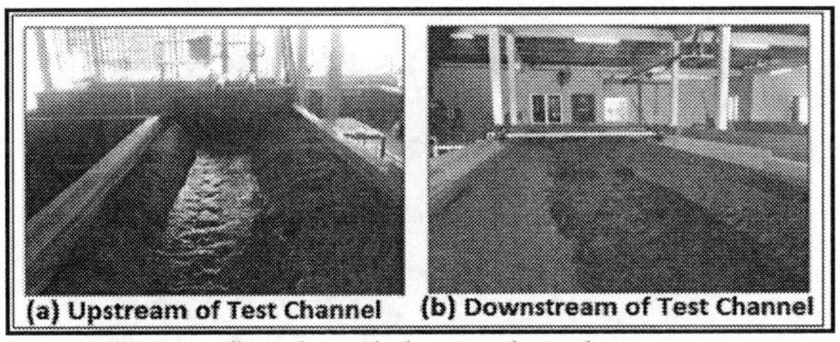

Figure 8. Water flow through the test channel.

Due to the occurrence of backwater in the channel length, only 4m length at the middle part of the test channel was used for testing. Velocities of test channel were measured by using flow meter. Three (3) different depths were used to measure flow characteristics in the test channel. These depths comprised 5cm, 10cm and 15cm. Three (3) readings for vertical velocity were

DOI: 10.1051/matecconf/20166808002

recorded by using a current meter and then average values computed for each cross section.

Afterwards the discharge, Q, was estimated by the velocity – area method, which was derived from Q = VA, where, V is the measured average velocity and A, being the cross sectional area. Then Manning's roughness, n, in Tables 3 and 4 were calculated using Manning's equation (6), which, although developed for conditions of uniform flow in open channels, may give an adequate estimate of the non-uniform flow which is usual in natural channels. The Manning equation states that:

$$Q = (AR^{2/3}S^{1/2})/n \quad (6)$$

where, Q = discharge (m^3/s), A = cross-sectional area (m^2), R = hydraulic radius (m) = A/P, P = wetted perimeter (m), and S = slope of gradient of the bed.

4 Results and discussion

4.1 Cross section profiles of test channel

Figures 9a and 9b show the cross sectional profiles of the test channel at different chains along the channel length. Also, it depicts the maximum level in which water may rise.

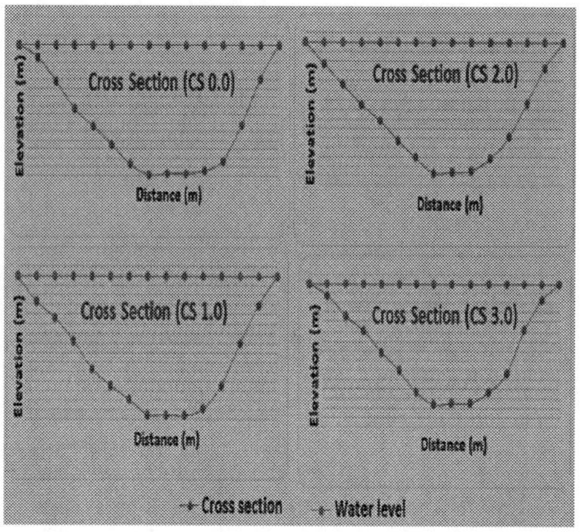

Figure 9a. Cross sectional profiles at 0 m, 1 m, 2 m and 3 m along the test channel.

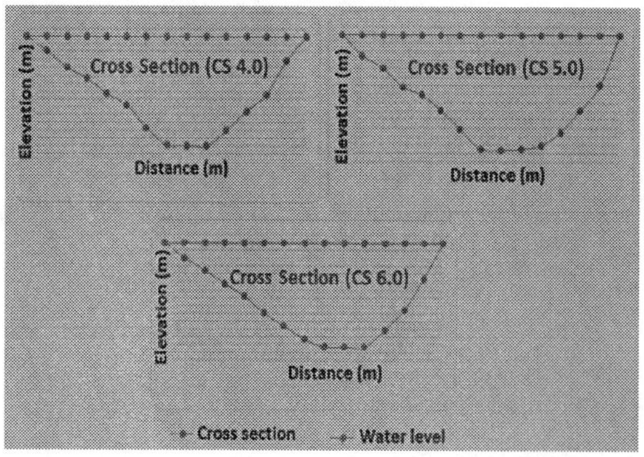

Figure 9b. Cross sectional profiles at 4 m, 5 m, and 6 m along the test channel.

4.2 Flow parameters

Starting with TRM test channel, from Table 3, the average vertical velocity for 5cm depth recorded was in the range of 0.6m/s to 1.2m/s. The average vertical velocity for 10cm depth recorded falls between 1.6m/s to 1.9m/s. While the average vertical velocity for 15cm depth recorded varies from 3.1m/s to 3.5m/s. Similarly, from Table 4, for ECB test channel, the average vertical velocity for 5cm depth recorded falls between 0.7m/s to 1.4m/s. The total average vertical velocity for 10cm depth recorded were within 1.5m/s to 2.7m/s. While the average vertical velocity for 15cm depth recorded varies from 3.2m/s to 4.1m/s.

Table 3. Total average vertical velocity and flow for TRM test channel.

Lining Material	Flow Depth (cm)	Cross Section (CH m)	Total Average Vertical Velocity, v (m/s)	Total Flow, Q (m^3/s)	Manning's n
TRM	5.0	0.00	0.772	0.0050	0.025
		1.00	0.647	0.0033	0.039
		2.00	0.760	0.0040	0.037
		3.00	0.920	0.0049	0.019
		4.00	1.172	0.0062	0.015
		Total	**4.270**	**0.0234**	**Mean 0.027**
	10.0	0.00	1.840	0.0206	0.018
		1.00	1.823	0.0183	0.020
		2.00	1.893	0.0173	0.021
		3.00	1.693	0.0157	0.024
		4.00	1.697	0.0179	0.028
		Total	**8.947**	**0.0898**	**Mean 0.022**
	15.0	0.00	3.160	0.0451	0.014
		1.00	3.470	0.0436	0.013
		2.00	3.400	0.0453	0.014
		3.00	3.857	0.0439	0.014
		4.00	3.453	0.0458	0.013
		Total	**17.340**	**0.2237**	**Mean 0.014**

Table 4. Total average vertical velocity and flow for ECB test channel.

Lining Material	Flow Depth (cm)	Cross Section (CH m)	Total Average Vertical Velocity, v (m/s)	Total Flow, Q (m^3/s)	Manning's n
ECB	5.0	0.00	0.783	0.0059	0.033
		1.00	1.007	0.0051	0.024
		2.00	0.780	0.0031	0.033
		3.00	0.782	0.0032	0.028
		4.00	1.347	0.0061	0.014
		Total	**4.698**	**0.0234**	**Mean 0.026**
	10.0	0.00	1.533	0.0179	0.024

	1.00	1.360	0.0117	0.026
	2.00	1.893	0.0130	0.018
	3.00	2.107	0.0121	0.018
	4.00	2.677	0.0164	0.012
	Total	**9.570**	**0.0711**	**Mean 0.019**
15.0	0.00	4.090	0.0683	0.013
	1.00	3.283	0.0479	0.017
	2.00	3.670	0.0526	0.013
	3.00	3.597	0.0471	0.015
	4.00	3.990	0.0515	0.012
	Total	**18.630**	**0.2673**	**Mean 0.014**

4.3 Shear stress for the channel lining materials

The flow depth of 15cm gives the highest values of estimated shear stress for both lining materials. The experiment conducted for TRM lining material gives 1.950 N/m^3 for bed and 1.575 N/m^3 for bank of test channel. While for ECB lining material gives 2.138 N/m^3 for bed and 1.721 N/m^3 for bank of test channel. The flow depth of 5cm gives the lowest values of estimate shear stress for both lining materials. Lining material of TRM gives the values 0.972 N/m^3 for bed and 0.769 N/m^3 for bank of test channel. While lining material of ECB gives the values 0.957 N/m^3 for bed and 0.766 N/m^3 for bank of test channel. Table 5 shows the summary of average shear stress for test channel bed and bank. Also, Figure 10 & 11 show the variations of shear stresses with the aspect ratio (B/y). As the aspect ratio decreases the shear stresses increases and vice versa in the both test materials.

Table 5. Average shear stress for test channel bed and bank.

Lining Material	Parameter: Flow Depth (cm)	B/y (Average)	τ_b (N/m^3)	τ_s (N/m^3)
TRM	5	8.071	0.972	0.769
	10	6.096	1.568	1.262
	15	4.861	1.950	1.575
ECB	5	7.525	0.957	0.766
	10	5.113	1.418	1.139
	15	4.467	2.138	1.721

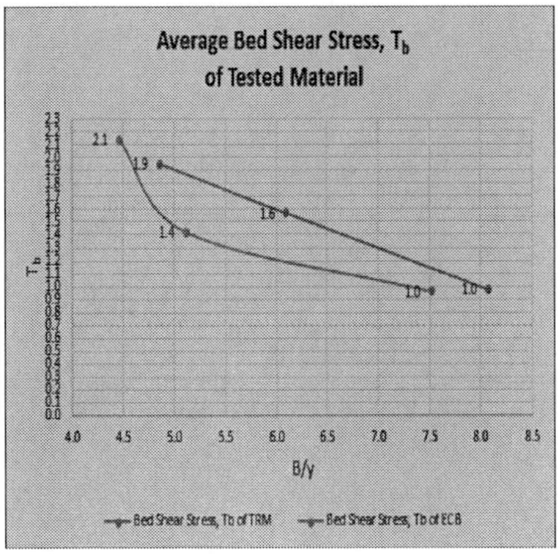

Figure 10. Average bed shear stress for tested material.

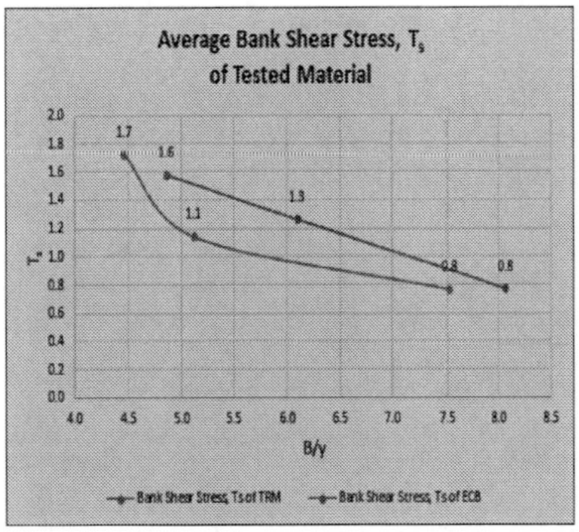

Figure 11. Average bank shear stress for tested material.

5 Conclusion

Within the limit of experimental error it can be concluded that both the average vertical velocities and estimated shear stresses of the different flow depths, show that the greater the flow depth, the higher would be the values of velocity and shear stress for both TRM and ECB materials respectively. Therefore, at low flow depths the hydraulic parameters obtained are within the allowable values and appropriate for designing a stable channel. Generally, TRM performed the best being having higher shear stress values compared to ECB.

Furthermore, investigation on combining TRM or ECB with green vegetation is required, as vegetation is more sustainable to erosion effects and eco-friendly to the environment.

Acknowledgements

The authors would like to acknowledge the short term grant from Universiti Sains Malaysia (Grant No. 304.PREDAC.60312035 / 1001.PREDAC.814225). The second author would also like to acknowledge Universiti Teknologi PETRONAS for the fellowship scheme to undertake the postgraduate study, and to as well the collaborative research with REDAC.

References

1. Department of Irrigation and Drainage Malaysia (DID). Urban Stormwater Management Manual for Malaysia, (2012)
2. Shahkolahi, A., Tadman, A. and Crase, J. An Economical Solution for Flood Protection and Channel Erosion Control. in Institute of Public Works Engineering Australiasia (IPWEA) Queensland Conference, (2014)
3. Fischenich, C. Stability thresholds for stream restoration materials, DTIC Document, (2001)
4. Evette, A., Labonne, S., Rey, F., Liebault, F., Jancke, O., Girel, J. History of bioengineering techniques for erosion control in rivers in Western Europe. Environmental Management, **43**, 6 972-984 (2009)

5. Donat, M. Bioengineering techniques for streambank restoration. A Review of Central European Practices. Vancouver, BC, Canada: Watershed Restoration Program. Ministry of Environment, Lands and Parks, and Ministry of Forests, (1995)
6. Li, M.-H. and K.E. Eddleman. Biotechnical engineering as an alternative to traditional engineering methods: A biotechnical streambank stabilization design approach. Landscape and Urban Planning, **60**, 4 225-242 (2002)
7. Riley, A.L. Restoring streams in cities: A guide for planners, policymakers, and citizens, Island Press, (1998)
8. Kilgore, R.T. and G.K. Cotton, Design of roadside channels with flexible linings hydraulic engineering Circular Number 15, (2005)
9. Witheridge, G. Erosion and Sediment Control – A Field Guide for Construction Site Managers: Catchments and Creeks, P. Ltd, Editor, (2010)
10. Fishenich, C. and H. Allen. Stream Management, Water Operations Technical Support Program. US Army Corps of Engineers, Environmental Laboratory. Waterways Experiment Station. Vicksburg, MS, (2000)
11. Sprague, C. Green Engineering: Design Principles and Applications using Rolled Erosion Control products. Part Two. Civil Engineering News, **11**, 2 (1999)
12. Saldi-Caromile, K., Bates, K., Skidmore, P., Barenti, J., and Pineo, D, Stream habitat restoration guidelines: Final draft. Washington Depts. of Fish and Wildlife and Ecology and the U.S. Fish and Wildlife Service. Editor Olympia, WA, (2004)

Design Procedure Enhanced with Numerical Modeling to Mitigate River-Bank Erosion

Mohamed Elhakeem[1], and Thanos Papanicolaou[2]

[1]*Abu Dhabi University, Abu Dhabi, P.O. Box 59911, UAE*
[2]*The University of Tennessee, Knoxville, TN, 37996, USA*

Abstract. In this study, the 2D Finite Element Surface Water Modeling System (FESWMS) is used to design barb structures to mitigate river bank erosion in a stream reach located on the Raccoon River near Adel, Iowa, USA just upstream of the US Highway Bridge 169. FESWMS is used also to access the barbs effect on the study reach. The model results showed that the proposed barb structures successfully reduced the flow velocity along the outside bank and increased the velocity in the center of the stream, thereby successfully increased the conveyance towards the core of the river. The estimated velocities values along the river-banks where the barbs exist were within the recommended values for channel stability design. Thus, the barb structures were able to reduce the erosion along the bankline.

1 Introduction

Barb (either spur or bendway weir) structures have been used in many streams to mitigate streambank erosion by redirecting stream flow away from the banks towards the core of the channel [1]. Yet, detailed guidelines for barbs design are limited due to the site specificity involved in most studies [1]. Therefore, hydraulic engineers must resort to physical and/or computational models to predict the effects of barb structures on riverine environment and base their design on the model outputs. Computational models can be adapted for different physical domains more easily than physical models [2]. In addition, computational models are not subjected to distortion effects like physical models when a solution can be obtained for the same flow conditions as those present in the field (i.e., complete similarity by having the same length scale in the three directions, and identical hydrodynamic and kinetic forces). However, computational models must be calibrated and verified against laboratory or field data [2].

Numerous commercially hydrodynamic models exist for various intended applications, which can broadly be classified into 1D, 2D and 3D models [2]. 2D-hydrodynamic models have been deemed to be ideal for simulating flow in stream reaches containing barb structures because they can resolve the large-scale turbulent flow structure and provide spatially varied information regarding the flow pattern around barbs [3]. In addition, 2D models require less data than 3D models for model calibration and verification, which make them more attractive [3]. The majority of the 2D-hydrodynamic models typically solve the depth-averaged continuity and Navier-Stokes equations using finite-difference, finite-element, or finite-volume schemes [2].

In this study we used the Finite Element Surface Water Modeling System (FESWMS) to design the barb structures and access their effect on a stream reach located on the Raccoon River near Adel, Iowa, USA just upstream of the US Highway Bridge 169. FESWMS is a commercially available 2D-hydrodynamic model developed by the Federal Highway Administration [4]. The model has a proven track record in modeling riverine environment and was successfully used to simulate flow in river bends where secondary currents were prevalent, and to study bank erosion in a straight reach of a stream [5]. It was also used to study fish habitat in a gravel-bed stream under different bed-roughness conditions and to study shallow water habitat around dikes in the Missouri River [6]. Detailed sensitivity analysis and error propagation studies have shown that FESWMS provides in general satisfactory results for streams of complex geometry and bed-roughness when the model is carefully calibrated against field measurements [7].

2 Methodology

The methodological steps needed to accomplish the study objective include: 1) a review of the essential criteria for designing barb structures; 2) description of the study site and collection of field data for model validation; 3) description of the important features of the 2D hydrodynamic model.

2.1 Barbs design guidelines

DOI: 10.1051/matecconf/20166808003

The general design guidelines of barbs (e.g., length, spacing) have been documented in many hydraulics textbooks [1]. The design criteria depend mainly on the purpose of the structures (e.g., bank erosion protection vs. navigation improvement) and are based on a fundamental understanding of the prevailing flow regions in the vicinity of the barbs. The length of barbs is primarily a function of the amount of flow needed to be redirected towards the core of the channel [1]. The maximum length of a barb is determined from the maximum permissible flow depth and velocity at the main-core region [1]. Typical design specifications recommended that the barb length to stream width ratio (L/B) is less than 0.5 [1]. For L/B values greater than 0.5, the flow becomes increasingly contracted causing excessive bed-scour in the main-core region and back water effects. The spacing between the barbs depends upon the structure length and typical design specifications recommended that the spacing to length ratio (S/L) is between 0.75 and 2.0 [1]. The majority of the existing studies have evaluated the design criteria (other than spacing) from investigating the flow pattern around a single barb. However, multiple barbs are often used to control bank erosion in a stream reach. Introducing a series of barb structures in a stream reach can have large scale effects by increasing the total flow resistance of the reach creating backwater effects [8]. The presence of barbs affects the water surface profile and the overall spatial distribution of the velocity in the main-core and the barb regions [1]. It is therefore important to simulate the entire series of barbs to understand fully the large scale effects of these structures on the hydrologic and morphologic characteristics of a stream [8].

2.2 Study site and field work

The study site is located on the Raccoon River near Adel County, Iowa, USA just upstream of the US Highway Bridge 169. Figure 1a shows the study site after the construction of the barbs. The study reach has a sandy bed of mild gradient with an average bed slope of 0.00061, and a meandering pattern with an average sinuosity of 1.8, producing deeply entrenched corners in silt-clay banks with height exceeding 3.0 m. The bridge causes excessive bank erosion and formation of a meander bend well-upstream of the bridge span. The observed river-bank erosion appears to be accelerating as the bridge acts as a control structure not allowing the downstream portion of the reach to adjust to changes upstream. Field measurements were performed to support the numerical simulations. The field work included detailed velocity measurements using Large Scale Particle Image Velocimetry (LSPIV) technique (Figure 1b), and collection of bathymetry data of the study reach using standard surveying practices and an Eagle FishElite® 480 sonar. LSPIV is an image-based approach [9] for measuring time-averaged, water surface velocities over large areas (< 225 m^2). It provides not only the velocity magnitudes, but also the overall flow pattern. This non-intrusive method, which is based on particle tracking velocimetry can provide large data sets of simultaneously measured velocity vectors and resolve for large-scale flow patterns with errors less than 10 % [9]. The velocity vectors can be determined using particle image velocity algorithms, which estimate the displacement of seeding particles through a sequence of recorded images. Eco-friendly wood mulch as the seeding material and video cameras with a resolution of 720 × 480 pixels mounted on a truck are often sufficient to capture the mulch movement at the free surface. The large sample size provided by the LSPIV technique allows for a better statistically significant comparison between the measured and predicted velocity values. Bathymetry data were collected using standard surveying practices and an Eagle FishElite® 480 sonar. The Eagle FishElite® 480 is a coupled sonar/GPS instrument capable of recording depth and corresponding positional data at a rate of 1.0 Hz with errors less than 5 % [10]. These measuring techniques allowed recording large samples of data to be used for computational mesh generation and model calibration and verification.

Figure 1. a) Study site after the construction of the barbs; b) the Large Scale Particle Image Velocimetry (LSPIV) technique.

2.3 Model description and calibration

FESWMS is part of the commercially available Surface water Modeling System (SMS) package version 10.2 Graphical Interface that combines a series of hydrodynamic/ sediment codes [4]. FESWMS solves the differential forms of the continuity and the momentum equations in the stream wise and transverse directions

DOI: 10.1051/matecconf/20166808003

using the Galerkin method of weighted residuals, providing water depth and depth-averaged velocity magnitude in x and y directions at each node in the grid [4]. The governing equations are written in the conservative form hence, the momentum is conserved along the streamline and the model is capable of capturing shock effects [11]. The conservative form was chosen because of its robustness in solving critical and transcritical flow fields under both low and high flow conditions by allowing dry-elements to exist within the computational mesh [11].

FESWMS solves the following equations simultaneously:

$$\frac{\partial z_w}{\partial t}+\frac{\partial q_1}{\partial x}+\frac{\partial q_2}{\partial y}-q_m=0 \tag{1}$$

$$\frac{\partial q_1}{\partial t}+\frac{\partial}{\partial x}\left(\frac{q_1^2}{d}+\frac{1}{2}gd^2\right)+\frac{\partial}{\partial y}\left(\frac{q_1 q_2}{d}\right)+gd\frac{dz_b}{\partial x}$$
$$+gn^2\frac{q_1\sqrt{q_1^2+q_2^2}}{d^{7/3}}\sqrt{1+\left(\frac{\partial z_b}{\partial x}\right)^2+\left(\frac{\partial z_b}{\partial y}\right)^2} \tag{2}$$
$$-2d\varepsilon_{xx}\frac{\partial^2\bar{u}}{\partial x^2}-\varepsilon_{xy}\frac{\partial}{\partial y}\left(\frac{\partial\bar{u}}{\partial x}+\frac{\partial\bar{v}}{\partial y}\right)=0$$

$$\frac{\partial q_2}{\partial t}+\frac{\partial}{\partial x}\left(\frac{q_1 q_2}{d}\right)+\frac{\partial}{\partial y}\left(\frac{q_2^2}{d}+\frac{1}{2}gd^2\right)+gd\frac{dz_b}{\partial y}$$
$$+gn^2\frac{q_2\sqrt{q_1^2+q_2^2}}{d^{7/3}}\sqrt{1+\left(\frac{\partial z_b}{\partial x}\right)^2+\left(\frac{\partial z_b}{\partial y}\right)^2} \tag{3}$$
$$-2d\varepsilon_{yy}\frac{\partial^2\bar{v}}{\partial y^2}-\varepsilon_{yx}\frac{\partial}{\partial x}\left(\frac{\partial\bar{u}}{\partial x}+\frac{\partial\bar{v}}{\partial y}\right)=0$$

where (1) is the continuity and (2) and (3) are the momentum equations in x and y directions, respectively. In the equations, t is time (s), d is water depth (m), ρ is water density (kg/m^3), g is the acceleration due to gravity (m/s^2), n is Manning's coefficient of roughness, z_w and z_b are water surface elevation and bed elevation above certain datum (m), q_1 and q_2 are the unit discharge fluxes (m^2/s) defined as $\bar{u}_d$ and $\bar{v}_d$, respectively, $\bar{u}$ and $\bar{v}$ (m/s) are the depth-averaged velocities of an element in the stream wise and transverse directions, respectively, q_m is the resultant inflow or outflow from that element (m/s), ν_{xx} and ν_{yy} are the normal components of the eddy viscosity (m^2/s) in the x and y directions, respectively, and ν_{xy} and ν_{yx} are the shear components of the eddy viscosity (m^2/s) applied to the $x-y$ plane.

FESWMS inputs for model calibration are the Manning's coefficient of roughness n, and the eddy viscosity ν. Model inputs must be assigned correctly to represent the physical processes occurring in the modeled reach and to produce accurate model predictions. Manning's n is an empirical coefficient that accounts for the total flow resistance caused by flow interaction with the boundary [12]. FESWMS utilizes Manning's n to account for momentum loss due to bed-shear, which may vary significantly in a stream reach in accordance to bed-bathymetry and roughness. The second input variable used by FESWMS is the eddy viscosity ν. Eddy viscosity accounts for flow resistance due to the internal shear stresses, or the Reynolds' stresses of the fluid incorporating the added energy dissipation due to turbulence in the flow [13]. Therefore, eddy viscosity is not a physical property of the fluid, but rather a turbulent characteristic of the flow. Because of the complex geometry of the study reach, distributed Manning's n and eddy viscosity values are used for model calibration (Figure 2).

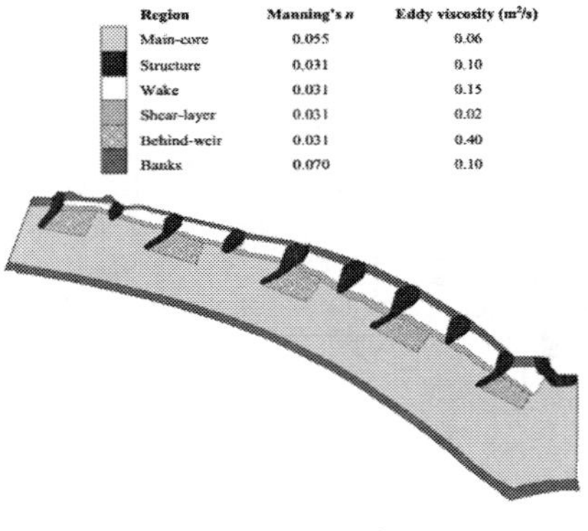

Figure 2. Spatially distributed Manning's n and eddy viscosity values over the study reach.

3 Results

3.1 Barbs design

Barbs (spurs and bendway weirs) were proposed for river-bank protection (Figure 3). The average length to width ratio (L/B) and average spacing to length ratio (S/L) of the structures are 1:3 and 2:1, respectively. Spurs was proposed to protect the bank line and force the river to the north, and bendway weirs to further redirect the flow from the backline towards the channel core. The major advantage of this system is correction of the incoming flow. The average length of the spurs and bendway weirs were 16 m and 32 m, respectively, with a top width of 2.5 m and side slope of 2:1. The spurs are perpendicular to the flow field, whereas the weirs have an angle of 65° with the flow field. The design is based on that the noses of the spurs were located above the streambed for most of the time during the year, and rise with a slope of 6:1, until they meet the bank elevation. However, the crests of the weirs extend horizontally from the elevation of the spurs noses toward the river centerline. Spurs noses and weirs crests were set at elevation 255.5, which corresponds to 2 year return period of the water flow depth. This elevation corresponds to 0.75 m above the streambed and discharge of 26 m^3/s. Top of the spur was set at constant elevation of 265.3, which corresponds to the bank-full condition. This elevation corresponds to 4.3 m above the streambed and discharge of 312 m^3/s. The size of the barb structures is based on the 100-year flood of the stream, which is 430 m^3/s.

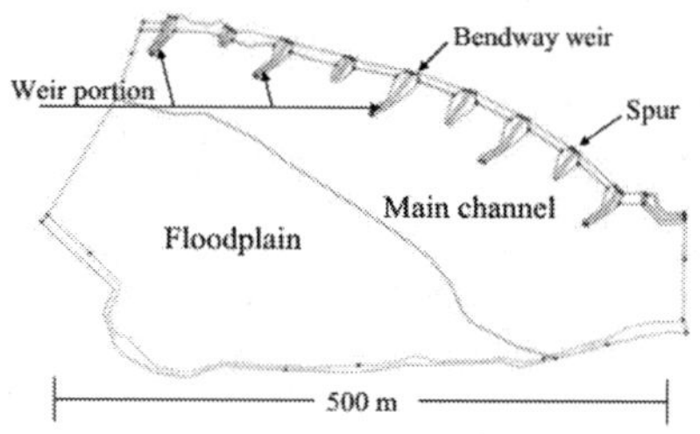

Figure 3. Plan view of the spurs and bendway weir arrangement.

3.2 Performance evaluation

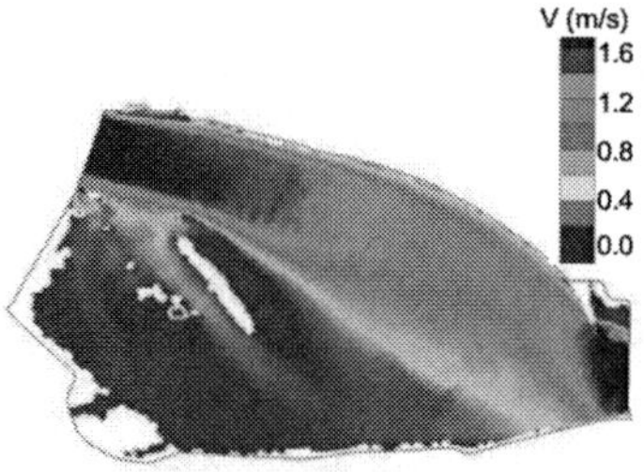

Figure 4. Flow condition with no barbs (velocity distribution).

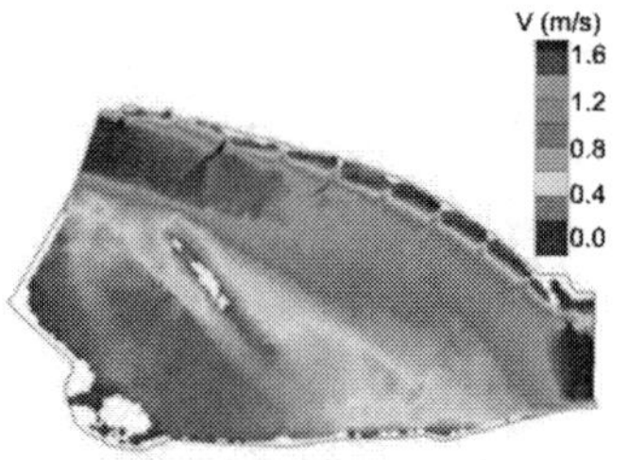

Figure 5. Flow condition with optimal number of barbs (velocity distribution).

The hydrodynamic model FESWMS was used to obtain the optimal number, location, and spacing between the barbs. The model inputs are: the stream bathymetry, values of Manning's coefficient n and eddy viscosity ν shown in Figure 2, flow rate at the upstream section Q = 312 m^3/s, and water depth at the downstream section y = 4.3 m. The bank-full flow condition (Q = 312 m^3/s; y = 4.3 m) was used to evaluate the overall performance of the barbs (spurs and bendway weirs) in controlling bank erosion. In performing the numerical simulations, the model was run with and without the barbs present. The flow condition in the case of no-barbs (Figure 4) was used as reference for comparison and to evaluate the barbs effect on the stream reach. The first simulation (not shown here) started with the 15 barbs (7 spurs and 8 bendway weirs) distributed equally over the study reach. If this barbs arrangement alters the flow condition by increasing the flow velocities on the opposite bank (compared with the no barb condition), the number of barbs in the next run is reduced until the barbs effect on the opposite river-bank has minimal on the flow condition. The model simulations (Figure 5) showed that the optimal number of barbs are 9 (4 spurs and 5 bendway weirs) with a spacing of 45-50 m between them (Figure 3). As can be seen from Figure 5 the barbs reduced the velocity dramatically near the north bank.

4 Summary

The 2D Finite Element Surface Water Modeling System (FESWMS) is used in this study to design the barb structures to mitigate river-bank erosion in a stream reach located on the Raccoon River near Adel, Iowa, USA just upstream of the US Highway Bridge 169. FESWMS results showed that the proposed barb structures successfully reduced the flow velocity along the outside bank and increased the velocity towards the center of the stream, thereby successfully increased the conveyance in the core of the river. The estimated velocities along the banks of the river where the barb structures exist were within the recommended values for channel stability design (velocity = 0.5 m/s). Thus, the barb structures were able to reduce the erosion along the bankline.

References

1. B. Przedwojski, R. Blazejewski, and K.W. Pilarczyk, River Training Techniques, A.A. Balkema: Rotterdam, Brookfield, Vt., (1995)
2. A.N. Papanicolaou, M. Elhakeem, G. Krallis, S. Prakash, and J. Edinger, J. Hydraul. Eng., **134**, 1 (2008)
3. M. Spasojevic, and F.M. Holly, Proc. 4th Inter. Conf. Hydroinformatics, The University of Iowa, Iowa, (2000)
4. D. Froelich, User's Manual for FESWMS Flo2DH: Two-Dimensional Depth-Averaged Flow and Sediment Transport Model, Release 3, (2002)
5. A.Z. Khan, and B. Barkdoll, Int. J. Comput. Eng. Sci., **2**, 3 (2001)
6. A.N. Papanicolaou, M. Elhakeem, D. Dermisis, and N. Young, River Research and Applications, (2010)
7. G.B. Pasternack, A.T. Gilbert, J.M. Wheaton, and E.M. Buckland, J. Hydrol., **328**, 1-2 (2006)
8. H. Azinfar, and J.A. Kells, J. Hydraul. Eng., **134**, 8 (2008)
9. J.D. Cruetin, M. Muste, A.A. Bradley, S.C. Kim, A. and Kruger, J. Hydrol., **277**, (2003)
10. T.D. Osting, CRWR Online Report 04-01, Center for Research in Water Resources, University of Texas, Austin, TX, (2004)
11. H.M. Chaudhry, Open Channel Flow, Springer, New York, (2008)
12. D.M. Hicks, P.D. Mason, Roughness Characteristics of New Zealand Rivers, New Zealand DSIR Marine and Freshwater Resources Survey, Wellington, NZ, (1991)
13. A.N. Papanicolaou, R. Hilldale, J. Eng. Mech., **128**, 9 948-960 (2002)

Research on Flow Pattern of Nitrogen Tetroxide Liquid in the Different Bend Radii Pipes

Pengfei Hao, Wenji Si, Hui Zhang, Ping Zhang, Shouti Yue

Department of Technique, Taiyuan Satellite Launch Centre, Taiyuan, Shanxi, China

Abstract. N_2O_4 is a common rocket fuel propellants, it has the characteristics of low boiling point and a large viscosity , the friction between viscosity fluids and pipeline dramatic leads to a huge sticky heat, therefore, the vaporization phenomenon often occurs in the pipeline, particularly in bending of the viscous heat. For this reason, the research of the different bending radii vaporized fluid conditions for optimizing the piping and precise the filling flow is significant. In this paper, the MIXTURE mixed flow model is used to achieve the numerical simulation the pipelines filling of the three different bending radii, it still have not solved the mass transfer problem between the different phases. Therefore, the custom functions are needed to define the mass transfer problems from the liquid phase to the vapor phase. Though the contrast among the volume phase cloud of six different elbow models , we have the following conclusions: 1) In the entire pipeline transportation, the distribution vaporization rate from the inlet pipe to the outlet pipe follows the distribution of the first increasing and then decreasing, the gas rates of the elbow area is highest; 2)Analyzing the sticky heat for different bend radii, we have the conclusion that the lowest bending vaporization the of the optimal radius is 0.45m. The above conclusions are drawn in good agreement with the actual law, can effectively guide the engineering practice, have important significance for the future design for the optimization of the fuel pipeline transportation.

1 Introduction

Nitrogen tetroxide (N_2O_4) has strong oxidation,its available oxygen can be up to 70% , it can be spontaneous in contact with amines, hydrazine, alcohol and many other fuels. Therefore it often used as bipropellant liquid propellant oxidizer. Currently, in the military aspects, unsymmetrical dimethyl hydrazine (UDMH) and N_2O_4 are combined at room temperature to raise the rocket chamber. N_2O_4 has huge sticky heat, therefore, the pressure in the pipe will produce viscous heat transport, the boiling point of the substance is 21.2C.Therefore, long-range transport in the pipeline is likely to occur vaporization phenomenon due to the viscous heat, thereby reducing the transportation flow throughout the pipeline, particularly in bending of the viscous heat. The main content of this paper is the optimal bend radius of the filling system, that find out that under any bend radii, the number of the vaporization due to the sticky heat produced by N_2O_4 is lowest.

Currently, the fluid researches are focused on low-viscosity fluids. Thus, the practical study does not need to consider the viscous heat problem. The main reason that the vaporization phenomenon occurs in the elbow is that fluid cavitation. In this research, there also exists some cavitation in elbow. However, compared with the vaporization due to the viscous heat, it can be negligible. The CFD numerical simulation method is adopted to study the optimal radius of the filling systems.

Zhan Han-hui containing [6], under different flow conditions ,the secondary flow (Dean Vortex) of the 90 elbow inside is simulated, the simulation process uses the $k \sim \varepsilon$ model with a swirl correction is calculated to obtain the pressure and speed distribution of the different flow regimes in flow direction of the cross-section;Shi Feng [7] etc., do the study $R_c/d = 1.69$ (R_c is the average radius of the bend , d is the Hydrodynamic diameter elbow) turbulent flow in the $90°$ bend based on the Standard $k \sim \varepsilon$ turbulence model, it is found when using the model to get the flow field velocity distribution compared with experimental data, in the bend $\theta = 45°$, both are in good agreement, in other words the calculation error is small, in the bend $\theta > 45°$, there are larger error. Ding Jue [8] etc., use 3 turbulence model(the Standard $k \sim \varepsilon$ turbulence model, RNG $k \sim \varepsilon$ turbulence model, Realizable $k \sim \varepsilon$ turbulence model),simulation a square cross-section three-dimensional turbulent flow in the $90°$ large curvature of the curved pipe, RNG $k \sim \varepsilon$ turbulence model and two-wall model are more applicable in the numerical simulation of turbulent considered strong curvature of the

DOI: 10.1051/matecconf/20166808004

bend.The above findings are not considered viscous heat production, therefore, to study the subject has some innovative.

2 Control equations

2.1 Multiphase flow mixing equation

This paper uses a mixed flow model for two-phase flow simulation,the mixed is mixed liquid and gas phase, it variable density and ignore the relative movement between the liquid and gas,flow field calculation just needs a control equation that is the mixed phase equations,gas phase volume distribution can be obtained by gas phase mass transport equations.

Mixed phase continuity equation:

$$\frac{\partial}{\partial t}(\rho_m)+\nabla\cdot\left(\rho\overline{V_m}\right)=0 \tag{1}$$

$$\rho_m=\rho_v\alpha+\rho_l(1-\alpha) \tag{2}$$

In the equation: ρ_m is the mixed phase density; V_m is the mixed phase velocity.

Mixed phase momentum equation:

$$\frac{\partial}{\partial t}\left(\rho\overrightarrow{V_m}\right)+\nabla\cdot\left(\rho\overrightarrow{V_m}\overrightarrow{V_m}\right)= \vec{F}-\nabla p+\nabla\cdot\left[\mu_m\left(\nabla\overrightarrow{V_m}+\left(\nabla\overrightarrow{V_m}\right)^T\right)\right] \tag{3}$$

Gas phase transport equation:

$$\frac{\partial}{\partial t}(\rho_m f)+\nabla\cdot\left(\rho_m f\overrightarrow{V_m}\right)=m_e-m_c \tag{4}$$

In the equation: f is the gas phase mass fraction; m_e is the vaporization source term ; m_c is the Condensation source term.

2.2 Energy equation

FLUENT Solving the following energy equation:

$$\frac{\partial}{\partial t}(\rho E)+\nabla\cdot\left(\vec{u}(\rho E+p)\right)= \nabla\cdot\left(k_{eff}\nabla T-\sum h_j\overrightarrow{J_j}+\left(\overline{\overline{\tau_{eff}}}\cdot\vec{u}\right)\right)+S_h \tag{5}$$

In the equation , k_{eff} is the effective thermal conductivity ; $\overrightarrow{J_j}$ is the diffusion flux of constituent j .The first three to the right of equation, respectively, due to thermal conduction, species diffusion, viscous dissipation arising energy transfer. S_h containing chemical reactions release (absorption) heat and any other user-defined volumetric sources.

$$E=h-\frac{p}{\rho}+\frac{u^2}{2} \tag{6}$$

For incompressible fluid:

$$h=\sum_j Y_j h_j+\frac{p}{\rho}$$

Y_j is the mass fraction of constituent j .

$$h_j=\int_{T_{ref}}^{T} c_{p,j}dT$$

T_{ref} is 298.15K

Equation contains the viscous dissipation term, which represents the heat generated by the viscous shear in the flow process. When using the segregated solver, the default energy equation does not contain this one in FLUENT (because generally ignored the heat generated by viscous).When Brinkman, Close to or larger than 1,the Viscous fluid heat can not be ignored:

$$Br=\frac{\mu U_e^2}{k\Delta t} \tag{7}$$

Δt is the temperature difference calculated within the region.

2.3 Mass transport model

The mixture model of multiphase flow to Fluent, still can not solve the problem in a different mass transfer between phases. So, we need to use a custom function (UDF) to define the mass transfer problems from the liquid phase to the steam phase. To do this, on the one hand we define the mass absorption phase in the steam continuity equation, on the other hand we define the equivalence of mass source term in the liquid continuity equation.

3 Mathematical model

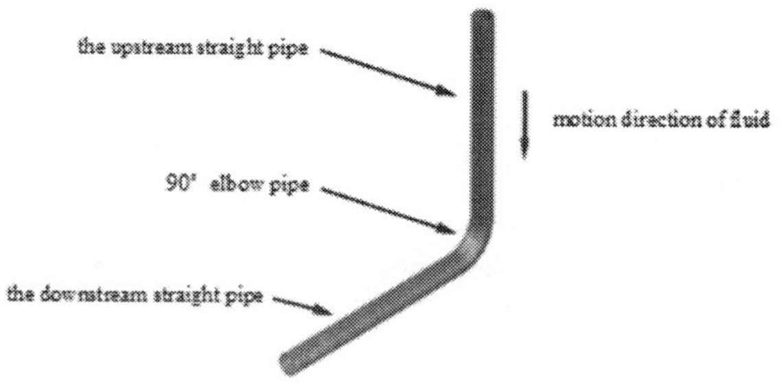

Figure 1. Mathematical model in the simulation region

Mathematical model consists of Upstream straight pipe, 90 degree elbow segment and downstream straight pipe three parts for the pipeline to transport, in order to ensure the model is comparable, upstream and downstream straight pipe sections were maintained at 1m, 90 degree bend segments selected six different radius to simulate, the diameter of the entire pipe is 0.1m,the radius 90 degree bend segments selected are 0.15m, 0.25m, 0.35m, 0.45m, 0.55m and 0.65m.

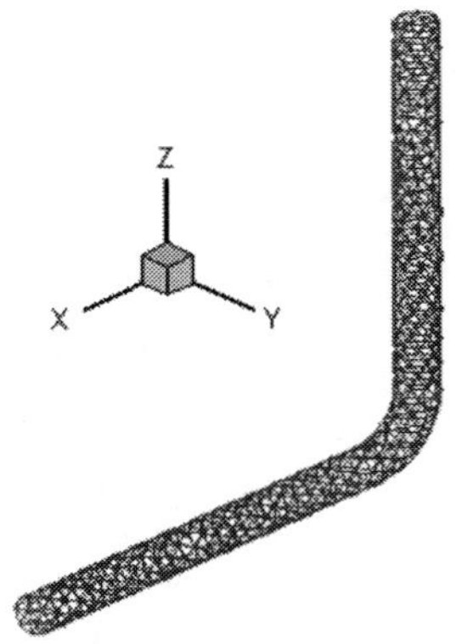

Figure 2. Computational domain and the mesh

To do the three-dimensional solid modeling using Gambit. The model handles the viscous heat problem,that is larger friction between the wall and the fluid, so have the need to encrypt boundary of the inner wall to the pipeline, the first layer thickness of the boundary is 0.001m,the thickness growth rate between the two layers is 1.2, in total four layers,the thickness of the boundary layer is 0.005368m.In order to make the numerical calculation facilitate,the entire mesh of the calculation model take the tetrahedral grid structure,the number of the mesh is 8,532,450,000 .

The main simulation object in this paper is N_2O_4, so, we need o modify material properties in the FLUENT,

Physical parameters	Value
Liquid Density（21.1℃, 101.325kPa）	1443kg/m3
Gas density（21.1℃, 101.325kPa）	3.40kg/m3
Specific heat capacity (gas, 101.325kPa)	Cp=831.5kJ/(kg•K)
Viscosity（101.325kPa，25℃）	0.0132mPa·S
Viscosity (liquid, 20℃)	0.42mPa·s
Thermal conductivity（101.325kPa，50℃）	0.16747W/（m·K）

This calculation model take the two-phase Mixture Model to simulate, the model is applicable to the calculation of two-phase flow due to vaporization phenomena, for cavitation, the interaction command in the model can solve, there need to fill the corresponding saturated vapor pressure and the surface tension coefficient of the fluid.For the calculation of two-phase flow, the problem of quality transmission still can not be solved, need to customize UDF programming to resolve.

The key program of C language:

$$if(\mathrm{C\_T(cell, mix\_th)} >= \mathrm{T\_SAT})$$

$$\begin{cases} m\_dot\_v = 0.1 \cdot C\_VOF(\mathrm{cell, pri\_th}) \cdot \mathrm{C\_R(cell, pri\_th)} \\ \cdot\, \mathrm{fabs(C\_T(cell, mix\_th) - T\_SAT)/T\_SAT;} \, dS[eqn] = 0.; \end{cases}$$

$$else$$

$$\begin{cases} m\_dot\_v = -0.1 \cdot C\_VOF(\mathrm{cell, sec\_th}) \cdot \mathrm{C\_R(cell, sec\_th)} \\ \cdot\, \mathrm{fabs(T\_SAT - C\_T(cell, mix\_th))/T\_SAT;} \\ dS[eqn] = -0.1 \cdot C\_R(\mathrm{cell, sec\_th}) \\ \cdot\, \mathrm{fabs(C\_T(cell, sec\_th) - T\_SAT)/T\_SAT;} \end{cases}$$

In which T_SAT is the Vaporization temperature, through the above C language programming, to realize the equal mass exchange of gas-liquid two-phase material ,otherwise, will arise the conserved problem to the continuity equation, cause unnecessary problems to the Calculation.

The import border is set to the entrance of the speed in the calculation ,the flow rate of the fluid in the pipe is 5m/s, we use outflow free flow boundary conditions in outlet .Take the no-slip boundary conditions to the wall, do discrete based on the finite volume method to equations,take the second-order upwind to convection term, use second-order central difference scheme diffusion, take *Simplec* algorithms to pressure-velocity coupling, unsteady discrete format is a first-order implicit scheme, the time step is 0.001s.

4 Numerical calculation

In order to study the sticky heat vaporization phenomenon of the pipelines better under different conditions bend radius, we have six different flow field bend radius numerical simulation calculations.

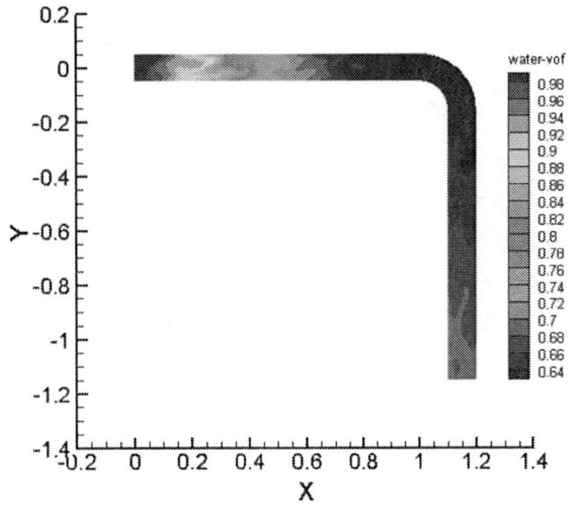

Figure 3-a. Bend radius 0.15m

DOI: 10.1051/matecconf/20166808004

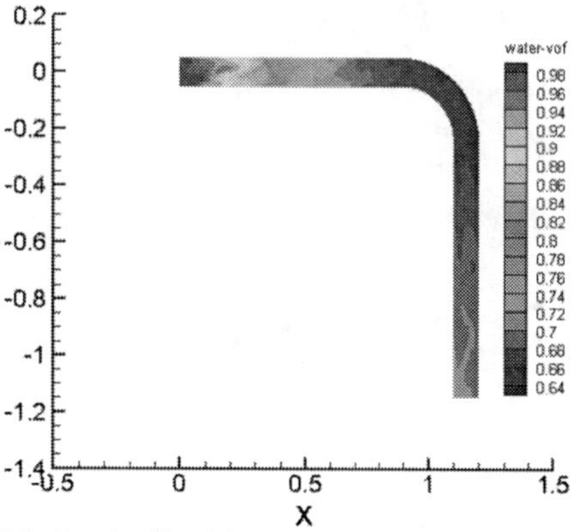

Figure 3-b. Bend radius 0.25m

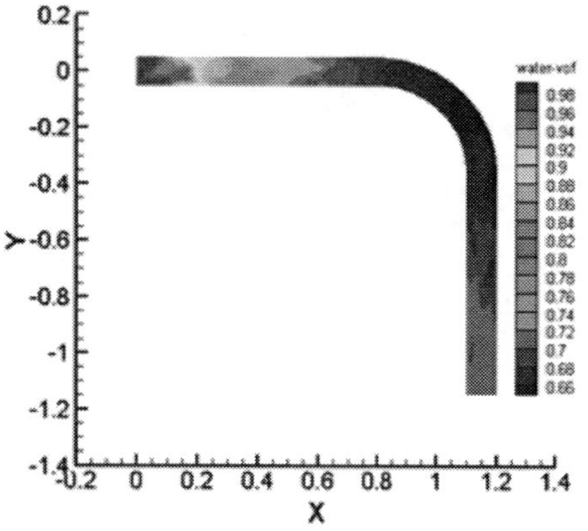

Figure 3-c. Bend radius0.35m

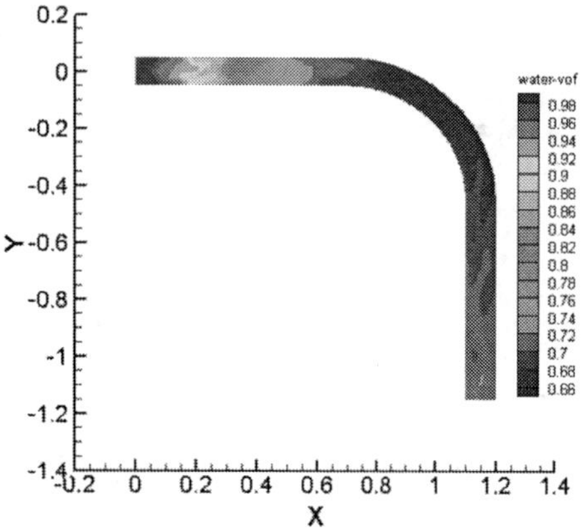

Figure 3-d. Bend radius 0.45m

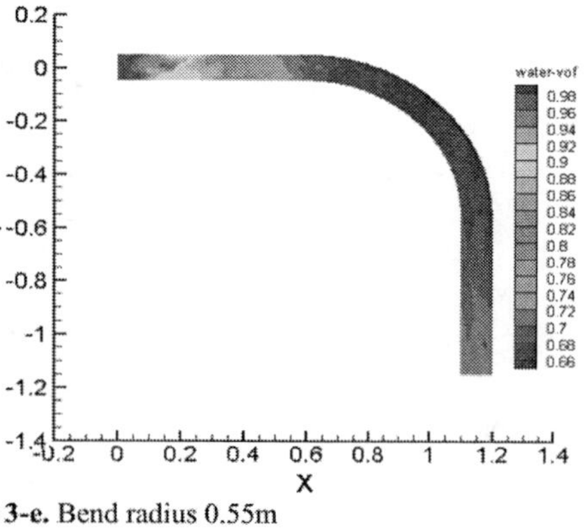

Figure 3-e. Bend radius 0.55m

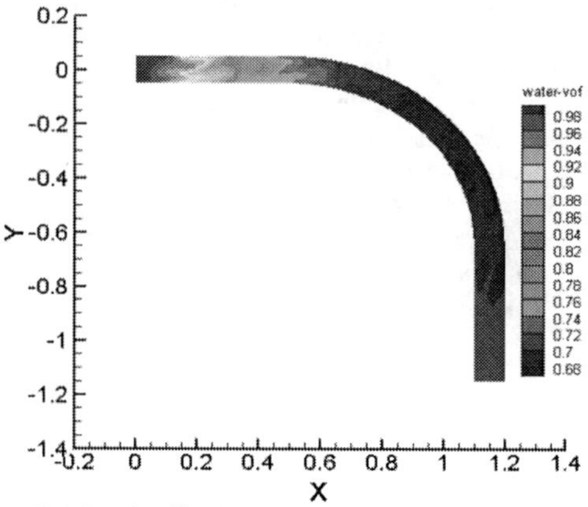

Figure 3-f. Bend radius 0.65m

From Fig. 3, the vaporization in elbow is more serious, the distribution vaporization rate from the inlet pipe to the outlet pipe follows the distribution of the first increasing and then decreasing. The main reason is: N2O4 fluid viscosity is greater, the intense friction between the high-speed flow wall, results in a viscous heat, achieves the vaporization temperature of the fluid, the fluid will occur vaporization in section, aeration pipelines to some extent reduce the total transport fuel, this brings some inconvenience to the rocket fuel filling cavities. The fluid in the pipe suffers the action of centrifugal force, The temperature produced by great force generating of the inner wall of the outer fluid pipe is higher, so as the serious vaporization.

As can be seen from Fig. 4, of the distribution of the pipe viscous heat of vaporization first decreases and then increases the with vaporization increasing of the bending radius, there exists optimal pipe radius of the minimum amount of the vaporization.The vaporization distribution of the elbow is related to the length of the elbow and the total length of the transmission pipelines, the longer distance of elbow the more serious of the turbulent fluid, however, the longer the elbow the total length of the entire pipe will be reduced,friction distance is shorter between the fluid and pipeline,viscous heat generation is reduced,in both factors simultaneously,there must exist a more reasonable bend radius.From the figure it can be drawn when the bending radius is 0.45m,viscous heat is the least.

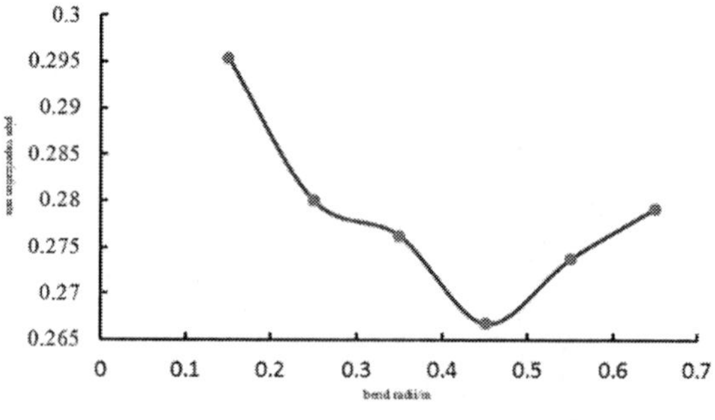

Figure 4. Vaporization distribution in the different bend radii pipeline

5 Conclusions

In this paper, the MIXTURE mixed multiphase flow model and modified $RNG\ k \sim \varepsilon$ turbulence model are used to simulate the three-dimensional model of the fuel filling pipe elbows. The mass transport problem can not be solved by this model, thus we have to write UDF mass transport equations. Six different transport pipeline bend radius of the fluid is simulated, we obtain the distribution of the two-phase flow at different radii bends and the same time, The pipeline and the pipeline serious vaporization zone vaporization rates are compared and analyzed, we have the following conclusions:

1) In the entire pipeline transportation, the distribution vaporization rate from the inlet pipe to the outlet pipe follows the distribution of the first increasing and then decreasing, the gas rates of the elbow area is highest;

2) Analyzing the sticky heat for different bend radii, we have the conclusion that the lowest bending vaporization the of the optimal radius is 0.45m.

The above conclusions are drawn in good agreement with the actual law, can effectively guide the engineering practice, have important significance for the future design for the optimization of the fuel pipeline transportation.

References

1. L. J. Xie, Si-yu, J. G. Yu, PIV Analysis of Inner Flow Field in90°Bending Duct of Circular-Section with Fore-End Valve, Journal of Shanghai Jiaotong University, **45(9)**, 1395-1399, (2011)
2. J. Ding, P. F. Weng, Numerical simulation of theoretical models & flow characteristics in 90°bending duct, Chinese Journal of Computational Mechanics, **21(3)**, 314-321, (2004)
3. Y. Kliafas, T. M. Holt, LDV measurements of a turbulent air-solid two-phase flow in a 90 degree bend, Expriments in Fluids, **5(2)**, 73-85, (1987)
4. K. P. Taylor, J. H. Whitelaw, M. Yiannes, Curved ducts with strong secondary motion: velocity measurements of developing laminar and turbulent flow, Journal of Fluids Engineering, **104,** 350-359, (1982)
5. Y. J. Zhao, Y. Y. He, Hydraulic characteristics of Z-type pipe combination with two similar rectangular bends, Hydraulic Engineering, **37(7)**, 778-783, (2006)
6. H. H. Zhan, H. Zhu, J. D. Chen, Numerical Simulation of secondary flow (dean vortices) in 90° curved tube, Boiler Technology, **41(4)**, 1-5, (2010)
7. F. Shi, Z. Xu, C. F. Ma, Turbulent Flow in Curved Duct with Large Curvature Numerical Computation and Measurement, Acta Aerodynamica Sinica, **8(4)**, 423-429, (1990)
8. J. Ding, P. F. Weng, Numerical Study on Three Dimensional Turbulent Separated Flow in Right-angled Curved Duct by Three Turbulent Models, Chinese Journal of Computation Physics, **20(5)**, 386-390, (2003)
9. ANSYS Inc. FLUENT User's Guide, Lebanon: Fluent Inc, (2003)
10. L. L. Wang, J. Lu, H. Zhu, Three-dimensional large eddy simulation of water wave using the dynamic coherent structures eddy model, Huazhong University of Science and Technology(Natural Science Edition), **39(1)**, 106-109, (2011)
11. R. L. Bass., E. B. Bowles, R. W. Trudell, et al. Modeling criteria for scaled LNG sloshing experiments, Transactions of the ASME, **107**, 272-280, (1985)
12. O. M. Faltinsen, A numberical nonlinear method of sloshing in tanks with two-dimensional flow, Journal of Ship Research, **22(3)**, 193-202, (1978)

MATEC Web of Conferences **68**, 09001 (2016) DOI: 10.1051/matecconf/20166809001
ICIEA 2016

Rice Grain Physical and Chemical analysis from Paddy Drying by Using a Solar Tunnel Dryer at Prathum Tani, Thailand

Krissadang Sookramoon[1] and Mananya Khamwachirapithak[2]

[1]*Faculty of Industrial Technology Vallaya Alongkorn Rajabhat University 1 Moo 20 Tambol Klongnueng Amphoe Klongluang PathumThani, Thailand*
[2]*Faculty of Science and Technology Vallaya Alongkorn Rajabhat University 1 Moo 20 Tambol Klongnueng Amphoe Klongluang PathumThani, Thailand*

Abstract. This paper presents the RD31 rice grain physical and chemical analysis from paddy drying by using a solar tunnel dryer. RD31 paddy was dried in a compact solar tunnel dryer, with a drying chamber volume in the size of 0.113 m^3 and 2.112 m^2 flat plate collectors. The experiment was performed on May 15, 2014 at Faculty of Industrial Technology Vallaya Alongkorn Rajabhat University Pathum Tani Thailand. The samples for testing were three varieties of Thai rice namely RD31 (from the farm), RD31 (dried in a solar tunnel dryer), and Prathum Tani 1 (from the market). The test consists of RD 31 dried by a solar tunnel dryer; the moisture content was reduced to about 14 wt. %. Then 3 varieties of rice were milled and sent to the laboratory for analyzing physical and chemical properties, repeated 3 times. The results showed that the average drying temperature was 67.75 °C. The paddy moisture reduction content was estimated from 49.96 to 14.64 MC (% w.b.) within 6 hours. All dimensions of RD31, dried by using solar tunnel dryer milled rice were closely related to those of the Prathum Tani 1, purchased from the market. The ratio of length to width of 3 samples: RD31 (raw rice), RD31 (dried in solar tunnel dryer), and Prathum Tani 1 (from the market) were 3.54, 3.47, and 3.44, respectively. 3 rice grain varieties were slender. The yield percentage was 60%, 75 %, and 70 %, respectively. Results from chemical test showed that the alkaline spreading value were 4-5, 3, and 2, respectively. The quality of the RD31 rice grain dried in a solar tunnel dryer passed the Thailand Ministry of Agriculture rice quality standard

1 Introduction

Thailand, one of the rice exporter countries in the world produces high quality rice such as RD31 Thai rice. Rice can be used for producing many products such as rice flour, rice oil, and many food applications. The quality of drying paddy is important for producing high quality rice. Rice quality is a combination of physical and chemical characteristics. Some studies of solar dryer for paddy drying have been reported. Excell and Kornsakoo [1] designed a cheap efficiently solar dryer for paddy drying namely AIT (Asian Institute of Technology) dryer in Thailand. It consists of the solar collector, paddy drying chamber, and the exhaust air chimney. Suharwadji and Takiyah [2] evaluated AIT dryer in Indonesia. 100 kilograms of paddy were dried in 76 h with a range of the moisture content of 22-10 % and concluded that there was no difference between this method and traditional sun drying method. On and Than [3] developed a paddy dryer with an intermittent batch in Malaysia. Centrifugal blower was used to move hot air from the solar collector to dry the products. Harahap and Mustadjab [4] tested a cabinet type solar dryer for paddy drying in Indonesia. 200 kg Paddy was dried from moisture content of 22 24% to 14-15 % within 3 days. Tongprasert *et al.* [5] reported the experiment of a forced convection paddy solar dryer. The dryer with a 3.74×4.48 m^2 flat plate solar collector and a drying bin showed that 1,000 kg of paddy could be dried in 1-4 days from a moisture content of 17.21-14%. The objective of this work is to investigate the RD31 rice grain physicochemical properties from RD31paddy drying by using a solar tunnel dryer and compare the product quality with RD31, a product of the tradition sun drying method (14 wt.% moisture content).

2 Materials and methods

2.1 The Solar Tunnel Dryer for paddy drying

This work focuses on rice grain physical and chemical analysis from paddy drying at Prathum Tani, Thailand. The paddy is sent to the solar tunnel dryer for drying. The paddy is loaded into the drying chamber of the solar tunnel dryer [6]. Hot air, which was obtained from 14 W, 1000 ft^3/min DC electric fan, is circulated and passed through the RD31 paddy in drying chamber and the paddy dry. The moist air is passed out through the exhaust after drying. The door is opened and RD31

 DOI: 10.1051/matecconf/20166809001

paddies are taken out of the tunnel and new batches of wet paddies are introduced into drying chamber for repetition. The solar tunnel dryer, is 2.99 meters long and 1.77 meters wide, with a flat plate solar collector area of 2.112 m^2 and a drying chamber volume of 0.113 m^3. The paddy drying test was started from 9:00 A.M. to 3.00 P.M. on May 15, 2014. The paddy samples were polished and sent to analyze the physical and chemical properties in a laboratory. The dried paddy was weighed for moisture content collection. The initial and final moisture contents of the product before and after each day of drying were determined. The solar radiation data and other meteorological data were obtained from the Mechanical Technology Workshop Station at Vallaya Alongkorn Rajabhat University Pathum Tani. Fig. 1 shows Solar Tunnel Dryer.

Figure 1. The solar tunnel dryer with paddy drying.

2.2 Samples preparation

3 milled rice grain samples: RD31 (raw rice from farm in Prathum Tani Province), RD31 (dried in a solar tunnel dryer), and Prathum Tani 1 (purchased from the market) with approximately 14 % moisture content were collected and prepared. This work was carried out in the Laboratory Department of Food Science and Technology in Vallaya Alongkorn Rajabhat University, Thailand. The grains were cleaned to remove all substances. The length, width and slenderness were measured using a digital vernier caliper (0.001 mm accuracy). Grain mass was measured with a sensitive electronic balance (0.001 gm sensitivity). After the determination of the dimensions, the grain density was determined using a beaker and electronic balance. The bulk density was determined with a weight (gram) per milliliter. Then the rice grain color measurement and chemical properties were measured.

2.3 Physical characteristics of rice grain

Grain quality physical examination generally includes dimensions, shape, color, percentage of the rice mixture, and density. Grain Dimensions were measured with a digital vernier caliper. Measurements are in millimeters. Grain dimensions are used to separate the rice into three categories; short, medium, or long grains.

1. Rice grain dimensions measurements

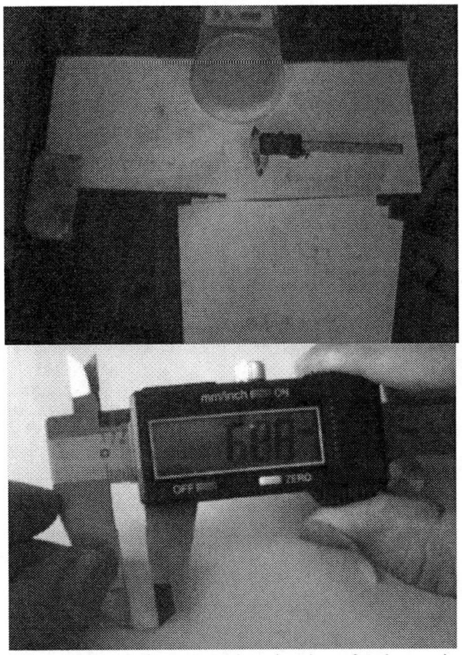

Figure 2. Equipment for measuring the size of a rice grain, including an electronic balance, a digital vernier caliper, and a digital camera.

2. Rice grain color measurement

The rice grain color meter was used to determine rice grain color. The obtained value was evaluated by using a Minolta chart as shown in Fig. 3 below.

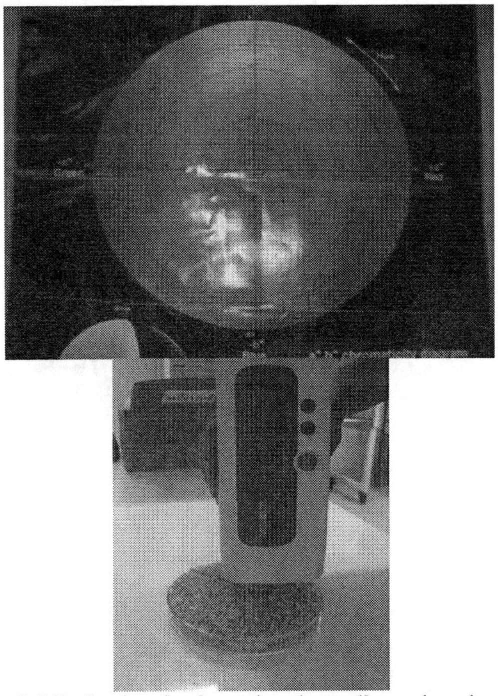

Figure 3. Minolta chart for rice grain color reading and a color meter for measuring the rice grain color.

 DOI: 10.1051/matecconf/20166809001

2.4 Chemical characteristics of rice grain (Alkali Spreading Value)

The purpose of this technique is to asset the gel type of the rice, which indicates the gelatinization temperature (60-70°C). The first step was preparing a solution of potassium hydroxide in a concentration of 1.7 wt. % (0.05%). Consider the dissolution of the grain in each seed spotted by a breakdown in Table 4.

3 Results and discussion

Table 1. Results of rice color evaluation

Order	Type of rice grain	Color Evaluation
1	RD31 (raw rice)	No.1 L70.0 a+1.5 b+13.4 No.2 L65.8 a+0.9 b+12.4 No.3 L68.9 a+1.1 b+13.8 **Ave. L68.23 a+1.16 b+13.2**
2	RD31 (dried in solar tunnel dryer)	No.1 L61.4 a+2.5 b+15.4 No.2 L61 a+2.2 b+15.9 No.3 L62.4 a+2.0 b+14.7 **Ave. L61.6 a+2.23 b+15.33**
3	Pratum Tani 1 (dried and purchased from the market)	No.1 L68.3 a+4.1 b+18.7 No.2 L64.9 a+0.6 b+19.3 No.3 L64.2 a+3.5 b+20 **Ave. L65.8 a+2.73 b+19.33**

L = brightness a = red color b = yellow

Table 2. The average length and width of a grain of rice

No.	Length (mm)				Width (mm)			
	1	2	3	Ave.	1	2	3	Ave.
RD31 (raw rice)	6.88	7.15	6.81	6.946	1.94	2.02	1.93	1.963
RD31 (dried in solar tunnel dryer)	7.05	6.76	6.8	6.87	2.07	1.87	2.01	1.983
Pratum Tani 1 (from market)	7.5	7.32	7.51	7.443	2.14	2.61	1.87	2.206

Table 3. Grain shape, length / width of a grain

Type of rice grain	Grain shape, length/width of a grain			
	1	2	3	Ave.
RD31 (raw rice)	3.55	3.54	3.53	3.54
RD31 (dried in solar tunnel dryer)	3.4	3.61	3.38	3.47
Pratum Tani 1 (from market)	3.5	2.8	4.02	3.44

RD31 (raw rice)	RD31 (dried by solar tunnel dryer)	Prathum Tani 1(dried/purchased from the market)
4-5	3	

Figure 4. The alkali spreading values

Table 4. The alkali spreading value

Type of rice grain	Alkali Spreading Value	Description
RD31 (raw rice)	4-5	The surface of the transverse or longitudinal seams grain and flour spread out of the seeds over a wide area.
RD31 (dried in a solar tunnel dryer)	3	Seeds, puffed rice and flour spread from some parts of the grain.
Prathum Tani 1 (purchased from the market)	2	Seeded rice crispiest

4 Summary

The physical and chemical analysis from paddy drying at Prathum Tani, using solar tunnel dryer was experimental study. The results clearly indicate that the average drying temperature was 67.75 °C. The RD31 paddy moisture content was reduced from 49.96 - 14.64 MC (% w.b.) in 6 hours. All dimensions of RD31, dried by using solar tunnel dryer milled rice were closely related to those of the Prathum Tani 1, which was purchased from the market. The ratio of length to width of 3 samples: RD31 (raw rice), RD31 (dried in solar tunnel dryer), and Prathum Tani 1 (from the market) were 3.54, 3.47, and 3.44, respectively. Three rice grain varieties were slender. The yield percentage was 60%, 75 %, and 70 %, respectively. Results from chemical test showed that the alkaline spreading value were 4-5, 3, and 2, respectively. The quality of paddy dried in this solar tunnel dryer has the same quality as a conventional drying technique, but could reduce the time of drying in one third.

Acknowledgement

The authors would like to acknowledge the Research and Development Institute of Rajabhat Vallaya Alongkorn University for financial support for this research.

References

1. Excell RHB, Kornsakoo S, A low cost solar rice dryer, Journal of Approp Technol, **5**(1) 23-24, (1987)
2. Suharwadji, Takiyah, Kamar Pengering Model EXELL, Technical Report, Lembaga Fisika Nasional, Bandung, Indonesia, (1983)
3. Ong KS, Than CF, A forced convection solar paddy dryer, Bulletin, July, Institution of Engineers Malaysia, (1982)
4. Harahap F, Suwono A, Abdurrachim Sapardi J, Prayudi, A Description and preliminary test results of a two ton solar paddy dryer. In Proceedings of regional Asia and Pacific workshop on the applications of solar energy in agricultural and post-harvest activities, Bandung, Indonesia, (1981)
5. Thanvi KP, Pandey PC Development of low cost solar Agricultural dryer for regions of India energy in agricultural dryer for arid regions of India, Journal of Energy Agri. **6,** 35-40, (1987)
6. Krissadang Sookramoon, Performance Evaluation of a Solar Tunnel Dryer for Paddy Drying at Prathum Tani, Thailand, Journal of App Mec and Mats, 799-**800**, 1455-1458, (2015)

Dynamics Analysis of Castor Hulling in the Process of Air-and-Screen Cleaning

Ruitao Gao, Yuhua Cao, Liangliang Yao and Hong Jin

Key Laboratory of Key Technology on Agricultural Machine and Equipment, Ministry and Education, South China Agricultural University, Guangzhou 510642, China

Abstract. The air-and-screen cleaning mechanism of castor hulling is analyzed in this paper. And the numeric expression equations of the castor dynamics analysis are established. The correlation of floating speeds vp, air speeds w, the direction angle of airflow velocity β, the direction angle of vibration δ, the friction angle φ, the slip coefficient, the fell coefficient and the jumping coefficient are analyzed.

1 Introduction

Castor is one of the world's top ten oil crops. The economic value of castor is very high, which is renewable green oil and can replace petroleum. There are many researches about agricultural materials hulling and cleaning structure, i.e. the effect of decorticating machine and the improving method [1], [2]. Existing researches primarily focuses on the mechanism of castor peeling and testing, the force analysis of castor in the process of shelling, the optimum design of key hulling parts and the peel rate and breakage rate testing [3].

However, there is few researches pay attention to the method and structure of castor cleaning except the aerodynamics analysis of air-and-screen cleaning mechanism of castor hulling [4]. In this paper, the relationships of floating speeds v_p, air speeds w, the direction angle of airflow velocity β, the direction angle of vibration δ, the friction angle φ, the slip coefficient, the fell coefficient and the jumping coefficient are analyzed.

2 Motion analysis of the shaker in the screening structure

The shaker in the screening structure does harmonic reciprocating motion driven by a crank mechanism [5]. The schematic diagram is showed in Fig. 1.

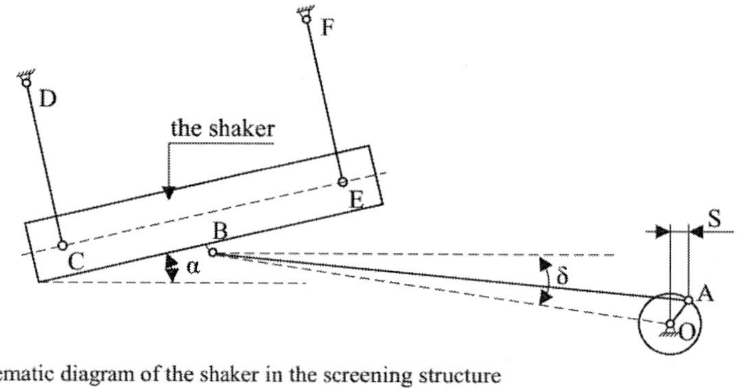

Figure 1. The motion schematic diagram of the shaker in the screening structure

The crank is OA. The linkage CD is parallel and equal to the linkage EF. That the direction of the center of the crank O connects with the hinge point B in the shaker is the direction of vibration. The angle between it with the horizontal direction is the direction angle of vibration δ. The direction angle of vibration is 0°. Compared with the crank OA, the length of the linkage CD, the linkage EF and the linkage OA is much longer. So, each point on the shaker is seen to do linear simple harmonic motion. Supposing the direction OB is the positive direction of vibration, and the castor upstream direction along the screen surface is the positive direction of relative movement. Then, the equations of the screening structure motion relationship are as follow:

DOI: 10.1051/matecconf/20166809002

$$\text{Displacements: } S = R\cos\omega t\,, \tag{1}$$

$$\text{Speeds: } v = -R\omega\sin\omega t\,, \tag{2}$$

$$\text{Accelerations: } a = -\omega^2 R\cos\omega t\,. \tag{3}$$

where R is the crank radius length, ω is the crank angular velocity, t is the time.

3 Dynamics analysis of the castor on the screen surface

The motion of the castor on the screen surface is complex. In order to simplify the motion analysis, the castor is assumed that it was only affected by the friction of the shaker surface F_f, the force of the air flow P, the inertial force Q, the gravity G and the supporting force N. the castor does the upward sliding, down sliding or jumping movement. The size and direction of the inertial force are determined by the size and direction of the acceleration of the castor. Therefore, before analyzing the state of motion of the castor on the shaker surface, the size and direction of it should be determined firstly. Then, the size parameters of the screening structure are selected.

4 The upward sliding of the castor

When the crank is located I or IV quadrant, $\omega t \in (0,\pi/2)\cup(3\pi/2,2\pi)$, the direction of acceleration a is the positive direction which is a leftward direction. Since the inertial force direction is the opposite direction of the acceleration direction, so the inertial force is a rightward direction. Therefore, the castor along the shaker surface has an upward sliding tendency. The castor force analysis is showed in Fig.2.

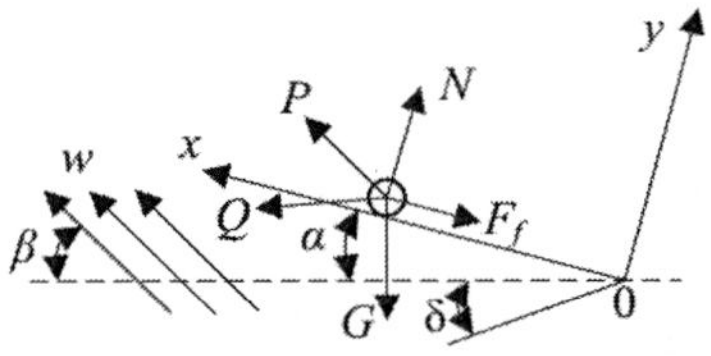

Figure 2. The castor upward sliding force analysis

The castor stress situation is as follows:

$$\begin{cases} m\,\mathrm{d}^2x/\mathrm{d}t^2 = Q\cos(\alpha+\delta) + P\cos(\beta-\alpha) - F_f - mg\sin\alpha \\ N + F\sin(\beta-\alpha) = mg\cos\alpha + Q\sin(\alpha+\delta) \\ F_f = fN = N\tan\varphi \end{cases} \tag{4}$$

where $Q = mR\omega^2\cos\omega t$, $P = mg\,w^2/v_p^2$, α is the shaker installation angle that is the angle between the horizontal direction and the shaker, v_p is the floating speed of the castor, w is air speeds, β is the direction angle of airflow velocity, δ is the direction angle of vibration, f is the coefficient of friction, φ is the friction angle, m is the mass of the castor particles and g is the acceleration due to gravity.

Therefore,

$$\frac{\mathrm{d}^2x}{\mathrm{d}t^2} = \frac{R\omega^2\cos\omega t\cos(\alpha+\delta+\varphi)}{\cos\varphi} - g\left[\frac{\sin(\alpha+\varphi)}{\cos\varphi} - \frac{w^2\cos(\beta-\alpha-\varphi)}{v_p^2\cos\varphi}\right] \tag{5}$$

When $R\omega^2\cos\omega t/g \ge \sin(\alpha+\varphi)/\cos(\alpha+\delta+\varphi)$, then $\mathrm{d}^2x/\mathrm{d}t^2 \ge 0$, and the castor sliding upwards along the shaker surface. Since $0 \le \cos\omega t \le 1$, setting $\mathrm{d}^2x/\mathrm{d}t^2 = 0$, then the critical condition of the castor upward sliding is presented below:

$$\frac{R\omega^2}{g}\cos\omega t = \frac{\sin(\alpha+\varphi)}{\cos(\alpha+\delta+\varphi)} - \frac{w^2\cos(\beta-\alpha-\varphi)}{v_p^2\cos(\alpha+\delta+\varphi)} \tag{6}$$

So if the castor is demand to sliding upwards, the acceleration ratio $R\omega^2/g$ should satisfy conditions:

$$\frac{R\omega^2}{g} > \frac{\sin(\alpha+\varphi)}{\cos(\alpha+\delta+\varphi)} - \frac{w^2\cos(\beta-\alpha-\varphi)}{v_p^2\cos(\alpha+\delta+\varphi)} \tag{7}$$

Supposing the slip coefficient is K_1, $K_1 = \dfrac{\sin(\alpha+\varphi)}{\cos(\alpha+\delta+\varphi)} - \dfrac{w^2\cos(\beta-\alpha-\varphi)}{v_p^2\cos(\alpha+\delta+\varphi)}$. Setting $K = R\omega^2/g$.

When $K > K_1$, then the castor sliding upwards along the shaker surface. Since $\omega = n\pi/30$ (n is the number of turns of the crank revolutions per minute), the critical rotating speed of the crank when the castor sliding down along the shaker surface is as below:

$$n_1 = (30/\pi)\sqrt{gK_1/R} \tag{8}$$

If $-1 \le \cos\omega t \le 0$, then $K < K_1$, the castor sliding down.

5 The down sliding of the castor

When the crank is located II or III quadrant, $\omega t \in (\pi/2, 3\pi/2)$, the direction of acceleration a is the negative direction which is a rightward direction. Since the inertial force direction is the opposite direction of the acceleration direction, so the inertial force is a leftward direction. Therefore, the castor has a down sliding tendency. The castor force analysis is showed in Fig. 3.

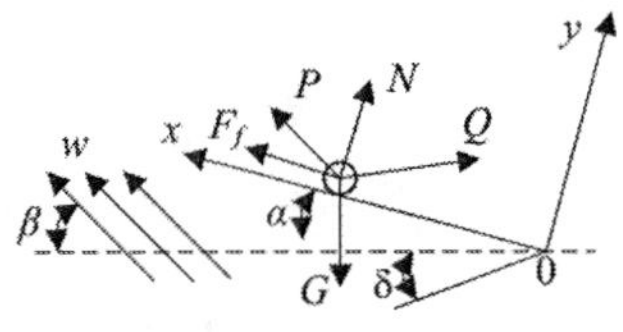

Figure 3. The castor down sliding force analysis

The castor stress situation is as follows:

$$\begin{cases} m\,\mathrm{d}^2x/\mathrm{d}t^2 = -Q\cos(\alpha+\delta)+P\cos(\beta-\alpha)+F_f-mg\sin\alpha \\ N+Q\sin(\alpha+\delta)+P\sin(\beta-\alpha)=mg\cos\alpha \\ F_f = fN = N\tan\varphi \end{cases} \quad (9)$$

$$\frac{\mathrm{d}^2x}{\mathrm{d}t^2} = -\frac{R\omega^2\cos\omega t\cos(\alpha+\delta-\varphi)}{\cos\varphi}+g\left[\frac{\sin(\alpha-\varphi)}{\cos\varphi}-\frac{w^2\cos(\beta-\alpha-\varphi)}{v_p^2\cos\varphi}\right]$$

Since $-1 \le \cos\omega t \le 0$, setting $\mathrm{d}^2x/\mathrm{d}t^2 = 0$, then the critical condition of the castor down sliding is presented below:

$$\frac{R\omega^2}{g} > \frac{\sin(\alpha-\varphi)}{\cos(\alpha+\delta-\varphi)}+\frac{w^2\cos(\beta-\alpha-\varphi)}{v_p^2\cos(\alpha+\delta-\varphi)} \quad (10)$$

Supposing the fell coefficient is K_2, $K_2 = \frac{\sin(\alpha-\varphi)}{\cos(\alpha+\delta-\varphi)}+\frac{w^2\cos(\beta-\alpha-\varphi)}{v_p^2\cos(\alpha+\delta-\varphi)}$. When $K > K_2$, then the castor sliding down along the shaker surface.

6 The jumping state of the castor

When the crank is located I or IV quadrant, the direction of acceleration is the positive direction which is a rightward direction. It is in the opposite direction of the supporting force. Under the gravity, the pressure could not be 0. Then, the castor is impossible to have jumping states when the crank is crank is located I or IV quadrant.

When the crank is located II or III quadrant, the inertial force Q equals $mR\omega^2\cos\omega t$. The direction of it is the same as the direction of the supporting force. So, the castor has a tendency to be thrown upward. Then, when the supporting force N is 0, the castor would be thrown from the shaker surface.
Therefore,

$$N = mg\cos\alpha + Q\sin(\alpha+\delta) - P\sin(\beta-\alpha) = 0 \quad (11)$$

$$mg\cos\alpha - mR\omega^2\,|\cos\omega t|\sin(\alpha+\delta) - mgw^2\sin(\beta-\alpha)/v_p^2 = 0 \quad (12)$$

where $|\cos\omega t| \le 1$.

So, the critical condition of the castor jumping is presented below:

$$K = \frac{R\omega^2}{g} > \left(\frac{\cos\alpha}{\sin(\alpha+\delta)} - \frac{w^2\sin(\beta-\alpha)}{v_p^2\cos(\alpha+\delta)}\right) \quad (13)$$

Supposing the jumping coefficient is K_3, $K_3 = \left(\frac{\cos\alpha}{\sin(\alpha+\delta)} - \frac{w^2\sin(\beta-\alpha)}{v_p^2\cos(\alpha+\delta)}\right)$. When $K > K_3$, then the castor is in jumping states. Then, the rotating speed of the crank can be calculated as follows:

$$n_3 = (30/\pi)\sqrt{gK_3/R} \quad (14)$$

From the above analysis, if friction angle φ, the shaker installation angle α, the direction angle of airflow β and the floating speed of the castor v_p can be determined, the critical air speeds w and the critical crank angular velocity ω could be calculated.

The states of the castor are depends on the relationship of coefficient K, K_1, K_2, and K_3. The states are analyzed as follows:

(1) If $K \le K_1$ *and* $K \le K_2$, the castor is in a relatively quiescent state.

(2) If $K_2 > K > K_1$, the castor only have the upward sliding state.

(3) If $K_1 > K > K_2$, the castor only have the down sliding state.

(4) If $K > K_1 > K_2$, the castor have both upward and down sliding states. But more castors have an upward sliding tendency.

(5) If $K > K_2 > K_1$, more castors have a down sliding tendency.

(6) If $K > K_3$, the castor is in jumping states.

7 Conclusion

In air-and-screen cleaning mechanism, the castor motion on the shaker is analyzed and the numeric expression equations of the castor motion are established. The relationships of the castor states with coefficient K, K_1, K_2, and K_3 are analyzed. In the future, the association curves of floating speeds v_p, air speeds w, the direction angle of airflow β, the direction angle of vibration δ, the friction angle φ, the coefficient K, K_1, K_2, and K_3 would be analyzed. The main factors that affect the cleaning rate and loss rate of mechanism would be identified.

Acknowledgements

The work was supported by the Science & Technology Foundation of Guangdong Province (No.2013B020313004).

References

1. L. Zhang, L. Gong, Status quo about shelling machine and discussing about improving methods, Food and Machinery, **22** (4), pp. 72-74, (2006)
2. C. Deng, D. Tao, Y. Gao, Research situation and development trend on grain sorting machinery, Journal of Agricultural Mechanization Research, **2**, pp. 5-7, (2005)
3. Y. Cao, C. Li, Z. Zhang, Y. Qing, L. Yao, Improvement design and test to key components of castor capsule hulling device, **28**(18), pp. 16-22, (2015)
4. L. Yao, Y. Cao, C. Li, Z. Fang, Dynamics property of air-and-screen cleaning mechanism of castor hulling, **12**, pp. 70-74, (2014)

DOI: 10.1051/matecconf/20166809002

5. C. Chen, Z. Wang, Q. Li, Analysis of aerodynamic properties of rape extractions, Transactions of the Chinese Society for Agricultural Machinery, **35**(5), p. 90-93, (2004)

DOI: 10.1051/matecconf/20166809003

Assessing Sustainable Rural Community Tourism Using the AHP and TOPSIS Approaches under Fuzzy Environment

M. Mujiya Ulkhaq, Pradita Y. Akshinta, Rizal L. Nartadhi and Susatyo Nugroho W. P.

Department of Industrial Engineering, Diponegoro University, Semarang 50275, Indonesia

Abstract. Tourism is currently a sector that is growing into an important and significant world activity. The development of an area where the tourist destination located in affects the growth of the tourism. In addition, the success of tourist destinations are influenced by their relative competitiveness; hence, they do compete each other to offer the best service to satisfy their customers. Rural tourism in Indonesia is believed as emerging business since there are abundant sites located in rural area that offers fascinating attractions to the visitors. This study aims to evaluate the rural tourism using sustainable indicators, namely, service quality, facilities, management system, and outcome. A combination of fuzzy AHP and TOPSIS are employed to select five rural tourism in Central Java Province. Result shows that service quality is considered as the most important attribute with weight of 28.6%, while Dieng is named for the excellent rural tourism. This finding might offer the service providers with valuable insights into the attribute that reflects customers' perceptions about rural tourism; also to position their services based on their competitors.

1 Introduction

Over the past decade, the convention industry has continued to grow on a global and regional basis. It is a multi-service industry that delivers huge benefits both directly and indirectly in various area including tourism, convention center, accommodations and transportation, restaurants, entertainment, communication, trading, and many more. It also has been recognized for its valuable economic contribution to tourism destinations and its substantial growth potential [1].

Tourism, as a part of convention industry, is one of the largest and fastest growing industries in the world [2]. It is commonly believed to be a viable means of raising the economic activity of such areas. Furthermore, the development of a tourism industry is widely considered to promote the destination's image, enabling the area to achieve other goals, such as business recruitment and retention [3].

According to the World Travel and Tourism Council, in 2014, tourism industry has total contribution to world's GDP of USD 7,580.9 billion—it is expected to grow more than 50% in 2025—and to employment of 276,84—it is expected to grow about 30% in 2025 [4]. In Indonesia, total number of foreign tourist arrivals in 2014 was 9.44 million, up 7.19% from the preceding year, meaning that the government target of welcoming 9.3 million foreign tourists last year was achieved [5]. Given the expected growth of tourism, there is a great optimism for rural tourism. Numerous researchers have acknowledged tourism as a potential economic development tool, particularly for rural communities [6]. It is progressively being used as a development strategy to improve the social and economic well-being of rural areas.

Rural tourism encompasses a huge range of activities and interests in farms, nature, adventure, sports, health, education, arts, and heritage taking place in countryside as a multi-faceted activity rather than farm-based tourism only [7]. Reference [8] further broaden the concept of rural tourism to include living history of countryside such as rural custom and folklore, local and family traditions, values, beliefs, and common heritage.

Rural tourism in Indonesia holds great potential in terms of sustainable rural development. Sustainable development for community tourism should intend to improve the residents' quality of life by optimizing local economic benefits, protecting the natural and built environment, as well as providing a high-quality experience for visitors [9], [10]. It should both provide a long-term economic linkage between destination communities and industries [11], and also minimize the negative effects of tourism on the natural environment and improve the socio-cultural well-being of the destination communities as well. It is getting increased recognition as a rural economic development tool. The government initiatives have sought to boost tourism as an economic substitute for traditional rural economic activities.

Rural tourism in Indonesia is quite emerging since the rural population is about 47% of total population in 2014 [12] and 31.5% of total land is agriculture [13]. Consequently, there are abundant tourist destinations located in rural area that embrace neighborhoods community to attract the visitors. This study aims to select the best sustainable rural community tourism in Central Java

DOI: 10.1051/matecconf/20166809003

Province. It may provide some insights to the officer or supervisor of the rural tourism on how patrons rate the rural tourism; hence enabling the officers to position their businesses based on their competitors and to discover the dimensions of sustainable rural community tourism that they have to improve.

This study utilized the combination of fuzzy analytic hierarchy process (AHP) and fuzzy technique for order performance by similarity to ideal solution (TOPSIS) to pursue the objective of the research. To be simple, the fuzzy AHP and TOPSIS approach extends the AHP and TOPSIS [14], [15] by combining it with the fuzzy set theory [16]. Despite of the advantages as seeking consistency and allowing decision makers to structure complex problems in the form of a hierarchy levels, AHP and TOPSIS has been criticized for its inability to handle the ambiguity of the concepts that are associated with human being's subjective judgment which are represented as precise numbers. However, in real life situations, the linguistic assessment of perceptions are fuzzy. Hence, it is more reasonable to represent it in terms to give interval judgments rather than fixed value judgments. The combination of fuzzy AHP and TOPSIS is widely used in the researches; see for example [17]–[19].

2 Research design

In this study, the attributes for evaluating sustainable rural community tourism were determined as four main attributes, i.e. service quality, facility, management system, and outcome. The attributes and their sub-attributes are based on research by [20]. However, they had been slightly modified since this study was conducted in Indonesia, especially in Central Java Province, which has different characteristics than the original research.

The first main attribute is service quality (SQ), which is actually an abstract construct that is difficult to define and measure, since the nature of the service is intangible, simultaneous, and heterogeneous. Despite of numerous researches that are studied about how to measure the service quality, there is no general agreement with regard to the content of its dimensions [21]. In this study, it does not utilize the dimensions of service quality used by several researchers, for example [22], [23], since the dimensions are developed in the hospitality service, such as restaurants, hotels, banking, health cares, etc., besides, the sub-attributes that are used in this study are accessibility (SQ1) and convenience (SQ2). The first can be defined as the ability for visitors to access the site location; it can be measured with the provision of site homepage, guidebook and map, and directional sign to the location within let say five kilometers. The second can be measured by ratio of householders who have checklists for bedding and able to pay using credit card.

The second main attribute is facilities (F), which consists of three sub-attributes: accommodations (F1), subsidiary facilities (F2), and environment (F3). Accommodation is measured by the ratio of rooms isolated from host family house, the ratio of rooms with flush toilet, and the ratio of rooms with shower bath. Subsidiary facilities mean meeting room with about a 20-person capacity and parking lots that hold over let say 20 cars. Environment can be measured by refuse disposal system in their community.

The third main attribute is management system (MS), which consists of four sub-attributes: community planning (MS1), collaborated community business (MS2), community management (MS3), and tourism business (MS4). Community planning can be divided into long-term community plan for last three years, fund-raising in their community, and amount of expert consultation for developing tourism. Collaborated com-munity business holds four criteria, i.e. community festival, ratio of house-holders with tourism management diary, value-added processing businesses in their community, and internet sales for their own products in their community. Community management can measured by minutes and account book for community council meetings, propaganda within last three years, organizing and participating in local tourism development council, and enacting community agreements. The last sub-attribute, tourism business, comprises four criteria, i.e. ratio of householders' participating in community tourism business, full-time worker to manage their tourism business, visitors' insurance, and constructing customer database.

The last attribute is outcomes (O), which comprises of two sub-attributes: satisfaction (O1) and income total sales (O2). Satisfaction represents residents' satisfaction and visitors' satisfaction. Income and total sales means increasing rate of a number of visitors, increasing rate of direct sales, number of one company one village linkage, and ratio of equity capital per government assistance.

The objective of this study is set to find out the best sustainable rural community tourism in Central Java Province, Indonesia. Bandungan, Baturaden, Dieng, Kopeng, and Sidomukti are selected because they could be considered as leading tourist attractions in Central Java Province. The map of those five is depicted in Fig. 1. Visitors can learn and explore the nature, the panoramic beauty of the scenery, as well as interact with the residents who are living there.

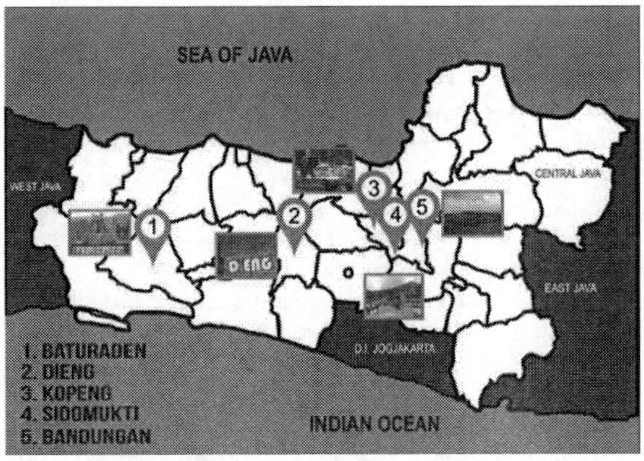

Figure 1. Map of objects of the study.

Bandungan is a potential tourist destination, which is located on the mountain slope at the Ungaran Mountain. It has a very beautiful natural scenery. Numerous distinctive vegetables, fresh fruits, and lovely flowers can be

DOI: 10.1051/matecconf/20166809003

found along the way. Bandungan is a paradise for nature lovers; there are many wonderful rides and challenging natures. Coffe café, spa, traditional temple called Gedong Songo, camp of rose, flower markets, and Rawa Pening, a beautiful tiny-lake, can be discovered in Bandungan. In addition, its beautiful panorama is usually exploited by photographers, or even by a couple for his/her wedding proposal's photoshoots.

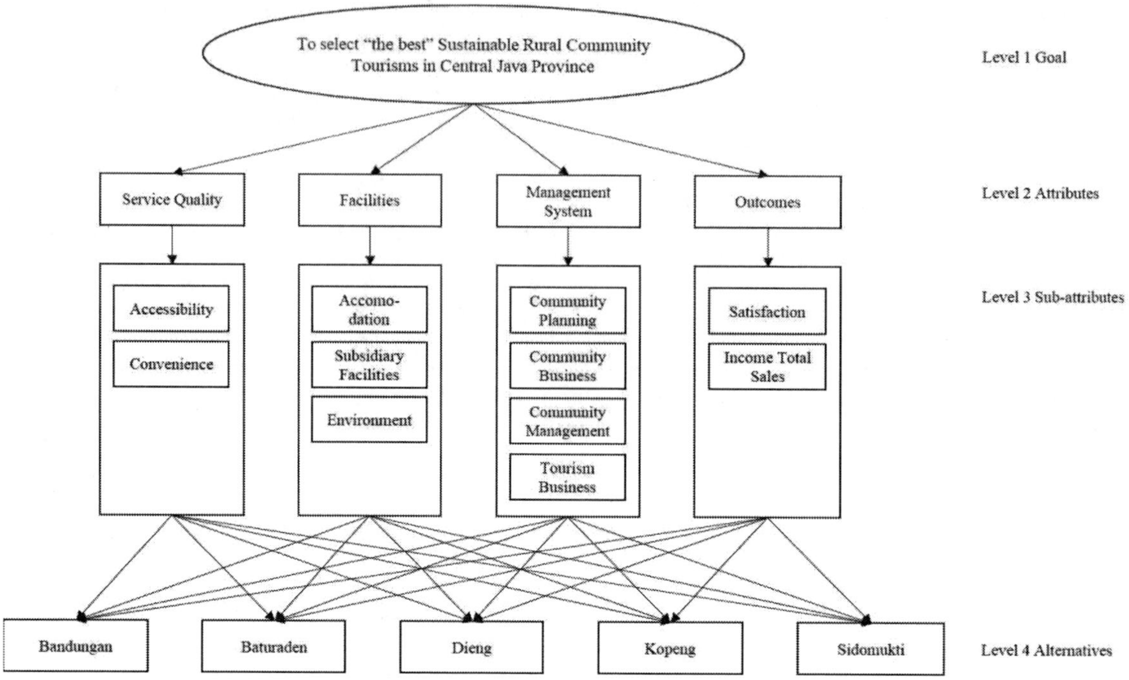

Figure 2. The evaluation framework of the sustainable rural community tourism model.

The next site is Baturaden. Five of eleven tourist destinations in Banyumas are located in Baturaden. It becomes a leading tourist district area in Banyumas since it is frequently visited by domestic or foreign visitors. It has been known as a tourist resort with a variety of exciting attractions with complete facilities supporting it. A plateau of Dieng is located at the northeast Banjarnegara, and could be regarded as the number two tourist destination in Central Java after Borobudur. There are Arjuna and Ghatotkacha Complex—two major characters of puppetry in Java legend, Bima temple, Dwarawati and Jalatunda wells, archeological museum, Balekambang lake, Merdada tiny-lake, Sikidang Candradimuka, and Sileri craters, as well as Bitingan hot spring waters. Located at the altitude of 2,093 meters above the sea level, Dieng plateau has fresh air temperature range from 15 to 20 degrees Celsius, and sometimes foggy when the weather is cloudy.

Kopeng is located in the Semarang area. It has several ideal locations that could be used by family as a recreation spot, camp, meeting, and seminar held by government or private institution. It also provides a camping ground for outbound activity. There are lodging of Kartika Travel, Songo Bannerman camping ground, and Mount Merbabu climbing spot. It presents gorgeous panorama in countryside combined with the landscape full of beautiful flowers and various fresh vegetables. Sidomukti is also located in the Semarang area precisely in the slope of Ungaran Mountain. Sidomukti nature parks is located at an altitude of 1,200 meters above the sea level. The open-for-public swimming pool is very attractive for visitors since it serves a beautiful view of the mountain. Sidomukti also provides an exciting adrenaline game for visitors with affordable cost.

In order to apply the fuzzy AHP and fuzzy TOPSIS methodology to the prioritizing of attributes for assessing the sustainable rural community tourism, the attributes abovementioned are initially structured into different hierarchy levels. The hierarchy of the decision model is shown in Fig. 2, where the goal is to select the best sustainable rural community tourism in Central Java Province. This hierarchy of attributes is the subject of a pairwise comparison of the methods being used. Data are collected from five experts who are very fond of traveling and have abundant experiences in visiting rural community tourism. The decision makers are asked to compare the elements on a pairwise basis in order to estimate their relative importance in relation to the element at the immediately preceding level, i.e. the attributes. A nine-point scale questionnaire is used to show the decision makers' judgment between options as equally, moderately, strongly, very strongly, and extremely favourable (or unfavourable).

3 Fuzzy AHP and fuzzy TOPSIS

There are various fuzzy AHP methods proposed by numerous authors to handle the comparison matrices. Among them, [24] is widely used due to its implementation simplicity to calculate relative weights.

A fuzzy number is a special fuzzy set $F = \{(x, \mu\tilde{A}(x)), x \in R\}$, where x takes values on the real line, R: $-\infty < x < +\infty$ and $\mu\tilde{A}(x)$ is a continuous mapping from R to the closed interval [0,1]. A triangular fuzzy number denoted as $\tilde{A} = (l, m, u)$, where $l \leq m \leq u$, has the following triangular-type membership function:

$$\mu\tilde{A}(x) = \begin{cases} \frac{x}{m-l} - \frac{l}{m-l}, & x \in [l,m], \\ \frac{u}{u-m} - \frac{x}{u-m}, & x \in [m,u], \\ 0, & \text{otherwise,} \end{cases} \quad (1)$$

where l, m, and u are lower, mid, and upper value of the support of M respectively, as well as m is the mid-value of $\tilde{A}$. In this study, first linguistic terms are used to represent the decision makers' assessments, and then triangular fuzzy numbers (TFN) are used for evaluations.

The four-step-procedure is given as follows:

1. Compare the performance score. Linguistic terms are used to indicate the relative strength of each pair of elements in the same hierarchy.
2. Construct the fuzzy comparison matrix $\tilde{A}$.
3. Solve the fuzzy eigenvalue. A fuzzy eigenvalue $\tilde{\lambda}$, is a fuzzy number solution to $\tilde{A}\tilde{x} = \tilde{\lambda}\tilde{x}$, where $\tilde{\lambda}_{max}$ is the largest eigenvalue of $\tilde{A}$ and $\tilde{x}$ is a non-zero fuzzy vector containing fuzzy number $\tilde{x}_i$.
4. The priority weight of each alternative can be obtained by multiplying the matrix of evaluation ratings by the vector of attribute weights and summing over all attributes.

After having weight of each alternative, fuzzy TOPSIS method is being employed. The steps are presented as follows:

1. Compute the fuzzy positive ideal solution (FPIS) and fuzzy negative ideal solution (FNIS) as follows:

$$\text{FPIS} = A^+ = (\tilde{v}_1^*, \tilde{v}_2^*, \ldots, \tilde{v}_n^*) \quad (2)$$

$$\text{FNIS} = A^- = (\tilde{v}_1^-, \tilde{v}_2^-, \ldots, \tilde{v}_n^-) \quad (3)$$

where $\tilde{v}_j^* = \max_i(v_{ij})$, $\tilde{v}_j^- = \min_i(v_{ij})$, $i = 1, \ldots, m; j = 1, \ldots, n$; and v_{ij} is weighted normalized matrix.

2. Compute the distance of each alternative from FPIS and FNIS (d_i^*, d_i^-):

$$d_i^* = \sum_{j=1}^{n} d_v(\tilde{v}_{ij}, \tilde{v}_j^*) \; ; \; d_i^- = \sum_{j=1}^{n} d_v(\tilde{v}_{ij}, \tilde{v}_j^-) \quad (4)$$

where $d_v(\tilde{v}_{ij}, \tilde{v}_j^-)$ is the distance measurement bet-ween two fuzzy numbers $\tilde{v}_{ij}$ and $\tilde{v}_j^-$.

3. Compute the closeness coefficient (CC_i) of each alternative:

$$CC_i = \frac{d_i^-}{d_i^- + d_i^+} \quad (5)$$

4. Rank the alternatives. In step 3, the different alternatives are ranked according to the closeness coefficient CC_i in decreasing order. The best alternative is closest to the FPIS and farthest from the FNIS.

4 Case study

The following is the application of fuzzy AHP and fuzzy TOPSIS to evaluate the sustainable rural community tourism in Central Java Province, Indonesia. The methods were employed based on the aforementioned attributes. First, the decision makers filled the pairwise comparison to express their preferences between the alternatives, attributes, and sub-attributes in a nine-point scale questionnaire. Their answers were then transformed into triangular fuzzy numbers.

The results of application of fuzzy AHP and fuzzy TOPSIS are alternative priority weights (APW) indicated the important of attributes of sustainable rural community tourism and the alternatives' priority referred to each attribute. The results can be seen in Table 1.

As the result from Table 1, service quality is regarded as the most important attribute with 28.9%. The next is management system with 24.4%, facilities with 24.3%, and the least important is outcome with 22.7%. This seems that the decision makers viewed the service given by the tourist destination as the main factors that have to manage to win a competition among their competitors.

The weights for sub-attributes of service quality are given as follows. Accessibility (SQ1) is believed to be the most important with the weight of 0.856, while convenience (SQ2) is considered the least important with 14.4%. The tourist destination with the best performance on service quality attribute is Dieng. For facilities attribute, environment (F3) is considered as the most important sub- attribute with weight of 34.6%. The next is accommodation (F1) with 33.8% and the last is subsidiary facilities (F2) with 31.5%. Sidomukti is believed to be the best rural tourism in accordance with facilities attribute by 21.6%.

The third attribute, i.e. management system, the sub-attribute which is thought to be the most important is community planning (MS1) with weight of 25.9%. Tourism business (MS4) and community management (MS3) are believed to be the second and third most important with weight of 25.6% and 24.6%. The least important sub-attribute is collaborated community business (MS2) with 24.2%. Dieng is regarded to be the best rural tourism in line with management system attribute.

The last attribute, i.e. outcome, satisfaction (O1) is considered as the most important sub-attribute with weight of 50.9%, and the other sub-attribute, i.e. income and total sales (O2) has weight of 49.1%. The decision makers chose Dieng to be the best rural tourism in performing outcome attribute.

Finally, the sustainable rural community tourism also were compared using the fuzzy TOPSIS. According to all the calculations, Dieng is considered as the best site with the value of relative approximation value versus ideal solution is 0.855. The next is Sidomukti with 0.615, Baturaden with 0.462, Bandungan with 0.456, and the last is Kopeng with the value of zero.

Table 1. Summary of APW for each attribute.

SQ	SQ1	SQ2			APW
Weight	0.856	0.144			0.286
Bandungan	0.200	0.203			0.201
Baturaden	0.201	0.210			0.203
Dieng	0.210	0.198			0.208
Kopeng	0.195	0.180			0.193
Sidomukti	0.193	0.210			0.196
F	F1	F2	F3		APW
Weight	0.338	0.315	0.346		0.243
Bandungan	0.207	0.206	0.178		0.197
Baturaden	0.195	0.202	0.215		0.204
Dieng	0.207	0.200	0.188		0.198
Kopeng	0.196	0.188	0.201		0.195
Sidomukti	0.195	0.204	0.218		0.205
MS	MS1	MS2	MS3	MS4	APW
Weight	0.259	0.242	0.246	0.256	0.244
Bandungan	0.204	0.198	0.206	0.197	0.202
Baturaden	0.198	0.206	0.191	0.181	0.195
Dieng	0.206	0.218	0.208	0.231	0.216
Kopeng	0.191	0.165	0.189	0.185	0.183
Sidomukti	0.201	0.213	0.206	0.205	0.207
O	O1	O2			APW
Weight	0.509	0.491			0.277
Bandungan	0.184	0.202			0.193
Baturaden	0.196	0.207			0.201
Dieng	0.224	0.198			0.211
Kopeng	0.196	0.183			0.190
Sidomukti	0.200	0.209			0.205

5 Conclusion and future research

The purpose of this study is to assess the performance of sustainable rural community tourism in Central Java Province, Indonesia, based on aforementioned attributes and sub-attributes. The result of this study which is shown in Table 1 is indicated that service quality is considered as the most important attribute with weight of 28.6 % and the least important attribute is outcome. It shows that service providers of rural tourism should focus more on service quality to perform satisfactory service.

The ranking of rural tourism based on calculation is Dieng for the first rank, Sidomukti for the second, Baturaden for the third, Bandungan for the fourth, and Kopeng for the last. It does not mean that Dieng provides a gorgeous service than the others. In fact, other rural tourism should improve their service considering with these attributes. This finding can provide the officers of the rural tourism with valuable insights into the attribute that reflects customers' perceptions about rural tourism; also to position their services based on their competitors.

There can be other method to evaluate the service of sustainable rural community tourism. It is highly recommended to utilize the other tools such as analytic network process [25] when the attributes are assumed correlates each other. Further research may be the application of these methods to evaluate other sustainable rural community tourism. Validating the scale in other site locations is an interesting area to be pursued.

References

1. L. Dwyer, *Economic contribution of convention tourism: conceptual and empirical issues* (Haworth, New York, 2002).
2. D.L. Edgell, *International tourism policy* (Van Norstrand Reinhold, New York, 1990).
3. D.L. Barkley, *Economic adaptation: alternatives for non-metropolitan areas. The potential for tourism development in non-metropolitan areas* (Westview Press, San Francisco, 1993).
4. World Travel and Tourism Council, *Travel and tourism: economic*, https://www.wttc.org/-/media/files/reports/economic%20impact%20research/regional%202015/world2015.pdf (2016).
5. Statistics Indonesia, *Tourism in Indonesia: record high number of foreign visitor arrivals*, http://www.indonesia-investments.com/news/news-columns/tourism-in-indonesia-record-high-number-of-foreign-visitor-arrivals/item5280.
6. G. Prosser, *Regional tourism research: a scoping study* (Centre for Regional Tourism Research, 2000).
7. B. Bramwell, B. Lane, *Rural tourism and sustainable rural development* (Channel View, London, 1994).
8. J. Pedford, *Seeing is believing: the role of living history in marketing local heritage* (Hisarlink Press, Enfield Lock, 1996).
9. D.B. Park, Y.S. Yoon, Tourism Manag. **30** (2009).
10. D.B. Park, Y.S. Yoon, M. Lee, J. Econ. Geo. Soc. **11**, 4 (2008).
11. H.C. Choi, E. Sirakaya, Tourism Manag. **27** (2006).
12. World Bank, *Rural population*, http://data.worldbank.org/indicator/SP.RUR.TOTL.ZS (2016).
13. World Bank, *Agricultural land (% of land area)*, http://data.worldbank.org/indicator/AG.LND.AGRI.ZS (2016).
14. T.L. Saaty, *The analytic hierarchy process: planning, priority setting* (McGraw-Hill International Book, Co., New York, 1980).
15. C. Hwang, K. Yoon, *Multiple attribute decision making: methods and application* (Springer, New York, 1981).
16. L.A. Zadeh, Info. Cont. **8** (1965).
17. F.R. Lima-Junior, L.C.R. Carpinetti, Int. J. Prod. Econ, **174** (2016).
18. Y.H. He, L.B. Wang, Z.Z. He, M. Xie, Eng. App. Artif. Intelligence, **47** (2016).
19. D.A. Wood, J. Nat. Gas Sci. Eng, **28** (2016).
20. D.B. Park, Y.S. Yoon, Int. J. Tourism Res. **13** (2011)
21. M. K. Brady, J.J. Cronin, J. Marketing, **65**, 3 (2001).
22. A. Parasuraman, V.A. Zeithaml, L.L. Berry, J. Retailing, **64**, 1 (1988).
23. J.J. Cronin, S.A. Taylor, J. Marketing, **56** (1992).
24. D.Y. Chang, Eur. J. Oper. Res. **95** (1996).
25. T.L. Saaty, *The analytic network process: decision making with dependence and feedback* (RWS Publications, Pittsburgh, 1996).

Charting the Research Course for Sustainable Aquaculture in Sabah, Malaysia

L. W. Vun, C. M. Payus and S. A. Mohd Ali

Faculty of Science & Natural Resources, Universiti Malaysia Sabah, 88400 Kota Kinabalu, Sabah, Malaysia

Abstract. Due to arising needs and demands, aquaculture is currently the fastest growing food production sector. In order to increase yield and yet to remain sustainable, the challenges would be to minimise impact on the environment and ecosystem services. Aquaculture activity contributes significantly to Malaysia and also the state of Sabah's economy and food security. Hence, the future changes in the environment as a result of rapid population growth and development would pose as threats to this industry in terms of quality, quantity and sustainability. Unforeseen environmental changes such as environmental pollution from other sources, climate change and the changes in policies would jeopardize the sustainability of this industry. In order to anticipate such impacts to the aquaculture activities, this paper set to chart a sustainable course for its development. Four important research courses were proposed: establishment of a sustainable framework, assessment of impacts of climate change, viability and vulnerability assessment due to future environmental changes and food security. Such findings would eventually allow the stakeholders to plan and manage the resources and aquaculture activities in such a way that foster sustainable food security and resilient aquatic ecosystems.

1 Introduction

Aquaculture is currently the fastest growing food production sector and it is expected to supply over half of the world's seafood [1]. With the arising needs and demands, Wilfart et al. [2] foresees that aquaculture will face four inevitable challenges: increasing cultivable areas without decreasing biodiversity or increasing water demand, improving food quality, producing ecosystem services and adapting to climate change. Thus, aquaculture faces the pressure to increase yield yet be sustainable.

In Sabah, aquaculture has great potential to contribute to the state's economy and food security, although it is not as significant as the palm oil industry, it is promoted as a promising commercial venture to meet consumer demand for seafood and support community livelihoods in general [3]. It is inevitable that there will be changes to the structure of the activity (from small scales to corporate), the coverage (from meeting domestic needs to meeting regional and global needs), the intensity, the approach (from conventional to innovative) and also the stakeholders (from individual to local to national to multiplayers).

Hence, aquaculture is vulnerable to the impact of climate change, future variability and changes in the environment as a result of rapid population growth. This would mean that aquaculture production in the state will be exposed to the threats of future environmental changes to its quality, quantity and sustainability.

Continued growth in aquaculture production is likely to set off intensification of production and this would bring about a range of resource and environmental problems. Yet at the same time, this sector is at risk due to possible future environmental changes, particularly as aftereffects of climate change and increased anthropogenic wrong doings. Climate change is only one among many environmental and anthropogenic stresses faced by aquaculture but is likely to complicate the process of achieving sustainability [4].

2 The need for a sustainable course

The first question that we pose in this scenario is how to ensure that this growing sector will incorporate sustainable practices and avoid some of the possible resource and environmental problems that have plagued the agricultural and livestock sectors in the past decades. The second question we ask is how the aquaculture sector can anticipate the impacts of future environmental change.

The answer to the first question have been aptly addressed by Klinger and Naylor [5] in a recent review, they have managed to identify some of the most promising pathways toward sustainable growth in the future of aquaculture sector. They raised three main issues: the technological innovations related to energy efficiency, the waste management matter and the genetic technologies for the enhancement of terrestrial aquafeed ingredients. They acknowledged that most innovative, productive, profitable, and environmentally sound systems are likely to vary by species, country, and policies.

In view of this, we attempted to propose measures to address the second question on charting a course for Sabah, so that the future of development in aquaculture in the state will be on a sustainable course.

Anticipating future environmental changes such as climate change and anthropogenic stresses will enable policy-makers and the stakeholders to be more aware and alert of the future challenges, and they could plan and manage the resources and activities in such a way that foster sustainable and resilient aquatic ecosystems. As this will benefit the aquaculture industries, and also provide goods and services at the national and even global levels, for example, through sustained food security and the conservation of biodiversity.

3 Charting a sustainable course

When attempting to meet sustainability, I firmly believe that sustainable development comes after taking into due account the future sustainable opportunities for the future generations. The simple idea would be as follows:

Resources – Future Sustainability and Opportunities = Sustainable Development

As clearly started by Serageldin [6] and in his own words: sustainability requires leaving to the next generation exactly the same amount and composition of natural capital as we found ourselves, and to substitute a more promising concept of giving future generations the same, if not more, opportunities than we found ourselves.

Thus, we attempt to address the future aspect first by developing a sustainable framework for aquaculture activities, and then follow by preparing the industry in anticipating future environmental changes.

Smil [7] in his book on *Global Catastrophe and Trends: The Next Fifty Years* remarked that any prediction of the future environmental changes could potentially be hampered by two incessant processes: the ever changing existing trends and also the shifting significance and concerns due to interactions of current and underestimated trends. However, Smil also mentioned that catastrophes and endings are also opportunities and beginnings. Taking these into consideration, we believe that a framework would serve as a basis for us to begin with (in moving towards sustainability) and also a platform for us to deal with any imminent negative environmental changes.

3.1 Sustainable framework for aquaculture

Developing a sustainable framework for aquaculture activity in Sabah will provide valuable understanding of the limit and provide the standards for the industry. The identified and developed suits of indicators will enable all stakeholders to know the rhythm of the aquaculture activities, symptoms and signs are good indicators that foretell possible problems and concerns.

Sustainability indicators also enables monitoring and reporting on the state of the aquaculture activities and relevant ecosystems locally and regionally. By taking into account a broad spectrum of sustainability perspectives and concerns, a comprehensive framework for the assessment can therefore be proposed, and results communicated to the decision making process. Although this does not necessarily offer a definitive judgement on sustainability, but present a holistic view, allowing recognition of trade-offs or compromises involved between conflicting sustainability objectives if any.

3.2 Impact of climate change on aquaculture

Understanding the impact of climate change on aquaculture is necessary on the research agenda as climate change is inevitable. Review of local and regional adaption and mitigation strategies related to aquaculture sector in view of climate change will allow stakeholders to anticipate such impacts [8].

This issue is also highly relevant to the formation of Malaysia's national climate policies and strategies. With the close of the recent COP21 (2015 Paris Climate Conference), the commitment to tackle climate change will again be renewed and new targets will be pledged. Thus, it is important to identify the impact of this on aquaculture as it posed as one of the most vulnerable economic sector, especially in developing countries like Malaysia. Of course the actual impact will be hard to predict, but due consideration can be given to the different impacts of climate change over a period of time, whether they are positive or negative.

3.3 Effects of environmental changes on aquaculture

Open-intensive aquaculture activities would have an impact on the water environment, nonetheless, this activity is susceptible to the changing environment, and such as the decline in water quality will affect the long term sustainability of the aquaculture activity. Thus, it is important to carry out long term and periodical monitoring to look at the potential sources of environmental pollutants that might threaten this activity, so that the management of this issues could be aptly addressed. Understand the risks of aquaculture activities in relations to nutrients salts, medication and also organic pollutants within the vicinity. This would enable the establishment of standards for continuous monitoring and food safety.

Once we understand the pollutants in the environmental, we would be able to implement useful measures, such as the 4S strategy (Stop, Slow, Simplify and Share) in dealing with environmental problems [9].

3.4 Aquaculture and food security

No matter how much we consider the issue of sustainability, ultimately it is the food security for the present and future generations that must be addressed.

The impact of aquaculture on the environment and local communities: issues on food security, food nexus, access to and management of common pool resources, could result in conflicts with exiting users and potentially acute social, political, and economic problems [10]. So, partnership is a must for this to be addressed.

3.5 Aquaculture and ecosystem health

Ecosystem health is closely related to aquaculture and this depends greatly on how the farmers, aquabusiness, aquaculture scientists, local communities and policy makers to understand the interrelationships and work towards tapping into what aquaculture activities can offer towards ecosystem health. Ecosystem health has been a neglected aspect for most development as the approach tends to drain certain services and sometimes even cause disservice. Aquaculture activities could well be a double edged sword as it could also provide beneficial environmental services if governed and practised in an ecosystem approach and an integrated manner. However, such ideal or sustainable use of goods and services and impacts on ecosystem integrity depend greatly on our understanding of the system dynamic. Further research is required to answer this because to achieve such a balance is very challenging.

3.6 The sustainable aquaculture framework

Thus, to achieve sustainability in this activity, having a sustainable framework would be a good start. As this would provide the minimal framework for all the stakeholders to begin their work with. Furthermore, an existing framework would serve as a precautionary step to move forward. Oversight in development would jeopardise the industry. And finally, a framework would help the stakeholders to take action on any future environmental changes that might have impacts on the industry.

Figure 1 depicts the framework proposed in this paper towards sustainable aquaculture. The box in the middle shows that farming activities and processing have an effect on the water quality, that is why technical options are important as they play a role in identifying the most effective and efficient strategy to reduce such impact. Dotted boxes show the important factors and players that will influence aquaculture activities, minimal environmental changes and healthy ecosystem services would provide the desired environment for the industry to thrive in.

Environmental policies changes (e.g. climate change policy) will also define the direction of the industry. The same goes for governance and best management practices as they would also shape the aquaculture activities. One last but not least is the partnership from the local communities. Taking all these into account would make a good alliance in bringing the industry towards sustainable aquaculture.

The viability of a "sustainable aquaculture" industry would depend on a collaborative and malleable process that allow all the stakeholders to adapt and communicate in order to establish, shift and deploy any policy goals [11].

From the business front, there is a need to experiment with new ways of conducting the business affairs, according to Williams [12], businesses that fail to take responsibility, embrace transparency and open up to new collaboration will increasingly seem out of touch in this green era. Thus, Williams has proposed a New Behavioural Contract that allows the businesses to try-out innovative and never-thought-possible ways of working together towards sustainability. A new sustainable paradigm shift for the businesses.

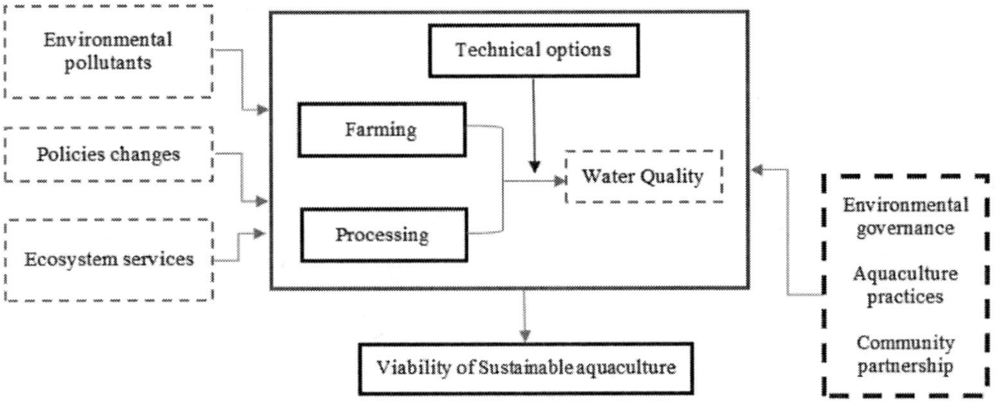

Figure 1. Framework towards sustainable aquaculture for Malaysia.

4 Conclusion and suggestion

By anticipating future environmental changes such as climate change and anthropogenic stresses will empower policy-makers and all stakeholders to be more alert of the future challenges, and they could plan and manage the resources and activities in such a way that foster sustainable and resilient aquatic ecosystems. This will not only benefit the aquaculture industries, but it will also provide goods and services at the national and even global levels, for example, through sustained food security and the conservation of biodiversity.

The proposed framework for sustainable aquaculture will help to organize and guide the planning to move forward while leveraging and incorporating all stakeholders' input as well.

With the presence of a framework, the organization would help to put everything into proper perspectives, remove any potential assumptions and sort out what is the best step(s) to go forward based on the best available knowledge and information about the industry and also the environmental dynamics.

The framework will also incorporate others' thinking and ideas. In this process, not only that the precautious principles are applied, it also could be an eye opener to realise that there are better ideas that could solve potential problems, such as local knowledge that could address the same issue without carrying out any time and cost consuming researches for that matter. With such inputs, they will also stimulate the thinking in others to propose ideas that are even better. Thus, this framework can also be viewed as a platform or starting point for the stakeholders to interact with one another and decide on the best solution.

Critical gaps in planning and application, policies and enforcement could be identified and filled, hence the implementation of adaptation and mitigation measures would be possible. Raised awareness and well managed aquaculture activities play important roles also in maintaining healthy and productive ecosystems and vice versa. Indeed, it is the economic returns that will drive and influence and farmers' production decision. However, through awareness, farmers could learn that sound environmental responsibilities actually make good business and create wealth and that pollution will lead to the outbreak of diseases and this is not business smart.

The two main purposes of sustainable aquaculture are for the sustained business for the farmers and also for the sustained and desirable healthy generations to come because the food source is meant for all but it is not just about feeding but staying healthy. Thus, the research course helps to move the industry towards that end.

In principle, the 4S approach would help: Stop, Slow, Simplify and Share [9]. Collectively, if all the stakeholders share the same approach, this would help to regulate the course for sustainable aquaculture to be always on the right track. For example, stakeholders united to *stop* the use of drugs, *slow* down the intensification that might cause the misuse of unhealthy seeds, *simplify* certain processes and *share* or advocate water recycling or treatment before discharge would promote a shared partnership towards environmental protection. Only in protecting the viability, vitality and diversity of the environment that we can ensure sustainability.

In propagating industrial and agricultural activities in the past, mistakes have been made because of ignorance, carelessness and the lack of restrain. As one moves towards a surging and challenging industry, we should embrace prudence and seek to create a desirable environment for us and also the future generations.

Charting the research course for sustainable aquaculture in Sabah, Malaysia is to be ahead of the *game*. We especially need committed proactive and aggressive management of resources for sustainability.

Acknowledgement

The authors thank the Ministry of Higher Education (MOHE), Malaysia, in providing the Niche Research Grant Scheme (NRGS) for this project. The authors also thank Isaac W. Witus, Weipeng Lee, Vincent Woon, Peirong Han, Nurul Fazlina for being a part of this project.

References

1. D. Cressey Nature, **458**, 398-400 (2009)
2. A. Wilfart, J. Prudhomme, J. P. Blancheton, J. Aubin, Journal of Environmental Management, **121**, 96-109 (2011)
3. SEDIA. http://www.sedia.com.my. (Accessed on 06.01.2016)
4. FAO, *Climate Change For Fisheries And Aquaculture. Technical Background Document*, Rome: FAO (2008)
5. D. Klinger, R. Naylor, Annu. Rev. Environ. Resourc., **37**, 247-276 (2012)
6. I. Serageldin, *Making Development Sustainable: From Concepts to Action*, World Bank (1994)
7. V. Smil, *Global Catastrophe and Trends: The Next Fifty Years*, MIT Press (2012)
8. FAO, Climate change implications for fisheries and aquaculture, FAO Rome (2009)
9. L. W. Vun, International Journal of Environmental Studies, **72**, 599-600 (2015)
10. B. A. Costa-Pierce, D. M. Bartley, M. Hasan, F. Yusoff, S. J. Kaushik, K. Rana, D. Lemos, P. Bueno, A. Yakupitiyage, Responsible use of resources for sustainable aquaculture, Global Conference on Aquaculture 2010, Sept. 22-25, 2010, Phuket, Thailand, Food and Agriculture Organization of the United Nations (FAO), Rome, Italy (2011)
11. E. A. Wolters, B. S. Steel, D. Lach, D. Kloepfer. Ocean & Coastal Management, **122**, 95-102 (2016)
12. E. F. Williams, *Green Giants: How Smart Companies Turn Sustainability Into Billion-dollar Businesses*, Amacom (2015)

 DOI: 10.1051/matecconf/20166810001

A Control Approach of Battery Energy Storage Systems to Reduce kW Demand

Mancheol Shin[1], Seongmin So[1] and Kernjoong Kim[2]

[1]Power21 Corp., Techno 3-ro, Yuseong-gu, Daejeon, Korea
[2]Dept. of Electrical Engineering, Chungnam Nat'l Univ., Korea

Abstract. This paper presents a control approach of battery energy storage systems to reduce kW demand under the two-part tariff pricing environment. The proposed algorithm is to minimize the maximum peak load using the forecasting load information during each interval for the charging and discharging periods of a storage system; the optimization of each interval is solved using the dynamic programming technique. The simulation result shows that it is greatly useful to reduce the maximum peak load as well as accomplish load leveling. This work will contribute to electricity cost savings.

1 Instruction

A two-part tariff pricing for electrical power has been adopted in most of the countries. It estimates for the rates of the connected kW demand and electrical kWh usage for electricity consumers. The first part of the two-part tariff considers maintenance and depreciation fees for electrical facilities that have been invested in providing consumers with electricity, and the second part considers fuel costs used for generating electricity.

Battery energy storage systems (BESSs) have been popular among industry areas because of its economic benefits and power quality enhancement [1]. In economic aspects of BESSs, peak load shaving or load leveling has been utilized in various sites such as factories, campuses, microgrids, etc. as a technique to reduce utility power consumption and energy costs during peak-load periods. In [2], a peak load shaving method in power grids with utility owned a BESS is proposed. Using energy bars and the utilization factor of the BESS, the total loss, purchased power and demand reductions are estimated. In [3], a dispatch method to enhance unbalanced low voltage on distribution grids is proposed using a BESS. It emphasizes the reduction in annual maximum demand and enhanced load leveling effect. In [4], an economic operation method in isolated islands is proposed. Using the operation plans of generators and a BESS, it shows the reduction of the annual system costs. Above-mentioned studies accomplish the peak shaving objective based the electrical kWh usage. It is also necessary to reduce the connected kW demand on the demand side in a two-part tariff pricing environment.

In Korea, the monthly electrical power price is calculated by adding the basic price based the kW demand and the electrical kWh usage price for industrial customers [5]. After finding the maximum peak load for December, January, February, July, August and September among the previous year including the current month, the basic price is calculated by multiplying the selected maximum peak load and the constant price. The candidates for the maximum peak load are calculated with the consumed load during each 15 minutes for the periods; most of the countries adopt 15 minutes as a certain amount of time to determine the maximum peak load to avoid a temporary peak power due to faults.

This paper proposes the peak load shaving algorithm to reduce the basic price based on dynamic programming and load forecasting. The kW demand is determined considering the state of charge (SOC), the target values of the BESS and the forecast load for the discharging and charging periods.

2 Proposed algorithm

2.1 Assumptions

There are some assumptions to simplify the proposed algorithm.

- An application for forecasting day-ahead electricity load is performed in the test system.
- Charging and discharging periods of a BESS during each day is decided.
- A meter which can measure electricity load supplied by the utility each minute has been installed in the test system.

2.2 BESS operation strategy

In order to ensure the life of a BESS, charging and discharging is composed of a 24-hour cycle, and it has a one charging period and one discharging period, respectively as shown in Fig. 1. The charging (or discharging) might be temporarily suspended during the charging period (or discharging period). Though the supplied power from the utility during each 15 minutes for a 24-hour cycle is lower than the past maximum peak load, the basic price is not lower; however, the electrical kWh usage price can be lower. Therefore, it is more effective in the energy cost that the BESS should be operated to reduce the supplied power from the utility as far as possible.

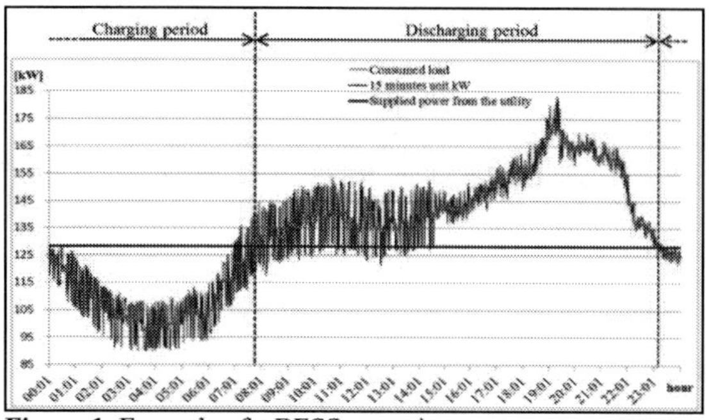

Figure 1. Example of a BESS operation.

2.3 Definition of the problem

2.3.1 Nomenclature

k — A 15-minute interval for the discharging (or charging) period within a 24-hour cycle ($k \in [k_1, k_{end}]$; k is one of the only discharging (or charging) period.)

k_{cur} — The current 15-minute interval

$P_{base}(k)$ — The supplied power from the utility during the interval k

$P_{LF}(k)$ — The forecasting load during the interval k; k is one of the future intervals.

$P_{E_{given}}$ — The power which the BESS could discharge after completing its charge within a 24-hour cycle

$P_{E_{dis}}(k)$ — The power which the BESS discharges during the interval k

$P_{E_{chg}}(k)$ — The power which the BESS charges during the interval k

$P_{Dif_{dis}}(k)$ — The power which the BESS will discharge during the interval k; k is one of the future intervals.

$P_{Dif_{chg}}(k)$ — The power which the BESS will charge during the interval k; k is one of the future intervals.

$P_{E_{SOC\min}}$ — The minimum value of the BESS' state of charge (SOC)

$P_{E_{SOC\max}}$ — The maximum value of the BESS' state of charge (SOC)

$P_{E_{dis\max}}$ — The maximum discharge value of the BESS; it is set by the operator.

$P_{E_{chg\min}}$ — The minimum charge target value of the BESS; it is set by the operator.

$P_L(k)$ — The consumed power from the load components during the interval k (i.e., in the discharging period, $P_L(k)$ is the sum of the supplied power and the BESS' discharged power, and in the charging period, $P_L(k)$ is the sum of the supplied power and the BESS' charged power.)

2.3.2 Objective function and constraints for discharging

As we consider discharging of a BESS, the objective is to minimize the supplied power from the utility. The objective function is formulated as:

$$\text{Min } P_{base}(k) \tag{1}$$

As a constraint, the sum of the past discharged power and future discharging power (i.e., the sum of differences between the forecasting load and the supplied power since the current interval k until the last interval) is less than or equal to the maximum discharging power as expressed as:

$$\sum_{k=1}^{k_{cur}-1} P_{E_{dis}}(k) + \sum_{k=k_{cur}}^{k_{end}} P_{Dif_{dis}}(k) \le P_{E_{given}} - P_{E_{SOC\min}} \tag{2}$$

The maximum discharge value of the BESS is constrained as:

$$\sum_{k=1}^{k_{cur}-1} P_{E_{dis}}(k) + \sum_{k=k_{cur}}^{k_{end}} P_{Dif_{dis}}(k) \ge P_{E_{dis\max}} \tag{3}$$

During the interval k, if the consumed load is greater than the supplied power, the difference between them will be the discharging power of the BESS; otherwise, the BESS will not discharge. This constraint is expressed as:

$$P_{E_{dis}}(k) = \begin{cases} P_L(k) - P_{base}(k) & if\ P_L(k) > P_{base}(k) \\ 0 & if\ P_L(k) \le P_{base}(k) \end{cases} \tag{4}$$

Finally, during the interval k, one of the future intervals, if the forecasting load is greater than the supplied power, the difference between them will be the discharging power of the BESS; otherwise, the BESS will not discharge. This constraint is expressed as:

$$P_{Dif_{dis}}(k) = \begin{cases} P_{LF}(k) - P_{base}(k) & if\ P_{LF}(k) > P_{base}(k) \\ 0 & if\ P_{LF}(k) \le P_{base}(k) \end{cases} \tag{5}$$

The above objective function and constraints should be fulfilled during each interval for the only discharging period (i.e., $k \in [k_1, k_{end}]$). Since the optimization is

DOI: 10.1051/matecconf/20166810001

performed at each interval, the problem can be solved using the dynamic programming technique [6].

2.3.3 Objective function and constraints for charging

In this section, we consider charging of a BESS. The objective for the charging is to minimize the supplied power from the utility. The objective function is the same as shown in (1).

As a constraint, the sum of the past charged power and future charging power (i.e., the sum of differences between the supplied power and the forecasting load since the current interval k until the last interval) is less than or equal to the maximum value of the BESS' SOC as expressed in (6).

$$\sum_{k=1}^{k_{cur}-1} P_{E_{chg}}(k) + \sum_{k=k_{cur}}^{k_{end}} P_{Dif_{chg}}(k) \le P_{E_{SOC\max}} \tag{6}$$

The minimum charge target value of the BESS is constrained as:

$$\sum_{k=1}^{k_{cur}-1} P_{E_{chg}}(k) + \sum_{k=k_{cur}}^{k_{end}} P_{Dif_{chg}}(k) \ge P_{E_{chg\min}} \tag{7}$$

During the interval k, if the consumed load is less than the supplied power, the difference between them will be the charging power of the BESS; otherwise, the BESS will not charge. This constraint is expressed as:

$$P_{E_{chg}}(k) = \begin{cases} 0 & if\ \ P_L(k) \ge P_{base}(k) \\ P_{base}(k) - P_L(k) & if\ \ P_L(k) < P_{base}(k) \end{cases} \tag{8}$$

Finally, during the interval k, one of the future intervals, if the forecasting load is less than the supplied power, the difference between them will be the charging power of the BESS; otherwise, the BESS will not charge. This constraint is expressed as:

$$P_{Dif_{chg}}(k) = \begin{cases} 0 & if\ \ P_{LF}(k) \ge P_{base}(k) \\ P_{base}(k) - P_{LF}(k) & if\ \ P_{LF}(k) < P_{base}(k) \end{cases} \tag{9}$$

The above constraints and equation (1) should be fulfilled during each interval for the only charging period. Since the optimization for the charging is performed at each interval during the charging period, this problem can be also solved using the dynamic programming technique.

3 Simulation studies

3.1. Test system environment

A factory site which owns a 300 kWh BESS and an electrical power meter is studied as shown in Fig. 2. The maximum and minimum values of the BESS' SOC are 300 kWh and 30 kWh, respectively, and the past maximum peak load of the site is 142 kW. The meter only measures the supplied power from the utility side to the demand side.

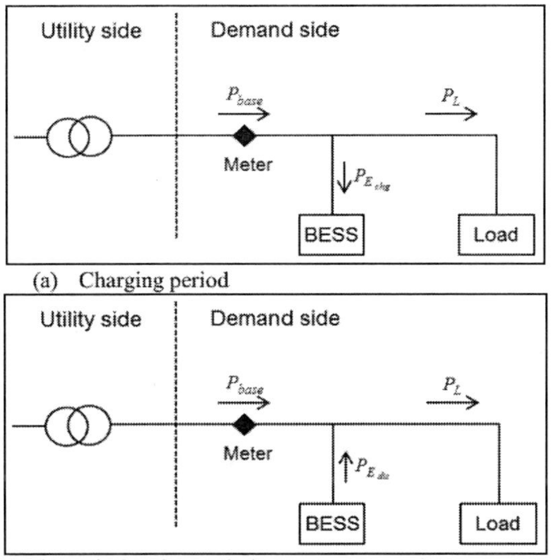

(a) Charging period

(b) Discharging period

Figure 2. Test system.

According to the two-part tariff pricing, the basic and the electrical kWh usage prices for the test system are given in Table 1; the kWh usage price is decided by the time of use (TOU) rate.

Table 1. Electricity prices

Basic price [\$/kW]	kWh usage price [\$/kWh]	
7	Low load (00:00 ~ 06:00)	0.05
	Middle load (06:00 ~ 09:00)	0.10
	Peak load (09:00 ~ 00:00)	0.17

3.2 Computer simulations

This section describes the simulation results for the charging and discharging periods of the BESS in the test system. Then the comparison of the system with and without the BESS is discussed.

3.2.1 Charging period

Figure 3 (a) shows the simulation result in the charging period. During the period from 01:00 to 05:27, the BESS charges about 60 kW, the consumed load is about 20 kW; therefore, the supplied power from the utility is about 80 kW. It is noted that the supplied power does not exceed the past maximum peak load and the current peak load, and the purely consumed load on the load follows the forecast load closely.

3.2.2 Discharging period

Figure 3 (b) shows the simulation result in the discharging period. The supplied power is calculated by the proposed algorithm. During the period from 06:00 to 07:00, the BESS does not either charge or discharge. The BESS starts the discharging from 07:00. After that time

DOI: 10.1051/matecconf/20166810001

the consumed load follows the forecast load closely, and the supplied power keeps almost constant value, 128.5 kW until 14:54. Subsequently, the consumed load rises than the forecast load until 19:30, and the supplied power increases to 133 kW in a unit step function. In the final period, the consumed load is lower than the forecast load until 24:00, and the supplied power keeps constant for a while and decreases to 127 kW afterward. From this discharging simulation result, the features of the proposed algorithm are observed as follows:

1) When the consumed load follows the forecast load closely, the supplied power keeps almost constant value.
2) When the consumed load tends to be higher than the forecast load, the supplied power increases not to exceed the maximum discharge power limit of the BESS.
3) When the consumed load tends to be lower than the forecast load, the supplied power decreases to maximize the discharge power of the BESS.

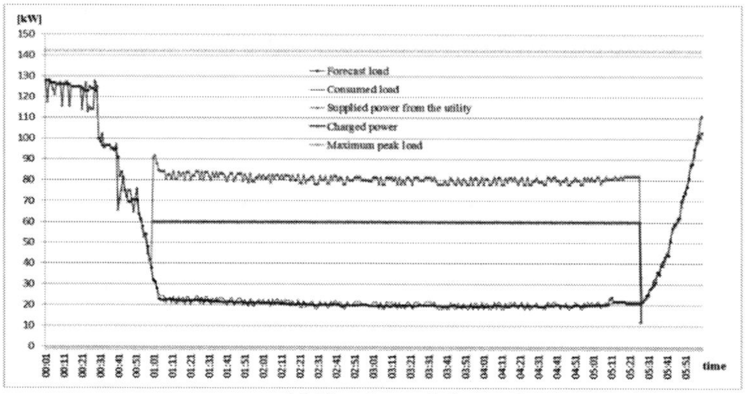

(a) Charging period

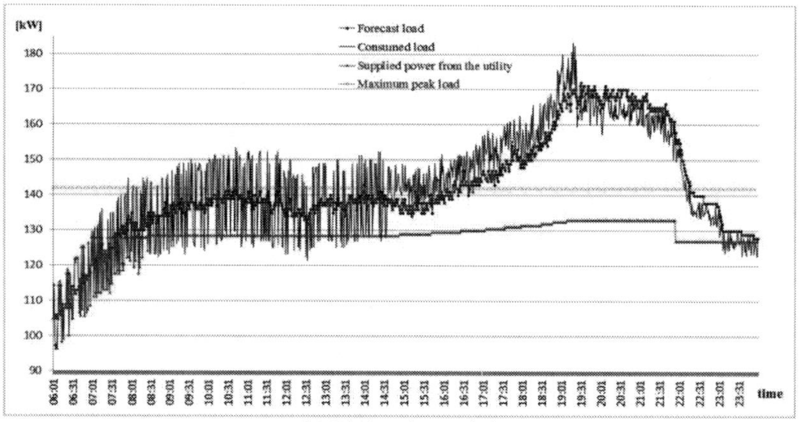

(b) Discharging period

Figure 3. Simulation result of the test system.

3.2.3 Comparison of the test system

The effectiveness of the BESS using the proposed algorithm is shown in Table 2. The maximum peak load is determined as 133 kW; therefore, the kW demand reduction is 26.5 % and kW demand price also decreases. In addition, the kWh usage price is reduced because of the TOU rate and the optimal operation of the BESS.

Table 2. Comparison of the test system with and without the BESS.

	Without BESS	With BESS
Peak load [kW]	180.9	133.0
kW demand reduction [%]	0	26.5
kW demand price [$]	1,266.3	931.0
kWh usage price [$]	421.0	390.4
kWh usage price reduction [%]	0	7.3

4 Conclusion

An algorithm to reduce kW demand under the two-part tariff pricing environment has been proposed. The kW demand is optimized depending on the difference between the forecast load and the real consumed load during the interval for the charging and discharging periods. The simulation results show that peak load shaving and the kW demand reduction have been achieved. Additionally, kWh usage reduction has been accomplished. In the next study, a research to reduce both kW demand and kWh usage using BESSs will be performed.

Acknowledgements

The authors would like to thank Power21 Corp. and KETEP. This research has been performed with the support of KETEP (No. 20151220100050).

References

1. K. C. Divya, Jacob Østergaard, Battery energy storage technology for power systems—An overview. Elsevier Electric Power Systems Research, **79**, 4 (2009): 511-520.
2. Azar Rahimi, M. Zarghami, M. Vaziri, S. Vadhva, A Simple and Effective Approach for Peak Load Shaving using Battery Storage Systems. In: Proc. of North American Power Symposium, (2013): 1–5.
3. Kalpesh A. Joshi, Naran M. Pindoriya, Day-ahead Dispatch of Battery Energy Storage System for Peak Load Shaving and Load Leveling in Low Voltage Unbalance Distribution Networks. In: Proc. of Power & Energy Society General Meeting, (2015): 1–5.
4. Enrique Lobato, Lukas Sigrist, Luis Rouco, Use of Energy Storage Systems for Peak Shaving in the Spanish Canary Islands. In: Proc. of Power & Energy Society General Meeting, (2013): 1–5.
5. Korea Electric Power Corporation, Electrical Power Supply Clauses. (Aug. 2015)
6. Richard E. Bellman, Stuart E. Dreyfus. Applied Dynamic Programming. Princeton University; (1962)

MATEC Web of Conferences **68**, 10002 (2016)
ICIEA 2016 DOI: 10.1051/matecconf/20166810002

The Research on Programmable Control System of Lithium-Bromide Absorption Refrigerating Air Conditioner Based on the Network

Lunan Sun[1], Lipeng He[2],Zihe Yin[1],Tianxun Li[1]and Shuai Guo[3]

[1]*Shandong Academy of Building Research, Jinan, 250031, China*
[2]*Shandong Hengxin Architectural Design co.,LTD, Jinan,250100, China*
[3]*Shanghai Advanced Research Institute, Chinese Academy of Sciences, Shanghai ,201210, China*

Abstract. This article regard the solar lithium-bromide absorption refrigerating air conditioning system as the research object, and it was conducting adequate research of the working principle of lithium bromide absorption refrigerating machine, also it was analyzing the requirements of control system about solar energy air conditioning. Then the solar energy air conditioning control system was designed based on PLC, this system was given priority to field bus control system, and the remote monitoring is complementary, which was combining the network remote monitoring technology. So that it realized the automatic control and intelligent control of new lithium bromide absorption refrigerating air conditioning system with solar energy, also, it ensured the control system can automatically detect and adjust when the external conditions was random changing, to make air conditioning work effectively and steadily, ultimately ,it has great research significance to research the air conditioning control system with solar energy.

1 Introduction

In the current, the global climate is becoming warmer and environmental pollution is more serious, government in countries around the world has put forward the measures and policy of saving the energy resources and reducing emissions, also actively researching the new technology to protect our earth's environment.

With the progress of the solar energy air conditioning technology, especially the application of automatic control technology and intelligent control technology is continuous improving, so that the solar energy air conditioners device will be widely used. The Solar energy air conditioning control system brings the field bus technology, network technology, virtual instrument technology and intelligent control algorithm together, not only it can reduce the labor intensity of operators, perfect working conditions, and improve the efficiency of management, but also can make the operation parameters of the unit work stability long-term within the reasonable scope,adapt to the development direction of modern industrial control system.

2 The working principle of the lithium-bromide absorption-type refrigerating machine

2.1 The Working Principle of the refrigerating machine

The single-effect lithium-bromide absorption refrigeration cycle is mainly composed of the following four cycle: hot water circulation, cold water circulation, circulation cooling water circulation and solution.

Firstly, the hot water circulation is that the water is warmed by the solar heating system, and was stored in the water tank, when the solar refrigeration systems is opening, the hot water driven pump must be opened, so the hot water have been stored in water tank was transported to generator in lithium bromide absorption refrigerating machine, to conduct heat exchange with the solution of lithium bromide and water. Secondly, the cold water circulation is the low temperature water system, which is taken place between the evaporator of lithium bromide-water chiller and the room radiators, its role is removing indoor cooling load of the room by transferring the cooling capacity in evaporator to the room. Thirdly, the cooling water circulation is extracting constant temperature water from outdoor cooling water, then transferring the cooling water to a refrigerator with the circulating pump to take away the waste heat inside the refrigerator. Finally, the solution circulation is that lithium bromide is refrigerating cold quantity inside the refrigerator.

From the above analysis, we can see that the hot water circulation is the dynamic system of lithium bromide-water absorption refrigerating machine to

DOI: 10.1051/matecconf/20166810002

promote the operation of entire refrigeration system. The solution circulation is the key of the refrigeration cycle, it connected the other three cycle to bring four cycle into an organic whole to finish the flow process of together, it is transferring the cold capacity in the high temperature source to the indoor outdoor low temperature environment, the working principle of lithium bromide absorption refrigerating machine and the operation principle is shown in Fig. 1.

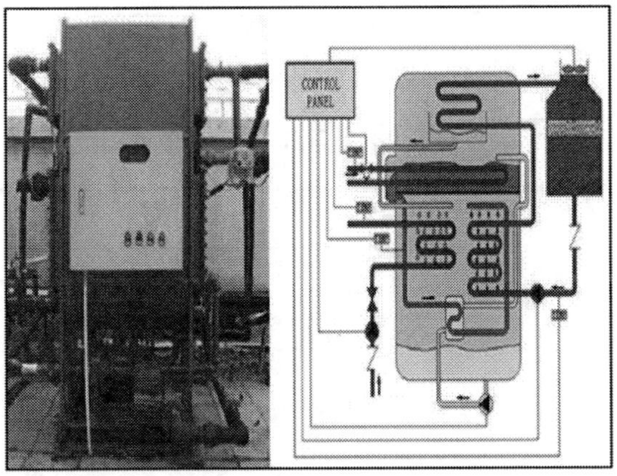

Figure 1. The working principle of lithium bromide absorption refrigerating machine and the operation principle.

2.2 The thermodynamic model of Lithium-bromide absorption refrigerating air conditioner

The new solar lithium-bromide absorption refrigerating air conditioner is main consist of additional the high-voltage generator, the high voltage generator, the low voltage generator, the high pressure absorber, the low pressure absorber, the condenser, the evaporator, the dilute solution storage and the cooling water storage reservoir, et al, when the water and solution is circulating flow in the equipment of all system, then, it is forming a thermodynamic model of air conditioning. And it was shown in Fig. 2.

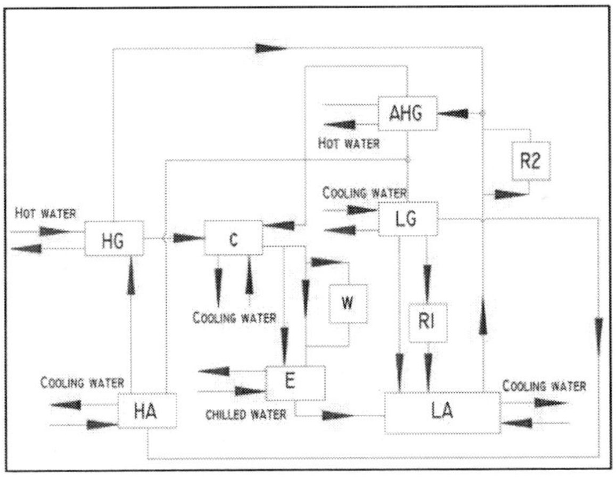

Figure 2. Thermodynamic model of air conditioning.

3 The PLC control system of lithium bromide absorption refrigerating air conditioner

When the system is under heating model, the thermal performance was tested and analyzed. Regarding existing problems and shortcomings of the whole system, the structure optimization analysis was conducted to seek for solutions to realize high efficiency thermal collection and application.

3.1 The control theory of Refrigeration system

According to the working principle of solar lithium bromide absorption refrigeration system, when it is working, the control system adjust the power switch, the circulating circulating pump, valve to realize the normal work of the four big circulation in system, the feedback data and information was mainly obtained through temperature sensor sensor, liquid level meter, flow meter, current meter, et al. The collector tubes absorb solar radiation to heat water, and hot water is transferring heat to lithium bromide solution through plate heat exchanger, according to temperature signals of the lithium bromide solution ,the high and low hair three-way valve will respectively the solution in double refrigerating machine of the high or low, when the heat source water inlet temperature is higher than the switch temperature 120 ℃(80 ~ 120 ℃) ,the valve will switch to high temperature generator, while lower than the switch temperature 5 ℃ (3~20℃), the valve will switch to low temperature generator, when the heat source temperature is lower than the water chiller, the pump must stop. Only through repeating the cycle operation, it can realize the normal work of the refrigeration system.

3.2 The programmable controller

Programmable Logic Controller is called PLC, it is the PLC program design of the control system that the design process of writing programs by users, program design process, also it is based on the instruction and combined with the requirements of the object and field signal of that the object in processes was controlled, then by comparing the control relay number of the PLC software to draw the ladder diagram of control system, then they have begun to appreciate the immediate benefit of interoper ability between heterogeneous platforms and programming languages, using the programming language for programming, the implementation team converts this abstract, high-level design into a detailed set of design models and the pro programming language implementation. According to the requirement of the control system, the OMRON (OMRON) CJ1 PLC what was from the Japanese company was chosen to finish the control system company CJ1 series programmable controller PLC, it has so many specific functions that supporting Ultra-high speed input, releasing the task of the CPU burden by owning independent motor position control unit and the structured text function block make floating-point

DOI: 10.1051/matecconf/20166810002

arithmetic programming more easily, so that it can meet the control demand of air-conditioning system .we can see the PLC from the following Fig. 3.

Figure. 3 The OMRON CJ1 PLC

3.3 The electric schematic diagram of refrigeration system

This system uses the three-phase 380V as power supply, air switch is master switch, which has the overcurrent and short circuit protection features.PLC power is 24V,and the function of each switch button are shown in following Table 1.

Table 1. The function of the switch in control system

Switch symbol	Function
1QA	Work start button
1TJ	The system stop button
1HK	The power switch of control circuit
1HD	The power indicator light
1RJ-6RJ	The heat protection relay of driving pump
1C-5C	The Circuit node switch
1D-6D	The pump motor in generator
D01-D09	The valve contactor
1DL	The mil malfunction Indicator Lamp
COM	The output terminals
GND	The SCT Service Circuit Terminal
X00-X13	The input interface of normally open contact
X14-X18	The input interface of normally close contact
U	A phase
V	B phase
W	C phase

3.4 The PLC electrical schematic diagram

According to the working principle of the lithium bromide absorption refrigeration system and system components symbol, which were described above account, so that we can determine the PLC electrical schematic diagram of lithium bromide absorption refrigerating air conditioner system, which is shown in Fig. 4.

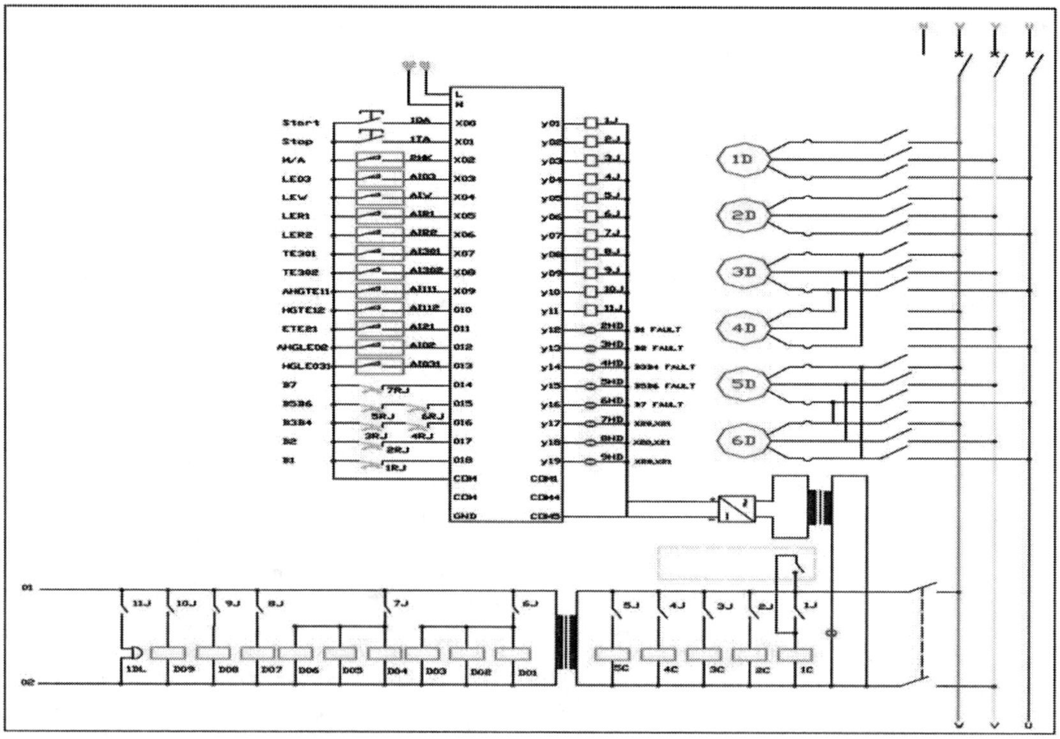

Figure 4. The PLC electrical schematic diagram of lithium bromide absorption refrigerating air conditioner system.

4 The field bus control system of lithium bromide absorption refrigerating air conditioner

Introduce the working principles of solar parabolic trough concentrating system and hot water single-effect lithium bromide absorption air conditioning parabolic trough type concentrating collector, respectively. At the same time carry out the theoretical researching works on thermal performance.

A control system has already described based on network used for solar energy .The control system can not only meet the requirement of solar energy, but also be applied in other industrial control system by using the control model. The system performs the function of running control, remote monitoring, and automatic control of heating air for solar energy control system.

The solar lithium bromide absorption refrigerating air conditioner system have a great of measurement parameters and control parameters, in order to design a reasonably control system, according to the control system parameters and performance, the field bus control system mainly includes the input and output module, the computer processing module, circuit module, remote control module, the storage system, the network communication module, the principle diagram of the module are shown in Fig. 5.

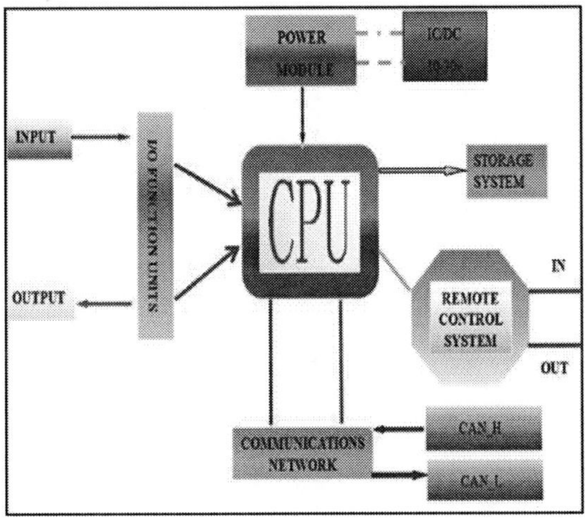

Figure 5. The field bus control system of lithium bromide absorption refrigerating air conditioner.

5 The control performance analysis of solar energy air conditioner system

In order to research the control performance analysis of solar energy air conditioner system, the outdoor dry bulb temperature change was regarded as a benchmark one day in a region, calculate the ourly cooling load and olar total radiation intensity in a company office conference room, the statistical results such as Table 2.

Table 2. A company meeting room statistics of the outdoor temperature, cooling load and the intensity of solar radiation

Time(h)	The Outdoor temperature (℃)	The sun total radiation (W/m^2)	Cooling load(W)
0	19.2	0.00	4808
1	19.0	0.00	4735
2	17.9	0.00	4663
3	17.4	0.00	4612
4	16.7	0.00	4517
5	16.1	0.00	4539
6	16.2	0.00	4567
7	15.6	25.00	4735
8	18.1	180.56	4984
9	20.0	391.67	5221
10	21.0	611.11	5460
11	23.0	800.00	5668
12	24.3	933.33	5826
13	26.0	988.89	6273
14	28.2	1000.00	6936
15	28.7	947.22	7801
16	28.8	825.00	8712
17	28.2	630.56	9399
18	28.2	405.56	9263
19	27.2	108.33	5354
20	25.2	19.44	5178
21	22.7	0.00	5056
22	21.7	0.00	4932
23	21.0	0.00	4910

6 Conclusion

In this paper, it focuses on the solution for the remote control system in solar energy lithium bromide absorption refrigerating air conditioner, and it based on field bus technology, this design scheme not only can effectively realize the function of the whole solar energy air conditioner operation system, but also, it can realize monitoring across the region to provide a more convenient and more efficient system solutions for staff through the network data interaction. It is this way that conducting a monitoring system based on networks can make the information transmission, monitoring, management, and other functions integrated into a modularization system to realize integration of management and control, all in all, it has great research significance to research the air conditioning control system with solar energy.

References

1. Zhang Yunjuan. Research on the Characteristic and Auto-control System of a Novel Solar Absorption Air-conditioning with an Energy Storage Unit[D].Wuhan: Huazhong University of Science & Technology.2005.4
2. M. Eck, E. Zarza, M. Eickhoff, et al. Applied research concerning the direct steam generation in parabolic troughs [J]. Solar Energy, 2003, 74: 341 - 351.

3. Xiande Fang, Baoyi Chen, Zhuyi Zheng. A study of the influence of solar radiation on the thermal performance of evaporators of heat pump system. Applied Thermal Engineering. 2003, **23**, pp. 1551-1557.
4. Inci Turk Togrul,Dursun Pehl Bvan,Cevdet Akosman. Development and testing of a solar air-heater with conical concentrator. Renewable Energy.2004, **29**.
 pp. 263-275.
5. A.Yattara, Y. Zhu, M. Mosa Ali. Comparison between solar single-effect and single-effect double-lift absorption machines. Applied Thermal Engineering.2003, Vol. 23, pp. 1981-1992.
6. Afif. Akel. Hasan, D. Yogi. Goswami, Sanjay Vijayaraghavan. First and second law analysis of a new power and refrigeration thermodynamic cycle using a solar heat source. Solar Energy. 2002, **73(5)**, pp. 385-393.
7. A. Syed, M. Izquierdo, P. Rodriguez, et al. A novel experimental investigation of a solar cooling system in Madrid [J]. International Journal of Refrigeration, 2005, **28**:859-871.
8. W.S.Duff,R.Winston,J.J.Gallagher, t,al. erformance of the Sacramento demonstration ICPC collector and double effect chiller [J]. Solar Energy, 2004, **76**:175-180.
9. T. Lin, X.M. Sui. Design of air conditioning system for an apartment hotel in Shanghai. Heating Ventilating & Air Conditioning, **38(6)**, (2008), pp. 98-102.
10. Doosam Song, Taeyeon Kim, Suwon Song, Suckho Hwang, Seung-Bok Leigh. Performance evaluation of a radiant floor cooling system integrated with dehumidified ventilation. Applied Thermal Engineering, **28**, August 2008, pp. 1299-1311.

Intelligent System for Continuous Steel Casting Based on Water Flow Control in the Secondary Cooling Stage

Gelu-Ovidiu Tirian [1], Ioan Filip [2] and Cristian Paul Chioncel [3]

[1]*Politehnica University of Timisoara/Department of Electrical Engineering and Industrial Informatics, Hunedoara, Romania*
[2]*Politehnica University of Timisoara/Department of Automation and Applied Informatics, Timisoara, Romania*
[3]*Eftimie Murgu University/Department of Electrical Engineering and Informatics, Resita, Romania*

Abstract. This paper presents an intelligent system capable of eliminating the surface flaws and casting rejects, to increase the quality of continuously cast products and implicitly to increase the productivity of the installation.This method contains a fuzzy controller which requires changes of the water flow value at the secondary cooling. Fuzzy system controlling in real time the distribution of the water flow rate according to the real situation in the installation and working as an adaptive. The simulations results show us that the control strategy proposed is a very efficient method for the water flow control in the secondary cooling stage.

1 Introduction

Nowadays, continuous casting provides the biggest part of semi-products for the metallurgical industry. The process is a very complex one, of high productivity, but, because of some technological issues it can generate quality defects in the wire or throw-outs by cracking the crust of the semi-product. This defects and throw-outs lead to lower quality of the semi-products, of the installation productivity and to economic losses. The continuous casting domain has developed and consolidated as an important research direction, concentrating collective research efforts of numerous specialists from the academic domain and from the metallurgical industry. The method of continuous casting consists of introducing the liquid metal with a well determined temperature in a cava shape which has the walls cooled inside with water named crystallizing, and the evacuation is made on the opposite side where is obtained the solidified chord [1-3].

A very important component of the continuous casting installation is the secondary cooling zone. The secondary cooling zone has the role to continue the wire cooling after it has emerged from the crystallizing and to assure the total solidification of the semi–product. It is considered "the heart" of a continuous casting and has the role of ensuring the quality of the material, the material surface shape and has to ensure a homogeneous cooling and a uniform repartition of the water on the materials surface [4].

The crust created in the crystallizing gives the shape of the molded section. In most cases the crust doesn't offer sufficient mechanical resistance to the action of ferro static pressure [5, 6]. To complete solidification and guidance in good conditions of the wire it is created the secondary cooling zone. This cooling is achieved by direct spraying pressurized water, through nozzles, it is able to cross the steam layer formed by evaporation and ensure continuous water-metal contact.

The secondary cooling can be made in different cooling environments. In practice, the water is used especially as cooling agent that is sprayed through nozzles (circular conical, conical ring and slot). In special cases it is added to the water compressed air for optimizing the automation sprayed water [7].

The heat from the wire is partially removed by heating the cooling water, but the most heat is extracted by water evaporation, even if the evaporation percentage of the sprayed water on the wires surface is in general lower than 20%. All the unevaporated water which flows from the wires surface are collected filtered and reused.

The importance of the problem from the scientific point of view is reducing the casting defects, increasing the productivity and increasing the quality of the continuous casting semi - products, and from the technological point of view is very important to assure the cooling programs, all kind of casting speeds and temperatures on which the products are casted. The secondary cooling is important because of the process that creates the crust, it is leading to a length adjustment technology so that surface effects to be minimized [8].

At this moment the cooling is done using some valves by using some classical PID loops which adjust the water flow depending on reference measure. Advantages of PID control are a relatively easy implementation, in general only three parameters P, I, and D to be set up and an automatic parameter tuning mechanism. The main drawback of PID is, however, its non–convenience for the control of non-linear systems and its use in systems with fast changes of process parameters. In [9] is described and discussed challenges and control issues of the PID use for the continuous cast-ing and also is analyzed limitations of PID control. In comparison to PID control the fuzzy logic offers better control performance, smaller, overshoots, faster response, robustness and stability [7].

In [10, 11] are present some solutions for control the continuous casting process with fuzzy logic, but these solutions not realize a uniform repartition of the water flow in the secondary cooling stage and not work like an adaptive system. Other authors [12-14] use fuzzy logic and neural networks for detecting the cracks in crystallizing and control the primary water cooling stage.

The goal of intelligent system proposes by us is to improve the classical adjustment system, the introduction of some adaptive components in the adjustment loops and some overall predictions over the continuous casting machine. The fuzzy system can be adapted to all types of continuous casting plants, because it does not require major hardware changes, and is practically grafted into the existing structure of the continuous casting plant.

2 The structure of the control system

In Fig. 1 we describe the modern pattern of the control system of the continuous casting. The neural network of crack detection delivers the logic signal "1" at the moment when a primary crack was detected; otherwise the output signal is "0". This information is used accordingly by a fuzzy decision system (FC) with two bases of rules, which analyze a set of data acquired from the process, triggering changes of cooling water flow rate (Δv) and of casting speed set points (Δq).

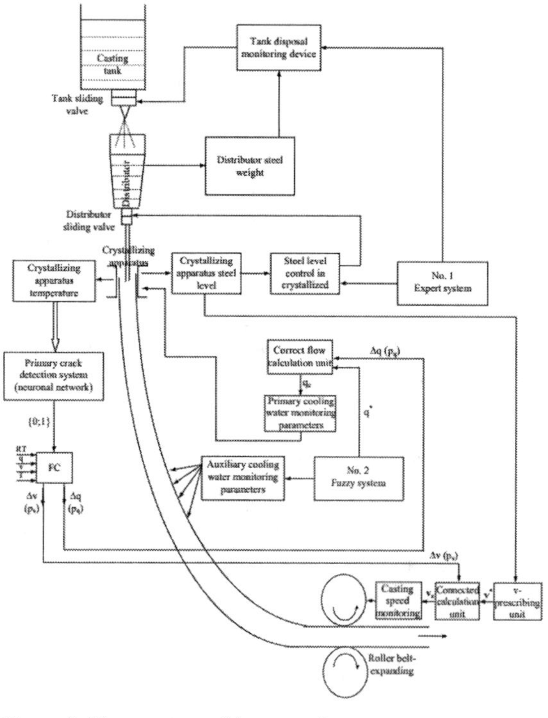

Figure 1. The structure of the control system.

Except for the technological risk (RT), which is set by the technological engineer, the other variables used as input data by the FC are received from the sensors already existing in the plant. The fuzzy decision system (FC) operates with two different rule bases, selected according to the signal (0 or 1) received from the neural network for the detection of cracks [13]. Fig. 1 shows that the two FC outputs (the casting speed correction Δ_v and the flow rate one Δ_q) are applied to calculation blocks that give the needed value for the correction of casting speed (v_c) and the flow rate(q_c), which, also are applied to the control systems of the respective parameters, existing in the plant. So, these values will replace the imposed values v^*, respectively q^* (control system initial set points) of the existent control system for both closed loops.

3 The control strategy proposed

The secondary cooling system is divided into 3 subareas such as: area 1 or foot roller area located on the crystallizing exit, area 2 and area 3, these areas are controlled individually. In general the sprayed water flow must decrease from the crystallizing in the direction of the cast. The sprayed water flows must be bigger in the superior side of the casting machine (area 0 and area1), for realizing the increasing of the semi-products crust and thus improve the resistance crust according to the effort. Lowering the cooling intensity in the casting direction must prevent that the surfaces temperature became too low in the straightening points. How we already presented the cooling is done using some valves by using some classical PID loops which adjust the water flow depending on reference measure. The goal of an intelligent system is to improve this classical adjustment system, the introduction of some adaptive components. Fig. 2 shows the block diagram of the proposed control strategy.

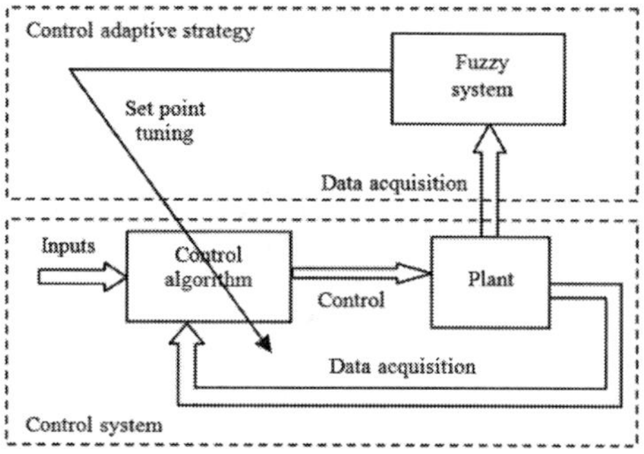

Figure 2. The block diagram of the control strategy.

From the technical point of view the changes made to the installation are not very important from the point of view of the execution and driving equipment, but are very important from the point of view of the control system, of the software which is used to control the system by introducing some fuzzy controllers, some predictive – adaptive systems or by introducing some standard models.

From the technological point of view the monitoring system must be rethought so that the technologist know which the ranges of the casting deviation for each recipe so he doesn't enter deviations that lead to different types of damage such as perforations of the semi-product, quality problems, problems related to the appearance of the casting surface and of the casting profile geometry.

The fuzzy system which will be created must also take into consideration the distance between the rings and the length of the curved wire. The installations geometry will reveal for a certain speed and casting recipe the needed quantity of the water to obtain the desired quality, this will make a spraying recipe which will give the quantity of water for every area, the values of the minimal flows needed for ensuring a minimal pressure in this areas (the minimal pressure for this areas is 2 bar) and implicitly the information needed to create the fuzzy systems rules which may take the needed measures in case the waters pressure is too low or too high, pressure which must be assured by a corresponding flow.

The classical systems are limited to a rigid repartition of the water flow in each area such as: 40% on the first area, 40% on the second area, and 20% on the third area.

An intelligent system has the capacity of eliminating this nowadays systems disadvantage by having the possibility of modifying in real time the water flow repartition for each area taking into consideration what happens in the installation.

4 Design of fuzzy controller

We will create a fuzzy controller having 3 inputs: speed, temperature and water flow and 1 output: Flow correction. Matlab design of fuzzy controller is present in Fig. 3.

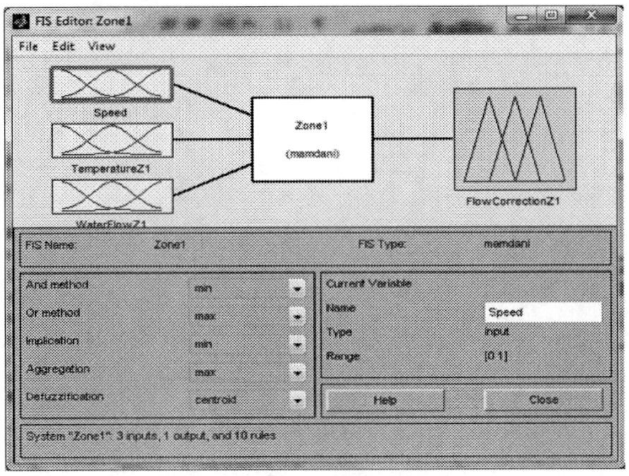

Figure 3. Block diagram of the fuzzy controller.

Input information

1) Speed (casting speed, [m/min])
Number of states: 5

States	Real domain [m/min]	Standardized values domain
Very small	0.7-0.8	0-0.2
Small	0.8-0.9	0.2-0.4
Medium	0.9-1.0	0.4-0.6
High	1.0-1.1	0.6-0.8
Very high	1.1-1.2	0.8-1

2) Temperature (steel temperature [^{0}C])
Number of states: 3

States	Real domain [^{0}C]	Standardized values domain
Small	410-610	0-0.33
Medium	610-810	0.33-0.66
High	810-1010	0.66-1

3) Water Flow (water flow [l/min])
Number of states: 5

States	Real domain [l/min]	Standardized values domain
Very small	900-940	0-0.2
Small	940-980	0.2-0.4
Medium	980-1020	0.4-0.6
High	1020-1060	0.6-0.8
Very high	1060-1100	0.8-1

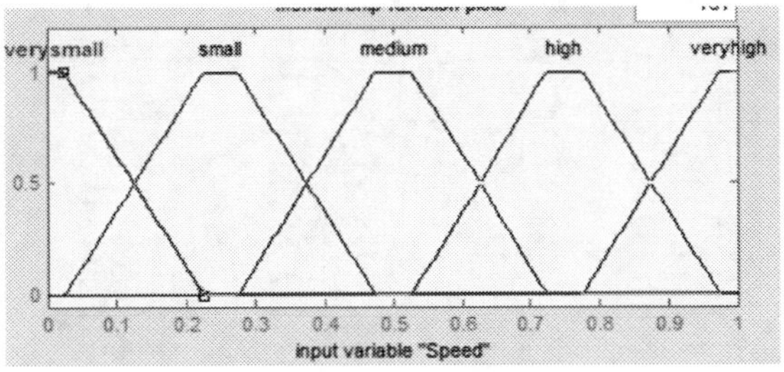

Figure 4(a). Membership function „Speed".

In Figs. 4(a), 4(b) and 4(c) we describe the membership functions for the input sizes.

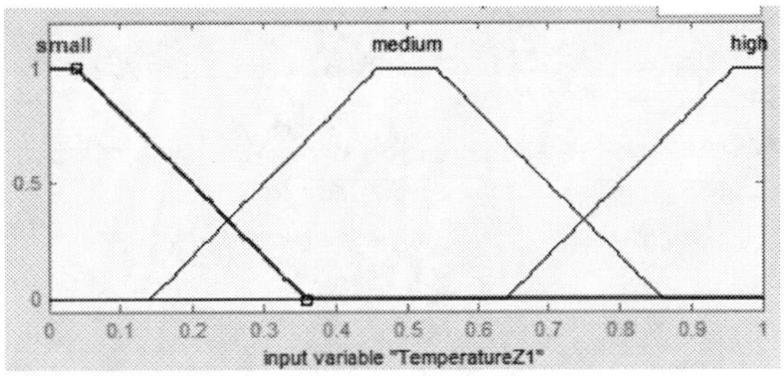

Figure 4(b). Membership function „Temperature".

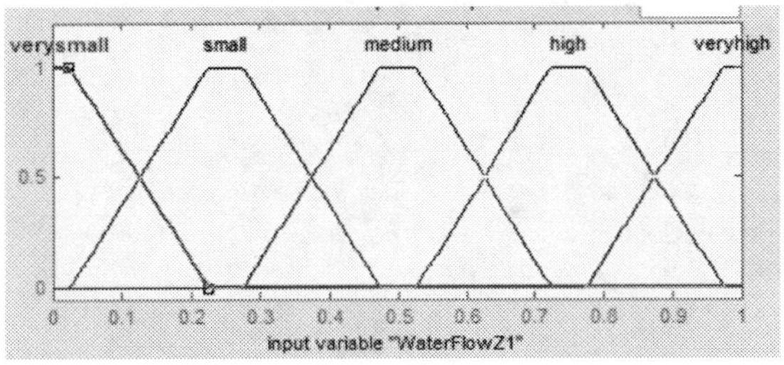

Figure 4(c). Membership function „Water Flow".

Output information

1) Flow Correction (water flow correction [l/min])
Number of states: 5

States	Real domain [%]	Standardized values domain
Verysmall	0-4	0-0.2
Small	4-8	0.2-0.4
Medium	8-12	0.4-0.6
High	12-16	0.6-0.8
Veryhigh	16-20	0.8-1

DOI: 10.1051/matecconf/20166810003

In Fig. 4(d) we describe the membership functions for the output size.

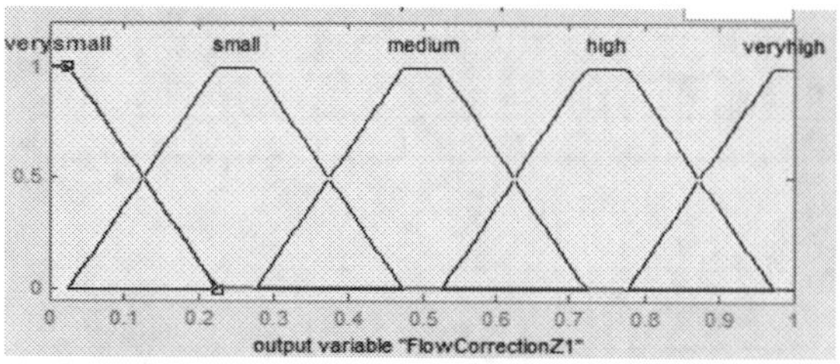

Figure 4(d). Membership function „Flow Correction".

Table of inference for some of the rules is given below:

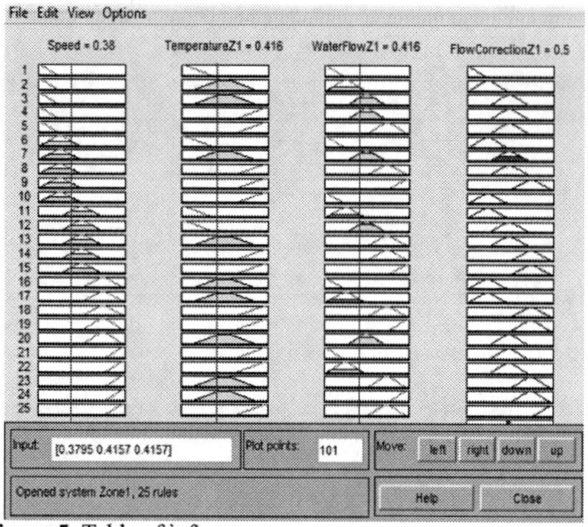

Figure 5. Table of inference.

Control surfaces obtained by simulations according to the block diagram from the Fig. 3 are listed below:

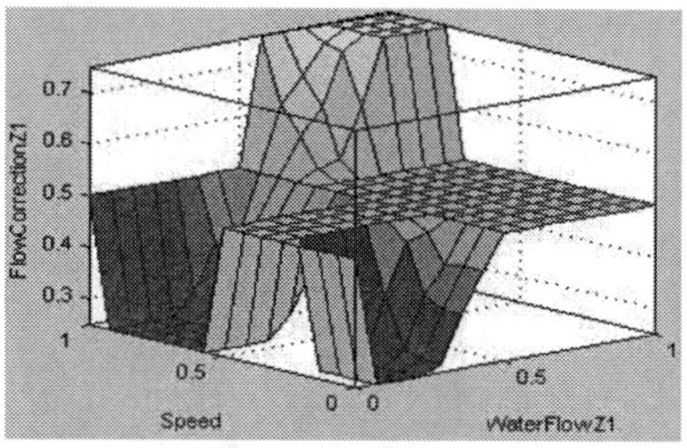

Figure 6. Control surface - Flow correction=f (speed, water flow).

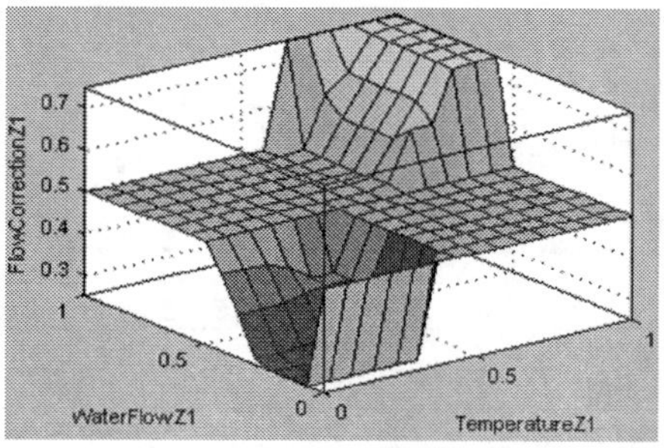

Figure 7. Control surface - Flow correction=f (water flow, temperature).

5 Validation of the simulated system

We have applied two different sets of data measured during the current process and stored into tables in case of validation of the simulated system operation, and for the input crack detection neural network. One of the sets refers to the situation when there are no cracks and the other one in case cracks occur.

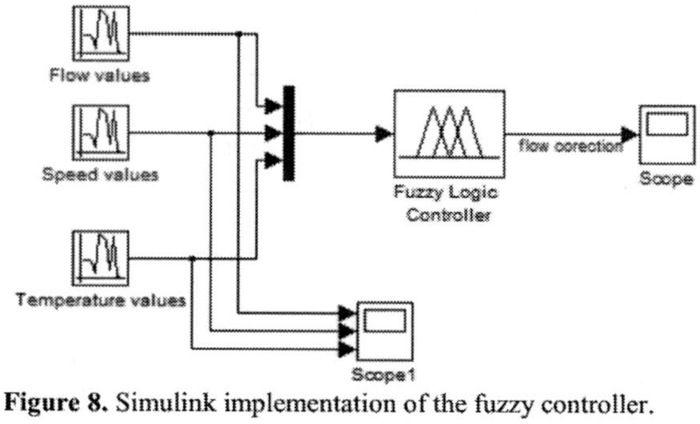

Figure 8. Simulink implementation of the fuzzy controller.

At the fuzzy controller input, there are several input values (flow, speed, temperature). These values are described in Fig. 8:

a) Time variation (120 seconds) of fuzzy controller input values;

b) Flow correction and new flow values.

Fig. 9 describes the situation during the first 30 simulation seconds, when both the cooling water flow and casting speed are low, the temperature is high. Speed correction is very low, hence required casting speed is almost unchanged.

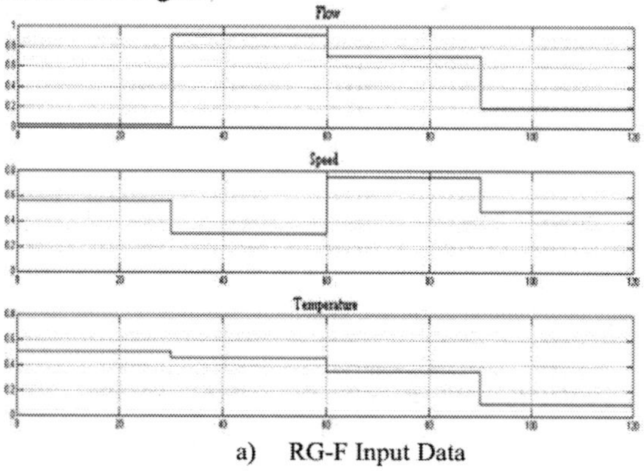

a) RG-F Input Data

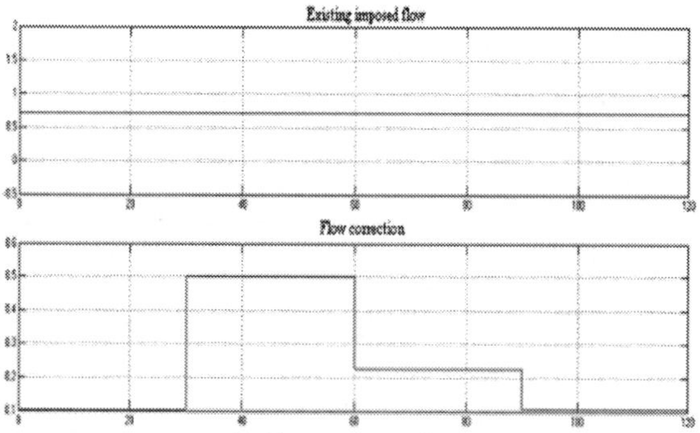

b) Output data - flow

Figure 9. Fuzzy controller validations.

During the whole time, cooling water flow increases significantly. We can see that the other two simulation rounds are similar. When analyzing all cases described in Fig. 9 we draw the following conclusions: fuzzy controller analyzes the input values and elaborates water flow corrections, according to the base sets; fore-writing block corrects all required values for flow, according fuzzy controller outputs. By simulating in Simulink, we have proved that all solutions are correct. Such simulation is made for performing a check out of the fuzzy system. During operation, all size values do not change so fast, hence some input values combinations are not that predictable. Once the system is implemented, the rules referring to such situations could be eliminated.

6 Conclusions

The main target of this paper was to develop an intelligent control system of the continuous casting process in the secondary cooling area capable of eliminating the surface flaws and casting rejects, to increase the quality of continuously cast products and implicitly to increase the productivity of the installation by controlling the water flow rate in the secondary cooling area. In order to achieve this objective, a solution was developed, meant to control in real time the distribution of the water flow rate according to the real situation in the installation and working as an adaptive system. The system was tasted by simulation and the results show us that the control strategy proposed is a very efficient method for the water flow control in continuous casting.

Acknowledgment

This work was supported by a grant of the Romanian National Authority for Scientific Research and Innovation, CNCS – UEFISCDI, project number PN-II-RU-TE-2014-4-1788.

References

1. E. Ardelean, et al., Simulation of continuous cast steel product solidification, Revista de Metalurgia **43** **(3)**, pp.181-187, (2007)
2. V. A. Efimov, *Casting and Crystallizing of Steel*, Ed. Technical, Bucuresti, (1986).
3. B. G. Thomas, J. T. Parkman, Simulation of thermal mechanical behavior during initial solidification, *Internat. Conf. On Thermomechanical Processing of Steel and Other Materials*, Wollongong, Australia, TMS, (1997)
4. B. G. Thomas, Modeling of the continuous casting of steel-past, present and future, *59th Electric Furnance Conference*, Phoenix, AZ, (2001)
5. G. O. Tirian, C. Pinca, C. Abrudean, C. Chioncel, Analytical identification, simulation and analysis of steel solidification process in crystallizer, Annals of University "Eftimie Murgu" Resita, XVIII, Nr. **1**, (2010)
6. M. Janik, H. Dyja, Modeling of three-dimensional temperature field inside the mould during continuous casting of steel, Journal of Materials Processing Technology, Volumes **157-158**, pp. 177-182, (2004)
7. J. Stetina, L. Klimes, T. Mauder, New challenges in modelling and secondary cooling control of continuous steel casting, *MATEC&2nd ESTEAD^{2015}*, Dusseldorf, Germany, pp. 1-8, (2015)
8. B. Kiflie, D. Alemu, Thermal analysis of continuous casting process, ESME *5th Annual Conference on Manufacturing &Process Industry*, Addis Ababa University, Ethiopia, (2000)
9. C. Furtmüller, L. D. Re, Control issues in the continuous casting of steel, *Proceedings 17 th World Congress Int. Feder. Autom. Cont.*, Seoul, Korea, pp. 700–7005, (2008)
10. L. Mugwagwa, et al. Design of a fuzzy controller for the continuous casting process, *International Journal of Engineering Science and Technology*, **5**, 1, (2013)
11. T. Mauder, J. Stetina, M. Masarik, On-line fuzzy regulatorfor continuous casting process, Metal 2013, Brno, Czech Republic, EU, (2013)
12. A. K. Bhattacharya, et al., Fuzzy diagnostics system for breackout prevention in continuous casting of steel, *Intelligent Control and Automation, WCICA 2004*, pp. 3141–3145, **4**, (2004)
13. G. O. Tirian, I. Filip, G. Prostean, Adaptive control system for continuous steel casting based on neural networks and fuzzy logic, Neurocomputing, pp. 236-245, **125**, (2014)
14. G. O Tirian, et al., Fuzzy system for implementing the cracks control during the continuous casting, *Proceedings of the 20th International Symposium*, **20**, 1, pp. 1661-1662, Austria, (2009)

A Roll Controlling Approach for a Simple Dual-Actuated Flapping Aerial Vehicle Model

Omar El-Farouk E. Labib, Sarah W. El-Safty , Mohamed H. El-Kouly , Mohamed M. Ramadan, Omar M. Shehata, ElSayed I. Morgan

Robotics and Autonomous Systems (RAS) Research Group, German University in Cairo (GUC), 5th settlement - New Cairo, 11432, Cairo, Egypt

Abstract. Aerial vehicles have been investigated recently in different contexts, due to their high potential of utilization in multiple application areas. Different mechanisms can be used for aerial vehicles actuation, such as the rotating multi-blade systems (Multi-Copters) and more recently flapping wings. Flapping wing robots have attracted much attention from researchers in recent years. In this study, a simple dual-actuated flapping mechanism is proposed for actuating a flapping wing robot. The mechanism is designed, simulated and validated in both simulation and experiments. A roll controlling approach is proposed to control the roll angle of the robot via controlling the speeds of both motors actuating each of the wings. The results achieved are validated experimentally, and are promising opening the door for further investigation using our proposed system

1 Introduction

In recent decades, Unmanned Aerial Vehicles (UAVs) has encountered a consistent ascent, specifically Micro Air Vehicles (MAVs) hold a unique sense of interest among engineers for both domestic and military applications. MAVs that fly utilizing flapping wings hold great potential for in- door surveillance and outdoor applications like photography and exploration. Progresses in the comprehension of flapping wing vehicles and flapping wing aerodynamic properties have incited a few endeavors to create flapping flight vehicles.

Endeavors to date are principally concerned with the attain- ability of these vehicles, as opposed to the advancement of their execution or the foundation of general outline stan- dards. Examinations have concentrated on amplifying thrust to-weight or minimizing power consumption. Work in the outline and advancement of individual vehicle subsystems incorporate endeavors to enhance stroke kinematics, wing shape and consistence, transmission proficiency and actuator execution.

The rest of this paper is organized as follows, section 2 presents an overview about research endeavors conducted by other researchers tackling the same problem. In section 3, the design and fabrication details are conducted, while section 4 presents the control and simulations conducted. Results are presented in section 5, and finally section 6 concludes this paper..

2 Literature review

In this section, investigation of related research efforts is conducted. This section is divided into two parts, The first will cover different designs investigated by researchers, and the second will be on the various control studies conducted.

2.1 Design review

To mimic the flapping wing movement seen in nature, particularly at small scales, MAVs have used just a few types of actuation effectively; piezoelectric and electromag- netic actuation. These sorts of actuation when combined with mechanical transmission subsystems, most popular being the ranges between 6-bar and 4-bar mechanisms [1] can produce improved stroke amplitudes and responding motion needed for flapping flight [2].

While the single actuator and wing is able to produce lift capa- ble of overcoming its own weight, this is only viable when the actuator is mounted rigidly to a fixture. To achieve free flight, the mean and instantaneous body forces and torques, or body wrench, delivered by a single flapping wing would prompt both lessened flapping sufficiency and unstable spinning. To make a powerful free-flight, extra actuators and wings must be added to counter this instantaneous torque [3]

In [4], a small butterfly-style flapping robots with wing veins and investigated their flight characteristics for different design parameters such as swept-forward wing angle and center of mass (COM). The experimental results showed that the body pitch angle was controlled by the swept-forward wing angle and by the relative positions of the COM and center of lift (COL).

In [5] the design and construction of a 2.6 gram electromag- netic actuator operated at resonance is presented. Independent bench tests for the coil and spring magnets were used to modify analytical models of the actuator, derived in detail in a parallel study. Wing kinematics and mean lift measurements were performed for the flapper operating at resonance, producing a lift-to-weight ratio for the actuator of over one at 24V.

In this study, it is desired to achieve a design that utilize 2 actuators to drive a mirrored mechanism, in order to be able to control each motor separately. This would be later used to experiment with the effects of variation in wing speeds on roll, using a developed test platform to isolate that degree of freedom which will be explained in details in the upcoming sections.

2.2 Control review

Several researchers investigated the problem of control of aerial vehicles. In [6] researchers have tried to investigate the potential of four design alternatives based on a piezoelectric driven passive pitch reversal wing. The models are compared based on their performance under an ideal linear quadratic reg- ulator (LQR) control scheme. The developed designs showed promising feasible controllable vehicles results.

While in [1], Liu, et al investgated the flight stability of a prototype flapping aerial vehicle. An integrated study of flexible wing aerodynamics and passive dynamic flight stability of the MAV by a combination of flexible wing kinematics and force measurements and computational approaches were conducted and simulated.

In this study, continuation upon previous researchers efforts is conducted in order to design and control a dual-actuated flapping wing robot, and control its roll-angle.

3 Mechanical design and fabrication

In this section, we will discus the process of reaching the final design to achieve our desired outcome, which will be explained in details in the coming subsections. The design iterations will be explained, the drawbacks and advantages of each design, and the fabrication process.

3.1 Designing Process

We stared initially using 2 servo motors, each connected to a wooden link. Both where fixed together using a wooden disc and suspended by a rod to view the effects on roll. The servo motors were slow and required us to change the direction of rotation to flap which was not efficient.

We decided to change the actuators to dc motors, and imple- ment a motion converting mechanism. In our first iteration of this idea, we developed a 6-bar mechanism. The system was manufactured using acrylic which is a commercial and light- weight material suitable for the flapping wing mechanism. The problem with the 6-bar mechanism was that it was too heavy and inefficient, The height of the mechanism was 26 cm.

A 4-bar mechanism was then designed, which greatly increased the compactness of the system. The final dimensions of the mechanism without the motors was 4.5 cm high, 9 cm wide and 3 mm thick. We mirrored the mechanism and made a full assembly with the 2 dc motors mounted on the final design. The final measurements of the mechanism without the motors was 4.5 cm in height, 9 cm width, and a thickness of 3 mm.

3.2 Design Analysis

Analysis of the mechanism was done twice; once without wings attached to the flapping links, and another using wings that we designed and will be disused in the wing design section. Tests were conducted on the wingless model, the wingless mechanism would act as a reference for testing different wing designs later if needed.

In the wingless tests, we first analyzed our mechanism experimentally using video tracking software called (Tracker) by (OSP) Open source physics.Track points were added to the tip of each flapping link,to be able to track its displacement in both X and Y axis, velocity components and magnitudes , and finally acceleration components and magnitudes. We tracked the mechanism starting from an operational voltage of 2 volts till we reached 12 volts, which is shown in the following graphs.

Mathematical analysis of the mechanism was done using (Matlab), the equations of the 4-bar mechanism were plotted using matlab and then compared with the video tracking software's results. The parameters of our mechanism are all sown in the following table.

Mechanism Parameters	
Parameter	Symbol
Ring radius	l_1
Connector	l_2
Flapper extension	l_3
Hinges Distance	l_4
Motor position	ϕ
Flapping angle	θ
Center of mass X component	P_x
Center of mass Y component	P_y

$$k1 = l4/l1 \quad (1)$$

$$k2 = l4/l3 \quad (2)$$

$$k_3 = \frac{l_1^2 - l_2^2 + l_3^2 + l_4^2}{2l_1 l_3} \quad (3)$$

$$A = \cos\phi \quad -k1 \quad -k2\cos\phi + k3 \quad (4)$$

$$B = -2\sin\phi \quad (5)$$

$$C = k1 \quad -(k2+1)\cos\phi + k3 \quad (6)$$

$$\theta = 2\arctan\frac{-B - \sqrt{B^2 - 4AC}}{2A} \quad (7)$$

DOI: 10.1051/matecconf/20166810004

3.3 Wing design and fabrication

To be able to test the effect of a flapping wing on our mechanism, we went for a minimalistic design. The reason being that we are investigating the effects of lift and thrust forces that are added to the system by the wings, and how they affect the roll. Takeoff with the current design is not the goal of this model, we are only using it as a testing platform to base our controller that will be discussed in details in the control and simulation section.

A symmetric wing was designed using using Computer aided design software (SolidWorks). The skeleton and connector are made up of acrylic. The skeleton is 25 cm long, 15 cm wide and 3 mm thick. The wings were made up of stencil paper as it is thin and durable.

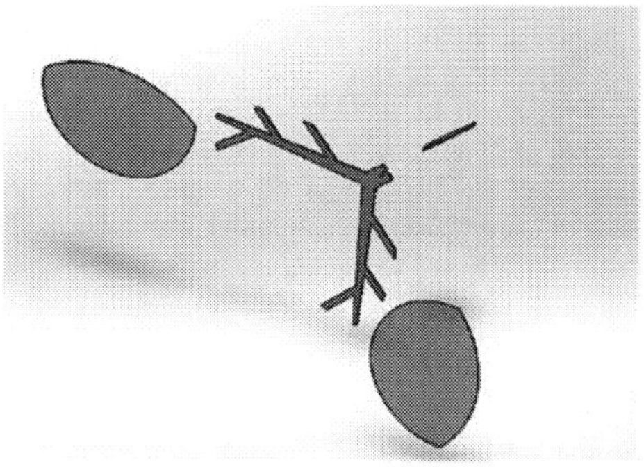

Figure 1. Wing assembly

3.4 Test bed

Finally a test bed was made to be able to run various test scenarios while implementing our controller model. The test bed was a simple design, a housing of dimensions 80 cm in height, 65 cm in length and width. Suspended in its interior are 2 supports to be connected to a rod that isolates roll only. The mechanism is given space and clearance to fully rotate at 360 degrees. as shown in Fig. 2.

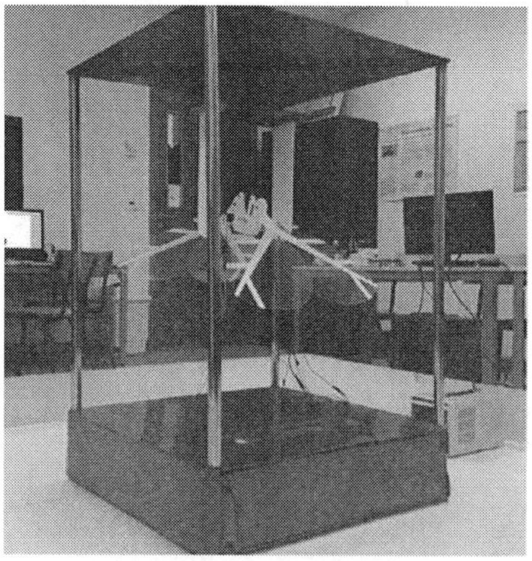

Figure 2. The testing platform : Bed and mechanism

4 Controllability analysis

The two DC-motors produce one effect that alter the forces and moments produced by wing motion. When the wing angle is changed, the resulting wing motion function remains a periodic function as shown in the design section regardless the speed of the motors. The difference is the frequency of which the moment arms of the wings change, hence, moments are altered. This effect can either be manipulated to change the moments of the whole system thus controlling roll as wanted, or manipulated to counteract external forces such as the counter torque mechanism of a helicopter.

Figure 3. The dual actuated flapping wing mechanism

4.1 Simulations

In the coming sub-sections, the three scenarios chosen for theory validation will be demonstrated. The first is stabilizing the model at a reference roll angle, while acted upon by external forces. The second is tilling the model to a desired angle. The third scenario utilizes the capabilities of our ap- proach, as we run a simulation of a predetermined sequence of commands. Simmechanics and Simulink were used to run simulations of the different scenarios. This was also done experimentally but only the second scenario was tested. The control of our model is primarily dependent on changing the DC-motor speeds while obtaining feedback from both the encoders and an inertial measurement unit (IMU).

4.2 Stabilization scenario

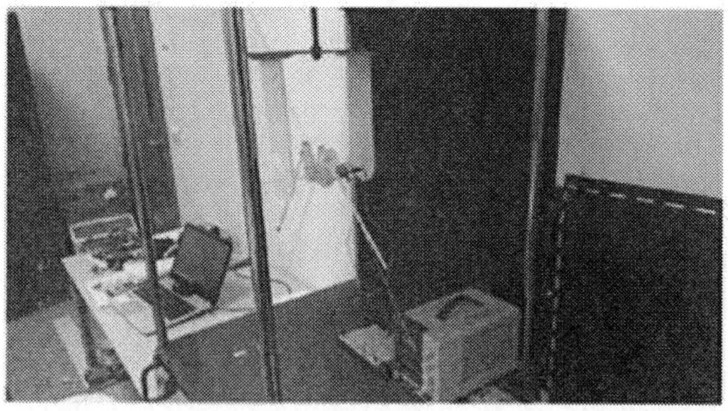

Figure 4. The first test scenario conducted

The stabilization scenario was conducted by adding a constant external force that would change the roll of the model. By using feedback from the roll angle, the angular velocity of each motor is changed to compensate this disruption. A PID controller is used for this closed loop, two scopes for the rpm of each motor separately and one scope for the roll of the model as can be seen in Fig. XX.

4.3 Desired angle scenario

In this scenario, a desired angle is given as an input and the speed of each motor is separately regulated using PID control till the desired angle is achieved by the model.

4.4 Predetermined sequence

The final scenario was a predetermined sequence of commands; stabilizing at reference angle 1 for 3 seconds then reference 2 for 2 seconds then back to reference 1 again indefinitely.

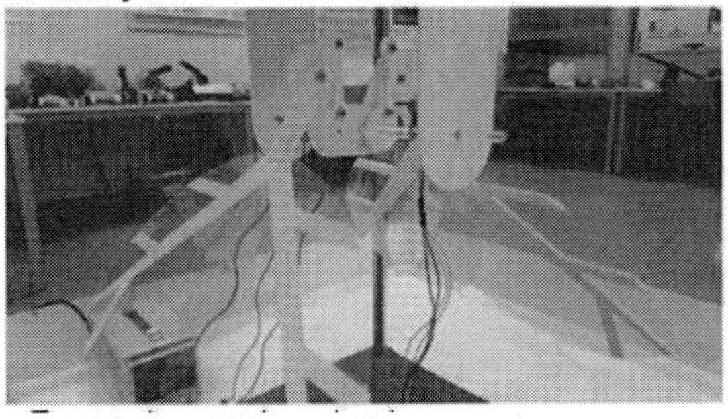

Figure 5. The second test scenario conducted

5 Results

The final design in our experimental modeling platform was able to produce sufficient moments and torques to change its roll efficiently and reliably. Three test scenarios were done, the first being the stabilization scenario which responded as expected giving satisfactory results but with some oscillations which can be overcome using different controllers such as fuzzy controller or tuning the PID values to minimize the overshooting. The second test scenario gave similar results but with less oscillations as there were no external forces acting upon the system. The third scenario obeyed the sequence given but the settling time was too long but can be refined by tuning the PID values.

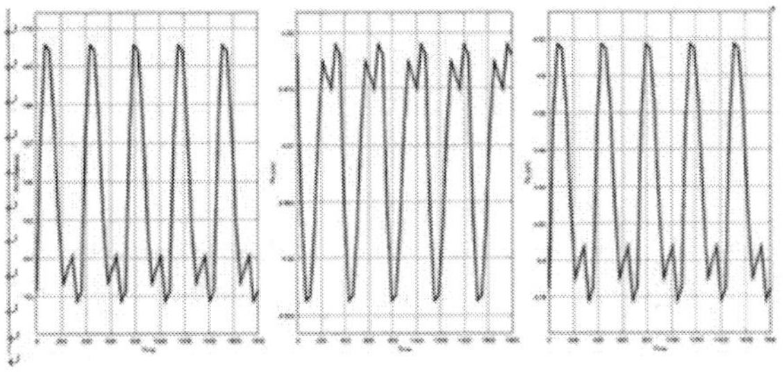

Figure 6. Phi, Px, and Py versus time

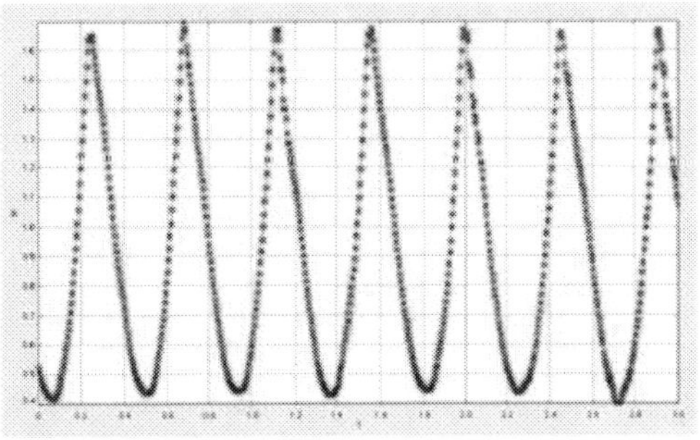

Figure 7. Angle at 8 volts

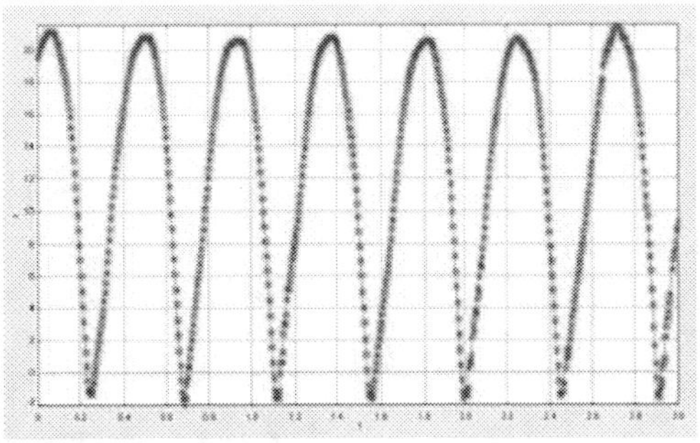

Figure 8. X position at 8 volts

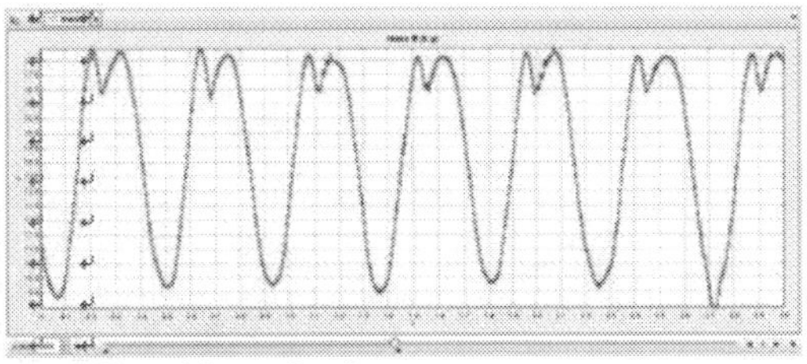

Figure 9. Y position at 8 volts

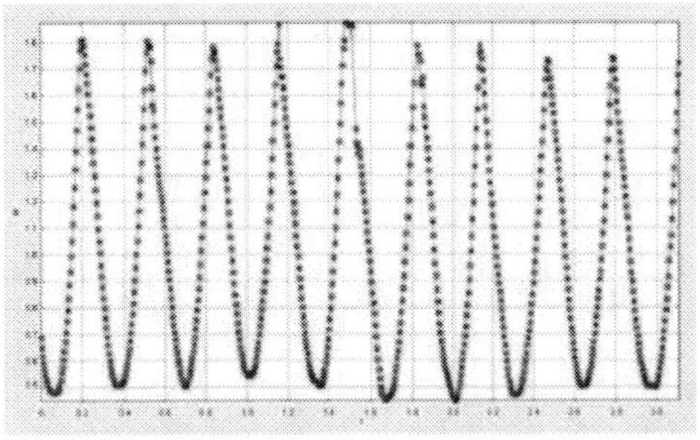

Figure 10. Angle at 12 volts

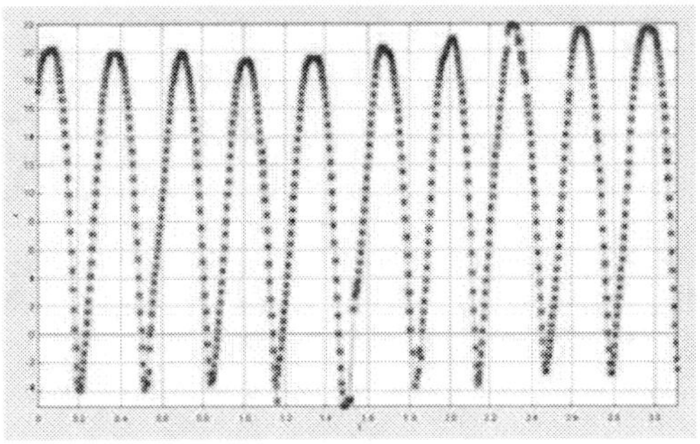

Figure 11. X position at 12 volts

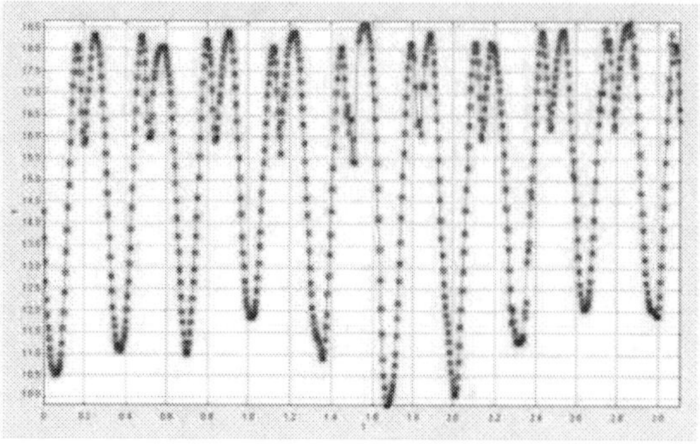

Figure 12. Y position at 12 volts

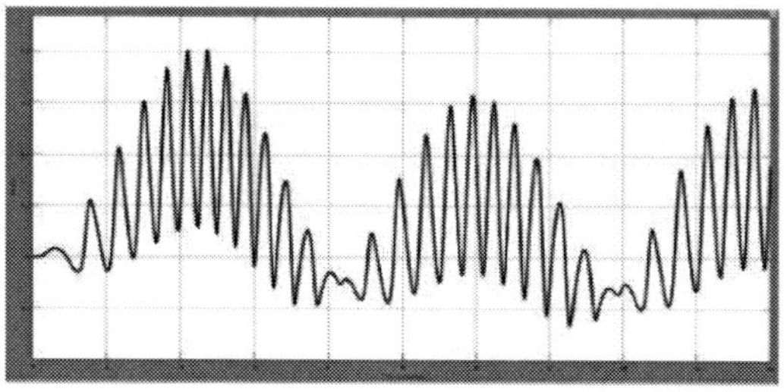

Figure 13. Scenario 1 Roll

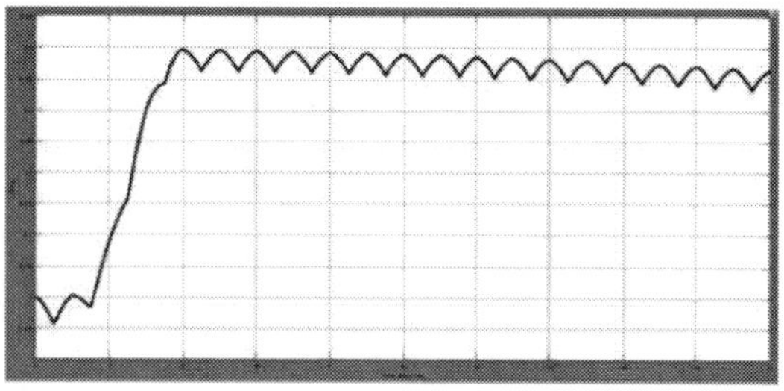

Figure 14. Scenario 2 Roll

6 Limitations

Many limitations were faced in the study; there was no variety in the commercially available components or materials. Moreover, the availability of DC motors with built-in encoders was limited. Also, the shaft encoders available were not compatible with the motors used.

7 Conclusion and future work

In this study we were aiming to tackle the stability problems which lead to short flight times of flapping wing MAVs. Using two actuators, one for each wing instead of a single one opens a lot of possibilities for the future of flapping flight. We designed a dual-actuated flapping MAV model which can be used as a testing platform for different roll control approaches. Which will help to build a better understanding of how to control this degree of freedom, and hopefully result in more stable flapping air vehicles and thus longer and more reliable ones. Our design was not aiming for take-off at this stage; however, due to the importance of roll in achieving longer flight times, the model is chosen to be as simple and general as possible to be easily tested.

For future work, using this platform to test various controllers and there effect on the system. Adding pitch control to the platform by using a method to manipulate it such as Festo's Smart bird's neck and tail actuation. Developing a smaller testing platform capable of takeoff using the same feedback method to further analyze the behavior.

Acknowledgements

The Authors would like to thank Professor Yasser Higazi Dean of the Information and Electronics Technology (IET)and Professor ElSayed Imam Morgan Dean of Mechatronics Engi- neering for there support and efforts to get these results out. It wouldn't have happened without you, thanks.

References

1. H. Liu, X. Wang, T. Nakata, and K. Yoshida, Aerodynamics and flight stability of a prototype flapping micro air vehicle, Complex Medical Engineering (CME), 2012 ICME International Conference on. IEEE, pp. 657-662, (2012)
2. T. Zhang, C. Zhou, and S. Su, Design and development of bio-inspired flapping wing aerial vehicles, Advanced Robotics and Intelligent Systems (ARIS), 2015 International Conference on. IEEE, pp. 1-6, (2015)
3. L. Hines, D. Colmenares, and M. Sitti, Platform design and tethered flight of a motor-driven flapping-wing system, in Robotics and Automation (ICRA), 2015 IEEE International Conference on. IEEE, pp. 5838-5845, (2015)
4. T. Fujikawa, Y. Sato, Y. Makata, T. Yamashita, and K. Kikuchi, Motion analysis of butterfly-style flapping robot for different wing and body design, Robotics and Biomimetics, 2008. ROBIO 2008. IEEE Inter- national Conference on. IEEE, pp. 216-221, (2009).
5. J. Roll, B. Cheng, X. Deng, et al., Design, fabrication, and experiments of an electromagnetic actuator for flapping wing micro air vehicles, Robotics and Automation (ICRA), 2013 IEEE International Conference on. IEEE, pp. 809-815, (2013)
6. L. L. Hines, V. Arabagi, and M. Sitti, Control performance simulation in the design of a flapping wing micro-aerial vehicle, Intelligent Robots and Systems (IROS), 2010 IEEE/RSJ International Conference on. IEEE, pp. 1090-1095, (2010)

MATEC Web of Conferences 68, 10005 (2016)
ICIEA 2016
DOI: 10.1051/matecconf/20166810005

A Hospital Medical Gas Monitoring System Based on Multi-mode Optical Fiber Ring Network

Wei Sun[1], Jian Wang[1], Li-Hong Gai[2], Jun-Jie Gu[2] and Li Li[1]

[1]School of Mechatronic Engineering and Automation, Shanghai University, 200072 Shanghai, China
[2]Environmental Engineering CO, LTD of Shandong Academy of Environmental Science, 250013 Jinan, China

Abstract. This paper proposes a hospital medical gas monitoring system based on optical fiber ring network aiming to solve the problems of current medical gas monitoring, for example, the function of acquisition points is single, data transmission is not real-time, and communication distance is short. This system can display real-time gas state data and alarm in time by using the multi-function regional monitoring alarms as monitoring nodes. Meanwhile, we use multi-mode optical fiber ring network to connect the monitoring nodes throughout the hospital because of a long-distance transmission. Experimental results show that the system has the features of multifunction, safety, reliability, convenience and utility.

1 Introduction

Medical gas project is known as one of the life support system, whose normalization, safety, reliability and reasonability directly related to the patient's life safety [1]. The function of medical gas monitoring system is providing real-time state data of medical gas system to device managers or medical staffs. Once the parameters are out of the normal range, it emits sound and light alarm signal. Most of the existing hospital medical gas monitoring systems use the method of installing the pressure or flow meters in the gas equipment as gas parameter acquisition node. This gas parameter acquisition node has only a single function, so it need artificial screening. When gas failure occurs, alarm occurs only in control table, which is unfavorable for medical staff to notice the gas breakdown in first time.

Currently, monitoring center and regional acquisition nodes of existing hospital medical monitoring systems communicate with each other through fieldbus. However, in the field environment, the effective transmission distance (such as RS485, CAN, etc.) of common fieldbus is generally only 300 or 400 meters, and their transmission rate is 100kbit/s or less. In addition, since most of major hospitals cover large area so that the gas region disperses in different locations of the hospital, which result in that most gas regions are far from the monitoring center, so the distance exceeds common fieldbus's effective transmission distance. Therefore, connecting monitoring center and gas religions with fieldbus cannot meet all the requirements in terms of transmission distance, transmission speed or data bandwidth. Compared with the fieldbus, multi-mode fiber supports longer transmission distance, it can support up to 2000 meters' transmission distance when used in 10Mbps and 100Mbps Ethernet. In this system, gas state data acquisition nodes of each regional monitoring subnet communicate with each other through fieldbus, regional monitoring subnets throughout the hospital are connected by multi-mode fiber ring network [2].

2 Overall system design scheme

The overall design scheme of the system shown in Fig. 1. The monitoring center is composed by the industrial personal computer (IPC), the printer and the switch. The switch is responsible for transforming the optical signals of multi-mode optical fiber ring network into electrical signals for IPC, at the same time, it transforms the electrical signal emitted by IPC into optical signals which can transmit in a multi-mode optical fiber ring network. Each gas region form a set of regional monitoring subnets in building, each floor place one or two regional monitoring alarm as regional monitoring nodes. Microprocessor is embedded in regional monitoring alarm, who not only can send the data measured by the sensors to the fieldbus, but also has function of real-time display and alarm.

The PLC with communication interface modules collects and processes data from fieldbus, then control the human-machine interface which we called regional monitoring panel to display the gas state data acquired from every acquisition node, RTU/TCP gateway achieve the function of mutual conversion between Modbus RTU and Modbus TCP protocols [3]. The medical gas supply stations such as negative pressure pump station, air pressure pump station and liquid oxygen station equipped with the same devices. PLC and monitoring center communicate via switches and multimode optical fiber ring network.

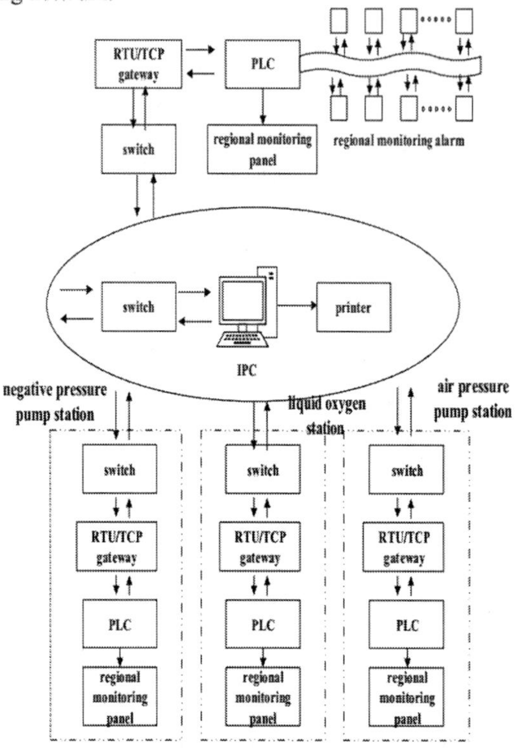

Figure 1. Overall system design scheme.

The IPC in monitoring center can monitor the hospital medical gas state of overall hospital, it has a function of historical data query, and the gas state data can be printed when needed. Regional monitoring panel can also set the upper and lower limit alarm values manually. Fieldbus and multi-mode optical fiber ring network two means of communication methods are used in this system in order to take full advantage of their advantages. In the process of monitoring, if the multi-mode fiber optic ring network breakage occurs somewhere for any reason, optical fiber ring network break into ordinary optical fiber line, while it is still capable of transmit optical signals which can ensure the normal communication [4].

3 Important modules of system

The hospital medical gas monitoring system based on multi-mode optical fiber ring network includes several important modules to ensure achieve the desired results, such as regional monitoring alarm and regional monitoring panel.

3.1 Formatting the title, authors and affiliations

Regional monitoring alarm is composed by modular splicing, each function module can display or monitor two kinds of gas data simultaneously. If you need to monitor multiple gases simultaneously, more function modules can be spliced together [5]. The hardware structure of each function module is shown in Fig. 2. It includes the modules of microprocessor (MCU), power supply circuit, clock circuit, input signal selection circuit, display driver, digital display, keys, DIP switch, buzzer, data storage circuit and fieldbus communication circuit.

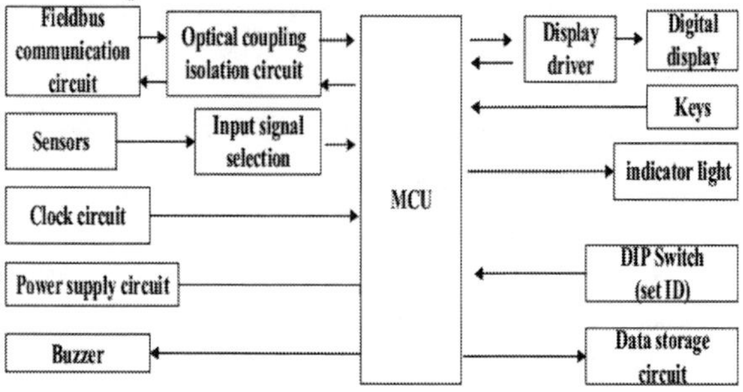

Figure 2. Hardware architecture of regional monitoring alarm.

Main software workflow of regional monitoring alarm is shown in Fig. 3, the MCU's analog channels receive the value of medical gas pressure, and its analog / digital (A / D) converter convert analog value, the converted data is stored after computed and processed by MCU, the display module calls the data and displays real-timely [6]. At the same time, communication module send data to the fieldbus. MCU compare the medical gas data with threshold value. If the data is higher than the upper limit, light on the indicator light "high", the buzzer sound the alarm; similarly, if the data is lower than the lower limit, light on the indicator light "low", the buzzer also sound the alarm. If the mute button is pressed when the buzzer is ring, the alarm sound pause for some time.

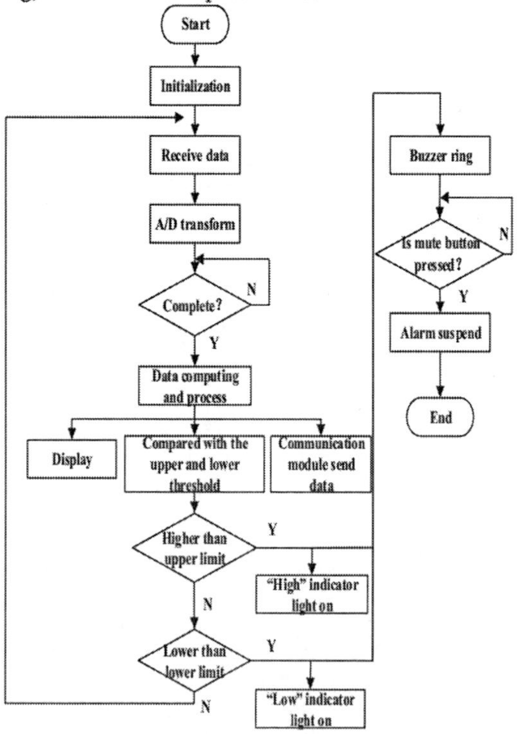

Figure 3. Hardware architecture of regional monitoring alarm.

DOI: 10.1051/matecconf/20166810005

3.2 Regional monitoring panel

The function of regional monitoring panel is monitoring all acquisition nodes of one of the buildings or medical gas supply station. If medical gas fault occurred in somewhere, regional monitoring panel sound the alarm, and what's more, it can show the fault position [7]. By operating directly on the touch screen, you can set the threshold value of the medical gas state data, and you can query historical data. All data processing and computing work is done by the PLC in this process [8]. The main workflow of PLC is shown in Fig. 4.

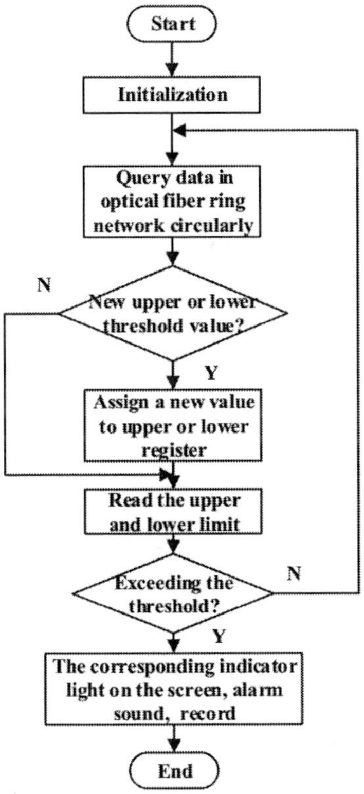

Figure 4. The main workflow of PLC.

4 Experimental test and analysis

In accordance with the system design scheme, the test samples we made are shown below (Fig. 5, Fig. 6 and Fig. 7). Fig. 5 shows the external and internal structure of the regional monitoring alarm while Fig. 6 shows the design of the regional monitoring panel and Fig. 7 shows the Human-machine interface of IPC.

In order to test the reliability of the system, we put the system in a large hospital which covers an area of 35000m2. The hospital has four gas usage buildings and three medical gas supply stations. We deploy 100 regional monitoring alarms, 8 switches, and 7 sets of regional monitoring panel in the experiment, and the total fiber length is 2km. During several months' experiment, the system alarmed twice. We confirmed that the medical gas does go fault when alarm occurred, and monitoring center has an accurate positioning of fault node. Every part of the system works properly throughout the course of the experiment.

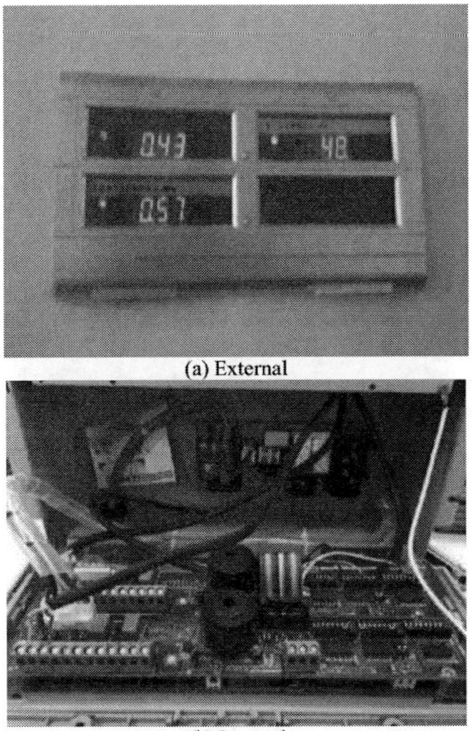

(a) External

(b) Internal

Figure 5. Regional monitoring alarm

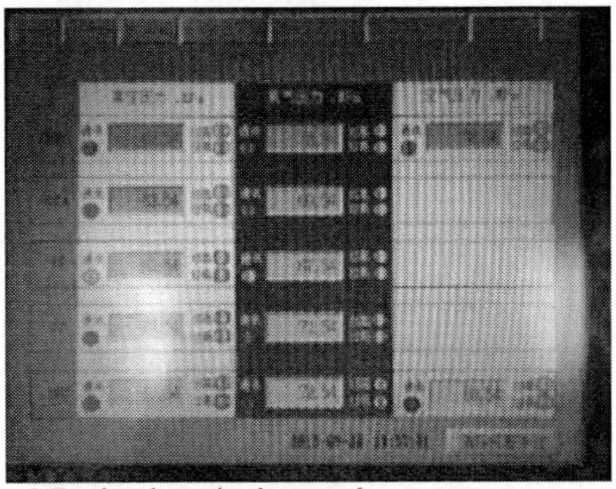

Figure 6. Regional monitoring panel.

Figure 7. Human-machine interface of IPC

5 Conclusions

This paper implements a hospital medical gas monitoring system based on multi-mode fiber optic ring network, and elaborates on the content of system structure, hardware and software design and so on. Regional monitoring alarms are placed in all medical gas monitoring nodes, regional monitoring alarms connect each other with fieldbus. Regional monitoring panel is placed in every building to monitor the entire building. Multi-mode optical fiber ring network connects all regional monitoring buildings and medical gas supply stations, the IPC of monitoring center monitor all monitoring nodes of the entire hospital. Two means of communication fieldbus and multi-mode optical fiber ring network exist in one system, in order to make best use of the advantages and bypass the disadvantages so that the medical gas monitoring is more accurate and more stable. In conclusion, the system has a high social value and economic value. The future work will further extend our system both theoretically and experimentally.

References

1. S. L. Xu, L. B. Tan, J. Pu, Medical Gas Pressure Monitoring System Base on Modbus Communication Protocol, Ind. Control Comput. **23**(8), 40-43 (2010)
2. N. J. Xu, Fault detection technique for wavelength division multiplexing passive optical network using chaotic fiber laser, Opt. Fiber Technol. **20**, 163-167 (2014)
3. Z. Y. Liu, Design of communication protocol transform between Modbus RTU mode and TCP mode, Sci. Technol. Eng. **13**(18), 1671-1815 (2013)
4. M. Grigoriu, M. C. Popescu, Study on Optical Fiber Insertion in Underground Telecommunication Networks Using Hydraulic Similarity, Procedia Technol. **22**, 637-644 (2016)
5. S. C. Xu, S. Q. Yin, R. Srinivasan, M. Helander, Proactive Alarms Monitoring using Predictive Technologies, Comput. Aided Chem. Eng. **31**, 1537-1541 (2012)
6. Q. S. He, AD conversion with high speed based on FPGA, Eletron. Des. Eng. **20**(8), 122-124 (2012)
7. D. H. Chace, Medical Monitoring and Diagnosis, Encyclop. Mass Spectrom. **9**, 305-311 (2016)
8. E. R. Alphonsus, M. O. Abdullah, A review on the applications of programmable logic controllers (PLCs), Renew. Sustain. Energy Rev. **60**, 1185-1205 (2016)

MATEC Web of Conferences **68**, 10006 (2016) DOI: 10.1051/matecconf/20166810006
ICIEA 2016

Based on Wide Area Environment Abnormal Behavior Analysis and Anomaly Detection Research

Lin Zhang[1] Xiaoping Li[2,]

[1] *Beijing University of Civil Engineering and Architecture, Beijing, China*
[2] *Beijing Institute of Technology, Beijing, China*

Abstract. Group anomaly identification and location is an important issue in the field of artificial intelligence. Capture of the accident source and rapid prediction of mass incidents in public places are difficult problems in intelligent video identification and processing, but the traditional group anomaly detection research has many limitations when it comes to accident source detection and intelligent recognition. We are to research on the algorithms of accident source location and abnormal group identification based on behavior analysis in the condition of dramatically changing group geometry appearance, including: 1) to propose a logic model of image density based on the social force model, and to build the crowd density trend prediction model integrating "fast and fuzzy matching at front-end" and "accurate and classified training at back-end"; 2) to design a fast abnormal source flagging algorithm based on support vector machine, and to realize intelligent and automatic marking of abnormal source point; 3) to construct a multi-view human body skeleton invariant moment model and a motion trajectory model based on linear parametric equations. The expected results of the research will help prevent abnormal events effectively, capture the first scene of incidents and the abnormal source point quickly, and play a decision support role in the proactive national security strategy.

1 Introduction

Some large-scale unexpected mass incidents occur more frequently in public area with the development of society and economy and growth of population density. These things will lead to serious consequences, if we can't capture the first scene and the source point quickly and do the rapid early warning.

The existing video surveillance system is lack of intelligence. Its real time performance of intelligent decision for group abnormal events is poor. It is always limited to fixed single camera which can only distinguish and determine some simple abnormal behavior and it is lack of recognition of the high level complex behavior of the human body. People need to find out the abnormal source from the massive video after the abnormal event occurs and this can't meet the real-time requirements of emergency management of major events. The existing monitoring system must be transformed from close to the Internet to improve the collaborative analysis of video surveillance system and the ability of group anomaly detection and localization and the working efficiency of the camera network. This paper is intended to study the detection of abnormal source points of the wide area group and the determination of the nature of the source points based on the understanding and analysis of the intelligent video, try to make breakthroughs on abnormal population model, abnormal source detection and location, the source of abnormal attribute analysis, model training and matching, camera collaborative scheduling key technology. In order to achieve population density trends prediction of according to the "fast fuzzy matching" and "accurate classification of training", group trend prediction should be combined with social force model and population density trend model should be built, anomaly source should be focused by multi stage split control strategy to capture the first scene of the abnormal events of the source point timely and predict the trend of group events rapidly. Rapid propagation of new abnormal group model, multi region and multi scene linkage real-time warning should be achieved by continuous training "group trend - the source of the object behavior attribute" relationship, determining attribute of abnormal source and pushing analysis on the attribute of the anomaly source. This will make a positive contribution to the early warning of public places.

2 Theoretical basis research

In view of the above questions, we analyzed the population density estimation research, intelligent video surveillance, moving object behavior analysis, and machine learning based on support vector machine and pattern classification methods.

2.1 Analysis of group event types

Intelligent video surveillance [1], become one of the hot research fields of artificial intelligence, has been transformed from the "passive monitoring" to "active control". Group of abnormal behavior recognition technology driven by requirements has attracted more and more attention of researchers. The U.S. Defense Advanced Research Projects Agency has set up a major project of visual surveillance VSAM which is mainly used for the analysis of crowd behavior in battlefield environment and civil video monitoring scene. Davies is the first to apply the crowd density monitoring system to the subway station in London using background subtraction and edge detection method to calculate the number of pixels. Ramin Mehran is the first to propose the method using the social force model to detect anomaly, use the optical flow method to extract the dynamic information of the image in the crowd and use Newtonian mechanics to characterize the individual and mutual movement. Saira Saleem Pathan is the first to use social entropy to quantify the uncertainty of potential field and use SVM to detect the crowd abnormal. However, the real-time detection of abnormal group still faces a series of technical problems, for example, the traditional video monitoring system still can't confirm the category of abnormal alarm event and information of abnormal source. The proper descriptive Characteristics of moving crowd must be found, a reliable motion model should be established, analysis and position of abnormal crowd should be realized in order to reduce the space dimension and improve processing speed.

2.2 Analysis of anomaly source location

At present, there isn't any research that studies the location of abnormal source and judgment of anomaly types, causes and attribution under the group abnormal environment. The logical relation between the abnormal area and the abnormal source should be established according to different group movement models because of complex variability of population movement, the motion model should be classified by mass samples of moving groups that are collected by the front end and a special group abnormal behavior model should be trained using SVM. The biggest advantage of SVM is that it does not need to reduce the dimension of data when it is classified and it has good performance in terms of convergence, training speed, generalization, and classification accuracy Rabaud and Belongie [2] proposed Kanade-Lucas-Tomasi tracking algorithm tracking the target by clustering similar features on the track. Chow estimated the population density based on global hybrid neural network learning algorithm. He got pixel statistics in image sequences, then estimated population density using neural network modeling after the study. Donatell found a method which could use SURF algorithm to extract the image feature points and use the support vector machine to train some selected samples, then count population density of scene through support vector machine. This paper proposes a feature representation based on the video image partition region density calibration and scalar, identifies the abnormal source in the abnormal group by semiautomatic and screens a more effective training set through continuous training and learning, then finds out the relationship between the population trend characteristics and the distribution of abnormal source points. At the same time it also realizes the all-time intelligent identification of abnormal source points using appropriate training machine kernel function and adopting moment invariant adaptive strategy considering the multi angle and multi modal changes in video. image.

2.3 Analysis of the behavior of abnormal objects

Behavioral attribute analysis has a wide range of applications, such as intelligent monitoring, human-computer interaction, training of athletes, video coding, etc. Many scholars have put forward a lot of behavior analysis algorithms to identify the complex event and individual behavior. Efros et al. poposed the characteristics of the optical flow of the low precision motion image using the fuzzy optical flow feature. Dalal et al. [3] analyze human using the differential optical flow histogram and gradient direction histogram as input features. Alberto et al. tried to recognize the human body posture using the fuzzy finite state machine. Vajda et al. used dynamic time warping method in motion matching and this method has the characteristics of time sequence, robustness and good performance. ZhiZhongused L-K algorithm to capture the optical flow information proposed by BruceD [4]. Lucas to get the image of the energy flow value and got an intuitive reflection of the dynamic changes of the target [5]. Billing et al. proposed two algorithms based on PSL (Predictive Sequence Learning) [6]. Lin et al. analyzed the principal component of video frames and then classified and identified the sequence of events, but the large amount of data of the image led to a very large amount of computation. Jacob combined the assumption of projection, context dependent reasoning, and structured HMM to identify the expected behavior in the GMT (Ground Motion Target Indicator) data. Crowd control system should be able to provide early warning of the occurrence of abnormal behavior in the video environment. Motion trajectory could be obtained by using the moving human tracking algorithm and the abnormal behavior could be estimated by extracting the length, pixel, covariance, motion degree and other information from the trajectory [7].The study of behavioral analysis mainly focuses on the behavior identification of a limited class of simple rules or abnormal behavior detection of single target in a particular scene [8], complex and large amount of calculation of the image processing should be carried out before behavioral analysis, this reduces the efficiency of the algorithm and it's difficult to achieve intelligent front-end equipment. Most of the algorithms can only achieve the static matching of single target behavior in the process of analysis and the capture efficiency is low.

3 Program architecture design

In order to analyze the group abnormal behavior and determine the abnormal source object attribute, we combine the social force model, the trend of group movement and density information, divide the Video frames into different region through the grid area scan, predict the speed and direction of population movement based on crowd time domain trajectory logic model and carry on the fuzzy matching of anomaly groups based on the crowd spatial domain density logic model, then give the predictive result by focusing on the source points and analyzing the source object attribute. The flowchart as shown in Fig. 3.1.

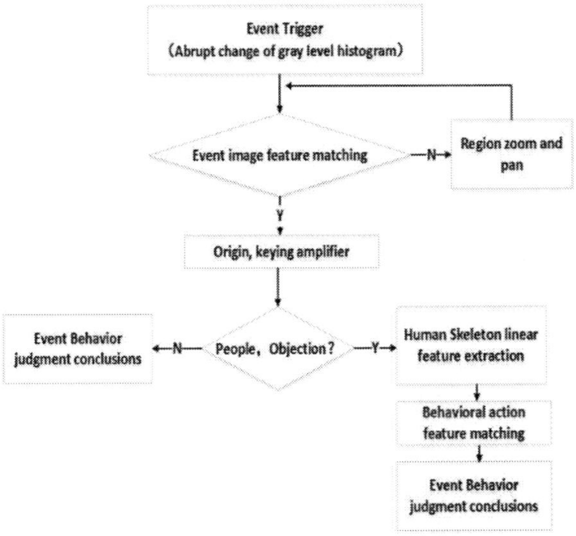

Figure 1. The process of the group abnormal behavior prediction and the source point attribute analysis.

4 Conclusions

Aiming at the problems of marking groups abnormal source and analyzing source behavior under intelligent video surveillance, the fast feature matching of the sudden mass events in public places and the first time to capture the abnormal object are realized by establishing of the model which is related to the change tendency of the group characteristic and the abnormal character of the source point; the fast algorithm designed should be focused on rapid detection of abnormal population in complex background, abnormal target location, Characteristic model of population movement tendency in the series of basic problems of abnormal behavior analysis, adaptive SVM training, analysis of abnormal source and image transmission coding protocol in wide area environment.

The suspected abnormal points is located through the early abnormal group area detection and locked combining with the lens multi-level collaborative scheduling algorithm, the model of "group individual" feature is established according to the change characteristic of abnormal group and the position information of abnormal point.

References

1. Zhang P, Zhang X, Li J, et al. The effects of temperature and salinity on the swimming ability of whiteleg shrimp, Litopenaeus vannamei[J]. Comparative Biochemistry and Physiology, Part A: Molecular & Integrative Physiology, 2007, **147(1)**: 64—69.
2. Vapnik V N.The Nature of Statistical Learning Theory[M].Springer,1995.
3. Dalal N, Triggs B. Histograms of oriented gradients for human detection[C]//Proceedings of the Conference on Computer Vision and Pattern Recognition(CVPR’05). 2005,1:886-893.
4. Bruce D, Lucas, Takeo Kanade. An Iterative Image Registration Technique with an Application to Stereo Vision[C]. Proceedings of the 7th International Joint Conference on Artificial Intelligence. 2006(11): 674-679.
5. Billing, E.A, Hellstro, Janlert, L.-E. Behavior recognition for Learning from Demonstration[C]. 2010 IEEE International Conference on Robotics and Automation (ICRA), 2010(5): 866-872.
6. Lin Chunli, Wang Kejun. A behavior classification based on Enhanced Gait Energy Image[C]. 2010 2nd International Conference on Networking and Digital Society (ICNDS). 2010(5): 589- 592.
7. Lee J J, Kim G J, Kim M H. Trajectory extraction for abnormal behavior detection in public area[C]//Emerging Technologies for a Smarter World (CEWIT), 2012 9th International Conference & Expo on. IEEE, 2012: 1-5.
8. Zhong Hua, Shi Jian-bo, Visontai M. Detecting unusual activity in video[A]. In: Proceedings of IEEE Conference on Computer Vision and Pattern Recognition[C], Washington DC, USA, 2004:819-826.

Framework for the Energetic Assessment of South and South-East Asia Fixed Chimney Bull's Trench Kiln

Niccolò Le Brun[1,2,a] and Christos N. Markides[2]

[1]*ThermoCrunch, 144a Albert Palace Mansions, Lurline Gardens, London, SW11 4DJ, United Kingdom*
[2]*Clea Energy Process (CEP) Laboratory, Department of Chemical Engineering, Imperial College London, South Kensington Campus, London, SW7 2AZ, United Kingdom*

Abstract. One of the major sources of fuel consumption and greenhouse gas emission in South and South-East Asia is brick manufacturing. One of the most commonly implemented technologies for brick manufacturing in this region is the fixed chimney Bull's trench kiln (FCBTK). This type of technology largely depends on manual labour and is very inefficient when compared to more modern technologies. Because the adoption of more advanced technologies is hindered by the socio-economical background, the much needed innovations in the brick sector are necessarily related to improving/modifying the FCBTK already operational. However, few scientific studies have been conducted on FCBTK probably due to the basic level of technological development. Such studies are however important to systematically and methodologically assess the challenges and solutions in FCBTK. In this study we develop a thermo-energetic model to evaluate the importance of the parameters pertained to FCBTK construction and operation. The prospective of this study is to build an initial thermo-energetic framework that will serve as a basis to investigate possible energetic improvements.

1 Introduction

Brick manufacturing in South and South-East Asia is a core economic sector which experienced a rapid expansion in recent years due to the rapid demographic growth and urbanization [1]. Notwithstanding the presence of countries with a developed brick manufacturing industry (e.g. Vietnam), most of the brick production relies on obsolete and inefficient methods which greatly contribute to the countries carbon footprint. In particular, fixed chimney bull's trench kiln (FCBTK) are one of the most implemented type of brick kiln in South and South-East Asia, accounting for more than 70% of the Indian total brick production [2], and are a considerable source of greenhouse gas emission.

FCBTK generally consists of an approximately 3 m deep, 8 m width toroidal trench connected with regular apertures to a fixed structure equipped with a fixed chimney. A sketch of a FCBTK is depicted in Fig. 1. The trench is partially filled with bricks and is then covered with insulation. During operation, green bricks are continuously added by batch in chambers from one end while fired bricks are removed from the other end. Coal (black arrows in Fig. 1) is provided from apertures at the top of the green bricks cluster in a so-called "firing zone", which is normally few meters long. The firing zone is continuously moved so that new green bricks are being fired. The kiln is designed so that the combustion air travels through the fired bricks cooling them; in this way the air is also pre-heated. In addition, the flue gases from the combustion zone pre-heat and dry the green bricks before being conveyed to the chimney. The air/flue gas path is achieved via closing/opening apertures in the fixed kiln structures.

Figure 1. Sketch of a fixed chimney Bull's trench kiln (FCBTK).

Even if FCBTK represents a major step forward compared to the previously adopted, movable chimney bull's trench kiln, they are still far from a satisfactory energy economy. Considering the size of the brick manufacturing industry and the corresponding fossil fuel consumption (second industrial sector for coal consumption in India [3]), new technological developments and best practices will be able to achieve a significant reduction in greenhouse gas emission along with substantially monetary savings [4]. The conversion of FCBTK to zig-zag bull's trench kiln (where the air is conveyed in a zig-zag pattern) is a good example of a

[a] Corresponding author: niccolo.le-brun11@imperial.ac.uk

relatively simple yet important innovation being able to increase both fuel economy and productivity. In order to further improve the efficiency of the FCBTK kiln and reach optimized energetic solutions, it is important to conduct systematic thermo-energetic studies. Apart from the works of Sameer and co-workers [2-3,5], such studies have been neglected in the past years due to the relatively undeveloped nature of FCBTK, which was probably regarded as less attractive than more advanced technologies. As part of this effort, this study proposes to develop a thermo-energetic framework over which the different challenges and the new solutions can be systematically studied.

2 Model

The thermal model of FCBTK was developed using a one-dimensional, steady-state, lumped approach where the brick kiln is considered as a continuous process in which the bricks and the surrounding material are progressively being heated by a fictitious stationary firing section, whereas in reality the fire is moved across the stationary bricks. To develop the model, the kiln was assumed as composed of three sections: a pre-heating, a firing and a cooling section. The equations for each section were derived by conducting a mass and energy balance in each section and using simplifying assumptions. The main equations and respective assumptions comprising in the model are listed and discussed in the following lines.

The pre-heating section was modelled as a counter-current heat exchanger where heat is transferred from the exhaust gas to the green bricks. The brick and gas temperatures were assumed to be uniform over the trench cross-sectional area. Because of the high water content in the green bricks, a constant mass flow-rate of bricks and flue gas could not be assumed during the heating and drying process of the bricks. Instead a spatially dependent mass flow-rate was adopted, as given by the following equation:

$$-\frac{d\dot{m}_g}{dx} = \frac{d\dot{m}_b}{dx} = S_{bg}k_{bg}\left(x_{sat}(T_b) - x_g\right), \quad (1)$$

where $\dot{m}_b$ is the mass of brick, S_{bg} is the available surface per unit of length between the bricks and the flue gas, k_{bg} is the mass heat transfer coefficient, x_{sat} and x_g are respectively the humidity ratio at the brick surface and in the flue gas. The temperature profile of the bricks and the flue gas in the heating section is found by conducting an energy balance on each stream:

$$(\dot{m}_b+\dot{m}_{sur})c_{p,b}\frac{dT_b}{dx} = S_{bg}h_{bg}\left(T_g - T_b\right) + S_{b0}h_{b0}(T_0 - T_b) - \frac{d\dot{m}_b}{dx}\Delta h_{vap} \quad (2)$$

$$\dot{m}_g c_{p,g}\frac{dT_g}{dx} = -S_b h_{bg}\left(T_g - T_b\right). \quad (3)$$

Here $\dot{m}_{sur}$ is the mass of the surrounding kiln bricks and soil. Neglecting $\dot{m}_{sur}$ in the model would add a major source of error; indeed as the fire is moved during the kiln operation, new sections of the kiln need to be heated up to temperature, therefore consuming additional energy. The other terms in Eqs. 2-3, namely $c_{p,g}$, $c_{p,g}$, T_b, T_g represent the heat capacity and the temperature of respectively bricks and air; h_{bg} and h_{b0} represents the brick/flue gas heat transfer coefficient and the brick/environment heat transfer coefficient; S_{b0} is the brick/environment surface area per unit of length.

The firing section was modelled by considering a uniform firing temperature T_f. The energy balance for the solid and gaseous phases can be written as

$$(\dot{m}_b+\dot{m}_{sur})c_{p,b}\left(T_{b,in} - T_f\right) - \dot{m}_b\Delta h_r + \dot{m}_a c_{p,a}\left(T_{a,in} - T_0\right) - \dot{m}_g c_{p,g}(T_f - T_0) - S_{b0}h_{b0}(T_0 - T_f) + \dot{Q} = 0 \quad (4)$$

where $\dot{Q}$ is the heat input rate, related to the amount (rate) of coal consumed. The term Δh_r is the enthalpy of reaction and accounts for the energy consumed by the chemical/physical processes during brick firing. The mass of the flue gases is found from the mass of air, $\dot{m}_a$, and the amount of fuel burnt, calculated from the knowledge of the heat input rate $\dot{Q}$.

The cooling zone is modelled similarly to the heating section, with the mass flow-rate of brick and the mass flow-rate of inlet air assumed constant. The energy balances for this section are:

$$(\dot{m}_b+\dot{m}_{sur})c_{p,b}\frac{dT_b}{dx} = S_{ba}h_{ba}\left(T_a - T_b\right) + S_{b0}h_{b0}(T_0 - T_b) \quad (5)$$

$$\dot{m}_a c_{p,a}\frac{dT_a}{dx} = -S_{ba}h_{ba}\left(T_a - T_b\right). \quad (6)$$

The model is completed by giving the expression for the heat transfer coefficients and the mass transfer coefficient, along with the equation governing the combustion process. The complete set of equations was solved in MATLAB using an in-house made algorithm. The algorithm iteratively calculates the heat input rate $\dot{Q}$ based on the temperature profiles, then finds the mass flow-rate of each stream and finally the temperature profiles. These steps are repeated until convergence. The model was checked for mesh independency and it was found that a 1,000 point spatial discretization gave acceptable results.

3 Results

The model was applied to a typical design of a FCBTK. The values of some parameters (such as the mass of surrounding bricks/soil material) were inferred from the study of Sameer [4] which reported a detailed analysis of a Bull's trench kiln. The main values of the model

DOI: 10.1051/matecconf/20166811001

parameters are hereby described. The kiln trench was assumed to have a height of 3 m and a width of 8 m, with an overall length of 130 m. The pre-heating, firing and cooling sections were assumed to have a length of respectively 40, 10, 80 m. The daily dry brick production was assumed to be 80,000 kg approximately corresponding to a production of 30,000 bricks per day. The baseline air flow-rate was assumed to be 80,000 kg/d of dry air. The amount of coal used varied according to the heat losses and the air flow-rate, with an indicative value of 3,500 kg/d. The coal was assumed to have a carbon ratio of 0.72 and a lower heating value of 22 MJ/kg. The values of brick-gas/air and brick-surrounding heat transfer coefficients were indicatively between 1 and 10 W/m^{2}K depending on the different conditions. The mass heat transfer coefficient was assumed to be 0.01 kg/m^{2}s while the geometric parameters, such as S_{ba}, were calculated based on the standard brick disposition in Bull's trench kilns.

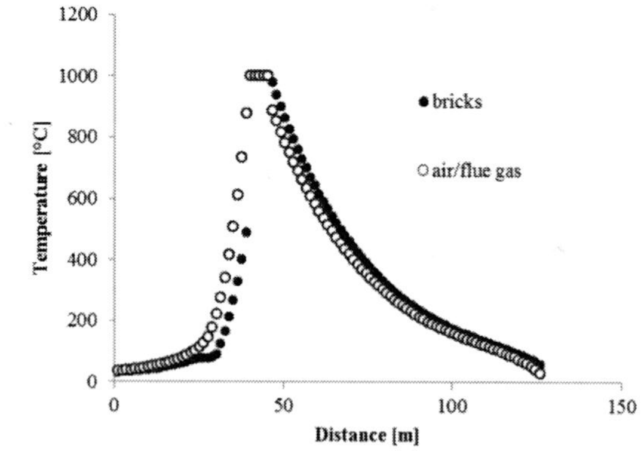

Figure 2. Temperature profiles across the kiln length. The bricks move from left to right in the figure whereas the air/flue gas from right to left. The kiln is divided in a pre-heating section (0 – 40 m), a firing section (40 – 46 m), and a cooling section (46 – 126 m).

The results regarding typical temperature profiles in the pre-heating, firing and cooling regions of the kiln are reported in Fig. 2. It is evident how the temperature of the cooling section decreases more slowly compared to the pre-heating section. The main reason for this behaviour is the lower heat capacity of the air flow compared to the mass of bricks and surrounding material. The mass of surrounding material shifts the heat capacity ratio of the two streams in counter-current flow (hot solid and cold air) from around 1 to more than 1 (around 1.7 in the present case). The lower the heat capacity ratio, the shorter is the distance over which the heat recovery from the hot solid to the cold air takes place. With a higher heat capacity ratio, due to the longer distance exhibiting higher temperatures, the heat losses to the environment from the cooling section are substantial, comprising the largest part of the energy loss. To further understand the effect of air flow-rate, Fig. 3 reports the specific energy consumption (SEC) for different values of excess air and brick-air normalized heat flux, equal to $S_{ba}h_{ba}$. As showed, increasing the excess air diminishes the fuel consumption, thus increasing the efficiency. Large amount of excess air (200 – 250%, corresponding to an equivalent fuel/oxidant ratio of 0.33 – 0.28) can achieve a 5 – 10% reduction in energy consumption. From the present discussion it follows that the use of a low equivalent ratio (in the range 0.5 – 0.25) is an intrinsic characteristic of Bull's trench kilns which allows for better heat recovery. Of course, too large amount would results in larger heat losses in the heating section and as sensible heat in the flue gas stream. In Fig. 3, there is an evident increase in SEC for too large amounts of excess air used. This situation was reported in the study of Majumdar *et al.* [6], in which the major source of heat loss was identified as the sensible heat in the flue gases (60%). An accurate tuning of the air flow-rate in the kiln is therefore considered essential to improve the performance of Bull's trench kilns.

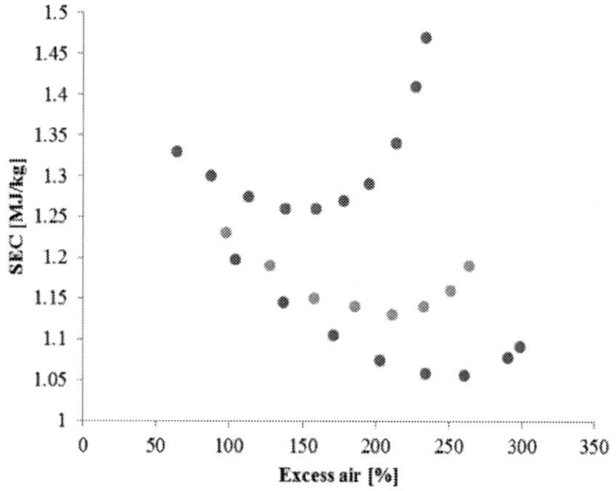

Figure 3. Specific energy consumption (SEC) as a function of the excess air percentage. The different colours refer to different values of $S_{ba}h_{ba}$: 150 W/K (blue), 225 W/K (green), and 300 W/K (red).

Another possibility to diminish the heat losses to the environment is to increase the brick-air surface and heat transfer coefficient. As showed in Fig. 3, the SEC reduction by increasing the value of $S_{ba}h_{ba}$ is substantial. This is the one of the main reasons for the increase in efficiency caused by the conversion from a standard brick arrangement to a zig-zag pattern. The zig-zag flow pattern of the air stream increases both the air-brick surface and the heat transfer coefficient. A higher value of $S_{ba}h_{ba}$ leads to a more efficient heat recovery, which in turn decreases the distance over which the temperature is maintained high, decreasing the heat losses to the environment. A zig-zag brick pattern however also increases the pressure losses in the kiln reducing the flow of air. A reduced flow of air could hinder the beneficial effect of the zig-zag pattern as mentioned in the previous discussion. Therefore, in order to fully exploiting the benefits of zig-zag kilns, it seems necessary to increase

DOI: 10.1051/matecconf/20166811001

the draft in order to compensate for the larger pressure drops. This can be done by adopting a fan or increasing the height of the chimney. Alternatively, the bricks could be placed in a zig-zag pattern with a less compact arrangement; in this case the pressure drop would be reduced at the expense of productivity.

References

1. Status and development issues of the brick industry in Asia, Research report, Food and agriculture organization of the united nations, Bangkok, (1993)
2. M. Sameer, D. Lalchandani, G. Malhotra, P. Bhanware, R. Uma, *et al.*, Brick Kilns Performance Assessment, Monitoring of brick kilns & strategies for cleaner brick production in India, Research Report, (2012)
3. M. Sameer, Evaluating energy conservation potential of brick production in India, Greentech Knowledge Solutions Pvt Ltd, Research report, (2013)
4. Introducing energy-efficient clean technologies in the brick sector of Bangladesh, Research report, Environment, climate change, and water resource unit, South Asia Region, (2011)
5. M. Sameer, Energy utilization in brick kilns, PhD dissertation, Indian Institute of Technology, Bombay, (2003)
6. Majumdar NC, Ahmed FU, Das K. Survey of thermal efficiency of Bull's trench kilns. Transaction of the Indian Ceramic Society **27**, 79–88 (1968)

A Transient Model for Simulating the Freezing Process of Molten-salt Coolants

Niccolò Le Brun and Christos N. Markides

Clea Energy Process (CEP) Laboratory, Department of Chemical Engineering, Imperial College London, South Kensington Campus, London, SW7 2AZ, United Kingdom

Abstract. Even though molten salts have many useful characteristic, especially as coolants for nuclear reactors, they are prone to freezing due to their high melting point. The solidification of the salt inside the piping system could cause structural damage and stop the flow of coolant with possible serious problems. Modelling the freezing process is therefore of primary importance for nuclear safety. In this study a quasi-steady thermo-hydraulic model has been derived and implemented to describe the transient freezing of molten salts. The partial differential equations describing the solidification/melting of the salt are solved numerically using a combination of standard explicit and implicit methods. Validation of the model is presented based on previous experimental studies for two separate cases.

1 Introduction

One of the main characteristic of a next-generation nuclear reactor is the integral safety of the design which has to withstand many different accident conditions. Molten salt reactors are currently being scrutinised as an alternative to conventional nuclear power plants because of their inherent safety [1]. In a molten salt reactor passive safety systems, which do not rely on human intervention and/or supply of power, can be used to assure the removal of decay heat under various critical scenarios. DRACS (Direct Reactor Auxiliary Cooling System) is a passive safety system currently being considered as a viable design component; it consists of a series of heat exchangers which are capable of removing decay heat by the natural circulation of different molten salts. A schematic of DRACS design outlining its main components is available in Fig. 1. The preliminary design of such a system features a two natural convection loops: a primary and secondary loop. In the primary loop the salt circulates through a pebble bed reactor and rejects heat to the secondary molten salt loop through the Direct Heat Exchanger (DHX). The secondary loop features a molten salt/air heat exchanger (NDHX in Fig. 1) through which the decay heat is ultimately dissipated to the environment [2]. Although very promising, DRACS can still be prone to failures and requires further study. A possible mode of catastrophic failure is the freezing of the salt within the NDHX, which would interrupt the flow of coolant in the secondary natural circulation loop. With the interruption of the secondary flow the heat removal capability of DRACS is compromised with possible severe consequences. It is therefore important to assess the risk of freezing in the piping system and heat exchangers of DRACS. Such an analysis can also be valuable for molten salt based, solar power plant in which freezing is also a possible dangerous occurrence during night-time [3]. The first step in such an analysis is to develop a simple yet accurate sub-model of molten salt freezing which can be incorporated in more complex models. In this study we present a 1-D transient freezing model which can reasonably capture the main features of the freezing process. The main equations describing the freezing process are explained in section 2 and are referred to a pipe section, which is the most common case encountered in real applications. The model was validated against experimental results which are reported in section 3. In section 4 the main conclusions of the study are drawn.

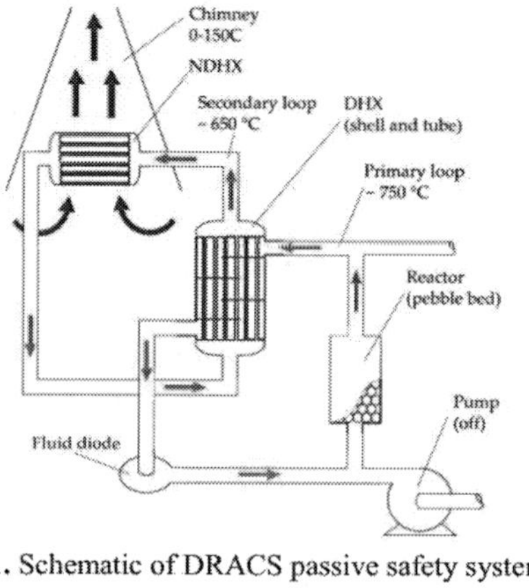

Figure 1. Schematic of DRACS passive safety system.

2 Model

The freezing process of molten salts inside a pipe is modelled by assuming that the salt solidifies homogenously along the pipe walls. In this case, a uniform layer of solidified salt is formed between the wall and the molten salt. This assumption could not hold in cases of weak adhesion forces between the frozen salt and the pipe wall. In this latter case the salt would solidified through the formation of slurry with a uniform temperature distribution. However the formation of a uniform solidified layer of salt is a more dangerous scenario and thus the mentioned assumption can be regarded as conservative from a safety point of view. The layer is assumed to form when the temperature of the wall drops below the melting temperature of the salt T_{melt}.

The main equation describing the freezing process where derived by considering a pipe section of length ds. The energy balance in the liquid phase in contact with a frozen layer of salt is:

$$-\frac{\partial T}{\partial t}+\frac{W}{A\rho}\frac{\partial T}{\partial s}=-\frac{h(T_{melt}-T)P}{A\rho c_p}. \quad (1)$$

Here P and A are respectively the perimeter and the cross sectional area of the pipe wet by the liquid phase (without considering the layer of solidified salt); h is the heat transfer coefficient; ρ and c_p are the density and specific heat capacity of the molten salt; T is the temperature of the salt and W the mass flow-rate. A similar energy balance is conducted on a section of the pipe wall to find the wall temperature. The equation is derived by assuming the quasi-steady-state condition for the temperature profile inside the solidified salt:

$$A_w\rho_w c_{p,w}\frac{\partial T_w}{\partial t}+\xi(m_s)\frac{\partial T_w}{\partial s}=-h_{ext}(T_w-T_{ext})P_{ext}-q_{sw}P_w. \quad (2)$$

Here the subscript "w" and "ext" refers to the wall and the external side of the pipe; q_{sw} is the heat flux to the wall from the layer of solidified salt and $\xi(m_s)$ is a function of the solidified mass of salt. The function $\xi(m_s)$ accounts for the energy lost or gained in the solidified salt during the transition between different quasi-steady-states caused by a change in T_w; its value is calculated from the steady state temperature profiles in the salt:

$$\xi(m_s)=\rho c_p\pi\left(\frac{R(m_s)^2-R_0^2}{2}\ln\frac{R(m_s)}{R_0}+R_0^2\right), \quad (3)$$

where R_0 is the pipe radius and $R(m_s)$ is the radius of the liquid cross section in a pipe in which a layer of solidified salt is present with a mass per unit of pipe length m_s. The interface between the solid and the fluid is assumed to be at the melting point of the material. The mass of solidified salt is found by conducting a mass balance over a pipe section through the following equation:

$$\left(\Delta H+(T-T_{melt})c_p+\psi(m_s,T_w)\right)\frac{\partial m_s}{\partial t}=h(T-T_{melt})P-q_{sf}P_w, \quad (4)$$

where q_{sf} is the solid-side heat flux at the interface between the fluid and the solidified mass of salt, ΔH is the enthalpy of fusion and $\psi(m_s,T_w)$ is a function accounting for the energy loss/gained during transitions from different quasi-steady-states caused by a change in the mass of solidified salt per unit length m_s. The expression for $\psi(m_s,T_w)$ was calculated as the first order term of the Taylor series of the energy differential between two different quasi-steady-states caused by a change in mass dm_s:

$$\psi(m_s,T_w)=c_p(T_{melt}-T_w)\left(\frac{1}{2}\ln\frac{R_0}{R(m_s)}+\frac{1}{2}\ln\frac{R_0}{R(m_s)}^2+\frac{1}{2}\ln\frac{R_0}{R(m_s)}^2\frac{R_0^2}{R(m_s)^2}\right). \quad (5)$$

Eq. 1-5 where combined with standard 1-D thermo-hydraulics equations. Here we just briefly summarized the computing algorithm for a piping network of the type described in Fig. 1. The general system of equations is solved using the forward time upwind space (FTUS) first order explicit scheme [4]. First the pressure drops are calculated (which are function of the mass of salt solidified), as well as the pressure head (based on the buoyancy caused by the spatial temperature distribution); successively the flow-rate in the piping system are computed by solving the corresponding momentum equations; the heat transfer coefficients (which depend on the local velocity profiles) are then calculated and, from them, the energy equations are solved to find the temperature profiles both in the liquid and in the wall; successively the values of $\xi(m_s)$ and $\psi(m_s,T_w)$ are calculated based on the mass of solidified salt from the previous step and the temperature of the wall (Eq. 3 and 5); finally the mass of salt solidified is calculated through Eq. 4.

3 Validation

The current approach to model molten salt freezing was validated against two experimental studies. The first case regards the work of McDonald and co-workers [5]. The authors monitored the freezing time of a pipe section full of water exposed to both natural and forced convection. The outside air temperature was varied between -6 and -26 °C. A thermocouple was placed at the centreline of the pipe to record the temperature of the water. The authors measured the time required for complete freezing by monitoring when the temperature of the water fell below the freezing point, indicating that the water was completely frozen and that ice cooling was taking place.

The experiment of McDonald *et al.* [5] was simulated by considering a pipe full of quiescent water at approximately 21 °C (the stated initial temperature in the experiments). The authors reported the values of the heat transfer coefficient on the outside of the pipe for every case, which were used as a boundary condition in the simulation. The liquid-side heat transfer coefficient was calculated by assuming Nu = 3.66, which is the limiting solution for stagnant fluid and a constant pipe wall temperature. Fig. 2 shows the experimental results along the prediction of the simulations. The error bars in the figures indicate the range of results obtained by changing the thermo-physical properties of water, air and ice over a reasonable range (±15 °C). A good agreement exists between the simulations and the experimental results, even though for relatively high outside air temperatures the experimental results are slightly overestimated. Considering the simplicity of the current approach and the possible experimental errors, the results are deemed satisfactory.

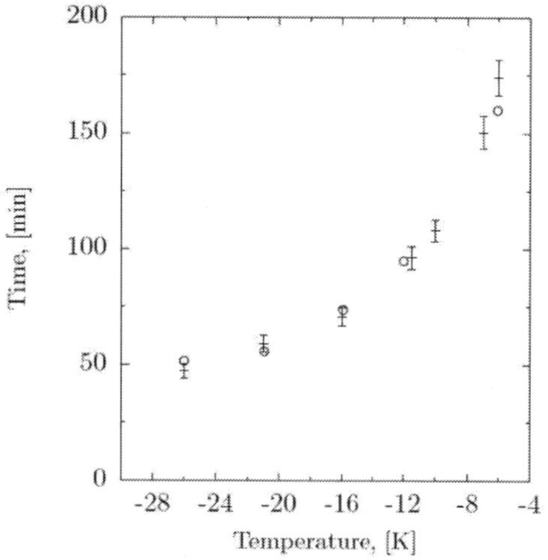

Figure 2. Comparison between the total freezing time measured by McDonald *et al.* [5] (circled dots) and the simulated time (bars) for a pipe full of water subjected to a flow of air.

The second case used to validate the model is the experimental study of Cheung and Baker regarding the transient water freezing in pipe flow [6]. A picture of the experimental setup, taken from the mentioned study, is shown in Fig. 3 and depicts a coiled tube fully immersed in liquid nitrogen. During an experiment, water is injected in the initially dry tube through the fast opening of a valve. The experiment aims at measuring the penetration length reached by the liquid before complete freezing occurs. The authors measured such penetration length for different initial temperatures of the water (ranging from 0 °C to 90 °C) and different average flow velocities inside the tube (ranging from 60 cm/s to 140 cm/s). To simulate the experiment of Cheung and Baker we considered a pipe partially immersed in liquid nitrogen, with a fixed wall temperature of -197 °C. This assumption can be justified by the very high heat transfer coefficient from the pipe wall to the boiling liquid nitrogen. At t = 0 the tube is considered partially full of water with an initial velocity V_0 and the rest of the tube is considered empty. A constant pressure head is imposed on the fluid. The average velocity of the water during an experiment is therefore a function of both the initial fluid velocity and the imposed pressure head. The simulation ends when a section of the cryogenic pipe becomes clogged. The penetration length is computed by calculating the amount of water which flowed in the cryogenic pipe over the total simulation time. As showed by Fig. 4, the difference between the experimental results and the simulations are within a 30% error which is more than satisfactory in the light of the fast dynamic of the processes involved. Some of the simulations, especially for higher initial temperature of the fluid, overestimate the experimental results. A possible explanation for such a pattern is the over-prediction of the heat transfer coefficient caused by adopting the Sieder-Tate equation when all the liquid is almost frozen.

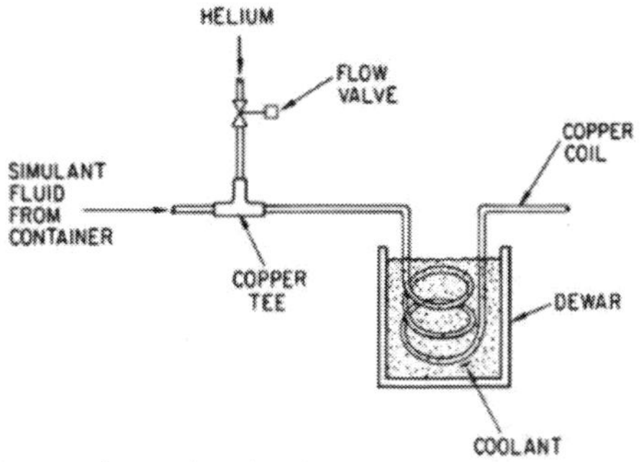

Figure 3. Experimental study of Cheung and Baker [4] used to validate the transient freezing model.

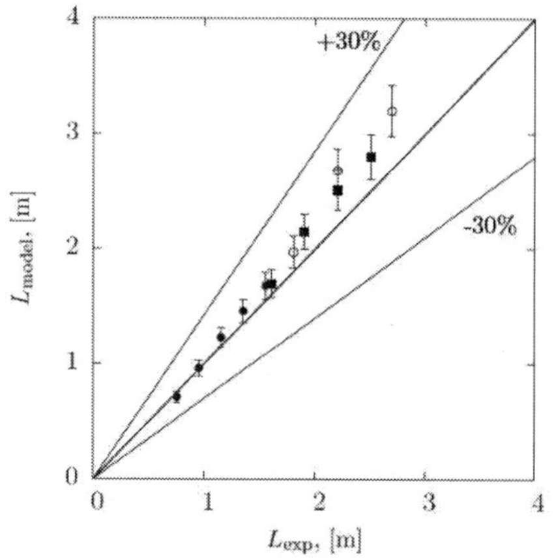

Figure 4. Comparison between the liquid penetration length measured by Cheung and Baker [6] (Lexp) and the penetration length simulated by the model (Lmodel) for different flow conditions.

Different symbols refer to different flow conditions obtained by varying the fluid initial temperature and velocity.

4 Conclusions

A thermo-hydraulic model was developed to simulate the freezing process of molten salts in piping systems. The model was developed by using a 1-D approach under the quasi-steady-state assumption. The model was successfully validated against two experimental studies. The approached mentioned in the present study can be used to assess the susceptibility of molten salt nuclear reactor and molten salt-based solar power plants to the freezing of the coolant, which is a major drawback hindering the development of these technologies.

References

1. J. Serp, M. Allibert, O. Benes, S. Delpech, *et al.*, Progress in Nuclear Energy **77**, 308-319 (2014)
2. Q. Lv, H. Lin, I. Kim, X. Sun, *et al.*, Annals of Nuclear Energy **77**, 115-128 (2015)
3. D. Kearney, U. Herrmann, P. Nava, B. Kelly B, *et al.*, J. Sol. Energy Eng **125**, 170-176 (2003)
4. R. Courant, E. Isaacson, M. Rees M, Communications on Pure Applied Mathematics **5**, 243-255 (1952)
5. A. McDonald, B. Bschade, E. Sullivan, R. Marsden R, *et al.*, Applied Thermal Engineering **73**, 142-153 (2014)
6. F. B. Cheung, L. Baker, Nuclear Science and Engineering **60**, 1-9 (1976)

MATEC Web of Conferences **68**, 12001 (2016) DOI: 10.1051/matecconf/20166812001
ICIEA 2016

Simulation and Optimization of Contactless Power Transfer System for Rotary Ultrasonic Machining

Xinwei Wang, Aimin Wang and Xiaolong Wang

Beijing Institute of Technology, School of Mechanical Engineering, 5 South Zhongguancun Street, Haidian District, Beijing, China

Abstract. In today's rotary ultrasonic machining (RUM), the power transfer system is based on a contactless power system (rotary transformer) rather than the slip ring that cannot cope with high-speed rotary of the tool. The efficiency of the rotary transformer is vital to the whole rotary ultrasonic machine. This paper focused on simulation of the rotary transformer and enhancing the efficiency of the rotary transformer by optimizing three main factors that influence its efficiency, including the gap between the two ferrite cores, the ratio of length and width of the ferrite core and the thickness of ferrite. The finite element model of rotary transformer was built on Maxwell platform. Simulation and optimization work was based on the finite element model. The optimization results compared with the initial simulation result showed an approximate 18% enhancement in terms of efficiency, from 77.69% to 95.2%.

1 Introduction

Rotary ultrasonic machining (RUM) is a non-traditional machining process that combines the material removal mechanisms of diamond grinding and ultrasonic machining. It comprises of a tool mounted on a rotary spindle that is attached to a piezoelectric transducer to produce the rotary and ultrasonic motion [1]. In RUM machines, the ultrasonic vibration power provided by the ultrasonic generator need to be transferred to the machine tool that rotating in a high speed. The conventional energy transfer method for RUM is via slip ring, which exists a lot of drawbacks and cannot be applied to a spindle rotating in high speed, because it would affect the machining accuracy and working efficiency of RUM [2].

Loosely coupled inductively power transfer (LCIPT) is a novel energy transfer technology that could achieve power transfer without physical contact. LCIPT owns many merits, such as reliability and minimum maintenance requirement. It is unaffected by dirt, dust, water or other chemicals and capable of working in high-speed rotation condition. LCIPT is widely used in areas, such as aerospace [3], medical electronics [4] and electric vehicle [5]. These all proved that LCIPT technology is perfect for rotary transformer. Therefore a contactless power transfer system, also named rotary transformer, using LCIPT technology was designed [6].

Though rotary transformer have been designed for years and used in many areas, optimization research and articles related to rotary transformer for URM are relatively scarce. This paper focused on optimization of the rotary transformer, by optimizing the gap between the two ferrite cores, the ratio of length and width of the ferrite core and the thickness of ferrite, to enhance its efficiency. The optimization work of rotary transformer was performed and validated through Maxwell platform based on a finite element model of rotary transformer of URM that had already been built in former work [7].

2 Principle of rotary transformer

Based on the principle of electromagnetic mutual inductance theory, the rotary transformer contains a set of pot ferrite cores with windings fitted into both of them (Fig. 1). The primary pot core is static, and the winding in it is connected to the ultrasonic generator, which produces electrical power of ultrasonic frequency. The secondary pot core is fixed to the rotary spindle (spinning when working) and connected to the ultrasonic transfer, and winding in it is induced to generate an alternating current of the same frequency as the primary winding to achieve the power transfer.

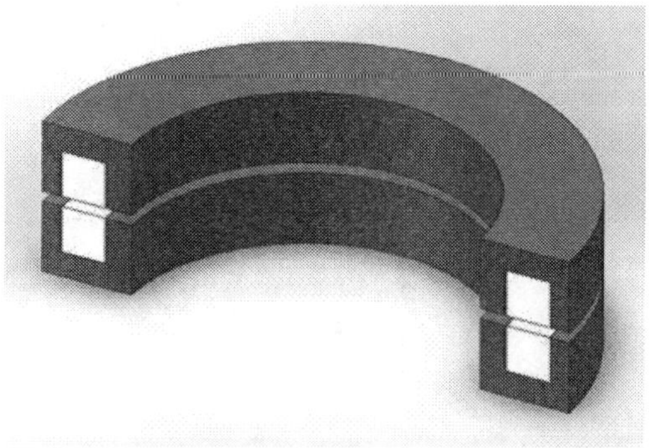

Figure 1. Contactless power transfer system for RUM

3 Simulation of rotary transformer

The simulation of the rotary transformer was carried out on the Maxwell platform. The transient magnetic analytical module of Maxwell was adopted to simulate the mutual inductance coupling model of rotary transformer. Firstly, the mechanical model of the RUM rotary transformer has been built with dimensions according to the (Fig. 2) and Table 1, and the core material was set to ferrite with a relative magnetic permeability of 2300. Then the mesh operations, primary and secondary windings, the boundary conditions and the later processing of the contours of the electric and magnetic field have been set. The ultrasonic-frequency (20 KHz) AC signal is assigned to the primary winding. It has been proved the spinning movement of the secondary ferrite core has little influence of the efficiency of the rotary transformer [6], so it is neglected in the simulation work.

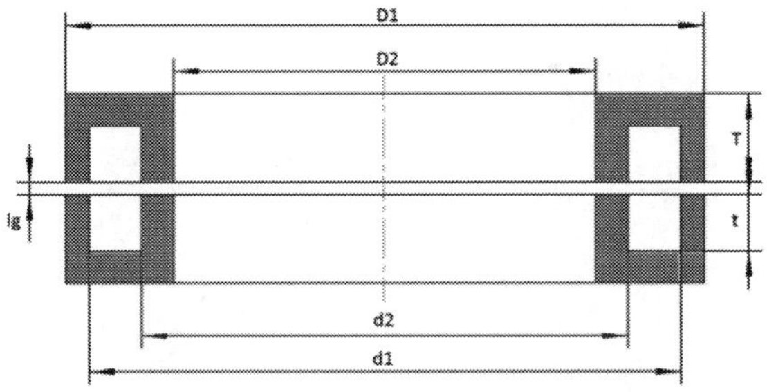

Figure 2. The dimensions of rotary transformer

Table 1. The initial dimensions of rotary transformer before optimization

D_1/mm	100
D_2/mm	66
d_1/mm	92.6
d_2/mm	76
T/mm	14
t/mm	9
lg/mm	1

The simulation results were showed in (Fig. 3) showed the relationship between the secondary winding (winding2) and primary winding (winding1). The simulation results of the rotary transformer were same with the theory analysis. The frequency of the secondary winding is the same as the primary winding, but the phase of the secondary winding is half cycle behind the primary winding, due to the principle of electromagnetic mutual inductance. The amplitude inducted current (secondary winding) I_2 is smaller than the input current (primary winding) I_1 because of the energy loss during the transfer process.

For a better understanding of the simulation results of the rotary transformer, the current density distribution of rotary transformer at 15 μs was showed in (Fig. 4). The result is corresponding with (Fig. 3) that the current density of the primary winding is higher than that of the secondary winding.

The magnetic field B of rotary transformer at 25 μs was showed in (Fig. 5). We can see that the distribution of magnetic fields B is relatively equal in the ferrite core at about 300 mT, which is the perfect working area for ferrite.

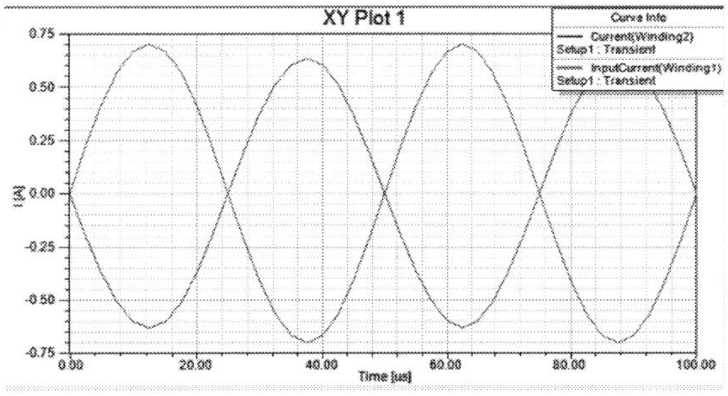

Figure 3. Simulation results for Winding2 under the ImputCurrent Winding1

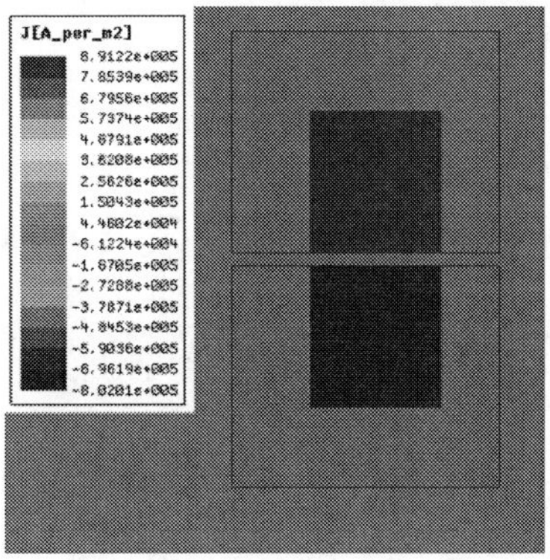

Figure 4. Current density distribution of rotary transformer at 15 μs

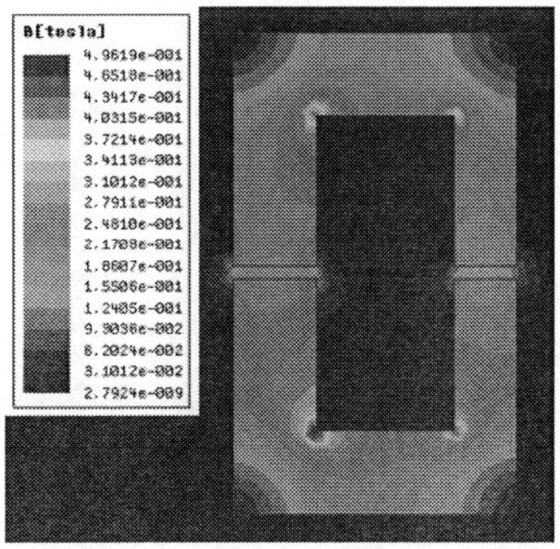

Figure 5. Magnetic field B of rotary transformer at 25μs

4 Optimization of rotary transformer

Based on the simulation results, the optimization work begins. The main parameters of the rotary transformer, including D_1, D_2, d_1, d_2, T, t, lg were all considered in the optimization work. The optimization work mainly considered three factors. First the gap between the two ferrite cores, then the ratio of length and width of the ferrite core and finally the thickness of ferrite. The initial dimensions of the rotary transformer are showed in (Fig. 2) and Table. 1. The transfer efficiency is defined as *E*,

$$E = \frac{I_2^2 Z_2}{I_1^2 Z_1} \quad (1)$$

where Z_1 and Z_2 is the impedance of the primary ferrite core and secondary ferrite core, and since they are identical $Z_1 = Z_2$.

4.1 Gap between two ferrite cores

The change of the gap between two ferrite cores *lg* from 0.5mm to 3mm has been showed in (Fig. 6). And the relationship between *lg* and transfer efficiency *E* is showed in (Fig. 7). With all the other parameters unchanged, but the air gap between the two ferrite cores lg changing from 0.5mm to 3mm, the efficiency of the rotary transformer reduced from approximately 86% to 55%. From this point of view, the larger the air gap, the lower the transfer efficiency.

This could be explained by the bigger the air gap the more amount of energy will loss to the air, so the less the transfer efficiency. However, due to the physical and machining restrictions, the air gap couldn't be too small. Therefore 0.8 mm was chose for the air gap of the rotary transformer, and the latter optimization work of the ratio of the length and width of the rotary transformer and the thickness of ferrite were based on this air gap.

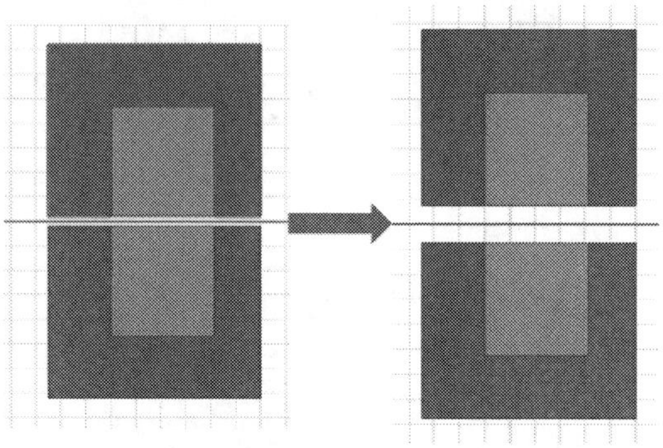

Figure 6. Optimization of the gap between two ferrite cores

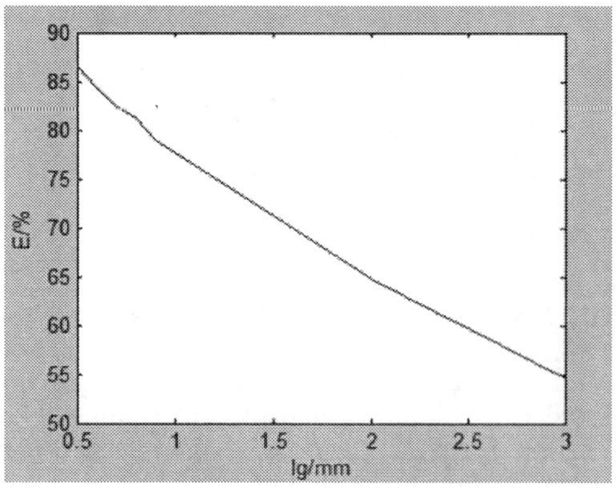

Figure 7. The relationship between lg and *E*

4.2 Ratio of length and width.

Define the ratio of the length and width as *R*,

$$R = \frac{D_1 - D_2}{2T} \quad (2)$$

Changing *R* from 1.2 to 2.8 by changing *T* (Fig. 2) of the rotary transformer while the length of the ferrite core D1, D2, d1, d2 (Fig. 2) was kept unchanged, like showed in (Fig. 8), the rotary transformer become more short.

The relationship between *R* and efficiency *E* was showed in (Fig. 9). We can see that with the increasing of *R* from 1.2 to 2.8, the efficiency *E* was also increasing, though with a decreasing acceleration, from 81% to 95%, respectively.

Due to this fact, it is better to design the rotary transformer in large length and small width, as long as there is enough space for the winding, and the rotary transformer could be suitable to fit in the spindle of the ultrasonic machining tool.

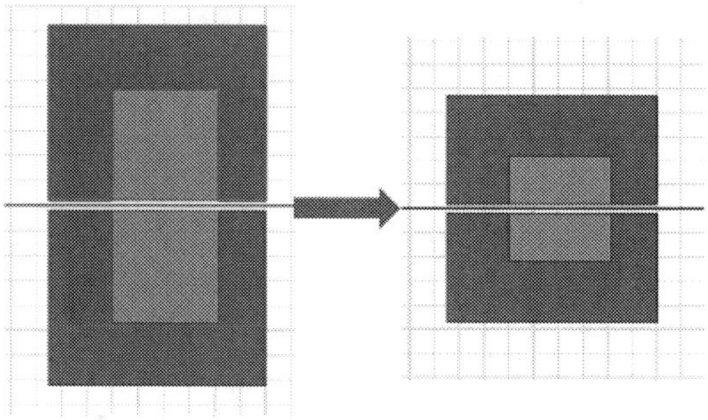

Figure 8. Optimization of ratio of length and width

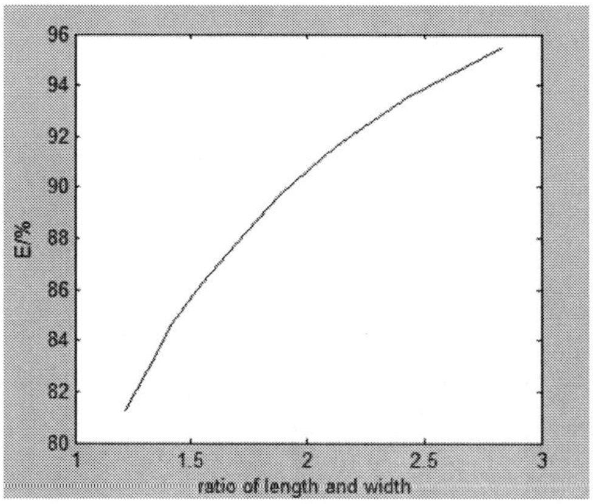

Figure 9. The relationship between ratio of length and width and *E*

4.3 Thickness of the ferrite cores

Researches have also showed that the thickness of the ferrite cores could help enhance the efficiency of the rotary transformer. The change of the thickness of the rotary transformer of from 5 mm to 12 mm was showed in (Fig. 10). The relationship between the thickness of the

DOI: 10.1051/matecconf/20166812001

ferrite core and the transfer efficiency E is showed in (Fig. 11).

With the thickness of ferrite increasing, the efficiency of the rotary transformer also increased, through with a decreasing acceleration. It could be explained by the theory that the thicker the ferrite is, the better it is at accumulating the magnetic power, which means less energy loss.

However, the ferrite of the rotary cannot be too thick for many reasons. First, the magnetic field B distribution is related to the thickness of the ferrite. The thicker the ferrite the smaller the magnetic field B, and there is a perfect working area for magnetic field B which is 300 mT [8]. Secondly, the size of the ferrite core should be suitable for it to assembling to the spindle of the ultrasonic machine, so the ferrite core cannot be too big. Finally, the ferrite cannot be fully used if it is too thick, so for the energy efficiency, the ferrite of the rotary should not be too thick.

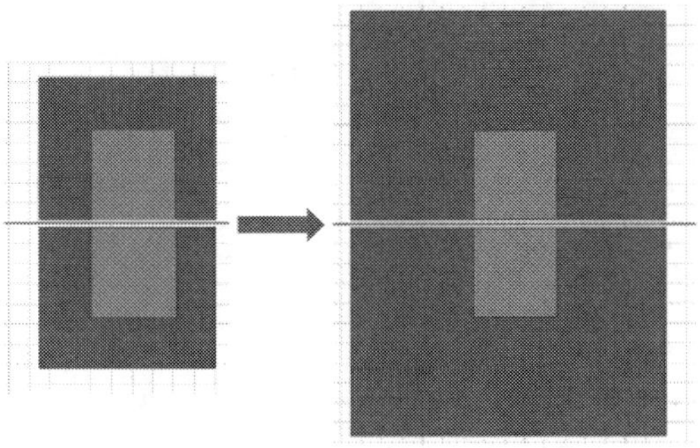

Figure 10. Optimization of the thickness of the ferrite

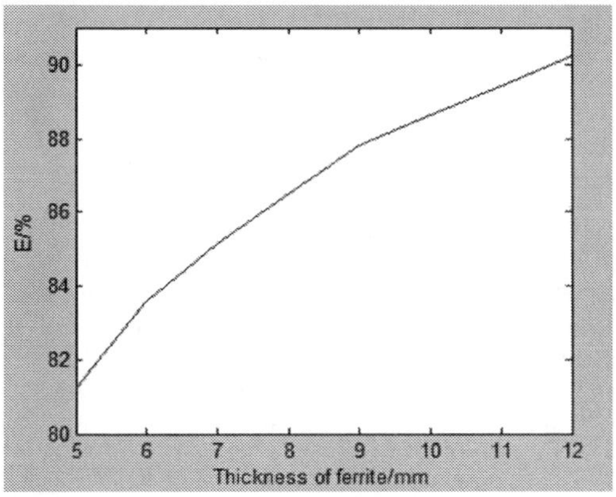

Figure 11. The relationship between thickness of ferrite and E

5 Conclusion

Based on the simulation result, we conclude the designed rotary transformer can be used to transfer the electromagnetic energy from the stationary primary windings to the rotating secondary windings, so the ultrasonic spindle can obtain stable energy in a high rotating state.

In the optimization part, three main factors that affect the transfer efficiency of the rotary transformer, gap between the two ferrite cores, the ratio of length and width of the ferrite core and the thickness of ferrite that affect the efficiency of the rotary transformer were all analysed on Maxwell platform. We concluded that the less the gap is, the bigger the ratio of length and width is, and the thicker the ferrite, the more efficient the rotary transformer becomes.

However, the size and physical conditions of rotary ultrasonic machine should also be taken into account such that the rotary transformer could be assembled onto the rotary ultrasonic machine. It means that the gap between the two ferrite cores is impossible to be zero (greatest efficiency), the ratio of length and width is impossible to be infinity and that the ferrite could not be too thick in the dimension control of rotary transformer.

Based on all these factors, the optimal parameters of the rotary transformer can be obtained. Suitable gap between the two ferrite cores, the ratio of length and width of the ferrite core and the thickness of ferrite could be chosen. The rotary transformer can greatly improve the transfer efficiency and work in the best state to meet the maximum power transfer principle, reducing the unnecessary electrical and magnetic energy loss, with power transfer efficiency up to 95%. The simulation results showed an excellent result for the rotary transformer optimization, though experiments have not been conducted yet, the experimental efficiency of rotary transformer is expected to be up to 90%.

References

1. N. Churi, Rotary ultrasonic machining of hard-to-machine materials, B. S., University of Mumbai, India, (2002).
2. L. P. Liu, Study on the key technologies of rotary ultrasonic machining based on accessing of the machine tool, Tianjin Univerosity, (2011).
3. K. D. Papastergiou and D. E. Macpherson, An airborne radar power supply with contactless transfer of energy-part I: Rotating transformer, IEEE Trans. Ind. Electron, 5-45: 2874-2884, (2007).
4. P. T. Theilmann, Wireless power transfer for scaled electronic biomedical implants, University of California, San Diego, (2012).
5. C. P. Choudhury, Modeling and analysis of an inductive power transfer system, Queen's University, December, (1989).
6. Z. L. Long, X. W. Lin, W. Yuan and J. G. Zhang, Modeling on a non-contact power transmission system in ultrasonic machining, Proceeding of the IEEE International Conference on Robotics and Biomimetics, ROBIO, Shenzhen, China, December (2013).
7. X. W. Wang, A. M. Wang, X. L. Wang, Modeling analysis of contactless power transfer system for rotary ultrasonic machining, submitted to Applied mechanics and materials, (2015).
8. H. Hao, FEM Analysis of Contactless Power Transmission System for Ultrasonic Assisted Machining, Dalian University of Technology, (2013).

The Influences of Fast Switching Series Compensation for Distance Protection

Qin Lijun, Guo Jing

North China Electric Power University, bei nong lu 2 hao, China

Abstract. This paper describes the technical principles of series compensation, and introduces the series compensation device structure, and finally got through simulation effects of rapid switching Series Compensation for distribution network distance protection. Distribution network is an important network connections and power grid users, from the distribution of electric energy in the power system, the strength of its transmission capacity directly determines the power user power quality is good or bad. However, since the electricity distribution network of long-term failure to get the attention of practitioners, its construction and the relatively slow pace of reconstruction, the aging serious, increasingly unable to meet the annual growth of electricity demand.The main problem with the presence of the grid are:Distribution network construction is lagging behind, power transmission grid capacity is insufficient. Network structure is irrational, power supply radius is too long, circuitous route power line terminal voltage is too low. Part serious aging distribution equipment and lines, the line number is small, backward technology, serious power loss. Reactive power shortage, power grid is not economical and low power quality. Therefore, how to improve the status of the distribution network has become the focus of power workers

1 Value of series compensation and compensation principle

1.1 Value of series compensation

By capacitive reactance series compensation to offset the line inductance, reducing the electrical distance between the power source and the load point, can solve a series of problems exist in the distribution network, its value lies in:

1) solve the supply radius distribution network end of the line is too long due to excessive voltage offset problems and improve power quality.

2) to solve the problem of heavy loss of distribution network and improve the economics of grid operation.

3) increase the transmission capacity of the line to avoid the reconstruction of the line [1-8].

1.2 Series technical principle of compensation Action series line voltage compensation to improve:

AC line, due to the presence of line resistance and reactance, in power through the load, will produce a voltage drop, the voltage is reduced. Series capacitor compensated by the capacitive reactance to counteract the line inductance reduces the voltage drop across the load current line reactance enhance the line voltage and reduces the phase angle difference between the terminal voltage line first.

Series compensation for reducing line loss effect:

When the load current flows through the transmission line will produce power loss on line resistance proportional to the square of its size and the current value. Since the series compensation capacitor of access, so that the line voltage increases, at constant load power load current is reduced, thereby reducing the power loss generated on the line.

Series Compensation for improving the transmission capacity of the line:

Limit the power line transmission power and line reactance inversely, if we can reduce the line reactance, it is possible to improve the stability limit and transmission power lines. Series compensation is used to compensate for line reactance, thereby enhancing the transmission capacity of the line [9-12].

2 Series compensation structure introduction

2.1 General structure type Series Compensation

Conventional Series Compensation Series Compensation based on the type of protection can be divided into the main spark gap series compensation devices and metal zinc oxide voltage limiter type (referred to as MOV type)

Series Compensation [13].

2.1.1 FSC spark gap device configuration Introduction

Early FSC bypassed by a spark gap device and bypass breaker components. When a short circuit fault line, the fault current to the rapid increase in the terminal voltage of the capacitor. When the spark gap felt transient voltage exceeds the trigger value, the gap flashover bypass capacitor. Since the spark gap does not have the ability to self-extinguish the arc prepared, so that by closing the bypass switch interrupter. After troubleshooting, bypass switches gate, series capacitor running again.

In order to fill the gap as a string through the main protection voltage capacitors, specifically divided into single space, double space and SIC nonlinear resistance band gap protection mode, the typical configuration is shown in Fig. 2.1.

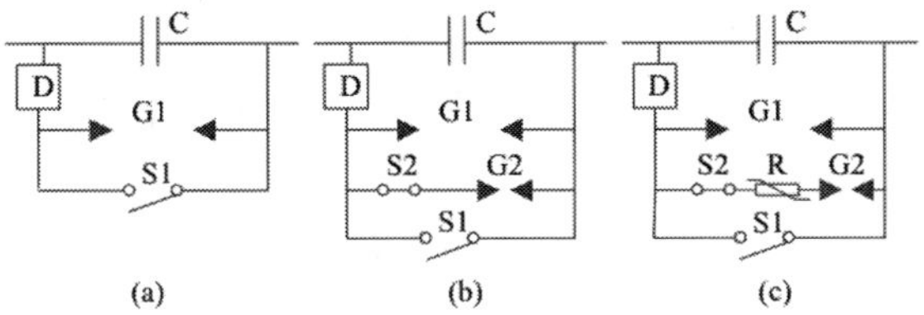

C—Capacitor banks D—Damping circuit R—SIC resistance
G1,G2—Spark gap S1,S2—Bypass Switch

Figure 2.1. Series Compensation spark gap.

For the gap compensation means in series, due to the instability of its action, after maintenance difficulties, short-circuit failure bypass and re-invest their time to reach 200 ~ 400ms, resulting in a greater impact on the protection line. And because of the complex structure of the spark gap, make huge volume series compensation means expensive.

2.1.2 Metal oxide type series compensation voltage limiter device configuration Introduction

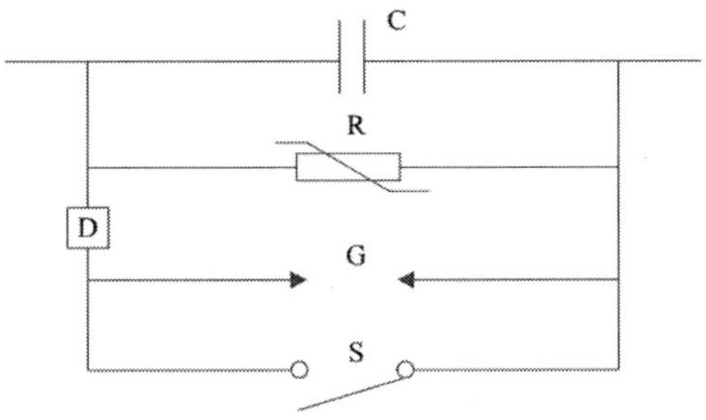

C—Capacitor banks D—Damping circuit R—ZnO resistance
G—Spark gap S—Bypass Switch

Figure 2.2. Metal oxide type voltage limiter Series Compensation.

Currently series compensation devices typically use their equipment from a metal oxide zinc bypass voltage limiter (MOV) and trigger gap composition, MOV as the main capacitor overvoltage protection, trigger gap as the backup protection MOV, which arrangement method shown in Fig. 2. 2.

MOV as the main protection series compensation capacitor bank, under steady state conditions showed a high impedance characteristics, the equivalent of an open state, then the capacitor in series with the circuit [14]. When a fault occurs after the capacitor voltage reaches its tolerance limit, MOV conduction bypass capacitor bank. After the fault is cleared, the capacitor back into operation.

2.1.3 Fast switching series compensation device configuration Introduction

Conventional series compensation device, because of the configuration of complex, bulky, expensive, input and output is relatively low, the device's own security, reliability and poor range of issues in high voltage distribution network has not been widely used.

Clearance is the application of rapid vacuum switch in the bypass device based on the fast switching series compensation device replace the spark that once the wiring diagram is shown below.

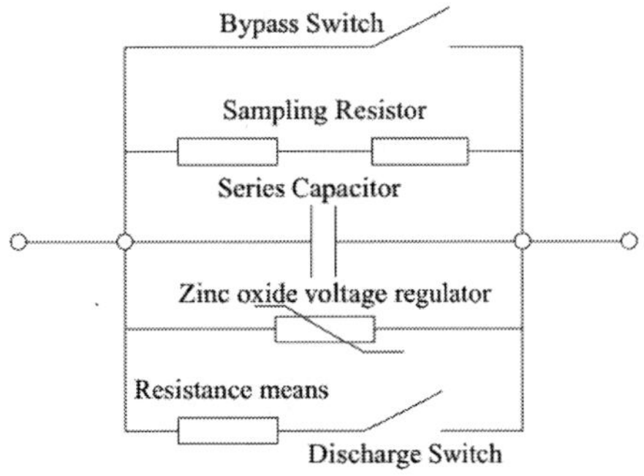

Figure 2.3. Fast switching Series Compensation Schematic.

Fast switching series compensation means each part of the main features:

1) Series capacitor banks: to compensate for line drop, improve voltage quality. According to the requirements of the rational allocation of capacitor compensation depth, the normal operation of the capacitor in series with the circuit;

2) Zinc oxide components: series compensation for limiting the voltage across the capacitor. Zinc oxide valve series and parallel composition, zinc oxide series voltage lower than the threshold voltage of the capacitor power lines during normal operation, zinc oxide does not move; when external short circuit, the voltage across the capacitor is limited to a low level, to protect the capacitors are not damaged;

3) The rapid vacuum discharge switch: for rapid release of the charge storage capacitor in series eddy current drive technology based on rapid development rapid vacuum circuit breakers. Normal operation Fast switching device is in the open state, for turning the capacitor discharge circuit when the line is short-circuited

[15]. Its closing time can be controlled at about 10ms, opening time can be controlled within 5ms. Rapid vacuum circuit breakers, combined with the short-circuited rapid identification technology to achieve a short-circuit in the wiring to series capacitors rapid short about 15ms, greatly shortening the duration of the over-voltage, so that the zinc oxide component It can greatly reduce the required capacity. Reduces capacitor operating safety protection and the zinc oxide component can capacity for research and development of series compensation device technical basis;

4) Discharge damper device: to limit the discharge current of the capacitor, the parallel discharge by the discharge resistor and an inductor formed. The discharge current is limited to within the permissible range, to prevent the occurrence of the discharge switch contact welding;

5) Sampling resistor: capacitor for detecting the operating state and provide electrical quantities to the device control unit to control fast switching and bypass switches open and close [16].

Fast switching series compensation capacitor protection device of its higher reliability, the device smaller and more affordable cost, so that the distribution network series compensation application possible.

3 Impact on the distribution network from the protection of fast switching type Series Compensation

Series capacitors added, undermines the ratio between the line impedance and measure the distance between the impedance of the mutation will take place before and after the series capacitors installed position so that protection can not correctly measure the fault distance, the impact on the distance protection [17].

Theoretical studies show that if can instantly bypass capacitor series compensated quickly, it is possible to weaken the influence of the capacitors on the line after a short circuit fault protection occurs [18]. In recent fast switching series compensation appears a good solution to the problem of bypass capacitors, the following simulation study by the fast switching type Series Compensation Capacitor to Distance Protection.

In this paper, 110kV radial distribution systems for the simulation object, failure time is set to 0.1 seconds, according to the operating characteristics of the fast start time is set after the bypass capacitor short circuit occurs 15ms. Series compensated line system wiring diagram is as follows:

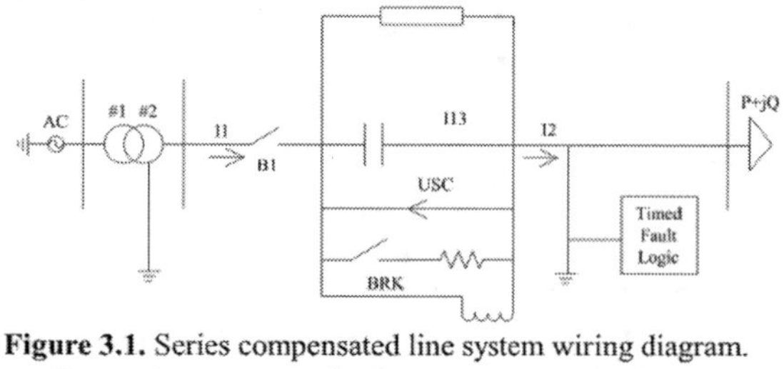

Figure 3.1. Series compensated line system wiring diagram.

Protection segment I Modeling and Simulation

Fault point is located within the protective action zone protection installation at the nearby

Series compensation device is mounted to the head-end line, series compensation capacitor circuit protective device side most likely to cause a short circuit fault repellent action, therefore, choose this as the point of failure, tested in the worst case, whether it will protect the series capacitor Impact [19]. A phase short circuit to ground fault, for example, the simulation results are as Fig. 3.2.

The simulation results show that when the series compensation device is mounted in the first end of the line, when the single-phase short-circuit ground fault circuit protection are not subject to influence fast switching series compensation deviceoccurs Series Capacitor line side, can accurately reflect the type of fault and action.Similarly available in any type of fault, fast switching series compensation did not affect the line protection [20-23].

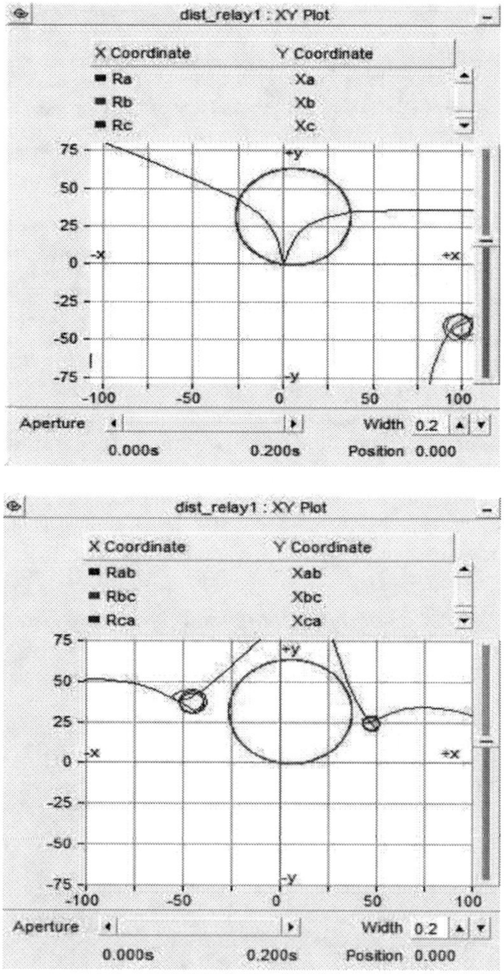

Figure 3.2. Line protection measure impedance trajectory A phase to ground fault occurs.

4 Conclusion

Due to limited space, this is no longer on the installation location and other types of capacitor protection simulation done carefully explained, only the simulation results are listed below:

Protection I segment:

1) The fault is located within the protective action zone to protect the installation place

Conclusion: no effect;

2) fault point is located outside the protective action zone

Located at the end of the fault line of this paragraph

Conclusion: Protection segment I protection extended to the distance of the line length;

Fault line located adjacent the first end of the series compensation capacitor series compensated line side

Conclusion: the degree of compensation and load carried on this line, but is usually not affected;

Protection II section:

1) located in the lower line fault protection stage O operation outside the area

Conclusion: no effect, the line segment of the section II grade trip protection does not occur;

2) The end of the line fault point is located in this paragraph

Conclusion: no effect, protecting II section to protect this segment line length

Protection III segment:

Conclusion: no effect

In this paper, the impact of series compensation on power relay to conduct a comprehensive study, and fast switching series compensated for the impact of the distribution network from the protection of the simulation experiments, the research has achieved initial results. Limited space, the article did not compensate for a series of low-voltage distribution grid current protection impact do in-depth research, this part of the work remains to be further in the future to perfect.

References

1. M. E. Moursi, A. M. Sharaf, K. El-Arroudi, Optimal control schemes for SSSC for dynamic series compensation, Electric Power Systems Resear, **784**, (2007).
2. X. Y. Li, Nonlinear controller design of thyristor controlled series compensation for damping inter-area power oscillation, Electric Power Systems Research, **7612**, (2006).
3. T. Z. Jiang, C. Chen, A design method of nonlinear optimal predictive controller for thyristor controlled series compensation, Electric Power Systems Research, **769**, (2005)
4. E. A. Leonidaki, G. A. Manos, N. D. Hatziargyriou, An effective method to locate series compensation for voltage stability enhancement, Electric Power Systems Research, **741**, (2004).
5. S. K. Gupta, N. Kumar, A. K Gupta, Controlled series compensation in coordination with double order SVS auxiliary controller and induction M/C for repressing the torsional oscillations in power systems, Electric Power Systems Research, **622**, (2002)
6. F. Huo, X. C. Xi, A. N. Poo, Generalized Taylor series expansion for free-form two-dimensional contour error compensation, International Journal of Machine Tools and Manufacture, **531**, (2011)
7. A. Strickholm, A single electrode voltage, current- and patch-clamp amplifier with complete stable series resistance compensation, Journal of Neuroscience Methods, **611**, (1995)
8. M. S. Widyan, Controlling chaos and bifurcations of SMIB power system experiencing SSR phenomenon using SSSC, International Journal of Electrical Power and Energy Systems, **49**, (2013)
9. L. A. T. Guajardo, A. C. Enríquez, Enhance performance for distance relays due to series capacitors in transmission lines. Electric Power Systems Research, (2013)
10. M. Biswal, Adaptive distance relay algorithm for double circuit line with series compensation, Measurement, **53**, (2014)
11. J. C. N. Martinez, K. Nieradzinska, O. Anaya-Lara, Dynamic series compensation for the reinforcement of network connections with high wind penetration, Energy Procedia, **53**, (2014)
12. A. Moharana, R. K. Varma, R. Seethapathy, Subsynchronous Impact of Series Compensation on Induction Generator Based Wind Farm, Electric Power Components and Systems, **4111**, (2013)
13. L. Y. Sun, J. Zhao, G. M. Dimirovski, Adaptive coordinated passivation control for generator excitation and thyristor controlled series compensation system, Control Engineering Practice, **177**, (2009)
14. O. M. Neto, D. C. Macdonald, Analysis of subsynchronous resonance in a multi-machine power system using series compensation, International Journal of Electrical Power and Energy Systems, **288**, (2006)
15. E. J. D. Oliveira, J. L. R. Pereira, I. C. S. Junior, A. N. G Paulo, Wheeling cost influence in hydrothermal dispatch and series compensation allocation, International Journal of Electrical Power and Energy Systems, **262**, (2004),.
16. J. Y. Zou, J. X. Chen, E. Y. Dong, Study of fast-closing switch based fault current limiter with series compensation, International Journal of Electrical Power and Energy Systems, **249**, (2002)
17. C. S. Indulkar, Self-excited oscillations in series and shunt compensation schemes of ac transmission systems, International Journal of Electrical Power and Energy Systems, **197** , (1997)
18. H. W. Lee, J. C. Lee, A flexible CPW series resonator using phase compensation method on polyimide film at 5.8 - GHz[J]. Microw. Opt. Technol. Lett., **571**, (2015)
19. Y. A. Safar, M. M. Saied, The feasibility of multiple-capacitor long line series compensation, Electric Power Components and Systems, **235**, (1995)
20. A. Y. Abdelaziz, M. A. El-Sharkawy, M. A. Attia, Optimal location of thyristor-controlled series compensation and static VAR compensator to enhance steady-state performance of power system

with wind penetration, Electric Power Components and Systems, **4318**, (2015),.

21. M. Aredes, C. Portela, E. L. Emmerik, R. F. Silva Dias, Static series compensators applied to very long distance transmission lines, Electrical Engineering, **862**, (2003)
22. X. Z. Duan, J. Y. Wen, S. J. Cheng, Bifurcation analysis for an SMIB power system with series capacitor compensation associated with sub-synchronous resonance, Science in China Series E: Technological Sciences, **522**, (2009)
23. E. Ulas, Hocaoglu, M Hakan,Yalcinoz, Tankut. Transmission line shunt and series compensation with voltage sensitive loads, International Journal of Electrical Engineering Education, **464** (2009)

An Optical Inspection System Development for Defects on Multi-surfaces of Chips

Ming-Fu Chen[1], Chih-Wen Chen[1], Chih-Chung Chou [1], Chih-Yen Chen [1] and Yen-Chi Chiu [2]

[1]*Instrument Technology Research Center, NARLabs, 30076 Hsinchu, Taiwan*
[2]*Department of Power Mechanical Engineering, National Tsing-Hua University, 30013 Hsinchu, Taiwan*

Abstract. We proposed a new architecture with multiple pick and place heads to suck inspected chips and to be moved synchronously for acquiring the images of different surfaces on different chips at their corresponding acquisition stations. The computation loading of defect inspection for each chip also can be distributed because of sequential image acquisition. Considering the different computing loading of 6 surfaces of a chip, image acquisition for the front and back sides of chips is allocated at the first and second stations, and then 4 edge sides of chips. The modification and calibration of position and orientation for each opto-mechanical device will be simpler and easier when the characteristics of inspected chips changed. Thus system can accomplish the inspection for 6 surfaces of each chip and determine its quality in real-time before that chip is put onto target wafer. Developed inspection system has advantages and contributions of real-time inspection with high throughput during chip sorting process, and easily adjustment and calibration for all opto-mechanical devices to inspectdifferent types of chips. The first phase of development was accomplished for an AOI system to inspect defects on 5 surfaces of chips. System has been integrated in customer's chip sorter for testing. The inspecting accuracy for chip backside is better than 95%.

1 Introduction

The thickness of 3DIC chips with circuit design both on front and back sides is thinner (<100μm) than common IC chips due to application requirements. The cutting techniques then become the critical factor of cutting quality for these thin chips. But only few semiconductor companies with advantaged technologies own these cutting technologies. Chipping and glue could exist on chip surfaces of front, back and edge sides after cutting, and might impact chip quality and yield rate of manufacturing. In order to ensure the product quality, chips have to be inspected to feedback chips quality of production lines. The manufacturing processes then can be improved as well. Image acquisition and defect inspection for 6 surfaces of each chip shall be completed during chip sorting process according to customer's requirements. And chips need to be allocated on target wafer or tray in real-time. Then customers can reduce the wastes and costs of packaging the unqualified chips.

The main referenced patent related to multi-surface inspection of chips and applications is US 7283253 B2 [1]. A complicated opto-mechanical architecture, shown as Figure 1, for simultaneously acquiring images of 5 surfaces of chips is proposed in that patent. Its architecture seems rather convenient. But in fact, there're many constraints used for inspection. For example, different lighting sources are need to accommodate the requirements of acquiring images for back and edge sides of chip and protruding features on images. Moreover, the system inspection performance will decrease substantially in operations because of the inspected chips have to be sunk into the structure of Opto-Mechanical Device (OMD).

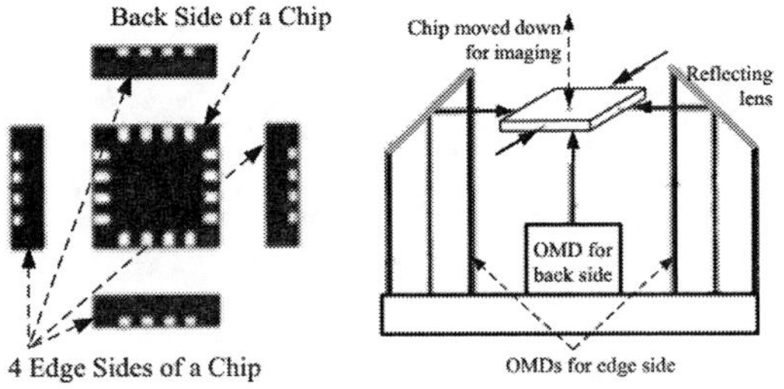

Figure 1. Diagram of an opto-mechanical device for images acquisition on 5 surfaces of chips in reference [1].

Owing to above limitations, most of patents and related applications were designed to extract its partial concepts and principles for developing new structures and applications. For instance, system consists of one OMD for acquiring image of chip back side and two OMDs for acquiring chip edge side. And there're 3 stations for multi-image acquisition of chips by using the multiple PP heads moved circularly. But inspected chips

DOI: 10.1051/matecconf/20166812003

still need to be sunk into the structure of OMDs, and the efficiency problem exists.

Here we proposed a new architecture consisted of multiple PP heads moved simultaneously to acquire images of corresponding chips at different positions for 6 surfaces of chips. The sequence of image acquisition for front and back sides of chips is disposed at 1st and 2nd stations considering that they need larger computation time. Then inspection system can complete the inspection for all surfaces and the quality determination of each chip before placed in GO/NG target wafer or tray. Therefore, the quality of chips can be accomplished in real-time during sorting process to achieve high efficiency of performance. The qualified chips then can be rearranged on target wafer for follow-up packaging and quality tracing.

2 System architecture and design

AOI system for detect defects on 6 surfaces of chip consists of 6 ODMs and their corresponding software modules, and infrastructure based on industrial personal computer (IPC) platform for system integration and operations. Developed AOI system was designed to inspect chip defects during sorting process with high inspecting performance.

The operations scenario and external interfaces have to be described as clearly as possible. Thus we can correctly define the system architecture, requirement specifications, function blocks and internal interfaces completely. The operation scenario and architecture of AOI system for 6 surfaces of chips can be schematically shown in Figure 2.

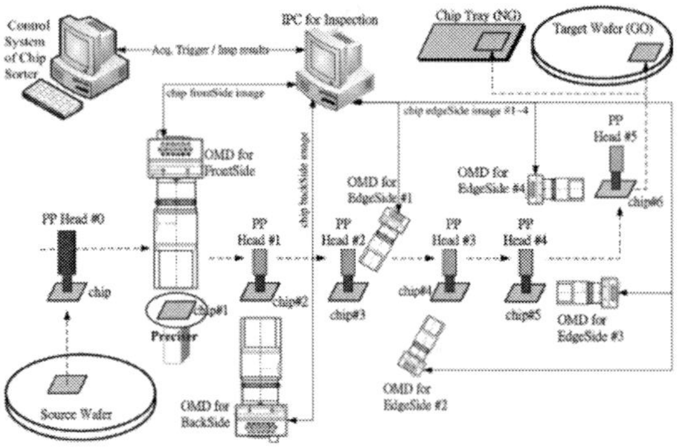

Figure 2. Operations scenario and architecture of AOI system for chips.

Control system of packaging machine sends the trigger signals of image acquisition for each OMD according to inspecting procedures. Each OMD immediately acquires and stores image for different surfaces of 6 chips simultaneously after receiving the trigger signal. Then 6 modules of inspection software conduct the image processing and defect inspection in real-time based on acquired raw images. A chip is located at different stations sequentially to acquire all images of its 6 surfaces. After finishing the image acquisition and defect inspection for the chip located at 6th station, the quality of the chip therefore can be determined. The inspection results for each chip are transmitted to control system. Then the qualified chip is rearranged on target wafer and unqualified chip will be put in tray of NG.

2.1 System requirement specifications and analysis

(1) Inspecting targets for 6 surfaces of chips

- Front and back sides of chip: Defect size of glue or contaminant on surface, and chippings at the rim of chip is greater than the inspecting criteria.
- Edge side of chip: Defect size of glue or chippings is greater than the inspecting criteria for 4 edge sides.

(2) Specifications for opto-mechanical devices

- OMD for image of chip front side: Spatial resolution for each pixel on image shall be better than 1.57 μm. Chip size of no greater than 7.2 x 6.8 mm can be inspected.
- OMD for image of chip back side: Spatial resolution for each pixel on image shall be better than 3μm. Chip size of no greater than 7.2 x 7.2 mm can be inspected.
- OMD for image of edge sides: Spatial resolution for each pixel on image shall be better than 3.45μm. Chip size of no greater than 7.2 x 5.6 mm can be inspected.

(3) Specifications for defect inspecting criteria

- Defects on front side: Defect size of greater than 8 x 8 μm for glue or contaminant shall be detected. Both the depth of chippings from chip boundaries and their corresponding width of greater than 21μm shall be inspected.
- Defects on back side: Defect size of greater than 10x10 μm for glue or contaminant shall be detected. Both the depth of chippings from chip boundaries and their corresponding width of greater than 21μm shall be inspected.
- Defects on edge side: Defect size of greater than 10 x 10 μm shall be detected for glue or chipping.

(4) Requirement specifications for inspection accuracy

- Error rate for inspecting front side of chip $< 5\%$.
- Error rate for inspecting back side of chip $< 10\%$.
- Error rate for inspecting edge side of chip $< 5\%$.

The AOI system for 6 surfaces of chip comprises 6 OMDs for front side, back side, 4 edge sides of chips, and their corresponding inspection software (S/W) modules, and an IPC platform. Each OMD includes the functions of illumination, image acquisition and storage. Each inspection software includes the modules of image preprocessing, chip image extraction and defect inspection. The system integration and tests for OMDs, software modules and packaging machine shall be accomplished to simultaneously fulfill all functions and performance of sorting and inspection for chips.

2.2 Design for opto-mechanical device

COTS products are adopted for all lenses and cameras of OMDs considering their quality, stable sources, and flexibility of update for critical modules. Key features of COTS critical modules for 3 types of OMDs are briefly described as below:

(1) OMD features for chip front side
- Camera: Size of CCD sensor is 5.5μm. Image resolution is 29M with data interface of camera link (CL). EureSys GrabLink Full card [2] is selected.
- Lenses: Magnification is 3.5. Resolution and DOF are about 1.2μm and 13μm respectively. Both of ring and coaxial lighting modules are designed.

(2) OMD features for chip back side
- Camera: Size of CCD sensor is 4.5μm. Image resolution is 25M.
- Lenses: Magnification is 0.5~2.0. Resolution and DOF are about 3μm and 500μm respectively [3]. Both of ring and coaxial lighting modules are included.

(3) OMD features for chip edge sides
- Camera: Size of CCD sensor is 3.45μm, and image resolution is 5M with GigE data interface.

2.3 System functional block and interfaces

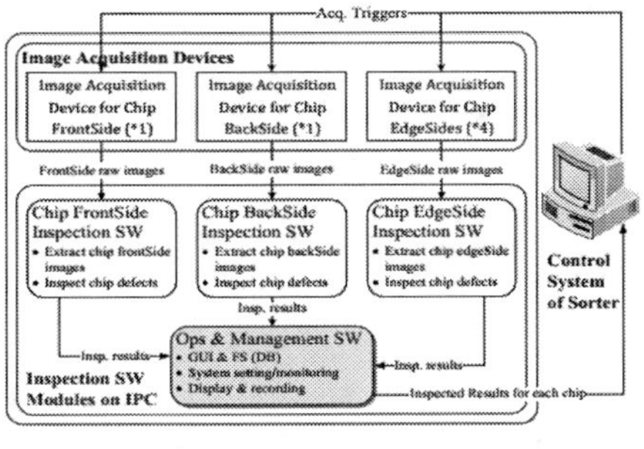

Figure 3. System Functional Block Diagram and Interfaces.

Developed inspection system for multi-surfaces of chip consists of following OMDs and software modules:

(1) OMD and S/W module for chip front side: The spatial resolution of acquired image is 1.57μm/pixel and FOV is about 10.32 x 6.88 mm.
Inspection S/W module real-time read images of provided by OMD via CL interface. Image processing by using methods of frequency domain analysis for image registration and template-based matching for defect detection are applied. Defective chips will be designated and reported for quality evaluation and classification if the size of defects is greater than criteria.

(2) OMD and S/W module for chip back side: Ring illuminating modules is used for acquiring the dark field image of chip backside. But coaxial lighting module is also designed for inspecting few defects of hard to be detected. The spatial resolution of image is 3 μm/pixel and FOV is about 15 x 15 mm.
Chip backside image of dark field is applied for inspection S/W module. Image enhancement is applied to stand out the defects on backside image [4].

(3) OMD and S/W module for chip edge side: The spatial resolution of image is 3.45 μm/pixel and FOV is about 7.3 x 6.1 mm. S/W module for inspecting defects of chip edge side, including glue and chipping.

S/W module for operations and management can provide the functions of system setting and monitoring, and graphic user interfaces (GUI). Inspection system was integrated with chip sorter. Qualified chips will be rearranged on target wafer, and the defective chips will be allocated in NG tray after chip sorter receives the inspection results of each chip. Figure 3 shows the functional block diagram of developed system.

3 Inspection methods for 6 surfaces of chips

3.1 Defect inspection for front side of chips

Inspection methods and flowchart for chip front side are depicted in Figure 4. Images are processed by registration [5, 6], including rotation, scale and position, based on the template chip image. Defects can consequently be detected by using background filtering and pattern matching [7, 8]. The inspection algorithm for front side of chip now is still being improved to achieve higher accuracy and performance of inspection.

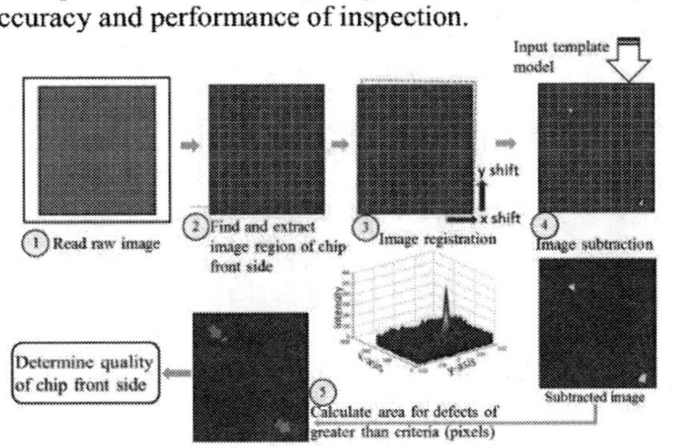

Figure 4. Inspection methods and flowchart for chip front side.

We can consider the defect inspection on images as the problem of background subtraction. The chip images with regular type or profile are regarded as "background". The region with intensity difference between inspected image and background image are the "foreground". We can therefore designate and determine these foregrounds as defects with features or size of greater than criteria [8]. The background model used as matching template here is generated by Gaussian Mixture Model (GMM) [9]. There're few variation of pixel intensity for designated point on acquired images due to the small change of illumination, the machine vibration and so on.

The update of background model includes the derivation of Gaussian model for template update. The pixels intensity of images under different time, , ,..., , can be described by Gaussian distribution respectively. Hence we have following equation:

$$P(I_t) = \sum_{K=1}^{K} \omega_{t-1,k} N(I_t; \mu_{t-1,k}; \sigma^2_{t-1,k}) \quad (1)$$

where is Gaussian distribution function, , and are mean value, variance and weighting factor respectively.

3.2 Defect inspection for back side of chips

The methods and flowchart for chip backside inspection are depicted in Figure 5. After the image of chip area is segmented, the dark field image of chip backside will be pre-processed by image enhancement to make defects more clearly on chip image. We then can inverse the gray-level of image to derive an inverted image. Defects of glue, contaminant and boundary chipping can be derived by blob/contour finding. The size of defects of greater than criteria will be designated and reported for quality check.

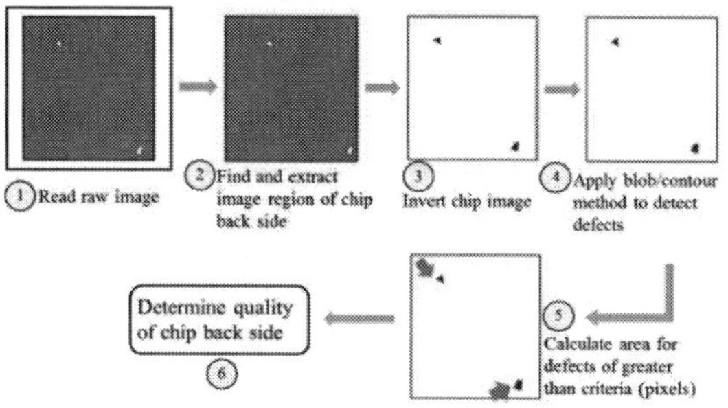

Figure 5. Inspection methods and flowchart for chip backside.

3.3 Defect inspection for edge sides of chips

Figure 6 is the flowchart and methods of defect inspection for chip edge sides. We designate the region of edge side of chip, and then extract the chip area [10, 11]. Thus the defects of glue and boundary chipping can be derived by using methods of blob/contour finding. And the defects of greater than criteria are designated and reported.

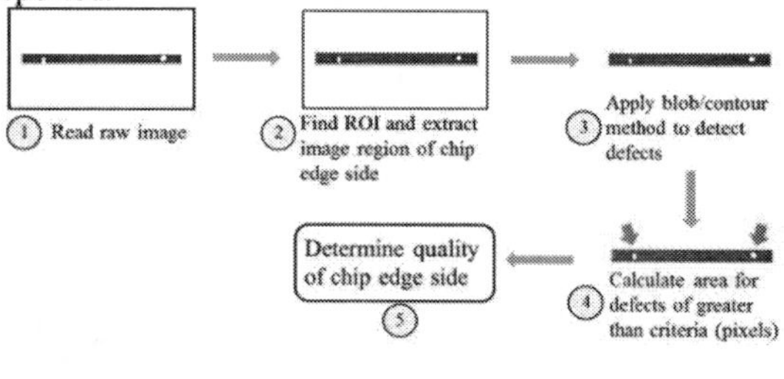

Figure 6. Inspection methods and flowchart for chip edge side.

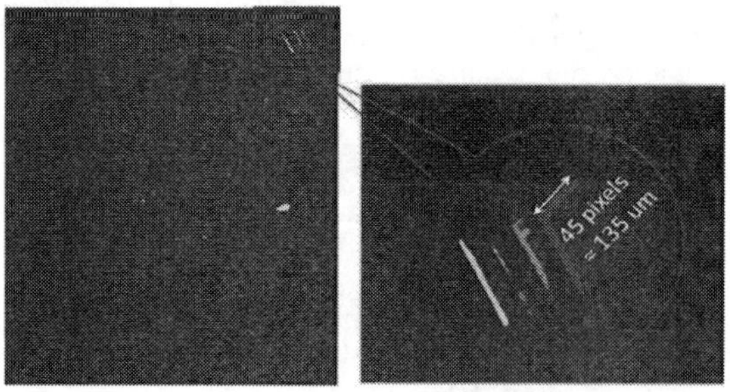

Figure 7. Defects size on chip back side detected by developed AOI system.

4 Experiments and discussion

The prototypical inspection system for 5 surfaces of chip has been implemented and integrated with chip sorters. The defects on chip backside are inspected by dark field image as shown in Figure 7. Defect size of inspected by developed AOI system is also depicted in Figure 7. Measured result for the same defect by microscopy is shown in Figure 8. The size error of defect on chip backside is within 1 pixel (3um) compared to the measured result by using microscopy.

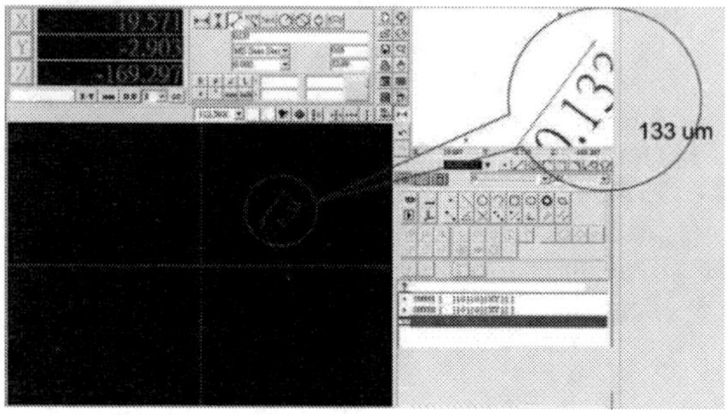

Figure 8. Defects size on back side of chip measured by microscopy.

Based on the preliminary test results of inspection, the accuracy of inspection for developed system is better than 95% for backside of chips. Now we continuously improve the image quality and inspection methods to try raising the performance and accuracy of inspection.

The defects size for an edge side of a chip, detected by developed system and measured by microscopy, are shown in Figure 9 and 10 respectively. The inspection error is within 1 pixel (3.45um) as well. But defects of glue and chipping can't be discriminated because of their reflective features on image are very similar. Cutting scars on images of chip edge sides is another problem to decrease the inspecting accuracy.

Figure 9. Defects size on chip edge side inspected by developed AOI system.

Figure 10.Defects size on chip edge side measured by microscopy

Figure11 is the inspection result for front side of a chip by our developed system. Defects on front side can be inspected successfully, but the inspection accuracy needs to be verified by using more images in the near future. And we still try to improve the inspecting algorithms for chip front side continuously for the moment.

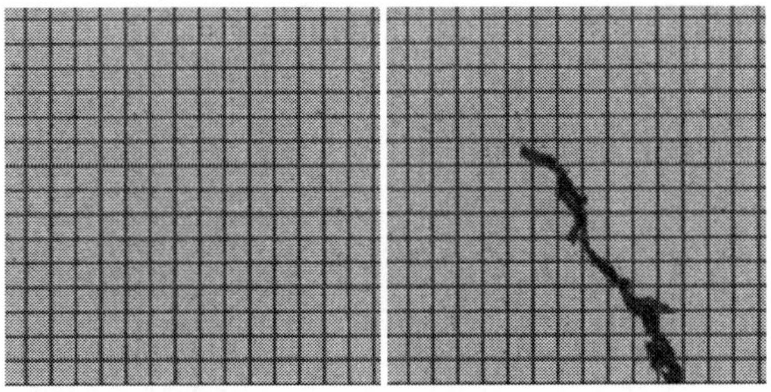

Figure 11. Inspection result for front side of a chip (left: good chip, right: inspected chip with defects)

5 Conclusion

An AOI system with new architecture of imaging devices for defects on multi-surfaces of chips was developed and presented in this work. System comprises 6 OMDs to acquire raw images for their corresponding surfaces on different chips, inspection S/W modules, and industrial PC platform. The first phase of developing an AOI system for defects on 5 surfaces of chips was accomplished. System has been integrated in customer's chip sorter for testing. The inspecting accuracy for chip backside is better than 95%. Defects with flat feature which is peeled off chip boundary can't be detected constantly. Therefore, bright-field images need to be applied to improve the inspection accuracy in our future work. Another challenge is the cutting scars on chip edge side that will interfere with inspection accuracy, and have to be handled continuously.

Comparing to other similar AOI system, developed system here with multiple PP heads to suck different chips for simultaneously acquiring images on 6 surfaces of chips has better system throughput and is easily adjusted for different types of chips. Chips can be inspected in real-time during sorting process for quality assurance to improve the production efficiency.

References

1. Aldo salvi, La ChauX de FOndS, Optical Device and Inspection, Patent No.: US 7,283,235 B2 (2007)
2. EureSys Corporation, Datasheet of EureSys Grablink Full frame grabber (2013)
3. Jos. Schneider Optische Werke GmbH, Datasheet of Schneider MRV 4.5/85 Lens (2007)
4. Intel Corporation, Reference Manual of Open Source Computer Vision (openCV) Library, version 1.0 (2001)
5. Gonzalez, R.C., and Woods, R.E., Digital Image Processing, Prentice-Hall, New Jersey (2002)
6. B. Srinivasa Reddy and B. N. Chatterji, An FFT-Based Technique for Translation, Rotation, and Scale-Invariant Image Registration, IEEE Transactions on Image Processing, Vol. 5, No. 8, 1266-1271 (1996)
7. Chi-ho Chan and Grantham K. H. Pang, Fabric defect detection by Fourier analysis, Industry Applications, IEEE Transactions, Vol. 36, 1267-1276 (2000)
8. Z. Zivkovic, F. van der Heijden, Efficient Adaptive Density Estimation per Image Pixel for the Task of Background Subtraction, Pattern Recognition Letters, vol. 27, no. 7, pages 773-780 (2006)
9. Z. Zivkovic, F.van der Heijden, Recursive unsupervised learning of finite mixture models, IEEE Trans. on Pattern Analysis and Machine Intelligence, vol.26, no.5, pages 651-656 (2004)
10. Heath, M., Sarkar, S., Sanocki, T. and Bowyer, K., Comparison of Edge Detectors: A Methodology and Initial Study, Computer Vision and Image Understanding, Vol. 69, No. 1, 38–54 (1998)
11. Ming-Fu Chen, Chih-Chung Chou, Chun-Chien Lien, Rui-Cian Weng, Development of an AOI system for chips with a hole on backside based on a frame imager, 7th International Symposium on Precision Mechanical Measurements (ISPMM), Xiamen, China, August 7-12 (2015)

Effect of Temperature and Vibration on Electrical Connectors with Different Number of Contact Cores

W. L. Song[1], J. T. Li[1], N. Wang[1], Z. L. Sun[1], M. J. Zhang[2], Q. Yang[1]

[1]*School of Mechanical Engineering and Automation, Northeastern University, Shenyang 110819, China*
[2]*Shenyang Xinghua Aero-electric Appliance Limited Liability Company, Shenyang 110142, China*

Abstract. In this paper, we presented the results from three related analysis performed by adopting the failure models, which provided an explanation of performance influencing factors caused by different number of contact cores, for the purpose of measuring the temperature change and deformation value, which were the factors causing contact failure. The failures were localized in contact parts of the connectors. Performed investigations included thermal analysis, modal analysis, harmonic response analysis and contact failure analysis. From the results of these simulations, related temperature and vibration analysis nephograms were got respectively. And the correctness of results of thermal analysis was verified by Fourier law. The research results of this paper provide a reference for thermal analysis and vibration analysis of electrical connectors, which is important for ensuring the reliability and safety of electrical connectors.

1 Introduction

The increase in the number of electronic and electric systems used in the aerospace industry necessitates a better performance for electrical connectors. The main failure mode is contact failure, and the temperature and vibration are the main reasons that determine the performance of electrical connectors, as under the action of environmental stress [1]. Both of the formation of fretting wear, the occurrence of transient fracture and the growth of oxide film, that caused by vibration and the oscillation of temperature will have adverse effects on the electrical connectors. Therefore, thermal analysis and vibration analysis of electrical connectors are necessary for meeting its performance.

Few related studies about contact failure of electrical connectors have been done. HAYASHI et al. [2] present that at least four contact points are necessary in order to decrease noise radiation from two independent experiments performed by adopting the Contact Failure Model. Hayashi et al. [3] study the effects of the contact distribution and the number of contact points on the common-mode current in the coaxial cable as well as on the current distribution among the contact points. LeFevre et al. [4] conduct two case studies for failure analysis of connector which provided a better understanding of the mechanisms for design and testing purposes. Zhou et al. [5] conduct a failure analysis for a kind of low power connector which adapted fretting simulation by vibrator, showed that connection resistance of connectors was raised up, even to ohms level.

However, the effect of temperature and vibration on the relationship between reliability and different cores of electrical connectors has not been reported.

The purpose of this work is to study the behavior of electrical connectors with different cores that used in the aerospace industry by basing on establishing three models with the same structure and dimension for the contact pin, insulator and shell, but with a different of number contact pins.

2 Thermal analysis of electrical connector

Thermal transmission results from the temperature difference between two objects or an object interior. The main thermal transmission mode of electrical connectors is internal thermal conduction, followed by the thermal convection formed from the shell exposed to air.

Table 1. Material and its parameters for electrical connectors.

component	material	thermal conductivity $[Wm^{-1}K^{-1}]$	Specific heat capacity $[kg\ J^{-1}K^{-1}]$	Elasticity modulus [GPa]	Poisson's ratio
Contact body	brass	122	380	106	0.324
Shell	constructional steel	43.53	460	106	0.4
insulator	PTEF	0.197	1.0	120	0.4

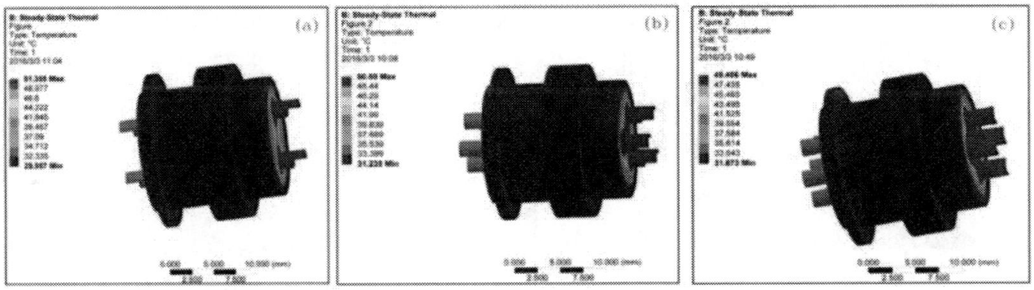

Figure 1. Temperature distribution (a is for 2 cores, b is for 3 cores, c is for 4 cores).

For steady-state thermal analysis, material property-thermal conductivity is only required. Set material properties as showed in Table 1 after the three models imported in Workbench ANSYS.

The working state of 25 °C is set for steady state thermal analysis of electrical connectors. The natural convection coefficient is 5.78 [w / (m^2·°C)], which is applied to the shell by the contact body with the rated current of 15 A, and the steady-state temperature distribution diagrams of the models are obtained showed in Fig. 1 (a), (b) and (c) respectively.

Transient thermal analysis with analysis time and time step set to 45 s and 5 s is conducted on the basis of steady state thermal analysis. When the internal heat generation load is applied, the transient maximum temperature of the models can be obtained by loading the load applied to the steady state analysis.

Fig. 1 clearly shows the steady-state temperature field distribution of electrical connectors. Computational results of temperature field distribution show that high temperature concentration existed at the root of core inserts and the shell near the cores. The maximum temperatures of model I, II and III are 51.36 °C, 50.59 °C and 49.40 °C respectively show that with the increase of the number of cores, the maximum temperature shows a decreasing trend.

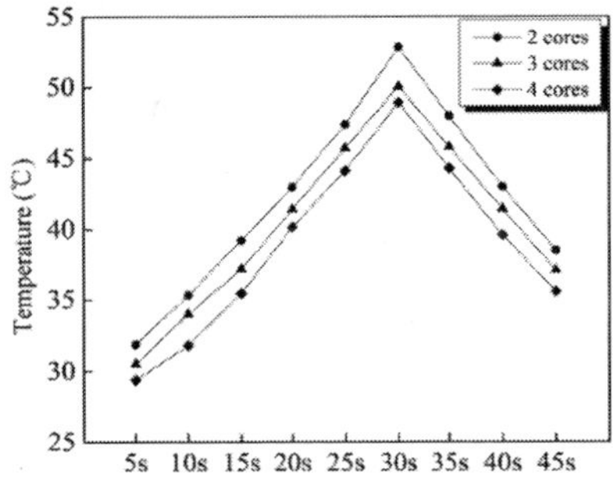

Figure 2. Maximum temperature (transient state).

Fig. 2 shows the transient temperature of the three models respectively. Observe carefully, the temperature of model I and II are higher than that of the model II and III respectively in every period of time, shows that with the core insert added, the heat dissipation performs better and better, which is consistent with the conclusion of steady state thermal analysis.

3 Vibration analysis of electrical connector

3.1 Modal analysis

The modal is the natural vibration characteristic of the mechanical structure. In recent years, the modal testing of structures has grown rapidly following developments in the hardware and software employed. In addition, its application is required as an auxiliary tool in design, structural modification, validation, model update and damage identification projects [6], [7].

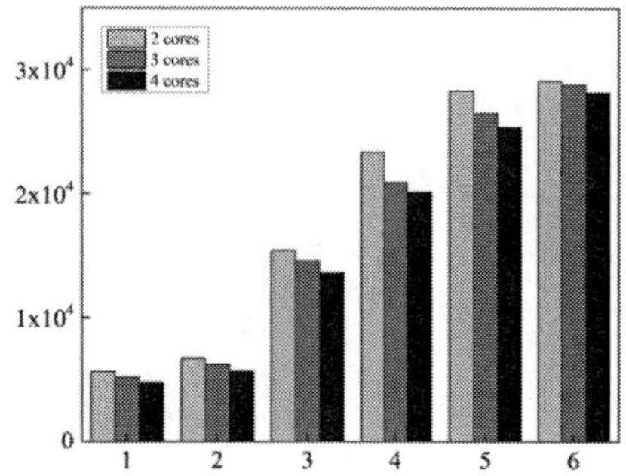

Figure 3. Comparison diagram of natural frequency for model I, II and III.

Import the models into Workbench ANSYS and set material properties shown in Table 1 respectively. The natural frequencies of models are obtained as indicated in Fig. 3.

Fig. 3 shows calculated natural frequencies for each model. The frequency response situations of the three models at the measurement points have clearly different natural frequencies. Analyze and conclude that, the natural frequency of the different orders increases with the increase of modal order, and the natural frequency of the same order decreases with the increase of contact cores. Therefore, the electrical connector with four cores has a better performance.

3.2 Harmonic response analysis

The harmonic response analysis is a time domain analysis method and therefore is of fundamental importance [8]. In the process of analysis, the steady-state of the structure is calculated only, and the transient vibration is not considered at the beginning of excited vibration. The acceleration load is used in this analysis.

According to the experiment condition of vibration stipulated in ANSI/EIA- -364-28E (electrical connector test method) and Test condition II , the frequency and accelerated speed are selected as 100 Hz and 98.1 ms^{-2}, 300 Hz and 98.1 ms^{-2}, 500 Hz and 98.1 ms^{-2} respectively in the simulation. Such three sets of parameters are used to carry out harmonic response analysis on the models [9], [10].

Now, the correctness of the simulation result can be verified by the vibration formula:

$$Z = A \cdot \sin(2\pi f \cdot t + \theta) \quad (1)$$

Second derivative of Equation (1) for time can be written as the form:

$$a = -4A \cdot \pi^2 \cdot f^2 \cdot \sin(2\pi f \cdot t + \theta) \quad (2)$$

where Z, A, f, θ and a are vibration displacement, maximum value of displacement Z, vibration frequency, initial phase and the vibration acceleration.

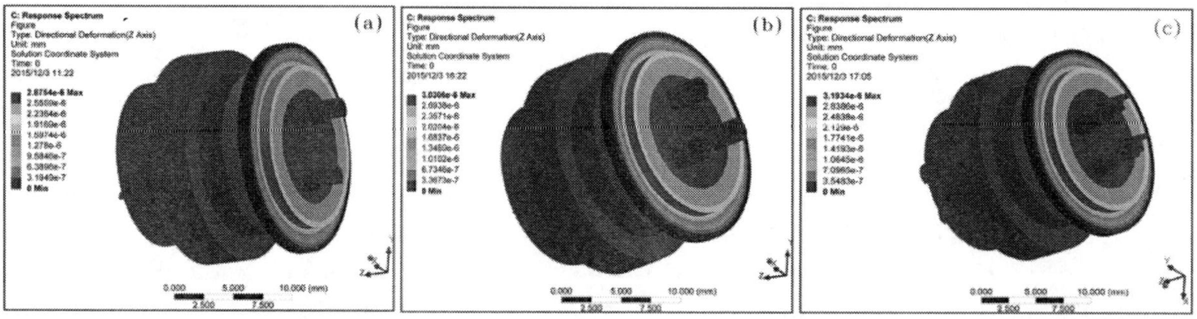

Figure 4. Deformation nephogram of harmonic response analysis (a is for 2 cores, b is for 3 cores, c is for 4 cores; a = 98.1 ms^{-2}; f = 100Hz).

An equation can be got in consideration of formula (2) and the acceleration amplitude specified in Test condition

II presented as follow:

$$4A \cdot \pi^2 \cdot f^2 = 98.1 \tag{3}$$

As can be seen, A is inversely proportional to f. The vibration Z will be small as substituting getting smaller amplitude affected by frequency's getting bigger to Equation (1). In modal analysis, the natural frequency and the maximum deformation of the electrical connector with the core number of 2, 3 and 4 are gradually reduced and gradually increased respectively, which is consistent with the results of ANSYS.

4 Conclusion

(1) Through steady-state analysis and transient temperature analysis, it is indicated that the heat dissipation increased gradually with the core number. The simulation results calculated by Fourier law coincide with the results of theoretical analysis that the electrical connector with 4 contact cores has the highest heat transfer rate and best heat dissipation performance. (2) By vibration analysis, it is concluded that with the increase of vibration frequency, the maximum deformation of the models in increasing. At the same frequency, the maximum deformation increased gradually with the number of cores.

Acknowledgements

This study was funded by the Fundamental Research Funds for the Central Universities (Grant No.N130303002), the Postdoctoral Science Foundation of China (Grant No.2014M561243), the Scientific Research Foundation for the Returned Overseas Chinese Scholars, State Education Ministry (50-5), and the Doctoral Scientific Research Foundation of Liao Ning Province (Grant No.20141004).

References

1. Y. V. Murty, Ency. Mater: Sci. and Technol.: 2483–2494 (2001)
2. Y. Hayashi, T. Mizuki, H. Sone, *IEEE Int. Symp. on EMC*: 209-212 (2009)
3. Y. Hayashi, S. P. Wu, J. Fan H. Sone, *IEEE Int. Symp. on EMC*: 743-747 (2010)
4. B. G. LeFevre, W. W. King, A. G. Hardee, A. W. Carlisle, K. B. Bradley, J. Lightw. Technol. **Vol. 11, No. 4**: 537-541 (1993)
5. Y. L. Zhou, N. Zeng, L. J. Xu, J. Stefan, J. Zhejiang Univ. Sci. A **Vol. 8, No. 3**: 384-392 (2007)
6. E. N. B. S. Júlio, C. A. S. Rebelo, D. A. S. G. Dias-da-Costa, Eng. Struct **Vol. 7, No. 11:** 3,468-3,477 (2012)
7. M. Lepidi, V. Gattulli, D. Foti, Eng. Struct **Vol. 31, No. 7**: 1,486-1,500 (2009)
8. L. Li, Y. J. Hu, X. L. Wang L. Lü, Mech. Syst. Signal Proc. **Vol. 42, No. 1-2**: 31–41 (2014)
9. D. L. Cronin, S. Adhikari, Eng. Mech. **Vol. 125, No. 12**: 1,372–1,379 (1999)
10. L. Li, Y. J. Hu, X. L. Wang, L. Ling, J. Finite Elem. Anal. Des. **Vol. 72, No. 19**: 21–34 (2013)

MATEC Web of Conferences **68**, 12005 (2016) DOI: 10.1051/matecconf/20166812005
ICIEA 2016

The Model and Quadratic Stability Problem of Buck Converter in DCM

Xiaojing Li [1, 2], Aiguo Wu [1]

[1] *School of Electrical Engineering and Automation, Tianjin University, China*
[2] *School of Electrical Engineering, Tianjin University of Technology, China*

Abstract. Quadratic stability is an important performance for control systems. At first, the model of Buck Converter in DCM is built based on the theories of hybrid systems and switched linear systems primarily. Then quadratic stability of SLS and hybrid feedback switching rule are introduced. The problem of Buck Converter's quadratic stability is researched afterwards. In the end, the simulation analysis and verification are provided. Both experimental verification and theoretical analysis results indicate that the output of Buck Converter in DCM has an excellent performance via quadratic stability control and switching rules.

1 Introduction

Hybrid systems (HS) is defined as a unitized dynamic system interacted by discrete and continuous parts. DC-DC converters are typical HS because the operation of each mode can be regard as the continuous dynamic subsystems and the turn-on or turn-off of power switch as the discrete dynamic subsystems.

Switched linear systems (SLS), an important type of HS, have attracted considerable attention in modeling, analysis and design. Quadratic stability is one of the important problems of SLS. This problem is more complex since it depends on the switching rules as the stability of all the subsystems.

Looking from the existing literatures [1]-[5], Lyapunov theories are the dominant approaches used in study of stability. For example, the stability of DC-DC Converters in CCM (Continuous Current Mode) was analyzed in [6]. The aim of this paper is to study the quadratic stability of DC-DC converters via using Lyapunov function based on the model in DCM (Discontinuous Current Mode).

2 Modeling of buck converter in DCM

2.1 Buck converter in DCM

The topology structure of Buck converter is shown in Fig.1. Assume all of the components are perfect. Three work modes of it in DCM are shown in Fig. 2.

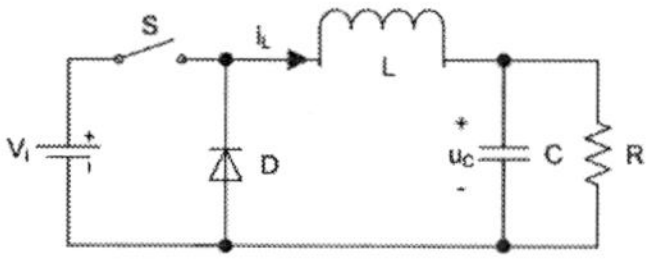

Figure 1. Buck converter.

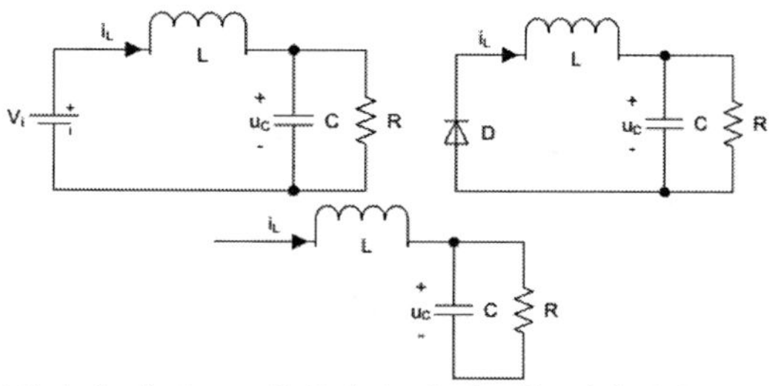

(a) Switch tube is on (b) Switch tube is off and diode is on (c) Switch tube and diode are all off

Figure 2. Buck Converter in DCM

2.2 SLS Model of Buck Converter in DCM

Supposing all of the components are ideal and the state vector is $x(t) = [\ i_L \quad u_C\ ]^T$,the output vector is $y(t) = u_C$, consequently, the state equations of Buck Converter in SLS model are[7]

$$\begin{cases} \dot{x}(t) = A_j x(t) + B_j \\ y(t) = C_j x(t) + D_j \end{cases} \tag{1}$$

The parameters matrixes of Buck Converter in DCM are expressed as follows according to Fig. 2.

$$\text{(a)}\ A_1 = \begin{bmatrix} 0 & -\dfrac{1}{L} \\ \dfrac{1}{C} & -\dfrac{1}{RC} \end{bmatrix},\ B_1 = \begin{bmatrix} \dfrac{V_{in}}{L} \\ 0 \end{bmatrix},$$

$$C_1 = [0 \quad 1],\ D_1 = \begin{bmatrix} 0 \\ 0 \end{bmatrix}.$$

(b) $A_2 = \begin{bmatrix} 0 & -\frac{1}{L} \\ \frac{1}{C} & -\frac{1}{RC} \end{bmatrix}$, $B_2 = \begin{bmatrix} 0 \\ 0 \end{bmatrix}$,

$C_2 = [0 \quad 1]$, $D_2 == \begin{bmatrix} 0 \\ 0 \end{bmatrix}$.

(c) $A_3 = \begin{bmatrix} 0 & 0 \\ 0 & -\frac{1}{RC} \end{bmatrix}$, $B_3 = \begin{bmatrix} 0 \\ 0 \end{bmatrix}$,

$C_3 = [0 \; 1]$, $D_3 = \begin{bmatrix} 0 \\ 0 \end{bmatrix}$.

So it can be seen that Buck Converter in DCM has three switched modes, (A_1 , B_1 , C_1 ,D_1), (A_2 , B_2 , C_2 ,D_2) and (A_3 , B_3 , C_3 ,D_3),which belong to three subsystems respectively, Σ_1, Σ_2 and Σ_3.

3 Quadratic stability of SLS

3.1 Propaedeutics

Given a stable equilibrium point $\bar{x}$. Since any other equilibrium point can be shifted to the origin via a change of variable $\tilde{x} = x - \bar{x}$, we can assume the switched equilibrium is the origin $\bar{x}=0$ without loss of generality.

Definition: if and only if there exist a matrix $P = P^T > 0$ and a constant $\varepsilon > 0$ such that for the quadratic function $V(x) = x^T Px$, we have $\dot{V}(x(t)) \leqslant \varepsilon x^T x$ along all system trajectories, the switched equilibrium $\bar{x} = 0$ can be said quadratically stable[8].

(a) when subsystem Σ_i is active,

$\dot{V}(x(t)) = \dot{V}_i(x(t)) = x^T(A_i^T P + PA_i)x + 2b_i^T Px$,

(b) at the switching point between Σ_i and Σ_j ,

$$\dot{V}(x(t)) = \sup_{\gamma\in[0,1]} \{\gamma \dot{V}_i(x(t)) + (1 \quad \gamma)\dot{V}_j(x(t))\}.$$

If we denote $A_{eq} = \Sigma_{i=1}^m \alpha_i A_i$ and $b_{eq} = \Sigma_{i=1}^m \alpha_i b_i$,where $0<\alpha_i<1$ and $\Sigma_{i=1}^m \alpha_i = 1$, a convex combination of the subsystems is defined as

$$\Sigma_{eq} : \dot{x} = A_{eq}x + b_{eq}. \tag{2}$$

3.2 Quadratic stability

Theorem 1: As for system (l), the point $\bar{x} = 0$ is a quadratic stable switched equilibrium if there exist $\alpha_i \in (0,1)$, $i =1,...,m$

$$\text{such that } \sum_{i=1}^m \alpha_i = 1, \tag{3}$$

$$A_{eq} = \sum_{i=1}^m \alpha_i A_i \text{ is Hurwitz,} \tag{4}$$

$$b_{eq} = \sum_{i=1}^m \alpha_i b_i = 0 \tag{5}$$

Since the convex combination A_{eq} in Eq. 4 is stable, there exist two positive definite symmetric matrices P and Q such that

$$A_{eq}^T P + PA_{eq} = Q. \tag{6}$$

According to Eq. 4, Eq. 6 can be rewritten as

$$\sum_{i=1}^m \alpha_i x^T (A_i P + PA_i)x = x^T Qx \tag{7}$$

From Eq. 5, we can also get the null term

$$2(\alpha_1 b_1 + \alpha_2 b_2 + ... + \alpha_m b_m)^T Px = 0 \tag{8}$$

Now, a new equation is obtained as follows

$$\alpha_1(x^T(A_1^T P + PA_1)x + 2b_1^T Px) + \alpha_2(x^T(A_2^T P + PA_2)x + 2b_2^T Px) + ... + \alpha_m(x^T(A_m^T P + PA_m)x + 2b_m^T Px) = x^T Qx \leqslant \varepsilon x^T x \tag{9}$$

where $0<\varepsilon \leq \lambda_{\min}$ and $\lambda_{\min}$ is the smallest positive real eigenvalue of Q. Then Eq. 9 is equivalent to Eq. 10 as follows.

$$\alpha_1(x^T(A_1^T P + PA_1)x + 2b_1^T Px + \varepsilon x^T x) + \alpha_2(x^T(A_2^T P + PA_2)x + 2b_2^T Px + \varepsilon x^T x) + ... + \alpha_m(x^T(A_m^T P + PA_m)x + 2b_m^T Px + \varepsilon x^T x) \leqslant 0 \tag{10}$$

4 Hybrid feedback switching rule

In this rule, a lower bound on the decay rate $\dot{V}_{\min}(x(t)) = -\varepsilon x^T x$ with $\varepsilon > 0$ is presented primarily, then the active subsystem is switched off only when it no more satisfies the required constraint.

The procedures of hybrid feedback switching rule is listed as follows.

(*initialization*) at time $t = 0$ activate the subsystem Σ_{i_0} with $i_0 = \arg\min\{(\dot{V}_i(x_0))\}$.

(*switching off rule*) if subsystem Σ_i is active and $x^T(A_i^T P + PA_i)x + 2b_i^T Px > -\varepsilon x^T x$,the system will switch to subsystem Σ_j with $j = \arg\min\{(\dot{V}_j(x))\}$.

(*equilibrium neighborhood rule*) if $\|x\| \leq \rho_{off}$ switching will be stopped until $\|x\| \geq \rho_{on}$ ($\rho_{off} < \rho_{on}$) .

5 Quadratic stability of buck converter

5.1 Coordinate Transformation of Buck Converter

Assume the anticipant output $\bar{x} = [\bar{x}_1 \quad \bar{x}_2]^T$ as the switched equilibrium, we use $\tilde{x} = x - \bar{x}$ for conversion of coordinates to transfer switched equilibrium to original point. The state equations after coordinate transformation is given by Eq. 11 and Table 1.

Table 1. Parameters matrixes of buck converter after coordinate transformation

Σ_1 :w=1, v=1 (Switch tube is on)	Σ_2 :w=1, v=0 (Switch tube is off and diode is on)	Σ_3 :w=0, v=0 (Switch tube and diode are all off)
$A_1 = \begin{bmatrix} 0 & -\frac{1}{L} \\ \frac{1}{C} & -\frac{1}{RC} \end{bmatrix}$	$A_2 = \begin{bmatrix} 0 & -\frac{1}{L} \\ \frac{1}{C} & -\frac{1}{RC} \end{bmatrix}$	$A_3 = \begin{bmatrix} 0 & 0 \\ 0 & -\frac{1}{RC} \end{bmatrix}$
$B_1 = \begin{bmatrix} \frac{V_{in}}{L} - \frac{1}{L}\bar{x}_2 \\ \frac{1}{C}\bar{x}_1 - \frac{1}{RC}\bar{x}_2 \end{bmatrix}$	$B_2 = \begin{bmatrix} -\frac{\bar{x}_2}{L} \\ \frac{1}{C}\bar{x}_1 - \frac{1}{RC}\bar{x}_2 \end{bmatrix}$	$B_3 = \begin{bmatrix} 0 \\ -\frac{1}{RC}\bar{x}_2 \end{bmatrix}$

$$\begin{bmatrix} \dot{\tilde{x}}_1 \\ \dot{\tilde{x}}_2 \end{bmatrix} = \begin{bmatrix} 0 & -\frac{1}{L}w \\ \frac{1}{C}w & -\frac{1}{RC} \end{bmatrix} \begin{bmatrix} \tilde{x}_1 \\ \tilde{x}_2 \end{bmatrix} + \begin{bmatrix} \frac{V_{in}}{L}v - \frac{1}{L}w\bar{x}_2 \\ \frac{1}{C}w\bar{x}_1 - \frac{1}{RC}\bar{x}_2 \end{bmatrix} \tag{11}$$

5.2 Switching control of buck converter

Buck Converter in DCM includes three subsystem, consequently there exist three coefficients α_1, α_2 and α_3 in the convex combination of Eq.3 .According to Eq.3, we have α_3=1-(α_1+α_2). Then the convex combination can be written as

$$A_{eq} = \begin{bmatrix} 0 & -\frac{\alpha_1 + \alpha_2}{L} \\ \frac{\alpha_1 + \alpha_2}{C} & -\frac{1}{RC} \end{bmatrix},$$

$$b_{eq} = \begin{bmatrix} \frac{\alpha_1 V_{in} - (\alpha_1 + \alpha_2)\bar{x}_2}{L} \\ \frac{(\alpha_1 + \alpha_2)R\bar{x}_1 - \bar{x}_2}{RC} \end{bmatrix}.$$

Assume $P = \begin{bmatrix} L & 0 \\ 0 & C \end{bmatrix}$, the Lyapunov function is

$$V(\tilde{x}) = \tilde{x}^T P \tilde{x} = \frac{1}{2}L\tilde{x}_1^2 + \frac{1}{2}C\tilde{x}_2^2 = \frac{1}{2}Li_L^2 + \frac{1}{2}Cu_C^2.$$

Obviously $V(\tilde{x})$ is an energy function. Then we get

$$A_{eq}^T P + P A_{eq} = \begin{bmatrix} 0 & 0 \\ 0 & -\frac{2}{R} \end{bmatrix} = -Q.$$

Where Q is semi-definite positive. Q can be chosen as a semi-definite matrix because $\dot{V}(\tilde{x})$ is not identically vanishing along all system trajectories for DC-DC converters. Let $\lambda_{\min}$ =2/*R* is the smallest positive real eigenvalue of *Q* and $0<\varepsilon \leq \lambda_{\min}$. So the regions of subsystems can be divided by hybrid feedback rule.

5.3 Simulation of buck converter

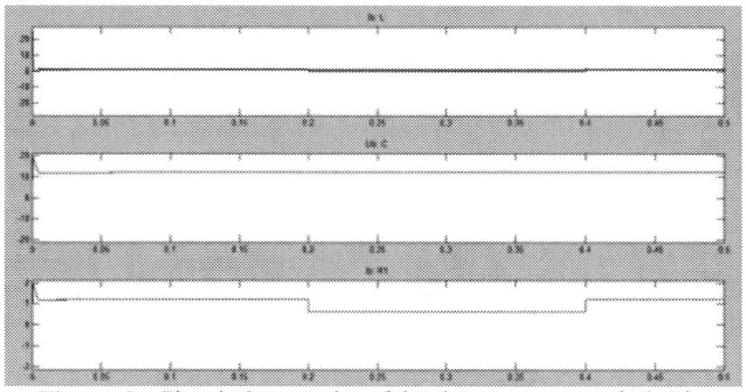

Figure 3. Simulation results of buck converter on hybrid feedback switching control rule.

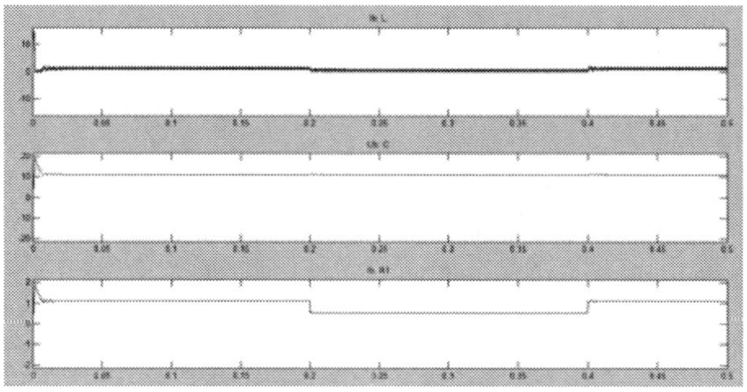

Figure 4. Simulation results of buck converter in open-loop control.

In this section, simulation results of Buck converter based on hybrid feedback switching rule are presented in Fig. 3(V_{in}=24V, *f*=20KHz, *R*=10Ω, *L*=400μH, *C*=750μF). Fig. 4 demonstrates the simulation results when Buck Converter is open-loop as a compared object.

The load resistor *R* has two salutations during the process of simulation, one is from 10Ω to 20Ω when t=0.2s, the other is from 20Ω to 10Ω when t=0.4s. According to Fig.4,we can see the output voltage reach its steady state with some overshoots and it spends more time to stabilize compared with the converter in Fig.3 when the load resistor changed. So the hybrid feedback switching rule appropriates good performance of transient and steady state dynamic responses.

6 Conclusion

In this paper, the model of Buck Converter in DCM, which is based on the concept and theory of SLS, is built first and foremost. Then the quadratic stability and switching rule of it are studied. Afterwards, simulation results are presented. This approach can also be extended to other converters and port controlled Hamilton linear switched system. The research results are beneficial to the development of nonlinear control strategy and the practical applications of power electronic systems.

References

1. Z Sun,D Zheng,On reachability and stabilization of switched linear control systems. IEEE Transactions on Automatic Control, Vol. 4(2001).
2. G Xie, D Zheng, L Wang: Controllability of switched linear system. IEEE Transactions on Automatic Control, Vol. 47(2002).
3. Xie, L Wang. Controllability and stabilizability of switched linear systems[J]. Systems&Control Letters, (2003).
4. Xuesheng Zheng,Chunwen Li, and Yuanjie Rong, "Hybrid Dynamic Modeling and Model Predictive Control for DC/AC Converter," *Transactions of China Electrotechnical Society*, vol.24,No.7(2009).
5. D Liberzon, A S Morse, "Basic problems in stability and design of switched systems," IEEE Control Systems Magazine, Vol.19, No.5(1999).
6. Yimin Lu, Bo Zhang, and Liyun Yin: Switched Affine Systems Modeling and Control of DC/DC Converters. Proceedings of the CSEE, Vol. 28(2008).
7. Paolo Bolzern and William SPinelli: Quadratic stabilization of a switched affine system about a nonequilibrium Point. Proceedings of the American Control Conference, (2004).
8. A.F. Filippov: Differential equations with discontinuous right-hand sides. Kluwer Academic (1988).

MATEC Web of Conferences **68**, 12006 (2016)
DOI: 10.1051/matecconf/20166812006
ICIEA 2016

The Design of Phase-Locked-Loop Circuit for Precision Capacitance Micrometer

Shujie Li [1], Xiang Deng [2]

[1] *School of Mechanical, Electronic and Control Engineering, Beijing Jiaotong University, Beijing, 100044, P. R. China*
[2] *Beijing Fei Na Ying Chuang Measurement, Control Technology & Instrumentation Co., LTD, Beijing, 100098, P. R. China*

Abstract. High precision non-contact micrometer is normally divided into three categories: inductance micrometer, capacitance micrometer and optical interferometer micrometer. The capacitance micrometer is widely used because it has high performance to price ratio. With the improvement of automation level, precision of capacitance micrometer is required higher and higher. Generally, capacitance micrometer consists of the capacitance sensor, capacitance/voltage conversion circuit, and modulation and demodulation circuits. However, due to the existing of resistors, capacitors and other components in the circuit, the phase shift of the carrier signal and the modulated signal might occur. In this case, the specific value of phase shift cannot be determined. Therefore, error caused by the phase shift cannot be eliminated. This will reduce the accuracy of micrometer. In this design, in order to eliminate the impact of the phase shift, the phase-locked-loop (PLL) circuit is employed. Through the experiment, the function of tracking the input signal phase and frequency is achieved by the phase-locked-loop circuit. This signal processing method can also be applied to tuber electrical resistance tomography system and other precision measurement circuit.

1 Introduction

When measuring the micro displacement with the capacitive sensor, in addition to the weak measurement signal, there are also other noise signals. In order to distinguish measurement signal from noise signals, method of modulating the useful signal is used. In this design, amplitude modulation is employed. At receiving end, modulated signal is demodulated, and the measured signal is eventually reduced. Phase sensitive detection circuit is used in this design. The system block diagram of this design is shown in Fig. 1.

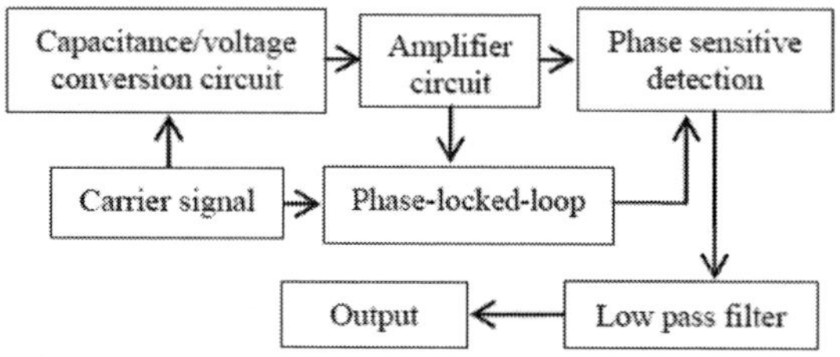

Figure 1. The system block diagram of measurement circuit.

The measured signal is loaded onto the carrier signal, after the capacitance/voltage conversion circuit, the corresponding voltage signal is obtained. In the process of signal transmission, phase shift will be occurred, due to the existing of resistors, capacitors and etc. In order to improve the accuracy of phase sensitive detection, PLL circuit is added. The function of the PLL is making the two signals of the input of phase sensitive detection with the same frequency and same phase. Thereby the output precision of phase sensitive detection and the accuracy of the measurement results will be improved.

2 The principle of PLL for demodulation circuit

Amplitude modulation is widely used because of simple circuit. Carrier signal is generated by the Wien bridge oscillator in this design. The general expression of the linear amplitude modulation signal is:

$$u_s = (U_m + mx)\cos \omega_c t \tag{1}$$

where: ω_c is the angular frequency of the carrier signal, U_m is the amplitude of the carrier signal, and m is modulation depth [1].

In order to obtain the measured signal, the demodulation circuit is needed. There are two ways to demodulation: the envelope detector and phase sensitive detector. In order to make the circuit have the ability to determine the phase and frequency of signal, improve the ability of anti-interference, phase sensitive detection circuit is used. In general, the block diagram of phase sensitive detection system consists of multiplier circuit shown in Fig. 2.

DOI: 10.1051/matecconf/20166812006

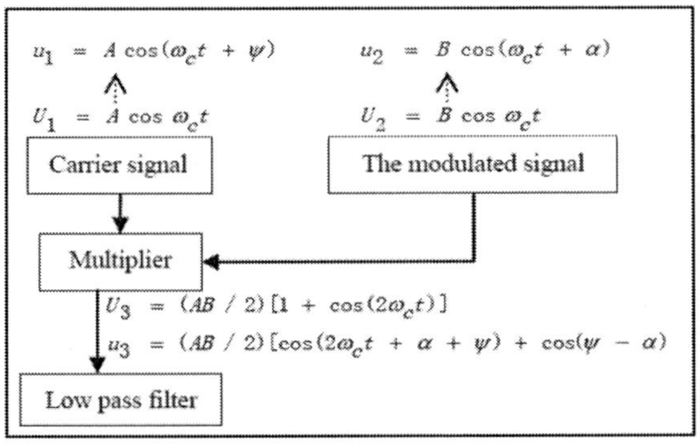

Figure 2. The block diagram of phase sensitive detection system.

Two inputs of the multiplier are carrier signal and the modulated signal. Under the ideal conditions, the carrier signal is written as:

$$U_1 = A\cos\omega_c t \tag{2}$$

where: A is amplitude of carrier signal.

The modulated signal is expressed as:

$$\begin{aligned} U_2 &= (U_m + mx)\cos\omega_c t \\ &= B\cos\omega_c t \end{aligned} \tag{3}$$

where: B is amplitude of the modulated signal.

The output of multiplier is:

$$\begin{aligned} U_3 &= AB\cos\omega_c t\cos\omega_c t \\ &= \frac{AB}{2}[1 + \cos(2\omega_c t)] \end{aligned} \tag{4}$$

In practice, the phase shift will occur in circuit. The phase shift of the carrier signal is ψ, the phase shift of the modulated signal is α, then:

$$u_1 = A\cos(\omega_c t + \psi) \tag{5}$$

$$\begin{aligned} u_2 &= (U_m + mx)\cos(\omega_c t + \alpha) \\ &= B\cos(\omega_c t + \alpha) \end{aligned} \tag{6}$$

Under such circumstances, the output of the multiplier is drawn as:

$$\begin{aligned} u_3 &= AB\cos(\omega_c t + \psi)\cos(\omega_c t + \alpha) \\ &= \frac{AB}{2}[\cos(2\omega_c t + \psi + \alpha) + \cos(\psi - \alpha)] \end{aligned} \tag{7}$$

According to formula (7), even if the high frequency signal is filtered by the low pass filter, the output is also related to the size of the phase shift. In order to eliminate the impact caused by the phase shift, the phase-locked-loop is added in this design.

The system block diagram of phase-locked-loop method is shown in Fig. 3 [3].The principle of PLL is using the phase difference between two signals to control the frequency of the output signal of the voltage controlled oscillator (VCO). Ultimately, the frequency of the output signal equals to the input signal, and the phase difference is zero or a fixed value. The input signal is described as:

$$u_i(t) = U_{im}\sin[\omega_i t + \varphi_i(t)] \tag{8}$$

where: ω_i is instantaneous oscillation angular frequency, and $\varphi_i(t)$ is instantaneous phase.

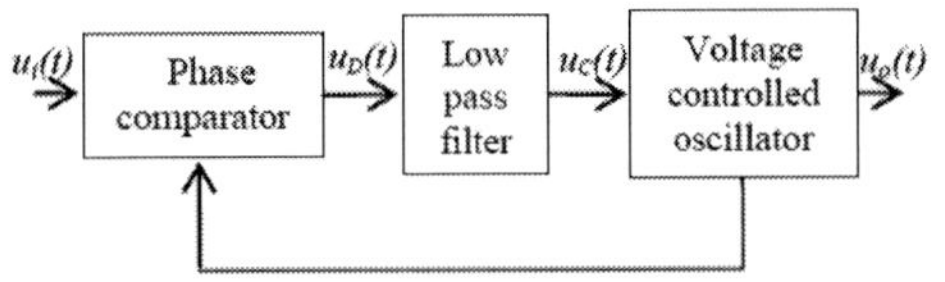

Figure 3. The system block diagram of phase-locked-loop method.

The output of the VCO can be obtained:

$$u_o(t) = U_{om}\cos[\omega_o t + \varphi_o(t)] \tag{9}$$

where: ω_o represents natural oscillation angular frequency, and $\varphi_o(t)$ is instantaneous phase.

The output of the phase comparator is [5]:

$$u_D(t) = Ku_i(t)u_o(t) \tag{10}$$

The output of the low pass filter is described as:

$$u_C(t) = \frac{1}{2}KU_{im}U_{om}\sin[\omega_i t + \varphi_i(t) - \omega_o t - \varphi_o(t)] \tag{11}$$

The relationship between the instantaneous frequency and phase is written as:

$$\omega(t) = \frac{d\theta(t)}{dt} \tag{12}$$

From formula (12), it can be deduced:

$$\theta(t) = \int \omega(t)dt + \theta_{d0} \tag{13}$$

Instantaneous phase difference is:

$$\theta_d = (\omega_i - \omega_0)t + \varphi_i(t) - \varphi_o(t) \tag{14}$$

After differential, we can get:

$$\frac{d\theta_d}{dt} = \frac{d(\omega_i - \omega_0)t}{dt} + \frac{d[\varphi_i(t) - \varphi_o(t)]}{dt} \tag{15}$$

When formula (15) is zero, the frequency of output equals to the input, and the phase difference is zero or a fixed value.

In this design, the phase-locked-loop method is applied to the demodulation circuit, and the system block diagram is shown in Fig. 4. From the above analysis, we know that u_4 and u_2 are signals with the same frequency and same phase. So the output of multiplier is written as:

$$\begin{aligned} u_5 &= BC\cos(\omega_c t + \alpha)\cos(\omega_c t + \alpha) \\ &= \frac{BC}{2}[\cos(2\omega_c t + 2\alpha) + 1] \end{aligned} \tag{16}$$

Then the output of low pass filter is:

DOI: 10.1051/matecconf/20166812006

$$u_6 = \frac{BC}{2} \tag{17}$$

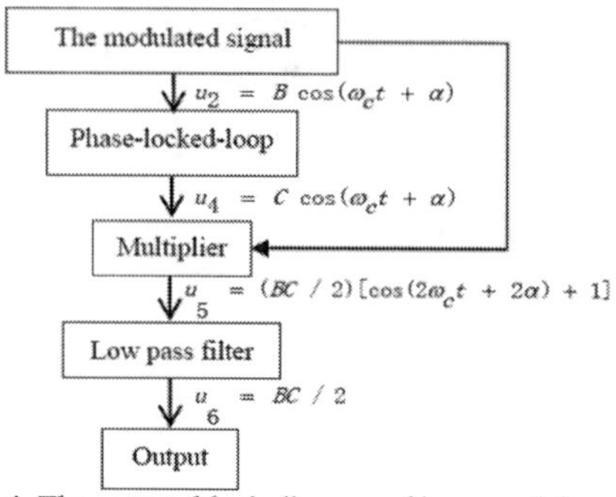

Figure 4. The system block diagram of improved demodulation circuit.

3 The parameters design of PLL circuit

CD4046 chip is employed in the PLL circuit design. The low pass filter of PLL circuit requires external connecting of resistors and capacitors [6]. The left picture of Fig. 5 shows the diagram of the designed PLL circuit.

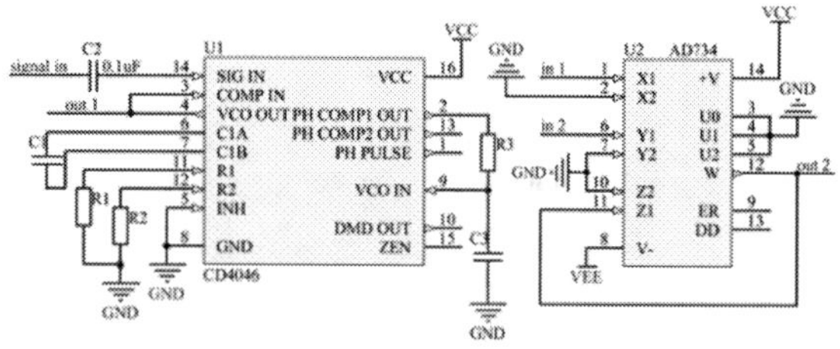

Figure 5. The circuit of PLL and multiplier.

In this design, R_3 and C_3 are used as consisting of low pass filter. The value of R_3 and C_3 are determined by the time constant of low pass filter. After the frequency capturing range of PLL is acquired, the value of R_1, R_2 and C_1 can be roughly calculated by the formula (18) and (19). Finally, the selected accurate value can be amended by experiment.

$$f_{\min} = \frac{1}{R_2(C_1 + 32pF)} \tag{18}$$

$$f_{\max} = \frac{1}{R_1(C_1 + 32pF)} + f_{\min} \tag{19}$$

In our testing platform, which consists of signal generator, oscilloscope and other devices, through the strictly experiment the selected value of R_1, R_2 and C_1 are achieved:

$$R_1 = 4.7\text{k}\Omega$$
$$R_2 = 82k\Omega$$
$$C_1 = 4400pF$$

Then:

$$f_{\min} = 2.75kHz$$
$$f_{\max} = 50.0757kHz$$

4 Experimental results and analysis

Fig. 6 shows the experimental results of PLL circuit. Signal 1 is the input. Signal 2 is the output of the PLL circuit. The input signal is generated by the sinusoidal signal generator. The output of PLL is square wave, and the value is only related to the power supply voltage. It can be seen that the phase difference between the input and output is almost zero. The frequency of the output equals to the input. And the output of low pass filter is only related to the amplitude of the modulated signal. From the calculation above, it can be known that the frequency range is 2.75 kHz~50.0757 kHz. Through the experiment, it is found that the circuit can track the signal of 10 kHz~40 kHz very well.

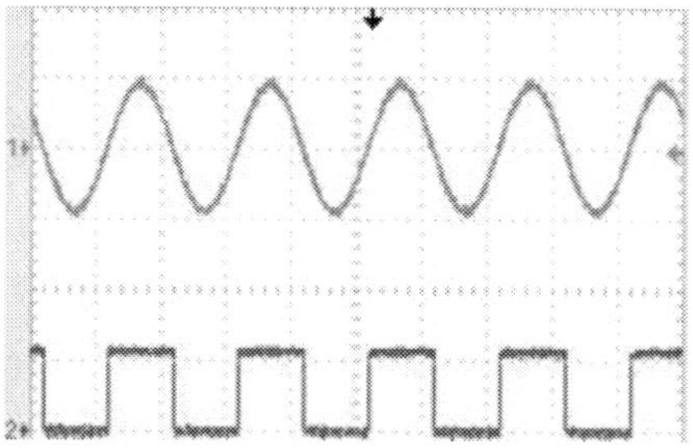

Figure 6. The experimental results of PLL circuit.

Multiplier AD734 is used in this design. It provides the transfer function $W=XY/U$. The circuit diagram of multiplier is shown in Figure 5. Figure 7 shows the experimental results of multiplier. Figure 8 is the output of low pass filter MAX297.

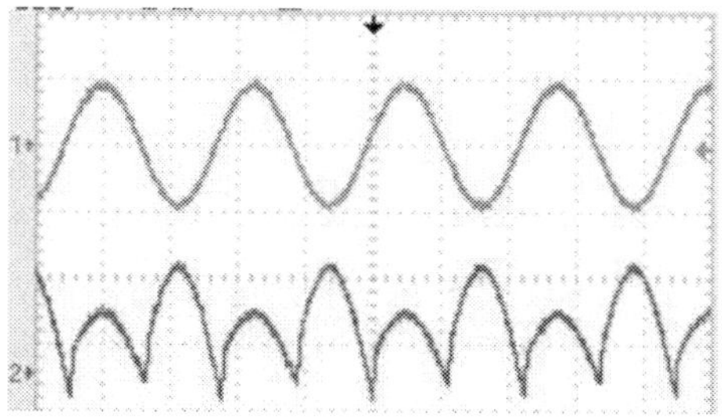

Figure 7. The experimental results of multiplier.

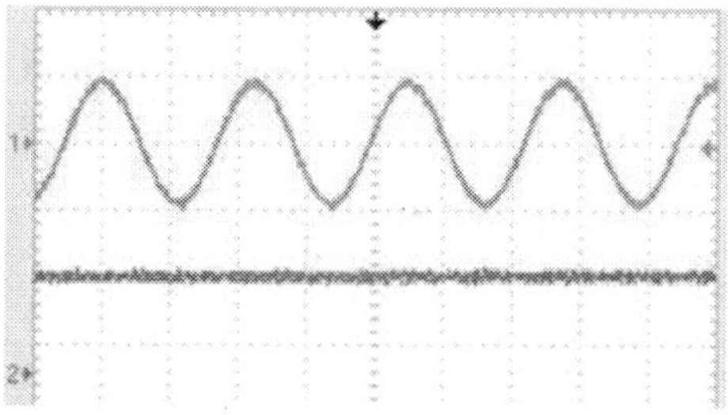

Figure 8. The output of low pass filter MAX297.

From the above experimental results, it can be seen that the applied PLL technology to the phase sensitive demodulation circuit makes the measurement circuit has high accuracy of measurement and stable phase tracking characteristics.

5 Conclusion

As long as the resistors, capacitors exist in the modulation and demodulation circuit, phase shift on the signal will no doubt occurs. And it cannot be eliminated quantitatively because the size of the phase shift is difficult to be measured in general circuit. In the phase-locked-loop circuit, the function of tracking frequency and locking phase are achieved through the feedback loop. After phase-locked-loop added in the phase sensitive detection circuit, the carrier signal is no longer used as reference signal for demodulation, but the output of the PLL circuit. Thus, the frequency and phase of two input signals of the multiplier can be same, and the output of the multiplier is only associated with the amplitude of the input signals. After such modification, the sensitivity of the system enhanced. It also makes the measurement results more accurate. Through the experiment, it is obvious that this method has strong practicability in the precision measurement of capacitance micrometer.

References

1. Zhang, G.X., "Measurement and Control circuit," China Machine Press (2001).
2. Zhu, Y.Q., Deng, X. and Cui, G.L., "Design of dynamic filter for precision vibration measurement," Journal of Electronic Measurement and Instrument, Vol. **23**, Issue 1, pp. 52-57(2009).
3. Zeng, Q.G., "PLL IC Principles and Applications," SHANGHAI SCIENTIFIC & TECHINICAL PUBLISHERS (2001).
4. Lei F, White M H.A preliminary study of the Coherent Phase Synchronous Oscillator (CPSO) for Phase-Locked Loop (PLL) applications[C]//Circuits and Systems (MWSCAS), 2013 IEEE 56th International Midwest Symposium on. IEEE, 2013: 908-911.
5. Zeng, S.Q., "The Application Design and Study of Phase-Locked Loop CD4046," Electronics Quality, Issue 1, pp.72-25(2012).
6. Zhang, K., Pu, J., "Low power consumption phase-locked loop 4046 applied research and simulation," Popular science & technology, Vol. **16**, Issue 6, pp. 124-126(2014).
7. Luo, Y., Zhang, Z. Y. and Yang, N., "Demodulation Technology Based on Phase-locked Loop," Acta Metrologica Sinica, Vol. **36**, Issue 4, pp. 432-435(2015).
8. Pontikakis B, Bui H T, Boyer F R, et al.A novel phase-locked loop (PLL) architecture without an analog loop filter for better integration in ultra-deep submicron SoCs[C]//Circuits and Systems and TAISA Conference, 2008.NEWCAS-TAISA 2008. 2008 Joint 6th International IEEE Northeast Workshop on. IEEE, 363-366(2008).

Research of Electrical Conductivity Measurement System for Mine Bursting Water Based on Dual Frequency Method

Zhou Mengran, He Chenyang, Liu Dong, Wu Xiaoyun

Department of Electric Engineering, Anhui University of Science and Technology, Huainan,China

Abstract. This paper presents a double frequency conductivity measurement method for measuring mine bursting water, to solve the capacitance effect of the conductivity sensor itself has the help. The core controller of the system is the single chip microcomputer ATMEGA128. This paper introduces the basic principle of the measurement of the existing problems and the dual frequency measurement method, and then introduces and analyzes the hardware. To test and analyze the collected data, the double frequency method is found to have good stability and accuracy in the measurement of the electrical conductivity of mine inrush water. It is proved that the method and the system design of the hardware circuit can accurately measure the electric conductivity of the mine inrush water source.

1 Introduction

Mine flood is one of the major hazards in coal mine production safety, mine water inrush occurred suddenly, flooding tunnels quickly, causing economic dangerous, and can causing casualties, it is possible to real-time recognition of mine water inrush sources, grasp the characteristics of underground water, the safety in production of mine has important significance. This design through the measurement of underground water conductivity to determine mine bursting water type and real-time control of water source changes, in order to solve the conductivity cell capacitance can not be measured. In this paper, by using dual frequency determination method to determine the sensor resistance capacitance network charge and discharge time constant, so as to effectively eliminate the influence of distributed capacitance, improving the measurement accuracy of the system greatly [1]-[4].

2 Principle of Measurement

Conductivity sensor measuring principle is to add certain voltage of two parallel plate electrodes into the solution to be measured in, in order to prevent the electrode polarization, usually has a high frequency AC voltage, through the current flowing between the measuring electrode to obtain conductivity. According to Ohm's law, the conductance of the conductor G is the reciprocal of the resistance R, and the resistance of the solution is directly proportional to the distance between the electrode and the L, and is inversely proportional to the cross-sectional area of the electrode A [5]-[7]. As the follow:

$$R = \rho\left(L / A\right) \tag{1}$$

In the follow, ρ is resistivity have 1cm long, resistance conductor cross-sectional area of 1C, the value of which is decided by the nature of the material itself.

The electrical conductivity can be expressed as the next follow:

$$G = 1/R = (1/\rho)\times(A/L) = K\times(1/J) \tag{2}$$

In the follow, K is the electrical conductivity, J is the electrode constant. From the above analysis, we can know that the resistance value of the solution can be obtained.

In the actual measurement, the conductivity sensor and measured the composition of the solution conductivity cell is a complex electrochemical system, there is C_m inter electrode capacitance between two electrodes, between the capacitor and resistor R_m in series, also lead electrodes exist lead capacitance C_n, between the capacitor and the resistance and capacitance composed of series parallel circuit[8-11]. Under normal circumstances, the value of lead capacitor is pF level, and capacitance between electrodes is μF level. $C_m >> C_n$. When two ends of the electrode add high-frequency alternating current, the capacitance C_m is very small, can be neglected [12], [13].

The source of mine bursting water is a solution of low resistance, so available high frequency AC signal as excitation source, the measurement system choice

0~5000Hz square wave signal, neglecting the influence of capacitance between the electrodes, conductance pool available Fig. 2 shows the equivalent circuit.

In the Fig. 3, R is the range switching resistance, U is the sensor output voltage, this voltage is amplified by the signal amplification circuit into the precision half wave rectifier circuit to filter negative voltage, and then for the microcontroller ADC sampling. Therefore, it can get the final signal sampled for DC output by half wave rectification.

Under the action of the sinusoidal signal with frequency ω, the RC parallel impedance is

$$Z = \frac{R(1 - jwRC)}{1 + w^2R^2C^2} \tag{3}$$

Mode of impedance is

$$|Z| = \frac{R}{\sqrt{1 + w^2R^2C^2}} \tag{4}$$

Phase angle is

$$\varphi = -\mathrm{arc}\,tg\left(wRC\right) \tag{5}$$

In parallel model of RC, when ωRC<<1, Z is similar to R. Conductivity meter is mostly used the AC conductance determination method, because of not considering the factors that the phase angle between the voltage and the current, so the measured resistance is used to replace the Z. However, due to the influence of the capacitor R, C is very large, it is clear that the measurement will bring a large error.

The effective value and peak value of the input and output sine wave are measured under the action of the sine signal, and the numerical value obtained by the formula calculation is the reciprocal of the impedance modulus 1/Z. When the frequency of the sinusoidal signal is 2ω, the mode of RC is

$$|Z| = \frac{R}{\sqrt{1 + 4w^2R^2C^2}} \tag{6}$$

When $t = w^2R^2C^2$, $r = |Z|/|Z|'$, have follows

$$r = \sqrt{\frac{1+4t}{1+t}} \quad \text{or} \quad t = \frac{r^2 - 1}{4 - r^2} \tag{7}$$

And have a follow

$$R = |Z| \cdot \sqrt{1+t} \tag{8}$$

To solve the actual water resistance R, there is no theory error, the complete elimination of the influence of AC conductance measurement method of capacitance effect on the precision of measurement, finally get water conductivity, the realization of the determination of conductivity of dual frequency. The equivalent circuit of the conductivity cell is shown in Fig. 1. It can be simplified equivalent circuits are obtained, as shown in Fig. 2.

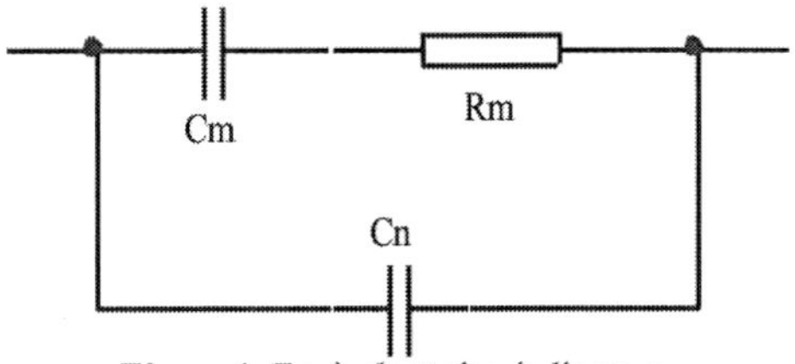

Figure 1. Equivalent circuit diagram.

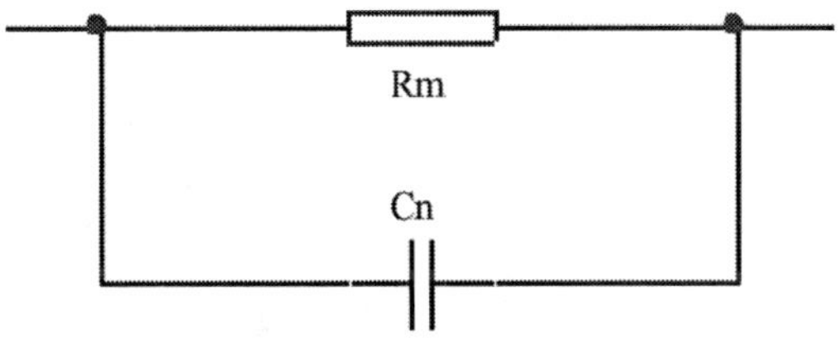

Figure 2. Simplified equivalent circuit diagram.

Thus dual frequency measurement principle diagram as shown in Fig. 3.

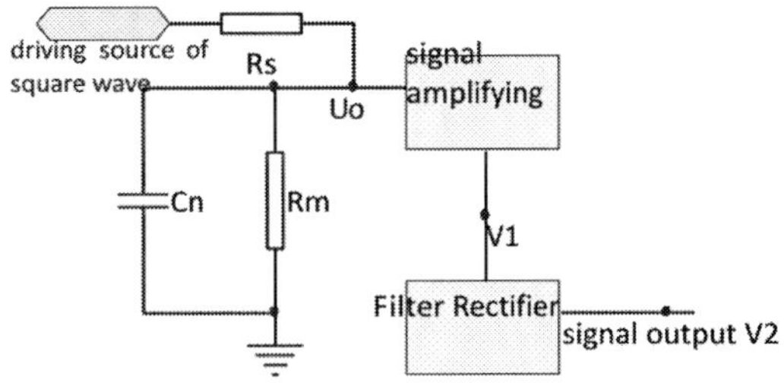

Figure 3. Principle diagram of dual frequency measurement.

3 Test Result Process & Analysis

The 5 water samples extracted from a mining area in Huainan were used as the system to test the water samples. The water samples were collected from 111302 goaf water, 111305 goaf water, 1 layer of sandstone water, 9 layer of sandstone water and ordovician limestone water.

Make 5 times to measure 5 kinds of water were recorded as D1-D5, then use the frequency determination model for the system of sampling data measuring and processing, also for the 5 measurements were recorded as S1-S5. In the end, the two sets of data are compared with the original value of water samples, as shown in Fig. 4.

From Fig. 4, after the measurement group is more obvious than dual frequency measurement model after measuring group of ordinary algorithm more accurate. Compared with the original electrical conductivity, the measurement results of D group were obviously not close to the original conductivity of the S group.

The measurement results of D group and S group were screened respectively, and a group of results that can represent the characteristics of each group were selected. The results and the original values are statistics in Table 1.

Table 1. Measurement result

	111302 goaf water	111305 goaf water	1 layer of sandstone water	9 layer of sandstone water	ordovician limestone water
D	1974.33	3626.13	5319.51	7012.84	4113.47
S	2201.04	3986.29	5894.96	7286.17	4352.28
O	2157.55	4024.53	5940.69	7338.52	4372.28

From Table 1 data shows, without taking into account the conductivity cell is the influence of distributed capacitance, the measured results and the real value has always been rather too small. Because existence of capacitor lead to the conductivity cell voltage across a resistor is rising less than the steady-state value of sake. Taking into account the effect of capacitance, the use of dual frequency measurement method can be clearly seen that the compensation effect, although there are errors, but the range is within ±70, require the measurement accuracy.

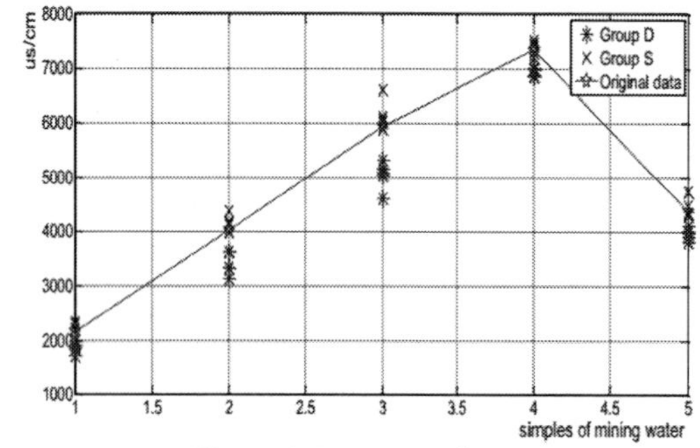

Figure 4. Data comparison.

4 Summary

System has been high-speed ATmega128 microcontroller to build hardware platform, combined with dual frequency determination of influence of distributed capacitance compensation method of sensor, not only improve the measurement accuracy in the system, and improve the system of data processing is fast, the system of mine water inrush early-warning ability can be improved.

Acknowledgements

Supportive fund, Project number. For example, 12th Five-Year key Project supported by the National Science and Technology(Grant No. 2013BAK06B01).

References

1. Liu Y J, Hao Q, Smalley J S T, et al. A frequency-addressed plasmonic switch based on dual-frequency liquid crystals[J]. Applied Physics Letters, 2010, **97(9)**:091101-091101-3.
2. Cai W, Liu B, Duan S, et al. An Islanding Detection Method based on Dual-Frequency Harmonic Current Injection under Grid Impedance Unbalanced Condition[J]. IEEE Transactions on Industrial Informatics, 2013, **9(2)**:1178-1187.
3. Trokhanova O V, Okhapkin M B, Korjenevsky A V. Dual-frequency electrical impedance mammography for the diagnosis of non-malignant breast disease.[J]. Physiological Measurement, 2008, **29(6)**:S331-S344.
4. Hu S, Wu K, Wang H, et al. Electrical conductivity measurement method in seawater desalination based on variable frequency excitation[C]// Electronic Measurement & Instruments, 2009. ICEMI '09. 9th International Conference on. IEEE, 2009:1-810 - 1-813.
5. Ji H, Chang Y, Huang Z, et al. Voidage measurement of gas–liquid two-phase flow based on Capacitively Coupled Contactless Conductivity Detection[J]. Flow Measurement & Instrumentation, 2014, **40.**
6. Cai-Ping L I, Zhang Y M, Zhang Z G. Design of monitoring and control system for water bursting in floor of coal seam[J]. Industry & Mine Automation, 2013.
7. Yan D, Xi Y, Pan L. Coal Floor Water Bursting Risk Evaluation System Based on Water Bursting Coefficient——A Case Study in Gujiao Mine Area, Xishan Coalfield, Taiyuan, Shanxi[J]. Coal Geology of China, 2010.
8. Guo M, Chen X, Liu W, et al. Application of technology combined GIS and transient electromagnetic technique in coal mine water burst prevention — A case of Sihe Coal mine[C]// International Conference on Geoinformatics. IEEE, 2011:1-5.
9. Wang J, Li X, Cui T, et al. Application of distance discriminant analysis method to headstream recognition of water-bursting source[J]. Procedia Engineering, 2011, **26(4)**:374-381.
10. Wei-Wei H U, Zhi-Yuan M A, Cao H D, et al. Application of Isotope and Hydrogeochemical Methods in Distinguishing Mine Bursting Water Source[J]. Journal of Earth Sciences & Environment, 2010.
11. Wu L. An Analysis and Countermeasure under Water Bursting of Xialei Branch Mine of Daxin Manganese Mine under Water Bursting[J]. Chinas Manganese Industry, 2010.

12. Wu Q, Yang L, Zhu B, et al. Application of Vulnerability Index Method in Coal Floor Water Bursting Evaluation in Zhaogezhuang Coalmine[J]. Coal Geology of China, 2009.
13. Hongqing Y U, Xing W. Reasons Analysis and Prevention Measures of Water Bursting From Hidden Subsided Column in Huangcun Branch Coal Mine 10111 Working Face[J]. Safety in Coal Mines, 2014.

MATEC Web of Conferences **68**, 12008 (2016) DOI: 10.1051/matecconf/20166812008
ICIEA 2016

Hardware Design of Tuber Electrical Resistance Tomography System Based on the Soil Impedance Test and Analysis

Shuyi Liu , Xiang Deng , Zili Jiang and Yu Tang

School of Mechanical, Electronic and Control Engineering, Beijing Jiaotong University, Beijing, 100044, P. R. China

Abstract. The hardware design of tuber electrical resistance tomography (TERT) system is one of the key research problems of TERT data acquisition system. The TERT system can be applied to the tuber growth process monitoring in agriculture, i.e., the TERT data acquisition system can realize the real imaging of tuber plants in soil. In TERT system, the imaging tuber and soil multiphase medium is quite complexity. So, the impedance test and analysis of soil multiphase medium is very important to the design of sensitive array sensor subsystem and signals processing circuits. In the paper, the soil impedance test experimental is described and the results are analysed. The data acquisition hardware system is designed based on the result of soil medium impedance test and analysis. In the hardware design, the switch control chip ADG508, the instrumentation amplifier AD620 and programmable amplifier AD526 are employed. In the meantime, the phase locked loop technique for signal demodulation is introduced. The initial data collection is given and discussed under the conditions of existing plant tuber and no existing plant tuber. Conclusions of the hardware design of TERT system are presented.

1 Introduction

Tuber electrical resistance tomography (TERT) originates from industrial ERT technology. TERT system is aimed to extend the application of the technology into agriculture, which could be used in monitoring the status of tuber growth, breeding new varieties of plant tubers and fine cultivation. Compared with the traditional ERT system, TERT system pays more attention to the image reconstruction quality rather than the data acquisition rate. Thus the hardware design of TERT data acquisition, especially in excitation and measurement strategies, is also different from the industrial ERT system [1].

The structure of TERT system showed as Fig. 1 contains: excitation current source module, sensitive array sensor subsystem, switching control module, signal conditioning module and computer of image reconstruction [2]. The data acquisition mode of the system is adjacent mode with four electrodes [3], which is a widely application data acquisition mode in ERT system. There are sixteen electrodes in one layer, and it will get boundary voltage values in adjacent mode [4].

The sensitive array sensor subsystem is 16 electrodes system in adjacent mode, which designed by our research group [5]. The container's diameter is 300mm, and with an electrode array of 6×16 electrodes on the inner wall. The material of the electrodes is titanium, which provides a good electrical conductive performance and corrosion proof. As the overall electrodes are manufactured by using wire-electrode cutting technique, there is no effect of welding stress and deformation. In addition, the electrodes are made by using chamfering dovetail sink whole process, and adhesive on the container with special glue, this made the device has a good water proof and immersion proof performance. The customized connect cable has very small impedance value and excellent anti-jamming performance.

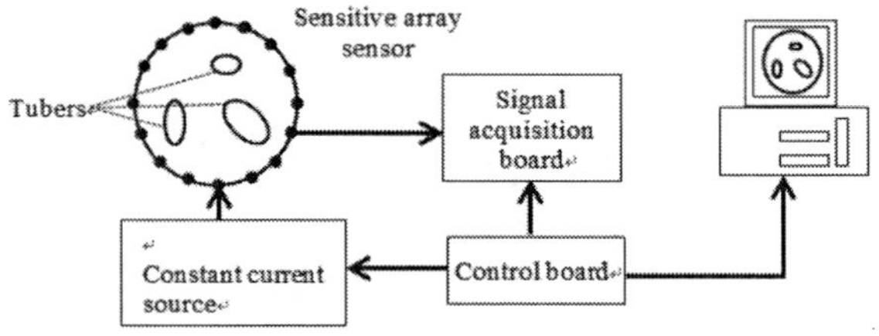

Figure 1. TERT system structure.

2 Soil impedance test and analysis

The measured objective field-the medium of soil in the TERT system is different from which of traditional ERT system. Its resistivity numerical is high, distribution is not homogeneous, and humidity impact, soil quality and other factors cannot be ignored [6]. So, in the hardware design of TERT data acquisition system, the soil impedance characteristics test and analysis is needed.

The test of soil impedances in different moisture content is completed. It gives a basic range of soil impedance and provides a reference in the hardware design. Then the voltage values between the electrodes are measured, this will provide a basic acknowledge of amplification ratio for the subsequent circuits design. The experimental instruments are HP-4284A impedance tester and the 34401A digital multimeter with 61/2 digital resolution. The picture of the test is shown in Fig. 2.

Figure 2. Soil impedance test.

2.1 Soil impedance test

In the Quasi-static field theory, it requires the medium of the field is homogeneous [7]. However, the absolutely homogeneous soil is difficult to achieve because soil multiphase medium is rather complexity. Before the soil impedance formal test the first thing is finding out the ratio of soil and water in saturated soil. The detailed measuring process is: the impedance of adjacent electrodes is continuously measured while water is adding into the container with sensitive array electrodes progressively; when the impedance value of adjacent electrodes is no longer change, which means the soil has reached the saturated state, and then the ratio of water is just about 10.71%. The next is the formal soil impedance test. The measured data in saturated soil is showed in Table 1. It can be drawn in curve graphs as shown in Fig. 3.

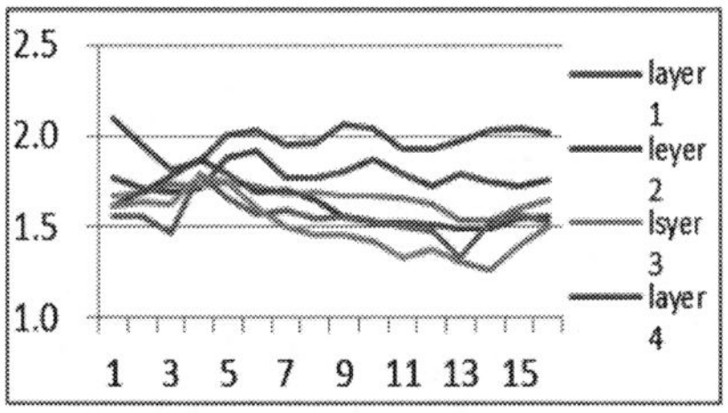

Figure 3. Impedance tested in saturated soil.

From Fig. 3, it can be seen that the resistance of saturated soil is between 2.1kΩ and1.3kΩ .And the soil impedance (resistance value) decreases according to the decreasing of layers' number. The distribution of soil impedance (resistance) is relatively uniform, which is much better than the dry soil, it reaches the requirement of quasi-stationary field theory.

2.2 Measurement of the boundary voltage between each two electrodes

The experimental test aims to get the range of boundary voltage output, and it can provide important guidance in the hardware design. The excitation current source for the test is 30mA, 5 KHz sine wave. The measured voltages between electrodes are shown in Table 2, and the drawn curves are shown in Figure 4.

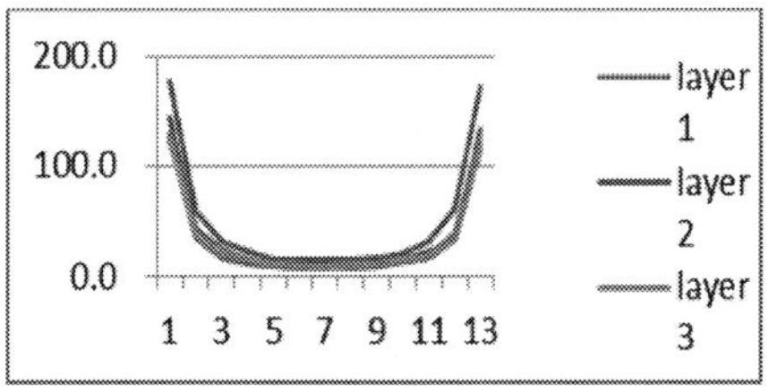

Figure 4. Adjacent electrode voltage change curves.

From the Figure 4, it can be seen that the tendency of voltage values change, i.e., it reduces first and then enlarge symmetrically, it meets the requirement of quasi-stationary field theory, and is in the range of more than a dozen mV and a few hundred mV. As is shown that the maximum voltage value is ten or fifteen times than the minimum one, programmable amplifier circuit should be designed in the data acquisition system.

3 Data Acquisition System Design

The hardware of TERT system is mainly including three parts: constant current source, one signal acquisition board for each layer, and one board of main control circuit. The constant current source can produce sine excitation current signal, and it is designed based on the impedance characteristic of the array electrodes sensing subsystem. There are six pieces of signal acquisition board; each board is used to collection the boundary voltage of one layer's electrodes. As the central control part, the main control circuit board is responsible for the switching process and data transfer control of the whole system [8]. Power supply is also provided on the main control circuit board. Overall system composition is shown in Fig. 5.

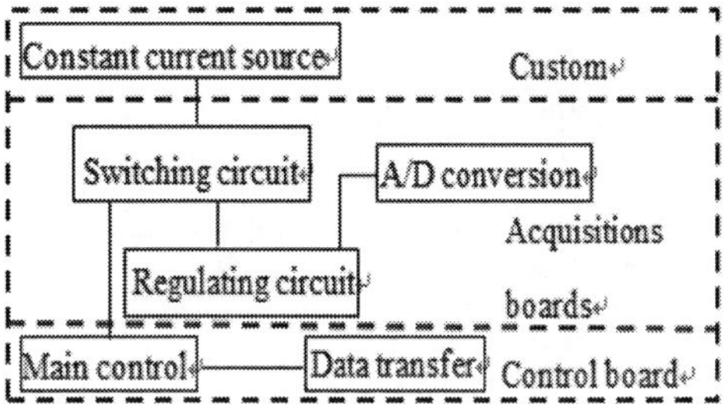

Figure 5. Signal acquisition board architecture .

3.1 Hardware design

3.1.1 Switching circuit

Because the soil impedance is greater than the traditional medium impedance of ERT system, three things should be taken into consideration. First, the intensity of excitation current single should be increased; second, in order to control the bigger excitation current single, the switching chip should withstand a higher voltage value; last, TERT system in data acquisition is more concerned with accuracy, reliability, and less noise ratio rather than speed. So the switching circuit is different from the traditional ERT system. In the switching process design, relays and switching chip ADG508, which is CMOS monolithic analog multiplexers with 8 channels, are employed. The switching circuit for one channel is shown in the Fig. 6.

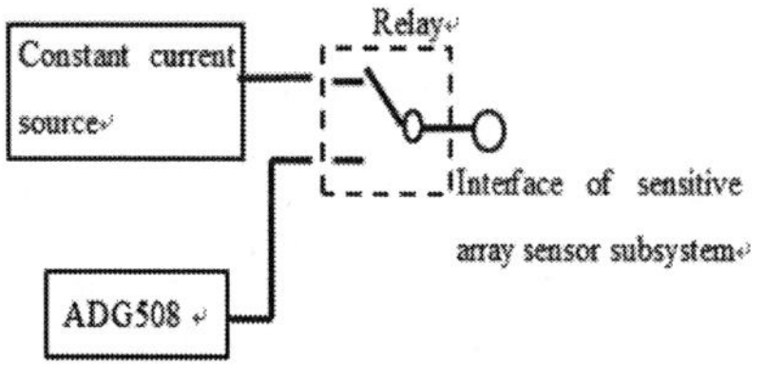

Figure 6. Switching circuit for one channel.

3.1.2 Amplifier and Filter

This part includes three parts: differential operational amplifier circuit, low pass filter circuit, and programmable amplifier circuit. Differential amplifier circuit is used to preliminary amplifier the voltage between electrodes and to make the two way signal input into one way signal output. The AD620, a three-transport discharge scale amplifier chip, is used as a 20 times the first level of amplification [9]. After the differential operational amplifier circuit, a fourth-order filter, which is consisted by two second-order Butterworth filter is used to removing the high-frequency noise introduced by the electronic switch, which always up to a few MHz. Then the AD526, single-chip software programmable gain amplifier, is used to amplify the signal. The AD526 can realize a gain of up to 16 multiples, which covered the multiple among the voltage values, obtained in one ring collection, and it can amplify the voltage values into a similar voltage range effectively.

3.1.3 Multiplication demodulation

In order to convert the AC signal into a DC signal the multiplication demodulation method is used in the system. The Schematic diagram of multiplication demodulation is shown in Figure 7.

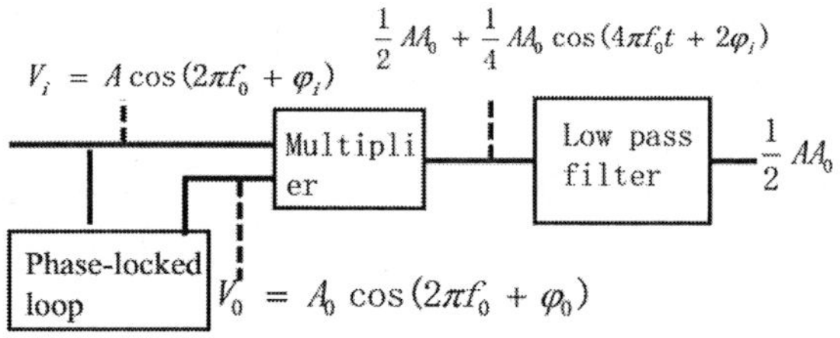

Figure 7. Schematic diagram of multiplication demodulation .

Suppose the demodulated signal is $V_i = A\cos(2\pi f_0 t + \varphi_i)$, the reference signal is $V_0 = A_0\cos(2\pi f_0 t + \varphi_0)$, and V_i multiply V_0 can get V :

$$V = \frac{1}{2}AA_0\cos(\varphi_i - \varphi_0) + \frac{1}{4}AA_0\cos(4\pi f_0 t + \varphi_i + \varphi_0) \quad (1)$$

where φ_i is the phase of the demodulated signal; φ_0 is the phase of the reference signal; f_0 is the frequency of both demodulated signal and reference signal.

MAX297 chip with an 8th-order switched-capacitor filter function is used to filter out the AC signal expressed in formula (1). In order to get a higher precision, a phase-locked loop is used in the system to insure V_0 has a same phase with V_i , this will make $\cos(\varphi_i - \varphi_0) = 1$ in formula (1) [10], and get the largest dc component.

3.2 Initial data collection test and results analysis

In order to determine the feasibility and rationality of the designed hardware of TERT system, the boundary voltages between each two electrodes are measured. The data collection test (only the third layer electrodes) is completed in the condition of existing plant tuber or not in the TERT system. The initial test results are shown in Fig. 8.

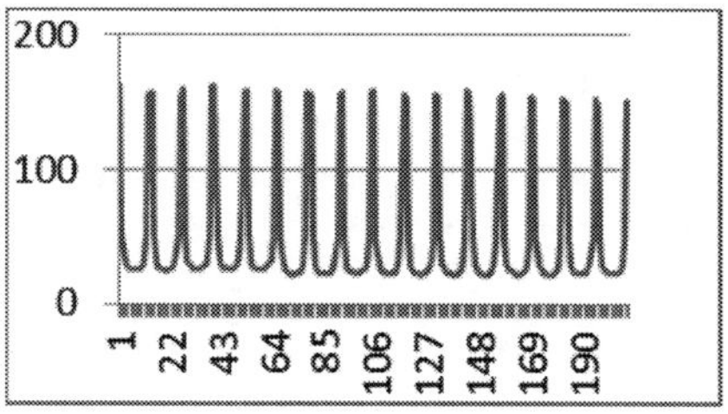

a. No exist plant tuber.

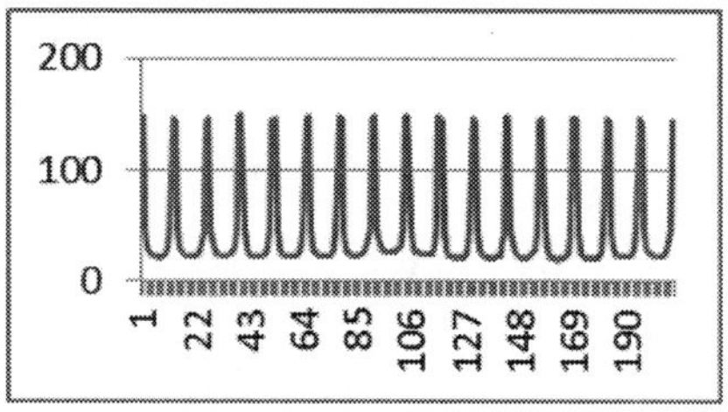

b. Exist plant tuber.

Figure 8. Boundary voltages test of TERT system in existing plant tuber or not.

From Figure 8, it can be seen that the boundary voltages data and curves are very similar to the traditional ERT system. So, after the accurate circuit amplifying, filtering, and signal multiplication demodulating, the image reconstruction of plant tuber by using specific imaging algorithm can be realized.

4 Conclusion

In the soil impedance experimental test and the hardware design, it can be drawn the following conclusions: the ratio of water is about 10.71% in the saturated soil of the test; the impedance value of saturated soil is between 2.1kΩ and 1.3kΩ ; the boundary voltage between each adjacent two electrodes is in the range of more than a dozen mV and a few hundred mV, and the maximum voltage value is ten or fifteen times than the minimum one. Based on the saturated soil impedance test, in the hardware design, the customized excitation current source with 30mA, 5 KHz sine wave and the switching circuit with both relays and switching chips are selected; based on the test of boundary voltage characteristics of the TERT system, the amplifier with a 20 times' three-transport discharge scale amplifier and a software programmable gain amplifier are employed; then a multiplication demodulation based on the phase-locked loop is adopted to transform AC signal into a DC signal. The soil impedance test provides the basis of hardware design, and the initial data collection test verifies the rationality of hardware design.

Acknowledgement

The authors would like to thank the National Natural Science Foundation of China (No.: 61271307).

References

1. Deng, X., Dong, Z., Chen, L., "Research on Excitation & Measurement Strategy of Tuber Electrical Resistance Tomography System", Applied Mechanics and Materials. ISSN: **1660-9336** (2014)
2. Gong, Y., Kong, D., Shang, F., "Design and implementation of wide-range dynamic ERT system based on FPGA", Application of Electronic Technique. **40(2):74-77** (2014)
3. Deng, X., Li, G., Yan, Z., Yang, W., "Theoretical study of vertical slug flow measurement by data fusion from electromagnetic flowmeter and electrical resistance tomography", Flow Measurement and Instrumentation **22: 272–278** (2011)
4. Gong, Y., Kong, D., Shang, F., "Design and implementation of wide-range dynamic ERT system based on FPGA", Application of Electronic Technique. **40(2):74-77** (2014)
5. Zhang, L., Wang, H., "Data Acquisition System of ERT Based on DSP and FPGA", Chinese Journal of Sensors and Actuators. **24(7):1011-1016** (2011)
6. Xiao, Q., Wang, H., Shao, X., "Optimization of finite element model with novel topology in ERT", Chinese Journal of Scientific Instrument. **35(6):1298-1305** (2014)
7. Du Plooy, R., Villain, G., Palma Lopes, S., Ihamouten, A., Dérobert, X., "Electromagnetic non-destructive evaluation techniques for the monitoring of water and chloride ingress into concrete: a comparative study". Materials and Structures**48.12 (1): 369-386** (2015)
8. Zhang, L., Wang, H., "Data Acquisition System of ERT Based on DSP and FPGA", Chinese Journal of Sensors and Actuators. 24(7):1011-1016 (2011)
9. Sun, H., "Pre-processing Module Circuit Design of Weak Signal Detection". Jiang Xi Science. **(4):598-601** (2015)
10. Zhang, K., Pu, J., "Low power consumption phase-locked loop 4046 applied research and simulation". Poplar Science & Technology. **(6):124-126** (2014)

Sustainability or Bust: Malaysian Home Buyers' Stated Preferences for Sus-Tainable Housing

A. Syahid, M. A. Tareq, and S. A. Zaki

Malaysia-Japan International Institute of Technology, Universiti Teknologi Malaysia, Jalan Semarak, 54100 Kuala Lumpur

Abstract. The lack of data on sustainable home buying behaviour in developing countries such as Malaysia is due to the absence of sustainable housing itself. However, it is still possible to solicit home buyers for their stated preferences and quantify its effects on housing demand. In this study, a sample of 300 responses to a Discrete Choice Experiment (DCE) on sustainable housing features was analysed using the "support.CEs" program. This study found that the addition of sustainable features; renewable energy generation, enhanced soundproofing and ventilation, energy saving features, and higher green area ratios significantly increase home buyer's willingness to pay (WTP) for sustainable housing.

1 Submitting the manuscript

The notion of sustainable housing, previously absent from public consciousness, is slowly coming to the fore in de-veloping countries such as Malaysia [1]-[4]. However, de-velopers are yet to catch up to this sentiment change [5], resulting in the unavailability of sustainable housing to match these changing attitudes.

Under-development in the past has blessed Malaysia with potential sustainability gains in energy efficiency and greenhouse gas emissions [6]. However, these gains have not realised because it is more effective to build sustain-ability into new projects [1], including sustainable housing, rather than retrofitting existing buildings. Because of this, it is important to get developers into building sustain-ably sooner rather than later, from the get-go instead of than as a bolt-on to previously less sustainable plans.

A major reason cited by developers for not building sustainable housing is the perceived lack of demand [5], which leads to the current catch-22 on Malaysian sustain-able housing: developers won't build without knowing if home buyers will buy and it is impossible to know if home buyers will buy unless developers build.

However, by adapting Discrete Choice Experiments (DCE), a technique used in non-market valuation [7], to evaluate Malaysian sustainable housing demand it is possible to know the Willingness to Pay (WTP) for the incorporation of sustainable features in homes amongst Malaysian home buyers.

2 Research methodology

2.1 Experimental design

As with all attribute-based methods for eliciting stated preferences, the one first steps in eliciting home buyer preferences for sustainable features in homes is to identify and describe the attributes that define the choices given to respondents [8]. For this purpose, the authors have resorted to the definitions for sustainable housing used in sustain-able building standards, namely the Green Building Index (GBI) [9] to suit the Malaysian home buying public. The authors reasoned that using established standards provides a ready baseline to define "sustainability" in real estate and enhance the applicability of this study's results.

The experimental design of the overall survey is based on the L^{MA} design[1] generated internally from the support.CEs program [11], where the experimental design is directly from the orthogonal main effects plan [12]. This experimental design is generally larger than most orthogonal main effects plan for DCEs [13], which the authors mitigate through effective separation of the choice sets into multiple blocks; subsets of choice sets.

Based on previous evidence, the authors have decided to construct the survey based on six choice sets of two options each containing six attributes of three or two levels, separated into six blocks. Figure 1 is a sample of a question in the survey, where respondents choose options 1, 2, or neither options to indicate their preference in sustain-able housing.

The authors have decided to incorporate pictorials and colour coding to allow the survey's attribute and attribute levels to be more easily understood. This follows the effects of "traffic light" system which enhances the

[1]The L^{MA} design is defined in Viney *et al.* as an experimental design, where L is the number of levels, M is the number of alternatives in each choice set, and A is the number of attributes for each alternative [10].

 DOI: 10.1051/matecconf/20166813001

visibility of pertinent information [14]. Also, having a more "game-like" survey technique, which include more visual rather than textual information, leads to a more enjoyable experience for respondents [15].

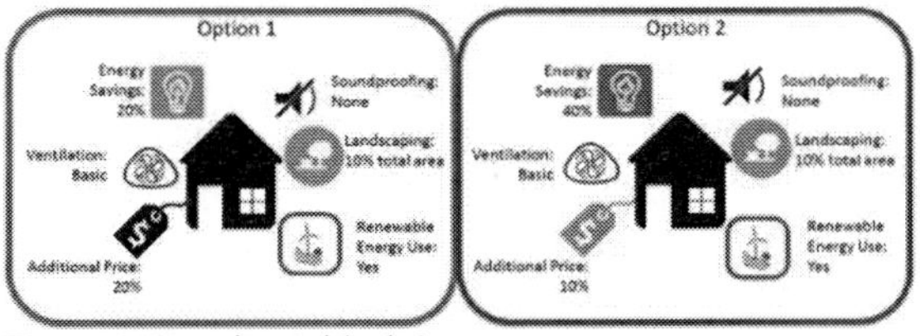

Figure 1. Question 1, block 1.

2.2 Theoretical background

The basis of many attribute-based methods such as DCEs is the assumption that agents would choose alternatives yielding the highest utility, also known as the Random Utility Theory [2] [17]. Assuming respondent i selects alternative j to maximise his or her utility. The utility from making the choice, U_{ij} can be decomposed into:

$$U_{ij}=V_{ij}+e_{ij} \quad (1)$$

where V_{ij} is the systematic component of the utility of respondent i from selecting alternative j, and e_{ij} is the stochastic component of the utility [18]. The systematic component of the utility is assumed to be as follows:

$$V_{ij}=ASC+b_1 EN_{ij}+b_2 SD_{ij}+b_3 VN_{ij}+b_4 LD_{ij}+b_5 R_{ij}+b_6 PR_{ij} \quad (2)$$

where ASC denotes an alternative specific constant for housing choices relative to the 'neither' option, where the systematic component of the utility for the option is normalised to zero. The definitions for other variables are tabulated in Table 1:

Table 1. Variable definition.

Coef.	IV	Type	Definition
b_1	EN_{ij}	Continuous	Percentage of energy saved from the respondent's energy consumption
b_2	SD_{ij}	Dummy	Enhancement of interior soundproofing
b_3	VN_{ij}	Dummy	Enhancement of indoor ventilation
b_4	LD_{ij}	Continuous	Percentage of development area set aside for landscaping and recreational uses
b_5	R_{ij}	Dummy	Production of renewable energy within the development area and usage of renewable energy in common areas
b_6	PR_{ij}	Continuous	Increase in price as a function of the respondent's perception of house price

[2]A further exposition of this theory and its applications to DCEs can be found in Alberini *et al.* [16]

2.3 Survey methodology

This survey was carried out online using Google Forms as its basis, which allowed for branching questionnaires that enabled effective separation of the choice sets into blocks that reduce the number of choice sets faced by respondents, which significantly reduce complexity [19] and lessens respondents' cognitive burden [20].

The preliminary sections of the survey include basic demographic information; age, income, education, and employment which allows the authors to separate the WTP for sustainable features against different demographic groups. Previous studies have shown that demographics affect WTP for sustainable housing features [21], [22], which the authors believe is also the case amongst Malaysian homebuyers.

The following section looks at the respondents' current housing situation, including home ownership, current type of house, and future housing purchase decisions. The most pertinent of these questions is whether future house purchases is for investment or own use, which has been shown to affect WTP for sustainable housing [23].

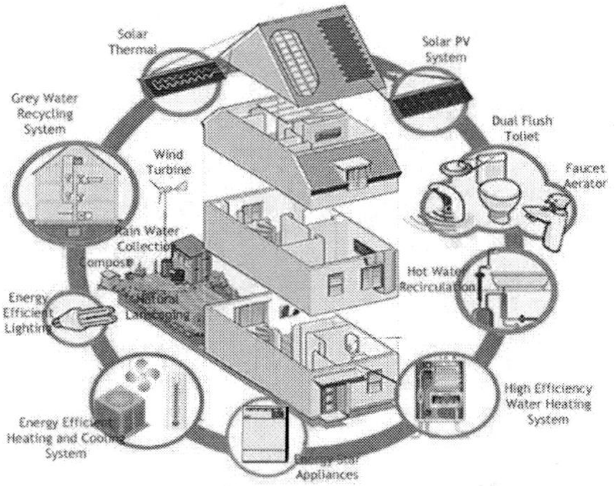

Figure 2. Illustrated list of sustainable features in homes.

This survey also includes a set of primer questions (see Figure 2) that is meant to shift respondents' thoughts towards sustainability and sustainable behaviour. The survey provides respondents with a list of sustainable features and sustainable behaviours [24] for them to acknowledge either knowing or performing.

While this section of the survey is not analysed further, the authors believe it is useful to get respondents who would not have otherwise thought of sustainability and sus-tainable features in homes to consider the breadth of possibilities in sustainable housing.

3 Results

3.1 Respondent demographics and home ownership

This survey sustainable home buying preferences from 50 respondents. A summary of respondent demographics is given in Table 2.

Table 2. Summary of respondents" demographics (number of respondents, n=50).

Category	n	(%)
Age		
18-25	6	(12%)
26-35	23	(46%)
36-45	12	(24%)
46-55	4	(8%)
>55	3	(6%)
Prefer not to say	2	(4%)
Education		
High school or lower	1	(2%)
Undergraduate Degree	24	(48%)
Postgraduate Degree or higher	24	(48%)
Prefer not to say	1	(2%)
Annual Income		
less than MYR 10,000	6	(12%)
MYR 10,000 - MYR 35,000	12	(24%)
MYR 35,000 - MYR 70,000	15	(30%)
more than MYR 70,000	12	(24%)
Prefer not to say	5	(10%)
Employment		
Unemployed	4	(8%)
Student	4	(8%)
White collar	34	(68%)
Blue collar	8	(16%)

Table 3. Summary of respondents" housing circumstances (number of respondents, n=50).

Category	n	(%)
Current house type		
Apartment/Condominium	21	(42%)
Detached/Semi-Detached	8	(16%)
Terrace/Town-house	20	(40%)
Prefer not to say	1	(2%)
Home ownership length		
Not home owner	22	(44%)
Less than 2 years	7	(14%)
Between 2 - 5 years	6	(12%)
More than 5 years	15	(30%)
First home		
True	8	(16%)
False	8	(16%)
Not applicable	14	(28%)
Unknown	20	(40%)
Planned future purchase		
No purchase intention	13	(26%)
MYR 100,000	2	(4%)
MYR 100,001-200,000	5	(10%)
MYR 200,001-300,000	6	(12%)
MYR 300,001-400,000	7	(14%)
MYR 400,001-500,000	9	(18%)
MYR 500,001-600,000	3	(6%)
MYR 600,001-700,000	1	(2%)
MYR 800,001-900,000	3	(6%)
MYR 900,000	1	(2%)

Following these questions, respondents are given a preamble on the main part of the survey and a choice of selecting one of six blocks prepared for this DCE. Each block consists of six questions each with two options and a null option.

Due to the respondent sampling methodology, it was expected that age groups, income, employment, and education demographics are skewed towards young, lower mid-income, white collar university graduates. Majid *et al.* [25] has shown that these demographic factors are major determinants to home ownership and that a previous study on first time home buyers in Syahid *et al.* [26] exhibit these same demographic characteristics.

From a home ownership perspective, this study also evaluated respondents' present housing situation, enquiring about their current house type, ownership status and length, as well as their future housing situation; their bud-get and purpose for future house purchases. These are summarised in Table 3.

Here, we can see that the most respondents are not home owners, but instead are renting the condominiums, apartments, terraces, or town-houses they are currently living in. This is symptomatic of the sampling methodology, which relies on the authors' personal acquaintances through social media channels. Most are willing to spend upwards of MYR 500,000 to purchase their future house.

While it would be interesting to see how demographics such as age and income affect home ownership, limited data prevents competition of this study against more through analyses such as in Tan [27].

3.2 Home buyer stated preferences

Based on a sample of 300 responses[3], the following esti-mations were run using the "support.CEs" program [11]. After the survey responses are encoded into a form legible by the program and supplementing it with previously generated survey design information, the commands are run to provide estimates for Malaysian home buyers' WTP for sustainable features using the conditional logit framework, shown in Table 4.

The results show the estimated coefficient; *coef*, its exponential; *exp(coef)* and standard error; *se(coef)*, z-, and p-values for each variable. The former three form the ba-sis for further WTP estimations while the latter indicate the results' statistical significance, which evidence all vari-ables to be statistically significant. Similarly, the p-value of the likelihood ratio test, which tests the null hypothesis of all coefficients being zero, also show that the estimation as a whole is statistically significant.

Table 4. Conditional logit estimation results.

	coef	exp(coef)	se(coef)	z	p
ASC	0.44	1.55	0.62	0.70	$4.80e^{01}$
No Renewables	-0.85	0.43	0.34	-2.5	$1.20e^{02}$
Enhanced Soundproofing	0.64	1.89	0.35	1.81	$7.10e^{02}$
Enhanced Ventilation	0.44	1.55	0.32	1.36	$1.70e^{01}$
Energy	0.06	1.06	0.013	4.47	$7.70e^{06}$
Landscaping	0.035	1.04	0.02	1.76	$7.80e^{02}$
Price	-0.03	0.97	0.019	-1.58	$1.10e^{01}$

Likelihood ratio test=93.9 on 7df, p=0 n=360, number of events=120

[3] Based on the sample calculation methodology highlighted in Aizaki et al. [12]: Number of respondents questions per respondent

DOI: 10.1051/matecconf/20166813001

Table 5. Marginal WTP estimations.

	MWTP	2.50%	97.50%
No Renewables	-27.802	-234.293	155.671
Enhanced Soundproofing	20.909	-94.779	166.089
Enhanced Ventilation	14.352	-93.299	154.339
Energy	1.829	-9.82	14.629
Landscaping	1.152	-6.594	11.176

method = Krinsky and Robb

Further goodness of fit tests are conducted using the "gofm()" function, with the output as follows:
Rho-squared = 0.36
Adjusted rho-squared = 0.30
Akaike Information Criterion (AIC) = 183.79
Bayesian Information Criterion (BIC) = 203.31
Number of coefficients = 7
Log likelihood at start = -131.83
Log likelihood at convergence = -84.90

The rho-squared and adjusted-rho squared values also indicate a good fit, since it lies between 0.2 and 0.4 [7]. Because this study does not include comparisons between models, the Akaike and Bayesian information criteria will not be analysed further.

The conditional logit estimates are then fed into the "mwtp()" function which calculates the average marginal WTP as well as the two-tailed 5% confidence intervals. The latter is produced using the Krinsky and Robb method [28], which calculates simulated WTP values, sorts these values, and selects values corresponding to the desired confidence levels.

All the marginal WTP estimations show the expected signs for the presence (or absence, as in the case of renew-able energy generation) of sustainable features. All fig-ures are shown as percentage increase or decrease in house prices.The confidence intervals mainly indicate that, given the simulated WTP values generated through the Krinsky and Robb method, that there is 95% confidence the results lie in between the values tabulated in Table 5.

Table 6. Marginal WTP in monetary terms.

	MWTP
No Renewables	-91312.89
Enhanced Soundproofing	68673.52
Enhanced Ventilation	47137.71
Energy	6007.17
Landscaping	3783.63

All figures given are in MYR

4 Discussion

The results indicate that for all sustainable features highlighted, Malaysian home buyers are willing to pay a premium to have these features included in their homes. The survey is worded in such a way to imply that these are new home purchases, therefore these results are not applicable for retrofitting of such features, which is shown to be a different kettle of fish altogether [29]. Unlike buying a new home, more factors need to be taken into account in sustainable retrofitting works [30], thus would require a different approach compared to the relative genericisation of the home buying process used in this study.

This study also chimes with previous works on Malaysian sustainable home buying behaviour. Jusoh *et al.* [2] showed that Malaysian home buyers are keen on sustainable housing features; CO_2 emissions, rainwater harvesting, natural ventilation, and landscaped areas, although the analysis does not show home buyers' WTP for these features individually. Bertrand [1] also gives evidence that the majority of interested sustainable housing buyers in Malaysia are willing to pay a 5% premium over conventional housing.

Because this study uses percentage increases in house prices to represent the cost of including sustainable features in lieu of an actual price [2], it is not possible to replicate real world price comparisons as seen in Park et al. [21]. However, taking a simple average of prices respondents are willing to pay for their next house; MYR 328,440, and using the marginal WTP results produced in this study (see Table 5), it is possible to show how much the average home buyer is willing to pay for the sustainable features highlighted in this study, as tabulated in Table 6. The fact that no studies have been done to compare WTP with market prices highlight the need to have such research conducted amongst Malaysian home buyers.

5 Conclusion

This study has shown that amongst Malaysian home buyers, there exists positive WTP for sustainable housing. This information is presented as percentage premia over the price of conventional housing, which could be useful for developers in deciding the limits they can charge for any one sustainable feature, given current house prices.

While sustainable premia in dollar terms is unclear based on this study, it would be interesting to see if the linear relationship between house prices and sustainable features implied in this study carries over when real prices are applied. Additionally, the respondents in this study are mainly highly educated urban dwellers within the Kuala Lumpur conurbation, will be overcome in future studies.

Acknowledgment

This research was supported by the Malaysian Ministry of Higher Education (MOHE) under the Research University Grant (4F267) project of Universiti Teknologi Malaysia.

References

1. L. Bertrand, *Can Developers Harvest the Benefits of Green Building While Reducing the Risks and Cost of Green Building Accreditation*, in *Conference on Sustainable Building South East Asia* (Kuala Lumpur,

2010), http://www.mgbc.org.my/Resources/Day1/Workshop C/Day1-Workshop C-GreenA Consultants Paper.pdf
2. Z. Jusoh, H. A. Rahim, E. Abu Bakar, A. Jamaluddin, R. Rasdi, Z. Zaini, OIDA International Journal of Sustainable Development, 4, 83 (2012)
3. T. H. Tan, The International Journal of Environmental, Cultural, Economic and Social Sustainability, **7**, 1 (2011)
4. T. H. Tan, *Assessing Green Home Performance: The Case Study from a Developing Coun-try*, in *Asia Pacific Network for Housing Re-search Conference 2013* (Kuala Lumpur, 2013), http://eprints.sunway.edu.my/207/1/Tan Teck Hong - Assessing Green Home Performance.pdf
5. N. Zainul Abidin, N. A. Yusof, H. Awang, International Journal of Environmental, Ecological, Geological and Mining Engineering, **6**, 55 (2012)
6. R. A. Begum, J. J. Pereira, Australian Journal of Basic and Applied Sciences, **4**, 5012 (2010)
7. J. J. Louviere, D. A. Hensher, J. Swait, *Stated Choice Methods: Analysis and Applications*, Cambridge University Presscam, Cambridge, (2000) ISBN 0521788307
8. T. P. Holmes, W. L. Adamowicz, in *A Primer on Non-market Valuation*, edited by P. A. Champ, K. J. Boyle, T. C. Brown, Kluwer, Dordrecht, (2003) ch. 6
9. Greenbuildingindex, Tech. Rep. March, Greenbuildingindex (Kuala Lumpur, 2011) http://www.greenbuildingindex.org/Resources/GBI Tools/GBI RNC Residential Tool V2.0 Final.pdf
10. R. Viney, E. Savage, J. J. Louviere, Health Economics, **14**, 349 (2005)
11. H. Aizaki, Journal of Statistical Software, Code Snippets, **50**, 1 (2012)
12. H. Aizaki, T. Nakatani, K. Sato, *Stated Preference* Methods *Using R*, Chapman & Hall/CRC The R Series (Chapman and Hall/CRC, Boca Raton, 2014), ISBN 978-1-4398-9047-9, http://www.crcnetbase.com/doi/book/10.1201/b17292
13. D. J. Street, L. Burgess, J. J. Louviere, International Journal of Research in Marketing, **22**, 459 (2005)
14. F. Mawad, M. Trías, A. Giménez, A. Maiche, G. Ares, Food Research International, **74**, 1 (2015)
15. K. Ogawa, S. Hattori, T. Fukushima, **2**, 18 (1993)
16. A. Alberini, A. Longo, M. Veronesi, in *Valuing Environmental Amenities Using Stated Choice Studies: A Common Sense Approach to Theory and Practice,* edited by B.J. Kanninen (Springer Netherlands, Dordrecht, 2007), vol. 8, ch. 8, pp. 203-227, ISBN 978-1-4020-4064-1, http://link.springer.com/10.1007/1-4020-5313-4, http://opus.bath.ac.uk/9816/
17. D. McFadden, in *Frontiers in Econometrics*, edited by P. Zarembka, Academic Press, New York, (1974)
18. H. Aizaki, T. Nanseki, H. Zhou, Animal Science Journal, **84**, 82 (2013)
19. D. A. Hensher, Journal of Transport Economics and Policy, **38**, 425 (2004)
20. J. Rolfe, J. Bennett, Ecological Economics, **68**, 1140 (2009)
21. M. Park, A. Hagishima, J. Tanimoto, C. Chun, Building and Environment, **60**, 225 (2013)
22. C. Marmolejo-Duarte, M. Ruiz-Lineros, Journal of European Real Estate Research, **6**, 63 (2013)
23. S. L. Heinzle, A. Y. Y. Boey, M. Y. X. Low, Urban Studies, **50**, 1970 (2013)
24. W. Poortinga, L. Steg, C. Vlek, G. Wiersma, Journal of Economic Psychology, **24**, 49 (2003)
25. R. Majid, R. Said, M. N. Daud, Journal of Surveying, Construction & Property, **3**, 1 (2012)
26. A. Syahid, M. A. Tareq, *The elephant in the room : Sustainability and housing demand in Asia*, (2015)
27. T. H. Tan, Habitat International, **32**, 318 (2008)
28. I. Krinsky, A. L. Robb, The Review of Economics and Statistics, **68**, 715 (1986)
29. L. Vlasova, K. Gram-Hanssen, Building Research & Information, **42**, 512 (2014)
30. B. Dong, C. Kennedy, K. Pressnail, Canadian Journal of Civil Engineering, **32**, 1051 (2005)

MATEC Web of Conferences **68**, 13002 (2016) DOI: 10.1051/matecconf/20166813002
ICIEA 2016

Application of the BIM Technology in the HVAC Design for an Office Building in Nanjing

Ren Lu-lu, Yu Yue-jin, Luo Lang

School of Energy and Mechanical Engineering, Nanjing Normal University, Nanjing, Jiangsu

Abstract: Based on the 2-dimensional design, this paper takes an office building in Nanjing as an example to use the BIM technology for the 3-dimensional optimization. With the detailed layout of pipelines, it sets the elevation to replan the original unreasonable designing. Then based on the BIM technology, it uses Navisworks to carry out the collision detection, puts forward the adjustment scheme, and gives types of invalid collision according to the practical experience. And finally it sums up some experiences, which provide optimization scheme for the future design.

1 Introduction

Building information modeling (BIM) is a kind of volume modeling founded by Autodesk company, which sets the preliminary scheme, the construction of the mid and the late operation and management of the project as a whole, and covers all parameters of the project (including construction information, equipment information, construction cycle planning, real estate operation and management, etc.). Now the BIM technology has been gradually changed from theoretical research to practical engineering application, and for construction projects, improving the quality of the solution and the accuracy of the design can reduce the unnecessary investment from all sides in the process of construction, shorten construction life cycle, save resources and reduce construction cost.

Based on an office building project in Nanjing, this paper uses BIM 3D modeling technology to redefine the elevation, to precisely arrange the pipeline and equipment, and to deepen the 2D design. Aimed at the "fight" of the previous pipeline, with the Navisworks software, this paper carries out collision detection, gives solutions to specific problems, and reduces the phenomenon of rework in the later construction, which truly has clear modes and is adjustable on the basis of visualization.

2 Engineering situation

This project is a six-storey office building, the building area of which is about 32680.7 m2 and the building height of which is 23.4 m. The office building is divided into two portions: the overground portion and the underground portion. The building area of the overground portion is 22200 m2, which is mainly used as offices, conference rooms, lecture theatres, and sports places and related facilities; the underground portion is used as garages and equipment rooms, the building area of which is 10000 m2. The architectural renderings of this project are shown in Fig. 1.

Figure 1. Architectural renderings

3 Deepening design of an office building in Nanjing based on BIM

3.1 2-dimensional design

3.1.1 Load calculation

The HVAC load of this project is calculated by Tengen software, and the calculated load and the load indicator of construction area of different seasons are shown in Table 1.

Table 1. Load calculation

	Summer	Winter
Calculated load	2758KW	1686KW
Load indicator of construction area	125W/m2	77W/m2

3.1.2 Scheme of the HVAC system

The project consists of a ground floor and 6 layers over ground, in which offices and conference rooms use the primary air fan-coil system, cooperation company offices, general classrooms, blueprinting rooms, printing rooms, reading rooms and the Network Center use the multi-split air fan-coil system, the sports field uses the all air system, and fire duty rooms use unit split air conditioners. The air-cooled liquid chillers and multi-split outdoor units are placed in the 6th floor roof and the 4th floor roof respectively, as shown in Fig. 2 and Fig. 3.

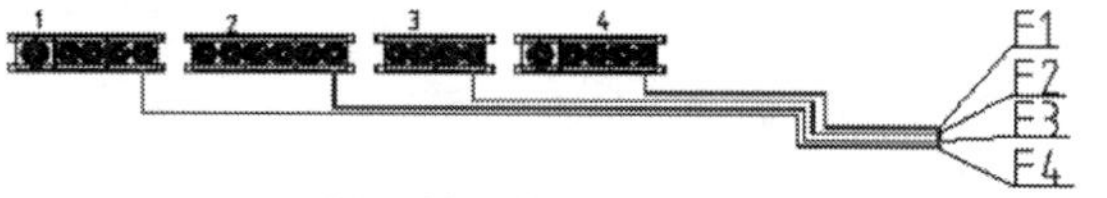

1-4 Multi-split outdoor unit

Figure 2. The conceptual drawing of outdoor units in the 4th floor roof

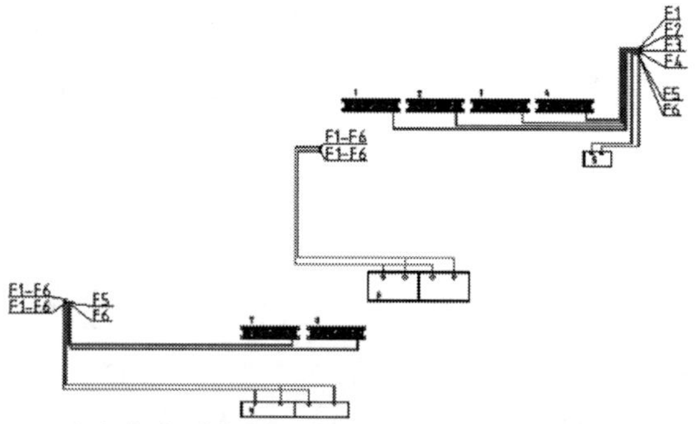

1-4, 7, 8 Multi-split outdoor unit 5, 6, 9
Air-cooled screw compressor water chiller

Figure 3. The conceptual drawing of outdoor units in the 6th floor roof

3.2 Deepening design based on BIM

3.2.1 Flow of the deepening design (as shown in Fig. 4)

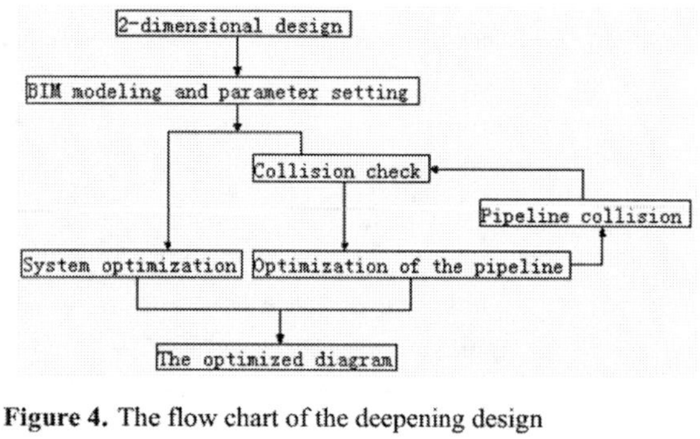

Figure 4. The flow chart of the deepening design

3.2.2 System optimization

There are many pipelines aside mechanical equipments, and because of the simple location layout of the 2-dimensional design, it cannot effectively define the height and lacks the concept of 3-dimensional space, leading to some unreasonable designs [1]. As shown in Figure 5, there are the air and exhaust system and the refrigerant system, including the heat pump air blower unit, the air supply pipe, the exhaust pipe, the condensate pipe, the refrigerant line, the air terminal, etc.

There are many pipeline elbows in the original design, leading to larger system resistance, so the system needs to be optimized. In the system optimization, we readjust the elevation of the original equipment, which is 300mm upwards, and make it closer to the floor, ensuring that there is enough space to arrange the pipeline below the equipment. Pipelines in the air supply pipe outlet and in the air return pipe inlet are respectively inclined 640mm and 430mm downwards, which ensures that the pipeline is below the beam, without a collision. There are 8 elbows in the original design of the air supply pipe and the exhaust pipe, the multiple local resistances of which make the system resistance larger; and in the optimization design of the air supply pipe and the exhaust pipe, there are 6 elbows, which reduces local resistances, and the length of the wind pipe becomes shorter, reducing the on-way resistance. The 3D modeling technology based on BIM optimizes the complex pipeline layout, reduces the systemic resistance, and provides an intuitive and effective design scheme for the construction.

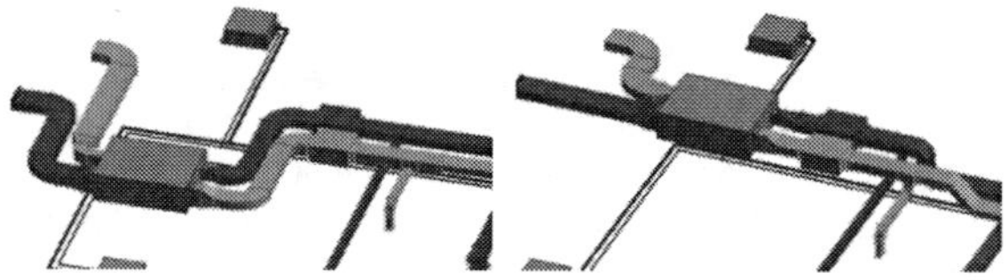

Figure 5. The original system design and the effect after adjustment

3.3 Optimization of the pipeline

Navis Works software is a 3D simulation software by Autodesk Inc., which can read the model data created by BIM 3D software after the extension of data interface, and can provide simulation and analysis for the model. The software consists of 7 function modules, in which the clash detective module can identify collision problems between the construction and the pipeline, and collision problems between pipelines [2]. Collision check is divided into effective collision and invalid collision [3], which need artificial screening.

3.3.1 Effective collision

There are numerous pipelines in the heating and ventilation design, including the air supply system, the air return system, the exhaust system, the circulation supply and return water system, the refrigerant system, the condensate system, the smoke control system, the cooling water system and other subsystems. How to effectively arrange the complex pipelines becomes a hard problem to solve for the 2-dimensional design. If the pipelines are placed improperly, there will be pipeline collisions at the construction site, which will cause the phenomenon of rework, resulting in waste of construction materials, efficiency decrease, extension of the project schedule and so on. However, based on BIM, the collision detection technology in NavisWorks software can effectively solve these problems.

The 3D models are established in the pre-construction, in which all pipelines are intuitively displayed, and the elevation of equipments and pipelines are set. The BIM models are firstly imported into NavisWorks software, and then each floor is checked as a unit for collision check, generating a list of collision points; and aimed at different collisions, it redefines the elevation, draws the pipelines, and finally uses different schemes to solve the problems. The pipeline collision and adjustment are shown in Fig. 6.

Figure 6. The pipeline collision and adjustment

3.3.2 Invalid collision

The so-called invalid collision is a "collision" appeared in the collision check, which is not because of the faulty design. The BIM technology in our country is in the initial stage of development, and due to the unskilled use of the software by designers, there will be errors in operation; the software itself exists some bugs, and the library family is not yet perfect, so invalid collisions appear in the collision check. These "collisions" need carefully artificial screening, and usually they don't need to be adjusted.

Figure 7. The disoperation

Fig. 7 shows an invalid collision, which is generated by a faulty operation in the drawing process. Fig. 8 is a "collision" produced by the software bug. When connecting the air terminal with the air pipe, it selects the property of the air duct as "connector connection", which will be a "collision" in the collision check; this situation is because of the own problem of the software, which doesn't need to be adjusted. Fig. 9 is a "collision" caused by no suitable connections. When the nominal diameter of the pipeline is larger and the bent downward slope is smaller during drawing downward inclined pipelines, it often meets with a situation which cannot be drawn out; the reason is that the library family of the BIM software is not yet perfect, and sizes of the air pipe accessory and the fitting are just specified, causing a great inconvenience to the design process. When it cannot find the right connection pipe during the drawing process, it will be a "collision" in the collision check, which is not the true sense of the collision and can be ignored.

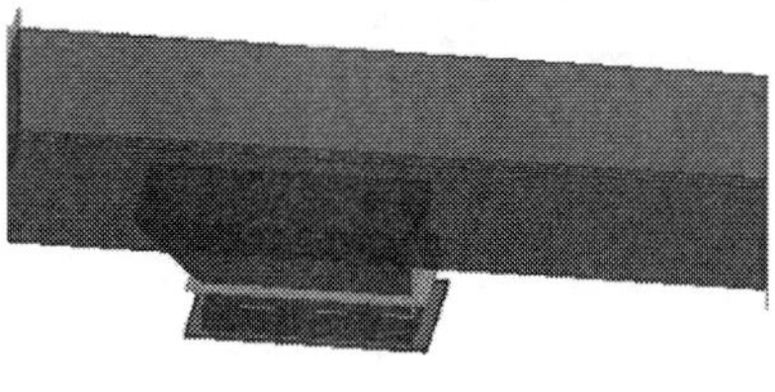

Figure 8. The software bug (connector connection)

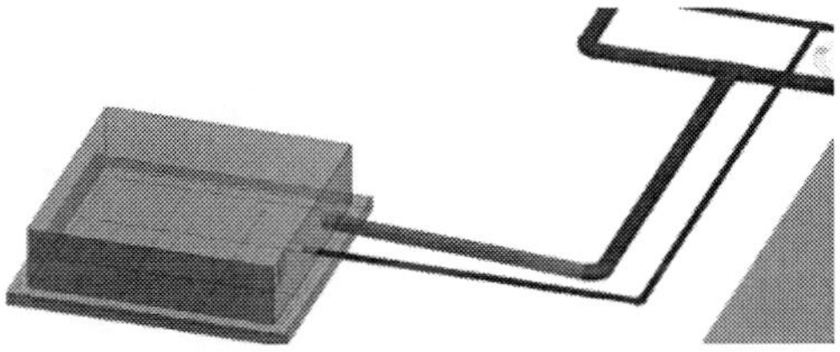

Figure 9. A "collision" caused by no suitable connections

4 Experiences and conclusions of the BIM technology

Compared with the 2-dimensional technology, the application of the BIM technology in HVAC design involves more information, visualization and digitization, and the HVAC design from 2D to 3D takes a revolutionary step. In the 3D models, there is all detailed information of the equipments and pipelines, including size, performance, material, elevation and so on, which greatly improve the design accuracy[4]. The NavisWorks collision check function based on BIM shows the collisions between pipelines, the pipeline and the equipment, the pipeline and the construction. At the design stage, the designers' quickly, precisely and effectively solving collision problems can effectively reduce the construction rework rate and shorten the construction period.

At present stage, the BIM technology is in the initial stage of development, and related standards are not perfect. The BIM technology is not used by the majority of designers. Parts of the domestic design units establish their own material library and library family, but they are only used by their own designers, not shared with the outside world, which restrict the development of BIM to some extent. Hope that the software can improve the library family, perfect the material library, improve the load calculation, combine with the energy consumption analysis software Energy Plus or indoor environment simulation software Fluent etc., make full use of its information advantages and finally achieve the purpose of building energy conservation [5].

References

1. W. J. Wang, Application of BIM Technology in Electromechanical Installation of a Nanchang Station

Building, Construction Technology, 44(12), 90-92, (2015)
2. N. N. Wang, Q. C. Wang, F. Wang, Key Technology in the Modeling and Application Process of Construction BIM. Journal of Dalian Nationalities University, 17(5), 495-499, (2015)
3. K. D. Liu, Y. C. Zhang, L. J. Chen, Research on Subway Station Pipeline Installation Collision Based on BIM Technolog, Journal of Civil Engineering and Management, 32(1), 53-58, (2015)
4. D. G. Dong, Y. D. Cai, J. Zhang, Application of BIM technique to HVAC design. Heating Ventilating & Air Conditioning, 43(12), 105-109, (2013)
5. Z. M. Pan, X. Y. Guo, Applied analysis of Revit MEP in HVAC engineering design, Fuiian Architecture & Construction, 11, 107-109, (2013)

Modeling Dubai City Artificial Channel

Mohamed Elhakeem and Mohamed El Amrousi

Abu Dhabi University, Abu Dhabi, P.O. Box 59911, UAE

Abstract. Dubai's new channel further enhances the urban-scape of the city offering new waterfront developments, transportation venues and diversified panoramas to the city. This paper performs a study to simulate the flow field in the proposed Dubai artificial channel using a 2D hydrodynamic model. The model predicts the flow depth and velocity in the channel, lagoons and bends. The model predictions show that the velocity is higher in the channel sections compared to the lagoons and bends sections. On the other hand, the water depth is lower in the channel sections compared to the lagoons and bends sections. Nonetheless, the velocities in the channel are within the accepted range that prevents boundary erosion and sediment deposition.

1 Introduction

Like several modern Arabian Gulf cities, Dubai is interested in creating landmarks to attract investors through display of its ability to create architectural spectacles that can be part of the World Genius Record. Among these landmarks are the artificial islands of Palm Jebel Ali and the Palm Jumeirah that shape Dubai new shoreline (Figure 1). The Dubai Palms increased the emirate's shoreline by a total of 520 km promoted the establishment of 100s luxury hotels, residential beach-side villas and apartments, marinas, water theme parks, restaurants, shopping malls, sports facilities and health spas.

Figure 1. Jumeirah hotel and residences around its artificial canals and lagoons.

Dubai's new waterfront developments in the form of artificial Islands, lagoons and water channels are all part of branding Dubai as part of the emerging network of modern global cities such as Kuala Lumpur and Singapore. The marketing power of Dubai as a brand name and lifestyle by EMAAR and NAKHEEL, and the instability created by the Arab Spring in cities like Cairo, Damascus that were shaped by multi-layers of history and collective memories have contributed further to Dubai as a city spaces of hope for the future [1].

The investment to create a main water channel from the boundaries of Diera onwards manifests the constant desire of Dubai authority to create new real estate properties along this channel as part of Dubai developments. The project includes digging of the water channel, lining of its boundaries and constructing a number of bridges to cross the channel. The project includes also the construction of 12 marine transit stations to enable convenient and effective water-transportation to link Dubai's several man made islands with the downtown. According to the Roads and Transport Authority (RTA), the channel will offer new means of public transportation for one million of Dubai residents and will become part of the mass transit network of Dubai [2]. The new channel (Figure 2) which is 2.8 km long, 80 m wide and 6 m deep, will connect the historic Khor Dubai to Al-Safwa shorefront passing through downtown of Dubai near Burj Khalifa area and its surrounding artificial lagoons [3]. The channel will receive its water from the sea and delivers it back to the sea. The new channel manifests how Dubai is constantly seeking to enhance the image and urban visibility of the city. Dubai's high-end shorefront developments reveal the essence of the presence of water not merely as a landscape feature but ideologically connecting Dubai's past as a pearl fishermen port city to a modern urban enclave of spectacle and richness.

Due to the lagoons and bends effects, the flow field in the proposed Dubai artificial channel would be of complex nature. This can be investigated by either physical models using experimental flumes or numerical models using computational fluid dynamics (CFD) methods. Numerical models are cheaper and offer more flexibility compared to physical models when modifications are needed to the channel shape. In addition, CFD simulations can provide detailed information of the flow field in the channel and lagoons under various hydraulic conditions [4]. The main objective of this study is to investigate the flow pattern in Dubai Channel using a 2D water surface model. The 2D water surface model can help in planning the landscape of the channel by minimizing the scour or sediment deposition at the channel entrance and exit and minimizing the eddy effects in the bends sections.

Figure 2. Dubai water channel project [3].

2 Methodology

Due to the complex landscape geometry of the Dubai Channel (Figure 3), the 2D Finite Element Surface Water Modeling System (FESWMS) is used to investigate the flow pattern in the channel. FESWMS is part of the commercially available Surface water Modeling System (SMS) package version 10.2 Graphical Interface that combines a series of hydrodynamic/ sediment codes [5]. The model was successfully used to simulate flow in stream with bends and lagoons similar to Dubai Channel [6].

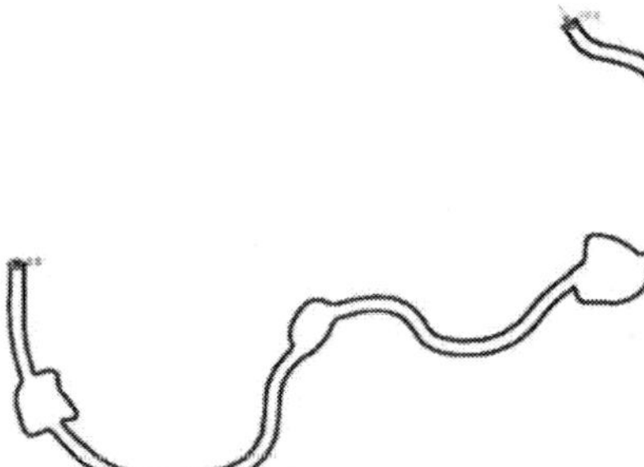

Figure 3. Proposed Dubai water channel layout.

FESWMS solves the differential forms of the continuity and the momentum equations in the stream wise and transverse directions using the Galerkin method of weighted residuals, providing water depth and depth-averaged velocity magnitude in x and y directions at each node in the grid [7]. The governing equations are written in the conservative form hence, the momentum is conserved along the streamline and the model is capable of capturing shock effects [4]. The conservative form was chosen because of its robustness in solving critical and transcritical flow fields under both low and high flow conditions by allowing dry-elements to exist within the computational mesh [4].

FESWMS solves the following equations simultaneously:

$$\frac{\partial z_w}{\partial t} + \frac{\partial q_1}{\partial x} + \frac{\partial q_2}{\partial y} - q_m = 0 \qquad (1)$$

$$\frac{\partial q_1}{\partial t} + \frac{\partial}{\partial x}\left(\frac{q_1^2}{d} + \frac{1}{2} g d^2\right) + \frac{\partial}{\partial y}\left(\frac{q_1 q_2}{d}\right) + gd\frac{dz_b}{\partial x}$$
$$+ gn^2 \frac{q_1\sqrt{q_1^2 + q_2^2}}{d^{7/3}} \sqrt{1 + \left(\frac{\partial z_b}{\partial x}\right)^2 + \left(\frac{\partial z_b}{\partial y}\right)^2} \qquad (2)$$
$$- 2d\varepsilon_{xx}\frac{\partial^2 \bar{u}}{\partial x^2} - \varepsilon_{xy}\frac{\partial}{\partial y}\left(\frac{\partial \bar{u}}{\partial x} + \frac{\partial \bar{v}}{\partial y}\right) = 0$$

$$\frac{\partial q_2}{\partial t} + \frac{\partial}{\partial x}\left(\frac{q_1 q_2}{d}\right) + \frac{\partial}{\partial y}\left(\frac{q_2^2}{d} + \frac{1}{2} g d^2\right) + gd\frac{dz_b}{\partial y}$$
$$+ gn^2 \frac{q_2\sqrt{q_1^2 + q_2^2}}{d^{7/3}} \sqrt{1 + \left(\frac{\partial z_b}{\partial x}\right)^2 + \left(\frac{\partial z_b}{\partial y}\right)^2} \qquad (3)$$
$$- 2d\varepsilon_{yy}\frac{\partial^2 \bar{v}}{\partial y^2} - \varepsilon_{yx}\frac{\partial}{\partial x}\left(\frac{\partial \bar{u}}{\partial x} + \frac{\partial \bar{v}}{\partial y}\right) = 0$$

where Eq. (1) is the continuity and Eqs. (2) and (3) are the momentum equations in x and y directions, respectively. In the equations, t is time (s), d is water depth (m), ρ is water density (kg/m^3), g is the acceleration due to gravity (m/s^2), n is Manning's coefficient of roughness, z_w and z_b are water surface elevation and bed elevation above certain datum (m), q_1 and q_2 are the unit discharge fluxes (m^2/s) defined as $\bar{u}_d$ and $\bar{v}_d$, respectively, $\bar{u}$ and $\bar{v}$ (m/s) are the depth-averaged velocities of an element in the stream wise and transverse directions, respectively, q_m is the resultant inflow or outflow from that element (m/s), ν_{xx} and ν_{yy} are the normal components of the eddy viscosity (m^2/s) in the x and y directions, respectively, and ν_{xy} and ν_{yx} are the shear components of the eddy viscosity (m^2/s) applied to the $x-y$ plane.

FESWMS inputs are the Manning's coefficient of roughness n, and the eddy viscosity ν. Model inputs must be assigned correctly to represent the physical processes occurring in the modeled reach and to produce accurate model predictions. Manning's n is an empirical coefficient that accounts for the total flow resistance caused by flow interaction with the boundary [7]. FESWMS utilizes Manning's n to account for momentum loss due to bed-shear, which may vary significantly in a stream reach in accordance to bed-bathymetry and roughness. The second input variable used by FESWMS is the eddy viscosity ν. Eddy viscosity accounts for flow resistance due to the internal shear stresses, or the Reynolds' stresses of the fluid incorporating the added energy dissipation due to turbulence in the flow [8]. Therefore, eddy viscosity is not a physical property of the fluid, but rather a turbulent characteristic of the flow. For isotropic flows in prismatic channels, a single eddy viscosity value is sufficient to describe the turbulence flow characteristics within the modeled reach.

3 Results

The appropriate size of the mesh elements can be estimated from the mesh Peclet number criterion, defined as the ratio of advection to diffusion occurring in an element [9]. This can be expressed mathematically as:

$$P_e = UR / \nu_t \quad (4)$$

where U (m/s) is the resultant depth-averaged velocity of the element ($U = \sqrt{\bar{u}^2 + \bar{v}^2}$), ν_t is the eddy viscosity (m^2/s) in the flow direction ($\nu_t = \sqrt{\varepsilon_{xx}^2 + \varepsilon_{yy}^2}$), and R is the maximum distance between two nodes on the element (m).

The mesh Peclet equation shows that U and ν_t determine the ideal element size R and thus the minimum eddy length that can be resolved. However, the element size is also a function of local bathymetry including bed roughness, and tailoring a mesh grid design to ensure a constant mesh Peclet number throughout the reach is laborious and unreasonable. It is therefore sufficient to keep the mesh Peclet number within the recommended range for all the elements in the computational mesh. Prior studies using FESWMS [10, 11] recommended a mesh Peclet number between 10 and 40 to ensure that: 1) momentum is dominated by advection; 2) eddy viscosity maintains flow consistency and prevents oscillation; and 3) proper amount of energy loss due to dispersion takes place in each element to account for microeddies, which are too small to be resolved in the mesh (i.e., those eddies that are smaller than the local element size). Keeping the mesh Peclet criterion within the recommended range and resolving for the ideal element size R minimizes the errors that can result from the spatial resolution of the computational mesh [12, 13, 14]. Figure 4 shows the size of the generated mesh elements in the channel, which satisfy the range of the recommended mesh Peclet number.

Although FESWMS possesses the capability of spatially distributing both Manning's n and eddy viscosity in the computational mesh, it is common to use averaged values for Manning's n and eddy viscosity. In prismatic channels under uniform flow conditions, an average eddy viscosity value can be assigned to the study reach as a function of depth and bed slope [15, 16]. In this study, it is assumed that the channel has a rectangular cross-section with a width of 80 m and variable width at the lagoon sections. The model inputs are: the channel bathymetry, Manning's coefficient n = 0.02, eddy viscosity ν = 0.01 m^2/s, flow rate at the upstream Q = 500 m^3/s, water surface elevation at the downstream y = 5.8 m. Figures 5 and 6 show the water depth in meter and the water velocity, respectively in the simulated channel. The minimum and maximum depths in the channel were 5.2 and 6.0 m, respectively. Velocity is the range of 0.3 to 1.2 m/s with higher values in the channel sections and lower values in the lagoons sections. The model predictions show that the water depth is lower in the channel sections compared to the lagoons and bends sections. On the other hand, the velocity is higher in the channel sections compared to the lagoons and bends sections. Figure 7 shows the velocity vectors in the eastside lagoon. It is clear from the figure that the model was able to mimic the expected flow field in the lagoon.

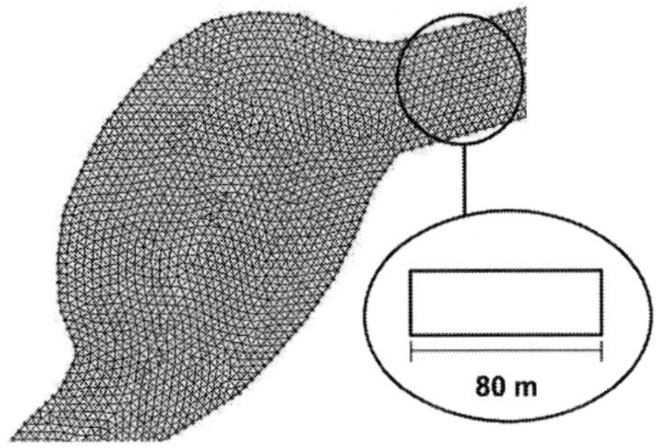

Figure 4. The size of the generated mesh elements for the middle lagoon section.

Recommended minimum velocity in rigid boundary channels should be in the range of 0.6 m/s such that sediment is not deposited, aquatic growth is inhibited, and sulfide formation does not occur, while recommended maximum velocity should be in the range of 6.0 m/s to avoid boundary erosion [17]. In addition, the Froude number $\left(F_r = U / \sqrt{g(z_w - z_b)}\right)$ should be in the range of 0.3 so that the water surface does not become rough, especially downstream of obstructions and in bends. From the model simulation, it can be seen that the velocities as well the Froude number in the Dubai artificial channel are within the accepted range recommended for rigid boundary channels.

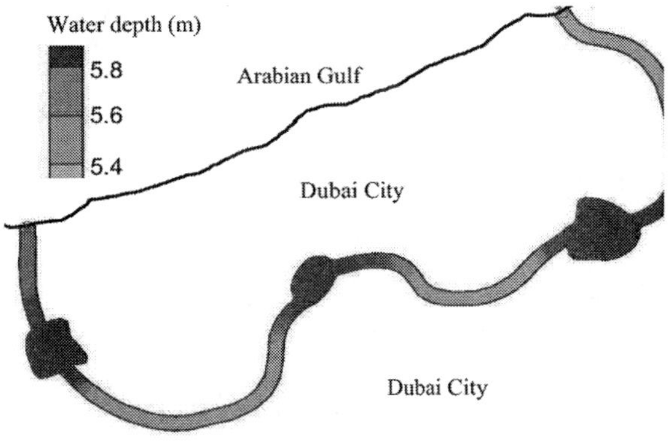

Figure 5. Simulation of the water depth in the Dubai city channel.

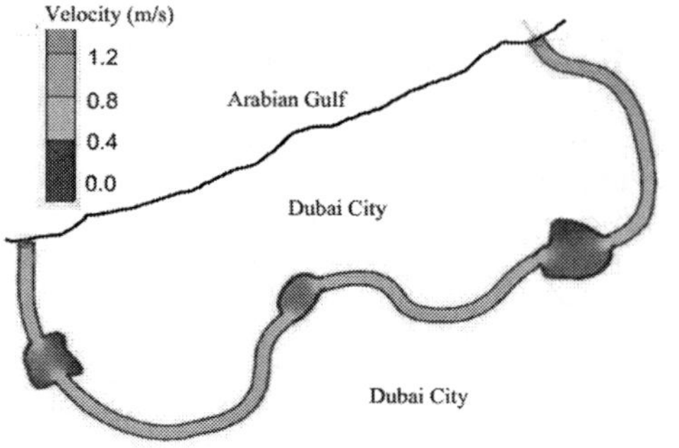

Figure 6. Simulation of the flow velocity in the Dubai city channel.

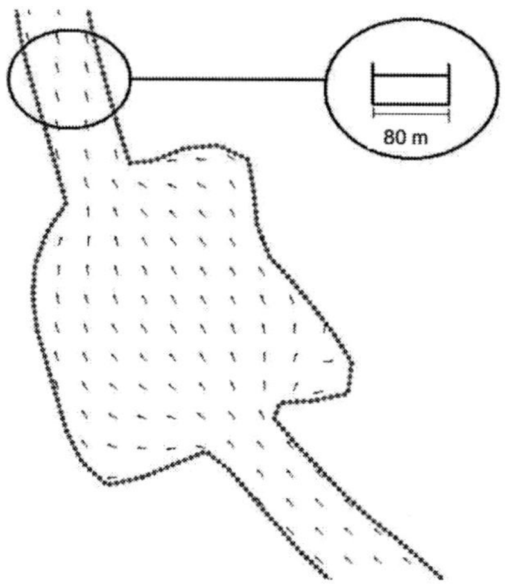

Figure 7. Flow direction in the eastside lagoon of the channel.

4 Summary

Dubai's new channel further enhances the urban-scape of the city offering new waterfront developments, transportation venues and diversified panoramas to the city. A study was performed to simulate the flow field in the proposed Dubai artificial channel using a 2D hydrodynamic model. The model predicts the flow depth and velocity in the channel, lagoons and bends sections. The model predictions show that the velocity is higher in the channel sections compared to the lagoons and bends sections. On the other hand, the water depth is lower in the channel sections compared to the lagoons and bends sections. Nonetheless, the velocities in the channel are within the accepted range that prevents boundary erosion and sediment deposition. In addition, the Froude number was in the range that averts the water surface from becoming rough, especially in the bends sections.

References

1. M. El Amrousi, P. Caratelli, S. Shakour, ACSA/AIK International Conference 2014: Open Cities, Seoul Korea, (2014)
2. Business Bay, Water Canal marine stations, Emirates 24/7, (August 15, 2015)
3. Business Bay canal extension set to energize Dubai's landscape, McClatchy-Tribune Business News, Washington, (October 2012)
4. H.M. Chaudhry, Open Channel Flow, Springer, New York, (2008)
5. D. Froelich, User's Manual for FESWMS Flo2DH: Two-Dimensional Depth-Averaged Flow and Sediment Transport Model, Release 3, (2002)
6. A.Z. Khan, B. Barkdoll, Int. J. Comput. Eng. Sci., **2**, 3 453-467 (2001)
7. D.M. Hicks, P.D. Mason, Roughness Characteristics of New Zealand Rivers, New Zealand DSIR Marine and Freshwater Resources Survey, Wellington, NZ, (1991)
8. A.N. Papanicolaou, R. Hilldale, J. Eng. Mech., **128**, 9 948-960 (2002)
9. J.C. Tannehill, D.A. Anderson, R.H. Pletcher, Computational Fluid Mechanics and Heat Transfer, 2nd Ed, Taylor & Francis Publisher, Philadelphia, (1997)
10. A.J. Miller, B.L. Cluer, Fluvial Processes in Bedrock Channels, AGU Monograph 107, AGU, Washington, D.C., 61-104 (1998)
11. M. Elhakeem, A.N. Papanicolaou, International J. River Basin Management, **6**, 1 63-77 (2008)
12. A.N. Papanicolaou, M. Elhakeem, and B. Wardman, Journal of Hydraulic Engineering, **137**, 1 75-89 (2010)
13. D.W. Crowder, and P. Diplas, J. Hydrology, **230**, 3–4 172–191 (2000)
14. R. J. Hardy, P.D. Bates, and M. G. Anderson, M. G. J. Hydrology, **216**, 124–136 (1999)
15. G.E. Freeman, Solving the dilemma: To wave or to oscillate? Opposing formulations of the shallow water equations in river modeling, Ph.D. dissertation, Texas A&M Univ., (1992)
16. D.R. Richards, Proc., 1990 National Conf. Hydraulics Engineering, H. H. Chang and J. C. Hill, eds., **2**, ASCE, Reston, (1990)
17. L.R. Mays, Hydraulic Design Handbook, McGraw-Hill, New York, (1999)

MATEC Web of Conferences **68**, 13004 (2016) DOI: 10.1051/matecconf/20166813004
ICIEA 2016

Strategies of Energy Efficiency Design in Traditional Kangbaiwan Mansion in China

Xiaoqing Song and Caili Zhang

Zhongyuan University of Technology, Zhengzhou, China

Abstract. The building sector is one of the highest energy consuming sectors in the world as well as in China, it is urgent to seek an energy efficiency way of sustainable architecture development. From the perspective of tradition, this paper focus on strategies of energy efficiency design that contained in excellent vernacular dwellings. On the basis of analyzing an example of Kangbaiwan Mansion, it illustrates the advantage of environment ecosystem, and summarizes the physical and cultural characteristic of its buildings, especially the climate-adapting overall arrangement and sustainable strategies of natural ventilation and passive solar gain, which can be a fertile source of modern energy efficiency architecture design as well as a proper way of inheriting the outstanding traditional culture.

1 Introduction

Facing to the environmental issues such as accumulating energy crisis and climate change in recent decades, which are not the only ecological concern that should on our list, it is becoming crucial and an inevitable option for sustainable development. Sustainability obtained attention from all walks of life, including building field.

Sustainable architecture in the past thirty years has experienced vigorous development, emerged in a number of important theories and research results. One of them is *ecological architecture*, which focus on the concept of man's bioclimatic approach and alternative energy strategy. Another is energy-efficient building which concentrates more on the appropriate technique adopted to lower the relative consumption of energy [1]. The further developed concept is defined as *green building*: The practice of increasing the efficiency with which buildings and their sites use energy, water, and materials, and reducing building impacts on human health and the environment, through better siting, design, construction, operation, maintenance, and removal--the complete building life cycle [2]. In the process, advanced new technology was adopted in the form of all kinds of equipment and installation. Energy efficiency and green building are equivalent to high technology, which has misguided many architecture designs. The comparison of energy-saving design became the competition of higher technology, regardless of the essence of saving.

With the deepening of cognition, architectural experts finally fixed their attention upon sustaining the compatibility of architecture, human beings and the nature, which was named *passive design* when related to the specific design strategy. Passive design is mentioned much more frequently to the solar usage aspect, which means some designed architectural elements that help gain more solar heating in winter and prevent from overheating in summer, by contrast with active electric solar gain systems.

The broader understanding of passive architecture design can be described as a cost-efficient way of coordinating relationship between artificial structures and the natural environment, which we found some of the principles just existed in the traditional architecture. It is no surprise when the essential of architecture is concerned. From its very beginning, the characteristic of architecture form has been closely related to the specific climate and topography in a certain location. That is the most basic principle appropriate not only in the ancient times but also in modern times. There are also many other strategies deserve review our vision of historical and traditional architecture, which should become a rich source for the creative design of sustainable architecture. "This was not done by the blind imitation and copy of its appearance, but through the conscious and deep understanding of esoteric principles of architecture." [3].

Taking a case study of China as an example, this article discusses the strategies of energy efficiency design in traditional architecture which can be the inspiration for modern sustainable architecture design.

2 The Kangbaiwan mansion in Gongyi city, Henan province of China

The Kangbaiwan (a millionaire with the last name Kang) Mansion was first established in the late Ming Dynasty (AD1368~1644) and maintained till the early part of the twentieth century over a period of more than 300 years. At its peak, the relatively self-sufficient estate covered an area of 240mu (1mu=0.0667hectares) and included not only a walled compound for the immediate family and nearby residences for other kin as well as retainers, but also

DOI: 10.1051/matecconf/20166813004

ancestral worship facilities, workshop areas, inns for travelers, stables for riding, pack, and work animals, livestock pens, a brick and tile-making kiln, a military camp, and a granary and stores complex [4]. Nowadays, with the traces of most of the ancillary structures, only the main parts of the mansions such as the walled terrace area exist and is an officially protected site in China.

Figure 1. Zoning map of building thermal design in China.

2.1 Geographic location and climate

The Kangbaiwan Mansion sits on the terminal of the Loess Plateau, along the Yiluo river, where there is the Yellow river and Mount Mangshan in the North and the Mount Songshan in the south. It's an advantaged geographical position. As a sentry, the Gongyi city next to the ancient capital-Luoyang. So, Gongyi(with the name Gongxian in the history) is the center of wars and a place with economic prosperity from the ancient times. The longitude of the city is 34°31'~34°52' North, its latitude is 112°49'~113°17' East.

Gongyi has a temperate and monsoonal climate with four clearly distinct seasons. It is cold in winter and hot in summer. The average annual temperature is 15.1℃, About 220 days in the year are frost-free. With the annual mean rainfall is 585 millimeters, the climate is dry (Fig. 1) [5].

2.2 The environment-adapting arrangement

From ancient times people have reacted to the natural environment and created a built space not only to provide protection from the vagaries of the weather but to express an understanding of the world [6]. The concept of "fronting water and with hills on the back" is one of the fundamental principles of the ideal environmental pattern for residential area in the architecture tradition of China. This is a true portrayal of Kangbaiwan Mansion's environment. Clinging to the southern slope of Mangshan mountain, The main dwelling houses of the Mansions follow the lie of steep cliff and make the main orientation facing South and East, which is the best orientation to gain more passive solar heating. All the buildings situated on a loessic terrace which is 83meters from north to south, 73meters wide and 12m~15m high from the ground. The ramp in the arched tunnel is the only access to the terrace which is surfaced by bricks. This ingenious measure provides a safeguard against both natural disaster such as flooding and invaders in troubled times (Fig. 2).

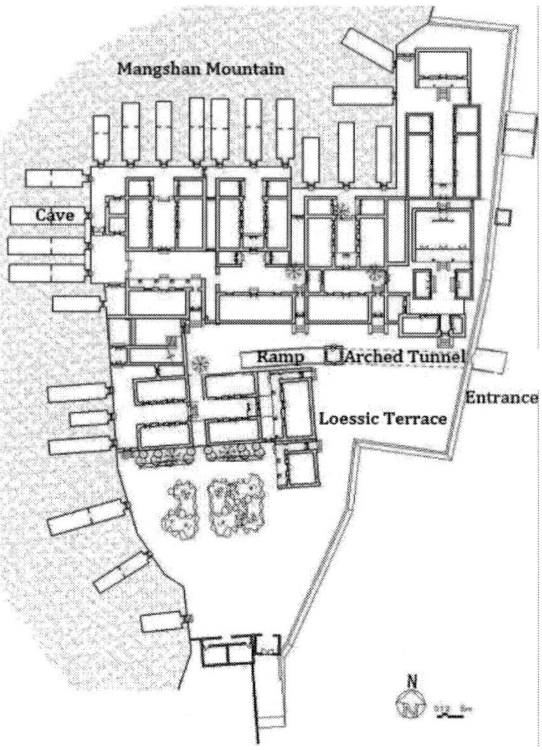

Figure 2. Plan of the main residential area of Kangbaiwan mansion.

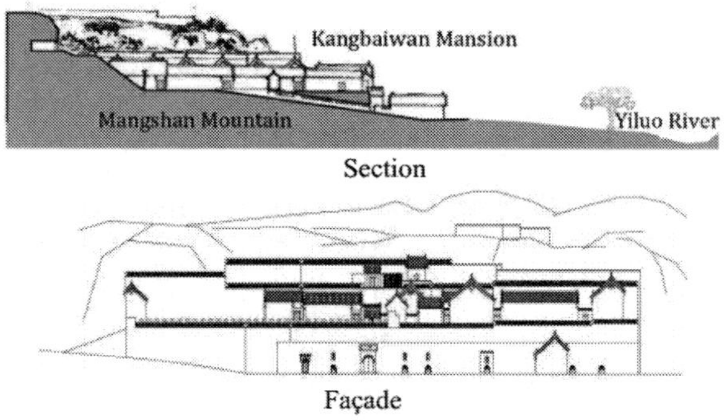

Figure 3. Environmental Sketch drawing of the main residential area of Kangbaiwan mansion.

To the east of the mansions,the Yiluo river flows pass from southwest to northeast joining into Yellow River about 15 kilometers away. Such geomatic and treasured site offers the residence area to against the chill wind from northwest in winter and guarantee adequate ventilation in summer due to the specific terrain of China which is high in west and low in east. It meets the concept of "harmony between man and nature" of the Oriental philosophy, as well as the spirits of sustainability in modern times (Fig. 3).

When it comes that there is no river available nearby, an alternative solution is a dug pool to fulfill the requirement of water supply for living and fire protection. As the style of

 DOI: 10.1051/matecconf/20166813004

life changed, the standards of treasured ideal environment for dwelling gradually turn from a physical environment-judge practice to a psychological understanding of living space.

2.3 The courtyard

The courtyard is the fundamental unit of architectural complex in China. According to the geographic latitude, the size of yards vary very much. In general, they are wide to obtain enough solar radiation in the high latitude area and narrow to provide more shading in the low latitude area.

As for the Kangbaiwan mansion, Most of Caves were dug as the space of main hall, which form a rectangular courtyard with conventional timber framed, slope roof buildings in other three sides (Fig. 4).

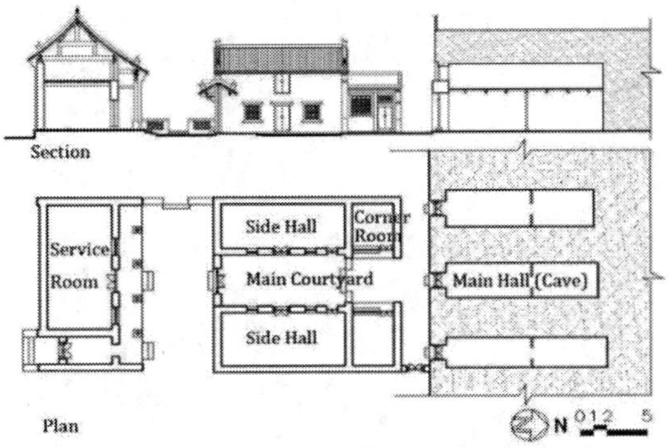

Figure 4. Typical courtyard unit of Kangbaiwan mansion.

With all the doors and windows face the courtyard, it becomes the central area of domesticity. Due to the latitude and mountainous topography of western part of Henan province, the mansion courtyards which only the central bay of the main hall can be seen from the bottom, is relatively narrow compared with the courtyard in the southern of the province which the whole facade of the main hall can be seen (Fig. 5) [7]. Based on the sunlight distribution analysis, it is easy to conclude that the scale of the narrow courtyard fulfill the basic request of solar heating, which was adopted merely followed the good tradition and builders' rich experience.

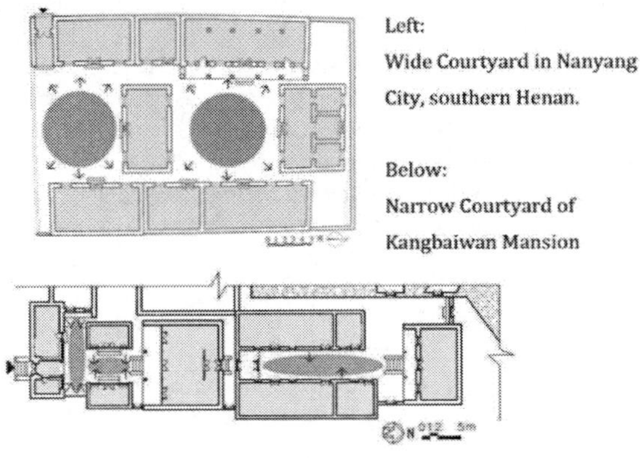

Figure 5. Comparation of two type of courtyard.

Other than that, the trees and flowers planted in yards offer not only a beautiful scenery but also a necessary shading in summer.

Such two or three courtyards range in line forming a central south-north axis which is an ordinary pattern of layout of chinese tradition architecture. The more living spaces are needed, the more such axial group of buildings are built. Most of yards conform with the regulation in Kangbanwan Mansion, while a few form an east-west axis due to the lie of cliff.

2.4 Strategies of energy saving in architecture

Many kinds of traditional rural buildings which came into existence after a long time of evolvement can adapt to the local climate and can achieve satisfying thermal comfort.

One of the most significant architectural character of Kangbaiwan mansion is the various architecture typologies that can be arranged in three classes as follows: the cliffside cave dwellings, simulated subterranean structures, and traditional post-beam structures.

Taking advantage of the special geological conditions and climate characteristics, cave-dwellings become the optimal choice for the residence living in the loess plateau area. They are relatively warm in the winter and cool in the summer, as well as cost-efficient and land-saving. The soil has been the main building material for thousands of years due to its great support force and the rainless climate. The arch structure gives full play to such local material and contributes much to cave dwelling. To give more support to the structure, both the facade and interior surface are laid by gray bricks.

The defects of cave-dwellings are the poor lighting and ventilation conditions. The significant feather of cliffside caves in the mansion is that they are two storeys. In addition, there is even a three-storey cave, which is the only one that has been known. Though the space of the second or third floor serve just as storages, the natural lighting and ventilation are improved because of the additional windows and the higher interior space (Fig. 6). The lime mortar painted on the interior surface makes the room bright and absorb humidity as well.

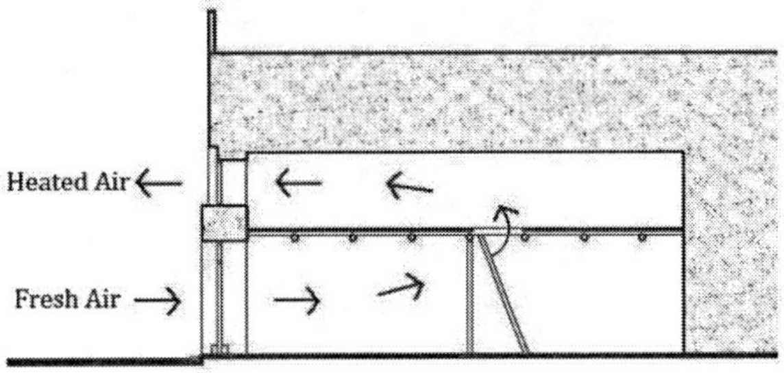

Figure 6. Ventilation analysis of two-storey Cave dwelling in Kangbaiwan mansion.

Such merits are inherited and developed in new type of cave-dwellings, which are performing well both in thermal conditions and energy saving (Fig. 7) [8]. Split-level space is designed to gain more sunlight in the room and increase the natural ventilation. In addition, an air exchange fan is installed to refresh the interior space. The creative element is a sunroom in front of the facade, which is integrated for solar heating, thermal insulation, ventilation and cooling.

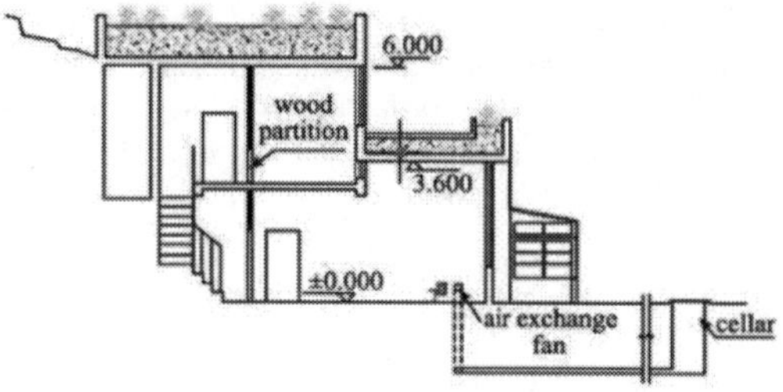

Figure 7. A new type of cave dwelling in Yan'an ,Shaanxi Province.

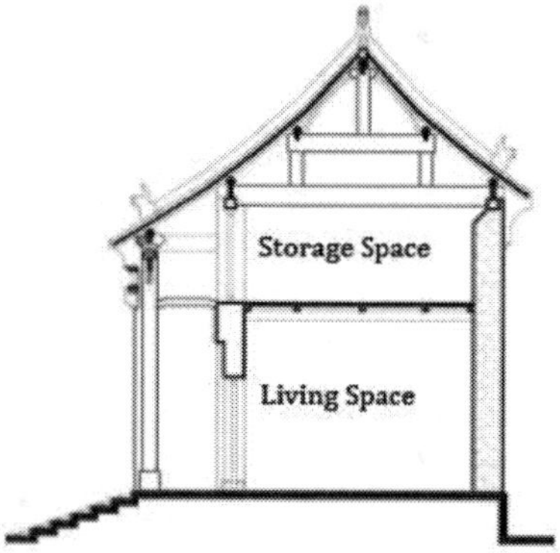

Figure 8. Storage space in the pitched roof building.

Figure 9. Walls meet the roof truss with some narrow gaps.

The same two-storey strategy was adopted in the traditional post-beam timber architectures in Kangbaiwan Mansion. The storage space under the Pitched roof not only increases the usable area but also improves the insulation performance as a transition area (Fig. 8). The porch in front of the main facade functions as a horizontal transition space mostly in the way of residents' daily activities. It serves as the signal between interior and exterior. Under the roof of the porch, walls meet the roof truss with some narrow gaps by which natural ventilation is promoted, so that the wooden frame can be prevent from decay (Fig. 9). For the main living space, chinese-feathered furnitures also play an important role in the interior comfort. Though designed more likely for privacy, for example, the elaborate alcove bed is an additional compensation of thermal comfort (Fig. 10) [9].

Simulated subterranean structures are created to continue the benefits of caves in the area without available loess cliffs. Just as the name, they simulated the cliffside cave-dwellings by constructing the arch-roof structures and covering with soil, which can also be interpreted as a way of inheriting the local customs and developing regional distinctive features. There are more simulated subterranean buildings than the cliffside cave dwellings in Kangbaiwan mansion because of flexibility.

Figure 10. Elaborate alcove bed in Kangbaiwan mansion.

3 Conclusion

Before high-tech devices are installed, Strategies of passive building design must be carefully examined and given precedence to act, then twice as much can be accomplished with half the effort. To be part of the movement to seek sustainable development to our troubled earth, many things can be do in the field of energy saving architecture. Looking back into the past, historic and traditional architectures become a fertile source of new sustainable ideas. The proper situation and layout related to a certain geographic location and climate, best orientation and windows arrangement related to effective way of obtaining natural lights and ventilation, various types of space designed to comfort the living space, all of above are the root aspects of energy efficiency design, which our designer should put into practice. Only these basic principles are adopted, shall the subsequent mechanical installation are meaningful.

From the discuss of the example of Kangbaiwan mansion, the key passive design aspects can be summarized as follows:

1) The layout of the site plan is corresponding with the certain geographical environment in the first place.

2) The scale of the narrow courtyards strike a good balance between fulfilling the basic request of solar heating and limited mountainous building site.

3) The architecture typology of cave dwellings and the second-floor storage spaces are climatic adaptable, energy saving and a good strategy to improve the interior comfort of the main living spaces.

4) Adjusting measures to local conditions, such as local materials and techniques, is the effective way to save the construction cost as well as inherit the local culture.

5) Trees are planted to provide seasoning shading and better the microclimate.

Acknowledgment

This work was supported by a grant from China Scholarship Council.

References

1. Yehao Song, The Overview of Development of ecological architectural theories in the West, Word architecture, **1**, 67-71 (1998)
2. Sam Kubba, Handbook of Green Building Design, and Construction, LEED, BREEAM, and Green Globes, Butterworth-Heinemann, Oxford (2012)
3. Javad Samadi, Utilizing the Central Courtyard of Traditional Architecture in Modern Architecture, Research Journal of Environmental and Earth Sciences. **6**, 3 (2014)
4. Ronald G.Knapp, Chinese Houses:The Architectural Heritage of a Nation, Tuttle publishing, Hong Kong, 146, (2005)
5. Drawn according to image from: http://aq.zhulong.com/info_wiki/read609337.html
6. Nalanie Mithraratne, Brenda Vale, Robert Vale, Sustainable Living: the Role of Whole Life Costs and Values, Elsevier Limited, Oxford, 4 (2007)
7. Manchang Zuo, Xianchen Bai, Vernacular dwellings in Henan province, China Building Industry Press, Beijin, 165 (2007).
8. Jiaping Liu, Xinrong Zhu, Liu Yang, Rongrong Hu, Exemplary project of green cave dwellings in Loess Plateau, Frontiers of Energy and Power Engineering in China, March, **4**, 1 (2010)
9. Ronald G.Knapp, Chinese Houses:The Architectural Heritage of a Nation, Tuttle publishing, Hong Kong, 151 (2005)

Risk Analyst: A New Player in the Construction Processes – Portugal

Nuno Dinis Cortiços

Centro de investigação em Arquitetura, Urbanismo e Design at Faculdade de Arquitetura, Universidade de Lisboa/Secção das Tecnologias de Construção, Lisbon, Portugal

Abstract. In Portugal, the established structure for the construction, does not work in favor of its quality! The construction sector, at the height of its financial strength, from 1980 to 2007, has always had the tendency to skid beyond the initial budgets. It is true that these were drawn up on a stage still under construction, for the sake of time management and to please the Project Owner (P.O.). When transiting to the construction site all was uncovered, and the P.O. had to deal with higher costs and unanticipated. The General Contractor (G.C.) took advantage of this weakness and would rework its proposal - dumping was a current practice, with the certainty of what would happen - what in fact only aggravated the situation. The designers were averse to change and conflicts would arise with the P.O. and G.C.. Given this the only entity considered to be independent, to mediate the situation was the Inspection. This, for its part, took the side of the strong players - P.O. and G.C. - in an attempt to balance the immediate savings of both, leading to systemic build failures with huge impact on the edified, which in turn affected the warranty period. As a rule the G.C. would refuse to do any repair - claiming that she had just built in accordance with the instructions - hiding behind the imposed economy and ignoring the projects - with the consent of the P.O.

1 Introduction

Portugal, not just today, but always, has presented contradictions in its economic growth. With considerable heritage legacy that consumes resources the institutions, which is not being used for the purpose for which it was built. In addition, some of that heritage is doomed to abandonment because, simply, there is no money available to meet the maintenance and function needs. Or, ultimately, operating with almost no maintenance, ensuring only the functioning, maintaining the space on the basis of the anomalies.

The technology that supports the production of papers to instruct the construction work is nowadays transversal to the various branches of knowledge, these parts do not always convey the whole of the decisions taken or the originally planned rigor, but its value is indisputable for the duration of the development processes. However, its use creates a distance between the operator and the end result, which originates difficulties in perceiving shape, durability and compatibility of the various solutions, of their products and materials. In the construction process everything happens in layers - with different times and natures, which are error enhancing factors - because each new layer will hide the previous, allowing only the observation of the latter, which impairs the detection of anomalies, the notion of longevity and performance of the all.

2 The problem

In the current framework, the Inspection has taken on a cross-cutting role in the control of the quality of the construction work, despite the systematic failures identified in this joint model between the P.O. and the G.C.. The Inspection takes a position in favor of immediate work economy - which favors the P.O.'s positions in collusion with the GM: the players in power.

The only entity obliged to inspect is more concerned with staying on the sidelines of the conflict, calling upon itself the mediator position. This position of neutrality is accompanied by: the defense of the established bureaucracy and legality; the preparation of documents for the various phases and their archival; that accumulates with the preparation of weekly meetings, their minutes and required approval; the monitoring of work (work fronts) in order to register the progress; and the approval of the measurement records. These last two matters are undoubtedly the most problematic - and the ones that are assumed as quality control mechanisms. Although these entities possess production mechanisms, established by practice and framed legally, they tend to ignore its legacy in favor of a less firm position, less interventionist but proven to be more profitable for itself, the P.O. and the G.C.

3 Possible answer

The answer to the problem detected implies the reorganization of the teams involved in the process, the Inspection will be confined to aspects which dominates qualitatively. The issues so far overlooked will be assigned to a new class of professionals, which shall bear the responsibility for assessing the constructive robustness: the Analysts. Aiming to address any gaps in monitoring the construction process, by examination and testing of the buildings.

In Project:

- with the evaluation of local conditions;
- controlling the solutions;
- checking proposals and coordinating building systems;
- ascertaining coordination, control and management of the different specialties;
- managing costs and budgets;
- evaluating new equipment and technologies;
- managing infrastructures;
- analyzing papers, preliminary to the work;

In the Construction Site:

- checking the coordination of the implementation process for the various work fronts;
- implementing tests and analyzing results;
- assessing the viability and sustainability (durability / performance);
- controlling the performance of the construction;
- development of testing proposals, on site, to validate the solutions;
- the execution of the certification of building systems and of the requirements of the spaces, general and specific.

In Warranty:

- monitoring claims, drawing conclusions and evolving in the way work is done.
- In particular, we suggest the following rate of responsibilities:
- check local conditions: sea air, prevailing winds, rainfall projections, groundwater, soil instability, accesses, pollution and noise levels;
- validate the microclimate of the architectural solution: shadow areas, wind confluences, exposure to the elements and thermal degradation in the coatings;
- legitimize the technical and constructive character of decisions on the drawing board, with particular attention to aspects related to the "envelope" (physics of the constructions);
- enable the constructive solutions proposed in the architectural framework and different specialties;
- recognize the comfort features when addressing the solutions and equipments with higher performance / sustainability;
- provide for the general coordination of architectural projects and specialties;
- program and implement tests to be performed on post-construction, even before the topcoats;
- check the technical solutions (architectural and from the specialties) to implement and anticipate durability tests;
- recognize technical documents: Specifications, Measurements, Quantities and Budgets;
- execute the report on the errors and omissions of the projects before the construction work;
- essay the buildings constructed in relation to: the "envelope" (the physical buildings); of architectural aspects (coatings and maneuver); and from the specialties, given the current durability;
- certify the performance of the spaces in functional expectations: collection of use and comfort data (acoustical, thermal, luminal, hygiene, safety and health promotion);
- prepare a report on the temporal durability of solutions, monitor the guarantee and maintenance;
- monitor the guarantee period in the perception of the performance of validated systems;
- complete the closure report listing a critical history of occurrences

The rendering of services ends with the expiration of the guarantee period.

Please note that in the system suggested Analysts act, both in design (prevention) and in construction work (validation), at the service of the Insurance Companies with the benefit for control and risk management. This format does not occur in any of the countries usually taken as a reference: Spain and France.

The assumption of those obligations does not intend, in any way, to empty of responsibilities the traditional players, contrariwise, the purpose is to evaluate the performance of these as to their degree of risk to the Insurance Companies, affecting the insurance tax to the enterprise, both funded by the P.O.. On our proposal the merit of those involved is quantifiable and differentiating, and the information is made available to the P.O., and will allow lower insurance premiums. Aiming for an improvement in the technical quality, counteracting rooted practices, backed by professional associations and with obvious gains for construction sector and final users. Serializing the information, by intervener, allows the quantification of experience which will act as a business card for a market progressively more demanding. In practice there will be the management of the curriculum of those involved, by participation: with the classification of the complexity of the solution; the relationship between initial and final budget; and quantification of risk (given by risk analysts).

The Insurance Companies, during the guarantee period, would report the performance of the solutions and construction quality. In event of a claim would act in the resolution of the damages, according to civil liability contracted, in favor of users and P.O.. Those, in turn, would provide the data to the Instituto de Seguros de Portugal (Insurance Institute of Portugal) that should assume two functions:

- the setting up and management of a database based on the professional register; and,
- publish a compilation of the most important claims, due to constructive anomalies.
- The advantage of such system lies in the knowledge available to professionals and future generations, with gains for the public in general.

The "regulation" of the market would be ensured by the Insurance Companies derived from work performed by the Analysts, trained risk assessors for the solutions proposed by the designers and the G.C. The options would be verified in the designing phase and would not reach the construction, without proper warning and report, ensuring a preventive control. The added value in establishing responsibility before the construction lies in the low cost of the resolution: it suffices to think of solutions "buried" beneath the surface quota.

The Insurance Companies would be responsible for providing data on the performance of the Contractors, if negative it should reverse on penalties in their business register managed by Instituto da Construção e do Imobiliário (Institute of Construction and Real Estate), and with direct impact on the Construction License. The advantage of this system is the additional security for the P.O. and the possibility to relativize the mandatory bank guarantees.

The buildings would be tested and any deficiencies corrected in the construction phase, immediately and without affecting adjacent areas and / or the quality of the whole.

The culmination of the whole process would by assuming responsibility for the analysis of results, as follows:

- Designer(s), with no contribution to the premium payment - performance evaluation and gradual registration on the basis of proven errors;
- Contractor(s), which share premium costs (percentage to be determined in terms of construction quality) - evaluation and testing of the built; and,
- Project Owner(s), co responsible in the premium costs (percentage to be determined on the basis of construction quality) - evaluation of the complexity of the operation.

The Guarantee System is based on assessments by Analysts, as service providers or employees of Insurance

Companies, and aims to ensure the quality of the end results, from the point of view of operators / users, in the allocation of responsibilities and risk classification.

4 Professional skills

The emergence of Analysts derives from some movements in recent history of the teaching of Architecture and Civil Engineering, especially in Europe, that support such need in the required primacy given to quality. A step forward in changing mentalities or a challenge to the construction sector: Technicians adjusted to a more efficient construction, in a world in rapid transformation - constant revolution of the complexities.

Nowadays, we deal with social, envIRONMENTAL AND energetic ISSUES in a way like never before.

The technical training of Analysts implies, inevitably, contact with various disciplines, such as the various branches of engineering; the science of construction; and, the basics of architecture. From the point of view of this research, aspects such as artistic, technological and scientific knowledge of architectural base; planning; the design; and, coordination between projects, are considered essential to the success of those, in a culture of quality and rigor to the ideal construction. And the knowledge of subjects, within the Insurance Activity, such as the control and perception of risk of operations; the effectiveness in the formulation and preparation of trials (before launching endeavor); the capacity to make assessment in situ; advising the coordination, between the various stages; and, finally, monitoring the guarantee period.

It is recognized that the Architect assumes, increasingly, more responsibilities for a process from which is distant, both in formation and in knowledge. And a deep knowledge in several areas with diverse natures and sometimes antagonistic.

In the current scenario, in higher education in Portugal, namely in Colleges - linked to the construction sector and facing difficulties in placing professionals in the market - it would be important to take responsibility for training technicians in response to the requirements and contemporary issues.

To promote the formation - the introduction of scientific knowledge (transverse to the sector) - for skilled technicians to support the P.O., Architects, Engineers, that would work as Analysts: evaluators of the design and the construction, by checking, suggesting and planning tests on matters that most affect the sector. Today, in the lead, we have problems with infiltrations, performance and overall assurance of the built. The tasks carried by those - in an extrinsic relationship to the design phase, work and guarantee - will enable the Designers ideal dedication to intrinsic aspects of the design.

Analysts would have an actual presence during the development phase, the construction work and, finally, on the premises, to evaluate the performance during the guarantee period.

The "arbitration", of the three appointed stages, requires industry analysis skills (technical-constructive) and economic and financial sustainability insights, to be able to readjust the architecture, the engineering, the models of management and operation of infrastructure. Something that does not happens in traditional courses, either in architecture or engineering.

In short, the training would combine different aspects of architecture and engineering, preparing graduates for an interdisciplinary analysis and high scientific level, that is, enabling professionals to perform highly complex technical tasks to control solutions. They could quickly achieve public recognition as the key element for the prevention and detection of constructive anomalies.

5 Academic context

Training in Construction Engineering, with a Master's Degree, would be dedicated to the constructive role, so critical to the rigor and robustness in construction (so popular among consumers / sector). An idea that develops from a natural movement, global and integrated into the legacy of the historical references of higher education, both from Architecture as modern Civil Engineering, tectonics vocation. The rationale for the need to create a new Degree is based on learning and research, legitimized by the demands identified by recent legislation and the need for greater liability on the projects and the constructive role. Situations increasingly more complex, detailed and interconnected in different specialties, with direct impact on the quality. The focus on training specificity and consequent functions could and should lead to greater liability of those involved. In the following stages:

The Degree should develop a kind of preparatory teaching with scientific basis, using subjects such as mathematics, geometry, physics, chemistry, computer science, property evaluation and management skills; without neglecting the constructive design of materials, history of human buildings, laboratory and constructive meaning, constructive composition techniques, technologies of tools and components, building systems in design, risk analysis, construction processes in construction work, evaluation methods, performance analysis, among others. At the end of this phase it is expected the introduction to the Construction Laboratory, the development of: development of theoretical analysis on integrated design, validation of elements in coordination, suggested tests, monitoring and tests on strands / built systems, validation and performance prediction; and,

The MSc (Integrated), in the fourth year, would be dedicated to scientific enrichment through the study of advanced and specialized disciplines. The conclusion of the learning process would go through technical and critical training in a practical professional domain and through monitoring of case studies. Preparing students for the elaboration of their Master's Thesis. The fifth year, the culmination of the qualifications, would be based on individual and specific choices through elective courses. The preparation of the Master Thesis will require the analysis of the design or the construction work or monitoring facilities, in close contact to the profession and its professionals, market and the construction sector overall. As usual, the thesis defense would happen before a jury composed of professors from various formations, ensuring a multidisciplinary and comprehensive analysis.

The purpose of this level of training involves perceive, identify, monitor and evaluate the building construction works. The contents, mostly of a practical nature, are supported by theoretical knowledge and practical methodologies, including experimentation, through workshops or accompaniments of construction work - in which students should develop and test their knowledge by collecting data, preparing reports and proposals by foreseeing performance - as well as communication and scientific research.

Training should produce qualified, aware and prepared professionals, for concrete and specific situations: analysis, critical construction project, organizing building activities, scheduling the construction process and risk management. Endowed with a holistic view of the entire development processes involving infrastructure, whether public or private.

Do not exclude the hypothesis of this Master occurring in continuity of the training for architects or engineers, in both cases adapted to their bases.

6 Conclusions

In the solution, which is presented in this paper, the Analysts are an entity with no legal standing in the Portuguese construction structure, small steps have to be taken to fill the gaps in quality control: the need for planning, preparation of tests and evaluation of constructive options in response to demanding solutions, and even to contemporary setbacks.

At the various stages of the construction process, Analysts would take on the task of collecting data and relevant information to the database - on the performance of those involved - reducing the risk in the following processes or even in future endeavors.

Monitoring the role of the different interveners is enough, from our point of view, to establish a culture of excellence, guided by merit and good performance in the conception of better building products with repercussions at various levels, with the active involvement of Professional Associations in defense their best professionals; while offering more consistent and adjusted tasks; promoting lifelong learning - legislative and technological - enhancing education, at all levels, with repercussions in the construction market.

The education system would be aware of the challenges students could face as professionals, their responsibilities and the best practices for excellence and performance, in favor of those who every day establish a relationship with the world around them.

A liability system based on a Insurance, guaranteed by the Insurance Companies, which attest to the quality of the property rather than to ascertain responsibilities by law: existing system that offers little assurance injured parties. We believe that this model, as with the other solutions offered by Insurance Companies, would commit those who systematically fail to their responsibilities. So, the recourse to the courts will be minimized, an option always so time consuming, and the continuous degradation of the property, with huge costs and highly complex repair: with the inevitable impact on those who manage and use buildings.

Therefore, we believe that a new articulation of processes (of the professionals involved and their work), despite the public and private interests, would serve to intervene, in time, to stop the addiction sources and control the trigger problems (mostly systemic), and yet ensure the earnings that motivate and encourage all those involved.

Acknowledgment

The author wish to thank Arch. Manuela Oliveira, Arch. Augusto Brandão and Eng. Engenheiro José Brito Subtil.

References

1. M. TAYLOR, H. HOSKER, *Quality assurance for building design*, Reino Unido, (1992)
2. L. Dias and M. Alves, *Organização e Gestão de Obras*, Instituto Superior Técnico, Departamento de Engenharia Civil e Arquitectura, Lisbon, (2002)
3. J. P. Couto, *A Deficiente Qualidade e Segurança na Construção Continua a Ser Determinante para a Falta de Competitividade* - Estudo em Curso Sobre os Factores de Competitividade do Sector da Construção, LNEC, Lisboa, (2006)
4. David I Cleland, William R. King, *Project Management Handbook*, Second Edition, John Wiley & Sons, (2008)
5. Moutinho and I. Vila, *Gestão e Organização de Obra*, Porto, Universidade Fernando Pessoa, Oporto, (2003)
6. J. Heizer and B. Render, *Operations Management*, International Edition, 7ª Edition, London, (2004)
7. Sycod*és* - Observatoire de la qualité de la construction – Qualité, Progressons Ensemble, Agence Qualité Construction (2006)
8. Sycod*és* - Les Indicateurs d'Évolution de la Qualité de Construtions, Agence Qualité Construction (AQC), Tableau de Bord (2008).

DOI: 10.1051/matecconf/20166813006

Effects of the Building Code on Construction and Design of Hong Kong Residential Buildings in the Colony

Wong Wah Sang

University of Hong Kong, Hong Kong SAR

Abstract. This paper discusses the effects of the Building Ordinance and its allied regulations to control buildings during the colonial period of Hong Kong and how these actually determined to certain extent the form of buildings especially for residential type of development. Various introduction of the building code at different times were used to illustrate the different effects and building forms created by the control in high density urban environment. Volumetric control and floor area control had been tried in the city with different construction of buildings developed. The paper concludes with the remark on possible new vision for building control in the modern high density city.

1 Background

1.1 Inroduction

The people of Hong Kong in the days of the colony (1841-1941, 1945-1997) had to live in an environment of high density due to the shortage of flat buildable land and the continued increase of population through immigrants coming from Mainland China. Hence, because of physical environment and political decisions, high density and even overcrowding was the common way of habitation. However, this led to problems of health and safety as evident with recurrent disasters of plague and fire spreading. Hence regulated control of the built environment by the government was a must and eventually determined the conventional form for the residential buildings.

The first type of control in the late ninetieth century looks at basic means of health and safety through both the building code and the lease to separate buildings and limit urban development. As building technology progressed in the twentieth century, the construction of high-rise buildings were then possible and the building control was focused at three-dimensional forms for sufficient light and open space. However, the building forms following the limitations of the ordinance are still evident even with more modern residential buildings.

1.2 A historical research analysis

To investigate into the relation between the physical buildings and the building code, the research methodology is to reveal all the ordinance and allied regulations issued by the Hong Kong government in the period of the colonial rule of Hong Kong and see how the physical forms such as building height, building features and floor plans of buildings have been affected by the legislative control. Due to the need of maximization of the developable space or floor area, the buildings have to conform to an imaginative mold set up by the building code. Hence, similarities in the external form and floor plans of buildings is the result of such building control.

Throughout the history of Hong Kong as a colony for about one and half centuries, the control of building development through various forms of Building Ordinances did occur. These ordinances have been changing because of the various human needs had changed, the advancement and development of building technologies and also the gradual cultural changes in the society. By revealing the different building code at various times and mapping it with the appropriate buildings at that time, the control on building form through the building code has been found to be obvious and evident to create physical architectural forms that have been distinctive of certain time and place.

Before going to the historical evolvement of the Building code, we have to see the most important background, i.e. the formation of high density in the colony.

1.3 Building in high density

According to the Census and Statistics Department, we notice that the land population density in Hong Kong at mid-2012 stood at 6620 persons per square kilometer. However, there are slopes and hills which remain green in Hong Kong so that the total built-up area accounts to less than one quarter of the total land area. Hence the actual average density should work out to be like 26,000 persons per sq. km. Kwun Tong, our most densely populated area, had 56,200 persons per kilometer [1]. Compared to the rest of the world, Manila has a density of approx. 43,000/sq. km., Paris approx. 22,000 and Cairo approx.18000.

Briefly, the answer to this when we trace back our history [10] can be in three folds: increase of population, lack of land supply and limited knowledge of building technology. Population at 26 January, 1841, when the Union Flag of Britain rose at Possession Point of Hong Kong Island was 7,450 but later the census of 1865 recorded a population of 125,504. However, land had never increased in the same rate with population and we can see the hilly Hong Kong Island had not encouraged any means of easy construction. Hence, low-rise buildings were mainly built and confined to the coastline or as merely supported by stilts on slopes. (Figure 1).

Figure 1. Potting Street in late 19th century.

People lived in the type of residential buildings called the "Tong Lau" [11] (figure 2) which were tenement houses with both Chinese and European characteristics. These are usually constructed of loadbearing brick walls supported timber and later concrete floors. Four storeys are the common limit of the building height. Each storey would be bounded by two side walls (called party walls) with windows only at front or back and then the long space is sub-divided into rooms with cooking and toilet facilities at the back. High density is created by the placing of these houses side by side, sharing the load-bearing wall to form row houses. Verandahs are common features which form extension of these buildings onto the pedestrian street. (Figure 3) The urban effect of this type of building can still be seen its five-meter modules of the previous façade along many streets with re-new developments. Fire and plague form two evils to this kind of residential accommodation as both can spread easily through the side-by-side houses.

2 The building code

2.1 Setting up of buildings ordinance

To look at the safety and health issues associated with such urban living, the first code was set up by the Hong Kong Government in 1856 [3] and was called the Ordinance for Buildings and Nuisance. Within this legislative control, terms for walls, storeys, buildings were defined and construction technique and materials were regulated. You may be surprised by the use of "Nuisance" in the ordinance. The nuisances mean the verandahs in poor state of care, buildings in ruinous and dangerous conditions, unauthorized building works, building parts erected with inflammable materials contravening the Ordinance etc. These are called nuisances because they can become causes of fire and unhygienic conditions. As this is written in the ordinance, it is probably a common phenomenon at those days.

Figure 2. "Blue house" at Wanchai.

Figure 3. Buildings at Shanghai Street.

Despite the control of high density through law, the demand for habitation continued as more people settled in Hong Kong from China such that plaque and fires did frequently occur. The issue for shortage of land supply was then partly resolved by reclamation – transforming land from the sea. And in 1868 the first coastal reclamation project (Praya Reclamation Scheme) [12] started to form at 1873 the present Des Voeux Road in Central. (figure 4) The British expert asked by the Colonial government to make recommendations on the urban situation in 1882 was Usbert Chadwick. Subsequently, the Ordinance was revised again and again. In the 1889 Buildings Ordinance, the limit of building height was set to 76 feet.

With continued increase in population, health was constantly a problem. Plaques broke out quite frequently so that in 1894, the Tai Ping Shan Resumption Ordinance gave authority to re-build 10 acres of the most unsanitary land. Health issues also led to the formation of the Public

Health and Buildings Ordinance in 1903 [4]. Here we have a definition for overcrowding of a domestic building written as "in excess of a proportion of one adult for every fifty square feet of habitable space or superficial area and five hundred and fifty cubic feet of clear and unobstructed internal air space".

In this 1903 Ordinance, the bye-laws list the construction of opium divans with standards of ventilation, drainage and opening hours. Yet in 1909, International Opium Conference was held at Shanghai to discuss the restriction on the sale of harmful drugs. Subsequently in 1910, all opium houses in Hong Kong were closed down.

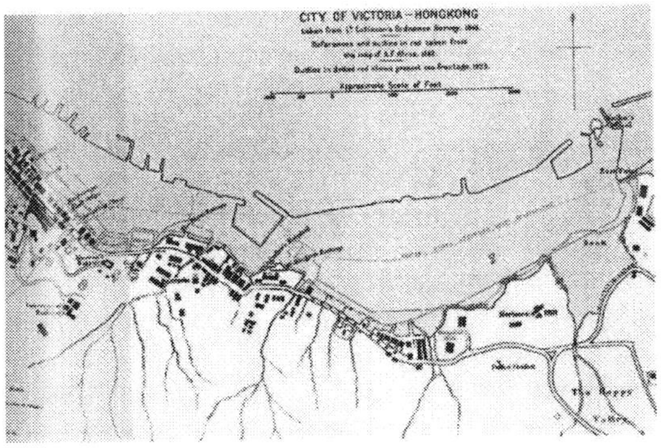

Figure 4. Map of Central in late 19th century.

Figure 5. The Hong Kong and shanghai Bank Headquarters in 1935

2.2 Constructin of tall buildings in twentieth century

At early twentieth century, reinforced concrete was introduced to the construction technology of Hong Kong. An example is the Hong Kong and Shanghai Bank (Figure 5) completed in 1935 October with a building height of 217 feet and 14 storeys. To cater for the technology advancement that brought increase of population density on a piece of land, the control of building was enacted as the 1935 Buildings Ordinance [5]. The main points of concern here are the minimum clear height of a floor to provide for sufficient lighting and ventilation, standard for the tread and rise of a stair to cater for fire escape, fire resisting door construction to hold up fire, minimum area for kitchen to secure a safe cooking place, minimum required opening for windows to allow for lighting and ventilation and provisions for open spaces for domestic (residential) buildings to allow adequate space around the building. To follow on this, the height of buildings was also controlled in relation to the adjacent street width. (Figure 6).

Following World War II was a period of intense building. The population was recorded to be 2.5 million at 1950. Fires got more serious and in 1953, a fire at the squatters of Shek Kip Mei (Figure 7) left 50,000 people homeless. Such were the background to bring about the establishment of the Housing Authority to look at public housing and the new Buildings Ordinance of 1955 to control private residential development. The statistics at 2012 reveals that 52.8% population lived in private housing, 29.7% public rental housing, 17.0% subsidized home ownership housing and 0.5% temporary housing. We can see both private and public sector play equal contribution to the residential living in Hong Kong.

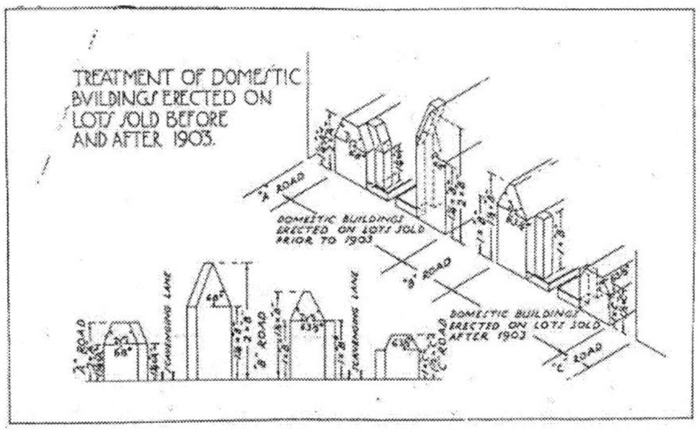

Figure 6. Control of building height based on street width.

Figure 7. Fire in Shek Kip Mei Xmas 1953.

2.3 Construction of buildings enforced by the ordinance

The high-rise era of Hong Kong began in the period of 50s to 70s with development of New Towns. At the same time, the very important Buildings Ordinance at 1955[6] established the basis of building development and control in the years following up till the present. The design and planning layout of buildings are controlled in terms of projections from buildings, building heights, volumes and open spaces, windows for lighting and ventilation, sanitation, staircases for fire escapes etc. Also, the construction of buildings was regulated with materials, structural loading, foundations, roofs, chimneys, fire-resisting construction, retaining walls, plumbing and drainage as well as lifts and escalators. (Figure 8).

Figure 8. Escalator in Man Yee Building of 60s.

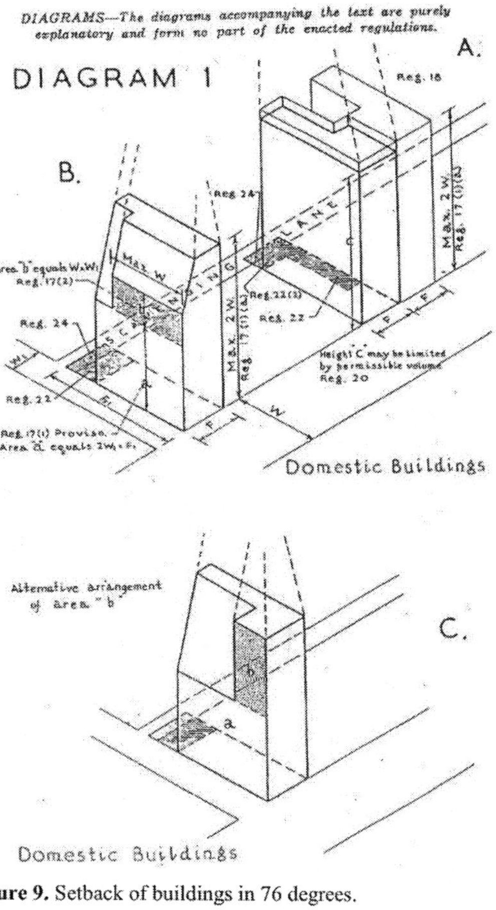

Figure 9. Setback of buildings in 76 degrees.

To clarify the control systems, the Buildings Ordinance was further divided into Building Regulations in three aspects of administration, planning and construction in 1956. The administration regulations take care of government procedures for approval of plans and permission for construction and occupation by users. The planning regulations oversee the design and layout of the buildings. The construction regulations look at the minimum standards of building materials and technology. And then subsequently regulations relating to demolition, sanitary provisions, private streets, refuse storage, lifts and escalators etc. were enacted.

The street adjacent to the building was made as a reference to development density control. The building bulk was related to the width of the street to allow adequate light and ventilation into the building. Briefly, it is based on "The wider the street, the higher the building." Setback of 76 degrees (ratio of 1:4) on the mass of the buildings was allowed and this molded the resultant building form like stepping pyramids (Figure 9). Then balconies and canopies over streets were also calculated with reference to the width of streets.

Restriction on the building bulk was also effected by the regulations on staircases and fire escapes. The riser height with maximum of 7 inches and the maximum of steps in one flight as 16 add up to 112 inches or 9 feet 4 inches. This is equivalent to the addition of 4 inches of structural slab plus about 9 feet as headroom of the internal space. Hence, if you need more headroom, you have to build more steps and an intermediate landing to maintain the maximum number of steps. Despite these high-rise developments, tenement houses still existed at that time and received more relaxed standard of control (Figure 10).

Figure 10. Post-war tenement houses.

A party wall is a common wall which supports two tenement houses or Tong-Lau. It also symbolizes ownership of the land because the site boundary actually lies in the middle of it. This poses a problem when the land owner decides to demolish the existing Tong-Lau, he cannot take down the party wall otherwise the adjacent Tong-Lau will collapse. And then subsequently, when the adjacent land owner wishes to demolish his building for development, he also cannot take down the party wall unless he got the agreement from all the individual owners of the new building next to him. As a result, these party walls become historical relics in the modern city.

2.4 Control of developable volume

An illustration of the control by volume for buildings can be seen in an example of a 1960s apartment building in Causeway Bay (Figure 11). The site area is approx.. 6200 sq. ft. and the resultant residential had got a ground floor of shops and then 13 floors of residential flats accumulating a total of 66000sq.ft. in floor area and approx. 645000 cubic ft. in volume. The building was a bulky development seemingly occupying the whole site and even more with projections of windows over the street and the upper floors with a 76 degree stepping form (Figure 12). This mode of development control actually generates the largest bulk of building.

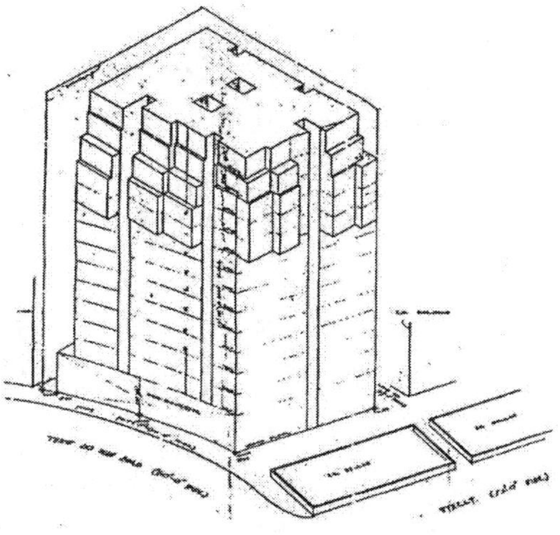

Figure 11. Apartment building in 60s.

Further categorization for building control can be seen in the 1962 Building (Planning) Amendments regulations. The notion of "plot ratio" and "gross floor area" were introduced. The gross floor area means the area enclosed by the external wall at each floor while the plot ratio is defined as the total gross floor area of a building divided by the area on which the building is erected. The plot ratio depends on the type of site which was defined in three classes. Class A means a site that abuts on one street; class B means a corner site that abuts two streets and class C means a corner site that abuts three streets or an island site surrounded by streets completely. (Figure 13) As a result of this, the more streets around you, the more you can build.

2.5 Building typology in the ordinance and control of building forms

Buildings were classified by the Ordinance into two types – domestic and non-domestic. One can imagine that domestic (residential) buildings with the population sleeping overnight will be more vulnerable to health and safety risks. The maximum plot for domestic buildings would work out to be 10 while that of non-domestic buildings is 15. Another important regulation clause to shape the urban environment is the permission of 100 % site coverage in the first 50 feet height of the building (Figure 14). The appearance for execution of this clause virtually cut off all potential open space in the ground floor and left the city with only open spaces in the form of roads for vehicular traffic like vertical concrete canyons.

Figure 12. Buildings with setback at top.

FIRST SCHEDULE

Percentage site coverages and plot ratios.

Height of building in feet.		Domestic buildings.						Non-domestic buildings.					
		Percentage site coverage.			Plot ratio.			Percentage site coverage.			Plot ratio.		
		Class A	Class B	Class C	Class A	Class B	Class C	Class A	Class B	Class C	Class A	Class B	Class C
Not exceeding	50	66.6	75	80	3.3	3.75	4.0	100	100	100	5	5	5
Over 50 feet not exceeding	60	60	67	72	3.6	4.0	4.3	97.5	97.5	97.5	5.8	5.8	5.8
Over 60 feet not exceeding	70	56	62	67	3.9	4.3	4.7	95	95	95	6.7	6.7	6.7
Over 70 feet not exceeding	80	52	58	62	4.2	4.6	5.0	92	92	92	7.4	7.4	7.4
Over 80 feet not exceeding	90	49	55	59	4.4	4.9	5.3	89	90	90	8.0	8.1	8.1
Over 90 feet not exceeding	100	46	52	55	4.6	5.2	5.5	85	87	88	8.5	8.7	8.8
Over 100 feet not exceeding	120	42	47.5	50	5.0	5.7	6.0	80	82.5	85	9.6	9.9	10.2
Over 120 feet not exceeding	140	39	44	47	5.4	6.1	6.5	75	77.5	80	10.5	10.8	11.2
Over 140 feet not exceeding	160	37	41	44	5.9	6.5	7.0	69	72.5	75	11.0	11.6	12.0
Over 160 feet not exceeding	180	35	39	42	6.3	7.0	7.5	66	67.5	70	11.3	12.1	12.6
Over 180 feet not exceeding	200	34	38	41	6.8	7.6	8.0	60	62.5	65	12.2	12.5	13.0
Over	200	33.33	37.5	40	8.0	9.0	10.0	60	62.5	65	15	15	15

Figure 13. Schedule of site coverage and pot ratio.

Figure 14. Podium of 100% site coverage.

Another control on building form was introduced in 1969 [7] as the "Street shadow area" through amendment on the building (Planning) Regulations. A formula of "F x W/2" was the limit to the street shadow of a building was formed by projecting the building like with a 76 degree sun onto the street. Here F is the width of the building frontage and W is the width of the street with the street shadow. The shadow was entirely mathematical and did not take into account the actual orientation of the sun. The result of this created buildings with the 76 degree sloping profile. This control created aesthetical effects generally not welcomed by the community and was later revoked. However, buildings (Figure 15) conforming to this control still remain.

Figure 15. High-rise buildings with sloping profile.

Residential prices in Hong Kong soared throughout the years so that through continuous experience with the real estate market and the restrictions from the Buildings Ordinance and regulations the most efficient residential floor plan is produced. (Figure 16) This ideal plan places all the common area into the central core of the building and usually divided into eight flats[2]. The staircase is an overlapping pair of staircases called the "scissors staircase". And the building of multi-storeys is basically an extrusion of the typical floor plan (Figure 17).

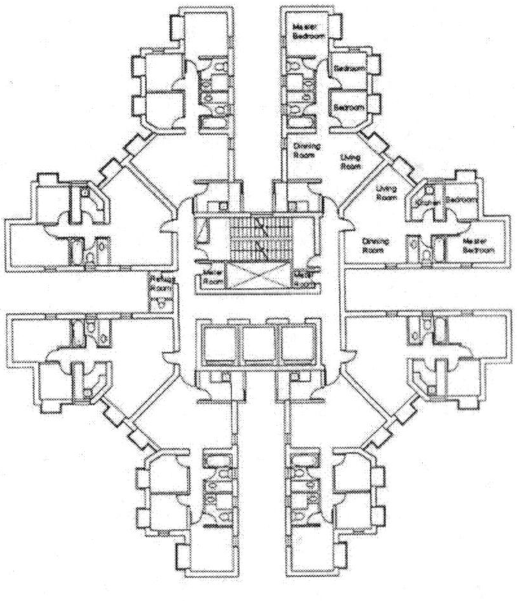

Figure 16. The floor plan with eight flats and central core.

Figure 17. Buildings as extrusion forms of the floor plan.

Figure 18. Typical building conforming to building code control.

The phenomenon of this floor plan can be revealed in further details. Projected windows called "bay windows" generally are assigned to each bedroom and living/dining room to enjoy the privilege of non-accountable floor area. Recesses on the floor plan called "re-entrants" are located to increase the length of building periphery as well as to provide windows for rooms especially the kitchen to satisfy the requirements for lighting and ventilation. And the whole tower may just rest on a 15m high podium to suit the property market or not. This type of plan[8] flourished all the years with high-rise residential building developments (Figure 18) and well carried into years after the return of the sovereign to China in 1997.

3 Final remarks and recommendation

Through the years of colonization in Hong Kong from early years with Tong Lau to the high-rise residential development in early 21st century, economy and efficiency has taken priority of development over other modern design parameters like sustainability, aesthetics, culture, computerization, etc. The control of development [9] is initiated through the government with the Director of the Buildings Department as the Building Authority. Initiation of building control rests with the government. The current methods of improvement and updating is still based on the basic 1955 Buildings Ordinance and regulations through modification, add-ons and amendments. As new technology from construction materials, building methods and computer engineering come about with social changes of working and living requirement, it looks that only with major change of the code can our living environment be varied and improved with a better holistic view of multiple issues.

This research study mainly has focus limited within the private residential developments in the history of colonial Hong Kong. Residential flats are called "domestic" in the building code while all other buildings are called "non-domestic". There is also a distinction between private and public residential buildings since private buildings are subject to control by the Buildings Ordinance and public residential buildings or housing are under the Housing Authority and do not require to be regulated through the formal procedure of the Buildings Department though they are still supposed to follow the building code. Hence forward research on this subject can be on public housing and non-domestic buildings to see how these building typologies have been influenced by the Building Ordinance and Regulations.

References

1. Buildings Department, "Building Development and Control in Hong Kong", Hong Kong Government, 27-53 (1999)
2. Chan, H W, Tang, BS, Wong, WS, "Density Control and the quality of living space: a case study of private housing development in Hong Kong", Habitat International, (2001)
3. Ordinance for Buildings and Nuisance in Hong Kong Gazette Note 12 of 1856
4. The Buildings Ordinance in the Hong Kong Government Gazette, (February 27, 1903)
5. The Buildings Ordinance in the Hong Kong Government Gazette, (April 12, 1935)
6. The Buildings Ordinance in the Hong Kong Government Gazette no. 68 of 1955.
7. The Buildings (Amendments) Regulations in the Hong Kong Government Gazette (1969)
8. Wong, WS, Legislative Control for Quality Buildings, HKIA Journal, **22**, 20-32 (1999)
9. Wong, WS, "Legislation and Safety of tall Buildings" from High-rise Living in Asian Cities, Springer, 81-112 (2011)
10. http://en.wikipedia.org/wiki/History_of_Hong_Kong_(1800s%E2%80%931930s)#Population
11. http://en.wikipedia.org/wiki/Tong_lau
12. http://en.wikipedia.org/wiki/Praya_Reclamation_Scheme

MATEC Web of Conferences **68**, 13007 (2016) DOI: 10.1051/matecconf/20166813007
ICIEA 2016

Developing Key Parameters for Green Performance of Partition Wall Blocks

GOH Cheng Siew[1] and NG Thomas Shiu Tong[2]

[1]*Heriot-Watt University Malaysia, Malaysia*
[2]*The University of Hong Kong, Hong Kong*

Abstract. To promote sustainable construction, it is important to consider green performance of construction materials throughout the life cycle. Selecting inappropriate materials could not only affect the functional performance but also preclude the achievement of green building performance as a whole. Green performance of construction materials has therefore been one of the primary considerations of green building assessment systems. Using partition wall blocks as an example, this paper examines green performance of building materials primarily from the cradle to gate boundaries. Nine key parameters are proposed for the green performance of partition wall blocks. Apart from environmental features, technical performance of partition wall blocks is also taken into consideration since it is the determinant of the lifecycle performance. This paper offers a roadmap to decision makers to make environmentally responsible choices for their materials of internal walls and partitions, and hence provides a potential sustainable solution for green buildings.

1 Introduction

The construction industry consumes 60% of raw materials globally [1] and it is the biggest consumers of materials by weight in US compared to other industries [2]. It brings major impacts on the depletion of finite resources, energy consumption, solid waste generation, pollution, and greenhouse gas emission [1, 3, 5, 6]. Furthermore, growing public awareness of environmental issues also lead to more attention to sustainable development in construction [4]. A green building paradigm adds significant impetus to the construction industry nowadays in order to reduce environmental impacts as well as improve the sustainability performance of the industry.

To develop green buildings, it is important to incorporate not only appropriate building design and technologies but also suitable building materials in the buildings. Wrong choices in materials specifications, suppliers, constructive technologies and management failure lead to a waste of material and handmade [2]. On the contrary, using appropriate construction materials can improve functional performance, aesthetics, and environmentally friendliness of buildings, in addition to a reduction of cost. As sustainability concept has grown in construction, there is a popularity of using green materials, which establishes the need for more sustainable solutions for construction materials.

A partition wall is an important component of buildings which divides space within a building into rooms or areas in order to enable more effective space utilization [1]. It can either be load-bearing or non-load-bearing, although in most cases it is non-load-bearing. The main functional requirements of partition walls are stability to self-weight, eccentric vertical loads and out of plane horizontal loads, visual, acoustic, and thermal and water resistance [5]. Moreover, contemporary internal partition walls need to take the following requirements into consideration: the speed of organizational and technological change; the increased number and complexity of services to be accommodated; the quality and aesthetic requirements; the need for acoustic separation of compartments; the flexibility for changing the area and organization of the internal spaces; thinner elements (to maximize the net areas); higher environmental life-cycle performance and optimal life cycle costs [6].

Environmental impacts of internal partition walls is attributable to the used material (including embodied energy and thermal properties) and the way the wall is built and maintained [6]. As described by Mateus *et al.* [6], partition walls have higher contribution to the overall life cycle impacts, as compared to other non-load-bearing elements. To determine key parameters of green performance, it is important to understand the variety of partition wall block, ranging from clay brick to hollow concrete block, autoclaved aerated concrete block, timber stud, lightweight sandwich membrane, fly-ash-lime-phosphogypsum hollow blocks, gypsum plasterboard, gypsum block and etc. [1, 2, 6, 7].

2 Partition wall blocks and types

Blocks are one of the important materials of walls and partitions and they have been used extensively as a popular choice in buildings. Hollow blocks or bricks gain the popularity in developing countries due to their low heat dissipation and less raw materials usage for forming cavities [8]. There are a great variety of partition wall blocks which are manufactured using different types of materials. They are also made in varying types, thickness, densities and sizes to accommodate for different demands. In general, their sizes are larger than standard bricks. Blocks are chosen because they can provide the better aesthetics and reduced costs by minimising the amount of cut units onsite. Grades of partition blocks are determined by the structural and functional performances of the partition, which are related to the level of activity in adjacent spaces, and the degree of care the partition is likely to receive [9].

2.1 Concrete blocks

Concrete blocks can be solid or hollow, with two or three cores or voids. The major materials of concrete blocks are cement, water, sand, gravel (quartz or stone) [10]. The common types of concrete blocks include dense block, lightweight foamed concrete blocks, and autoclaved aerated concrete (AAC) blocks. The shapes and sizes of most common concrete blocks have been standardized to ensure uniform building construction.

2.2 Gypsum blocks

Gypsum blocks are commonly made of gypsum plaster, water and additives such as vegetable or wood fiber for better strength. Gypsum blocks do not require sub-structure for erection. Instead of regular mortar, gypsum-based adhesive is used as the bonding agent.

2.3 Clay blocks

Clay blocks include honeycomb clay blocks and unfired clay blocks. Honeycomb blocks are formed with a cellular structure of vertical perforations. The perforating structure of honeycomb block reduces material use while giving better thermal performance [10]. For unfired clay blocks, they have excellent environmental credentials because approximate 30% of materials can be made from alternatives, recycled or secondary resources.

2.4 Hemp blocks

Hemp blocks (hempcrete) are blocks produced by blending hemp with a lime and cement binder. Hemp is a fast growing annual plant which is cultivated using different methods in different countries. They are manufactured using a cast or spray-on technology [10]. Hemp blocks are hygroscopic, thus offering a degree of humidity control.

3 Green performance of blocks

Selection of appropriate building materials is an important phase in the design process because it impacts not only future functionality of buildings but also their environmental performance [16]. This paper examines green performance of wall blocks principally from the perspective of cradle-to-gate boundaries, even though the impact after the product end-of-life is also taken into account. Cradle-to-gate boundaries evaluate the environmental impacts of materials during the production process, which embeds raw material extraction and processing, processing of secondary material input, transport to the manufacturer, manufacturing, waste processing and disposal of final residues during manufacturing phase [16]. Nine key parameters are identified for assessing green performance of partition wall blocks. They are 1) raw materials, 2) hazardous substances, 3) durability, 4) thermal insulation, 5) sound insulation, 6) fire resistance, 7) energy consumption, 8) waste management, and 9) reusability and recyclability.

3.1 Raw materials

There is an increasing trend to use by-products, recycled or waste materials to manufacture partition wall blocks. Vasconcelos *et al.* [5] proposed to use FGD gypsum (a synthetic industrial by-product), regranulated expanded cork, textile fibers and citric acid (setting time retarder) in manufacturing innovative blocks. Kumar [7] adopted fly ash, lime and gypsum to produce bricks and hollow blocks. Meanwhile, Wu *et al.* [8] used shale, fly ash, brick (from building rubbish) and waste paper pulp to produce fired shale hollow blocks. Recycling of those wastes and by-products by incorporating them into building materials is a practical measure to the pollution problem as well as to the conservation of depleting building materials [3].

Apart from waste materials, hemp (a renewable source) is also used as the raw material for partition wall blocks. Hemp is a robust annual crop with remarkable rate of growth, and hemp-lime wall has been demonstrated to be positive to the environment [11].

The use of waste materials and renewable sources (e.g. hemp) should not jeopardise the functional performance of blocks significantly. Several studies have been carried out to demonstrate that the addition of those materials would not affect the strength of blocks significantly [7, 19]. For instance, in the work of Kumar [7], hollow blocks made with high fly ash content can obtain the compressive strength of above 3.5MPa after 120-day curing, which can replace the burnt clay hollow bricks easily.

3.2 Hazardous substances

When more industrial by-products or wastes are used in manufacturing partition wall blocks, it is of concern to take hazardous substances into consideration. Hazardous substances and heavy metals shall be excluded from the use of production of partition wall blocks. If it is impossible to completely eliminate those substances, specific limits of hazardous substances/heavy metal shall apply for their use during manufacture as well as their presence in the final products.

3.3 Durability

A partition wall is expected to be serviceable for decades with relatively little maintenance. They need to be strong from frost action, salt crystallisation, biological agencies attack, chemical, aggressive environment and weather [7, 13]. To be durable, blocks shall acquire the minimum compressive strength to withstand the loads.

3.4 Thermal insulation

A maximum heat transfer coefficient has to be fulfilled by partition walls such as walls separating heated from non-heated zones, in order to promote more efficient indoor climate control systems [6]. Using high thermal resistance materials reduces the demand of electrical energy for cooling/heating purposes considerably [12]. For instance, introducing holes, cavities or air gaps in concrete blocks can improve the thermal insulation properties of materials since the thermal conductivity of concrete is higher than the thermal conductivity of air [12].

3.5 Sound insulation

Sound insulation is a significant attribute in designing internal walls to ensure there is minimum airborne sound transmission resulted from occupancies and traffic noise nuisance [13]. Although airborne sound insulation depends on the superficial mass, light and thin partition technologies, with appropriate design can perform better than conventional heavyweight wall elements [6].

3.6 Fire resistance

Partition wall blocks should be incombustible and provide a minimum period of time of fire protection as specified in building regulations [13]. Partition blocks with poor fire safety performance put the users and occupancies in danger. A partition wall block that excels environmentally is definitely not a good alternative if it fails to achieve the required fire properties.

3.7 Energy consumption

Lower environmental impacts, light weight and lower energy consumption used in composite wall systems gain growing interest in construction [14]. The embodied energy is the amount of energy consumed due to the raw materials extraction and manufacturing process to produce a finished product [1]. Therefore, the energy consumption (or embodied energy) can be used as an indicator to calculate the emission burden associated with the construction materials or components [1].

As indicated by Broun and Menzies [1], the embodied energy and air emissions associated with construction materials during their extraction, processing and manufacture take the largest percentage of total embodied energy and air emissions in buildings. Therefore, a significant reduction in energy use during the production can considerably improve the environmental performance of buildings. Therefore, blocks shall be designed and manufactured without much use of fossil fuel and electricity. The manufacturing process should be energy conservative, being devoid of sintering or autoclaving [7].

3.8 Waste management

The waste generated by each building solution can be managed as construction and demolition (C&D) waste [2]. C&D waste is about 13% to 50% of solid waste entering landfills in Australia, Finland, Germany, The Netherlands, United Kingdom and The United States [17, 18]. Wastes generate severe impact on the malodours air factor, climate change, land use, marine aquatic eco-toxicity and others [3]. To reduce waste, it is important to decrease the amount of waste generated as well as adopt an effective system for managing unavoidable waste produced [18]. Leiva *et al.* [20] suggested proper quantification and classification as the important steps of waste management to determine the type and amount of C& D waste available for recycling analysis. To achieve that, a waste management policy covering manufacturing operations shall be implemented.

3.9 Reusability and recyclability

The importance of achieving flexibility and increasing adaptability to uses and changeable needs is emphasised [6]. Emmanuel [18] highlights the importance of recyclability and reusability in assessing environmental suitability of wall materials since construction waste has a residual value and can be avoidable. Blocks can be designed to facilitate assembling and disassembling to prevent the production of huge amount of waste. The merits of recycling and reuse is also taken into consideration to reduce the embodied energy and pollution imposed by blocks after their end of life.

4 Discussion

Material selection is a key step to attain green performance of buildings. Using building materials with lower embodied energy and carbon emission is one of the approaches to reduce building embodied energy and carbon dioxide emission [1, 15, 16]. Research [16, 18] has been carried out to measure green performance of buildings using embodied energy and carbon dioxide emission.

Although environmental consideration such as carbon footprint or embodied energy is important, green performance should not be evaluated solely on the basis of environmentally friendliness. It should take into account physical-mechanical properties and technological requirements of the evaluated building materials. Emmanuel [18] advocated that the green performance of building materials should be held up against the technical and functional performance as well as economic reality. Environmental performance of building materials should not be developed at the cost of economy and society. If technical requirements are overlooked, the lifetime of materials could be significantly shortened.

By using partition wall blocks as an example, this paper examines green performance by considering their environmental factors as well as technical requirements. Durability, sound insulation, thermal insulation and fire resistance properties are identified as key parameters to ensure "all-round green performance" of blocks. Since wall block is a building material, it is favourable to reflect the environmental impacts incurred within the cradle-to-gate boundaries.

Cradle-to-gate assessment is a partial product life cycle assessment from the raw materials extraction to the factory gate. It is a useful tool for examining the environmental impacts of raw material production by considering the upstream flow of energy, fuel and raw materials.

With the identified key parameters, suppliers and manufacturers could improve their product development towards greener performance in construction. In the meantime, this paper also assists industry stakeholders and end-users in making greener choices for internal walls and partitions. The demand for and supply of greener building products in built environment can thus be promoted.

5 Conclusions

Choosing green materials in buildings can add values to the overall building design and sustainability performance. This paper examines the green performance of blocks principally from the cradle-to-gate boundaries. Nine key parameters are developed to assess green performance of partition wall blocks. In addition to environmental consideration such as energy consumption and waste management, the fulfillment of functional requirements of wall blocks are also essential for green performance. This paper provides a roadmap in guiding decision makers to make environmentally responsible choices for their materials of internal walls and partitions, and hence offers a potential sustainable solution for green buildings. Further research can be carried out to determine innovative design for making the blocks more sustainable, durable, flexible and recyclable.

References

1. R. Broun, _ G. F. Menzies, Life cycle energy and environmental analysis of partition wall systems in the UK, Procedia Engineering, **21** (2011) 864-873
2. A. _ández-García, V. Ibáñez-Forés, and M. D. Bovea, Eco-efficiency analysis of the life cycle of interior partition walls: a comparison of alternative solutions, J. Cleaner Production (2015)
3. K. Condeixa, A. Haddad, D. Boer, Life cycle impact assessment of masonry system as inner walls: a case study in Brazil, Constr Build Mater **70** (2014) 141-147
4. M. _, J. Braet, and A. Audenaert, Life cycle assessment in the construction sector: A review, Ren Sust Energ Rev **26**, 379-388 (2013)
5. G. _, *et al.* Innovative masonry blocks for _ walls, presented at XXXVII IAHS World Congress on Housing (2010)
6. R. _ *et al.*, Sustainability assessment of an innovative lightweight building technology for partition walls-Comparison with conventional technologies, Build Environ, **67** (2013) 147-159
7. S. _, Fly ash–lime–phosphogypsum hollow blocks for walls and partitions, Build Environ **38**, 2 (2003)
8. J. Wu *et al.*, Mechanical and thermal tests of an innovative environment-friendly hollow block as self-insulation wall materials, Constr Build Mater **93** (2015)
9. _ and Construction Authority, Performance specifications for buildable internal partition, Information on http://www.bca.gov.sg/publications/BuildabilitySeries/others/bswall.pdf
10. _, Blocks: Concrete masonry and its alternatives, Information on http://www.greenspec.co.uk/building-design/blocks/
11. K. _, and A. Miller, Life cycle greenhouse gas emissions of hemp–lime wall constructions in the UK, Resour Conserv Recy **69** (2012) 1-9
12. K. S. Al-Jabri *et al.*, Concrete blocks for thermal insulation in hot climate, Cement Concrete Res, **35**, 8 (2005) 1472-1479.
13. E. A.W. Hendry, Masonry walls: materials and _, Constr Build Mater **15**, 8 (2001) 323-330.
14. P. Samani *et al*, A sustainability assessment of advanced materials for novel housing solutions, Build Environ, **92** (2015) 182-191
15. Y. Saadah, and B. AbuHijleh, Decreasing CO2 _ and Embodied Energy during the Construction Phase Using Sustainable Building Materials," Int J Sust Build Technol Urban Dev, **1**,2 (2010) 115-120.
16. A. Estokova and M. Porhincak, Environmental analysis of two building material alternatives in structures with the aim of sustainable construction, Clean _ Environ Policy, **17**, 1 (2015) 75-83.
17. B. A. G. Bossink, and H. J. H. Brouwers, Construction Waste: Quantification and Source Evaluation, J Constr Eng M ASCE **122**, 1 (1996) 55-60.
18. R. _, Estimating the environmental suitability of wall materials: preliminary results from Sri Lanka, Build Environ **39**, 10 (2004) 1253-1261.
19. S. C. Kou, B. J. Zhan, and C. S. Poon, Properties of _ wall blocks prepared with fresh concrete wastes, Constr Build Mater **36** (2012) 566-571.
20. C. Leiva, Solís-Guzmán J, Marrero M, García Arenas C. Recycled Blocks with Improved Sound and Fire _ Containing Construction and Demolition Waste. Waste Manage **33**, 3 (2013) 663-671.

MATEC Web of Conferences **68**, 13008 (2016) DOI: 10.1051/matecconf/20166813008
ICIEA 2016

Enlightenment of Qilou Street Space Intelligence to Pedestrian System Design in Commercial District

Jing Chen[1], Fang Wang[2] and Jiaping Liu[3]

[1]*Xi'an University of Architecture and Technoloy Landcape Architecture Post-doctoral research station, Xi'an 710055, China*
[2]*Xi'an University of Architecture and Technoloy Rural and urban planning Post-doctoral research station, Xi'an 710055, China*
[3]*Xi'an University of Architecture and Technoloy college of Architecture, Xi'an 710055, China*

Abstract. Aiming at those generally existing problems in urban commercial pedestrian districts, through the design methods of extracting separation of man from vehicle in traditional Qilou pedestrian street space, continuous pedestrian space with shelter, and the behavior paths with combination of commercial activities, this thesis sets about from the behaviors of pedestrians in walking state, and finds that the traditional Qilou pedestrian street space can satisfy pedestrians' physiological and psychological demands in shopping better. At the same time, combining current urban commercial districts developing demands, and the successful experience from traditional Qilou, this thesis also proposes relevant improvement measures. The improvements in design means are expected to improve pedestrian environment for, and achieve the commercial values of these districts, as well as the win-win of environment improvement and commercial activity development.

1 Introduction

Urban commercial districts often located in some important core district of city where many people will gather together. The get-together of large numbers of pedestrians causes a series problems, such as disordered site, traffic disturbance, bad environmental quality, etc., and all these problems have been presented intensively in pedestrian space system of commercial districts. For a long time, the pedestrian system design of urban commercial districts includes two classifications: The first is the pedestrian system space with primary of complete pedestrian streets. This kind of pedestrian system can separate traffic flows out of pedestrian system, which can relieve the pressures to pedestrian streets by motor vehicles, and enhance the security of pedestrian streets. However, it also brings negative effects that block unban road traffic, increase pressures of those roads around pedestrian streets, and also influence nearby residents' going-out. At the same time, as some regions are deeply influenced by local special climatic conditions, efficiency of some pedestrian streets actually cannot be given full play. The second is the common pedestrian streets in general commercial districts, which has no significant differences with those ordinary urban streets, and it is only the sidewalk outward of commercial buildings in the space lay. This kind of pedestrian street is short of professional design, and except can satisfy essential traffic function, has few attractive effects to pedestrians by pedestrian space. Therefore, the values of commercial buildings cannot be expressed through pedestrian streets. In allusion to above two circumstances, we will proceed with analyzing commercial activities of pedestrians, and it is crucial to excavate successful experience of traditional commercial pedestrian streets.

2 Analysis of pedestrians' behavior psychology in commercial pedestrian districts

2.1 Security

The security of commercial pedestrian districts is a primary problem concerned by every waling shopper. *The Global* Situation *Report of Road Safety 2013* reveals that the number of people who die of road traffic accidents annually in the world reaches up to 1.24 million, among 270,000 of which are pedestrians, accounting for 22% of the total death toll for road transportation safety. The security problems of commercial pedestrian roads mainly concentrate upon coexistence of people and vehicles, signal lamp faults, unreasonable street crossing safety facilities set, unreasonable connection of pedestrian walkway and public transportation facilities, etc. As no consideration of growth rate of motor vehicles are taken into the design of some commercial districts, and at the later stage, roadways was broadened by occupying sidewalk blindly, it is easy to give rise to the complexion of people and vehicles coexistence. While some street roads are excessively wide and the waiting time for automobiles is too short in signal lamp design, all will lead to condition of snatching path by pedestrians and automobiles. It is quite dangerous for the disabled, the aged with mobility problem and children with weak transportation safety awareness.

2.2 Comfort

The comfort problem of commercial pedestrian districts is an important issue in pedestrian system design. At present, most are open type of pedestrian systems in commercial districts, and pedestrians mainly proceed along sidewalk. Two problems often will be neglected in the design of such open type of pedestrian system: 1. Behavior comfort of users. For the open type of pedestrian system, it can satisfy pedestrian comfort demands in specific time periods (such as early morning or dusk), while it cannot satisfy essential usage requirements with the effects of burning sun and intense fall within considerable time. 2. Psychological comfort of users. The pedestrian system design without support of building and environment is faced with some problems, such as hot weather, noisy surroundings, dusts all over the sky, waste water overflowing, excessively serried pedestrians, etc. Even though this pedestrian space can satisfy peoples' basic usage requirements, it cannot meet their sensory demands.

2.3 Convenience

Due to the uncertainty of pedestrians' behaviors, the behavioral purposes of those who are in a same road can be different. While in most cases, the diversity of people's pedestrian habits doesn't be taken into consideration by the pedestrian system in commercial districts, and the walkways are only developed into the means for traffic. For how to pass through urban streets conveniently and safely, how to enter into public traffic system conveniently and safely, how to distinguish traversing pedestrian stream and staying pedestrian stream in a same district in case of pedestrian congestion, and how to care for the disabled and those with mobility problems, this system still has no sufficient consideration. Some relevant data indicate that, if only passenger foot-bridge and tunnel are available for pedestrians to cross streets, 24% of pedestrians will give up street crossing, and 39% of them will reduce times of crossing streets (48% for the aged). Likewise, the arbitrary occupation of sidewalks for the blind, unreasonable public facilities set and several factors also bring more incommodities to the service of pedestrian space.

2.4 Affinity

Except a small number of specific commercial pedestrian streets, the business interface analysis along the streets in the pedestrian systems of most urban districts is relatively random, and business interfaces don't have a strong affinity to pedestrians. Business interface and urban pedestrian system are two separated systems, and buildings are separated with sidewalks by outer walls. The pedestrians usually pass by a hurried glance under the burning sun, except they have certain purposes to commercial space. Business interface and sidewalks lack a kind of bias relation.

2.5 Randomness

When pedestrians enter into commercial districts, most of them won't have obvious purpose. But with the attractions of commodities, advertisements, promotion performance and other outer commercial activities, or the external disturbances from weather, motor vehicles and construction activities, pedestrians often will not proceed upon the path schemed in original commercial pedestrian streets, but will select their routes upon their own subjective ideas.

2.6 Experience

In today's rapid development of E-commerce, co people hope to actively engage in it to acquire distinctive feelings and various integrity experiences, as well as make pleasant memories that are difficult to be forgotten. Consumers are not content with the single consumption experiences of "Window Shopping", but desire to enhance experiences through all-around participation of visual sense, taste sense, smell sense and auditory sense, to make full sense of urban cultural connotation.nsumption activities have not been limited to the simple service modes, like "home delivery service", and now

3 The intelligence of traditional qilou street to answer commercial activities

Qilou was a typical urban street and architectural form at the end of 19th century and the beginning of 20th century. The extensive development in China's South-East coastal areas for nearly 100 years explains that Qilou buildings and streets have a strong geographical adaptability. As the typical residential office building types, Qilou usually presented to core commercial districts cities and towns then. Although the commercial industry and building type of these commercial districts had a great difference with today's urban commercial districts, the consumer psychology of those pedestrians in commercial districts in recent years was similar. The extraction of Qilou street pedestrian space intelligence will be conducive to recognize the relations between human behaviors and pedestrian space.

3.1 Explicit space of separation of man from vehicle

Although at the beginning of constructing Qilou, the quantity of automobiles was much less than that of today, while urban planning, the relations between Qilou sidewalks and urban main streets has been taken into consideration. For example, in the 1920s, *Guangzhou city Municipal Office take arrest building 15 feet wide arcade charter* issued by Guangzhou City had the stipulations that the streets of 100 feet (about 30m) can be constructed with 20-feet-width (about 6.1m) Qilou, and 80 feet (about 24.4m) streets can be constructed with 15-feet-width (about 4.6m) Qilou. The traffic layout which divided urban traffic trunk road and footpath will be helpful to separation of pedestrian and vehicles. Through the research to domestic existing Qilou streets, we can find that: the main reasons Qilou street separating pedestrian and vehicles are as following: (1) There's 10~20 cm height difference between roadway and

 DOI: 10.1051/matecconf/20166813008

walkways, it limits clearly on the border between roadway and walkways. (2) Due to narrow footpath, if the car park at footpath, it will hinder the pedestrian traffic. (3) Continuous colonnade at footpath will hinder car for entering. (4) Peddlers who are along the outside of Qilou Street (near the roadway) will hinder vehicles closing. To sum up, Qilou building has separated pedestrians and motor vehicles clearly and ensured pedestrian traffic safety in the walking process (Fig. 1).

Figure 1. The ways of separation of man from vehicle.

3.2 Continuous sheltered pedestrian space

Another wisdom of Qilou Street lies in that it forms continuous semi open space through continuous side-by-side buildings. This kind of continuous shelter space ensures pedestrians can pass in the internal space no matter in hot sun or in the wind and rain. It changed the situation that people have to hide inside the building for lacking of covering measures and it provides more possibilities for pedestrians to do outdoor activity under special climate conditions.

3.3 Behaviors linked with commercial activities

Qilou building as a residential and commercial amphibious model building, its business development has strong link with traffic layout. (1) Pedestrian space of Qilou Street has shortened the distance of pedestrians and shops. Influenced by external climate and traffic, mostly pedestrians walk on walkway less than 3m. Often many pedestrians are walking by stores, it's better for them to find and realize what's in it. Interface along the street adopts the open stance. Most of Qilou commercial buildings use the completely open mode. Pedestrians can understand internal operating conditions at a glance passing by and easily enter in. those enclosed, limited gateway of commercial space cannot compare with it. (3) Commercial industry is developed in linear form rather than centralized distribution, thus it can avoid the strong purposeful centralized shopping mode usually at mall and bring positive influence to development of experiential business model (Fig. 2).

4 The way integrated traditional wisdom of qilou and commercial pedestrian system

4.1 Add continuous pedestrian space with shelter

For pedestrians, physical and psychological comforts are the first issue should be met during shopping. however, in the contemporary urban business district, buildings and roads are often disjointed and lack necessary linking, so pedestrian will pick up comfortable period (such as evening), or at the day time with purpose to enter certain mall rather than stay under strong sunlight pedestrian path. Enhance, in order to increase stream at daytime and extend their stay, it's necessary to stop the influence to pedestrian system by vehicles and transform interface along the street of original commercial buildings. The transformation mainly involves two aspects: (1) transformation on interface along the street. Back and leave some space of original commercial space interface to realize better pedestrian space with shelter. (2) Add vestibule with shelter between building and building. Due to the limitation of current requirements for fire protection and actual construction situation, it's difficult to realize continuous of hundreds of meters pedestrian street space with ceiling. In hence, it should link the pedestrian space by setting of vestibules appropriately (Fig. 3).

Figure 2. Open and closed commercial interface.

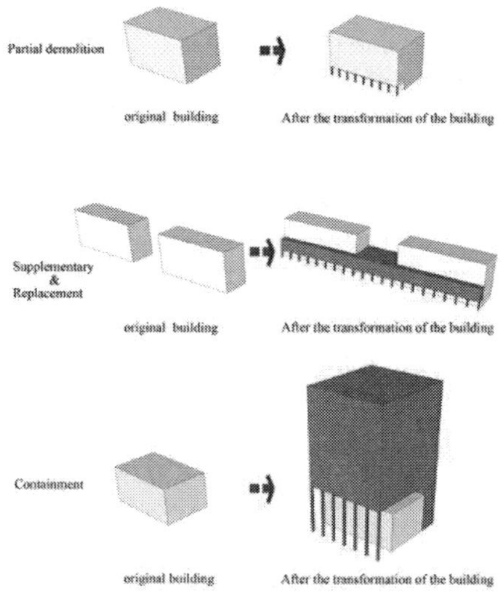

Figure 3. Transform ways of existing buildings.

4.2 Broaden the scale of walking space

Influenced by rules and regulations and its building structure system,traditional Qilou building usually has a spatial scale at width of 2.1~3m and height of 4~5m. This kind of space can satisfy the need of 2~3 streams of people. Due to the narrow space, it's difficult to increase people, other service facilities and to expand using function. Of course, modern commercial development requires a lot of pedestrians and at the same time, the progress of structural technology can offer the guarantee. Assuming that the standard column grid used by current market is 8 m, one column space back to street will form the 8 m wide pedestrian street space. Under this space, it not only can increase more service facilities, but also attracts more people to commercial interface. Meanwhile, it's possible for the vertical transport facilities with this width. For example, increase the height of Qilou building to 2 or 3 floors, it can lead pedestrians to floor 2 or 3 with escalator, thus it will increase the passenger flow of intervention (Fig.4).

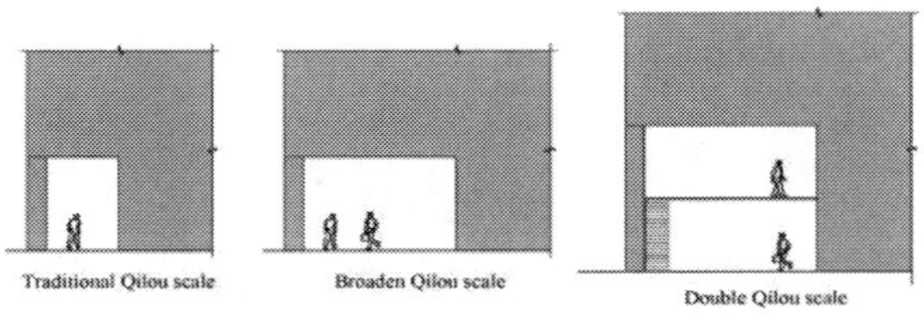

Figure 4. Broaden the scale of walking space.

4.3 Perfect external connection way of pedestrian system

Semi open space of Qilou Pedestrian Street and city vehicles system are isolated and interpenetrative. On the one hand, vehicles have been isolated from pedestrian system and on the other hand, pedestrians can easily enter into motorway, it provides lots of conveniences for linking of public facilities and pedestrian system. For example, people want to take public transportation, they have to walk through the bike lanes and then wait at the narrow waiting room. Under sun and rain, it often cannot meet the needs of a large number of passengers. While in Qilou Pedestrian Street, its width and continuous space can accommodate lots of passengers. At the same time, Qilou Pedestrian Street with a certain width can offer convenient access for metro.

4.4 Perfect the service facility in pedestrian system

Another important supporting factor is the perfect service facilities. Service facilities can be divided into service facility for rest, hygiene service facility, orientation and informational services and service facilities for pleasure and sightseeing (Table 1). In addition of those people who with purpose of passing through, most of pedestrian with weak objective need facilities to meet the needs of comfort or novelty.

5 Conclusion

Qilou pedestrian street space as the rudiment of modern commercial pedestrian system developed in China, despite its size, functions, forms still cannot meet the needs of commercial development in urban core areas, but it starts from mental shopping need of pedestrian and fully takes requirements of pedestrian mental activity into consideration and satisfy it through the way of building, the successful experience is worth studying. Extracted experience from traditional Qilou pedestrian space can be integrated into demand of current business district development, through improvement of design means, it can meet the psychological demand of pedestrians in pedestrian shopping also maximize the interests.

Table 1. Service facility in pedestrian system.

Item	Content	Function
Service facility for rest	seat	Meet the needs of rest and waiting
Hygiene service facility	Trash, waste recycling	Keep the street hygienic; recycle garbage can be recycled;
Orientation and informational services	road sign, house number, billboard	Hint the information of places, roads and traffic site
Service facilities for pleasure and sightseeing	Sculpture, sketch, greenery landscape, display space	Beautify road environment, show regional culture and strengthen the business promotion.

Acknowledgements

The research described in this paper was supported by science and technology talents fund of Xi'an University Of Architecture And Technology（RC1301）& the National Natural Science Foundation of China (Grant No. 51408470, 51221865,51178369).

References

1. The Chinese people's political consultative conference news.1/4 of traffic accident death toll is pedestrian in China [EB/OL]. http:// cppcc.people.com.cn/n/2013/0508/c34948-21401476.html, 2013-05-08. [In chinese]
2. Cao Guozhang, Yang Xiaoxiang. Study on walking street culture in the experience economy stage. Huazhong Architecture, 2010(11):163-165 [In chinese]
3. Huang Sujuan. Guangzhou Qilou's regulations and practice in the period of the republic of China. Academic Research, 2013(3):123-133 [In chinese]
4. Chen jing. Haikou Qilou regional adaptability model research [D]. Xian: Xian university of architecture&technology, 2012:164-165 [In chinese]

MATEC Web of Conferences **68**, 13009 (2016) DOI: 10.1051/matecconf/20166813009
ICIEA 2016

Building Rehabilitation with Dry and Wet Systems – Embodied Water Comparison

Mónica Macieira and Paulo Mendonça

Lab2PT, Architecture School of the University of Minho, Guimarães, Portugal

Abstract. Fresh water is a scarce resource. Its availability, essential for life, is being conditioned by several menaces, mostly from human origin. Studies on water saving strategies in building construction and what procedures should be adopted for its efficient use can respond to part of these problems, taking into account the relevance of construction activities, responsible for almost 40% of the overall environmental impact in Europe. Buildings rehabilitation assumes an important role in the construction activities of many European countries. The aim of this paper is to compare the water demand between dry and wet construction systems used in interior partition wall technologies existing in south European climates, specifically on Portuguese territory. This paper focus on the advantages of the dry and lightweight partition walls and it is expected that it may contribute for the reduction of water use in the construction industry. The evolution of interior partition walls, namely by its weight and embodied water analysis, is presented and discussed. Apart from the conclusion that dry construction systems, used in the rehabilitation of existing buildings, can contribute to a significant reduction of water demand, this analysis intends to also reveal on how much these systems can reduce other environmental impact indicators in relation to the wet conventional solutions. Dry construction strategies allows saving material and energetic resources, increasing flexibility and reducing wastes, what constitutes a relevant response to the growing concerns over environmental issues and incertitude on economic development.

1 Water use in the construction industry

Until 2050 it is expected a world population growth, mostly in cities of poor and under development countries. This fact will contribute to the increase of environmental pollution and to resources management problems. However, in the developed countries, the population is decreasing and policies are focused on a significant decrease on the pollutant emissions. Knowing that building construction is responsible for a significant percentage of the world consumption of fossil fuels and the emissions of greenhouse effect gases, it doesn't make sense to continue investing in new construction with so many buildings unoccupied or sub occupied. The rehabilitation of buildings has assumed an important role in the construction activities panorama of developed countries. According to published estimations, this activity represents in Europe about 45% of the total volume of investment [1-3]. Building rehabilitation will therefore be a preponderant factor to achieve sustainability.

Although the demand for water consumption has increased significantly over the last century, the infrastructure capacity to harvest adequate water to serve the Earth's growing population has not grown accordingly, and in many parts of the world, water scarcity is becoming a critical issue. In addition to the daily operation of buildings, water is used in the production of building materials.

In Europe, the construction industry is one of the sectors that consume more fresh water [4]. Water consumption practically accompanies the entire lifecycle of a building material, ranging from the extraction of raw material to the demolition phase at the end of building's life cycle.

The building's water consumption management can be realized at three levels: (1) selection of materials and components with low embodied water; (2) selection of more efficient sanitary appliances and devices; (3) provide rainwater harvesting and water reuse solutions.

Table 1. Water used in an office building during its life cycle of 50 years [5].

	Building life cycle phases		
Resource	Construction	Use	Deconstruction
Water (m^3)	3.000	228.000	0

There are few studies about water consumptions in buildings during its complete useful life. One available case study corresponds to an office building in Finland, with a gross area of 15.600m2, constituted by three towers with five floors each. According to Junnila [5], this building had its water consumption distributed over the different life cycles phases as indicated on Table 1.

At Table 1 it can be seen that during the building use phase occur the largest consumption, corresponding to approximately 99% of the total water used. Residential buildings present consumptions at the same or higher order of magnitude [5]. However, even if corresponding to just 1% of the overall number, the water consumption during the construction phase is still of 192 litre per square meter of gross area, which gives a notion of the significant water demand during the construction phase.

For the use phase is usually considered a time period of fifty years, while the construction phase has a significantly lower period of about one year. Thereby, dividing the amount of water used during the use period by fifty, corresponding to the time which it was used, a value of 4 560 m3 / year was obtained.

2 Dry construction

2.1 Contribute to the reduction of water use

With an increasing fresh water scarcity it is necessary to review the currently adopted construction systems. For example, the construction based on concrete mortars, generates additional water waste on the construction sites. Water consumption in masonry wall systems uses approximately 500 litres per square meter of gross area, much more than in dry systems as can be verified at the following topics.

2.2 Advantages for buildings rehabilitation and consequent environmental impact reduction

Construction and mainly the housing sector consume big quantities of materials, what result on big quantities of waste according to Mateus *et al* [6]. Therefore for an efficient construction system it is necessary to use prefabricated elements that facilitate assembling and disassembling.

The growing use of dry construction systems in the building sector is due to the many advantages of these systems, primarily their short construction time, economy, superior building functional performance when compared to masonry and concrete systems with the same thickness [7]. Dry systems allow achieving solutions that are: lighter, flexible, quick and easy assembled, don't require much available space on site, and are easily transportable to construction site – important aspects to safeguard on rehabilitation interventions.

Assuring comfort and energy efficiency during use phase is one important goal for sustainable construction. According to Friedman [8] such aim can be accomplished through, among other aspects, the rethink of indoor space, with increased flexibility.

2.3 Lightness

Dry construction systems can be added to existing buildings in order to improve specific properties, an aspect that is especially important for infill development tasks (extra storeys, expansion, extensions) and alterations. The low weight of dry construction systems means that loadbearing components can be sized more economically than would be the case with a fitting out scheme involving masonry and concrete. A clear reduction in mass and at the same time better sound and thermal insulation properties are readily achievable in wall systems (partitions, external walls, facades).

2.4 Flexibility.

A great portion of housing designs are not flexible in use and therefore they are not adjustable to the constant updating of life styles and variations on the composition of the households. The conventional practice uses heavyweight and static partition walls and therefore it is difficult to rethink the use of the indoor spaces. This problem reveals to be especially important in the refurbishment phase of existing buildings, but also when new buildings are designed.

According to Tichelmann & Pfau [9] these building elements were underestimated in the past. They also perspective that the vast majority of buildings erected between 1950 and 1995 will become unusable in the long term and will become increasingly difficult to let or sell. The small-format interior layouts acceptable in those days are no longer popular with actual users and buyers. Changing room sizes means an expensive intervention in the building.

The use of dry systems in lightweight construction, optimized for their particular functions, is generally associated with a gain in floor space and greater flexibility of usage – their intrinsic properties which were underestimated in the past.

Flexibility of housing spaces can increase the life time of the building, extending materials life-cycle [10]. Flexibility can be achieved through neutral and polyvalent spaces – passive flexibility – or through mobile elements that allow the transformation of the housing unit – active flexibility. The second strategy can produce daily, seasonal or annual transformations of the housing unit that increase its adaptability to the user and its changeable needs. This allows saving energy and resources, since people can use their home during larger periods of time and in a more adequate way to seasonal climate variations.

Innovation in building technology and products should be conducted to increase flexibility. One of these innovations is the distinction between the structural elements and the other elements of a building [11]. A reduction in the use of load-bearing interior partitions and increase in the number of pre-manufactured components provide the opportunity for greater choice and adaptability in design and construction [9].

3 Case study

3.1 Water consumption of interior partition wall systems

Interior partition walls are thin elements built to divide the indoor space into rooms or other compartments. The importance of these building elements is reflected in the global construction cost, estimated by Köning *et al* [12] to be around fifteen percent in office and administration buildings of European countries. In the non-loadbearing group of technologies used on the buildings construction, these represent 41% of the environmental impacts, as can be seen on Figure 1.

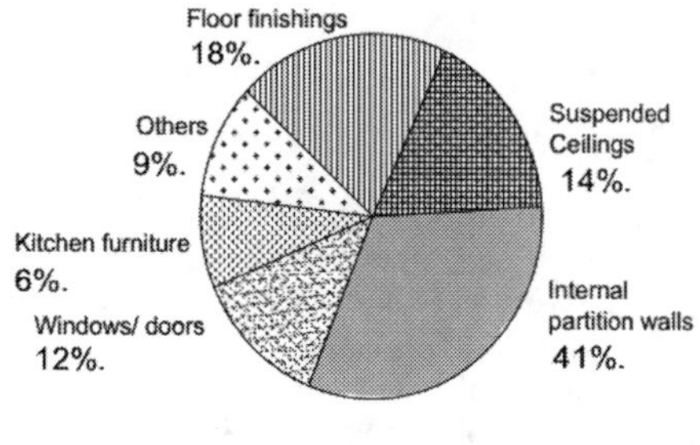

Figure 1. Environmental impacts from materials use on non-load bearing construction elements of a typical house over 60 years (adapted from Addis and Schouten [11]).

Generally, partition walls are non-load bearing. In terms of structural resistance, these elements need only to be strong enough to support themselves under normal conditions of service. In general, lightweight solutions present less embodied energy, less energy costs with transport to the building site and can be designed with smaller assembly fittings [13].

Many of the commonly used materials for interior partition walls use water as a resource for their manufacture. The following table lists some materials commonly used in construction of interior partition walls, with the corresponding amount of water demand in its manufacture according to their weight. It presents the amount of water which will have its quality drastically reduced as a result of the production process. The water recycling or purification was not taken into consideration.

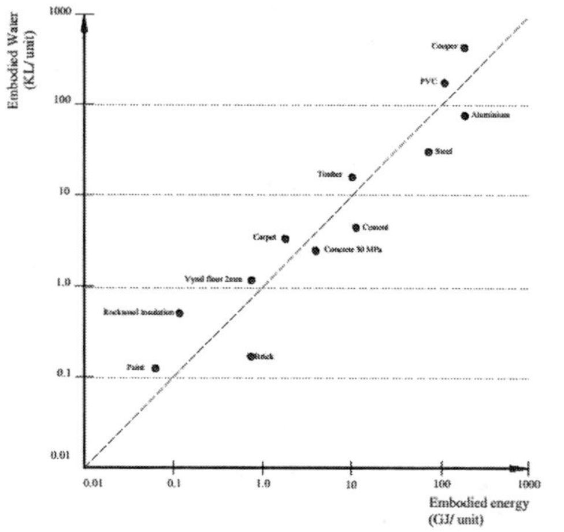

Figure 2. Embodied Energy and Embodied Water (adapted from Graham *et al* [14] and Vassig, Ozer and Spiegelhalter [15]).

The analysis of Table 2 makes possible to conclude that the careful selection of materials on the design phase is important for the water consumption reduction strategies on building construction. Materials such as galvanized steel consume too much water in its manufacturing process (3400 litres/kg). Moreover, the production of hemp fibres and straw presents a null use of water.

In European Palaeolithic period, the use of animal skins supported by large bones or wooden poles to build interior dividing walls in caves constituted an important innovation for future developments of exterior artificial dwellings, where the structural supporting element is independent from the covering layer [19].

Table 2. Water demand on the production phase of materials in function of their weight.

Material	Water demand (litres/kg)
Galvanized steel	0,07
Local pine wood**	4,2
Rockwool**	32,0
Ceramic solid brick**	1,9
Ceramic hollow brick**	1,4
Polyvinylchloride sheet**	511,9
Mortar**	3,3
Plasterboard*	0,2
Cork*	25,0
Cellulose (loose fill 100% recycled)**	21,0
Recycled textiles (matting glued with polyester)*	6,0
Polyurethane foam(PUR)**	3,5
Polyvinylchloride coated Polyester membrane***	1,4
Straw/ reed*	0,0
Canes/reed*	0,0
Plaster with sand**	0,2

* Source: [16]; ** Source: [17]; *** Source: [18].

By analysing the evolution of internal partition walls, shown at Figure 3, it is possible to verify that an initial tendency for dry and lightweight solutions evolved to wet and heavyweight solutions. However, nowadays it can be verified a tendency to the return of dry and lightweight solutions that can include easier construction/deconstruction design principles.

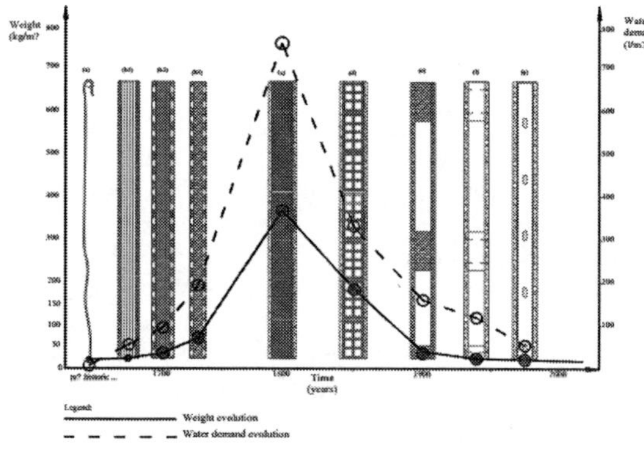

Figure 3. Evolution of weight and water demand on Mediterranean partition systems: (a) animal skin coating vegetable poles or bones; (b1) mat canes filled with reed; (b2) Wattle and daub filled with straw; (b3) wattle and daub with planks; (c) Solid Brick; (d) Hollow Brick; (e) Wooden board with timber frame; (f) Plasterboard panel with timber frame; (g) Plasterboard panel with light gauge steel frames. Scheme adapted from Mendonça and Macieira [20].

3.2 Mediterranean climate - interior partition walls

This study takes as reference the constructive solutions of existing dividing walls in Mediterranean countries, and specifically in Portugal.

The most common solutions that can be considered as vernacular [21] are: sun-dried blocks of soil (adobe) - characteristics of clay-rich zones, essentially in the south regions of Alentejo and Algarve, and the wattle and daub (Figure 4(c)) widespread through all the country, called as "tabique" (in all Portugal) and as "taipa de fasquio", (in the north region of Minho). This last one is made up of wooden planks placed vertically or in diagonal, over which are placed horizontally small strips of trapezoidal section, in which is adhered the mortar. Other solution combines timber frame filled with canes/reed. However, due to the loss of tradition in the techniques employed and consequent rising cost of hand labour, these solutions gave way to other wall types, such as the massive brick (Figure 3(c)), which later derived to the hollow brick wall (Figure 4(d)).

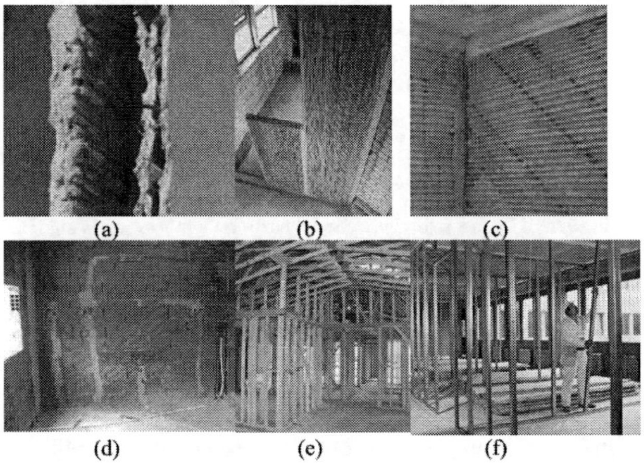

Figure 4. Non-loadbearing partitions walls in Portugal: (a) Mat canes in Lagos, Portugal; (b) wood frame filled with reed/straw; (c) Wattle and daub partition in Guimarães, Portugal; (d) Hollow brick; (e). Timber board with wood frame; (f) Plasterboard with light steel gauge frame.

The use of ceramic hollow brick in non-structural masonry became increasingly common, either by economic as well as ease of implementation reasons. With thicknesses, without finishing plaster, varying between 7 and 22cm, but more frequently with 11cm in interior walls, this system became the preferred solution and still remains in present times. Even if new and more efficient solutions are appearing in the market, most of them cannot compete with hollow brick, not only for the economic cost, but also due to the conservative mentality of the diverse agents present on the construction sector.

One exception is a solution that has been experiencing a great development since the 70's, the plasterboard panel. First it had a timber structure filled with a honeycomb cardboard in which were fixed the plasterboard plates (Figure 4(e)). Later, around the 90's, the wood structure was replaced by light steel gauge frames (Figure 4(f)). This type of partitions can be built with 51 × 76mm or 51 × 100mm studs normally spaced between 400 to 600mm, depending on the type and thickness of the wall finishing used.

4 Embodied water and environmental assessment

The environmental impact assessment of an internal partition wall solution can be made through several indicators, such as: the Embodied Energy (EE); the Global Warming Potential (GWP), the Acidification Potential (AP), the Chemical Oxygen Depletion (COD), the Photochemical Ozone Creation Potential (POCP), the Wastes generated in production process, the Coefficient of Heat Transmission [and the Embodied Water (EW).

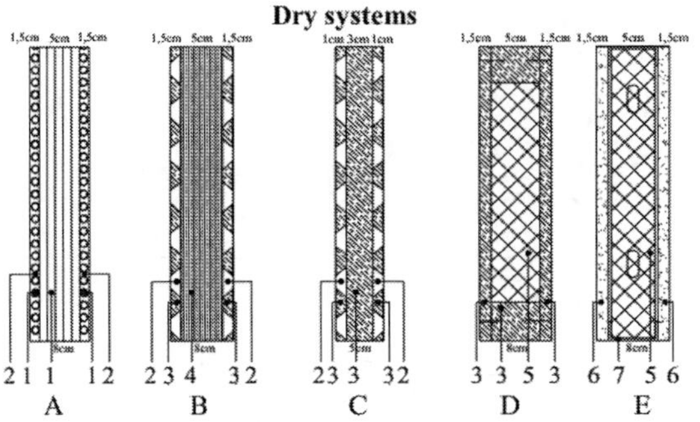

Figure 5. Transversal sections of dry interior partition wall systems analysed. Legend: 1. Canes/reed; 2. Plaster with sand; 3. Local pine wood; 4. Straw/ reed; 5. Rock wool; 6. Carton plasterboard; 7. Light gauge steel frame; and 12. PVC coated Polyester membrane.

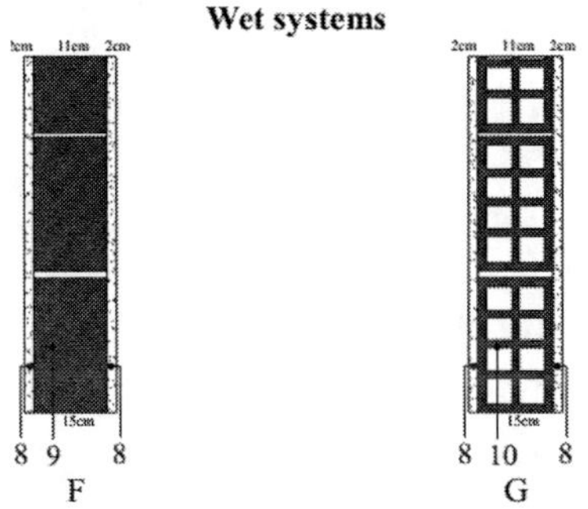

Figure 6. Transversal sections of wet interior partition wall systems analysed. Legend: 8. Mortar; 9. Ceramic solid brick; 10. Ceramic hollow brick.

The Embodied water is the amount of water used to manufacture materials. Water cost in the manufacture of goods and services required by construction may be significantly diverse depending to regional differences on water availability.

This study compares seven different partition walls commonly used in Mediterranean climates, five dry solutions and two wet solutions. Figure 5 presents the dry solutions: mat canes filled with reed (Figure 5A), timber frame filled with straw (Figure 5B), wattle and daub (Figure 5C), wooden board with timber frame and rock wool insulation (Figure 5D), plasterboard partition wall with light steel frame structure and rock wool insulation (Figure 5E).

Dry solutions are lightweight, allowing a lower embodied energy and other more favourable environmental

 DOI: 10.1051/matecconf/20166813009

impact indicators. Compared to wet and heavyweight solutions, such as hollow brick and solid brick walls, dry and lightweight solutions allow also easier deconstruction/reuse scenarios.

The values presented on Table 3 are based on bibliographical references [6, 22, 23]. The functional unit selected was one square meter of solution.

Table 3. Environmental impacts of the dividing wall systems analysed.

	dry					wet	
Impact categories	**A**	**B**	**C**	**D**	**E**	**F**	**G**
Specific Weight (kg/m^2)	17,5	28,0	39,0	19,0	30,4	386	179
EW [a] (l/m^2)	31,4	83,7	146,7	114,6	56,6	793,4	330,0
EE [b] kWh/m^2	37	28	14	13	62	457	140
GWP [c] (g/m^2)	971	2462	4182	2331	14387	69646	23218
AP [d] (g/m^2)	8	20	35	30	114	723	217
COD [e] POCP (g/m^2)	105	214	229	28	2211	6325	2340
Generated Waste [f] (g/m^2)	162	323	698	1012	2552	30725	6052
U value [g] W/m^2.ºC	0,6	0,7	1,4	0,6	0,7	2,3	1,8

[a] EW (Embodied Water);
[b] EE (Embodied Energy);
[c] GWP (Global Warming Potential) in equivalent CO_2 grams;
[d] AP (Acidification Potential) in equivalent CO_2 grams;
[e] COD (Chemical Oxygen Depletion) and POCP (Photochemical Ozone Creation Potential) in equivalent NO_x grams;
[f] Generated waste in production process;
[g] Coefficient of Heat Transmission [23].

5 Results and discussion

The aim of this research was to investigate the water demand in the production of interior partition wall systems for building rehabilitation.

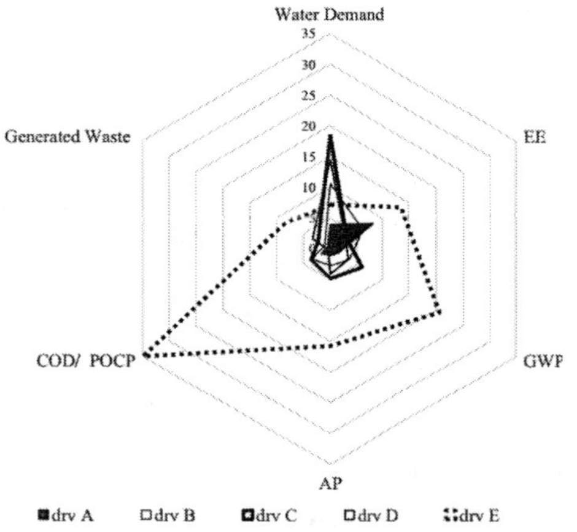

Figure 7. Comparison between dry partition wall systems analysed. The overall best solution presents the lower polygon area and the lower value for each indicator.

There is a considerable amount of water used in the materials production phase and in building construction phase. The highest values verified were for the wet and heavyweight systems.

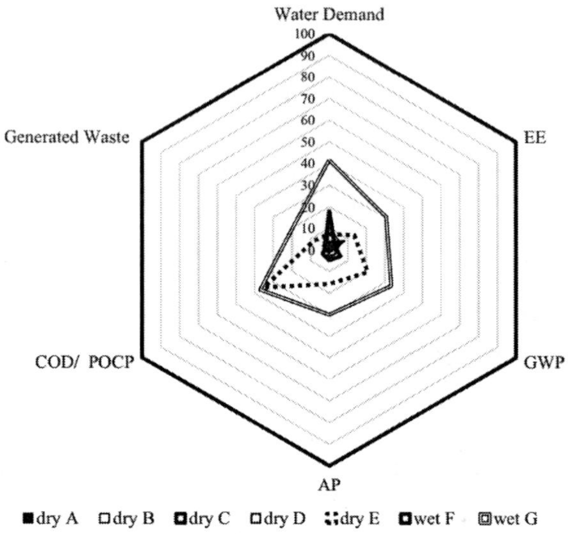

Figure 8. Comparison between all partition walls systems analysed. The worst solution presents the higher polygon area and the higher value for each indicator.

The highest value was 793.4 litres per square meter of gross floor area. These findings suggest that the selection of elements and materials has a great impact on a building's water demand during the construction phase. This research allows selecting the partition wall solutions that might be more sustainable in terms of water use.

The aim of the analysis was to provide multiple criteria for decision-making according to the values of different impact categories for several building materials.

From the results presented in Table 3, comparing the dry and the wet constructive systems, it is possible to conclude that the embodied water as well as other environmental impact indicators presents lower values in dry solutions. The worst solution in all indicators is the solution F. Between dry traditional Mediterranean solutions, like A B, C and D, the best solution is A except in EE indicator. However, solution D, wooden board with timber frame filled with rock wool, present the best values of EE and COD/POCP indicators, but it presents the second worst level of Water Demand and Generated Waste, when compared with other dry systems. Analysing the Figure 7, it can be seen that solution with worst overall impact is solution E and the best is A.

From the analysis of the solutions presented it can be concluded that to achieve a solution with low environmental impact levels it is necessary to reduce the quantity of materials used, especially those resulting from more industrialized and higher energy demanding production processes, with higher EE, such as light gauge steel frame or rock wool. For example, reed and cork present a significantly lower EE than Rockwool.

The dry and mechanically detachable wall systems have a smaller impact on the building's environmental impact, while the on-site construction systems present a significantly higher impact.

Policy focused on reducing water consumption should consider the use of dry and lightweight systems.

6 Conclusions

Dry systems of interior partition walls are certainly a wiser option in many situations as these can be more flexible and even portable, in some cases. The lower quantity of materials used in these lightweight solutions allow lower water use as well as other more favourable environmental impact indicators.

The reduction of material inputs to the minimum is a way to achieve higher eco-efficiency in a building and thus open the way to the development of efficient interior partition systems. The need for more sustainable construction, using innovative solutions has also motivated this study and the development of materials and construction technologies using lightweight materials.

Efforts for reducing the construction industry water consumption need to be focused up and downstream of the actual construction process. However, the complexity of the supply chain makes it difficult to ascertain precisely where efforts should be focused. A demand-side view is required to solve the problem of scarce water supplies. The drivers of demand, both direct and indirect, need to be managed through all stages of production.

In conclusion, the buildings rehabilitation with dry and lightweight systems constitute an important contribute to sustainability. These systems allow buildings to be adaptable in order to remain updated within the contemporary framework of sustainability.

Acknowledgments

The first author wish to thank FCT (Fundação para a Ciência e Tecnologia – Portugal), MCE (Ministério da Educação e Ciência – Portugal) and ESF (European Social Fund) for supporting the research fellow with the reference SFRH/BD/104891/2014. The second author wish to thank the financial support from the Project UID/AUR/04509/2013 by FCTMEC by national funding and, when applicable, FEDER co-financing under the new PT2020 partnership agreement.

References

1. Euroconstruct. *European construction: Market Trends to 2012 in* 68th Euroconstruct conference. KOF and ETH, Zurich; November 26-27 (2009)
2. C. Richard and C. Schulz. *Energy efficiency refurbishments: Principles, details, examples.* Detail Green Books, Munich (2013)
3. C. Schittiich. *Rehabilitation: Reconversion, Extensions, Reconversions.* Edition Detail, Munich (2006)
4. F. Jürgen. Water use in industry. Eurostat Statistics in focus 14/2014 (2014)
5. S. Junnila and A. Horvath. Life-Cycle Environmental Effects of an Office Building. Journal of Infrastructure Systems, 157-166, (2003)
6. R. Mateus, S. Neiva, L. Bragança, P. Mendonça, and M. Macieira. Sustainability assessment of an innovative lightweight building technology for partition walls – comparison with conventional technologies. International Journal Building and Environment. doi:10.1016/j.buildenv.2013.05.012. (2013)
7. P. Mendonça and M. Macieira. ADjustMEMBRANE: Innovative Lightweight Adjustable Partition Wall System. International Journal of Architectonic, Spatial, and Environmental Design. **7**, 2, 39-50 (2013)
8. A. Friedman. The Adaptable House: Designing Homes for Change. McGraw Hill Professional, (2002)
9. K. Tichelmann and J. Pfau. *Dry Construction – Principles: details and examples.* Edition Detail, Munich, 7-9 (2007)
10. L. Bragança, R. Mateus and H. Koukkari. Building Sustainability Assessment. Sustainability. Vol. **2**(7), July, p2010-2023 (2010)
11. W. Addis and J. Schouten. *Principles of design for deconstruction to facilitate reuse and recycling.* CIRIA, London, **38**, 39 (2004)
12. H. König *et al. A life cycle approach to buildings - Principles, calculations and design tools.* Detail Green books, Munich, **66** (2011).
13. EPA (Environmental Protection Agency), SBA (Science Advisory Board). Reducing Risk: Setting Priorities and Strategies for Environmental Protection, EPA, Washington, DC, United States, SAB-EC-90-021, (2000), pp. 13-14.
14. T. Graham, M. McCormack, L. Palmowski, and R. Fay. *Embodied water of construction.* Royal Australian Institute of Architects, BEDP Environment Design Guide, **58**, 1-8 (2011)
15. S. Vassig, E. Ozer and T. Spiegelhalter. *Best practices in Sustainable Building Design.* J. Ross publishing (2011)
16. B. Berge. *The Ecology of Building Materials.* Architectural Press, Bath (1999)
17. I. Bribián, A. Capilla, A. Usón. Life cycle assessment of building materials: Comparative analysis of energy and environmental impacts and evaluation of the eco-efficiency improvement potential, Building and Environment **46,** 1133–1140 (2011)
18. ASTM International.. CFFA SPPR PVC Membrane EPD, United States (2013)
19. M. Macieira. *Architectural membranes: potential application in the interior of buildings.* Dissertation for Master degree in Architecture. University of Minho, Portugal (2012)
20. P. Mendonça and M. Macieira. Interior partition walls in Mediterranean climates - lightweight versus heavyweight. International Conference "Sustainable Environment in the Mediterranean Region: from Housing to Urban and Land Scale Construction", Naples, Italy (2012)
21. F. Pinho. *Paredes de Edifícios Antigos em Portugal* (in Portuguese), 2nd edition, LNEC, Lisbon (2000)
22. P. Mendonça. *Living under a second skin - Strategies for Environmental Impact Reduction for Solar Passive*

DOI: 10.1051/matecconf/20166813009

Constructions in Temperate Climates. Doctorate Thesis in Civil Engineering, University of Minho, Portugal (2005)

23. C. Santos and L. Matias. *Coeficientes de transmissão térmica de elementos da envolvente dos edifícios* (in Portuguese), IT
24. E50, LNEC, Lisbon, 43-50 (2007)

MATEC Web of Conferences **68**, 13010 (2016) DOI: 10.1051/matecconf/20166813010
ICIEA 2016

Cost Parameters of Green Residential Buildings in Pune, India

Anshul P Gujarathi [a]

B.N. College of Architecture, Pune. Karvenagar, Pune, India

Abstract—Various studies have been performed on cost impacts and impact parameters in Green Buildings against the conventional ones. Most of the studies include commercial buildings prioritizing energy and water related impacts while fewer studies are done on residential scale. This paper attempts to study the residential side of Green.

Green certification and rating programs define the greenness quotient for a project with their predefined parameters. These parameters and their standards can be linked directly for their cost impact. In India, Pune is one of the pioneer cities with respect to the number and coverage of green building footprint. All the pertinent ratings of the country appreciate its prominent presence in Pune city with the advantages of applicable municipal incentives.

This paper aims to determine and analyse the cost incremental factors for the certified green residential buildings in Pune against the conventional buildings. It will also determine the incremental percentage or the monetary value which impact the project cost.

For the study, certified green projects from Pune are considered with at actual data pertaining to the project. The analysis is aimed for the overall effective resultant of cost incremental or decremental factors with the cost benefit analysis. Further a factor analysis is carried out for a survey conducted for the buyers with their approach to green buildings and cost associated.

1 Introduction

The evolving concepts of sustainability today, are no more limited to a typical segment of people like green building consultants or the authorities preaching them. The thoughts have penetrated much deeper and every individual is aware of depleting resources and global warming disadvantages. From an office going individual to a house wife; from a school going child to a grandfather; all realize the importance of saving of water and resources. The global awareness spread has positively impacted overall sustainability scenario in the country and benefitting the blessed city of Pune. Pune claims the status of one of the leading real estate destinations in India [18]. Due to Mumbai proximity and its strategic location in the state, it is spreading on all sides for development.

With the advantage of educational hub there has always been a good demand for jobs in the city. With IT & ITES development the demand was higher which inflicted upon the residential demand in the main town. Industrial development along with the ancillaries manufacturing units majored the growth of the Pimpri Chinchwad [PCMC] region thus further increasing the residential demand. Studies by major giants in real estate industry also claim that Pune real estate market has expanded by 16.6% in the time span of one year i.e. from June 2011 to 2012. [19]. Prices have escalated by 14% with massive price gain in areas under Gram Panchayat. [12]

The resource strain inflicted by the construction industry alone, dedicated stakeholders play an important part in forward progress. Many organizations and associations have been created to help the building industry embrace sustainability both practically and theoretically stressing on the actual implementation of green building practices in construction [12]. With intend of branding and / or expediting the associated intricate processes, green certifications are being adopted by almost every developer in Pune city. Association of Municipal tax rebates have further added to the intent of going green, giving monetary gains to the developer and the end user. The degree of greenness is proportionally linked with the tax benefits which help in pushing the stakeholders going with premium ratings. On the other hand, there are few reservations realized with green buildings amongst which the cost factor remains at the acme. [19]

This study will enlist the parameters responsible for cost incremental or decremental factors which are

[a] Corresponding author : anshulpgujarathi@gmail.com

additionally associated due to a green rating. The post occupancy or operational stage is not considered for study.

2 Environmental parameters - A brief: 2000 – 2013

The parameters for environment friendly constructional practises have been always a part of the bye laws and the literature of corporations. The environmental status reports have been a strong proof for the same. The parameters were never highlighted as a separate entity but always a part of the integrated construction process. The need for a recognised methodology and degree of greenness was realized and so the concept of green ratings was introduced in the country [2]. International protocols and growing awareness in the global scenario also created the need in the Indian market. The rating tool offered a similar concept which was associated with project branding and measurement of green status. The concept of green buildings picked up pace after year 2000 in India [7]. The authority organisations presented their voluntary assessment systems to the citizens which were propagating the green construction practices from design to operational phase. For its proliferation several offers and benefits including monetary benefits were attached to the rating systems. The globally recognised rating LEED made a way in India with its Green Homes rating which was further modified to Indian context by Confederation Of Indian Industry [CII]. The rating was termed as IGBC Green Homes, where IGBC is Indian Green building Council clearly highlighting the localised sense. Eco Housing India, IGBC Green Homes and Griha are the ratings that were presented to the city of Pune during the past decade. Environmental parameters influence the distribution, abundance and activity of animals and plants. Local meteorological conditions such as air temperature, rainfall or sunlight may affect the behaviour of terrestrial organisms, and water current, dissolved oxygen, suspended material and river bed topography may influence aquatic species. [12]-[19]

3 Green building rating systems

In India, there are rating systems like LEED, IGBC, Eco Housing and GRIHA. They have a predefined set of criteria that have intent towards direct applicable environment aspect with points against each of them. Few points are mandated while others are voluntary as per the choice of the project. Buildings are required to fulfill the defined criteria and achieve a certain number of points to be certified. In addition to these rating systems, there are codes that are applicable to defined projects. These codes are not mandated till date but might be a mandate to go with in near future. Energy Consumption Building Code [ECBC] has a direct implication with measurable energy savings. With all these guidelines there is National Building Code [NBC] which provide instructions on the energy consumption of buildings. All buildings in India need to comply with the guidelines set up by ECBC and the NBC.

We can define Green Buildings as structures that ensure efficient use of natural resources in construction. Use of eco friendly building materials, savings in water & energy and other resources with minimal generation of non-degradable waste are key parameters for green buildings. Further to the basic systems in the defined green way, technology support enhances the greenness quotient of the project. Technologies prove very helpful to help measure the actual savings that reflect in the system. Examples like efficient cooling systems with sensors that can sense the heat generated from human body and automatically adjust the room temperature, prove much more beneficial than only limiting to system installations, further saving energy. The same applies to the lighting systems too resulting in saving of energy against lighting. Green buildings have a smarter lighting system that automatically switches off when no one is present inside the rooms. Simple technologies like air based flushing system in toilets that avoids water use by 100%, use of energy efficient LED's and CFL's instead of conventional incandescent lamp, new generation appliances that consume less energy, and many other options help in making the buildings green and make them different from conventional ones.

In Pune city, all the four rating systems are predominantly noted. Due to the applicable benefits to the region, the city experiences a large green building footprint in the country with LEED, IGBC and GRIHA combined footprint.

4 Green building parameters

There are several parameters on which a building is evaluated before being awarded with a final certification. The overall parameters can be categorized as following major heads:

4.1 Efficiency of site selection and planning

The site selection and planning section addresses the ecology and environment concerns related to residential construction and site development activities including the design of landscape.

4.2 Water management

Efficient water and waste water management for reducing water usage demand on fresh water sources is the prime focus. The residential sector is one of the significant water consumers, after agriculture. The certification parameters encourage water usage in a self-sustainable manner through reduce-recycle-reuse concept.

4.3 Energy efficiency

Reducing demand of conventional energy by optimization of building design and structure is the prime focus. Incorporating renewable sources of energy and renewable energy based systems such as solar water heater to reduce the use of conventional energy are evaluated and applied in the project under this head.

4.4 Materials

Efficiency in selection of sustainable materials for construction and the technologies considered is evaluated.

4.5 Indoor environmental quality

Enhancement of indoor environment quality like indoor air and thermal quality is evaluated including daylight parameters.

4.6 Innovation

The credits that are performing exemplary in few of the above criteria can avail additional points under innovations. Also if the project attempts any beneficial activity that is not mentioned in the rating but that can help to prove positive benefit to the environment they can claim additional points under this head.

4.7 Audit & maintenance

Optimization of operation and maintenance.

Table 1. Combined weight age of criteria

FOCUS AREA	IGBC	GRIHA	Combined
Building Planning & Designing (AC/Non AC)	15%	12%	14%
Energy Efficiency (Lighting , HVAC, Renewable)	13%	12%	13%
Construction Materials	12%	12%	12%
Indoor Environmental Quality	11%	11%	11%
Waste Management	11%	10%	11%
Innovations	9%	9%	9%
Water Efficiency	7%	7%	7%
Social Consideration	7%	7%	7%
Landscaping	6%	6%	6%
Site Selection	4%	6%	5%
Preservation of Natural resources on Site.	4%	5%	5%
Efficient Transportation System	1%	3%	2%

To compare green and conventional buildings, major parameters that will be under consideration include those areas where the Green element shall make an impact. There are many parameters which will have common considerations as per the sanctioning mandates or by the environmental committee suggestive mandates. Only those parameters that are additional for green building will be accounted for this study [1]-[4].

4.8 Case study analysis

A brief survey was carried out for a sample of 150 where the preference of the developers and buyers was asked for the rating to be opted for while doing a green building.

Looking at the popularity of the ratings in the building sector in private and government sector both, IGBC and GRIHA prove to make a mark in the Indian industry. Similarly for the city of Pune both these ongoing ratings are prominent and exceed the footprint than other ratings.

We can thus study the two prominent ratings in Pune, i.e. IGBC and GRIHA; for the preferential weight age catered to each of the criteria as defined by Green norms. As per the defined categories of the ratings, the following table is identified with the weight age in terms of points given to each criterion under that category.

From the above tables it can be realized that the prevalent rating systems in India give most priority to the Building design and comfort parameters. Average high priority is for the materials and water and waste management is given considerable weight age. Significant points lie with the main credits but the supporting credits are designed in such ways that they have great influence on main points. With their minor number they can assume greater impact on main points further adding green quotient to the project. Further detailed analysis for these parameters is carried out on the basis of their cost impact against each of the criteria.

Cost impactive parameters for green and conventional buildings

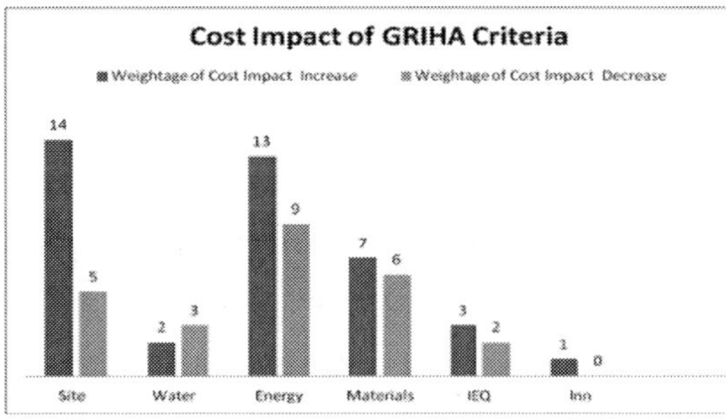

Figure 1. cost impact – Griha.

In residential sector for most preferred ratings of IGBC & GRIHA many projects can be identified that have reached a benchmark for study. Considering the parameters of conventional and green buildings, a detailed study is carried out for GRIHA and IGBC criteria to understand the impactive credits for the entire project. The cost impact shall only be considered for first costs impact, either increasing or decreasing cost for conventional against green building [3].

The overall study has been divided in two parts. The first part of the study is the identification of the variables of the cost parameters for conventional against green. Whilst the second part is the actual set of certified buildings are studied for the impacted cost. Here the detailed checklist for both rating systems is analyzed for the first cost impact that is probable for both the conventional and green [5].

Below analyses helps us to identify how the cost increment or decrement is impacting the green against the conventional case. This analysis is made against each sub part of the criteria for both ratings and segregated under heads as mentioned in the figures below.

It is noted that the parameters related to site and energy show a significant increase in cost parameters with green attempted criteria whilst there is also a significant decrease observed with the parameters of materials specifically under the structural head. With the increase in cost there is a high scope for cost decrease also with the attempt of right technology and correct material selection. Simultaneous decrease is also observed under the water and site credit with the responsible parameter of landscape.

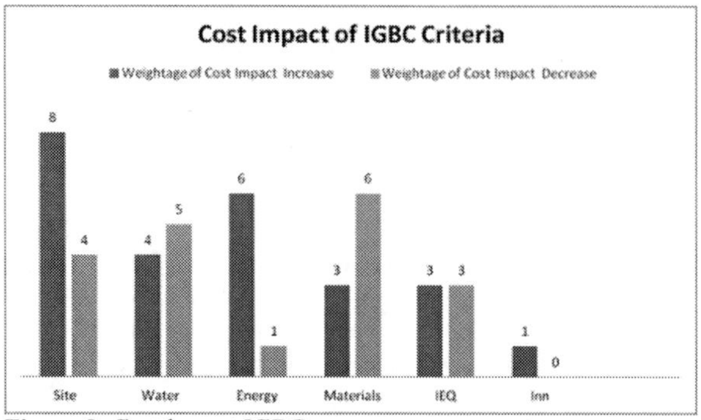

Figure 2. Cost impact IGBC.

4.9 Ranking of the activities based on their costs

For each of 15 sites, activities were arranged in the descending order based on the cost the activities. The top ten activities were assigned the ranks. Then the average rank was computed for each activity. The top ten expensive activities of each site were identified. The activities were rank based on the proportion of sites listed a particular activity to the total sites carried out that activity. E.g. Solar water heater was provided on nine sites and for all those sites this activity was identified as top ten expensive activities.

5 Cost impact parameters - discussion

An important limitation of all economic research on this topic is the absence of data directly linking specific resources and investments in construction to measures of energy efficiency or sustainability. Some evidence has been gleaned from experiments in construction and the subsequent operation of actual green buildings, but these are based on very small samples.

Research on climate change suggests that small improvements in the sustainability of the existing building stock can have large effects on energy efficiency in the economy. Increased awareness of global warming and the extent of greenhouse gas emissions in the real estate sector have increased attention to green building. [1]-[8] Measured attributes of sustainability and energy efficiency are incorporated in property rents and asset prices, and this seems to persist through periods of volatility in the property market. These developments will affect the existing stock of noncertified office buildings [1]. The findings already suggest that property investors attribute a lower risk premium for more energy-efficient and sustainable commercial space. Rated buildings may provide a hedge against higher energy prices, but also against the shifting preferences of both tenants and investors with respect to environmental issues. Increasing market awareness of climate change and rising energy costs can only increase the salience of this issue for the private profitability of investment in real capital.

Green Buildings have shown a gradual growth in past few years and related thoughts like increase in cost due to green parameters have often created the hindrance for the promotion of green buildings [4].

The above study gives us the variables for cost impact on both positive and negative grounds.

The Cost Positive Impacts – These set of variables prove beneficial when green compared to conventional case. The overall savings shall be anticipated with these criteria.

a) Choosing a brown field or a contaminated site might prove indirectly beneficial as it can be attained at a lower cost compared to market price.
b) Existing amenities can prove beneficial to the end users and give them an indirect benefit of fuel savings.
c) Use of AAC blocks or light weighted blocks for structural construction. This will reduce the overall load on structure thus anticipating huge savings in material and structural work.
d) Landscape norms and methods have a cost positive impact.
e) Reuse of materials and recycling will reduce the use and need for virgin materials and prove cost efficient.
f) Use of low energy materials can reduce the first cost of purchase.
g) Harvesting rain water can prove beneficial as the overall water use can be reduced thus reducing the cost associated with that.
h) Reuse of treated water for flushing can reduce the need of fresh water and new water use.
i) Proper site management can save 5% to 10% of material wastage thus saving on associated material cost.
j) Use of water based enamel for paints prove less costly compared to oil based.

The Cost Negative Impacts – These set of variables prove cost impactive when green compared to conventional case. The overall savings shall be anticipated at the later stages in the long run but for first costs they prove costly.

a) As per design and norms, dedicated parking and reduced hard areas in landscape may affect the overall area distribution that can be directly associated to the cost.
b) Site related maintenance activities and labour camp involve regular check and efforts. This is an ongoing cost to the project.
c) Use of mechanical and electrical appliances including pumps and motors as per the prescribed norms prove costly. Here the long term savings anticipated are not considered though they prove long terms beneficial in terms of both running and maintenance.
d) Specialty fixtures for water and accessories claim to be costly.
e) Envelope and the fenestration materials as prescribed by the rating norms.

f) Features like electric charging points and changing rooms or toilets may prove cost intensive.
g) Use of specialty products like certified wood or hybrid generators, etc. are cost intensive.

The overall comparison of green and conventional cannot be justified for cost increment or decrement unless actual projects are studied for. Since every project has a separate design brief and differently prescribed amenities, the comparison can never be on common grounds unless common case created for the same. For further study certified buildings were commonly clubbed for IGBC and GRIHA categories and separately studied for the parameters and further clubbed together with mean for analysis.

6 Cost analysis – case studies

The identified variables in the above study were further applied to a group of certified buildings which were studied for comparison of their own conventional and green case for cost impacts. As a standard the conventional case was considered as the case without the green implication for the same project in similar situation. Thus cost impact was taken nullified for that particular credit and the cost for conventional was considered less than the value against the green case. Similarly the cost positive areas were considered for the value which was reduced from the conventional case. The overall impact is the resultant value that is notified in the table. The cost data collected for 15 sites were analyzed in great detail to understand the difference in conventional and green cost of the construction. Also, the activities which contribute in these differences were studied and presented in the following section.

Table 2 presents the details of the cost data based all 15 sites under study. The conventional cost of the sites ranged 1,021 – 1,402 Rs./sq.ft. with the mean Rs. 1,238 [± 113]. Whereas, the mean green cost of the sites were Rs.1,253 [±117] per sq.ft. It was observed that the green cost of the sites were higher than the conventional cost. On an average there was an increase of 14.80 Rs./sq.ft in green cost, which accounted for 1.2 % increase over conventional cost.

However, the mean percent increase in green cost over conventional cost was highest for medium projects [1.6 ± 0.6] followed by small projects [1.3 ± 0.2]. Percent increase in green cost for large projects was only 0.6 [± 0.3], which was less than half of the increased observed in small and medium projects [Figure 5]. The difference in green cost of projects across the groups was statistically significant as indicated by the test statistics of non-parametric Kruskal Wallis Test.

Table 2. Average building conventional cost and green cost

Statistics	Conventional cost [Rs./sq.ft.]	Green cost [Rs./sq.ft]	Difference in cost [Rs./sq.ft]	% increase in cost
Mean	1,238	1,253*	14.80	1.2
Standard deviation of mean	113	117	13.10	1.0
Range	1,021 – 1,402	1,026 – 1,419	1.25 – 53.1	0.1 – 4.23

n=15
Non-parametric analog of paired t-test i.e. Wilcoxon Signed Ranks Test statistics [z=3.408, p=0.001] indicated that the difference between conventional cost and green cost was statistically significant.

For every typology the premium rated project shows higher increase in cost while the projects with two star or so show very negligible cost raise. If proper factor for maximum cost contribution is identified for the premium rated projects against the points earned definitely the gap can be bridged and the cost can be nullified maintaining the same point cadre.

7 Factor Analysis

Table 3. Factor tests.

KMO and Bartlett's Test

Kaiser-Meyer-Olkin Measure of Sampling Adequacy.		.675
Bartlett's Test of Sphericity	Approx. Chi-Square	2798.714
	df	276
	Sig.	.000

The opinions of the buyers about the green buildings were collected using 24 items questionnaire. The response format for each statement was the standard Likert 5-point scale. A factor analysis with varimax rotation was used to create the factor structure of the 24 items included in the questionnaire. Factor analysis was carried out using the maximum likelihood method with pair wise deletion.

Data adequacy for the spiritual climate inventory was checked with the KMO test, the value of which is 0.675 [Table 3]. A value is higher than 0.5 which indicated that the data is sufficient for factor analysis. Chi square value 2798.714 at 276 degrees of freedom was significant at less than 0.0001 level of significance. The significant value of Chi Square test indicates the acceptability of eight factor solution. The eight factor solution proposed by them holds valid on the current sample.

The factors were rotated by an orthogonal transformation [varimax] to achieve a simpler structure with greater interpretability. Data were aggregated in three factors with Eigen values greater than 2 and explained about 47.7 % of the variance. Items with Factor loadings > 0.55 were used to define factors [Table 4]. The factor score coefficients were estimated by the regression method, which produces uncorrelated scores with a mean of 0 and a standard deviation of 1.

Table 4. Factor loading matrix.

ITEMS	COMPONENT		
	1	2	3
Green building saves energy & resources at individual level.	*.878*	.081	.053
Today's residential complexes should be mandatorily "Green".	*.854*	.024	-.193
Pay back calculation in green building is essential.	*.829*	-.281	-.120

Studies related to COST of green buildings can help the reservation towards the green approach	*.782*	.010	.013
"Green building "refer to a] Energy & Cost saving b] Social welfare c] Environment protection	*.761*	.004	-.089
Green buildings help for climate impact mitigation in the long term.	*.718*	.439	-.067
COST one of the major hindrances that prevent the developers from taking a green approach.	*.705*	.182	.204
Green buildings benefits to overall society, a region & improves the quality of life.	*.658*	.207	-.125
Early initiation of green strategies result in COST benefitting structure	*.645*	.019	.191
Municipal Corporations should incentivize Green buildings	*.580*	.096	.074
Maintenance of Green residential buildings are costly compared to conventional ones.	*-.746*	.260	.065
Green buildings help in gaining monetary benefits to the end-users i.e. residents, in the long run.	.115	*.600*	-.080
Green building materials can influence the cost of the project negatively.	.013	*.581*	.436
COST is the utmost important factor for a green building.	.566	*-.686*	-.027
Payback calculation is reliable to a satisfactory extend.	.016	*-.724*	-.152
Services play a very important role in Green buildings and can save on huge resources.	.119	-.248	*.745*
Green building certification programs define a totality green building.	.042	.081	*.655*
The implementation of green features is cost intensive.	-.040	.029	*.646*
Green project should be preferred for stay than conventional one.	-.174	.269	.538
Experience of stakeholders with Green buildings result in COST saving.	.541	.248	-.528
Energy efficiency & Water Savings are related to COST savings.	.540	.300	.463
Green Building Certification guidelines improve the environment quality of the project.	.295	.492	-.104
Green Growth is related to Economic growth in real estate	-.417	.231	-.232
Overall awareness & green education is the key for the Green project	.553	.044	-.244

The factors extracted from the analysis were label based on the items which loaded [loading >0.05] on the factor. Internal consistency of these three sub-scales was assessed by calculating Cronbach's alpha. High values for Cronbach's alpha obtained during analysis indicated good internal consistency of the items in the scale.

Table 5. Factor details Cronbach's alpha.

Factor no.	Factor Label	alpha
Factor 1	Cost Conscious	0.837
Factor 2	Environment Conscious	0.673
Factor 3	Quality of life	0.636

8 Factor scores

The factor score coefficients were estimated by regression method, which produces uncorrelated scores with a mean of 0 & a standard deviation of 1.

Factor scores are composite variables which provide information about an individual's placement on the factor[s]. Higher the scores, higher will be the agreement with the factor.

The distribution of the respondents on three factors is given in following table.

Table 6. Factor outcomes.

No.	Factors	Frequency	Percent
1	Cost Conscious	64	42.1
2	Environment Conscious	49	32.2
3	Quality of life	39	25.7
	Total	152	100.0

9 Conclusion

Green Buildings can be made cost neutral with the right implementation of strategy at the correct aspect. Maintaining the balance between the cost raise and cost decrease a neutral approach can be attained where in a premium rating at no additional cost can be realized. The benefits on the life cycle performance are the added advantage for the project with long terms savings. Every activity associated with green is for the wider cause of the sustainable parameter and always proves beneficial to every stakeholder directly or indirectly.

A sufficient economic return on energy-efficiency investments is crucial for the sustainable development of the green building industry. The concern of environment and sustainable development has been increased recently. These problems force the countries to adopt a number of policies that enhance energy efficiency and apply baseline parameters in accordance with international standards. Green building has now become a forefront of sustainable development in this century that takes the responsibility for balancing long-term economic, environmental and social health. It offers an opportunity to create environmentally efficient buildings by using an integrated approach of design.

References

1. Cost of Green Revisited : Davis Langdon, Re-examining the Feasibility and Cost Impact of Sustainable Design in the Light of Increased Market Adoption, (July 2007)
2. The Economics of Green Building: Piet Eichholtz, Nils Kok, John M. Quigley, USGBC LEED Journal, (August 2010)
3. What Does Green Really Cost: Peter Morris, Davis Langdon, The Green issue Feature, PREA Quarterly, (Summer 2007)
4. Managing Costs of Green Buildings: Geof Syphers, Mara Baum, Darren Bouton, Wesley Sullens, State of California's Sustainable Building Task Force, the California State and Consumer Services Agency and the Alameda County Waste Management Authority, (October 2003)
5. Costing Green: A Comprehensive Cost Database and Budgeting Methodology, Lisa Fay Matthiessen, Peter Morris, Davis Langdon, (July 2004)

6. Environment report highlights the negative side of Pune growth story
7. Radheshyam Jadhav, TNN, 04.58am IST (Jul 27, 2011)
8. Tradition and modernity subtly synchronised in Pune
9. Presidency Pune District, **XVIII**, Parts I, II, III, Pune.
10. Government of India, Gazetteer of Bombay State – District (1954)
11. Series Vol. XX, Pune.
12. A Study of Residential Housing Demand in India; Online at http://mpra.ub.uni-muenchen.de/9339/
13. MPRA Paper No. 9339, posted 29. June 2008 02:55 UTC
14. City of Pune, An M.E. Thesis submitted at Government College of Engineering, Pune.
15. Sonawani, A., Haravalele Pune, Purva Publications, Pune (1998)
16. Speiregen, P., Urban Design: The Architecture of Towns and Cities, McGraw-Hill Book Company, New York (1965)
17. Pandya, Y, Housie, Architecture + Design, **XVI**, 5 (1999) pp. 43-47.
18. Pune Municipal Corporation, Development Plan of Pune, Pune Municipal Corporation, Pune (1991)
19. www.pcmc.gov.in

The Research on Project Cost of Housing Industrialization

Ling Fu, Jun Lu and Xiaojuan Wang

Engineering Economy and Cost Analysis, Chengdu University ,610106,ShiLing Zhen, Longquanyi Qu ,Cheng du, Sichuan China

Abstract. The difference between housing industrialization and traditional building are development model and process flow. The way of measurement and valuation for housing industrialization has changed. This paper compares the cost of housing industrialization with traditional building. The purpose is to border the cost of housing industrialization and lay a theoretical basis for housing industrialization cost. It figures out the boundary and constitution of construction project cost under the industrialization pattern before the large-scale development of the housing industrialization. Only figure out the intension and extension of the project cost can we effectively make a scientific contrast of the project cost. It makes a qualitative analysis of the constitution and characteristics of the project cost in the broad sense. It offers reference for the accurate calculation of the project cost of housing industrialization and provides development suggestions.

1 Current status of the housing industrialization cost research

The concept of housing industrialization has been introduced to China for more than 10 years. In order to promote the housing industrialization, many local governments have introduced a series of preferential policies; real estate developers, constructors and equipment suppliers have taken an active part in the practice of specific projects. As a result, model projects and model bases of industrialization have been emerged in succession suddenly. But the true meaning of housing industrialization is not be popularized in folk. From the perspective of industrialization development in other countries, industrialization is the inevitable outcome of the development of construction industry to a certain stage, but why China delays the promotion? There are two main reasons: one is the lack of perfect and systematic supporting policies of housing industrialization; the other one is the higher project cost of housing industrialization. About how to reduce the project cost of housing industrialization, the industry has many suggestions. For example, Xu Yuanming points out that the high costs of residential industrialization is mainly from the construction costs, such as the division of upstream and downstream of the industry chain, non forming the scale; high logistics costs; improper structure selection and the lack of skilled personnel and so on [1]. Gu Mingwang proposes that housing industrialization is made up of price A (the costs of raw materials and machinery) and price B (the labor costs); the obtained architectural value consists of value C (the value of service function) and value D (the time value); the production mode of traditional architecture is 1A+1B=1C+1D. The effect of housing industrialization is to reduce the expenditure of price A and price B, but to increase the value profits of C and D [2]. Han Xiaokui puts forward that the costs budget should be done before the design scheme is determined. The evaluation of the main costs of prefabricated components of housing industrialization is mold sharing costs and transportation costs. During the design process, it is necessary to compare the effect of different schemes on the reuse rate of mold and transportation efficiency as well as improve the product costs-effective [3]. Through the cases comparison, Wang Shuang finds that PC components production and installation fee can widely increase the project cost of housing industrialization. The prefabricated rate in China is only 15-20% while the increment of construction and installation fee is up to 20%-50%. Though the prefabricated rate of North America and Japan are up to 50%-75%, the cost is being controlled effectively. Based on this comparison, a series of improvement measures are put forward [4]. By means of the principle of life cycle, Cheng Fenfen divides the costs of housing industrialization into three parts: the construction costs, using costs and removal costs. Through comparing with cases, it can draw out that with the enlargement of architectural area, the saving costs of the life cycle will be more and more and the advantage of industrialization cost will become more apparent [5].

Considering opinions above, the author thinks the top priority is neither argues whether the project cost of the housing industrialization is lower or that of traditional on-site building is lower nor proves that there are many ways to reduce the project cost of housing industrialization. We should figure out the boundary and constitution of construction project cost under the industrialization pattern before the large-scale development of the housing industrialization. Only figure out the intension and extension of the project cost can we effectively make a scientific contrast of the project cost. Under the current bill valuation mode, it is suggested to increase the branch sub-item projects which are corresponding to the housing industrialization, and make relevant adjustments in order to welcome the promotion of housing industrialization better.

2 SI system of Housing industrialization

The housing industrialization is the fundamental change of the construction mode, which is related to the overall strategic mission of China's construction industry. SI system is a sustainable development system formed in the development process of housing industrialization. It separates the building Skeleton and building Infill, and realizes the standardized production of components. The building with SI system has the characteristics of long service life, energy saving and environmental protection, fast construction speed, variable house layout and so on. Though it is hard to separate the Skeleton from the Infill and construct respectively, the idea of SI system can be tried to use in engineering measurement and valuation of housing industrialization: it should separate the Skeleton from the Infill, and then lists items and valuates respectively.

Table 1. The comparison on the cost between industrialized building and traditional building.

Name of Cost	Industrialized Building (the symbol * means it is different from the traditional building)	Traditional building	Notes
1.Engineering cost	1.1 Acquisition expenses of the equipment and apparatus	1.1 Acquisition expenses of the equipment and apparatus	Acquisition expenses of the elevator, central air-conditioning.
	*1.2 Acquisition expenses of prefabricate and parts		Prefabricates and fillers such as nonbearing wall, kitchen, bathroom, storeroom, integrated system of equipment and pipes are produced in factory and transported to the spot to be assembled through logistics.
	*1.3 Civil and erection cost	1.2 Civil and erection cost	Narrow project cost, mainly including the engineering cost of support system, the engineering cost and measure cost of the filled system, and other project cost, fees, and taxes.
2. Other fees for engineering construction	2.1 Fees for construction land	2.1 Fees for construction land	Land-transferring fees, removal compensation and so on. Beijing, Shenzhen and other places have introduced preferential land policies about industrialized buildings.
	2.2 Fees for feasibility research report and geological exploration	2.2 Fees for feasibility research report and geological exploration	Consulting fees for consultation units.
	*2.3 Designing fees for supporting system and filler system	2.3 Designing fees	The supporting system (body of structures, bearing carriers, and public parts) has universality as the designing subject of the first stage; the filler system meets the individual requirements of clients as the deepened designing subject.
	*2.4 Integrated agile general and administrative fees	2.4 Management fees for construction(management expenses of construction unit, main contractor management expenses, cost of superintendence, fees for the quality supervision)	The whole process for processing industrialized buildings needs integrated agile resource integrator, concerns with a long industrial chain, including the early financing, land acquisition, calling for bid and bidding, product designing, sales, procurement of material, production equipment and parts, site operation, assembling, and after-sales service, etc,. It is suggested to outsource the non-core parts and lay emphasis on the control over the design, marketing, and integrated management.
	*2.5 Fees for product design and development	2.5 Research and experiment expenses	In the early period of industrialized building, many parts and technologies are introduced from the foreign countries. With the extension of new technologies, the imported parts gradually transit into domestic parts so the research and development expenses tend to decline.
	2.6 Fees for site preparation and temporary facilities	2.6 Fees for site preparation and temporary facilities	The wet construction in the industrialized building construction site declines a lot. The difficulty of first party to manage in the construction site decreases so fees of this item tend to decline.

	*2.7 Maintenance Funds	2.7 Insurance expenses of the project(not including the insurance which has been listed into the management fees of the engineering enterprise)	The maintenance funds should include the management fees, maintenance fees, and the fees for changing the filler in public parts after the construction products being used. The life cycle of the filler varies from 5 to 30 years. Suppose the life of industrialized building is 50 years, the filler should be changed for once at least.
	2.8 Fees for environmental assessment	2.8 Fees for environmental assessment	There are little wet constructions in industrialized buildings, which responds to the principle of "4 saving 1 environment protected", meeting the requirements of green environmental protection buildings
	*2.9 Government review and approval fees	2.9 Government review and approval fees(project application fees and civil air defense fees in different places)	The traditional government review and approval fees for construction mainly includes the backup fees of infrastructures, fees for quality supervision of project, fees for the construction of civil defense fees in different places. Government should give favors in the early period of industrialized buildings. Till June, 2015, there were more than 30 provinces and cities introduced the supportive policies for industrialization. Beijing and Shenzhen have connected the favor with the reward of the building area
	*2.10 The guide fees for introduced technologies	2.10 Fees for introducing technology and apparatus and other fees	The fees for industrialized technology and parts being introduced from the foreign countries are not included in the acquisition expenses of parts, such as the expenses for the buyer personnel to study abroad.
3. Budget reserve	3.1 Basic budget reserve	3.1 Basic budget reserve	The change of the industrialized construction sites decrease and the basic budget reserve will decrease too.
	3.2 Budget reserve for increasing price	3.2 Budget reserve for increasing price	The industrialized construction period shortens, and the budget reserve for increasing price will decrease.
4. Interest			The industrialized construction period shortens so the interest will decrease.

3 Project cost constitution of housing industrialization

3.1 Constitution of traditional construction project cost

The traditional project cost is divided into the broad sense and narrow sense. In the broad sense, the project cost is standing on the angle of the construction side or the investment side. It includes all expense generated in the decision-making stage from the generation of the project conception to set up the project, the implementation phase from design, bidding and tendering, construction to final acceptance of construction, and the operational use phase. It is called the fixed assets investment of construction project. In the narrow sense, it is standing in the perspective of project transactions. The contract price at the stage of biding and tendering, the construction and installation fee at the construction stage and the settlement price are the manifestation modes of the project cost in the narrow sense. Due to the long industry chain of housing industrialization, it is incomplete to only stand on the angle of the transaction part or take the construction price at any stage into consideration. From the perspective of resource integration, the paper analyzes the total fixed assets investment of housing industrialization.

3.2 Constitution of Housing industrialization Project Cost

From the angle of construction site, it can be found that the wet construction work on site of housing industrialization reduces while the assembling process, transportation and lifting workload increase; during the life cycle, the Infill system needs one or many replacements; the parts module installation in the kitchen and bathroom replaces the tedious site construction process and so on. From the entire development process, the development mode of housing industrialization is summarized as the preparation of specification---technology research and development---products design---marketing promotion---the second design---production---transportation logistics---installation and after sales service and so on. These unique process flows and operation modes of housing industrialization bring greater impact to the constitution and valuation ideas of traditional architecture project cost. In order to build a suitable valuation mode of housing industrialization and lay a good foundation for the promotion of housing industrialization, the paper tries to analyze the constitution of the housing industrialization project cost and compare with the traditional architecture project cost. Please see Table 1.

4 Discussion the effect of the project cost on industrialization mode

From Table 1, it can be seen that compared with the traditional architecture, the project cost under the industrialization mode has the following characteristics:

(1) The purchase cost of prefabricated component and the parts is the biggest price difference between housing

industrialization and traditional architecture. The fee of investing and building a plant is huge, it may cost the manufacturing enterprises of prefabricated component and parts about 200-400 million yuan [6]. The ex works price of prefabricated component and parts depends on the factory output and quantity of service projects. The service items are limited in the early stage of industrialization. The transportation fee depends on the distance from the factory to designated location at the construction site. Due to the large size of prefabricated component and parts, the purchase cost is usually high.

(2) The change of valuation thinking of construction and installation cost. According to the specialty, the valuation thinking of traditional construction and installation fee is divided into architecture engineering, decoration engineering and installation engineering and so on. After introducing the SI system of housing industrialization, it calculates based on the construction fee of Skeleton and Infill. List items belonged to the decoration engineering, installation engineering and non bearing architecture engineering will translate into the form of prefabricated component, parts or parts set, and then bring into the construction fee of Infill. Due to the combination of several construction procedures, and the use of massive finished and semi-finished products, the process of measurement and valuation of construction and installation fee can be simplified.

(3) The labor cost reduces but the machinery costs increases in the construction and installation fee. During the housing industrialization, the workload of cast-in-place concrete has been reduce greatly, so the labor costs and material costs of Skeleton template have been decrease greatly. The application of plant-prefabrication can improve the building quality, and it is good for energy saving and environmental protection. At the same time, it increases the fee of plant-prefabrication, mechanical service charges of field lifting and assembling. However, in the recent years, the labor cost in China rises sharply. In the long run, the labor cost saving has more advantages than the machinery installation fee in the field.

(4) The reserve ratio of other maintenance fee in the engineering construction will increase largely. In the SI system of industrialization, the life expectancy of Infill system is about 5-30 years, so it needs to be replaced or repaired at that time. The replace fee of internal filling parts can be suffered by the user, while the replace fee of the public part and filling parts of external wall should be included in the expenditure of the project cost in the broad sense. As far which expenditure should be included, it should wait for confirmation. In order to avoid developers pass the buck, it can establish the special maintenance fund or transfer the risk in the form of purchasing insurance.

5 Suggestions of the project cost development under the industrialization mode

The development and promotion of housing industrialization is in accord with the idea of sustainable development and green building in China. Professor Bao Jiasheng ever put forward that "Architecture is not just the solidification of music; it should be a dynamic and sustainable development space form." The construction project cost should cooperate with the promotion of industrialization actively:

(1) Establish valuation basis suitable for the construction engineering fee of housing industrialization. The current 2013 code for valuation with bill was prepared for the traditional architecture which lacks the corresponding list items and measure programs of parts and prefabricated component. The code is revised every five years, so it should take the projects of housing industrialization into consideration and make an adjustment of the division of the original branch and sub-item projects. The sales tax of parts and prefabricated component had been paid when it sold; the counting base of fees and taxes is suggested to deduct this part of the cost in order to avoid repeat counting. The clear scope definition of construction project cost and suitable valuation basis can provide the uniform specification and platform for the accurate accounting and comparison of the project cost between housing industrialization and traditional architecture.

(2) Cost personnel should keep pace with the times to study industrialization knowledge and BIM technology. There are large differences of the operation mode and the process flow of the field between the industrialization and traditional architecture, so cost personnel should learn its construction process and operation mode in advance. The core technology of BIM is a huge database formed by the 3D model. Its information can be used to not only prepare the project cost, but also manage the whole process valuation. These characteristics are fit in with the management model which has the long industry chain of housing industrialization, many participating enterprises and needs integration. With the development of computer technology, BIM will become the hot sharing model between the designer, constructor, and cost personnel in the architectural industry. Currently, the cost calculation software widely used in China is mainly based on the development of 2D platform; the traditional cost method is based on the calculation of construction process; the 3D dynamic of BIM makes models based on the building structure function. Whether it exports the BIM model to external software or uses the BIM as the calculation platform directly, it will bring challenge to the traditional valuation. Housing industrialization and BIM technology will bring great change to the architectural industry in China.

(3) The working emphasis of the cost consulting company and cost personnel should transfer to the active dynamic control from the passive calculation. In the traditional valuation work, it spends a lot of time and efforts to contrast the calculation measurement and control measurement of drawings, and most of them are doing the job passively. Under the industrialization mode, the use of most prefabricated components and parts and BIM technology lead to the simplification of calculation and valuation process, and some works even hand over to the software. Cost personnel can make more time to analyze cost indexes initiatively, actively control factors which may cause the increase of the project cost, treat the change and claimant in time which affects the project

cost, make preparation for settlement and final accounts and so on in order to realize the transformation from the counter of the project cost to the cost manager.

6 Conclusions

The paper makes definition and comparison of the constitution of the project cost of housing industrialization in the broad sense. Due to the limited experience, the frame structure of the project cost refers to the current valuation method. Though it doesn't make a quantitative index analysis by combining with specific construction projects, it makes a qualitative analysis of the constitution and characteristics of the project cost in the broad sense. It offers reference for the accurate calculation of the project cost of housing industrialization and provides development suggestions.

Acknowledgement

This paper is sponsored by 2015 Soft Science project of science and technology bureau of Cheng Du (GN: 2015-RK00-00042-ZF).

References

1. Yuanming Xu. *The cost analysis of housing industrialization. Construction economy*, **5**(2014):21-24.
2. Mingwang Gu. *Discussion the relationship between housing industrialization and economy. House industrialization*, **3**(2013):15-20
3. Xiaokui Han. *The thinking of practice on housing industrialization. ZheJiang Construction* , **2**(2014):59-61
4. Shuang Wang. *The cost comparison between the housing industrialization and the traditional construction. Construction and budget*,**7**(2014):26-29
5. Fenfen Cheng. *The research of housing industrialization promoting mechanism,Journal of Anhui Institute of Architecture&Industry*, **3**(2013):61-66
6. Tao Wang. *Comparative study on the cost between housing industrialization and traditional construction methods.Journal of Engineering Management*,**4**(2015):49-53

MATEC Web of Conferences **68**, 13012 (2016)
DOI: 10.1051/matecconf/20166813012
ICIEA 2016

Structural Model for the Effects of Perceived Indoor Work Environment on Sick Building Syndrome and Stress

Nor Hazana Abdullah[1], Nor Aziati Abdul Hamid[1], Muhamad Shahrul Amirul Shaif[2], Alina Shamsuddin[1], Eta Wahab[1]

[1]*Faculty of Technology Management, UTHM, Malaysia*
[2]*Ye Chiu Metal Smelting, Johor, Malaysia*

Abstract. Sick Building syndrome (SBS) and stress have a prevalent influence on organizational productivity and competitiveness. Unhealthy employees not only tend to have high medical leaves but also low productivity due to ailments and discomforts. Studies that investigate the effects of indoor work environment on Sick Building Syndrome (SBS) have yielded mixed results while their effect on stress has not been empirically established. Furthermore, studies that simultaneously investigate both SBS and stress are almost non-existent. Thus, this study aimed to study the effects of perceived indoor work environment on SBS and stress and the link between SBS and stress. A cross-sectional survey participated by 598 employees from various industries was conducted from September to October 2015. Data were analyzed using Partial Least Square Structural Equation Modeling (PLS-SEM) to assess both the measurement model and the path structure. The results suggest that indoor work environment has significant yet the weak effect on SBS while it has no effect on stress. However, SBS has a strong significant relationship with stress. The implication of this study on the importance of conducive indoor work environment is discussed with suggestions for future studies.

1 Introduction

Sick Building syndrome (SBS) refers to a constellation of symptoms experienced by employees who are working in the structural confinement of buildings. Such symptoms include a headache, nose and throat irritation, dry cough and itchiness without known causes/illnesses [1]. Stress, on the other hand, refers to "mechanism whereby the human body attempts to adapt to the environment". Failure to be adaptive to stress has dire consequences at individual and organizational levels [2]. Studies have shown that SBS and stress not only has an impact on productivity [3] [4], but also on job satisfaction [5] [6] and other organizational outcomes. In a nutshell, SBS and stress have a prevalent influence on organizational effectiveness. Unhealthy employees not only tend to have high medical leaves but also low productivity and commitment to work due to ailments and discomforts.

Among predictors claimed to affect SBS and stress is the indoor work environment [7]. As most employees spend the majority of their time at the workplace, the indoor work environment might not only influences their physical wellbeing but also their psychological states. Kogi [8] found that combined environmental exposure and indoor air quality are one of the top emerging issues of occupational and environmental health among fifteen Asian-Pacific countries being surveyed. Moreover, temperature, particularly heat, is ranked first as the most important occupational health problem which is aligned Hole and Pande's finding [9]. Statistics from the Social Security Organizations of Malaysia also shows an upward trend where the number of occupational diseases has increased from 194 cases in 2005 to 3002 cases in year 2014.

Studies that investigate the effects of indoor work environment on Sick Building Syndrome (SBS) have yielded mixed results while their effect on stress has not been firmly established. Moreover, majority of these studies are no longer recent and their findings need to be re-affirmed. Furthermore, studies that simultaneously investigate both SBS and stress are almost non-existent. Consequently, this study aimed to study the effects of perceived indoor work environment on SBS and stress and the link between SBS and stress using a more robust analysis via the Structural Equation Modeling.

2 Literature review

Discussion on related works of Indoor Work Environment, Sick Building Syndrome, and Stress is partitioned in subsequent sub-sections to reflect corresponding hypotheses formulation.

2.1 Indoor work environment and sick building syndrome

Indoor work environment refers to ambient environmental conditions which include air temperature and movement, relative humidity, and respirable. Indoor work environment, particularly, indoor air quality, could negatively impact employee's physical health such as asthma exacerbation, respiratory allergies and complications [10]. In terms of monetary returns, improving indoor air quality is claimed to earn potential annual savings and productivity gains of at least 29 billion and could reduce absenteeism up to USD400 per employee [11]. Employees who are exposed to various types of particles and gaseous pollutants [12] [13] tend to increase their health risks which consequently affect their productivity and performance. Various studies have shown that indoor environment conditions increase the prevalence and risk factors of SBS regardless of types of building [14], air-conditioned rooms/buildings [12] and age of the buildings [15].

For example, Tarcan and Varol [16] in their study involving 375 individuals working in 25 hospitals found that indoor air quality is highly predictive of SBS and building general sufficiency level. Abdel-Hamid [3] found that poor ventilation and high temperature are predictive of SBS among office workers. Zamani et al. [17] claimed that increasing the ventilation rates, ventilation effectiveness and reducing indoor air pollutant could reduce SBS although risk factors of SBS were different between old and new building. Similarly, Bholah et al. [18] found that mechanical ventilators induce higher SBS symptoms compared to naturally ventilated buildings. Hidayah et al. [19], concurred when they found that ventilation and accumulation of possible contaminants within indoor environment exacerbates SBS symptoms. It is evident that majority of studies have been focusing on indoor air quality which justify its inclusion in this study. Based on these literatures, it is hypothesized that indoor work environment is significantly related with SBS.

2.2 Indoor work environment and stress

Theoretical underpinning on how indoor work environment influences stress could be traced back from the seminal work of Lazarus and Cohen [20], Evans [21] and a few others [22] on environmental stress. According to Evans, stress is a function of variation in environmental quality. Since environmental conditions are inevitable and enduring, the extent of their interferences with optimal human functioning and coping processes could lead to psychological discomfort. Lazarus and Cohen [20] categorized 'daily hassles' which refers to persons' irritating daily experience that are stable and repetitive as one of important stressors. Badayai [23] reasoned that working environment such as temperature, air and noise are stress-provoking stimuli that influence employees' psychological processes, produce negative affection, reduce motivation and social interaction. Thus, prolonged exposure to such environmental stressors would affect the stress level of the employees.

Rashid and Zimring [7], in their framework, suggested that indoor work environment may elicit stress "by the ways in which it affects individual and/or workplace needs.". For example, if an employee perceives that he/she need comfortable temperature to work, absence of such need would induce stress. In their review of the massive literature on the impact of indoor environment and stress in both office and healthcare setting, they further concluded that empirical supports on the direct linkage of indoor environment – stress is lacking and remains a knowledge gap. Woo and Postolache [24] argued that work environment is closely related to mood disorders especially for those who suffer seasonal affective disorder (SAD). SAD sufferers who are exposed to heat during summer might fall into depression cycle. In addition, hot temperature not only induces perspiration and dehydration but also increases toxicity risks from psychotropic medications used to treat mood disorders. In Malaysia, Makhbul [25] explored the effect of indoor air quality, lighting, acoustics, furniture and tools and building general environment on academician's emotional health and found that only building general environment and workplace ergonomics are significantly related with emotional health. Therefore, it is hypothesized that indoor work environment is significantly related to stress.

2.3 Sick building syndrome and stress

The dynamic interaction between SBS and stress is still unclear. Crawford and Bolas [26], in their review, contended that stress is correlated with SBS but cautioned the causal direction. They argued that whether stress is the predictor of SBS or vice versa is debatable as most studies are cross-sectional in nature. Ooi and Goh [27], on the other hand, reasoned that despite the substantial change of work environment for the past years, the SBS cases have not abated. Thus, it is logical to assume that SBS is induced by stress rather than the work environment. This is especially true when employees have to cope with increasing workload, a higher pace of work with less autonomy. In their survey involving 2160 employees, they found the prevalence of SBS among employees who reported high levels of physical and mental stress. As studies scrutinizing this direct causal effect are limited, this study postulates that SBS and stress is related where SBS is regarded as the risk factor of stress in line with the stress model.

3 Methods

This was a cross-sectional survey research using questionnaires as data collection method. Unlike the majority of previous studies which utilized actual measurement of indoor work environment such as temperature and humidity levels, this study used employees' responses to measure all the three constructs following the suggestion of Hedge and Erickson[28].

3.1 Samples and procedure

Five hundred ninety-eight employees from twenty companies participated in this study. Selection of employees in each company was based on random sampling procedure while the selection of companies used convenience sampling. Since research participation in not encouraging, the use of convenience sampling is seen as the best option. Participating companies were located in the state of Selangor and Johor, Malaysia. The majority of the respondents worked in manufacturing companies (74.7%) while the rest worked in service-oriented companies. Male respondents were slightly higher at 51.8% compared to female (48.2%) with the majority of respondents aged between 20 to 30 years old (55%) followed by those aged 31-40 years old (32.1%), 41-50 years old (11.5%) and above 50 years old (1.3%). Majority of respondents (62.2%) had worked between 1-5 years, 21.2% had worked 6-10 years, 10.4% has worked 11-15 years while the rest had worked more than 16 years. In terms of total hours of working, 65.9% had worked between 40-49 hours per week, 16.9% worked less than 39 hours per week, and 17.1% worked more than 50 hours per week.

3.2 Measures

Adapted Cornell Office Environment Survey (short form) was used to measure Sick Building Syndrome (7 items) and Indoor Work Environment (7 items) [28]. Measurement of stress was developed based on the most common symptoms of occupational stress (7 items) which include insomnia, increase heart rate, lack of appetite, burnout, stomach ache, fatigue, and anxiety as shown in Table 1.

Table 1. Questionnaires items.

Construct	Item	Description
Indoor Work Environment	B1	Air temperature too cold
	B2	Air temperature too warm
	B3	Too little air movement
	B4	Air too dry
	B5	Unpleasant odour in air
	B6	Air too stale
	B7	Air too dusty
Sick Building Syndrome	C1	Irritated, sore eyes
	C2	Sore, irritated throat
	C3	Hoarseness
	C4	Stuffy, congested nose
	C5	Excessive mental fatigue
	C6	Headache across forehead
	C7	Unusual tiredness, lethargy
Stress	C8	Insomnia
	C9	Increase heart rate
	C10	Lack of appetite
	C11	Burnout
	C12	Stomach-ache
	C13	Fatigue
	C14	Anxiety

Two items from were taken out from the analysis during the assessment of measurement model which was C1 for Indoor Work Environment and C7 from stress since their loading is below 0.5 following the suggestion by Hair et al. [29].

3.3 Analysis

PLS-SEM Version 2.0 [30] was used to analyze the data. SEM is a second-generation multivariate data analysis method that combines factor analysis and multiple regressions. Although there are two types of SEM which are covariance-based SEM (CB-SEM) and partial least square SEM (PLS-SEM), the use of PLS-SEM in this study was more appropriate since it is exploratory in nature [31]. The use of SEM enables model testing that is more robust compared to conventional regression modelling. Prior to examining the structural model for hypotheses testing, a measurement model was assessed to determine constructs' convergent and discriminant validity.

4 Results

Table 2. Items loadings and reliability.

	Items	Loadings	M	SD	AVE	CR*	CA
IE	B2	0.724	2.600	1.105	0.569	0.887	0.847
	B3	0.662	2.669	1.019			
	B4	0.716	2.771	0.994			
	B5	0.802	2.664	1.123			
	B6	0.838	2.574	1.159			
	B7	0.771	2.686	1.135			
SBS	C1	0.733	2.227	1.049	0.571	0.888	0.849
	C2	0.809	2.324	1.013			
	C3	0.768	2.105	0.994			
	C4	0.769	2.557	1.125			
	C5	0.742	2.763	1.082			
	C6	0.708	2.513	1.085			
	C8	0.711	2.528	1.073			
	C9	0.744	2.647	1.061	0.580	0.906	0.879
STR	C10	0.740	2.124	1.063			
	C11	0.798	2.042	1.067			
	C12	0.750	1.997	1.091			
	C13	0.811	1.753	1.010			
	C14	0.774	2.005	1.104			

IE: Indoor Environment
SBS: Sick Building Syndrome
STR: Stress

Table 2 shows the mean, standard deviation (SD), composite reliability (CR), Cronbach Alpha, and AVE and loadings of each item for each construct. The measurement model shows that Indoor Environment, Sick Building Syndrome, and Stress had adequate reliability since the values of CR and AC were all above 0.7 [31]. Convergent validity of each construct was also sufficient as each factor loading and average variance extracted exceeded the recommended value of 0.5 [29]. Table 3 shows that the AVE value of each construct is also higher than the squared correlations between the latent variable and all other variables which indicate that discriminant validity is achieved [32].

Table 3. Discriminant validity.

	Correlations		
	IE	SBS	STR
Indoor Environment	**0.754**		
Sick Building Syndrome	0.435	**0.756**	
Stress	0.367	0.743	**0.762**

Notes: * Calculated using Fornell and Larker's (1981) method
Diagonals (in bold) represent the average variance extracted while other entries represent the squared correlations.

Table 4 and Fig. 1 show path coefficients of the structural model. The indoor work environment had a significant relationship with Sick Building Syndrome but not with Stress (*t*<1.96). Moreover, SBS had a significant strong positive relationship with stress. To conclude, hypotheses 1 and 3 were accepted and hypothesis 2 was rejected.

Table 4. Path coefficient.

Hypothesis	Relationship	Std. Beta	SE	t-value
H1	IE -> SBS	0.435	0.038	11.392**
H2	IE -> Stress	0.055	0.034	1.606
H3	SBS -> Stress	0.719	0.025	28872.000**

** p< 0.01

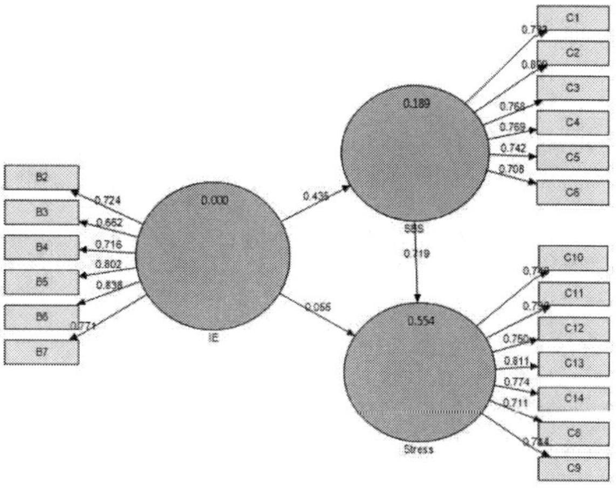

Figure 1. Path coefficient.

5 Discussion and implications

This study implies that further scrutiny is needed to investigate the dynamics of indoor work environment-SBS-stress linkages. It is evident that indoor work environment is an important consideration to reduce SBS but not stress. This finding contradicts with the theory of environmental stress [21], [22] where work environment has enduring features that influence whether or not stress is produced.. However, the strong relationship between SBS and stress demand further investigation. Perhaps, a concept of spill over could adequately explain the strong link of SBS and stress. When SBS become prevalent, it interferes or spill over to their physiological health. This explains the strong relationship of these two constructs. Future studies might need to dwell further on this interaction and provide a framework explaining their connection.

This study is not without limitation. As the nature of this study is cross-sectional, causality inference is cautioned. Furthermore, all measures are based on employees' perceptions rather than actual measurement of the indoor environment, SBS and stress. However, it is important to note that self-rating predicts better compared to objective measurement [33]. Furthermore, subjective measures are also more widely use and easy to assess.

Another limitation of this study is that it was conducted based on limited studies available and direct causal assumptions. Thus, future studies might investigate whether SBS mediates the relationship between indoor work environment and stress.

References

1. R. Runeson-Broberg and D. Norbäck, "Sick building syndrome (SBS) and sick house syndrome (SHS) in relation to psychosocial stress at work in the Swedish workforce," *Int. Arch. Occup. Environ. Health*, vol. 86, no. 8, pp. 915–922, 2013.
2. W. A. J. W. Yahaya, S. N. J. Ahmad, and M. Z. M. Zain, "Application of Persuasive Multimedia to Raise Stress Awareness among the Secondary School Students," *IERI Procedia*, vol. 3, pp. 105–113, 2012.
3. M. a Abdel-Hamid, S. A Hakim, E. E. Elokda, and N. S. Mostafa, "Prevalence and risk factors of sick building syndrome among office workers.," *J. Egypt. Public Health Assoc.*, vol. 88, no. 2, pp. 109–14, 2013.
4. P. K. Wilke, W. H. Gmelch, and N. P. Lovrich, "Stress and productivity: Evidence of the inverted U function," *Public Product. Rev.*, vol. 9, no. 4, pp. 342–356, 1985.
5. S. N. Kamaruzzaman, C. O. Egbu, E. M. A. Zawawi, S. B. A. Karim, and C. J. Woon, "Occupants satisfaction toward building environmental quality: structural equation modeling approach," *Environ. Monit. Assess.*, vol. 187, no. 5, 2015.
6. K. Fairbrother and J. Warn, "Workplace dimensions, stress and job satisfaction," *J. Manag. Psychol.*, vol. 18, no. 1, pp. 8–21, 2003.
7. M. Rashid and C. Zimring, "A Review of the Empirical Literature on the Relationships Between Indoor Environment and Stress in Health Care and Office Settings," *Environment and Behavior*, vol. 40, no. 2. pp. 151–190, 2008.
8. Kazutaka Kogi, "Current Problems-Emerging Issues in Occupational and Environmental Health," *Environ. Manag. Heal.*, vol. 8, no. 5, pp. 167–169, 1997.
9. J. A. Hole and M. Pande, "Worker productivity, occupational health, safety and environmental issues in

thermal power plant," in *Industrial Engineering and* Engineering *Management, 2009. IEEM 2009. IEEE International Conference on*, 2009, pp. 1082–1086.
10. J. Sundell, "On the history of indoor air quality and health.," Indoor *Air*, vol. 14 Suppl 7, no. Suppl 7, pp. 51–58, 2004.
11. P. Wargocki, "Productivity and Health Effects of High Indoor Air Quality," in *Encyclopedia of Environmental Health*, 2011, pp. 688–693.
12. B. F. Yu, Z. B. Hu, M. Liu, H. L. Yang, Q. X. Kong, and Y. H. Liu, "Review of research on air-conditioning systems and indoor air quality control for human health," *Int. J. Refrig.*, vol. 32, no. 1, pp. 3–20, 2009.
13. J. A. Bernstein, N. Alexis, H. Bacchus, I. L. Bernstein, P. Fritz, E. Horner, N. Li, S. Mason, A. Nel, J. Oullette, K. Reijula, T. Reponen, J. Seltzer, A. Smith, and S. M. Tarlo, "The health effects of non-industrial indoor air pollution.," *J. Allergy Clin. Immunol.*, vol. 121, no. 3, pp. 585–91, 2008.
14. J. C. Vischer, "The Concept of Workplace and its values to Managers," *Calif. Manage. Rev.*, vol. 49, no. 2, pp. 1–18, 2006.
15. K. Engvall, C. Norrby, J. Bandel, M. Hult, and D. Norback, "Development of a Multiple Regression Model to Identify Multi-Family Residential Buildings with a High Prevalence of Sick Building Syndrome (SBS)," *Indoor Air*, vol. 10, no. 2, pp. 101–110, 2000.
16. E. Tarcan, E. Sait Varol, and M. Ates, "A qualitative study of facilities and their environmental performance," *Manag. Environ. Qual. An Int. J.*, vol. 15, no. 2, pp. 154–173, 2004.
17. M. E. Zamani, J. Jalaludin, and N. Shaharom, "Indoor air quality and prevalence of sick building syndrome among office workers in two different offices in selangor," *Am. J. Appl. Sci.*, vol. 10, no. 10, pp. 1140–1147, 2013.
18. R. Bholah, I. Fagoonee, and A. H. Subratty, "Sick building syndrome in Mauritius: Are symptoms associated with the office environment?," *Indoor Built Environ.*, vol. 9, no. 1, pp. 44–51, 2000.
19. A. Norhidayah, L. Chia-Kuang, M. K. Azhar, and S. Nurulwahida, "Indoor Air Quality and Sick Building Syndrome in Three Selected Buildings," *Procedia Eng.*, vol. 53, no. 2010, pp. 93–98, 2013.
20. R. S. Lazarus and J. B. Cohen, "Environmental Stress," in *Human Behavior and the Environment: Current Theory and Research*, I. Altman and J. F. Wohlwill, Eds. New York: Spectrum, 1977, pp. 89–127.
21. G. W. Evans, *Environmental Stress*. Cambridge: Cambridge University Press, 1984.
22. S. Cohen, Gary W. Evans, D. Stokols, and D. S. Krantz, *Behavior, Health, and Environmental Stress*. New York: Springer, 1986.
23. A. R. A. Badayai, "A Theoretical Framework and Analytical Discussion on Uncongenial Physical Workplace Environment and Job Performance among Workers in Industrial Sectors," *Procedia - Soc. Behav. Sci.*, vol. 42, no. July 2010, pp. 486–495, 2012.
24. J. Woo and T. T. Postolache, "The impact of work environment on mood disorders and suicide: Evidence and implications," *Int. J. Disabil. Hum. Dev.*, vol. 7, no. 2, pp. 185–200, 2008.
25. Z. M. Makhbul, "Workplace Environment Towards Emotional Health," *Int. J. Acad. Res. Bus. Soc. Sci.*, vol. 3, no. 1, pp. 183–195, 2013.
26. J. O. Crawford and S. M. Bolas, "Sick building syndrome, work factors and occupational stress," Scand. *J. Work. Environ. Heal.*, vol. 22, no. 4, pp. 243–250, 1996.
27. P. L. Ooi and K. T. Goh, "Sick building syndrome: An emerging stress-related disorder?," *Int. J.* Epidemiol., vol. 26, no. 6, pp. 1243–1249, 1997.
28. A. Hedge and W. A. Erickson, "A Study of Indoor Environment and Sick Building Syndrome Complaints in Air-Conditioned Offices: Bencahmarks for Facility Performance," *Int. J. Facil.* Manag., vol. 1, no. 4, pp. 185–192, 1997.
29. J. F. H. Jr, W. C. Black, B. J. Babin, and R. E. Anderson, *Multivariate Data Analysis*, 7th ed. 2006.
30. C. M. Ringle, S. Wende, and A. Will, "SmartPLS 2.0.M3.," *Hamburg: SmartPLS, http://www.smartpls.de.*, 2005. .
31. J. F. Hair, M. Sarstedt, T. M. Pieper, and C. M. Ringle, "The Use of Partial Least Squares Structural Equation Modeling in Strategic Management Research: A Review of Past Practices and Recommendations for Future Applications," *Long Range Plann.*, vol. 45, no. 5–6, pp. 320–340, 2012.
32. J. Henseler, C. M. Ringle, and M. Sarstedt, "A new criterion for assessing discriminant validity in variance-based structural equation modeling," *J. Acad. Mark. Sci.*, 2014.
33. S. Cohen, T. Kamarck, and R. Mermelstein, "A Global Measure of Perceived Stress," vol. 4.

MATEC Web of Conferences **68**, 14001 (2016) DOI: 10.1051/matecconf/20166814001
ICIEA 2016

An Analysis of Energy and Carbon Intensities and Exergy Efficiency in Natural Gas Combined-Cycle Power Plants: A Case Study of Thailand

Yada Bunyamanid[1], Nantamol Limphitakphong[3], Orathai Chavalparit[1,3] and Thanapol Tantisattayakul[2]

[1] *Department of Environmental Engineering, Faculty of Engineering, Chulalongkorn University, Bangkok 10330, Thailand*
[2] *Faculty of Science and Technology, Thammasat University Rangsit Campus, Bangkok 12121, Thailand*
[3] *Research unit of Environmental Management and Sustainable Industry, Faculty of Engineering, Chulalongkorn University, Bangkok 10330, Thailand*

Abstract. The objectives of this study were to estimate energy and carbon intensities of natural gas combined-cycle power plants and to improve the energy efficiency of power plants through an exergy analysis. Three plants in Thailand were evaluated as case studies in this research using data collected in 2013 – 2014. The results showed that the average carbon and energy intensities of three Combined-Cycle Power Plants were 0.436 kg CO_2/kWh and 7.995 GJ/MWh, respectively. The range of the energy intensity in this study was slightly lower than that of other studies; however, the results of the carbon intensity were in the same range. The result of the exergy analysis illustrated that Heat Recovery Steam generator (HRSG) has the greatest exergy destruction due to the wide temperature range. To summarize, natural gas combined-cycle power plants require less energy and emit less carbon compared to other technologies or fuels. The exergy efficiency of HRSG was 60%, with an exergy destruction rate of 66.88 MW. One approach to improve the efficiency of power plants is to improve the efficiency and performance of HRSG.

1 Introduction

Global warming, which is caused by increasing greenhouse gas emissions, is mainly the result of human activities. Scientific studies have found that greenhouse gas emissions have increased abruptly over the last few decades. Increasing attention has been paid to finding methods for reducing greenhouse gas emissions. In Thailand, for example, policies related to climate change have been established, such as the climate change model scheme of 2015-2050; its main target is to reduce energy consumption and greenhouse gas emissions, mainly in the energy and transportation sectors [1]. The sector that accounts for the highest emissions in Thailand is the energy sector, with 69.6% of the total emissions in 2000. Within the energy sector, the energy industry has contributed the greatest amount of greenhouse gas emissions (approximately 41.7%) [2]. Furthermore, in 2014, according to the report of the Energy Policy and Planning Office, Ministry of Energy of Thailand, the electricity generation industry accounted for the highest emissions (40.1%) of the total energy sector emissions [3]. The indicators commonly used in many industries for evaluating the energy efficiency and environmental impact in terms of greenhouse gas emissions of plant operations are energy intensity and carbon intensity, respectively. Studies and reports have been conducted regarding the energy and carbon intensities in various industries, including the steel [4, 5], petrochemical [6], and electrical generation [7-9] industries.

Approximately 80% of electricity generation in Thailand currently depends on fossil fuels, which includes electricity generated from Combined-Cycle Power Plants (CCPPs) and thermal power plants, which account for 64.85% and 13.36% of the total installed capacity, respectively [10]. The CCPP is an energy efficient form of power generation that applies both Brayton (gas turbine) and Rankine (steam turbine) cycles to obtain higher thermal efficiency in comparison with individual steam or gas turbine cycles [11]. The gas-fired generation in California improved thermal efficiency more than 17% during 2001-2013 by introducing new CCPPs to the portfolio of natural gas power plants [12]. The analysis of carbon intensity of the electricity generation industry (fossil fuel power plants) in Thailand during 2004 - 2009 showed that CCPPs yielded the lowest carbon intensity of 0.426 $kgCO_2$/kWh compared to using other technologies or fuels [7]. The analysis of energy intensity in 2014 also presented consistent results in that natural gas CCPPs had the lowest energy intensity of 7.970 GJ/MWh, whereas coal thermal power plants had the value of 10.48GJ/MWh [9]. In the U.S.A., the average carbon intensity of coal power plants, natural gas power plants and natural gas CCPPs were 0.915 ± 0.0008, 0.5494 ± 0.0011 and 0.436 ± 0.0014 $kgCO_2$/kWh, respectively [8], which is in line with the results in Thailand. Due to the higher thermal efficiency and lower environmental impact than other types of fossil fuel power plants, CCPPs are attractive in the power generation field. Energy use improvement can contribute in both energy and environmental aspects.

An exergy analysis is a potential tool for the energy efficiency improvement of many energy engineering applications [13]. Exergy is the maximum work or available work that can be obtained from different forms of energy in relation to a given environment [14]. Exergy analysis applies the first and second laws of thermodynamics to determine the maximum performance of the system and identify the main plant component of exergy destruction to determine potential performance improvement opportunities of power plants. Studies regarding the application of exergy analysis have been conducted to determine energy efficiency improvement potential. I. H. Aljindi performed energy and exergy analyses of a steam power plant in Jordan. The system components include boiler pump heaters, turbines and condensers. The study revealed that the maximum exergy destruction occurred in the boiler system, followed by the turbine and the forced draft fan condenser at 77%, 13%, and 9%, respectively. The exergy destruction in a boiler system can be reduced by preheating the combustion air and reducing the air–fuel ratio [15]. S. C. Kamate et al. applied the exergy method to analyze the optimal steam inlet condition for the turbine in a cogeneration power plant [16]. R. Saidur et al. performed energy and exergy analyses of industrial boilers in Malysia and found that the combustion chamber is the major contributor for exergy destruction, followed by the heat exchanger [17]. An energy analysis of gas turbines with an air bottoming cycle was conducted by M. Ghazikhani et al. The result showed that the regenerator has the largest contribution in the exergy destruction of the air bottoming cycle, and the recovery of the exhaust exergy can increase the cycle performance [18]. The exergy and greenhouse gas emissions analyses of an integrated Organic Rankine Cycle (ORC) with a biomass combustor for different cases of operation was performed by F. A. Al-Sulaiman et al. The result showed that 1) the main two sources of exergy destruction were the biomass combustor and the ORC evaporator, and 2) the tri-generation (power, heat, and cooling) case can increase the exergy efficiency and, consequently, considerably reduce the carbon intensity [19].

For CCPPs, the exergo-environmental optimization of the Heat Recovery Steam Generator (HRSG), which is an important component and connection between the gas turbine and steam turbine cycles of a CCPP, was estimated through energy and exergy analyses. The results showed that for the given HRSG, most exergy destruction occurred in the high-pressure evaporator, and increasing the HRSG inlet gas temperature up to 650°C could increase the thermal and exergy efficiencies of the cycles [20].

In this study, the energy and carbon intensities of natural gas CCPPs in Thailand were analyzed. The results were compared to other studies. Moreover, exergy analysis of the plant components, including the HRSG and steam turbines, was conducted to analyze the energy improvement potential.

2 Methodology

To evaluate the energy and carbon intensity, the data of three natural gas CCPPs during 2013 - 2014 were collected and calculated. The energy intensity was calculated based on the data of energy used in the sample power plants, and the carbon intensity was calculated based on the IPCC guidelines for national greenhouse gas inventories [21]. The scope of emissions in this research includes direct (scope 1) and indirect (scope 2) emission sources. Moreover, the data of the processes in the plant analysis were also collected to perform the exergy analysis.

Three sample CCPPs with a total capacity of 1,648 MW (1,468 MW, 90 MW and 90 MW) were considered in this research. The evaluation indicators and calculation methodology are detailed in the following section.

2.1 Energy Intensity

The energy intensity is the total energy consumption per unit of energy generated. Energy intensity can be expressed as follows:

$$EI = \frac{EC}{P} \tag{1}$$

where EI is the energy intensity (GJ/MWh), EC is the total energy consumption (GJ), and P is the total power generated (MWh).

The total energy consumption is defined as a product of activity and net calorific values. The activity data include the amount of fuel consumed and the amounts of purchased electricity and steam. EC can be calculated as follows:

$$EC = \sum(AD \times NCV) \tag{2}$$

where AD is the activity data (unit), and NCV is the net calorific value (GJ/unit of activity).

2.2 Carbon Intensity

This research considered two sources of carbon dioxide emissions that were mentioned above. The emissions (E_{CO2}) can be expressed by the product of activity and the carbon dioxide emission factor as follows:

$$E_{CO_2} = AD \times EF \tag{3}$$

where AD is the activity, and EF is the carbon dioxide emission factor (kgCO_2/unit of activity). The activity data include the amount of fuel consumed in the case of direct emissions as well as the amount of purchased electricity and steam in the case of indirect emissions.

The total carbon dioxide emission is defined by direct and indirect carbon dioxide emissions as follows:

$$E_{CO_2,total} = E_{CO2,direct} + E_{CO2,indirect} \tag{4}$$

where $E_{CO_2,total}$ is the total carbon dioxide emission (kgCO_2), $E_{co2,direct}$ is the direct carbon dioxide emission caused by fossil fuel combustion in the plants, and

$E_{CO2,indirect}$ is the indirect carbon dioxide emission caused by using electricity and steam from outside the plants.

The carbon intensity is the carbon emission per unit of electricity generated. It can be expressed as the following equation:

$$CI = \frac{E_{CO_{2,total}}}{P} \tag{5}$$

where CI is the carbon intensity ($kgCO_2/kWh$), $E_{CO_{2,total}}$ is the total carbon dioxide emissions, and P is the total electricity generated.

2.3 Exergy Analysis

An exergy analysis is a potential tool in the analysis of design and performance improvement of power plants. The exergy balance equation for a control volume system can be presented as follows:

$$\dot{E}x_Q + \sum_i \dot{m}_i ex_i = \sum_e \dot{m}_e ex_e + \dot{E}x_W + \dot{E}x_D \tag{6}$$

$$ex = (h - h_0) - T_0(s - s_0) \tag{7}$$

where $\dot{E}x_Q$ is the exergy of the heat transfer rate (kJ/s), $\dot{E}x_W$ is the exergy rate of work (kJ/s), $\dot{E}x_D$ is the exergy destruction rate (kJ/s), ex is the specific exergy (kJ/kg), $\dot{m}$is the mass flow rate (kg/s), h is the specific enthalpy (kJ/kg), s is the specific entropy(kJ/kg), T is temperature (K), i is the inlet, e is the outlet, and 0 is the ambient condition.

The reference conditions used in this study were P_0= 1.01 bar and T_0 = 298.15 K.

Figure 1 presents the schematic diagram of the sample CCPP. The combined cycle facility consists of one gas turbine (GT), combustion chamber (CC), air compressor (AC), HRSG, gas GT air cooler condenser and cooling tower, two generators(Gs) and three steam turbines (STs). In this research, the exergy analysis was performed with the HRSG, high pressure steam turbine (HP ST), intermediate pressure steam turbine (IP ST) and low pressure steam turbine (LP ST).

2.3.1 Heat recovery steam generator (HRSG)

The exergy destruction rate of the HRSG ($\dot{E}x_{D,HRSG}$) can be calculated as follows:

$$\dot{E}x_{D,HRSG} = \sum \dot{E}x_{HRSG,i} - \sum \dot{E}x_{HRSG,e} \tag{8}$$

where $\dot{E}x_{D,HRSG}$ is the exergy destruction rate (kW), $\sum \dot{E}x_{HRSG,i}$ is the total exergy inflow rate (kW), and $\sum \dot{E}x_{HRSG,e}$ is the total exergy outflow rate (kW).

The exergy efficiency of the HRSG (ψ_{HRSG})can be calculated as follows:

$$\psi_{HRSG} = \frac{\sum \dot{E}x_{H_2O,e} - \sum \dot{E}x_{H_2O,i}}{\dot{E}x_{g,i} - \dot{E}x_{g,e}} \tag{9}$$

where $\dot{E}x_{HRSG}$ is the exergy rate of the HRSG (kW), $\dot{E}x_{H_2O,e}$ is the exergy rate of water or steam (kW), and $\dot{E}x_g$ is the exergy rate of gas (kW). The subscript i refers to the inlet and e refers to the outlet.

2.3.2 Steam turbine (ST)

The exergy destruction rate of ST ($\dot{E}x_{D,ST}$) can be calculated as follows:

$$\dot{E}x_{D,ST} = \sum \dot{E}x_{ST,i} - \sum \dot{E}x_{ST,e} - \dot{W}_{ST} \tag{10}$$

The exergy efficiency of ST (ψ_{ST}) is as follows:

$$\psi_{ST} = \frac{\dot{W}_{ST}}{\sum \dot{E}x_{ST,i} - \sum \dot{E}x_{ST,e}} \tag{11}$$

where $\sum \dot{E}x_{ST,i}$ is the total exergy inflow rate of steam into the ST(kW), $\sum \dot{E}x_{ST,e}$ is the total exergy outflow rate of steam from the ST (kW), and $\dot{W}_{ST}$ is thework obtained from the ST (kW).

According to equations 8 to 11, the exergy destruction and exergy efficiency of the HRSG and each ST can be calculated by the equations in Table 1. The subscript numbers in the equations refer to the schematic diagram in Figure 1.

Table 1. Exergy equation of each component.

Component	Equation	
HRSG	$\dot{E}x_D$	$(\dot{E}x_1 + \dot{E}x_2 + \dot{E}x_3 + \dot{E}x_4 + \dot{E}x_5)$ - $(\dot{E}x_6 + \dot{E}x_7 + \dot{E}x_8 + \dot{E}x_9 + \dot{E}x_{10})$
	ψ	$[(\dot{E}x_7 + \dot{E}x_8 + \dot{E}x_9 + \dot{E}x_{10})$ - $(\dot{E}x_1 + \dot{E}x_2 + \dot{E}x_3 + \dot{E}x_4)] / [\dot{E}x_5 - \dot{E}x_6]$
HP ST	$\dot{E}x_D$	$\dot{E}x_{11} - \dot{E}x_{15} - \dot{W}_{HP,ST}$
	ψ	$\dot{W}_{HP,ST} / (\dot{E}x_{11} - \dot{E}x_{15})$
IP ST	$\dot{E}x_D$	$(\dot{E}x_{12} + \dot{E}x_{13}) - \dot{E}x_{16} - \dot{W}_{IP,ST}$
	ψ	$\dot{W}_{IP,ST} / [(\dot{E}x_{12} + \dot{E}x_{13}) - \dot{E}x_{16}]$
LP ST	$\dot{E}x_D$	$\dot{E}x_{14} - (\dot{E}x_{17} + \dot{E}x_{18}) - \dot{W}_{LP,ST}$
	ψ	$\dot{W}_{LP,ST} / [\dot{E}x_{14} - (\dot{E}x_{17} + \dot{E}x_{18})]$

3 Result and Discussion

3.1 Energy Intensity

The average energy intensity of the three sample CCPPs are illustrated in Figure 2. The result showed an overall average energy intensity of 7.995 GJ/MWh. The major source of energy consumption was the natural gas combustion process, which was equal to 99.97% and 99.80% of the total energy consumption in 2013 and 2014, respectively. Compared with other research on energy intensity for combined cycles with natural gas fuel, it was observed that the energy intensity of this study was slightly lower than the other research [9, 12], as shown in Table 2.

3.2 Carbon Intensity

The carbon intensity of three sample CCPPs was evaluated, and the results of 2013 and 2014 are demonstrated in Figure 2. The average carbon intensity was 0.436 $kgCO_2/kWh$. The major source of carbon emissions was the natural gas combustion process, which accounted for 99.67% and 99.55% in 2013 and 2014, respectively. Compared with other research on carbon intensity for combined cycles with natural gas fuel, it was observed that the carbon intensity of this study was in the same range as other research [7-8], as shown in Table 2.

A comparison of energy intensity and carbon intensity of natural gas CCPPs in this research with the values of other technologies and fuel types from other research studies is shown in Table 2 [7-9, 12]. The result illustrates that the CCPPs in this study were operated with relatively high energy efficiencies to present. However, the reduction potential in terms of energy and carbon intensity can be assessed by determining the energy efficiency improvement potential through exergy analysis.

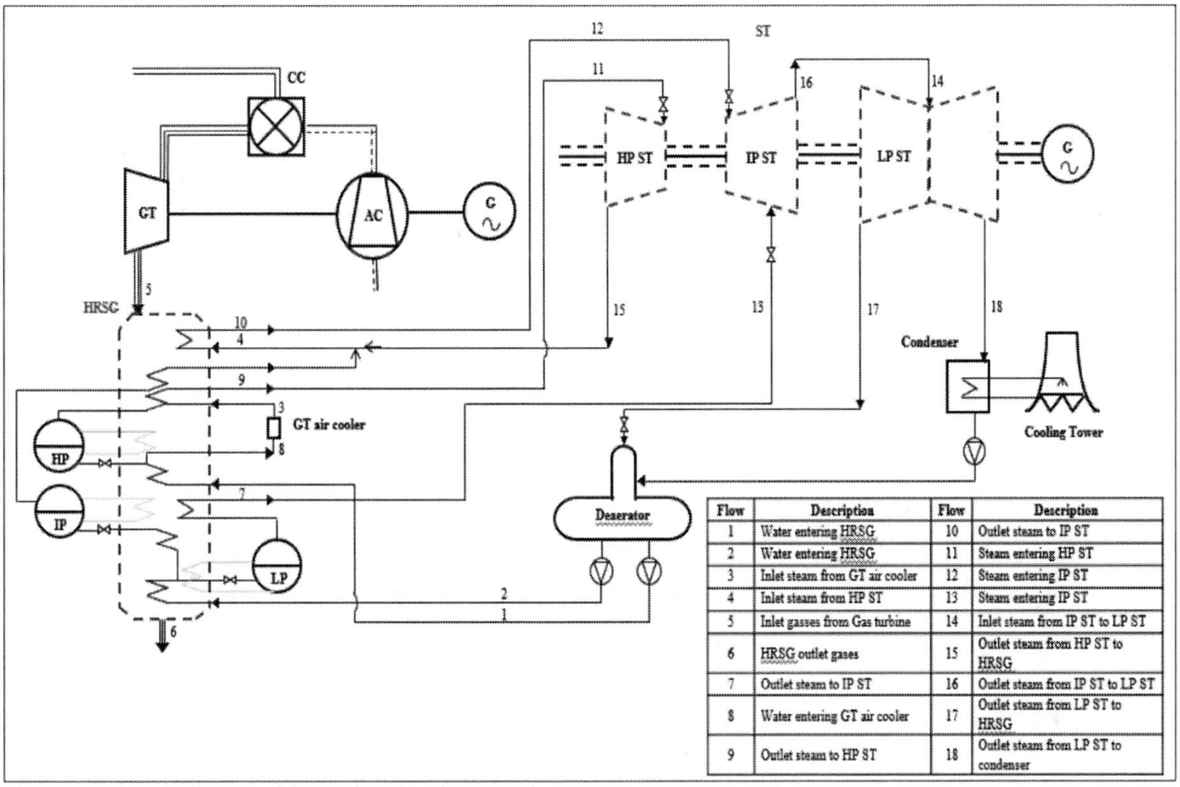

Figure 1. Schematic diagram of the combined cycle power plants.

Table 2. Comparison of carbon intensity and energy intensity.

Technologies	Fuel types	Energy Intensity (GJ/MWh)	Carbon Intensity ($kgCO_2/kWh$)	References
Combined cycle	Natural Gas	7.995	0.436	This research
		-	0.426	[7]
		-	0.436	[8]
		7.970	-	[9]
		9.037	-	[12]
	Natural Gas and Diesel	-	0.469	[7]
Thermal	Natural Gas Fuel oils and Diesel	-	0.622	[7]
	Natural Gas and Fuel oils	-	0.600	[7]
	Coal/Lignite	-	0.914	[7]
		10.48	-	[9]
Gas turbine	Diesel	-	0.970	[7]
Diesel engine	Diesel	-	0.842	[7]

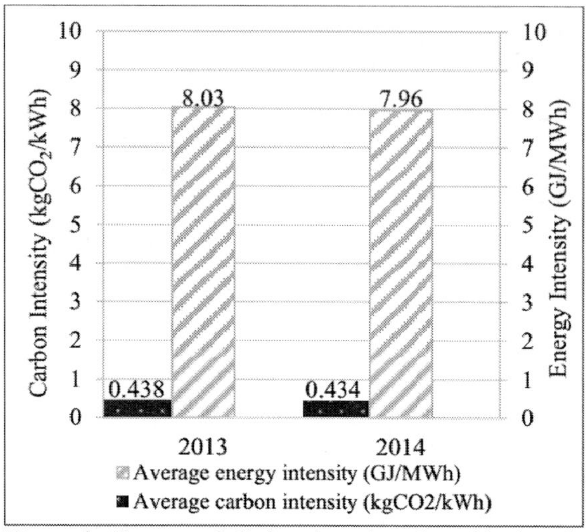

Figure 2. Energy and Carbon intensity of combined cycle power plants.

3.3 Exergy Analysis

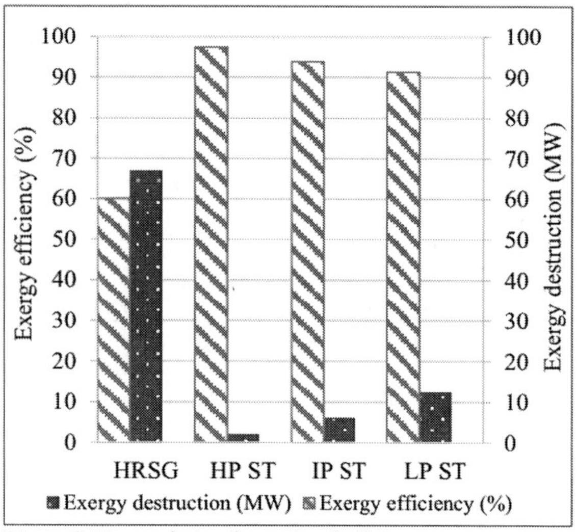

Figure 3. Exergy destruction rate and efficiency.

The exergy destruction and efficiency of the HRSG and three steam turbines were analyzed by the equations in Table 1 using the reference pressure and temperature of 1.01 bar and 298.15 K, respectively. The results are shown in Figure 3. The result revealed that the HRSG was the major contributor for exergy destruction, followed by the LP ST, IP ST, and HP ST, with values of 66.88 MW, 12.33 MW, 5.94 MW, and 1.88 MW, respectively. In addition, the HP ST had the highest exergy efficiency, followed by the IP ST, LP ST and HRSG, with values of 97%, 94%, 91%, and 60%, respectively. This indicates that the HRSG has more energy efficiency improvement potential than the steam turbines. The steam turbines had a high exergy efficiency of more than 90%, which suggests that the loss of available energy is less than 10%. However, the loss of available energy in the HRSG was approximately 40% of the total inflow available energy (exergy). One approach that can be used to improve the thermal and exergy efficiencies of HRSG is to increase the HRSG inlet gas temperature [20]. The improvement of performance of this cycle will consequently reduce the energy and carbon intensities.

4 Conclusion and recommendation

Natural gas combined-cycle power plants provide the lowest carbon and energy intensities compared to other technologies and other types of fossil fuels. This research aimed to evaluate the energy and carbon intensities of natural gas combined-cycle power plants and to determine the potential to reduce energy and carbon intensity. According to our findings, the average energy and carbon intensities of combine-cycle power plants in Thailand were 7.995 GJ/MWh and 0.436 kg CO_2/kWh, respectively. The potential of carbon intensity and energy intensity reduction was determined by an exergy analysis. The results of the exergy analysis revealed that the HRSG had the greatest exergy destruction contribution due to the wide range of temperature, followed by the LP ST and IP ST, whereas the HP ST produced the lowest exergy destruction. In addition, the HP ST had the highest exergy efficiency, followed by the IP ST, LP ST and HRSG. The exergy destruction rate of the HRSG was 66.88 MW, with an exergy efficiency of 60%. Therefore, improving the thermal efficiency of the HRSG can improve the efficiency of power plants. One approach to improve the thermal and exergy efficiencies of the HRSG is to increase the HRSG inlet gas temperature. Improving the performance will result in lower energy use and emission reduction.

Acknowledgement

This research has been supported by National Research University Project, Office of Higher Education Commission (WCU-58-015-EN).

References

1. Office of Natural Resources and Environmental Policy and Planning, climate, *change of model scheme 2015 – 2050*, (2015)
2. Joint Graduate School of Energy and Environment, *Second National Communication*, (2010)
3. Energy Policy and Planning Office, Ministry of Energy,*Energy Statistics of Thailand 2014*, (2013)
4. Department of Alternative Energy Development and Efficiency, *The analysis of indicators of energy efficiency in the steel industry*, (2015)
5. S. Juntueng, S. Towprayoon, S. Chiarakorn,Resour Conserv Recy, **87**, 46-56, (2014)
6. P. Kanchanapiya, N. Limphitakphong, C. Pharino, O. Chavalparit, Greenhouse Gas Measurement and Management, **4**, 161-177, (2014)
7. S. Patumsawad, *Carbon Intensity of Power Industry (Fossil fuel power plant)*, (2010)

8. J. Gouw, J. A. d., D.D. Parrish, G.J. Frostl, M. Trainer , *Reduced Emissions of CO_2, NO_x and SO_2 from U.S. Power Plants Due to the Switch from Coal to Natural Gas with Combined CycleTechnology*, (2014)
9. Electricity Generating Authority of Thailand, *SEPA Conference 2014 : Driving for Excellence*, (2014)
10. Electricity Generating Authority of Thailand, *National Grid*, (2015)
11. Y. A. Cengel, M. A. Boles, *Thermodynamics : an engineering approach 7st ed.* (Mcgraw-Hill, 2011)
12. M. Nyberg, *Thermal efficiency of gas-fired generation in California :2014 updated*, (2014)
13. I. Dincer, M. A. Rosen, Exergy, environment and sustainable development (Elsevier Sci., 2013)
14. T. J. Kotas, *The Exergy Method of Thermal Plant Analysis* (Paragon Publishing, 2012)
15. I. H. Aljindi, Appl Therm Eng, **29**, 324-328 (2009)
16. S. C. Kamate, P. B. Gangavati, Appl Therm Eng, **29**, 1187-1194, (2009)
17. R. Saidur, J. U. Ahamed, H. H. Masjuki, Energ Policy, **38**, 2188-2197, (2010)
18. M. Ghazikhani, I. Khazaee, E. Abdekhodaie, Energy, **72**, 599-607, (2014)
19. F. A. Al-Sulaiman, F. H. amdullahpur, I. Dincer, Appl Therm Eng, **31**, 439-446, (2011)
20. A. G. Kaviri, M. N. M. Jaafar, T. M. Lazim,H. Barzegaravval, Energ Convers Manage,**67**, 27-33, (2013)
21. IPCC, *IPCC Guidelines for National Greenhouse Gas Inventories Volume 2: Energy*, (2006)

A Study of Energy Intensity and Carbon Intensity from Olefin Plants in Thailand

Wasamon Chuapet[1], Nantamol Limphitakphong [2], Thanapol Tantisattayakul[3], Premrudee Kanchanapiya[4] and Orathai Chavalparit[1,2]

[1]*Department of Environmental Engineering, Faculty of Engineering, Chulalongkorn University, Bangkok, 10330, Thailand*
[2]*Research unit of Environmental Management and Sustainable Industry, Faculty of Engineering, Chulalongkorn University, Bangkok 10330, Thailand*
[3]*Faculty of Science and Technology, Thammasat University Rangsit Campus, Pathumthani, 12121, Thailand*
[4]*National Metal and Materials Technology Center, 114 Thailand Science Park, Pathumthani, 12120, Thailand*

Abstract. This research is aimed to estimate energy consumption and the greenhouse gases emission from olefins production process in Thailand and to analysis its correlation between energy intensity and carbon intensity. The results of five olefin case study plants showed that direct energy use of olefins production was about 87% of total energy consumption. The greatest amount of energy demand was supplied for fuel combustion in manufacturing processes. The olefin plants using gas feedstock could operate with less amount of energy consumed as the average energy intensity (EI) of 23.88 GJ/ton of olefin produced, whereas plants of olefin produced from steam cracking of liquid feedstock had the average EI of 33.21 GJ/ton of olefin produced. Greenhouse gas emission from olefin plants using gas feedstock were consistent with the result of EI. The carbon intensity (CI) from olefin plants using gas feedstock (0.98 tCO_2-eq./ton of olefin produced) was lower than the plants using liquid feedstock (2.01 tCO_2-eq./ton of olefin produced). The T-test function used to reflect the correlation between EI and CI emphasized that the carbon intensity was significantly correlated with the intensity of energy ($r = 0.9564$ and $p = 0.05$).

1 Introduction

The main cause of global warming is the emission of greenhouse gases from industry sector [1], mainly due to burning fossil fuels to produce energy and products. As a result of climate change at the global scale, changes in both physical and biological system [2] regarded as a major issue in all regions have been concerning very closely as seen from The twenty-first session of the Conference of the Parties (COP21) in Paris (2015 United Nations Climate Change Conference, COP21). The goal of limiting the average global temperature not to exceed 2 degrees Celsius from pre-industrial levels was agreed by 195 nations through the cooperation action and funding from developed countries to encourage stronger action by developing countries [3]. IEA states that the industry sector requires high energy use and emit a great amount of greenhouse gases. In 2005, the proportion of global energy consumption from industry sector was up to 33% of total final energy consumption particularly in petrochemical industry about 34 EJ [4]. The proportion of greenhouse gas emissions from industry sector was accounted for 21% of total global emissions, whereas 23% of that emissions contributed from petrochemical industry [5] due to the industrial use of fuel oil and natural gas as raw material and fuel in the production process. In 2006 furthermore, the demand of energy and raw materials in petrochemical industry was about 10% of total global final energy consumption, representing 30% of total final energy consumption in the industry sector [6].

Upstream petrochemical industry is the industry that produces three product groups including alkane group, olefin group, and aromatic group. Main product of the upstream industry is olefin group, including ethylene, propylene, mixed-C4, butadiene and butene-1. These products are used as raw materials for intermediate and downstream petrochemical industry. For example, ethylene is the feedstock for Polyethylene plastic beads such as low density polyethylene (LDPE) and high density polyethylene (HDPE), propylene is used in manufacturing of polypropylene (PP) plastic beads which is the downstream petrochemical products used in producing plastic items such as food-containing clear plastic bags and films [7].

In Thailand, the petrochemical industry normally uses steam cracking of gas feedstock (e.g., ethane and propane) or of liquid feedstock (e.g., naphtha). This process, mostly used worldwide currently, consists of the 'cracking furnace' which is the first unit drawing feedstock such as ethane gas, naphtha, LPG, propane, butane or mixed raw material into the cracking process obtaining compounds with smaller molecules. Then, the furnace transfers these materials to the temperature-reducing and pressure-increasing unit which are the two most energy-consuming units. From these two units, the materials are then transferred to heat exchanger and separated into a variety of petrochemical products and by-products [6]. This process consumes energy approximately 2-3 EJ with the amount of the greenhouse gas emissions approximately 180-200 ktCO_2-eq. [8].

There are some previous studies on olefin production. For example, steam cracking of ethane in U.S. required energy about 23.50 GJ/ton and emitted the greenhouse gas emissions by 0.90 tCO_2-eq./ton of ethylene [9-10]. For the production using naphtha as feedstock, the energy demand was approximately 31.5 GJ/ton and greenhouse gas emissions by 2.47 tCO_2 emission/ton of ethylene [9, 11].

Studying mitigation options to reduce energy consumption and greenhouse gas emissions is important to help untie global problems. IEA reported that industry sector can reduce either the energy consumption (25-37 EJ/year) or greenhouse gas emissions (1.9-3.2 GtCO_2/year) by improving energy efficiency or using best practices technology [4]. Whether implementation of best available technologies energy efficiency measures, switching to fuels with low carbon element and recycling materials to reduce direct greenhouse gas emission in 2050 was approximately 30% [12].

The main objective of this study was to evaluate the energy consumption in term of energy intensity and greenhouse gas emission in term of carbon intensity from the olefin production process in Thailand. An influence of different feedstock or fuel on energy intensity (EI) and carbon intensity (CI) was discussed. The findings of this study would be an alternative approach leading to achieving the global goal of reducing greenhouse gas emissions as one of the solution of global warming.

2 Methodology

This study is divided into two sections; the assessment of EI and CI from the olefin production process using different feedstock types and their correlation. Annual data of 2012–2014 were collected from five olefin plants to represent olefins manufacturing of Thailand. The scope of work was focused only on olefin production processes as shown in Figure 1.

2.1 Olefin plant case studies

According to collected data, five olefin plants in Thailand could be divided into two groups by type of feedstock; the first two plants using gas feedstock and the last three plants operating with liquid feedstock. Description of feedstock and fuel types each plants is shown in Table 1.

2.2 Energy consumption and energy intensity

The estimation of needed amount of energy consumption in olefin production process is divided into 2 parts; direct and indirect energy, which total energy consumption can be calculated by equation (1) as follows;

$$EC = EC_{Direct} + EC_{Indirect} \quad (1)$$

where EC is the total energy consumption for olefin process (GJ), EC_{Direct} is the needed amount of energy the needed amount of energy indirectly consumed in the process (GJ).

Table 1. Feedstock and fuel types of olefin production plants

Olefin Plant	1	2	3	4	5
Feedstock type (G:L)	100:0	100:0	70:30	20:80	0:100
Fuel used for direct consumption (FG:FO)	100:0	90:10	100:0	100:0	80:20

Note : G is Gas feedstock (i.e. Ethane)
L is Liquid feedstock (i.e. Naphtha)
FG is Fuel gas
FO is Fuel oil

The direct energy means the energy obtained by burning fuel in the manufacturing process such as fuel gas and fuel oil, while the indirect energy refers to the use of electricity and steam in the process. Both of that can be calculated by the equation (2) as follows;

$$EC_{Direct/Indirect} = \sum(FA \times NVC) \quad (2)$$

where $EC_{Direct/Indirect}$ is the amount of direct/indirect energy (GJ), FA is the quantity of energy (unit) and NVC is the net calorific value of fuel type (GJ/unit)

The assessment of the needed amount of energy consumption is presented in terms of energy intensity, which refers the needed amount of energy consumption in one unit manufacture as expression in equation (3);

$$EI = EC/P \quad (3)$$

where EI is the energy intensity of olefins (GJ/ ton of olefin produced), EC is the total annual energy consumption for the production process (GJ/year) and P is the total amount of olefin produced a year (ton of olefin produced /year)

2.3 Greenhouse gas emissions and carbon intensity

For estimating the amount of greenhouse gas emission from the olefin production process, three greenhouse gases; carbon dioxide, methane and nitrous oxide are mentioned. The calculation in this section refers to IPCC (2006) [13] and API (2009) [14] guidelines. To clearly identify the emission sources, two key issues are separately considered; direct and indirect emissions. According to Figure 1, direct emission refer to the emissions from carbon dioxide separation from gas feedstock (1), de-carbonization in the furnace (2), fuel combustion (3), methane leaking from olefin production process, fugitive, water treatment (4), and flare (5), whilst indirect emission means the emission related to consuming indirect energy (6).

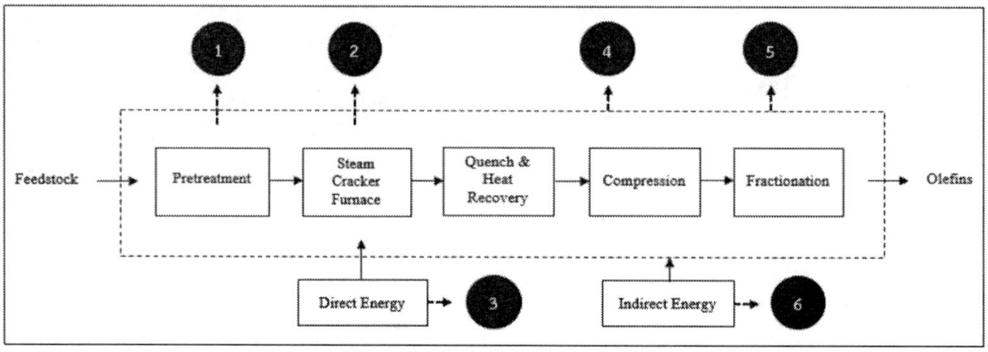

Figure 1. Olefin production flow.

The amount of greenhouse gases emitted by in each particular activities can be calculated by the equation (4) and (5) as follow;

$$GHG_{Emission} = AD \times EF \quad (4)$$

where $GHG_{Emission}$ is the amount of greenhouse gases emitted from all activities (t GHG), AD is the activity data of relevant olefin production process (unit) and EF is an emission factor of greenhouse gas emissions (t GHG/unit), by using the value obtained from the manufacturing plants. In case of the EF is unavailable from the plants, the value will be taken from IPCC (2006) [13].

$$E_{CO_2 eq.} = GHG_{Emission} \times GWP \quad (5)$$

where $E_{CO_2 eq.}$ is the total amount of greenhouse gases emission from olefin production process (tCO_2-eq.), $GHG_{Emission}$ is the amount of all greenhouse gases emitted from the process (t GHG) and GWP is the value of global warming potential of each greenhouse gases (tCO_2-eq./t GHG)

The evaluation of the greenhouse gas emissions was presented in terms of carbon intensity, which represents the amount of greenhouse gas emission in one unit of olefin produced, as expression in equation (6);

$$CI = E_{CO_2-eq.}/P \quad (6)$$

where CI is carbon intensity of olefins (tCO_2-eq./ ton of olefin produced), $E_{CO_2-eq.}$ is the total amount of greenhouse gas emissions from the process annually (tCO_2-eq./year) and P is the annual olefins production (ton of olefin produced /year).

3 Results and discussion

3.1 Effect of feedstock on EI of olefins

The estimation of energy intensity of olefin production using different types of feedstock was presented. Olefin produced from steam cracking of gas feedstock mostly uses ethane as feedstock, while that steam cracking of liquid feedstock uses only naphtha or mixed with other substances such as ethane, LPG and NGL as feedstock as shown in Figure 2.

An average EI from steam cracking of gas feedstock during 2012-2014 was approximately 23.88 GJ/ ton of olefin produced, whilst the average from steam cracking of 30%, 80% and 100% liquid feedstock were 30.25, 31.23 and 38.14 GJ/ ton of olefin produced respectively. Obviously, the key factor influencing the energy demand for production process is feedstock type. Olefins production using liquid feedstock requires greater amount of energy than using steam cracking of gas feedstock because the more number of carbon atoms in naphtha (5-12 atoms) absolutely needs the greater energy for breaking down the carbon-carbon bonds of long chain hydrocarbons into the simpler ones than olefins produced from steam cracking of ethane which consists of only 2 carbon atoms. Moreover, using steam cracking of liquid feedstock requires devices in separating process to separate the product and its by-product [8]. Consequently, the energy demand of electricity increased for such the process. Comparing the energy demand through energy intensity varied in portion of liquid feedstock, operating of 80% liquid feedstock consumed higher energy about 3% than that of 30% liquid feedstock, while using 100% liquid feedstock required greater amount of energy than operating of 80% liquid feedstock around 22%.

Olefins production requires direct energy consumption from 86-97% of the total amount of energy, which is higher than the use of indirect energy (3-24%) as shown in Figure 3, because the use of direct energy mainly contribute for heat raising and cracking process while using indirect energy for any supplement processes. IDB studied greenhouse gas emission from new petrochemical plants reported that the requirement of fossil fuel, steam and power for producing olefins were approximately 60%, 35% and 5% respectively [6].

3.2 Effect of feedstock on CI for olefins plants

As shown in Figure 4, carbon intensity of olefins production using gas feedstock was lower than that using

 DOI: 10.1051/matecconf/20166814002

liquid feedstock. Greenhouse gas emissions during

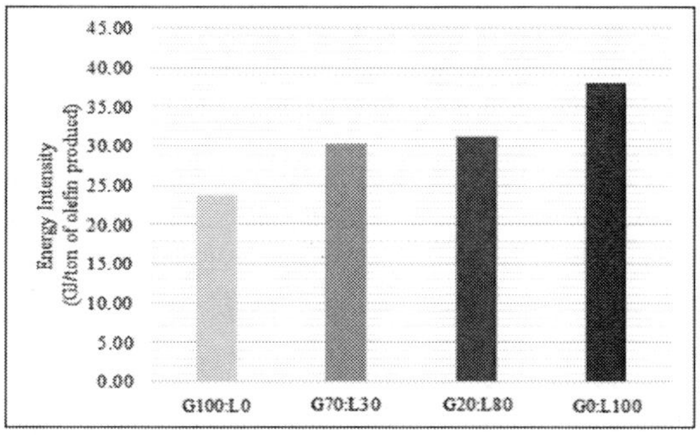

Figure 2. Effect of feedstock types on energy intensity in olefin production process.

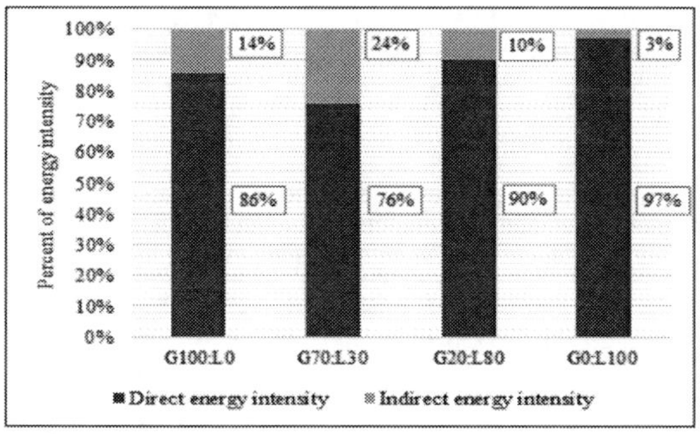

Figure 3. The proportion between direct and indirect energy consumption from the olefin production process.

2012-2014 from olefin production process using gas feedstock had an average of 0.98 tCO_2-eq./ton of olefin produced, whereas the average emission using steam cracking of 30%, 80% and 100% liquid feedstock were 1.77, 1.75 and 2.52 tCO_2-eq./ton of olefin produced respectively. Similar to the result of energy consumption, steam cracking of liquid feedstock (i.e. naphtha) which contents more number of carbon atoms emitted greater emissions than using gas feedstock. Moreover, using liquid feedstock gives the yield of ethylene (30%) less than using gas feedstock (80%) [8]. Therefore, different type of feedstock obviously have an effect on different carbon intensity of olefin produced.

In addition, the type of fuel used in the process were regarded as another important factors as it had influenced on CI of olefins. Using fuel oil for combustion process resulted in greater emissions than using only fuel gas because the coefficient of greenhouse gas emission of fuel oil is higher the coefficient of fuel gas. Therefore, the plants consumed mixed fuel (fuel oil with fuel gas) provide higher EI than that operating with fuel gas solely.

As shown in Table 2, it was clearly confirmed that different types of fuel and feedstock affect the energy demand and the greenhouse gas emissions from olefin production process. On the one hand, steam cracking of liquid feedstock provides greater value of EI and CI than that of gas feedstock for all types of fuel; on the other hand, using mixed fuel gives higher point of EI and CI than using single fuel for both feedstock types. For that reason, producing olefins via steam cracking of gas feedstock with fuel gas for combustion process would be an appropriate option both in term of energy and environment.

3.3 The correlation of EI and CI from olefin production

Olefin production process consumes high energy and emits a large amount of greenhouse gases. According to the analysis of the correlation between energy intensity and the carbon intensity via T-test, it has been found that carbon intensity and energy intensity has a positive correlation significantly ($r = 0.9564$ and $p = 0.05$). Furthermore, their relationship could be explained by the linear regression with R^2 of 0.9147 as shown in the Figure 5.

3.4 Comparison of energy intensity and carbon intensity with other research

A comparison of energy intensity and carbon intensity from olefin production in Thailand with other research is shown in Table 3 and Table 4. Based on types of feedstock, the energy demand of olefins from our finding was a bit higher than the result reported by other researchers, which meant there probably are some possibilities for increasing the energy efficiency used in upstream petrochemical industry. Note that, the findings of other research accounted for best practice technology (e.g., catalytic cracking process), which was not the same technology studied in this research.

Generally, the carbon intensity of our findings was the emissions from steam cracking of gas feedstock in this study was slightly less than the value reported by IPPC [13]. It might be because olefin plants in Thailand used cleaner energy (e.g., natural gas) for combustion process, while IPCC reported the average worldwide emissions that used a various kind of fuel.

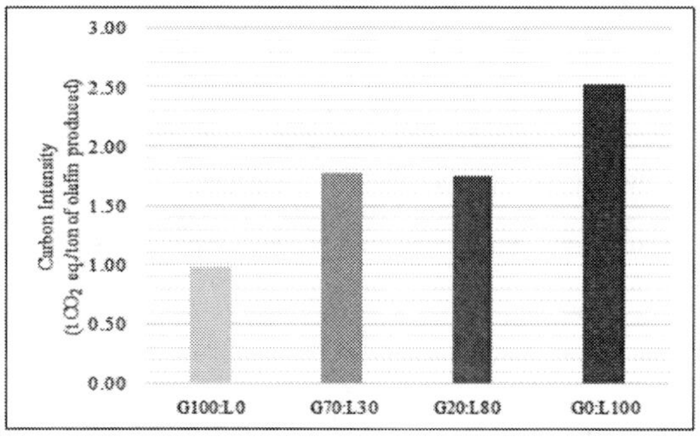

Figure 4. An average carbon intensity of olefin production in Thailand.

Table 2. Comparison of EI and CI of olefin produced from different types of fuel.

FG:FO	Gas feedstock		Liquid feedstock	
	100:0	90:10	100:0	80:20
EI (GJ/ton of olefin	21.04	26.71	30.74	38.14

produced)				
CI (tCO_2-eq./ ton of olefin produced)	0.79	1.17	1.76	2.52

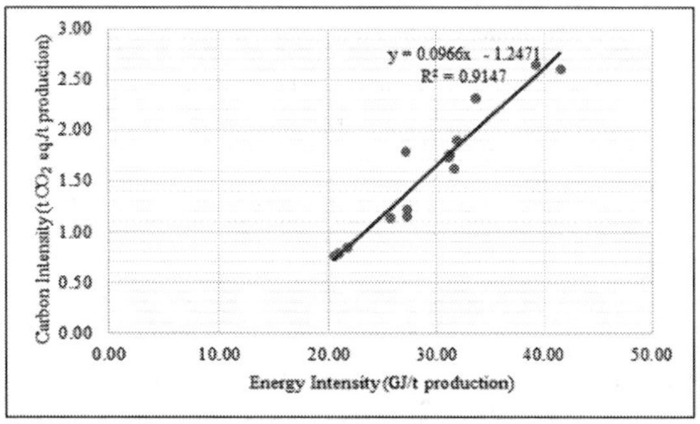

Figure 5. Correlation between energy intensity and the carbon intensity.

Table 3. A Comparison of energy intensity.

Feedstock type	Research	Source	
		[8]	**[15]**
Gas	23.88	17.00-21.00	16.00
Liquid 30%	30.25	26.00-31.00	16.50
Liquid 80%	31.23		
Liquid 100%	38.14		

Table 4. A comparison of carbon intensity.

Feedstock type	Research	Source	
		[13]	**[16]**
Gas	0.98	1.385	0.950
Liquid 30%	1.77	2.324	1.730
Liquid 80%	1.75		
Liquid 100%	2.52		

4 Conclusion

This research is aimed at estimating the energy intensity and carbon intensity from olefins production in Thailand. The annual data was collected from five olefin plants for three years (2012–2014). Our findings illustrated that olefins production via steam cracking of liquid feedstock consumed a larger amount of energy and emitted greater greenhouse gas emissions than that using gas feedstock. Olefins production required direct energy rather than indirect energy because it was used in combustion process for cracking carbon-carbon bonds. A majority of greenhouse gas emissions from olefin production was accounted to direct emissions mostly produced by fuel combustion process as well. The combustion process with a single type of fuel consumed less energy for heat raising and molecules cracking than that using mixed fuel. The correlation analysis of energy and carbon intensity via T-test clearly confirmed that energy demand for producing olefins in term of EI had a positive correlation significantly with its greenhouse gas emissions in term of CI ($r = 0.9564$ and $P = 0.05$). In conclusion, either type of feedstock or fuel have influenced on both energy and carbon intensity of olefins. Recommendation for new olefin plants or improving existing plants, therefore, would be supporting a plant using steam cracking of gas feedstock with a single clean fuel for combustion process which will be beneficial both in term of energy and environment.

Acknowledgement

This research has been supported by National Research University Project, Office of Higher Education Commission (WCU-58-020-CC)

References

1. N.N.N. Ahmad, Procedia Soc. Behav. Sci., **172** (2015)
2. S. Zeng, Y. Lan, J. Huang, Int. J. Greenh. Gas Control, **3** (2009)
3. UNFCCC, *Conference of the Parties Twenty-first session*, (2015)
4. IEA, *Worldwide Trends in Energy Use and Efficiency*, (2008)
5. K.A. Baumert, T. Herzog, J. Pershing, *Navigating the Numbers Greenhouse Gas Data and International Climate Policy*, (2005)
6. T. Benchaita, *Greenhouse gas Emission from New Petrochemical Plants*, (2013)
7. DEDE, *A Study on Energy Efficiency Index in petrochemical Industry*, (2007)
8. T. Ren, M. Patel, K. Blok, Energy, **31** (2006)
9. E. Worrell, D. Phylipsen, D. Einstein, N. Martin, *Energy use and energy intensity of the U.S. chemical industry*, (2000)
10. P. Suding, *Chemical Plant GHG Emission*, (2013)
11. Y.F. Huang, Y.C. Lin, J.T. Yang, Energy Policy, **38** (2010)
12. IEA, *Energy Technology Perspectives*, (2015)
13. IPCC, *Guidelines for National Greenhouse Gas Inventories, Volume 3*, (2006)
14. API, *Compendium of greenhouse gas emissions methodologies for the oil and natural gas industry*, (2009)
15. IEA, *Energy and GHG Reductions in the Chemical Industry via Catalytic processes*, (2013)
16. M. Neelis, M. Patel, K. Blok, W. Haije, P. Bach, Energy, **32** (2006)

 DOI: 10.1051/matecconf/20166814003

Policy Analytics, Household Informedness and the Collection of Household Hazardous Waste

Kustini Lim-Wavde [1], Robert J. Kauffman [1] and Gregory S. Dawson [2]

[1] *School of Information System, Singapore Management University, Singapore*
[2] *W.P. Carey School of Business, Arizona State University, United States of America*

Abstract. Proper collection of Household Hazardous Waste (HHW) is an important action to support environmental sustainability. We investigate the role of *household informedness*, the degree to which households have the necessary information to make utility-maximizing decisions, as they relate to participation in HHW collection programs. We find two factors that influence household informedness: the provision of public education about HHW and environmental quality information. We conducted an empirical study on HHW collection in California to obtain statistical evidence on the effect of these factors on the amount of HHW collected. The findings of this *policy analytics study* improve our understanding of how household informedness influences household decision-making in participating in HHW collection programs. This study is useful in the guidance it offers to devise new information policies to maximize households' participation in HHW collection program.

1 Introduction

Household informedness is the degree to which households have the necessary information to make utility-maximizing decisions in their daily activities, such as managing their home waste. It can also play an important role in the management of *Household Hazardous Waste* (HHW), particularly in optimizing their collection. Our study is intended to apply policy analytics methods and econometrics related to HHW management in a way that considers the role of household decision-making to improve the household waste management process.

Policy analytics emphasizes value-driven assessment of data for problems involving multiple stakeholders. It offers a framework for: identifying key issues, defining policy objectives, designing and testing policy, evaluating the effectiveness of policy implementation, and adjusting policy [1]. In this research, we analyze household data on HHW collection to obtain evidence on the impact of household informedness for household-level decisions.

In some parts of the world, local governments have organized various HHW collection programs that complement regular trash collection. These collection programs include HHW permanent facilities, load checking of hazardous materials in trash before disposal in landfills, door-to-door programs, mobile facilities, HHW recycling facilities, curbside facilities, and one-day collection events [2]. However, the success of these collection programs relies on voluntary participation of households because they are required to identify the HHW materials, and also to segregate, store, and transfer them to the collection facilities.

A recent review on HHW collection reported that the amount of HHW collected is only about 0.12% to 1.88% of total *Municipal Solid Waste* (MSW) [3]. This estimate is based on data from previous studies conducted in 20 European countries, several states in the U.S, and Mexico, Canada, Greenland, Japan, India, Pakistan, Hong Kong, and Nepal. These numbers do not include HHW that is disposed of improperly, for example, kept in storage rooms, poured into drains or disposed together with trash. The common belief is that the amount of HHW collected may be greater when more households are encouraged to participate in collection programs.

As we improve household informedness, households are likely to make better decisions to optimize their utility with respect to household waste. This, we believe, should lead to beneficial effects to increase the amount of HHW diverted from general trash, and consequently also improve environmental quality. Previous studies attempted to explain household-level decision-making related to their management of waste. Kinnaman and Fullerton [4], for example, developed a household utility maximization model to understand household's disposal and recycling decisions. The authors' model assumes that each household maximizes its utility subject to a budget constraint. Such decisions depend largely on the price or cost of disposal, recycling, or collecting HHW. Sidique et al. [5] adopted this model and added an environmental quality factor in the utility function: households also benefit from environmental quality.

To our knowledge, no researchers have studied the role of household informedness within a utility maximization framework that addresses the collection of

HHW. This study focuses on a key question related to this. How does household informedness influence household decisions in terms of the amount of HHW collected? We will identify and analyze two important factors that influence household informedness in HHW management: the provision of public education about HHW and the environmental quality information.

2 Household informedness

2.1 HHW public education

Public education about HHW has long been a part of waste management. We include this as an important part of household informedness because it can decrease the costs of household decision-making. It usually provides information about the potential hazards of corrosive, toxic, reactive or ignitable materials found in common household products. These include batteries, oil-based paints, latex paints, motor oils, cleaning products, pesticides, insecticides, and electronic devices. It also provides information on how to properly store, handle, and dispose these leftover materials. For example, paint cans have to be sealed before storing them in dry areas.

Similarly, pesticides also need to be stored in dry areas in their original containers with their labels intact. The leftover liquids should not be poured into the gutter, drain or put together with regular trash that usually goes to landfills. If they are not handled correctly, the contaminants can either seep into groundwater or be carried off by storm water run-off and eventually pollute public drinking water [6]. Without this kind of information households may pollute the water or land in their neighborhood with those hazardous materials.

This public education also provides information on a range of facilities, programs and events to guide households about the disposal of their HHW. HHW collection often follows certain schedules and depends on the type of facility or program type. HHW permanent facilities may receive HHW all year round, while temporary facilities may only accept waste on certain days of the week. One-day collection events also may be held. Some materials, for example, leftover paints, are accepted by local retail stores in the U.S.

Another objective of public information on HHW is to decrease the use of hazardous materials in households each day. This is achieved by providing information about alternative non-toxic or non-hazardous products that can substitute for the ones that are commonly used; for example, using baking soda with white vinegar to replace chemical oven cleaner. In this way, public education can be used as a *source control* measure that aims to reduce the generation of HHW at the source [2].

An effective public education program does not only create brochures or posters with such information. In implementation, there are many barriers to changing household-level behavior. A good education program manager will use various strategies to educate households, such as getting involved with them, giving feedback and motivating people, using friends and relatives, finding change agents and role models in the community, and presenting information effectively [7].

Due to the different types of information and methods to deliver it, it is difficult to measure the effects of public education on HHW collection. Empirical researchers often use *proxies* to quantify conditions that cannot be directly measured. Sidique et al. [5] used three-year cumulative recycling education expenditures to proxy for recycling awareness in Minnesota counties. They applied cumulative expenditures instead of yearly expenditures because of the cumulative effects of recycling education on recycling activities. They found that public education expenditures in recycling had a positive effect, increasing the recycling rate.

In our study, we use HHW public education expenditure as a proxy of household informedness through HHW public education. We believe that this will increase the amount of HHW collected in the short term. While it can act as source control, it will take a longer time for households to change their consumption behavior than what can come from participating in HHW collection programs to dispose of their HHW.

2.2 Environmental quality information

Fishbein [8] defined *environmental quality* as the quality of life within it. It can have different values for different stakeholders, such as conservationists, industrialists, general public, and so on. So it is usually governed by environmental policy in a nation. In our study, we adopt this definition for environmental quality: "varied characteristics such as air and water purity or pollution, noise, access to open space, and the visual effects of buildings, and the potential effects which such characteristics may have on physical and mental health" [9]. An example of environmental quality information is the level of contaminants in water, land, or air, and whether it exceeds the limit that poses a threat to human health as determined by the national environmental agency. This information can be delivered to the public in various formats and channels. A country's national environmental agency usually provides this information via memoranda, mobile applications, or web pages.

In our study, we use direct public notifications to measure household informedness related to environmental quality information. This kind of information is targeted directly to households, so it is most likely to affect their informedness levels related to HHW. Mobile applications or website information may also affect informedness; however, it is difficult to assess whether households have accessed information this way, because their behavior is unobservable.

Previous empirical studies have found that public notification, information disclosure, and advisory announcements related to environmental quality have significant effects on the behavioral changes of targeted consumers and households. For example, Shimshack et al. [10] found that well-informed consumers reduced their canned fish consumption in response to the U.S. Food and Drug Administration's (FDA) mercury advisories. Bennear and Olmstead [11] also found that information disclosures on violations in drinking water quality in the form of annual *consumer confidence reports* for people in Massachusetts reduced total violations between 30% and 44%. Zivin et al. [12] also studied the impact of water quality violation information in Northern California and Nevada, but they focused on the costs of avoidance behavior. They estimated the public's willingness-to-pay to avoid water quality violations. In economic terms, when the cost of avoiding water violations is more than the cost of disposing HHW properly via collection programs, households will prefer to participate in HHW collection programs.

Our study uses data on annual consumer confidence reports that contain water quality violations in California. We hypothesize that there is a relationship between perceptions of water quality and the amount of HHWs collected or disposed illegally. As a result, such information disclosure to households will create public awareness about unacceptable water quality to motivate them to participate in HHW collection programs.

3 Data and empirical model

3.1 Data: CalRecycle HHW collection

We explore and analyze HHW collection data published by California's Department of Resources Recycling and Recovery (CalRecycle) from 2004 to 2012. CalRecycle manages the waste handling and recycling program in California. It conducts annual mandatory survey to collect HHW data from public agencies that manage HHW within their jurisdiction in the county. These agencies submitted their data via CalRecycle Form 303 for the period of July 1 in the first year to June 30 in the following year [13]. The HHW collection report is available to public. It contains information on the reporting agency, report cycle year, material category and type, HHW program type, and the weight of the waste in pounds. We aggregated the data at the county level by summing up the weight of the collected waste in the same county, report cycle year, material category, and program type. We did so because we aimed to understand household waste collection behavior at the county level.

The second data source is the U.S. Census Bureau's American Community Survey [14]. From this source, we obtained California county and demographic characteristics, such as population, density, median age, average household size, rental household rate, unemployment rate, and percent of high school graduates.

Our study uses the U.S. dollar value of HHW grant awards for HHW public education programs as a proxy for HHW public education efforts. The CalRecycle HHW grant database data were obtained were the same in this respect. Similar to Sidique et al. [5], we also assume that public education may have cumulative effect to the following years. Thus, we use the three-year cumulative HHW grant awards for public education. Since the value of the grant awards is not only specifically spent for public education programs, we use a dummy variable instead of continuous variable for HHW public education, with 1 for public education, and 0 if there was none.

For environmental quality information, we use the presence of *Maximum Contaminant Level* (MCL) *violations* in the drinking water. The data were obtained from the annual compliance report by the California Department of Public Health. We have the violation counts of MCL in the report; however, we decided to use a 0/1 dummy variable to indicate whether there were any violations because the number of MCL violations may not represent the severity or health risk in household members' perceptions. Under the Safe Drinking Water Act (SDWA) in the U.S., water system companies are required to notify the affected households promptly, whenever there is a violation of drinking water quality regulations. So these data can be good proxy for household informedness through environmental quality information.

Table 1. Description of variables.

VARIABLE NAME	VARIABLE DESCRIPTION
County	County
ReportCycle	Report cycle year (2004-2012)
HHWCollectionQ	Quantity of HHW collected in pounds
HHWMaterial	Categorical variable for HHW material categories: flammable and poison; inorganic and organic acid; inorganic and organic base; oxidizers, peroxides, oxidizing acid and base; PCB-containing; reclaimable; asbestos; universal waste; others
HHWProgramType	Categorical variable for collection program types: permanent facilities, mobile facilities, temporary or periodic facilities, door-to-door (residential) programs, curbside programs, load checks, others
EduHS%	Population over 25 years old with high school diploma in percentage
UnemployRate%	Unemployment rate in percentage
Density	County density in thousand square feet per capita
Population	County population in million people
MedAge	Median age in year
Rental%	Rental household rate in percentage
AvgHHSize	Average household size
D_HHWEduCum	Dummy variable to indicate the presence of cumulative 3-year HHW grant dollar that is used for projects with public education or information
D_MCLviolation	Dummy variable to indicate the presence of Maximum Contaminant Level (MCL) violations in public drinking water

We combined the above data into a multidimensional panel with county as an *individual dimension*, and program type and material category as *group dimensions*, with report cycle year as the *time dimension*. A few counties did not have demographic characteristics data in the American Community Survey in some years. So there are some missing values in the panel. After omitting the rows with missing values, there are 7,936 data points covering 39 counties, 11 HHW material categories, and

10 HHW program types from the report cycle years 2004 to 2012. We used R and its statistical functions for the data processing and analysis [15].

3.2 HHW collection model

We model the HHW collection quantity as the function of household informedness and demographic characteristics variables. The panel data contain multiple dimensions so we employ *fixed effects* or *Least Squares Dummy Variable* (LSDV) regression. We control for the unobserved time-invariant effects of program types and material categories using *HHWProgramType* and *HHWMaterial* dummy variables. County characteristics variables are used to control for county variation. Our model allows for exogenous effects of statewide changes over time. This is controlled by using a dummy variable, *ReportCycleYear*. In this way, there will be no remaining heterogeneity in the error term due to unobservable effects from program type, material category, and statewide changes over time. We applied a natural log transformation on the dependent variable *HHWCollectionQ* because its distribution is skewed and its standard deviation is much larger than its mean. The econometric model is written as follows.

$$\ln(HHWCollectionQ_{cpmt}) = \beta_0 + \beta_1 D\_HHWEduCum_{ct} + \beta_2 D\_MCLviolation_{ct} + \beta_3\, EduHS\%_{ct} + \beta_4 UnemployRate\%_{ct} + \beta_5 Population_{ct} + \beta_6 Density_{ct} + \beta_7 MedAge_{ct} + \beta_8 Rental\%_{ct} + \beta_9 AvgHHSize_{ct} + \sum_{p=1}^{8} \theta_p HHWProgramType_{pct} + \sum_{m=1}^{11} \gamma_m HHWMaterial_{mct} + \sum_{t=1}^{9} \delta_t ReportCycleYear_{ct} + \varepsilon_{cpmt} \quad (1)$$

The subscripts in Equation (1) are defined as follows: c = county; p = program type; m = material category; t = report cycle year. The description of the variables is given in Table 1.

4 Results and discussion

4.1 Results

Table 2 provides the coefficient estimates of our model. The coefficient estimate for the dummy variable *D_HHWEduCum* is negative and significant. This result shows evidence that HHW public education is associated with decreased amount of HHW collection. This may indicate that HHW public education is effective in controlling HHW at the source, so it may influence households to reduce the use of products containing hazardous or toxic materials. This is not what we expecte, but it is interesting, and deserves comment.

The coefficient of the dummy variable *D_MCLviolation* is positive, but not significant. This result suggests that the effect of water quality information is not salient in increasing household participation in HHW collection programs. Households that receive information on MCL violations in drinking water, in our view, ought to perceive the quality of the environment to be low, and subsequently increase the amount of HHW collected and may be discouraged from disposing of HHW improperly. More observational data on environmental quality or other related factors may be necessary to pursue this aspect of our study further.

Table 2. Estimation results.

VARIABLES	COEF.		SE
Intercept	8.24	***	1.37
D_HHWEduCum	-0.22	***	0.05
D_MCLviolation	0.03		0.07
• **County Characteristics**			
EduHS%	1.05	*	0.62
UnemployRate%	-0.04	***	0.01
Population	0.26	***	0.01
Density	0.11	***	0.01
MedAge	-0.04	***	0.01
Rental%	-3.23	***	0.59
AvgHHSize	-0.54	***	0.19
• **Program Type**			
Door-to-Door Program	-0.08		0.14
HHW Permanent Facility	3.38	***	0.14
Load Check	0.20		0.15
Mobile Facility	0.27		0.22
Other	0.27	*	0.15
Recycle-only Facility	1.05	***	0.16
Temporary Facility	1.53	***	0.14
• **HHW Material**			
Aerosol Containers (UW)	0.84	***	0.21
Asbestos	-0.65	***	0.11
Base	0.26	***	0.08
Electronic Devices (UW)	4.24	***	0.09
Flammable and Poison	3.48	***	0.08
Other	1.27	***	0.08
Oxidizer	-0.83	***	0.09
PCB-containing	-0.93	***	0.09
Reclaimable	3.77	***	0.08
Universal Waste (UW)	1.86	***	0.08
• **Report Cycle**			
2005	0.33	***	0.09
2006	0.46	***	0.09
2007	0.51	***	0.09
2008	0.58	***	0.09
2009	0.73	***	0.11
2010	0.83	***	0.13
2011	0.79	***	0.14
2012	0.62	***	0.13

Note. Model: fixed effects; dep. var.: ln (*HHWCollectionQ*). Base case indicators for the categorical variables were not estimated. The base cases are *Curbside Program* for **Program Type**, *Acid* for **HHW Material**, and *2004* for **Report Cycle Year**. Adj. $R^2 = 0.63$; SE = 1.704; 7,901 d.f.; signif.: *** < 0.01, ** < 0.05, * < 0.10.

4.2 Discussion

More information may not necessarily give better impact on household's behavior. Shimshak et al. [10], in this study on the impact of mercury advisories, found that the ability to assimilate information (due to higher education) and access to information (via newspapers or other media) were important determinants in the response to the advisory. In our study, we have the percentage of county population who graduated high school as a proxy for a household's ability to assimilate information, but we have no data that can proxy for access to information yet.

Shimshak et al. [10] also found that information advisories may produce unintended spillover effects to non-targeted people. In our case, the MCL violation information in one county may also affect people in the neighboring counties. Thus, it makes it difficult to ascertain the precise effects of public information advisories. We are continuing to pursue this as a measurement issue as a result.

Further, the econometric model in this study has not included the potential simultaneity of demand for HHW collection and demand for *Municipal Solid Waste* (MSW) disposal. The factors that affect MSW generation (e.g., disposal unit pricing) may influence HHW collection, and at the same time, factors that affect collection of HHW (e.g., HHW programs or education) may also affect the amount of MSW. Hong [16] developed a *simultaneous equation model* based on the household utility maximization model that explicitly accounts for simultaneity between waste generation and recycling. We are exploring the development of this kind of extension to our work, especially the details of the equations and the data requirements. We will report on this aspect of our work-in-progress at the conference.

We also have not corrected the estimation results to deal with possible *policy endogeneity*. If policy choices are purposeful actions, then it is necessary to control for the factors that lead to policy decisions [17]. Policy-making in waste management may also be done to meet certain collection or recycling targets. In Hong's [16] model, the garbage unit pricing system was viewed as endogenous and had effects on both MSW generation and recycling. Kinnaman and Fullerton [4] also posited that policy choices were influential: they modeled garbage collection fees and curbside recycling as endogenous variables. They employed *two-stage least squares* (2SLS) regression to correct for the effects of endogenous policy choices. This kind of correction is useful to prevent over or under-estimating the effect of the policy.

Overall, our research gives preliminary evidence on our assertion that household informedness ought to affect household decision-making in HHW management. Our estimation results showed a negative association between HHW public education and the amount of HHW collected. We interpreted this in a positive way though: this may mean that HHW public education discourages households to use products with hazardous materials – q source control effect in other words. At the same time, we had rather weak statistical evidence to support our belief that information about maximum contaminant level violation in drinking water may encourage more household participations in separating, collecting, and delivering their HHW to appropriate HHW programs or facilities. This aspect of our empirical research is not yet conclusive, so we have more to do.

Further research work also is necessary to claim that there are *causal effects* of household informedness on HHW collection. We plan to develop a more complex and realistic simultaneous equation model to address the simultaneity of HHW collection and solid waste generation. The new model should consider endogenous policy choices, such as disposal pricing, HHW programs, and public education programs. We further intend to gather more data on environmental quality information that may have influences on household waste handling behavior. More HHW data from other states or other countries are required to generalize our findings on the effects of household informedness.

References

1. A. Tsoukiàs, G. Montibeller, G. Lucertini, V. Belton, Policy analytics: An agenda for research and practice. EURO J. Dec. Proc., **1**, 115-134 (2013)
2. H. F. Lund, *McGraw-Hill Recycling Handbook.* 21.1-21.29, McGraw-Hill, New York, NY (2001)
3. V. J. Inglezakis, K. Moustakas, Household hazardous waste management: A review, J. Envir. Mgmt., **150**, 310-321 (2015)
4. T. C. Kinnaman, D. Fullerton, Garbage and recycling with endogenous local policy, J. Urb. Econ., **48**, 419-442 (2000)
5. S. F. Sidique, S. V. Joshi, F. Lupi, Factors influencing the rate of recycling: An analysis of Minnesota counties, Res. Cons. Recyc., **54**, 242-249 (2010)
6. U.S. Environmental Protection Agency, Proper disposal of household hazardous wastes, Washington, DC (2014)
7. A. Frahm, D. Galvin, G. Gensler, G. Savina, A., Moser, Changing behavior: Insights and applications. Local Hazardous Waste Management Program in King County, King County Water Pollution Control Division, Seattle, WA (1996)
8. G. W. Fishbein, The real meaning of environmental quality, Am. J. Public Health, **59**, 2130 (1969)
9. European Environment Agency, Environmental quality (definition), Copenhagen, Denmark (2015)
10. J. P. Shimshack, M. B. Ward, T. K. M. Beatty, Mercury advisories: Information, education, and fish consumption, J. Envir. Econ. Mgmt., **53,** 158-179 (2007)
11. L. S. Bennear, S. M. Olmstead, The impacts of the "right to know": Information disclosure and the violation of drinking water standards, J. Envir. Econ. Mgmt., **56**, 117-130 (2008)
12. J. G. Zivin, M. Neidell, W. Schlenker, Water quality violations and avoidance behavior: Evidence from bottled water consumption, Amer. Econ. Rev., **101**, 448-453 (2011)
13. CalRecycle, Annual reporting: Household hazardous waste collection information, Sacramento, CA (2015)
14. U.S. Census Bureau, American Community Survey, Washington, DC (2012)
15. R Development Core Team, R: A language and environment for statistical computing, R Foundation for Statistical Computing, Vienna, Austria (2008)
16. S. Hong, The effects of unit pricing system upon household solid waste management: The Korean experience, J. Envir. Econ. Mgmt., **57**, 1-10 (1999).
17. T. Besley, T. C. Anne, Unnatural experiments? Estimating the incidence of endogenous policies, Econ. J., **110**, F672-F694 (2000)

Effects of the Coastal Park Environment Attributes on Its Admission Fee Charges

Erda Wang[1], Jiawei Zhu[1] and Yang Yu[2]

[1]*School of Business Management, Dalian University of Technology, No.2 Linggong Road, Dalian, Liaoning, 116024, P. R. China*
[2]*Yang Yu, Liaoning University of International Business and Economics, P.R. China*

Abstract. In this paper, we investigate the effect of those recognized nature-and-activity-based attributes on the level of park's admission fee charges using a panel data of 29 coastal recreation parks in Dalian city of China. A total of seven different Hedonic pricing model specifications are used in the estimating process. The results indicate that a numerous attributes have statistically significant effects ($\alpha \leq 0.10$) on the level of park admission fee charges. In terms of the economic valuation, the marine sightseeing results in the highest value of Marginal Willingness to Pay (MWTP) of \$6.4 as its quality rank improves to a designated higher level. As expected that the park congestion has a negative effect on the MWTP (-\$0.47) and overall park's rankings have a positive effect (\$0.05) on park's MWTP. However, many recreation activities accommodated by the park sites exhibit a relatively weak effect on the park entrance fee charges. One possible reason is perhaps owing to the single admission package fee policy adopted by the park management..

1 Introduction

As a result of a rapid tourism development in China, the nature-based park recreation demand has increased dramatically in recent years. Coastal parks endowed with a variety of marine and terrestrial features such as the sea water, mass of sand, and sunlight, so-called 3S resources, have become very important tourism attractants attracting millions of tourists to take their vacation and enjoy their leisure times every year not only in China but across over the entire world. Statistical data indicates that over 80 percent of tourism revenue is attributed to the coastal tourism in the United Sates. In the Europe, there are more than 3 million of international coastal arrivals and 2.5 million of domestic coastal arrivals, respectively, per annum (Needham & Szuster, 2011; Onofri & Nunes, 2013). By contrast, more than 950 million of the Chinese tourists choose coastal cities as their favorable holiday locations which accounts for 36% of the total tourism turnout in 2013 (National Bureau of Statistics[NBS], 2014a). However, in comparison with other types of tourism destinations, coastal parks sites bear unique non-substitutable ecological features. As such, they are extremely susceptible to human activities (Singh, 2015; Economou & Mitoula, 2013).

This research tries to fill in that gap by carrying out assessing on the economic importance of those nature-and-activity-based coastal attributes in Dalian coastal park areas. In the process, we take the following steps: (i) a comprehensive dataset is synthesized through field surveys of 29 coastal parks, which cover most coastal parks in Dalian city of China. (ii) The established dataset is used to estimate the relationship between park admission fees and their related attributes. Alternative model specifications are established using the control and interaction variables; (iii) the implicit price or tourist's marginal willingness to pay toward each attribute is estimated using the statistical regression model results. The rest of the paper is arranged as follows: Section 2 provides a brief introduction to the study area and data sources. Section 3 gives a detail description for the research methodology and the hedonic model formulations. Section 4 elaborates the model results, and Section 5 discusses procedures used to estimate the visitor's marginal willingness to pay for each specific attribute following by conclusions.

2 Literature review

From a recreation product, visitors can derive pleasure and satisfactions from consuming both use and non-use resource values embedded in coastal park resources. According to Lancaster's theory (Lancaster, 1966), all sorts of products can be understood as a bundle of attributes which exhibit some implicit characteristics without any exception. In this sense, actual utility accrued to a recreation trip is primarily determined by the park attributes such as amenity, clean sea water, sea animal species, recreation activity such as sea-animal shows, as well as associated with infrastructure facilities, etc. (Hasegawa, 2010; Alegre & Sard, 2015).

The Hedonic Price Method (HPM) is commonly used to value individual attribute embedded in a product and the revealed value information can help firm's product design and marketing strategy as well. Along the same line, the HPM has long been used to estimate the value of environmental goods such as air quality. HPM is frequently utilized in the field of real estate industry to identify the relationship between property price and local environment quality. Since then, the hedonic method has been widely used in valuing goods and services that cannot be priced directly via market transactions (Zygmunt & Gluszak, 2015). However, most of the studies are focused on price-making for an individual recreation site or activity, for example, price-making for a specific recreational resort and a designed holiday package (Alegre et al., 2013), second home (Saló et al., 2012), a restaurant (Yim et al., 2014) and a hotel room (Schamel, 2012). Certainly, when tourists encounter more inconveniences and/or unpleasant experiences caused by dirty surroundings and crowding conditions, etc., they must result in negative effect on related coastal park admission fee charges, thus causing the park's revenue losses (Taylor, 2008).

3 Study area and data sources

3.1. Study area

Dalian is located in the east coast of Eurasia and the southern tip of Liaodong Peninsular in the Northeastern China, which faces seas with three directions and one way to the continent. Surrounded by mountains and hills, Dalian has 2,211-kilometers of coastline, of which 139-kilometers of the coastal bank is suitable for tourism development. Currently, there are 36 coastal parks and 29 coastal beaches in total in Dalian area. Due to lack of comparative coastal resources in the neighboring areas, the Dalian coastal line becomes a unique asset for its costal tourism development. That is why it's been long viewed as competitive edge for Dalian's regional economic development.

3.2 Data sources

The panel data is collected from 29 coastal parks in Dalian city. According to the data documented in the Dalian statistical books, only 5 years of data (2010-2014) are available for analysis. To synthesize the data as best as we can, we search two well-known websites managed by related travel agencies (www.ctrip.com) and (www.qunar.com). The statistical books of the coastal tourism resources are published by the Bureau of China National Statistics, China National Tourism Administration, CNTA, 2008-2014, State Oceanic Administration (2014).

4 Research method

4.1 Framework of empirical model

The empirical analysis is based on the hedonic pricing model with an underlying assumption that the level of the admission fees charged reflects overall quality of the park's attributes and visitors' preferences. Thus, the park's admission fee can be treated as a function of both nature-based and activity-based characteristics.

$$P = P(x_m, z_n) \tag{1}$$

To be consistent with the panel data features, the Eq. (1) can be expressed as Eq. (2), where P_{it} denotes admission fee charged in a coastal park in period t, $x_{m_{it}}$ presents a vector of the nature-based characteristics, and $z_{n_{it}}$ denotes a vector of activity attributes; subscript i stands for an individual coastal park and t stands for different time periods beginning from the first quarter in 2010 to the fourth quarter in 2014.

$$P_{it} = P(x_{m_{it}}, z_{n_{it}}) \tag{2}$$

Considering the ranking status of each costal park achieved and the number of visitors to attract each year, we add two control variables (ranking and crowding) into the model specifications as shown in Eq. (3). $P(\cdot)$ now is a linear function of the parameters, $C_{ranking}$ and $C_{crowding}$ are the two control variables that have potential effects on the level of the park admission fee charges.

$$\ln P_{it} = P(x_{m_{it}}, z_{n_{it}}, C_{ranking}, C_{crowding}) + \mu_{it} \tag{3}$$

5 Model results and discussion

5.1 Comparison of alternative model specifications

All the Hausman tests of the models indicate that the random effect is not statistically significant at a conventional statistical significance level (0.05), whereas the fixed effects get to the statistically significant level of 0.05 or lower listed at the bottom of Table 3. Thus, the fixed effect becomes the priority for following analysis work. Additionally, the Durbin-Watson d statistic falls in the range of 1.7-1.9, approximately being equal to '2' except for model specification #6 which has the smallest value of d statistic of '1.561' among all-inclusive model specifications. We believe this is still acceptable. Table 1 shows the estimated results of all the 7 different model specifications based on Eq. (3). Specification #1 is the base model which uses a semi-log function where both the nature-based and activity-based characteristics are included as explanatory variables. The results indicate that most nature-based attribute variables are significant ($\alpha<0.05$) except for "sand". Crowding phenomenon has a long been one of the most important problems in all tourism parks in China, especially during the time of the nation's holidays. In order to capture this effect, a control variable of crowding is added to Specification #2 and it results in a significantly negative coefficient ($\alpha<0.05$), suggesting that congestion indeed has negative effect on traveler's willingness to pay or the park admission fees. In addition, if a park is located in a better coastal environment and at the same time offers better quality services, it will naturally have a better chance to be ranked an 'A' class than a park which lacks of those features. Therefore, a park ranking variable is added to Eq. (1) in order to examine whether the park ranking status has any markedly effect on the admission fee. It turns out that the ranking variable is not statistically significant (see Specification #3). Nevertheless, as both variables of crowding and park ranking are added to Eq. (1), the ranking variable becomes statistically significant (see Specification #4). This suggests that the park ranking status does have a markedly effect on park admission fees. Also, comparison on model results between Specification #1 and Specification #4, the adjusted R2 increases from 0.8187 to 0.8977, implying that the two control variables indeed add extra explanatory power to the park admission fee charges.

Table 1. OLS Estimates of different HPM specifications.

Specification	**#1**	**#5**	**#6**	**#7**
Acre	0.0184 ***	0.0314 ***		0.0197 **
Width	0.0032 **	0.0032 ***		0.0028 **
Sunshine	0.0190 **	0.0183 **		0.0098 ***
Degree	(0.0251) ***	(0.0318) **		(0.0362) **
Rain	(0.0252) *	(0.0253) **		(0.0171) **
Air	0.0113 **	0.0154**		0.0208 ***
Water	0.0408 **	0.0421 **		0.0170 ***
Condition	0.0107 **	0.0104 ***		0.0181 ***
Aquarium	0.0008**		0.0006*	0.0101**
Sea-Animal Show	0.0001**		0.0020 **	0.0113**
Sightseeing Boat	0.0067 ***		0.0092 ***	0.0071 ***
Diving	0.0027*		0.0027*	0.0005**
Crowding	-	(0.0327) ***	(0.019) ***	(0.0300) ***
Location Dummy	-	**	**	**
Time Dummy	-	**	**	**
_con	8.0875***	9.1843***	5.3834*	8.7235***
Adjusted R^2	0.8187	0.8079	0.6982	0.7902
AIC	5925.44	5810.76	5831.1	5760.26
DW	1.801	1.749	1.561	1.937
Hausman Test (Prob)	0.003	0.000	0.000	0.001

*p<0.1, **p<0.05, ***p<0.01

Table 2. Percentage impact and marginal willingness to pay of attribute variations.

Variable	**Coded**	**Coefficient**	**Variation**	**MWTP: USD (Mean= $15.65)**
Nature-Based	Acre	0.0197 **	1.97%	0.31
	Width	0.0028 ***	0.28%	0.04
	Sunshine	0.0098 ***	0.98%	0.15
	Degree	(0.0362)**	-3.62%	-0.57
	Good air	0.0208***	2.10%	0.33
	Water	0.0170***	1.70%	0.27
	Condition	0.0181***	1.81%	0.28
Recreation-based				
Attributes				
	Sea-Animal Show	0.0113**	1.13%	0.18
	Sightseeing Boat	0.0071***	0.71%	0.11
	Diving	0.0005**	0.05%	0.01
	Sightseeing Lift	(0.0007)	–	–
Control Variables	Crowding	(0.0300)***	-3.00%	-0.47
	Ranking	0.0032*	0.32%	0.05

5.2 Result and discussion

Overall, the seven model specifications are estimated. The main purpose of making such multitude model estimations is to enhance robustness and comprehensiveness of the analysis. However, only four model results (#1, 5, 6#, and #7) are presented in Table 1 due to the limited page space. One of the most informative analyses using Hedonic price model is to carry out price elasticity estimate (see Table 2). According to Table 1, most estimated coefficients for the nature-based attributes are positive. Each individual attribute's MWTP is calculated by employing the sample mean of entrance fee of $15.65. If the coastal park has a long coastline, its admission fee can be raised by 0.86% higher than without holding everything else remained constant; marine sightseeing can make the admission fee

4.19% higher as its quality ranking is enhanced by 0.1 point, and similarly, a better sea water quality (p=0.000) can increase the entrance fee by 1.71% over the base; marine physical condition (p=0.000) is able to contribute the entrance fee by 1.81% higher, and seashore sightseeing can raise entrance fee by 0.44% higher from the base.

Additionally, the mountain existent on the site can raise the entrance fee by more than 0.43%. Furthermore, 1% enhancement of air quality (p=0.000) can increase the park admission fee by 2.1%. By contrast, beach sand quality appears to have no statistically significant effect on park admission fee, which may be attributed to the fact that most beaches and coastal banks in the areas present either cobblestones or lower quality sand. The low variable variation contributes to the insignificant coefficient estimation result.

6 Conclusions

In this paper, we investigate the effects of both the nature-based and activity-based attributes of a coastal park on its admission fee charges and explore their sequential order to the parks' admission charges using a panel dataset of 29 coastal recreation sites in Dalian city of China. In the process, a total of seven different Hedonic pricing model specifications are estimated in order to demonstrate the robustness and comprehensiveness of the analysis with each model specification involving with different combinations of various explanatory factors.

The estimated results suggest that most sea-related nature attributes such as park size, length of beach, beach width, sunshine, air quality, water quality, etc. exhibit a positive and statistically significant effect on the park admission fee charges except for temperature, number of rainy days, and type of sand. The most important sea related attribute is the marine sightseeing, with the marginal willingness to pay being able to increase by \$6.40 (\$0.64/0.1 point) as the quality of the marine viewing is improved to one level higher.

Most water related attributes bring about a considerable positive effect on the entrance fees. Furthermore, both congestion and park ranking attributes exert statistically significant effect on park admission fees, suggesting that the park authority should pay more attention to the parks' carrying capacity management. Regarding the activity-based attributes, most of them impose statistically significant ($p<0.1$) effect on the park's entrance prices. However, those non-water related facilities end up with statistically insignificant results. These may suggest that there is no use for adding those none water-related recreation activities or facilities such as kid playground and fun fair to the coastal parks because they may play a role in detract not adds to the visitors' satisfaction (Loomis & Richard, 1997; Mueller et al., 2009).

One caveat deserves to mention regarding the overall significance of those assessed attributes in terms of their MWTP for park's admissions. Clearly, the contribution of each nature-based attribute to park entrance fee is underestimated by the model as most attribute changes make a relatively small contribution toward the change of park entrance fee, i.e., less than 3 percent except for temperature (Degree) and condition of marine (Marine). These underestimations are mainly due to the fact that the current entrance fee charging mechanism is under control of the city's government. As such, the coastal parks are treated as public goods, leading to the undervalue use of the resources. Certainly, from an economic efficiency viewpoint, the city government should reassess present coastal park admission charging system. A nature value-based fee charging system should be encouraged in order to enhance the optimal resource uses, thus to benefit both coastal environment amelioration and visitors' welfare.

This study has limitations in the following two aspects. First, the study did not address coastal tourism market equilibrium issues in Dalian and surround marketing areas. Due to the short of data sources this study did not carry out market equilibrium condition tests prior to constructing the Hedonic pricing models. Nevertheless, the property market equilibrium status test (or coastal tourism market in this study) is generally required in using HP model. Second, as described earlier, at present the coastal park admission fee charging system in Dalian may not fully reflect the true value of the park. Generally speaking, the admission in all parks in Dalian city is undercharged, which could potentially bring about downward biased MWTP estimates for all those attributes analyzed in this study. To improve the quality of this study, the future research may want to expand the study territory by including all the neighboring coastal park areas surrounding Dalian city, so that a larger dataset can be used in estimation, which will ultimately be able to improve the estimate results as more variation for each individual attributes is available in modeling process. In addition, as more marketing mechanism is adopted in determining the level of the entrance fee charges, the MWTP estimate for each considered attribute will be more accurate.

References

1. J. Alegre, M. Cladera, M. Sard, Tourist areas: Examining the effects of location attributes on tour-operator package holiday prices. Tourism Management, **38**(0), 131-141, (2013)
2. J. Alegre, M. Sard, When demand drops and prices rise. Tourist packages in the Balearic Islands during the economic crisis, Tourism Management, **46**(0), 375-385, (2015)
3. China National Tourism Administration [CNTA], The Yearbook of China Tourism. China Tourism Press: Beijing, China (Total 7 books in Chinese), (2008-2014)
4. Dalian Tourism Administration [DTA] (2010a), The Romantic City of Dalian- official brochure. (2010b) Dalian Bohai Sea Tourism Resources Survey. Liaoning Lijie Consultant Ltd: Liaoning, China. Dalian Tourism Industry Development Program-official brochure (total 3 books in Chinese), (2014)

5. A. Economou, R. Mitoula, Management of natural resources and protection of the coastal urban area of Glyfada. Land Use Policy, **35**(0), 204-212,(2013)
6. H. Hasegawa, Analyzing tourists' satisfaction: A multivariate ordered probit approach. Tourism Management, **31**(1), 86-97, (2010)
7. J. A. Hausman, Specification tests in econometrics. Econometrica: Journal of the Econometric Society, 1251-1271, (1978)
8. K. J. Lancaster, A new approach to consumer theory. The journal of political economy, 132-157, (1966)
9. J. B. Loomis, G. W. Richard, Recreation economic decisions, comparing benefits and costs, second edition, Venture Publishing, Inc., 101-109 (1997)
10. J. Mueller, J. Loomis, A. González-Cabán, Do repeated wildfires change homebuyers' demand for homes in high-risk areas? a hedonic analysis of the short and long-term effects of repeated wildfires on house prices in southern california, Journal of Real Estate Finance & Economics, **38**(2), 155-172, (2009)
11. National Bureau of Statistics of the People's Republic of China, (2010a-2014a), China Statistical Yearbook, (2010b-2014b). China City Statistical Yearbook, (2010c-2014c), Dalian Statistical Yearbook, China Statistics Press: Beijing, China (total 15 books in Chinese)
12. M. D. Needham, B. W. Szuster, Situational influences on normative evaluations of coastal tourism and recreation management strategies in Hawai'i, Tourism Management, **32**(4), 732-740, (2011)
13. L. Onofri, P. A. L. D. Nunes, Beach 'lovers' and 'greens': A worldwide empirical analysis of coastal tourism. Ecological Economics, **88**(0), 49-56, (2013)
14. A. Saló, A. Garriga, M. Vila, J. M. Sayeras, Differences in seasonal price patterns among second home rentals and hotels: Empirical evidence and practical implications, Tourism Economics, **18**(4), 731-747, (2012)
15. G. Schamel, Weekend vs. midweek stays: Modeling hotel room rates in a small market. International Journal of Hospitality Management, **31**(4), 1113-1118, (2012)
16. A. Singh, Managing the environmental problem of seawater intrusion in coastal aquifers through simulation-optimization modeling. Ecological Indicators, **48**(0), 498-504, (2015)
17. State Oceanic Administration People's Republic of China, *China Marine Statistical Yearbook*, China Ocean Press: Beijing, China, (In Chinese), (2014)
18. L. O. Taylor, Theoretical foundations and empirical developments in hedonic modeling. In Hedonic Methods in Housing Markets, (pp. 15-37). Springer New York, (2008)
19. E. S. Yim, S. Lee, W. G. Kim, Determinants of a restaurant average meal price: An application of the hedonic pricing model, International Journal of Hospitality Management, **39**(0), 11-20, (2014)
20. R. Zygmunt, M. Gluszak, Forest proximity impact on undeveloped land values: A spatial hedonic study. Forest Policy and Economics, **50**(0), 82, (2015)

Simulating Water Quality of Wei River with QUAL2K Model, a Case Study of Hai River Basin in China

Yali Zhang[1,2a], Huige Yang[1] and Zhisong Wang[3]

[1] *Institute of Resources and Environment, Henan University of Agriculture, Zhengzhou, 450002, China*
[2] *Key Laboratory of Soil Erosion Process and Control in the Loess Plateau of Ministry of Water Resources, Zhengzhou, China*
[3] *Industrial Development Company, Henan University of Agriculture, Zhengzhou, 450002*

Abstract. QUAL2K model is used to simulate the water quality of Wei River and the verification results are analyzed in this paper. For the actual situation of pollution in Wei River Basin, COD, ammonia nitrogen and total phosphorus are chosen as indexes of water quality prediction. The degradation coefficient of them, being one important water quality parameter, is calculated as 0.2428, 0.1019 and 0.2554, respectively. Conduct parameter setting and compared with the measured values combining comprehensive one-dimensional water quality model formula, with the measured data, then make adjustments to get new degradation coefficient with the values are 0.3631, 0.2046 and 0.2545, respectively. The simulation results are verified and showed that the predicted values fit with the measured ones well, the relative errors of COD, ammonia nitrogen and total phosphorus concentration are within 10%, and the simulation effect is good.

1 Introduction

Water quality simulation is important to master the variation law of pollutants in water body, reduce the risk of water pollution and its loss, and ensure the water safety and water environment quality. Research on water quality model is mainly in Europe and other developed countries where theoretical foundation and applied research of the model has been very mature [1]. While it starts relatively late, and is still in the exploratory stage of theoretical research and applications in China. QUAL2K model that evolved in QUAL2E is one of QUAL series models developed by the United States Environmental Protection Agency and applies to small and medium-sized dendritic rivers which mix well [2-4]. In order to solve some water quality problems, Seok Soon Park et al. [5] applied the above-mentioned two models to the Nakdong River, Korea and found simulated results of QUAL2K model was relatively more consistent with the measured values. Guo et al. [6] discussed water quality trends of the middle and lower reaches of Hanjiang River by QUAL2E and QUAL2K model with BOD as the predictor, and the results indicated that QUAL2K reflected the field data better than QUAL2E. Kalburgia et al. [3] conducts a study using the water quality model QUAL2K to develop the BOD-DO model and evaluation of the results for a 50 km stretch of river, and the results show that the values predicted by the model are in close agreement with measured values.

Wei River is the tributary of Hai River in China. It is the key of water pollution remediation area that required by *the 12th Five Year Plan* in China. According to the local *Environmental Quality Bulletin*, the water quality of Wei River was severe pollution from 2009 to 2014. Therefore, to provide a scientific basis for reducing water pollution risk and its losses, this study chooses a typical reach of Wei River and uses QUAL2K model for its water quality simulation.

2 Introduction of QUAL2K Model

QUAL2K model belongs to comprehensive one-dimensional water quality model and mainly simulates the main stream of rivers [2]. It is suitable for modeling the migration and transformation of BOD, N, P and other pollutants in well mixed dendritic river. The simulation with the elements as the basic unit, is a steady flow balance for each model element, namely:

$$Q_i = Q_{i-1} + Q_{in,i} - Q_{out,i} \tag{1}$$

where Q_i (m^3/d) is outflow from element i to i+1; Q_{i-1} (m^3/d) is inflow from element i-1; $Q_{in,i}$ (m^3/d) is inflow from point and non-point sources; $Q_{out,i}$ (m^3/d) is outflow from point and non-point sources of element i; there exists for any water quality components.

$$\frac{\partial C}{\partial t} = \frac{\partial\left(A_x E_x \frac{\partial C}{\partial x}\right)}{A_x \partial x} - \frac{\partial(A_x \bar{u} C)}{A_x \partial x} + \frac{dC}{dt} + \frac{S}{V} \tag{2}$$

where C(mg/L) is the concentration of a constituent; x(m) is the distance; t(s) is the time in days; A_x (m^2) is the river cross-sectional area from x; E_x (m^2/s) is the dispersion coefficient; $\overline{u}$ (m/s) is the average flow velocity; S(mg/L) is the external sources or sinks of the constituent; V(m^3) is the volume of each calculation unit. The four on the right side of the equation are respectively diffusion, advection, component reaction and the external sources or sinks of the constituent.

The model mainly simulates the water quality of the river main stream. The import tributary is regarded as a point source, and non-point source and its outlet are seen as a line source and line outlet. Firstly the river is divided into a series of steady nonuniform reaches according to the water quality, and then the reaches are divided into a number of calculation units. It conceptualizes the simulation river as a series of well mixed, end-to-end units formed through transport and diffusion by dividing reach units. Each unit is an ideal mixing reactor and the smallest calculation unit, and single unit calculate the corresponding equation based on the law of substance balance during the simulation. Among them, the substance balance includes water convection and dispersion. Not only will the substance of each unit increase or decrease along with the external water into or output, but also the concentration will rise or lower because of internal biological effect. Simulation generally includes the following basic steps: 1. material collection and field investigation; 2. divide reach and unit; 3. determine the model parameters; 4.validation and application of the model [2].

3 Modeling process for WEI River

3.1 Study area introduction

Wei River originates in the Taihang Mountain of Shanxi Province, with a total length of over 900km and length of 399km in Hebi range. The catchment area is totally 15830km^2. Its northwest is mountain with an area of about 6280km^2 and southeast is plain with an area of approximately 9010 km^2. There are many kinds of water pollutants and with wider range of pollution. We selected Chaiwan - Wuling which is the typical provinces controlled sections of the Wei river, with total length of 38.17km. There are two general monitoring sections, Xunxian bridge and Tunzi bridge with total length of 38.17km. Chai Wan to Xunxian bridge is 13.9km, Xunxian bridge to Tunzi bridge is 8.44 km, and Tunzi bridge to Wuling is 15.83 km. The entire river slopes slightly from south to north without significant pollution source, and pollutants are easily well-mixed in the lateral and vertical.

3.2 Basic data and river course generalization

When using the model, data needed to input mainly include river boundary conditions, reach data, meteorological data, model parameters, point and non-point sources pollutants, etc., so required data generally include watershed hydrology and spatial information, weather data, initial conditions, boundary conditions, node data of each reach monitoring sections and monitoring data that the model parameters calibration and verification require. The output includes calculation of the input data, temperature data, hydraulic and water quality data, modeling water quality components and variation curve in water [2,4]. According to the actual situation of Wei River, March, April and May in 2013 were determined as the model parameters calibration period and May 2012 as the verification period. We select COD, NH_3-N and TP as simulation pollutants, and the length of the calculation unit is 2km, the whole with a total of 19 (Table 1).

Table 1. The division of Wei River reach and calculation units.

Reach number	The starting point	Length (km)	Distance from the upstream boundary(km)	Calculation unit
1	Chaiwan-Xunxian bridge	13.9	13.9	7
2	Xunxian bridge -Tunzi bridge	8.44	22.34	4
3	Tunzi bridge -Wuling	15.83	38.17	8

3.3 Parameters calibration of the model

3.3.1 The boundary conditions

Take the parameters of Weihui estuary monitoring sections in the upstream water bifurcation as the boundary conditions of water quality simulation, which include design flow, design temperature, the initial concentration of COD, NH_3-N and TP and the relevant hydraulics parameters. According to the annual monitoring data, use the average annual flow as a design flow, take the mean of pollution indicators average measured concentration nearly 10 years in Wehui monitoring sections as the initial concentration of water quality modeling, and use the average temperature of the monitoring period as design temperature, in Table 2.

Table 2. Water quality and quantity parameters of source.

Temperature (℃)	Flow (m³/s)	COD (mg/L)	NH_3-N(mg/L)	TP (mg/L)
17.53	7.07	29.33	8.71	1.79

3.3.2 Water quality parameters

The degradation coefficient' determination at home and abroad are mainly data projections, experience estimation method, inverse model estimation method, water mass tracing valuation method and laboratory simulation method, etc. According the one-dimensional river water quality model[7], the degradation coefficient can be calculated with formula 3 (Table 3).

$$K = 86.4(\ln C_0 - \ln C \frac{\overline{u}}{L}) \tag{3}$$

where C(mg/L) is pollutants concentration monitoring values of next section; C_0 (mg/L) is pollutants concentration of previous section; $\overline{u}$ (m/s) is the average flow velocity; L(km) is the distance from next section to previous section; $K(d^{-1})$ is degradation coefficient.

Table 3. The initial value of degradation coefficient.

Degradation coefficient (d^{-1})	Chaiwan-Xunxian bridge	Xunxian bridge - Tunzi bridge	Tunzi bridge-Wuling	Mean
K_{COD}	0.3362	0.2704	0.2218	0.2428
K_{NH3-N}	0.1188	0.0816	0.1053	0.1019
K_{TP}	0.2586	0.2341	0.2735	0.2554

3.3.3 Hydraulic parameters

In view of the property of the existing materials, the flow curve formula and Manning formula demand for data, flow curve formula was used to describe hydraulics characteristics [8-11]. Thus, the hydraulic parameters are mainly related to the flow curve parameter and the diffusion coefficient K.

1) The flow curve

QUAL2K model assumes steady stream hydraulic characteristics, namely $\partial Q / \partial t = 0$, and hydraulic characteristics of each river meet the following form:

$$\overline{u} = aQ^b \qquad H = cQ^d \tag{4}$$

where Q (m^3/s) is flow; $\overline{u}$ (m/s) is the average flow velocity; a, b, c and d are the experience coefficient gotten by the velocity curve of drainage and water drainage curve, where b and d must meet the Table 4, and the sum of the two is not greater than 1.

Table 4. The parameter values of the flow curve.

Formula	Letter	Calibration value	Range of values
$\overline{u} = aQ^b$	b	0.43	0.4-0.6
$H = cQ^d$	d	0.45	0.3-0.5

Sectional area and width of the river can be gotten by formula 5.

$$A_x = Q / \overline{u} \qquad B = A_x / H \tag{5}$$

Then, the surface areas and volumes of reach units can be calculate by formula 6.

$$A_s = B\Delta x \qquad V = BH\Delta x \tag{6}$$

In this study, using the measured data of multi-sectional flow, flow velocity and depth, the values of a, b, c, d by the least squares method are 0.3, 0.28, 0.15, 0.45, respectively.

2) The diffusion coefficient

There are many methods to determine diffusion coefficient, and the study adopted the formula of QUAL2K model [2, 4].

$$E_x = 3.82Kn\overline{u}H^{5/6} \tag{7}$$

where E_x is the Longitudinal dispersion coefficient; K is the Longitudinal diffusion constant, and the empirical average is 5.93; n is Manning roughness coefficient; $\overline{u}$ is the average flow velocity; H is the average depth.

Wei River bend is with narrow and deep river channel, so Manning coefficient of it is relatively large. Contrasting with the common channel Manning coefficient, its value is ultimately determined 0.05.

4 Solution and verification of the model

4.1 Solution of the model and parameter adjustment

We model the water quality using the partial differential equations and solve it by the improved Euler Finite difference Scheme [2-4]. The solution process is omitted here, and the measured values are compared in Table 5.

Table 5. Simulation and measured values of parameter initial values.

Section	COD(mg/L)			NH3-N(mg/L)			TP(mg/L)		
	CV	MV	RE (%)	CV	MV	RE (%)	CV	MV	RE (%)
Xunxian bridge	28.85	25.01	15.35	10.45	9.27	12.73	0.93	0.84	10.71
Tunzi bridge	30.31	23.87	26.98	11.89	9.43	26.09	1.02	0.86	18.60
Wuling	26.89	24.12	11.48	10.12	9.06	11.70	0.95	0.71	33.80

Note: CV is Calculated Value; MV is Measured Value; RE is Relative Error; AE is Absolute Error.

It shows that the relative errors and the measured values are larger, especially Xunxian bridge-Tunzi bridge. It is advised to decrease the river longitudinal diffusion coefficient and to increase pollutant degradation coefficient. After repeated modification, the parameter values are as follows (Table 6).

Table 6. The determination results of model parameters.

Model parameter	Chaiwan-Xunxian bridge	Xunxian bridge - Tunzi bridge	Tunzi bridge - Wuling	Mean
$K_{COD}(d^{-1})$	0.4507	0.4839	0.1547	0.3631
$K_{NH3-N}(d^{-1})$	0.1277	0.2079	0.2782	0.2046
$K_{TP}(d^{-1})$	0.2588	0.2346	0.2701	0.2545

4.2 Analysis of the modelling results

Use routine monitoring data in May 2012 for model validation and the results are shown in Table 7. It shows that, except that the COD concentration simulation value at Tunzi bridge is slightly larger than the measured value, and the measured values of the other sections are less than the calculated values. The calculated and measured values of COD are with the same trends over reach distance, with a high correlation. There is little change in water quality from Chai Wan to Xunxian bridge, and it is deteriorating from Tunzi bridge to Wuling.

The measured values of NH_3-N concentration are slightly less than the calculated ones. Different from COD, the maximum error between NH_3-N model calculated and measured values is Tunzi bridge. The smallest error is in Wuling section with the relative error of 1.36% and the absolute error of 0.15.

The concentration calculated and measured values of TP have the same trend with NH_3-N, that pollutant concentrations increases from Xunxian bridge to Tunzi bridge and begins to decline after reaching a peak in Tunzi bridge.

In a whole, the relative errors of the concentrations of COD, NH_3-N and TP are within 10%. The calculation value of each water quality indicator significantly correlated to the measured ones. It indicates that the parameter values are relatively reasonable, and the model has high authenticity and reliability.

Table 7. Concentrations' comparison of calculated value and measured one for different river sections.

Section	COD(mg/L)				NH_3-N(mg/L)				TP(mg/L)			
	CV	MV	RE(%)	AE(mg/L)	CV	MV	RE(%)	AE (mg/L)	CV	MV	RE(%)	AE(mg/L)
Xunxian bridge	27.81	27.27	1.98	0.54	11.55	11.39	1.40	0.16	0.81	0.8	1.25	0.01
Tunzi bridge	27.55	28.36	-2.86	-0.81	11.65	11.45	1.75	0.2	0.85	0.82	3.66	0.03
Wuling	32.33	30	7.77	2.33	11.18	11.03	1.36	0.15	0.68	0.73	-6.85	-0.05

Note: AE is Absolute Error.

5 Conclusion

In this paper, QUAL2K model is used to simulate and verify the water quality of Wei River, which COD, ammonia nitrogen and total phosphorus are chosen as indexes of water quality prediction. Through the parameter optimization, mainly degradation coefficient, the modeling results show that the predicted values fit with the measured ones well. The relative errors of COD, ammonia nitrogen and total phosphorus concentration are all within 10%, and the effect is good. QUAL2K model relying Microsoft Office Excel interface, equipped with relatively simple VB programming, is easy to handle and learn. Meanwhile, to determine the water quality parameters is particularly important in the course of solving the model. To promote the results' rationality, we make appropriate adjustments of the parameters, mainly degradation coefficient. In the future, we should conduct more studies on the methods to optimize the parameter of the model for different river, especially for those without enough data.

Acknowledgment

The authors sincerely acknowledge the financial support by the key project of *three major* on the philosophy and social science of higher education in Henan Province, (Project No. 2014-szzd-27), key laboratory fund of soil erosion process and control in the Loess Plateau of Ministry of Water Resources, (Project No. 2013-03), and the National Nature Science Fund of China, (Project No. 40801102). The authors are grateful to the local administration for their generously providing research data.

References

1. J. H. Cho, S. R. Ha, Parameter optimization of the QUAL2K model for a multiple-reach river using an influence coefficient algorithm. Science of the Total Environment, **408:**1985-1991 (2010)
2. S. C. Chapra, G. J. Pelletier, H. Tao, *QUAL2K: a modeling framework for simulating river and stream water quality, version 2.07. Documentation and Users Manual*, Medford, MA: Civil and Environmental Engineering Dept., Tufts University, (2007)
3. P. B. Kalburgia, R. N. Shareefaa, U. B. Deshannavar, Development and Evaluation of BOD–DO Model for River Ghataprabha near Mudhol (India), using QUAL2K. I. J. Engineering and Manufacturing, **1**: 15-25 (2015)
4. R. K. P, Seockheon L, R K S, et al., Application of QUAL2Kw for water quality modeling and dissolved oxygen control in the river Bagmati. Environmental Monitoring and Assessment, **125**:1-3 (2007)
5. S. S. Park, Y. S. Lee, A water quality modeling study of the Nakdong River, Korea, Ecological Modelling, **152**:65-75 (2002)

6. Y. Guo, Y. Wang, Hydrochemical modeling and prediction of the middle and lower reaches of Hanjiang river——Application of QUAL2K model, Safety and Environmental Engineering, **10**:4-7 (2003)
7. R. M. Wright, A. J. Mcdoneii, In-stream deoxygenation rate prediction. Journal of the Environmental Engineering Division, **105**:323-333 (1979)
8. Z. Li, Study on automated monitoring, forecasting and emergency disposal of water pollution in Fushun section of Hun river Basin, Shenyang Agricultural University, (2013)
9. T. Ying, Application of QUAL2K model in water resources management and protection of Daling river basin. Water Resources & Hydropower of Northeast China, **05**:52-53 (2011)
10. L. Yang, J. Qian, Y. Wu, et al., Water quality optimization scheme for Qinhuai river based on QUAL2Kmodel, Water Resources Protection, **03**:51-55 (2013)
11. C. Ye, Progress of studies on water environmental mathematical model, Advances in Environmental Science, **01**:74-81 (1993)

MATEC Web of Conferences **68**, 14006 (2016) DOI: 10.1051/matecconf/20166814006
ICIEA 2016

Problems and countermeasures on the safety of rainwater harvesting for drinking in China

Liu Laisheng, Liu Linghua, Wu Leixiang, Wu Jiapeng and Huo Weijie

China Institute of Water Resources and Hydropower Research, Beijing 100038, China

Abstract. China is increasingly confronted with serious water shortages, so rainwater harvesting and utilization have gradually received attention with advantages such as accessibility, simple operation and low cost. The harvested rainwater can be used for drinking, irrigation, municipal greening, etc., and when applied for drinking, the demand for water quality is highest. Most existing researches have put their focus on improving the accumulation of rainwater, but there is a lack of in-depth studies on how to enhance the quality of rainwater. Based on the above considerations, this paper, by summing up the situation of rainwater harvesting and utilization in China, has systematically analyzed the system components of rainwater harvesting for drinking, i.e., the consisting units of rainwater harvesting, delivering, processing and accumulating. It also explains the sources of rainwater pollution and how to deal with it. Considering that the water harvesting system can make a contribution to the society and public welfare, this paper proposes a framework of participatory management for projects of rainwater harvesting for drinking.

1 Current practice of rainwater harvesting and drinking in china and abroad

1.1 Current practice of rainwater harvesting and drinking in foreign countries

In countries outside China, "rainwater utilization" is usually referred to as "rainwater management". From the beginning of the 1980s, the world began to explore how to utilize rainwater resources. At present, many countries in the world have recognized the value of using rainwater, and adopted a variety of technologies, equipments and measures for rainwater harvesting, application, control and management. In recent years, different scales of researches and applications on rainwater utilization and management have been carried out in more than 40 countries and regions such as the United States, Canada, France, India, Israel and Japan. In Australia's rural and suburban areas where houses are near, round sumps made of corrugated steel plates are commonly established to collect the rainwater from roofs. In many parts of Kenya in Africa, UNDP and the World Bank take rainwater storage tanks as an essential part of their Projects on Rural Water Supply and Sanitation. Later, such technique is spread to Botswana, Namibia, Tanzania and other places, leading to the development of African's rainwater harvesting project. In Latin America, Mexico and Brazil have also made a prevailing utilization of rainwater: In Mexico, there is relative mature system for rainwater harvesting on Chiapass plateau, and the system consists of aluminum roofs, trapezoidal underground pools, pool filters, pumps, etc.; In northeastern Brazil, poor people who lives in the semi-arid regions near the equator have built more than 2,000 water storage tanks by fence cement, precast concrete panels, lime and bricks under the fund of Canada and other international organizations.

Many countries and regions have introduced corresponding technical manuals, specifications and standards. They developed a series of laws and regulations on the utilization of rainwater as a support, e.g., Germany's Federal Water Act, Building Regulations and Local Regulations, and they all take the form of legal provisions to strengthen the protection of the natural environment and the sustainable use of water, such as the rule of requiring the owner of new projects to dispose and utilize rainwater.

1.2 Current practice of rainwater harvesting and drinking in China

China's water resource is averaged as 2300m³ per capita, only 1/4 of the world's per capita consumption. Such number is ranked 109 in the world, so China was listed as one of the world's 13 water-anemia countries. As rainwater harvesting benefits so much, the development and utilization of rainwater resources has achieved considerable development in recent years. Broadly speaking, rainwater harvesting and utilization has the following advantages: 1). Reducing Investment. China's rural and mountainous area is so vast that water users are relatively scattered, therefore long-distance water pipelines are needed for centralized water supply, which hugely increases the construction costs. 2). Saving energy. Long distance results in large water head loss, so it requires higher pump head with huge energy consumption. 3). Conserving water. The centralized water supply by existing plants in rural areas is mostly sourced from surface water or groundwater, such as rivers, lakes and reservoirs, so rainwater accumulation can effectively reduce the extraction of surface water and groundwater. 4). Reducing the erosion by rainstorm. Distributed rainwater harvesting can lessen the intensity of runoff and reduce the erosion by rainwater to the runoff-collecting surface, thereby reducing soil erosion.

In China, rainwater harvesting and utilization is mainly applied to the following types of water-deficient areas: 1). Water-deficient areas with a lack of resource. In arid and semi-arid areas such as Gansu and central Ningxia, the rainfall is only 200-400mm while the evaporation is over 1000mm, and no surface water and groundwater sources are available in these areas; 2). Seasonal water-deficient areas. In China's Fujian, Guizhou, Hubei and other hilly areas, there is abundant rainfall but with large seasonal variations, so water-deficiency still occurs in winter and spring; 3). Water-deficient areas with exploitation difficulty. In the southwest mountainous areas, the underground water is deeply buried because of deep valleys everywhere, which results in unbearable exploitation costs; 4). Water-deficient areas with poor water quality. In these areas are featured by poor water quality which is sourced from brackish water, fluoride water, high-arsenic water, etc. In recent years, China has constructed water cellars, tanks, ponds and other miniature water conservancy projects in the above water-deficient areas as an effective solution to the problem of water shortage, but there are still problems as follows: 1). Most projects are spontaneously constructed by farmers, so rainwater cannot be effectively collected due to the deficiency of harvesting brought by unprofessional design of projects. 2). The amount of rainwater harvesting is insufficient due to inappropriate layout of projects and a lack of unified technical guidance and planning; quality concerns also occur during the construction of projects, e.g., problems of cellars range from leaking to collapsing. 3). Lack of protection and pollution control measures in areas of rainwater resources. 4).Short of supporting policies. 5). Lack of professional technical atlas, standards and norms. 6).Drinking water is not safe for users due to a lack of effective rainwater treatment facilities.

2 System components of rainwater harvesting for drinking

The system of rainwater harvesting for drinking mainly includes four component units that respectively related to processes of harvesting, storing, processing and delivering. Figure 1 shows the system components of rainwater harvesting for drinking.

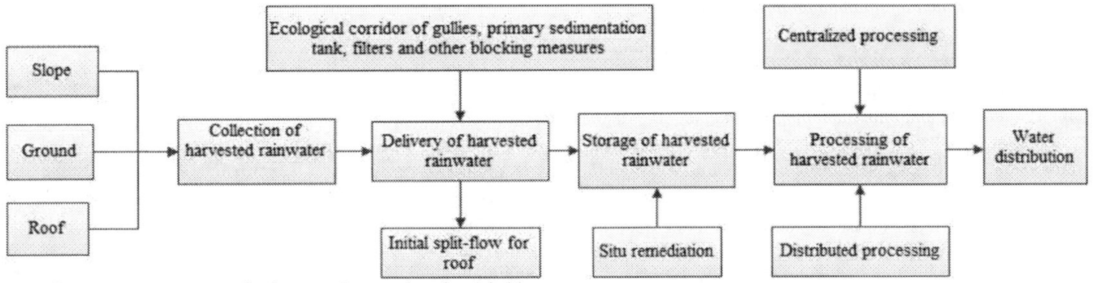

Figure 1. System components of rainwater harvesting for drinking.

2.1 Collecting units for rainwater harvesting

2.1.1 Types of runoff-collecting surfaces

Rainwater Harvesting and Utilization Project takes it as a priority to solve the problem of rainwater harvesting and storage. A rational, efficient and high-quality collection of rainfall runoff is regarded as the main objective of rainwater harvesting. In present China, roofs are mainly adopted as runoff-collecting surfaces, and some other areas also take hardened concrete patio floors to fulfill this function. runoff-collecting surfaces basically include three types: existing runoff-collecting surface, specialized runoff-collecting surface and natural runoff-collecting surface. 1). Existing runoff-collecting surface refers to the existing roofs (tile roof, concrete surface or mortar surface); 2). Specialized runoff-collecting surface refers to the hardened concrete and compacted solid in courtyard, threshing ground and other unused venues around houses; 3).Natural runoff-collecting surface refers to the natural slopes that used as runoff-collecting surfaces. At present in China's rural areas, existing and specialized runoff-collecting surfaces are most frequently adopted, and the surface area is based on average annual rainfall, rainfall characteristics, water demand planning and water consumption criteria.

2.1.2 Material of runoff-collecting surface

The material of runoff-collecting surface reflects one of the key technologies for projects of rainwater harvesting and utilization. In northwest China, a variety of ways are adopted to deal with runoff-collecting surfaces for rainwater harvesting projects, e.g., concrete, HEC solidified soil, AAM cement soil, compacted cement soil, cement tile, compacted 37 lime earth, plastic film, glass-fiber linoleum, chemical material handling and original soil rammed, etc. Concrete is one of the most commonly used material for runoff collecting with an efficiency of above 80%. HEC solidified soil functions as a curing agent that increases soil strength and impermeability. It takes industrial waste as its raw material, and is carefully mixed with core materials and other components to become a kind of cementing materials without clinker.

2.2 Delivery units for harvested rainwater

Different scholars give different definitions on initial discarded rainwater. Geiger (1987) defines it as "the largest proportion of pollution load in initial runoff" [1]; Saget (1996) defines the initial discarded as "30% of the initial runoff that contains 80% of pollution load"[2]. Factors with an impact on initial discarded rainwater include the followings: rainfall duration, rainfall amount, rainfall intensity, interval between two rainfalls, types of runoff-collecting surfaces, pollution sources and extent, etc. Figure 2 shows a typical attenuation of concentrations of pollutants in harvested rainwater. As the rain continues, the concentration of pollutants in rainwater is gradually reduced from L0 until the end of rainfall. Ding Kunlun, et al (2011) made a study on roof-harvested rainwater in some rural area located in Beijing suburb. The results show that, from the start of rainfall, the concentration of pollutants in roof-harvested rainwater showed a gradual downward trend. The initial rainfall had obvious scouring action. When the amount of rainfall reached 2-3 mm, 90% of contaminants in rainwater could be effectively removed and permanganate index was reduced by 40% [3]. According to the "Technical Specifications on Rainwater Utilization Project for Architecture and Community" (GB50400-2006), if the amount of rainfall is less than 10mm in Beijing, 2mm of the initial roof-harvested runoff contains more than 70% of the pollutants. The amount of initial discarded rainwater is determined in accordance with the permanganate index, suspended solids, chroma, etc. in rainwater, which are actually measured on the underlying surface. When there is no reference of measured data, 2-3mm of runoff depth could be adopted as the amount of initial discarded rainwater for roof-harvested rainwater, and for ground-harvested rainwater, the depth will be 3-5mm [4].

Foreign countries have an early study on devices for initial discarded rainwater with already approved mature products, while China has a relatively late start that confines its choice to the commonly used volumetric pool for discarded rainwater. Chen Gang et al. (2013) divided the devices for initial discarded rainwater into active ones (without mechanical power) and mechanical ones (using mechanical power) according to the discarding principles. Of which, the active devices are subdivided into volumetric devices, infiltration wells, and small-large tubes in line with the implementation of discarding rainwater. Since the volumetric discarding adopts an empirical coefficient method, the contamination of underlying surface often differs with seasons in practical engineering applications, but the volume of discarded rainwater has been fixed, so if the underlying surface has heavy contamination, the amount of discarded rainwater will not be enough. As a result, at the end of rainwater discarding, the pollutants in follow-up rainwater cannot be cleared. Yang Xiao et al (2010) designed a flap-typed separator for initial rainwater which makes use of hydraulic action to automatically flip the flap and in turn discard the initial rainwater [5]. Huang Yongqiang et al (2010) designed a set of hydrocyclone separator, which applies the combined action of gravity in cyclone separator and centrifugal force generated by vortex. The device can effectively remove the suspended solids, particulate nitrogen and particulate phosphorus. At the maximum, over 70% of SS can be removed [6]

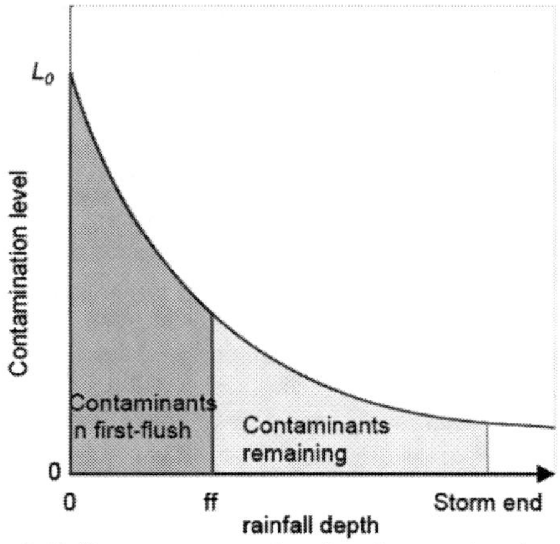

Figure 2. Pollutant concentrations in rainwater that change with rainfall duration.

2.3 Treatment units for harvested rainwater

In China's rural areas, residents are relatively scattered. Subject to the socio-economic and geographical conditions, conventional centralized water-plants cannot be spread in rural areas. Limited to economic conditions, knowledge levels, drinking-water safety awareness and other factors, usually farmers only make a coarse filter before storing the rainwater and adopt the process of precipitation after the storage. They rarely buy advanced devices to further treat rainwater. The so-called "coarse filter before storing rainwater and deposition after the storage" means that: Before rainwater enters into storage facilities, there are trash racks grit chamber and sand filter pool and other facilities set at water inlets to remove impurities like coarse sediments and weeds in rainwater; after the rainwater enters into storage facilities, it will be deposited and disinfected in a simple way by the added ash (potassium carbonate), alum (aluminum sulfate), "pestilence killer", etc.

Currently in rural areas, filtration technology is frequently used to treat drinking water, and ceramic filter is one of such common devices. This device has an advantage that it can be reused after the adhered contaminants are removed by a brush when being blocked, but the size of filter pores is decisive for the removal of contaminants, and such size is subject to the pressure of supplied water. With the development of technology, products of ceramic filter have been widely used in cities where tap water is available. However, as a drawback, ceramic filter does not apply to those rural single-household family without access to water pressure. In addition, the filter has a certain service life, and it has to be frequently replaced if water quality is poor, and that will burden water users with unbearable costs. The sand-filtering technology applied to centralized water plants can be used in rural areas after being downsized. This technology can make use of existing materials with less costs. Although most suspended solids can be removed by it, further chemical sterilization or pasteurized process have to be adopted since bacteria in the effluent are still out of limits. In recent years, more urban water users begin to use small-scale integrated water purification equipment, such equipment adopts ceramic filter, PP cotton, activated carbon, resin fiber as media for filtering, or uses nanofiltration, microfiltration, reverse osmosis and other deep processing techniques. Such device has better effects, but costly and complex to operate, and the charges for water and electricity generated during operation and backwash hinder its application in rural areas. The technology of slow biological filtration has been piloted in rural areas of Hubei, Fujian, Sichuan, Guangxi and other places in China and received recognition by the majority of water users because of its simple operation and maintenance, low cost of water system, and such water can be safely drunk without applying chemical disinfection for the effluents.

2.4 Storage units for harvested rainwater

Rainwater storage system mainly includes water cellars (assembled, bottle-shaped, cylindrical, spherical, tureen-typed, kiln-typed, cup-typed water cellars), ponds, tanks, flood pools, ponds and other water storage facilities. The most widely used rainwater storage facilities are mortar-surface water cellars, concrete pools and other masonry pools. In North China, mortar surface is usually applied for water cellars, and reservoirs are concrete-pouring (brick reservoir tends to crack in winter); while in some rocky areas of southern China, stones are easily accessible, so reservoirs are mainly built by stones. The above rainwater storage facilities are generally built around residential areas near farmers' houses for harvesting rainwater from roofs, patios and yards rainwater to meet the needs for domestic water. The volume of rainwater storage facilities is determined by the probability of water supply considering the water inflow and consumption. In loess regions of northern China, the volume of mortar-surface cellar is generally $30m^3$, and that of concrete pool is generally between 30-$50m^3$; while in rocky areas of southern China, the volume of masonry tank is generally above $50m^3$, but some can be up to about $100m^3$.

3 Protection to avoid pollution of rainwater

The cellars, tanks, trenches and other small projects for rainwater harvesting in China have played an important role in solving the problem of water shortage, but there is not enough attention attached to water quality. In the process of landing and runoff, rainwater runoff is easily polluted by atmospheric deposition, surface-sourced fertilizers and pesticides, solid debris and waste on runoff-collecting surface (garbage, animal manure, etc.) and other pollutants caused by human activities, so the water quality is worrisome. If being drunk without treatment, such rainwater would produce great harm to humans and animals. Pre-filtration system is capable of intercepting some large particles of stones, twigs, leaves, but small particles and bacteria attached there as well as organic matters will penetrate filters and enter into reservoirs. Extensive survey data indicate that, water-borne typhoid, dysentery, hepatitis and diarrhea are still the main intestinal infectious diseases in rural China. If people drink untreated rainwater directly, they could suffer from health risks including bacterial diarrhea (Salmonella and Campylobacter), bacterial pneumonia (Legionella), botulism (Clostridium), Giardia and Cryptosporidium diarrhea.

Pollution of water in reservoirs comes mainly from the processes of harvesting, using and storing water. The main protective measures include: 1) Protective measures against pollution during rainwater harvesting. i). Protection for the source of harvested rainwater. Area of runoff-collecting surface should be protected from the pollution of livestock, manure, garbage and stacking wood. Around the runoff-collecting surface, there should be facilities to prevent rainwater entering from other convergence areas. ii). Cleaning of rainwater harvesting structures. The scope of cleaning involves rainwater-harvesting surfaces and tanks, sink drains, grit chambers and filters. iii). Rainwater discarding devices for initial stage. If economically permitted, rainwater discarding devices for initial stage could be installed. Otherwise, the reservoir inlet should be blocked at the beginning of rainfall. It can be opened for water storage after the rainwater of initial 3-5min is discarded. 2) Precautions against pollution during storage. i). Reservoir protection. Reservoirs should adopt underground or semi-underground structure to reduce water temperature. They can be covered to reduce external contamination, avoid sunlight, and prevent algae from growing within. Bottom sediments should be cleaned regularly to reduce secondary pollution to water by pollutants in sediments. ii). Disinfection by chemical agents. In case of high turbidity during flood season in summer, simple disinfection and sedimentation should be applied for water by adding chemical agents like alum (aluminum sulfate), lime, basic aluminum chloride. 3) Measures against the pollution during getting and drinking water. i). If economically permitted, it would be better to use a hand pump or submersible pump to directly draw water into a storage tank/bucket; otherwise, the traditional way of getting water should be improved, and water should be taken by a fixed bucket to avoid pollution coming from ground, crowd and livestock. For temporary storage tank/bucket in kitchen, it should be covered to keep away insects, rats and mosquitoes. ii). Adopting the way of supplying water by quality degrades applied in cities, roof-harvested rainwater and courtyard/slope rainwater are stored in different reservoirs to avoid cross-contamination. Roof-harvested rainwater is for drinking, while patio/slope rainwater is for livelihood using. iii). Construct two reservoirs or more to alternately use them for water storage, so there would be settlement time for reservoirs to prevent the relatively clear water from secondary disturbance.

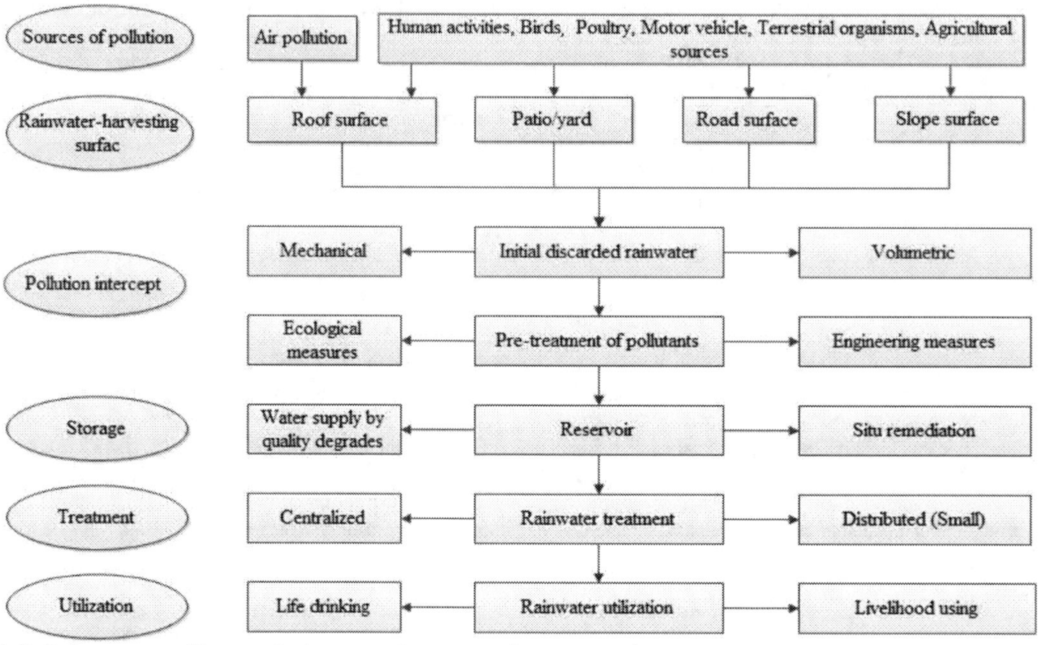

Figure 3. Pollution sources of harvested rainwater and processes of treatment units.

4 Maintenance and management on the system of rainwater harvesting for drinking

The property is owned by individuals as to the single-household decentralized engineering of rainwater harvesting for drinking. Due to the limit of knowledge, farmers are short of specialized skills for system construction, initial pollutant interception, further rainwater treatment and subsequent operation and maintenance. Therefore, the safety of distributed rainwater harvesting for drinking will be subject to the problems of how to technically guide these water users to improve the efficiency of rainwater harvesting system, how to maximize the interception of pollutants in rainwater and how to ensure a stable effect of further rainwater treatment system.

As a way of water supply in rural areas, the small centralized system of rainwater harvesting for drinking is beneficial for the society and public. Domestic and international practice shows that, the mechanism of participatory management runs a successful way to safeguard an effective supply of rural public goods. Under the framework of participatory mechanism, government should change its role to create a favorable environment for participatory management, so that enterprises and water-using associations will assume the responsibility of water supply. Government is responsible for the protection of water sources and public advocacy & training. Enterprises provide management, operation and maintenance according to contracts signed with the water-using communities. Water-using associations are in charge of collecting water bills. In this way, a framework of interaction for sustainable development will be constructed as shown in Figure 4.

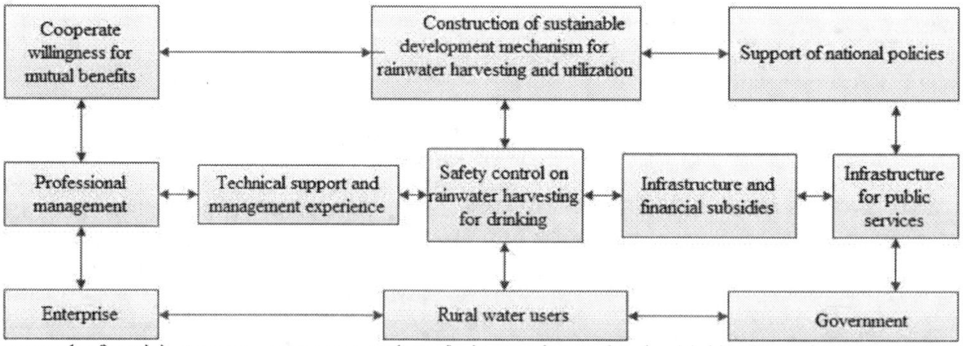

Figure 4. Framework of participatory management on project of rainwater harvesting for drinking in rural areas.

5 Conclusion

With China's social and economic development by leaps and bounds, the contradiction between social-economic growth and the carrying capacity of water resources and water environment becomes so increasingly prominent that it has aroused great attention of the country. In the "Views on Implementing the Most Stringent Water Management System" (National issue [2012] No.3) issued by the State Council, "three red lines" are explicitly put forward to administer the development and utilization of water resources, control the efficiency of water use, and prevent pollution in water functional areas. In current China, there is no new water source for development while existing sources are suffering from external contamination. Worse still, some sources are gradually shrinking and facing the risks of depletion along with increased imbalance between supply and demand. Confronting with problems like water scarcity, increasing water demand and deteriorating water quality, it would be of great significance to consider how to build a water conservation culture, maintain sustainable use of water resources, restore a healthy cycle of natural water, further tap the potential of rainwater resource for utilization, improve the efficiency of rainwater harvesting, and work on the research and development of appropriate technology for rainwater harvesting.

Acknowledgment

We would like to thank the National Science & Technology Pillar Program during the 12th Five-year Plan Period (2012BAJ21B03) for the financial support to this study.

References

1. W. Geiger, Flushing effects in combined sewer systems, Proceeding of the 4th Int.Cont.on Urban Drainage, Lausanne, Switzerland, (1987), 40-46
2. A. Saget, G. Chebbo, J.-L. Bertrand-Krajewski, The first flush in sewer systems, Water Science and Technology, **33**(9), 101-108, (1996)
3. K. Ding, W. Xie, W. Sun, Research on water quality and initial discarded amount of roof rain-off, China Rural Water and Hydropower, **1**:85-88 (2011)
4. Engineering technicalcode for rain utilization in buiding and sub-district, (GB50400-2006), Beijing: China Building Industry Press, 2006.
5. X. Yang, J. Zhang, T. Li, J. Wu, Design of initial rainwater split-flow instrument by reversing board[J]. Journal of Water Resources & Water Engineering, **21**(6):121-124 (2010)
6. Y. Huang, T. Wu, Initial Split-flow/Hydrocyclone/Eco-floating Bed for Treatment of Stormwater Runoff, China Water & Wastewater, **26**(11):1-4 (2010).

Study on Relationship between Seasonal Temperatures and Municipal Wastewater Pollutant Concentration and Removal Rate

Shaoxiong Yuan[1,2], Guangqing Huang[1,2], Wenyin Chen, Zhanghe Chen[3] and Xiaoling Yin[2]

[1]*Guangzhou Institute of Geochemistry, Graduate School of the Chinese Academy of Sciences, 510640 Wushan, Guangzhou, Guangdong, China*
[2]*Guangdong Open Laboratory of Geospatial Information Technology and Application, Guangzhou Institute of Geography, 100 Xianlie Middle Rd, 510070 Yuexiu, Guangzhou, Guangdong, China*
[3]*College of Life Science, South China Normal University, 510631 Tianhe, Guangzhou, Guangdong, China*

Abstract. In this study, the temperatures, pollutant concentrations and other indicators of municipal wastewater influent and effluent were tested for 7 months in 6 constructed wetland microcosms; the hydraulic retention time is 2 days. The results indicated that for both influent and effluent, there was a highly significant negative correlation ($P<0.01$) between the temperature and the pollutant concentrations, there was a significant difference ($P<0.05$) between seasonal temperatures, and the pollutant concentrations in summer and autumn were significantly different from those in winter ($P<0.05$). Furthermore, a regression analysis of pollutant concentration (y) based on changes in water temperature (x) in different seasons was performed. The analysis revealed that the relationship has the form '$y = a - bx + cx^2$', that under certain circumstances, pollutant concentrations can be calculated based on the temperature, and that the concentrations of NH_4-N, Total Phosphorus (TP) and Soluble Reactive Phosphorus (SRP) had a significantly negative correlation with their removal rate ($P < 0.01$). However, seasonal temperature clearly did not have a direct impact on the pollutant concentration, and some studies have indicated that the different manners in which urban residents use water as the temperature changes may be the real reason that the pollutant concentrations of municipal wastewater vary with seasonal temperature. Furthermore, when designing and operating constructed wetlands, the impact of the changes in pollutant concentrations generated by seasonal temperature should be fully considered, dilution and other means should be taken to ensure purification.

1 Introduction

From 1991 until 2011, the dominant Constructed Wetland (CW) research topics included water, nutrients, plants, and flow [1]. In developing countries, most researchers focus on the applications of CWs for removal of various contaminants from wastewater [2]. In most of the research regarding CW treatment of municipal wastewater, the purification effect is the main topic [3]-[7]. Many CW scholars have recognized the influence of temperature on the treatment effect, water temperature affected denitrification rates [8], the percentage reduction in coliforms was lower during winter compared to all other seasons [9], the removal rate of soluble phosphorus is lower than 30% after November [10], the removal rates of DON, NH_4^+-N and TN were significantly temperature-dependent [11], segmented regression analysis identified a breakpoint at 14.2°C for wastewater temperature that caused variations in TN and NO_3-N concentration reduction performances [12]. *Hydrilla verticillata* grew slower in 15°C than in 25°C and 30°C, and displayed lower TN and TP removal rates [13]. Because the treatment wetlands are solar powered ecosystems, Solar radiation will affect the temperature of the water and soils in the wetland, this cause changes in microbial activity and vegetation cycle, which in turn creates changes in water quality improvement [14]. Still not so many scholars have considered the seasonal relationships between temperature and pollutant concentrations, the relationships between pollutant concentrations and the removal of pollutants from municipal wastewater.

In this study, 6 constructed wetlands were constructed, and the temperature, pollutants, and pollutant removal rate were measured. We hoped to determine the relationships among those indexes and develop a more reasonable explanation for some of the problems of municipal wastewater purification due to seasonal changes.

2 Materials and methods

The constructed wetlands are located in Guangzhou City, Guangdong Province, China, South China Normal University Biological Park, within the latitude range of 23°8'29.89"- 23°8'0.94"N and longitude range of 113°21'-113°21'1.11"E in a subtropical monsoon zone. Six adjacent pools, each 3.5 meters long, 3 meters wide and 0.5 meters deep, were constructed. As this study tried to explore the relationship between temperature and the pollutant concentration, only *Eichhornia crassipes* and *Hydrilla verticillata* were planted, the field studies did not involve any vertebrate, endangered or protected species.

Six constructed wetland, designated C1 through C6, were divided into 3 types of microcosms, each with 2 replicates. *Eichhornia crassipes* was planted in C1 and C2, Hydrilla verticillata was planted in C4 and C5, and C3 and C6 were used as control.

The wastewater used in the experiment was an equal mixture of municipal wastewater from the teacher village buildings and tap water. When the sewage and tap water were evenly mixed, 5 water samples were taken from each of the pools, near the 4 corners and the middle of the constructed wetland. The ammonia nitrogen (NH_4-N), Total Phosphorus (TP), Soluble Phosphorus (SRP), Chemical Oxygen Demand (COD), and biochemical oxygen demand for five days (BOD_5) were measured immediately using standard methods [15]; the Air Temperature (AT) and Water Temperature (WT) were measured at the same time (the influent samples), and 2 days later, the temperatures and water pollutants were measured in the same manner (the effluent samples).

An analysis of variance was performed on all data. Duncan's new Multiple Range Test (DMRT) was performed to test for significant differences between seasons. The Pearson correlation coefficient was used to test for correlations among the seasonal temperatures, sewage pollutants, and percentage removal of pollutants. DMRT, the Pearson correlation analyses, and regressions were performed using IBM SPSS; Figures were made by IBM SPSS & Origin 9.

3 Results

3.1 Relationships between influent temperature and pollutants

The influent AT and WT exhibited a significantly negative correlation with the influent NH_4-N, TP, SRP, COD, and BOD_5 (Table 1). From summer to winter, both the AT and WT declined, whereas most of the pollutant concentrations increased (Figure 1). There was a significant difference among seasonal temperatures ($P<0.05$). The average AT in the 3 seasons differed from each other by 6-7°C, and the average WT differed by approximately 7°C. The COD content in autumn was approximately 40% greater than that in summer, whereas the COD content in winter was approximately 100% greater than that in autumn. Other pollutants did not exhibit a significant difference between summer and autumn, but the concentrations of pollutants in summer and autumn were significantly different from that in winter ($P<0.05$). The concentrations of NH_4-N, TP, SRP and BOD_5 in winter were approximately 160%, 104%, 103% and 76% greater, respectively, than the corresponding concentrations in summer and autumn.

The Concentration-Water Temperature Curve (C-WT) was drawn according to the change in concentration of pollutants at different WTs. A regression analysis was performed to obtain the equation and correlation (Figure 2). The general formula used was $y=a-bx+cx^2$, where $a>0$, $b>0$, and $c>0$, 'y' is the concentration of some pollutant (mgL- 1/molL-1), and 'x' is WT (°C). Table 2 showed the equations of the influent concentration of pollutants and water temperature curve. The results indicated that when the WT was less than 30°C, the pollutant concentrations of the water decreased as the temperature increased; when WT was greater than 30°C, the pollutant concentrations of the water increased as the temperature increased.

None of the indicators of water inflow in the three types of treatment exhibited any significant difference from each other.

Table 1. Correlations among Air Temperature (AT), water temperature (WT) and sewage pollutants of influent samples.

		Influent Pollutant Concentrations				
		NH_4-N	**TP**	**SRP**	**COD**	**BOD_5**
AT	Pearson Correlation	-.832**	-.604**	-.495**	-.785**	-.532**
	Sig. (2-tailed)	.000	.000	.001	.000	.000
WT	Pearson Correlation	-.838**	-.627**	-.514**	-.794**	-.519**
	Sig. (2-tailed)	.000	.000	.000	.000	.000

AT: air temperature; WT: water temperature; **: Correlation is significant at the 0.01 level (2-tailed).

Table 2. Equation of the influent sample pollutant Concentrations and Water Temperature (C-WT) curve.

	Influent sample C-WT	R^2
C_{NH4-N}-WT	$y=74.368-4.631x+0.76x^2$	0.863**
C_{TP}-WT	$y=6.624-0.389x+0.006x^2$	0.506**
C_{SRP}-WT	$y=9.649-0.657x+0.012x^2$	0.595**
C_{COD}-WT	$y=403.384-24.187x+0.396x^2$	0.767**
C_{BOD5}-WT	$y=98.187-5.417x+0.089x^2$	0.329

**. Correlation is significant at the 0.01 level. C-WT: concentration-water temperature curve; R^2: coefficient of determination.

3.2 Relationship between effluent temperature and pollutants

The effluent AT and WT still exhibited a significantly negative correlation with the effluent NH_4-N, TP, SRP, COD, and BOD_5 (Table 3). The effluent AT and WT and the concentration of pollutants changed in a manner similar to the water inflow, i.e., the concentration of pollutants generally increased as the temperature declined (Figure 3). The effluent AT, WT and NH_4-N exhibited significant differences among seasons ($P<0.05$). The AT in summer was approximately 5°C greater than that in autumn, the AT in autumn was approximately 13°C greater than that in winter, and the WT values in the three seasons differed from each other by 8-9°C. The NH_4-N increased by approximately 160% on a seasonal basis from summer to winter. The concentrations of other effluent pollutants did not exhibit any significant difference between summer and autumn, but the concentrations of pollutants in summer and autumn were significantly different from those in winter ($P<0.05$). The concentrations of TP, SRP, COD and BOD_5 in winter were approximately 190%, 250%, 120% and 160% greater, respectively, than the corresponding concentrations in summer and autumn.

Table 3. Correlations among AT, WT and pollutant concentrations of effluent samples.

		Effluent Pollutant Concentrations				
		NH_4-N	TP	SRP	COD	BOD_5
AT	Pearson Correlation	-.912**	-.744**	-.686**	-.802**	-.580**
	Sig. (2-tailed)	.000	.000	.000	.000	.000
WT	Pearson Correlation	-.882**	-.647**	-.607**	-.670**	-.384*
	Sig. (2-tailed)	.000	.000	.000	.000	.012

AT: air temperature; WT: water temperature; *: Correlation is significant at the 0.05 level (2-tailed); **:Correlation is significant at the 0.01 level (2-tailed).

The C-WT curve was drawn according to the change of concentration of pollutants at different effluent WTs. A regression analysis was performed to obtain the equation and correlation (Figure 4). Table 4 showed the equations of the effluent concentration of pollutants and water temperature curve. The results are similar to the C-WT curve of water inflow, and the trends in the concentrations of all pollutants reversed at approximately 30°C.

Table 4. Equation of the effluent sample pollutant concentrations and water temperature (C-WT) curve.

	Effluent sample C-WT	R^2
C_{NH4-N}-WT	$y=43.277-2.444x+0.036x^2$	0.857**
C_{TP}-WT	$y=4.618-0.274x+0.004x^2$	0.527**
C_{SRP}-WT	$y=4.271-0.264x+0.004x^2$	0.477
C_{COD}-WT	$y=252.773-14.975x+0.246x^2$	0.627**
C_{BOD5}-WT	$y=69.907-4.659x+0.082x^2$	0.393

C-WT: concentration-water temperature curve; R^2: coefficient of determination; **: Correlation is significant at the 0.01 level.

The effluent AT, WT and NH_4-N did not exhibit any significant difference among the three types of treatment, but the concentrations of other pollutants exhibited a significant difference between Eichhornia crassipes microcosm and control microcosm ($P<0.05$).

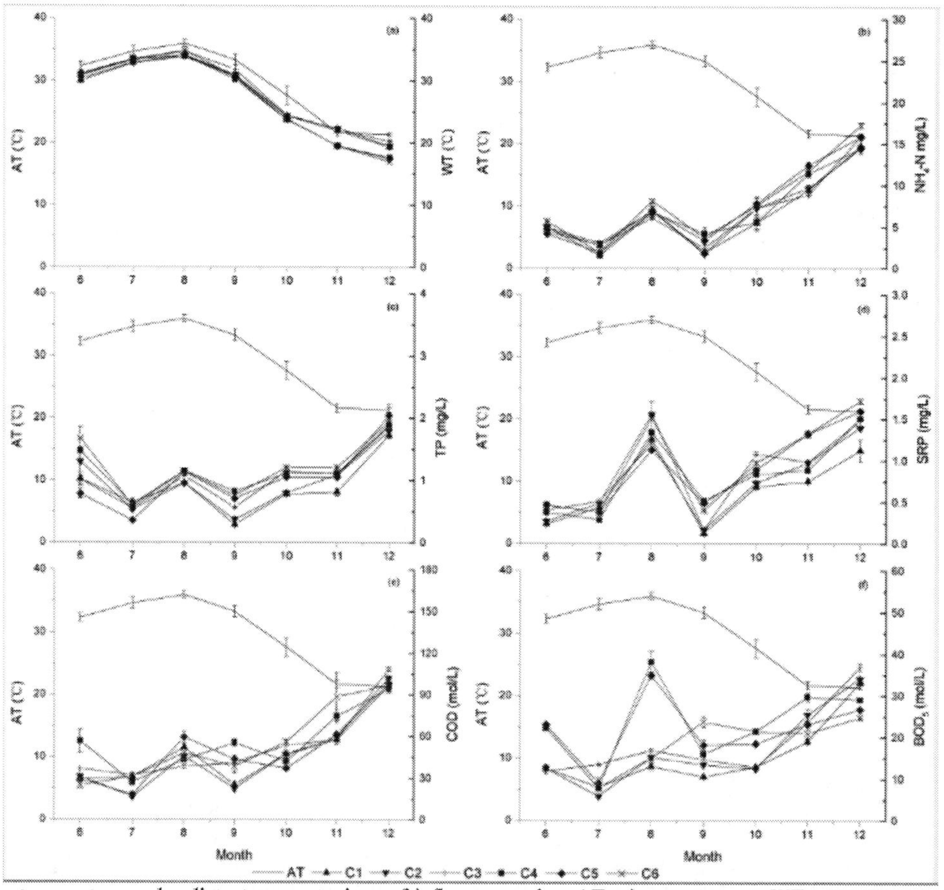

Figure 1. The temperature and pollutant concentrations of influent samples. AT: air temperature; WT: water temperature; C1-C6: Constructed wetland 1-6; (a) Relationship between air temperature and water temperature; (b) Relationship between air temperature and NH_4-N concentrations; (c) Relationship between air temperature and TP concentrations; (d) Relationship between air temperature and SRP concentrations; (e) Relationship between air temperature and COD concentrations; (f) Relationship between air temperature and BOD5 concentrations.

DOI: 10.1051/matecconf/20166814007

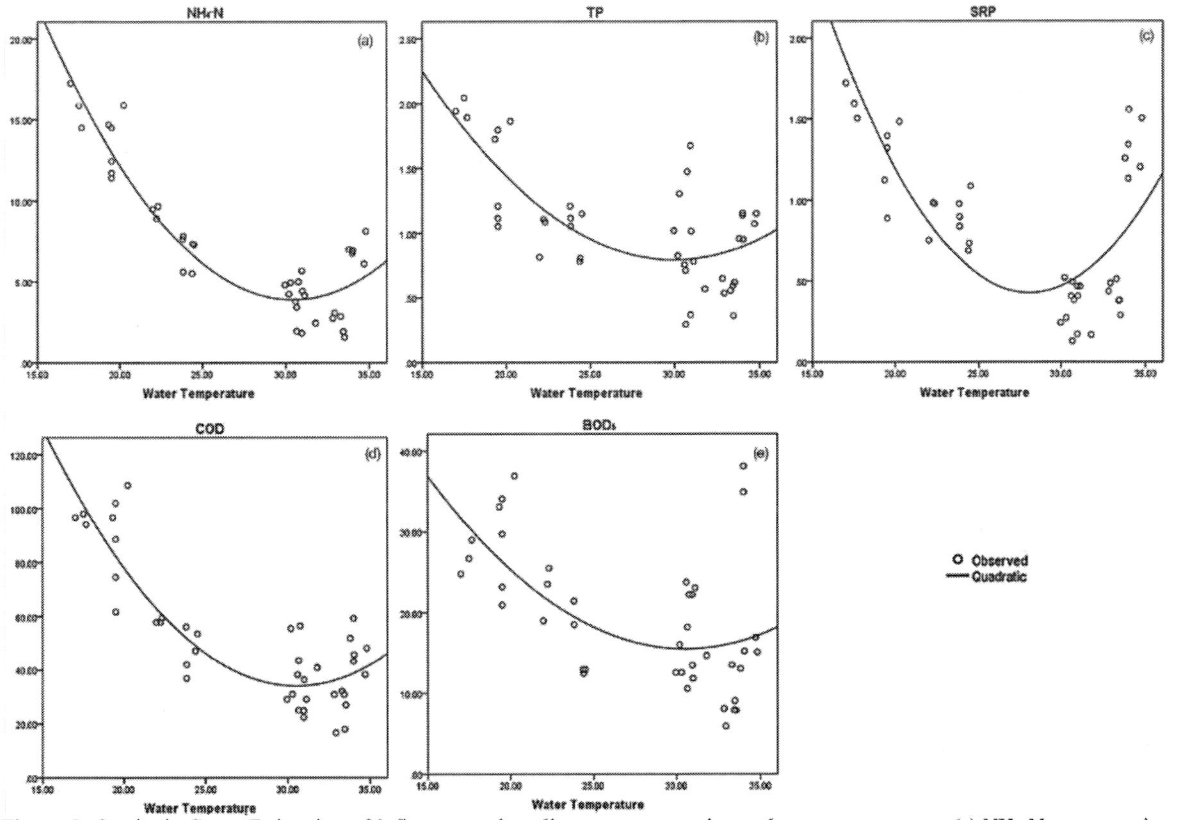

Figure 2. Quadratic Curve Estimation of influent sample pollutants concentration and water temperature. (a) NH_4-N concentration-water temperature curve; (b) TP concentration-water temperature curve; (c) SRP concentration-water temperature curve; (d) COD concentration-water temperature curve; (e) BOD_5 concentration-water temperature curve.

3.3 Relations among average temperature, pollutants and its removal rate

Similar to the findings of many scholars, the average temperature was correlated with the removal rate of some pollutants [11]-[13]. In this experiment, the hydraulic retention time of municipal wastewater is 2 days. The average AT and WT were significantly correlated with the removal rate of the NH_4-N ($P<0.01$), along with the removal rate of the TP and SRP ($P<0.05$), but were not correlated with the removal rates of COD and BOD_5 (Table 5). The average temperatures in the three seasons differed by 7-8°C from each other, exhibiting a significant difference ($P<0.05$). The removal rates of the COD and BOD_5 in the three seasons were not different from each other. The removal rates of NH_4-N in autumn and winter were significantly less than that in summer ($P < 0.05$). The removal rate of the TP and SRP in winter was significantly less than that in summer ($P<0.05$), but there was no significant difference between neighbouring seasons.

NH_4-N, TP and SRP exhibited a significantly negative correlation with their removal rates ($P<0.01$); the concentrations of other pollutants were not significantly correlated with their removal rates.

The average temperatures in the three types of treatment were not significantly different, but Eichhornia crassipes microcosm exhibited significantly greater removal rates of NH_4-N, COD and BOD_5 compared with Hydrilla verticillata microcosm and control microcosm, and it had significantly greater removal rates of TP and SRP compared with control microcosm. There was no significant linear relationship between the average temperature and removal rates of the pollutants.

Table 5. Correlations among the average AT, WT and removal rate of pollutants.

		Removal Rate				
		NH_4-N	TP	SRP	COD	BOD_5
AT	Pearson Correlation	.670**	.349*	.350*	-.046	-.062
	Sig. (2-tailed)	.000	.024	.023	.773	.696
WT	Pearson Correlation	.695**	.367*	.352*	-.080	-.119
	Sig. (2-tailed)	.000	.017	.022	.616	.451

AT: air temperature; WT: water temperature; *: Correlation is significant at the 0.05 level (2-tailed); **: Correlation is significant at the 0.01 level (2-tailed).

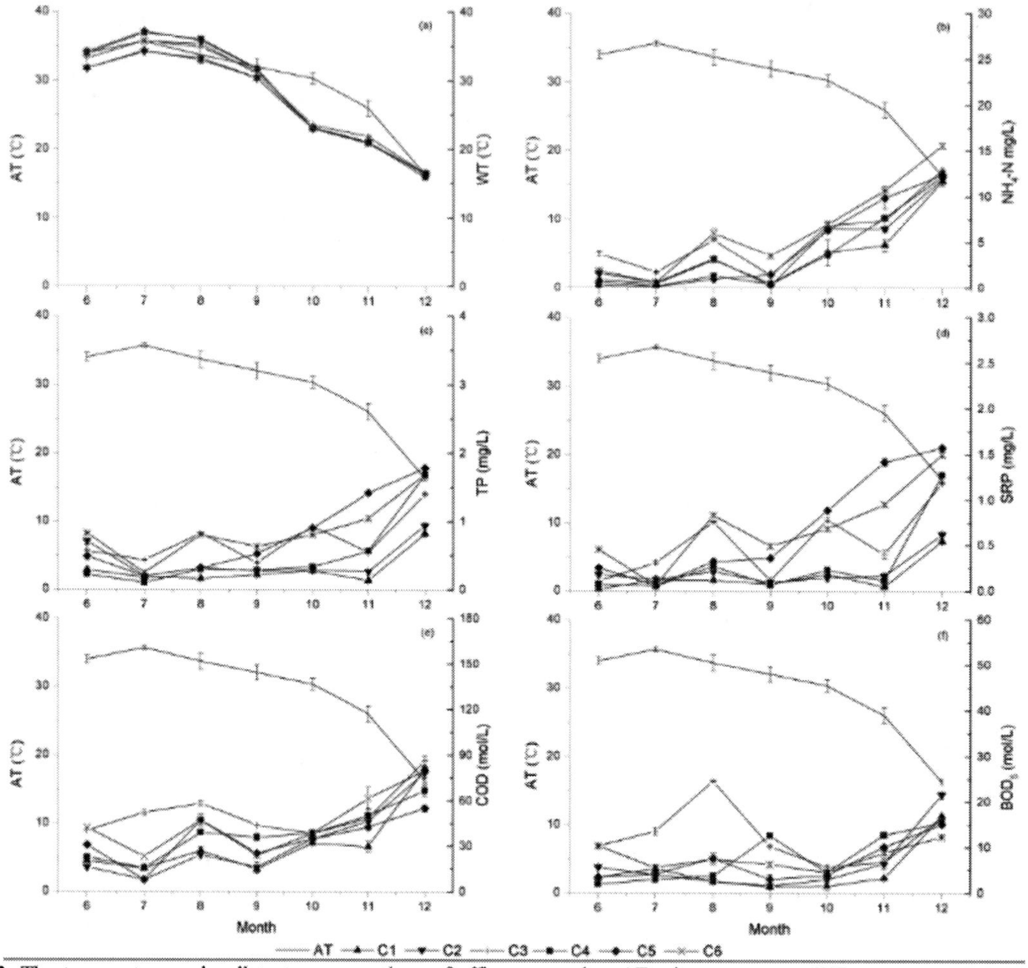

Figure 3. The temperature and pollutant concentrations of effluent samples. AT: air temperature; WT: water temperature; C1-C6: Constructed wetland 1-6. (a) Relationship between air temperature and water temperature. (b) Relationship between air temperature and NH_4-N concentrations. (c) Relationship between air temperature and TP concentrations. (d) Relationship between air temperature and SRP concentrations. (e) Relationship between air temperature and COD concentrations. (f) Relationship between air temperature and BOD_5 concentrations.

4 Discussion

Seasonal temperature influences the purifying effect of wetlands because of the effects of temperature on the growth of plants and microorganisms in wetlands [11], [16]-[18], but it does not have a direct impact on the concentrations of pollutants. Wei and Wan studied the characteristics of ammonia nitrogen pollution in an inner river branch and demonstrated that the concentration of pollutants in municipal wastewater is related to seasonal change [19]. They believed that the seasonal change in the concentration of ammonia nitrogen in the river branch was related mostly to the seasonal change in the quantity of water resources. In this study, municipal wastewater was extracted on days without rainfall to prevent the diluting effect of the rainwater. We observed that water consumption in the Guangzhou is greatly influenced by the season [20]. Along with the increase in seasonal temperature, the water consumption in July and August is relatively high. Compared with the months with low water consumption, high levels of water consumption in July and August may dilute the pollutants. As a result, the experiment demonstrated that the concentration of pollutants decreased as the temperature increased. This is reflects the different manners in which urban residents use water as the temperature changes, which may be the real reason that the concentration of pollutants in municipal wastewater varies with seasonal temperature.

The experimental results indicated that some concentrations of pollutants might have a negative correlation with their removal rates when constructed wetlands are used to treat municipal wastewater. Therefore, in addition to the influence of seasonal temperature on the treatment efficiency, which can be related to the growth of plants in the wetland, we should also pay attention to the influence of seasonal changes in the concentration of sewage on the treatment efficiency.

In this paper, the average temperature was correlated with the removal rate of some pollutants, this's quite similar to the findings of many scholars. Mechanism of the changes in the removal rate of pollutants have been widely discussed, removal of pollutants in CWs mainly base on the plant absorption and microbial decomposition, the temperature will affect the growth of plants and microbial, as temperature varies with the seasons, removal rate of pollutants changes in different seasons [11]-[13], [16]-[18]. Seasonal changes have a greater effect on NH4-N removal rate, mainly due to the removal is based on the action of microorganisms, temperature decrease resulting the microbial growth rate and activity decrease, especially nitrifying bacteria [21]. Kuschk et al [22] showed that, nitrification is subject to temperature limitations in all seasons, denitrification significantly limited by seasonal temperature of less than 15 ℃. Temperature also affected the transformations of glycogen and polyhydroxyalkanoates, and the activities of exopolyphosphatase and polyphosphate kinase activities, so the phosphorus removal rate will be affected by seasonal temperature [23].

In this study, tap water was used directly to dilute wastewater before the experiment for the purpose of preventing the treatment efficiency from being reduced when excessive concentrations of pollutants stop the growth of plants in the wetland. Therefore, several questions remain. For example, when we use large constructed wetlands to treat wastewater, in low-temperature seasons, is it necessary to reduce the damage of high concentrations of pollutants to the plants in the wetlands and improve the treatment efficiency in wetland via dilution? If necessary, what type of water should be used for dilution? Is it feasible to dilute wastewater by adding natural water during operations? An in-depth investigation is needed to eliminate the influence of seasonal changes in the concentrations of pollutants on the efficiency of treating sewage using constructed wetlands.

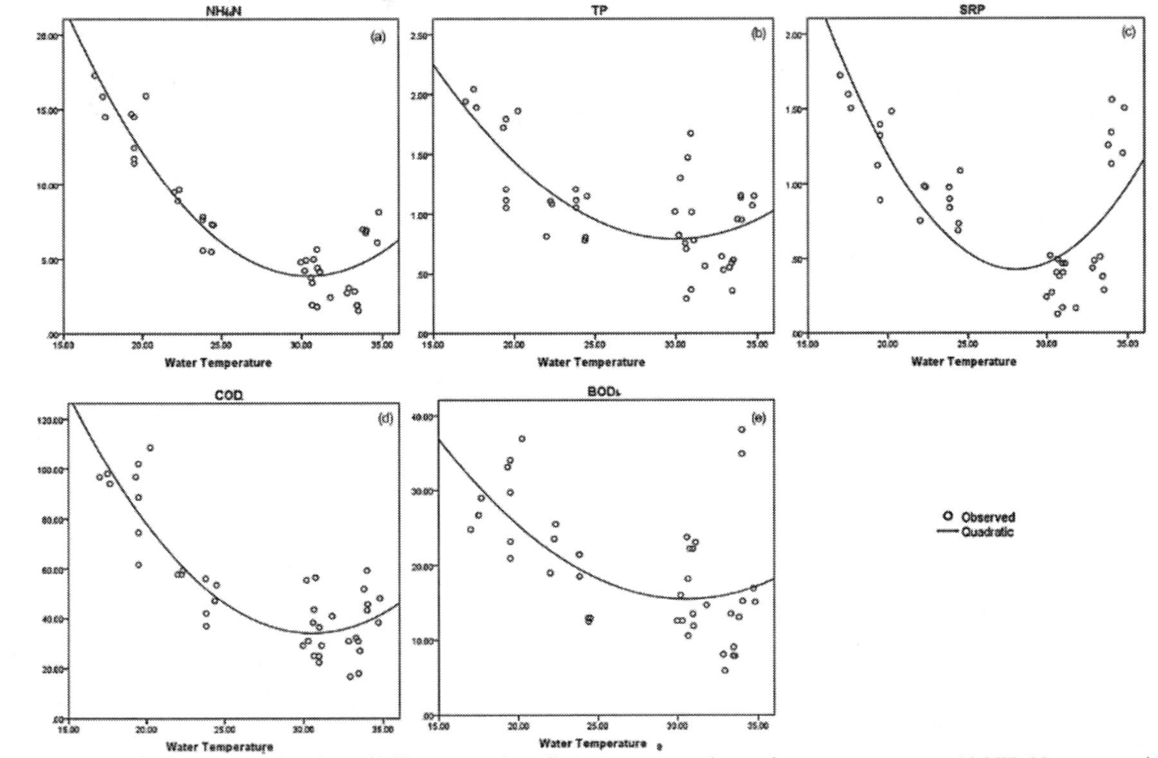

Figure 4. Quadratic Curve Estimation of effluent sample pollutants concentration and water temperature. (a) NH_4-N concentration-water temperature curve; (b) TP concentration-water temperature curve; (c) SRP concentration-water temperature curve; (d) COD concentration-water temperature curve; (e) BOD_5 concentration-water temperature curve.

5 Conclusion

In the experiment, seasonal temperature exhibited a significantly negative correlation with the concentration of pollutants in the water inflow of the municipal wastewater and the concentration of pollutants after the wastewater was treated for 2 days ($P<0.01$). The concentration of wastewater pollutants increased as the temperature declined from summer to winter. In contrast with the small difference in the concentration of pollutants between summer and autumn, the concentration of pollutants in summer and autumn was significantly different from that in winter ($P<0.05$). The average temperature was significantly correlated with the removal rates of TP, SRP and NH_4-N but was not significantly correlated with the removal rates of COD and BOD_5. NH_4-N, TP and SRP exhibited a significantly negative correlation with their removal rates ($P<0.01$), but other pollutants did not exhibit any significant correlations with their removal rates. Regression analysis of pollutant concentration (y) based on changes in water temperature (x) in different seasons was performed, and found that they have the formula '$y = a - bx + cx^2$', under certain circumstances, pollutant concentrations can be calculated by temperature.

Acknowledgment

This work was supported by the National Natural Science Foundation of China under grant numbers 30470346 and 41271029, the Major Science and Technology Production Planning Initiate Project of Guangdong Academy of Sciences, China, under grant number zdccyd201304, the Science and Technology Planning Project of Guangdong Province, China, under grant numbers 2013B060500067, 2013B060500068, 2013B060500072, and 2015B070701020.

References

1. W. Zhi, G. Ji, Constructed Wetlands, 1991-2011: A Review of Research Development, Current Trends, and Future Directions, Science of the Total Environment, **441**(0):19-27, (2012)
2. D. Q. Zhang, K. Jinadasa, R. M. Gersberg, Y. Liu, W. J. Ng, S. K. Tan, Application of Constructed Wetlands for Wastewater Treatment in Developing Countries – A Review of Recent Developments (2000–2013), Journal of Environmental Management, **141**(0):116-131, (2014)
3. C. X. Song, Mechanism Research and Application of Artificial Wetland for City Sewage Treatment, Dalian: Dalian Jiao Tong University, (2004)
4. Y. M. Cao, Study on Domestic Sewage of City Treatment by Artificial Wetlands, Technology of Water Treatment, **34**(9):26-29, (2008)
5. J. Vymazal, L. Kröpfelová, Removal of Nitrogen in Constructed Wetlands with Horizontal Sub-sureface Flow: A Review, Wetlands, **29**(4):1114-1124, (2009)
6. S. X. Yuan, M. Wang, W. Y. Chen, Z. H. Chen, J. M. Hu, H. J. Li, F. P. Lin, Eichhornia Crassipes and Nonwoven Fabric Filter for Eutrophic Water Decontamination, Journal of Freshwater Ecology, **25**(3):487-494, (2010)
7. K. Kelvin, M. Tole, The Efficacy of a Tropical Constructed Wetland for Treating Wastewater During the Dry Season: The Kenyan Experience, Water Air Soil Pollut, **215**(1):137-143, (2011)
8. P. A. Bachand, A. J. Horne, Denitrification in Constructed Free-water Surface Wetlands: Ii. Effects of Vegetation and Temperature, Ecological Engineering, **14**(1):17-32, (1999)
9. A. Zdragas, G. C. Zalidis, V. Takavakoglou, S. Katsavouni, E. T. Anastasiadis, K. Eskridge, A. Panoras, The Effect of Environmental Conditions on the Ability of a Constructed Wetland to Disinfect Municipal Wastewaters, Environmental Management, **29**(4):510-515, (2002)
10. J. G. Zhao, L. N. Liu, Z. H. Chen, Study on removal rate of pollutants and plant growth in subsurface and surface flow constructed wetlands, Ecologic Science, **25**(1):74-77, (2006)
11. J. Huang, W. Cai, Q. Zhong, S. Wang, Influence of Temperature on Micro-environment, Plant Eco-physiology and Nitrogen Removal Effect in Subsurface Flow Constructed Wetland, Ecological Engineering, **60**(0):242-248, (2013)
12. A. Mietto, M. Politeo, S. Breschigliaro, M. Borin, Temperature Influence on Nitrogen Removal in a Hybrid Constructed Wetland System in Northern Italy, Ecological Engineering, **75**(0):291-302, (2015)
13. W. Y. Chen, S. X. Yuan, H. J. Li, F. P. Lin, Q. Li, J. M. Hu, Z. H. Chen, Influences of temperature and light intensity on the growth and pollutant removal by Hydrillaverticillata, Acta Ecologica Sinica, **30**(10):2639-2646, (2010)
14. R. H. Kadlec, Chemical, physical and biological cycles in treatment wetlands, Water Science and Technology, **40**(3):37-44, (1999)
15. The Environmental Bureau of the State, Methods for Water Analysis, 4th ed., Beijing: Environmental Science Press, (2002) 210-285.
16. C. S. Akratos, V. A. Tsihrintzis, Effect of Temperature, HRT, Vegetation and Porous Media on Removal Efficiency of Pilot-scale Horizontal Subsurface Flow Constructed Wetlands, Ecological Engineering, **29**(2):173-191, (2007)
17. Z. Q. Yan, M. Liu, X. Y. Wu, Y. Liu, M. T. Pan, J. H. Wang, L. T. Cai, Q. Y. Li, Z. H. Chen, Influence of Temperature on the Growth and Nutrient Removal of Five Submerged Macrophytes, Ecological Science, **33**(5):839-844, (2014)
18. A. Mietto, M. Politeo, S. Breschigliaro, M. Borin, Temperature Influence on Nitrogen Removal in a Hybrid Constructed Wetland System in Northern Italy, Ecological Engineering, **75**(0):291-302, (2015)
19. J. T. Wei, J. M. Wan, Research on the Characteristic and the Source of Ammonia Nitrogen in Inner-river, Environmental Monitoring in China, **22**(2):62-65, (2006)
20. Y. Wang, X. W. Wu, G. Xu, Analysis of Change Law of Water Consumption in Guangzhou City, Journal of Economics of Water Resources, **25**(2):43-45, 83, (2007)
21. S. H. Wang, W. Wang, Y. Yu, Study on the Operating Characteristics of Subsurface Flow Constructed Wetland, China Water& Wastewater, **19**(4):4-11, (2003)
22. P. Kuschk, A. Wießner, U. Kappelmeyer, et al., Annual Cycle of Nitrogen Removal By a Pilot-scale Subsurface Horizontal Flow in a Constructed Wetland Under Moderate Climate, Water Research, **37**(17):4236-4242, (2003)
23. H. B. Chen, D. B. Wang, X. M. Li, Q. Yang, K. Luo, G. M. Zeng, Temperature influence on biological phosphorus removal induced by aerobic/extended-idle regime, Environ. Sci. Pollut. Res. Int., **21**(9):6034-43, (May 2014)

DOI: 10.1051/matecconf/20166814008

Leveraging Renewable Energies in Distributed Private Clouds

Christian Pape[1,a], Sebastian Rieger[1] and Harald Richter[2]

[1]*Department of Applied Computer Science, Fulda University of Applied Sciences, Fulda, Germany*
[2]*Department of Informatics, Division Computer Engineering, Clausthal University of Technology, Germany*

Abstract. The vast and unstoppable rise of virtualization technologies and the related hardware abstraction in the last years established the foundation for new cloud-based infrastructures and new scalable and elastic services. This new paradigm has already found its way in modern data centers and their infrastructures. A positive side effect of these technologies is the transparency of the execution of workloads in a location-independent and hardware-independent manner. For instance, due to higher utilization of underlying hardware thanks to the consolidation of virtual resources or by moving virtual resources to sites with lower energy prices or more available renewable energy resources, data centers can counteract their economic and ecological downsides resulting from their steadily increasing energy demand. This paper introduces a vector-based algorithm for the placement of virtual machines in distributed private cloud environments. After outlining the basic operation of our approach, we provide a formal definition as well as an outlook for further research.

1 Introduction

Cloud infrastructures and the underlying virtualization technologies are building the foundation of modern data centers. These paradigms also offer potential for reducing the energy consumption of data centers representing most of their ongoing operational costs. In this paper, we seize an opportunity to increase the energy-efficiency of data center operation by introducing a vector-based algorithm to support virtual machine placement decisions. After a brief introduction of related work in Section 2, we outline the basic operation followed by the formal definition of our algorithm. Further, we evaluate our approach and discuss the impact of migration costs in Section 4. Finally, we give an outlook on future work in Section 5.

2 Related work

The placement of virtual resources in modern cloud-based environments is subject of current research. A project called CÆSARA [1], introduced an algorithm for the energy efficient placement of virtual machines by estimating the server's energy consumption based on the running virtual machines' characteristics. The cost of virtual machine migration operations and their energy consumption itemized by the different type of data center equipment, is described in [2]. In [3], a utility is described that allows to distribute virtual machines considering the migration cost and also a basic analysis of migration cost and the impact of live-migration on the running application is outlined. A distributed algorithm for placing virtual machines in large cloud environments is outlined in [4]. The basic approach is that each server knows the CPU load of the other physical servers and each server tries to comply with an upper and lower threshold for the CPU load and initiates the migration of virtual machines when these thresholds are violated. Also, the underlying mathematical challenges like the set-partitioning [5], [6] and bin-packing [7]-[9] problems are still subject of current scientific studies and research. There also exist vector-based approaches for VM placement, but these mostly focus on intra data center placement of VMs on physical machines (PM). In [10] a vector-based methodology to model VM resources and to place VMs on PMs is introduced. Another vector-based constraint programming approach is described in [11]. A routing centric placement algorithm is introduced in [12] and describes a combined optimization approach for data center traffic and VM placement. Furthermore, the communication demand is focused in [13] for VM placement. In contrast to the listed publications, our approach also incorporates the use of renewable energies and their fluctuating characteristics concerning availability and pricing.

3 Inter-DC energy-aware placement of virtual machines

This section outlines the basic idea and functionality of our algorithm. In this context the iterative approach of the algorithm causes its complexity to be relatively low compared to bin-packing or set-partitioning algorithms. Thus, each single iteration will lead to a better overall topology. The algorithm is run continuously, though in

[a] Corresponding author: Christian.Pape@informatik.hs-fulda.de

DOI: 10.1051/matecconf/20166814008

reality a delay or pause between individual runs might be reasonable, especially, if a predefined threshold of migrations across multiple runs has not been reached. This limits the resources and management traffic generated by the execution of the algorithm. The algorithm consecutively considers the optimal placement for each virtual machine with respect to its network flows, corresponding relationships among them and connections to external clients.

3.1 Energy-efficient placement considering renewable energies

The algorithm used in this paper is a vector-based approach to optimize scheduling and placement decisions in private clouds. In this context the dimensions of the used vector space specify the characteristics of virtual resources as entities in distributed data centers.

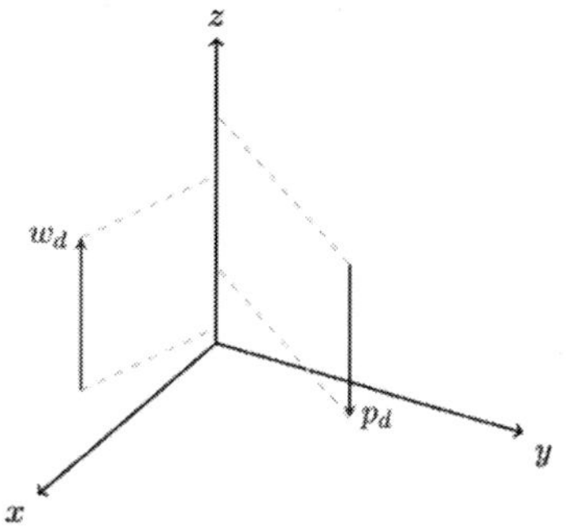

Figure 1. Example of the 3-dimensional vector space V.

To illustrate the basic operation of the algorithm, we will just use an example with three dimensions here. The dimension x and y depict the geographical location of the entity and the dimension z the availability of renewable energy sources for this data center site or, more generally speaking, location.

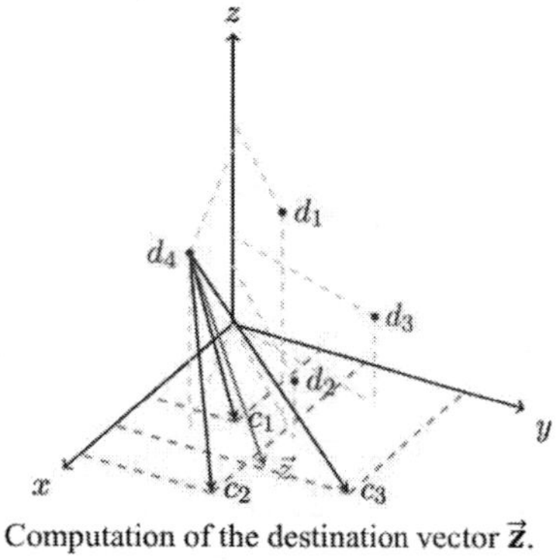

Figure 2. Computation of the destination vector $\vec{\mathbf{z}}$.

The example shown in Figure 1 shows the data center w_d representing a site with available wind energy and p_d depicting a site with available photovoltaic energy. The shown arrows indicate the modification of the positional vector of the data centers w_d and p_d over the time span from summer to autumn. Figure 2 illustrates the computation of a destination vector. Thereby d_1, d_2, d_3 and d_4 represent data centers and c_1, c_2 und c_3 clients with a uniformly distributed volume of communication. The destination vector in this example will be computed for a virtual machine currently executed in data center d_4. In this case the destination data center d_2 is chosen for the migration since this data center location has the shortest distance to the destination vector $\vec{z}$. The algorithm can be adapted to consider the network location instead of a geographical location, e.g., by including weights regarding the latency or other quality of service metrics between the data center and the clients.

3.2 Definitions

Let V be a finite-dimensional vector space over $\mathbb{R}$ with the dimension v and P the set of properties with $\beta_P: P \rightarrow V$ the function to map properties of P to the vector space V. Furthermore, let m be a virtual machine defined by the tuple $m = (i_m, P_m)$ with i_m a unique identifier of the virtual machine and $P_m \subset P$ the set of the associated properties. Moreover, let $D = \{(i_d, c_d \in \mathbb{N}^+, M_d)\}$ be the set of available data centers with i_d the unique identifier of the data center, M_d the set of virtual machines currently executed at this data center and c_d the capacity of this data center, so it holds $|M_d| \leq c_d$. The complete set M_{all} of all virtual machines is defined as follows

$$M_{all} = \bigcup_{(i_d, c_d, M_d) \in D} M_d \tag{1}$$

Of course, each virtual machine $m \in M_{all}$ is restricted to only be executed at one data center for any moment in time. This means

$$\begin{gathered}\forall i_d \in D, j_d \in D \mid i_d = (i_{d_i}, c_{d_i}, M_{d_i}),\\ j_d = \left(i_{d_j}, c_{d_j}, M_{d_j}\right),\\ i_d \neq j_d : M_{d_i} \cap M_{d_j} = \emptyset\end{gathered} \tag{2}$$

Additionally, we define a function $\beta_D: D \rightarrow V$ by

$$\beta_D(x_d \in D) = \vec{v}_{x_d} \tag{3}$$

so that $\vec{v}_{x_d}$ is the positional vector of the data center x_d. One more function $\beta_M: M_{all} \rightarrow V$ is used to map a virtual machine to the positional vector of the data center within it is executed and is defined by

$$\begin{gathered}\beta_M(m \in M_{all}) = \beta_D(d) \mid d \in D,\\ d = (i_d, c_d, M_d),\\ m \in M_d\end{gathered} \tag{4}$$

Also, we define N as the set of network flows by $N = \{(s \in P, d \in P, n \in \mathbb{N}^+)\}$ so that s is a property of the communication source, d a property of a communication destination and n the metric over a defined observation time span. Furthermore the euclidian distance for V is used for the distance function $\delta: V \times V \rightarrow \mathbb{R}$ as follows:

$$\delta((x_1, \dots, x_v) \in V, (y_1, \dots, y_v) \in V) = \| x - y \| = \sqrt{(x_1 - y_1)^2 + \dots + (x_v - y_v)^2} \quad (5)$$

Moreover, we define the following helper functions:

- The function $\sigma: M_{all} \rightarrow D$ maps each virtual machine to the data center it is currently executed in and is defined by:

$$\sigma(m \in M_{all}) = d \mid d \in D, d = (i_d, c_d, M_d), m \in M_d \quad (6)$$

- The function $\theta: M_{all} \rightarrow N$ maps each virtual machine to the set of network flows with matching properties:

$$\theta(m \in M_{all}) = \{(s, d, n) \in N \mid m = (i_m, P_m), s \in P_m\} \quad (7)$$

- The function $\phi: M_{all} \rightarrow \mathbb{N}^+$ defines for a virtual machine the sum of metrics of the associated network flows. It is defined by:

$$\phi(m \in M_{all}) = \sum_{(s,d,n) \in \theta(m)} n \quad (8)$$

- The function $\beta_A: P \rightarrow V$ maps a property to a vector. Respectively, for a property of a virtual machine the positional vector of the associated data center is given by:

$$\beta_A(p \in P) = \begin{cases} \beta_M(m), & \text{if } \exists m \in M_{all}, m = (i_m, P_m): p \in P_m \\ \beta_P(p), & \text{otherwise} \end{cases} \quad (9)$$

3.3 Continuous placement algorithm sequence

For the correct operation of the algorithm at least one data center of the set of data centers D must have spare capacity available. This means:

$$\exists (i_d, c_d, M_d) \in D \mid c_d > |M_d| \quad (10)$$

The sequence of the algorithm is as follows:

1) We define $T = M_{all}$
2) While $T \neq \emptyset$:
   a) We choose an arbitrary $t \in T$ with $t = (i_{m_t}, P_{m_t})$ and define $T = T \backslash t$. Furthermore, we set $t_d = \sigma(t)$ with $t_d = (i_{d_t}, c_{d_t}, M_{d_t})$ and define

$$D_t = \{(i_{d_x}, c_{d_x}, M_{d_x}) \in D \mid c_{d_x} > |M_{d_x}| \vee \exists m \in M_{d_x}: \phi(m) < \phi(t)\} \cup \{t_d\} \quad (11)$$

For the case $D_t = \{t_d\}$ we can examine the next virtual machine and start over with step 2 of the algorithm.

   b) Now we can compute the destination vector

$$\vec{z} = \beta_M(t) + \sum_{(s,d,n) \in \theta(t)} \frac{n}{\phi(t)} (\beta_A(d) - \beta_M(t)) \quad (12)$$

   c) We choose the destination data center $z_d = (i_{d_z}, c_{d_z}, M_{d_z}) \in D_t$ with the shortest distance to the destination vector $\vec{z}$ so that

$$\nexists y_d \in D_t, y_d = (i_{d_y}, c_{d_y}, M_{d_y}), y_d \neq z_d: \delta(\beta_D(y_d), \vec{z}) < \delta(\beta_D(z_d), \vec{z}) \quad (13)$$

For the case $z_d = t_d$ the virtual machine is already executed in the optimal data center, so we can examine the next virtual machine and start over with step 2 of the algorithm.

   d) For the case $|M_{d_z}| = c_{d_z}$, for the given destination data center $z_d = (i_{d_z}, c_{d_z}, M_{d_z}) \in D$, we choose the virtual machine with the lowest communication amount and move this machine to the nearest data center $s_d = (i_{d_s}, c_{d_s}, M_{d_s})$ with available capacity:
      i) We determine a virtual machine $x \in M_{d_z}$ so that $x = (i_{m_x}, P_{m_x})$ which holds

$$\nexists y \in M_{d_z}, x \neq y: \phi(y) < \phi(x) \quad (14)$$

      ii) Now, we identify $s_d = (i_{d_s}, c_{d_s}, M_{d_s}) \in D$ so that $c_{d_s} > |M_{d_s}|$ which holds

$$\nexists y_d \in D, y_d = (i_{d_y}, c_{d_y}, M_{d_y}), c_{d_y} > |M_{d_y}|, y_d \neq s_d: \delta(\beta_D(z_d), \beta_D(y_d)) < \delta(\beta_D(z_d), \beta_D(s_d)) \quad (15)$$

      iii) We define $M_{d_z} = M_{d_z} \backslash x$ and $M_{d_s} = M_{d_s} \cup \{x\}$
   e) Now, $|M_{d_z}| < c_{d_z}$, so we move the virtual machine t to the destination data center z_d. Finally, we alter the sets $M_{d_t} = M_{d_t} \backslash t$ and $M_{d_z} = M_{d_z} \cup \{t\}$.
3) The algorithm has now examined each virtual machine of M_{all}. After a new representative set of network flows is collected, we can start over with step 1 and a new iteration.

As outlined above, the algorithm starts over after all virtual machines were processed. Due to the online/iterative nature of this algorithm new network flows, historical data or even changes in the availability of renewable energy sources will be taken into account. This means, that this approach optimizes the overall topology continuously over time. It is obvious that the temporal resolution of the available data and its rate of change has to be considered when running the algorithm in a real private cloud environment to limit the amount of resources necessary to run the algorithm. Also, oscillations of virtual machines between data centers (e.g., due to similar clients or properties) should be prevented, e.g., using a hysteresis based on former placements of the virtual machine.

4 Evaluation of optimal VM placement and migration costs

The algorithm to optimize the placement of virtual machines described in the previous section is based on the minimization of operational characteristics of the VMs. In the outline of the algorithm described above, energy costs for the data centers hosting the VMs and the distance between the data centers and the clients are minimized. However, regarding the energy efficiency of the approach, the additional costs for the implementation of the algorithm have to be taken into account. While the energy consumption of the algorithm itself can be limited, e.g., by decreasing its resolution using longer intervals, the migrations resulting from the execution of the algorithm lead to additional energy costs. These migrations costs can be divided in direct and indirect costs. Direct migration costs arise from the effort to carry out the migration, i.e. using compute, storage and network resources across multiple data centers and links between them. Indirect costs are formed by the consequences of the migration and impacts on the operational characteristics of the virtual machine. For example, such indirect costs can result from a higher latency for some clients after the migration. The algorithm presented above does not consider possible service level requirements regarding the latency of all clients in favor of its primary goal, to benefit from the lowest energy price. Such constraints could be integrated in the optimization by using them as additional metrics of the virtual machines. This can be described as a the problem to minimize

$$C_{all} = C_{op}(M_{all}) + C_{alg}(M_{all}) + C_{mig}(M_{all}), \quad m \in M_{all}, m = (i_m, P_m), \quad \forall i_p \in P_m | P_{min} \leq i_p \leq P_{max} \tag{16}$$

whereas C_{all} are the overall energy costs to be optimized, $C_{op}(M_{all})$ the operational costs of the data centers based on the the properties and contraints of the virtual machines defined in P, $C_{alg}(M_{all})$ the costs to run the algorithm and the direct (mig_d) and indirect (mig_i) costs for the resulting migration and placement decisions of the algorithm $C_{mig}(M_{all}) = C_{mig_d}(M_{all}) + C_{mig_i}(M_{all})$. The minimization of C_{op} is subject of several projects and related work regarding the energy efficiency and power management of data center infrastructures. As stated above, C_{alg} can be limited by increasing the interval between iterations of the algorithm and its implementation. C_{mig_i} is primarily minimized by the defined constraints. C_{mig_d} is influenced by the migration technique used to transfer the virtual machines' resources between data centers. Based on related work in this area, in [2] we presented results from a simulation to leverage fluctuating renewable energies in northern, southern and central Germany. Also, we implemented an extension for OpenStack to use migrations to enhance the energy efficiency in distributed private clouds. Besides leveraging renewable energies, the testbed also consolidated the virtual machines across different data center sites to lower C_{op}. By integrating the algorithm described in this paper in this testbed and the previously presented simulation studies, also the placement of new virtual machines can be included in the optimization. This also includes the possibility to spawn multiple instances of a service provided by a virtual machine, e.g., to address the constraints regarding the latency between the service and the clients, hence increasing C_{op} in favor of a reduced C_{mig_i} and resulting lower C_{all}. This can solve situations in which the algorithm would migrate a virtual machine to a new site that is too far away from some connected clients and hence would violate predefined service level constraints (as, e.g., implemented by content delivery network and cloud providers).

5 Conclusions and future work

The algorithm described in this paper can be used to evaluate the use of renewable energies to enhance the overall energy efficiency across multiple data centers. Based on our previous research, we are evaluating to use the algorithm in simulations as well as integrate it in an extension for the OpenStack Nova scheduler, that we developed to leverage renewable energy sources and power management in distributed private cloud infrastructures. The scheduling extension can also consider the costs for placing or migrating virtual resources across multiple public or community clouds. Such hybrid cloud environments, can benefit from the energy efficiency and low energy prices in distant, supposedly public, clouds to lower the energy cost for workloads that are safe to be transferred to a third-party cloud provider. To benefit also from short-term fluctuations in the availability of renewable energy sources, e.g., on a daily basis, new live-migration and placement techniques for virtual resources have to be developed. We're currently working on container-based migration and service placement. This lightweight virtualization solution facilitates the transfer of the current state of the virtual resource and the underlying storage. Initial tests have shown that the amount of data that needs to be transferred is far less compared to full-sized virtual machines. However, the network virtualization, especially the data plane performance, and the effort to checkpoint and restore containers under load is still a challenge compared to existing and well-tested virtual machine live-migration techniques.

References

1. D. Versick, D. Tavangarian, in 6. DFN-Forum Kommunikationstechnologien, **31** (2013)
2. K. Spindler, S. Reissmann, R. Trommer, C. Pape, S. Rieger, T. Glotzbach, International Journal On Advances in Internet Technology, **8** (1 & 2), 13–28 (2015)
3. A. Verma, P. Ahuja, A. Neogi, Proceedings of the 9th ACM/IFIP/USENIX International Conference on Middleware, ser. Middleware '08. Springer-Verlag New York, Inc., 243–264 (2008)

4. X. Wang, X. Liu, L. Fan, X. Jia, Mathematical Problems in Engineering, (2013)
5. D. Levine, *A Parallel Genetic Algorithm for the Set Partitioning Problem*, Meta-Heuristics, Boston, MA, Springer US, 23-35 (1996)
6. A. C.-S. Lin, International Journal of Electronic Commerce Studies **4**, 33-46 (2013)
7. A. Beloglazov, R. Buyya, Proceedings of the 2010 10th IEEE/ACM International Conference on Cluster, Cloud and Grid Computing, ser. CCGRID '10. IEEE Computer Society, 577-578 (2010)
8. T. Carli, S. Henriot, J. Cohen, J. Tomasik, *A packing problem approach to energy-aware load distribution in clouds*, arXiv.org (2014)
9. L. Shi, J. Furlong, R. Wang, IEEE Symposium on Computers and Communications (ISCC), 9-15 (2013)
10. M. Mishra, A. Sahoo, IEEE CLOUD, 275–282 (2011)
11. H. N. Van, F. D. Tran, J.-M. Menaud, CIT, 357–362 (2009)
12. J. W. Jiang, T. Lan, S. Ha, M. Chen, M. Chiang, INFOCOM, 2876–2880 (2012)
13. O. Biran, A. Corradi, M. Fanelli, L. Foschini, A. Nus, D. Raz, E. Silvera, CCGRID, 498–506 (2012)

Comparing Administration of Hospital Wastes Using Social Network Analysis

Mustafa Ali[1], Wenping Wang[1] and Nawaz Chaudhry[2]

[1]*Department of Management Science & Engineering, Southeast University, Nanjing, P.R. China*
[2]*College of Earth & Environmental Sciences, University of the Punjab, Lahore, Pakistan*

Abstract. Waste management requires interaction between different players. It is important to map theses interactions to identify key players. This study utilizes social network analysis to compare the organization of waste management activities at two different hospitals. One of these hospitals was a large public hospital whereas the other one was a small private hospital. The structure and characteristics of the waste management network varied at both hospitals. On the whole the network at the larger hospital was institutional and the one at the smaller hospital was social in nature. At the larger hospital communication between any two members on average required three degrees of freedom. Moreover, its waste management network was characterized by the presence of three brokers whose absence could split the network into smaller disparate components. At the smaller hospital the waste management activities suffered from a lack of involvement from the paramedic staff. Thus its waste management network was more compact and communication between most of the network members did not require an intermediary. In both hospitals, a feedback mechanism between the municipal sanitary staff and hospital management was conspicuous by its absence. Hence issues such as an incidence of hazardous items mixed in the general waste could not be communicated to the hospital directly. The key positions identified in this study can be used for behavioral changes related to waste management among the concerned hospital staff.

1 Introduction

Health care activities can result in different kinds of waste. Mismanagement of these wastes can result in environmental and occupational health risks. It is important to analyze healthcare waste management (HWM) activities as this can help us gain a clearer picture of the situation on the ground and identify areas of improvement [1]. Most of the studies in existing literature focus only on the waste generation rates [2], behavior regarding waste management activities [3] or compliance with regulations [4]. Such studies fail to account for the complex interactions between the different people involved in waste management activities. With advancements in data mining softwares it is now possible to map these interactions in a graphical format [5]. This study involves an investigation of the waste management structure at two hospitals in Gujranwala, a major city of Pakistan. Here we employed Social Network Analysis (SNA) [6] as this approach leads to more information than afforded by organizational flow charts. It provides a clear advantage in identifying the influencers and opinion leaders in an institutional setting. Such positions can be exploited for monitoring, evaluation and change, if necessary. Existing models to investigate healthcare waste management structures employ techniques such as system dynamics [7], total quality management [8], life cycle analysis [9], etc. Few studies employing SNA on HWM can be found in the existing literature. Those that do exist, focus on the socio-economic interactions between a wide array of stakeholders [10]. Examining the internal waste management architecture at the hospitals is not their goal or purpose. Hence this study aims to fill a gap by focusing on the internal waste management interactions at the hospitals from a SNA perspective.

2 Methodology

The study was conducted between September 2014 and January 2015 at the two hospitals. One of these (H1) was the largest and the only public hospital in the city. The other hospital (H2) was a small private hospital with 37 beds. The latter represented the majority of the hospitals in the city which were also small and privately owned. The hospital staff involved in waste management activities were interviewed regarding their daily interactions. The composition of the staff was determined after visiting the hospitals and interviewing staff members and hospital administration regarding their daily interactions. Consequently at H1 the relevant staff included members of the hospital waste management team, the hospital sanitary workers, the nurses, the municipal sanitary workers, the scavengers and a

DOI: 10.1051/matecconf/20166814009

commercial firm charged with incineration of the medical wastes. At H2 these staff consisted of the Admin Manager, the Waste Manager, the Matron, ward boys, sanitary staff, the commercial firm and a scavenger. Each of these staff members was asked about her/his communications with staff from the same or other hospitals or clinics, laboratories, municipality, NGOs, scavengers, the local waste management company and the commercial firm. The study was conducted in the form of a questionnaire survey. Most of the sanitary staff were illiterate and hence they were directly interviewed by the first author. The response items included the names of the people, the frequency of interactions, the direction of communication between them and their category. The networks were constructed and analyzed using Gephi 0.8.2 [11].

3 Results & Discussion

Fig. 1 and 2 represent the social network diagrams of waste segregation, collection and transportation activities at H1 and H2 respectively. The size of each node represents its relative influence and the thickness of the links/edges between two nodes represents the frequency of their interactions. For simplicity purposes the networks have been considered to be undirected.

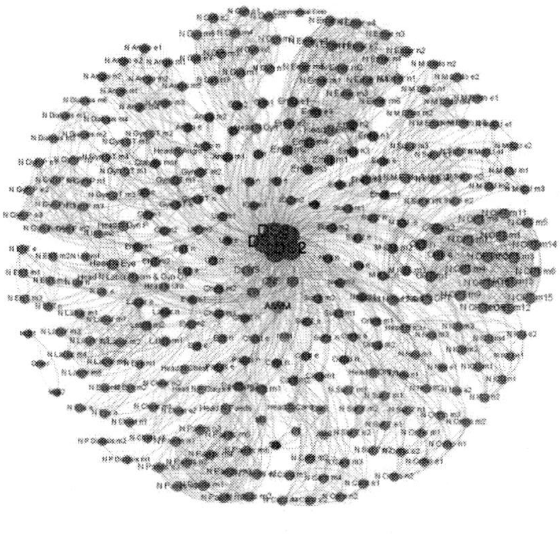

Figure 1. Waste management network at H1.

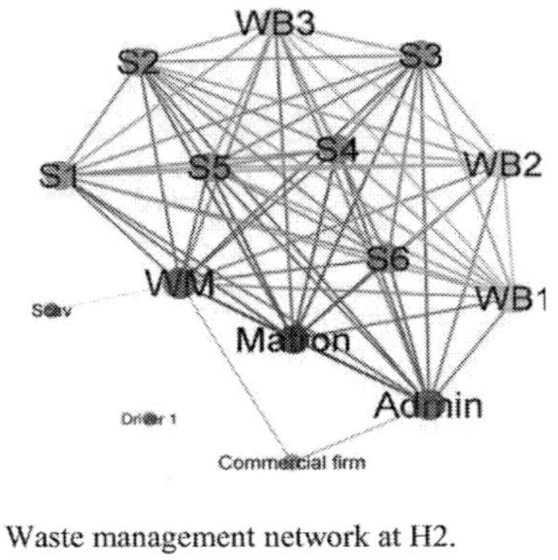

Figure 2. Waste management network at H2.

Both hospitals had constituted a waste management team in compliance with national regulations regarding safe management of hospital wastes. At H1 this team consisted of the medical superintendant, the deputy medical superintendant, a waste management in-charge, an assistant waste management in-charge, a doctor from the operation theatre, the environmental officer and the assistant medical superintendant. Waste segregation in each ward was conducted by the nursing staff whereas its collection was carried out by a sanitary worker specially assigned for this work. The disposal of the medical waste was carried out by a commercial firm. A lack of connections between the commercial firm and the waste management team or the sanitary supervisors at H1 represents a shortcoming in this network. Transportation of the general waste from the wards was carried out by the hospital sanitary staff. This waste was usually thrown around a municipal waste container by the hospital sanitary staff. The municipal sanitary staff, later, loaded it inside the container. The filled container was towed away to an open landfill site by a municipal waste truck, where it was subsequently emptied. The link between the driver of the municipal truck and the municipal sanitary staff represents the second component in the network at H1. No formal link existed between the two components. This highlights a lack of any feedback mechanism between the hospital management and the municipal sanitary staff. Hence any issues such as the presence of hazardous items in the general waste could not be directly communicated to the hospital. At H2 the waste management team consisted of the Admin Manager, the Waste Manager and the Matron. Source segregation of the hospital waste wasn't practiced and hence as opposed to H1, the nurses did not play any role in the waste management activities here. Moreover, there was no clear distinction between the roles of the ward boys and the sanitary workers. The male sanitary workers also acted as the ward boys while the female sanitary staff only handled sanitary operations. In Figure 2 the three ward boys at H2 are represented as WB1, WB2 and WB3 while the five female sanitary staff members are represented as S1 through S5. At H2, the waste was transported by the hospital sanitary staff to two wall mounted drums in a side street. A scavenger collected the wastes from the drums each morning and carried them to the nearest municipal container on his donkey cart. The waste from this container was collected twice a week by a municipal truck, represented by driver 1 in Figure 2. This waste was also thrown in the open land fill site. The municipal waste collector did not have any link with other members of the network at H2 and thus it is represented as an independent component. Table 1. provides the descriptive statistics [12] for the HWM network at H1 and H2.

The Table 1 shows that at H1 every member of the waste management network was connected to almost 13 people on average while at H2 the average degree was 9. The network density shows that the interactivity in the network at H1 was low as only 4.5% of all possible connections were actually realized. In contrast the network at H2 was highly social in nature as 65.7% of all possible connections were in existence. The diameter reveals that, on average, each person at H1 was 4 steps removed from her/his colleagues whereas at H2 the average distance was just 2. The average path length indicates that at H1 on average it would only take 3 ties to go from one member to another in the network. The average length at H2 was only 1. The average clustering coefficient at H1 had a value of 73% as compared to a value of 96.7% at H2. This shows that network members at H2 could relatively easily approach their colleagues. In essence, this information demonstrates that the network at H1 was institutional rather than social in nature. Moreover, a few powerful nodes existed that could control the flow of information in both directions. At H2 the network was highly social in nature as all nodes could easily approach each other. The removal of any node at H2 could not break the network into independent segments. Table 2. ranks the relative influence of the different nodes according to the metrics of degree, centrality and clustering coefficient [13]. Here, the degree of a node represents the number of direct connections it has with rest of the nodes in the network. Similarly, closeness centrality shows how far a person is from the others in a network, betweenness centrality shows the frequency with which a node forms a bridge between two others and Eigen vector centrality measures the centrality of a node relative to the centralities of its neighbors. The clustering coefficient is a ratio between the number of connections a node has and the total number of possible connections it could have. Thus clustering coefficient shows if the node is utilizing the full potential of its position by connecting to as many other nodes as possible.

Table 1. Descriptive statistics for hospital waste management network.

Measure	H1	H2
No. of nodes	296	15
No. of edges	1978	69
Average degree	13.36	9.2
Minimum	1	1
Maximum	94	10
Network density	4.5%	65.7%
Diameter	4	2
Average path length	3.07	1.24
Average clustering coefficient	73.2%	96.7%

The Table 2 shows that at H1 the sanitary supervisors and deputy supervisors had the most influential and central position in the network. On the other hand the commercial firm and the nurses working in the private dialysis ward (Nurse Pvt Dialysis) were the farthest away from others in the network. Among the waste management team only the assistant waste management in-charge (AIWM) had a high degree and betweenness centrality. It was learnt that the position of AIWM was held by a nurse. Thus she represented a vital link between the management, the nurses and the sanitary supervisors. Interestingly the sanitary workers in the operation theatre (Sanitary OPT) also had some of the highest degrees in the network. However they did not have high centralities relative to the others. The sanitary workers in the ICU ward (Sanitary ICU) and members of the waste management team were linked with all the possible connections they could have and thus they had the highest clustering coefficients in the network. At H2 the waste management team members had the highest degrees and network centralities. The scavenger and the commercial firm were the furthest away from other members of the network. Except for the Waste Manager, the Admin Manager and the scavenger every node in the network had a clustering coefficient of 100%. Thus most of the members had realized all the possible connections they could have.

Table 2. Ranking of nodes in hospital waste management network.

Ra nk	Degree	Centrality			Clustering Coefficient
		Closeness	Between ness	Eigen Vector	
(H1)					
1	Supervis or 1	Nurse Pvt Dialysis	Supervis or 1	Dep Supervis or 2	Sanitary ICU
2	Supervis or 2	Nurse Pvt Dialysis	Supervis or 2	Supervis or 1	Sanitary ICU
3	Dep Supervis or 1	Commerc ial Firm	Dep Supervis or 1	Supervis or 2	Sanitary ICU
(H2)					
1	Waste Manager	Scavenger	Waste Manager	Waste Manager	Matron
2	Admin Manager	Commerc ial Firm	Admin Manager	Admin Manager	Ward Boy 1
3	Matron	Matron	-	Matron	Ward Boy 2

4 Conclusions

This study represents the organization of waste management activities at two hospitals. One of these, H1, had a centralized structure Lack of formal connections between the commercial firm and the hospital waste management team was discovered as one of the weaknesses of this network. At H2 the network was highly social and thus not susceptible to division in the absence of a node. The absence of source segregation at H2 displayed a weakness in the network structure as it eliminated the role of the nursing staff in waste management activities. Lack of connections between the municipal sanitary staff and the hospital management at both hospitals also represented a weakness as there was no alternate feedback mechanism between these components. This study can be used by managers for the formation of a more resilient waste management structure which includes all relevant actors. Policy makers can use this study as a stepping stone for further research into the organization of infectious waste management at different healthcare facilities.

Acknowledgment

The writers would like to acknowledge the support extended by Major Program of the National Social Science Foundation of China [12&ZD207].

References

1. B. Sapkota, G. K. Gupta, D. Mainali, Impact of intervention on healthcare waste management practices in a tertiary care governmental hospital of Nepal, Bmc Public Health **14**, (2014)
2. R. Gai, C. Kuroiwa, L. Xu, X. Wang, Y. Zhang, H. Li, *et al.*, Waste Manag Res **27**, 336-42(2009)
3. P. Lakbala, M. Lakbala, Knowledge, attitude and practice of hospital staff management, Waste Manag Res. **31**, 729-32 (2013)
4. M. A. Patwary, W. T. O'Hare, M. H. Sarker, An illicit economy: scavenging and recycling of medical waste, J Environ Manage **92**, 2900-6 (2011)
5. J. N. Hall, S. Moore, A. Shiell, Assessing the congruence between perceived connectivity and network centrality measures specific to pandemic influenza preparedness in Alberta, Bmc Public Health **10**, (2010)
6. L. Chauvel, A. K. Leist, Social Epidemiology, In: Wright JD, editor. International Encyclopedia of the Social & Behavioral Sciences (Second Edition). Oxford: Elsevier, (275-81) (2015)
7. M. Chaerul, M. Tanaka, A. V. Shekdar, A system dynamics approach for hospital waste management. Waste Manag **28**, (442-9) (2008)
8. M. Askarian, P. Heidarpoor, O. Assadian, A total quality management approach to healthcare waste management in Namazi Hospital, Iran. Waste Manag **30**, (2321-6) (2010)
9. S. R. Soares, A. P. Finotti, V. P. da Silva, R. A. Alvarenga, Applications of life cycle assessment and cost analysis in health care waste management, Waste Manag **33**, (175-83) (2013)
10. M. Caniato, T. Tudor, M. Vaccari, Understanding the perceptions, roles and interactions of stakeholder networks managing health-care waste: A case study of the Gaza Strip, Waste Manag **35**, (255-64) (2015)
11. M. Salter-Townshend, A. White, I. Gollini, T.B. Murphy, Review of statistical network analysis: models, algorithms, and software, Statistical Analysis and Data Mining **5**, (243-64) (2012)
12. J. Holt-Lunstad, *Social Integration, Social Networks, and Health*. Wright JD, editor. Oxford: Elsevier; (365-70) (2015)
13. J. Scott, *Social network analysis*: Sage, (2012)

MATEC Web of Conferences **68**, 14010 (2016) DOI: 10.1051/matecconf/20166814010
ICIEA 2016

Statistically Filtering Data for Operational Modal Analysis under Ambient Vibration in Structural Health Monitoring Systems

Liang Wang, Ruixiang Song, Yubin Wu, Wencheng Hu

Beijing Municipal Institute of Labour Protection, 55 Taoranting Road, Xicheng District, Beijing 100054, China

Abstract. Operational modal analysis (OMA) is prevalent in large structure modal identification for that it asks for output measurements only. To guarantee identification accuracy, theoretically, OMA data need to be a random process of Gaussian white noise (GWN). Although numerous OMA applications are found in practice, few have particularly discussed the data distribution and to what extent it would blur the modal judgement. This paper presents a method to sieve segments mostly obeying the GWN distribution out of a recording. With a windowing technique, the data segments are evaluated by the modified Kurtosis value. The process has been demonstrated on the monitoring data of two case study structures: one is a laboratory truss bridge excited by artificial forces, the other is a real cable-stayed bridge subject to environmental loads. The results show that weak randomness data may result in false peaks that would possibly mislead the non-parametric modal identification, such as using the Frequency Domain Decomposition method. To overcome, cares on selecting the optimal segment shall be exercised. The proposed method is verified effective to find the most suitable data for modal identification of structural health monitoring systems.

1 Introduction

Operational modal analysis (OMA) has been extensively applied in modal tests of practical structures due to its convenient output-only merit. In bridge monitoring projects, for example, a well-known example is the WASHMS system which has mounted more than 800 sensors to monitor and identify the operational condition of three landmark bridges in Hong Kong [1]. Also, Shama et al. [2] conducted ambient vibration tests on North Grand Island Bridge in West New York for the FE model validation. Brown john et al. [3] acquired the OMA data of the Humber Bridge in 2008 and studied the modal parameter variation from an old dataset obtained decades ago. The OMA techniques are also safe and suitable for the modal test of historical architecture. Relevant applications include an ancient Cypriot aqueduct built in 1747 [4], a Tunisian temple built in16th century [5], [6], and an Egyptian Al-Sultaniyamirarets built in 1337 [7].Moreover, the rapid increase of the research in theory and applications of this area has motivated the creation of the International Operational Modal Analysis Conference (IOMAC) since 2005.

During an OMA test session the input is natural excitation such as traffic, wind, and other natural loads, and all OMA methods attempt to extract low noise information from the measured output. Popular approaches include frequency domain decomposition (FDD), covariance/data driven stochastic subspace identification (SSI-COV/-DATA), poly-least squares complex frequency domain (p-LSCF), maximum likelihood (ML) identification, and instrumental variable (IV) method etc. [8].

No matter whether a method is in time or frequency domain, one assumption of the OMA technique must be fulfilled: the responses of the underlying structure excited by natural loads are Gaussian White Noise (GWN). In literature, however, many have simply assumed that the data of adequately long time recording is supposed to be a white noise. Few have further investigated the measured data's randomness and its influence to modal identification results. In order to check the distribution characteristics of the output measurement and help analysts select the data segments that most likely have reliable identification results. To meet this need, the paper proposes a statistical index of the modified kurtosis to judge data randomness. The data selection process based on the index is illustrated by the ambient vibration data acquired a laboratory truss bridge (Case-study I) and a practical cable-stayed bridge (Case-study II). In addition, a windowing judgement strategy is adopted to locate the optimal data segment for subsequent analysis. In the study FDD method has been employed to determine the resonance of the structure. The comparative results showed that the data with weak randomness may affect the modal identification judgment, and the proposed method is straightforward to implement. Optimal data segments are filtered out in both case studies.

2 Frequency domain decomposition

FDD is a non-parametric frequency domain method extended from the basic frequency domain (BFD) technique[9] where resonance is subjectively determined by identifying spikes in singular value decomposition spectra. Consider the response $y(t)$ of a time-invariant, light damped structure being decomposed into different modes Φ in modal coordinates $q(t)$ [10]

$$y(t) = \Phi q(t) \tag{1}$$

The correlation between the response $y(t)$ of the current and τ steps afterwards is given by Eq.(2)

$$\begin{aligned} C_{yy}(\tau) &= E\left[y(t+\tau) y(t)^T \right] \\ &= E\left[\Phi q(t+\tau) q(t)^H \Phi^H \right] = \Phi C_{qq}(\tau) \Phi^H \end{aligned} \tag{2}$$

where $(\bullet)^H$ denotes the conjugate transposition of a vector or matrix and E[•] the mathematical expectation. Fourier transformation of Eq.(2) gives

$$G_{yy}(\omega) = \Phi G_{qq}(\omega) \Phi^H \tag{3}$$

where $G_{qq}(\omega)$ is the spectrum matrix of the modal coordinates. $G_{yy}(\omega)$ is the spectrum matrix of the response that can be estimated from the auto-power spectral density and the cross-power spectral density by the Welch's method [11]. Singular value decomposition (SVD) of the measured response spectrum is written as

$$\hat{G}_{yy}(\omega) = VSV^H \tag{4}$$

It is noticed that Eq.(4) has a similar form as Eq.(3). Therefore, the diagonal matrix S is an estimate for modal coordinates $G_{qq}(\omega)$ while the corresponding column in V is an estimate for Φ. If the input is a GWN in time and space, the estimated modal coordinates are uncorrelated, and the columns are orthogonal to each other. Under this context, once the SVD is calculated, mode shapes are estimated at the frequencies of singular vector peaks, and the frequencies associated are natural frequencies. This is because the modes at these peaks are most likely independent and the modal estimation is reliable. However, it is worth noted that such modal estimation would be invalid if the input fails to follow a GWN distribution. Modal identification using the FDD method is illustrated in Fig. 5 and Fig. 8 as examples.

3 Kurtosis-based data selection

Having noticed the importance of the "quality" of an acceleration time-series in terms of statistical distribution, checking the Kurtosis value of the recording is helpful to select the most optimal segment out of standalone measurement. Statistically, Kurtosis measures the peakedness of the probability density function (PDF) of a real-valued random variable, and is the fourth moment of the data or population scaled to the fourth power of its standard deviation [12].

$$\gamma(x|\mu,\sigma) = \frac{E\left[(x-\mu)^4\right]}{\sigma^4} \tag{5}$$

where μ and σ is the mean and the standard deviation of series x, respectively. In theory, Kurtosis of a normal distribution is equal to three. In this paper, instead, a modified Kurtosis described in Eq. (6) is used.

$$\gamma^*(x|\mu,\sigma) = \frac{E\left[(x-\mu)^4\right]}{\sigma^4} - 3 \tag{6}$$

If a dataset is of GWN, the modified Kurtosis should be zero and it is ideal for modal identification. However, since data acquired from the real-world are always subject to environmental unknown errors, even the data recorded for a long period may not render a zero Kurtosis. Practically the analyst has to select the segment with the minimal Kurtosis to represent the structure's current condition. For this reason, the research adopts a windowing strategy to select the most appropriate data segment in terms of its Kurtosis value.

To begin, detrend the raw time-series data and divide it into overlapping segments in adequate length. A minimum segment length, for instance, suggested by SVS [13] is T_{min}=1000/f_{min} where f_{min} is the lowest natural frequency of interest of the underlying structure. In the whole time-series, the segment seg_i overlaps with its neighbouring counterparts by w per cent. An overlapping factor is set to control the interval of moving "segment windows". The total number of segments equals to

$$nseg = \frac{L - segL}{(1-w)segL} \tag{7}$$

where L is the length of raw time-series and $segL$ the length of each segment.

Next, calculate the modified Kurtosis for each segment and plot a Kurtosis time-history graph for all the segments. Since most often a test session is likely to have more than one channel, we use an averaged Kurtosis index to check the randomness. Then the optimal data segment is selected and used for subsequent processes.

4 Case-study I: Laboratory truss bridge

A laboratory steel truss bridge shown in Fig. 1 is a scaled model established to study structural health monitoring systems [14]. OMA tests were conducted and the FDD method was adopted to obtain modal properties. Ambient natural loads were artificially simulated by powerful floor fans blowing in random directions. Due to the limited number of sensors, eight sensor layouts were set to reach all the locations as depicted in Fig. 5 (LVT3~8 as example). For some layouts two sessions were recorded, such as LVT1-1, LVT1-2, ... LVT6-1, LVT6-2.Each session consists of 9 channels. Among the layouts two channels (CH1 and CH2) were fixed while the other is roving. Upon the completion of the test, all layouts were combined to profile the mode shapes of the structure.

Figure 1. Configuration of the laboratory truss bridge model.

In practice, randomness assumption is often valid if a time-series has a small Kurtosis and the data is suitable for post-processing. Fig. 2 shows the result using Kurtosis to check the raw data of the model. It is found that Kurtosis for most layouts is less than 1.0 except for LVT8 which has a relatively high value. Hence,further data selection for this layout is required.

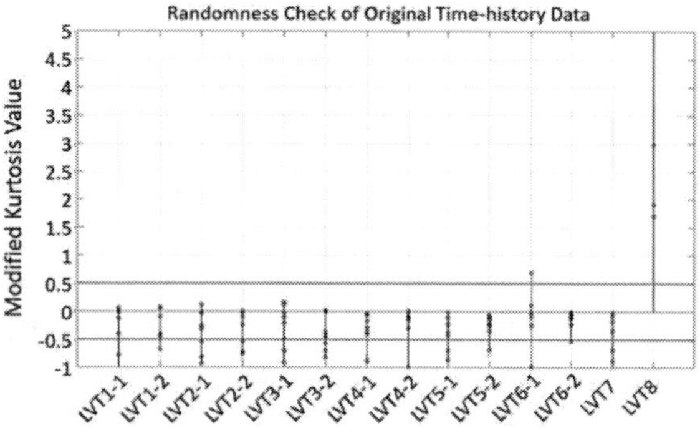

Figure 2. Kurtosis check of acceleration time-history.

4.1 Data selection and the windowing strategy

To apply the OMA technique, adequate length of data of the complete 450 seconds is necessary. In this case, pre-analysis indicated that the lowest natural frequency is 6.5 Hz, which leads to a minimum segment length (*segL*) of 1000/6.5=154 seconds. The sampling frequency is 883.4Hz which is adequate for the interested frequency up to 200Hz [15]. The overlapping ratio *w* was 95%.Take layout LVT3-2 as an example, Fig. 3 highlighted the segment with the minimum Kurtosis value equal to 0.3. Here the plotted Kurtosis was the average of all channels and the blue dotted line located the beginning point of the optimal segment. In the bottom time history, outliers are found in the beginning where the segments' Kurtosis was very high, which means that impacts during a test have substantially affected the randomness.

In addition, Table 1 lists the modified Kurtosis of the complete time-series against the optimal segment for all roving sensor layouts. Among them, for LVT 2-2, 3-2, and LVT8 Kurtosis of the best segment is 20% less than that of the whole data session.

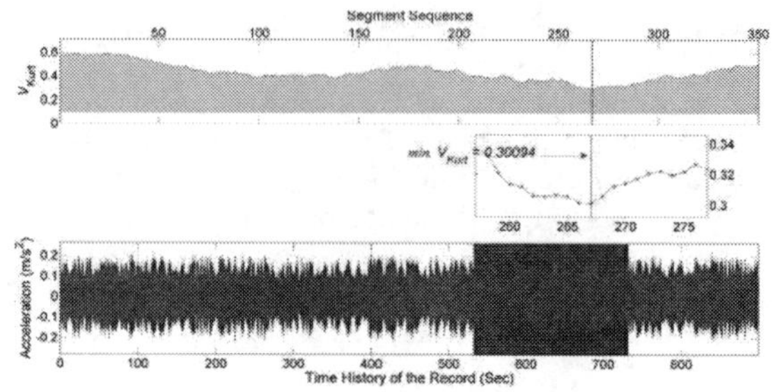

Figure 3. Time-history of data's Kurtosis of layout LVT8.

Table 1. Modified Kurtosis of whole and optimal data segment.

Test Session	Modified Kurtosis		
	Whole	Optimal	Error
LVT1-1	0.448	0.463	3.2%
LVT1-2	0.342	0.301	12.0%
LVT2-1	0.471	0.459	2.5%
LVT2-2	0.386	0.295	23.6%
LVT3-1	0.411	0.390	5.1%
LVT3-2	0.468	0.301	35.7%
LVT4-1	0.339	0.329	2.9%
LVT4-2	0.236	0.220	6.8%
LVT5-1	0.382	0.369	3.4%
LVT5-2	0.300	0.277	7.7%
LVT6-1	0.380	0.324	14.7%
LVT6-2	0.148	0.123	16.9%
LVT7	0.494	0.480	2.8%
LVT8	8.156	0.050	99.4%

In order to further reveal the influence of the data randomness, the SVD spectra used in FDD technique made by the data segment of the lowest and highest Kurtosis of layout LVT8 were compared in Fig. 4.

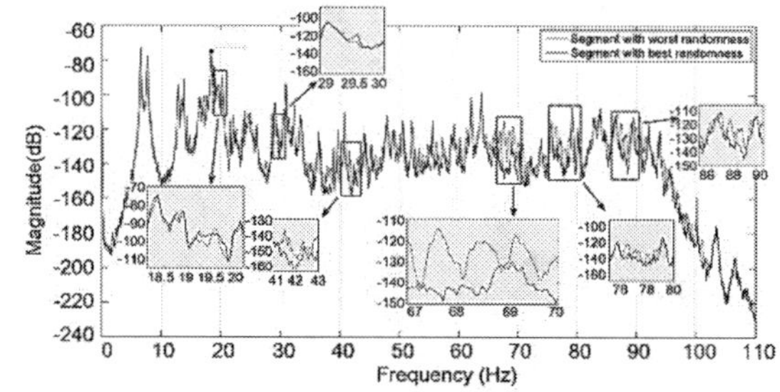

Figure 4. SVD spectra made by best/worst data segments in terms of randomness.

In Fig. 4, obvious discrepancy has been found under 20Hz, and unmatched peaks of the segments with the highest and the lowest Kurtosis existed at many frequency points. Some peaks disappeared or shifted in the spectrum. The comparison indicates that careless selection of the data segment, especially where the unwanted impacts are involved, may lead to inaccurate resonance judgment in the non-parametric OMA methods.

5 Case-study II: Cable-stayed bridge

Another case-study was the application of the selection method on a 25-year cable-stayed bridge in north China.

This SHM-instrumented cable-stayed bridge is regarded as a benchmark structure as shown in Fig. 6 [16]. 16-channel of acceleration are monitored at the sampling frequency of 100 Hz. In the study, a 24-hour dataset of "Damage status 1" were selected to execute the randomness check.

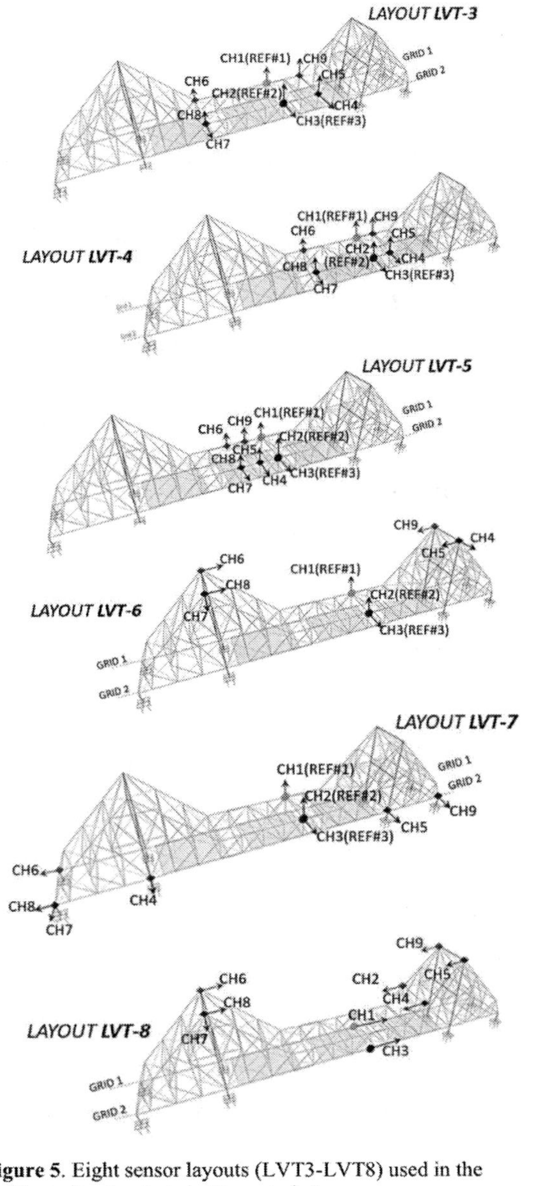

Figure 5. Eight sensor layouts (LVT3-LVT8) used in the ambient vibration tests in Case-study I.

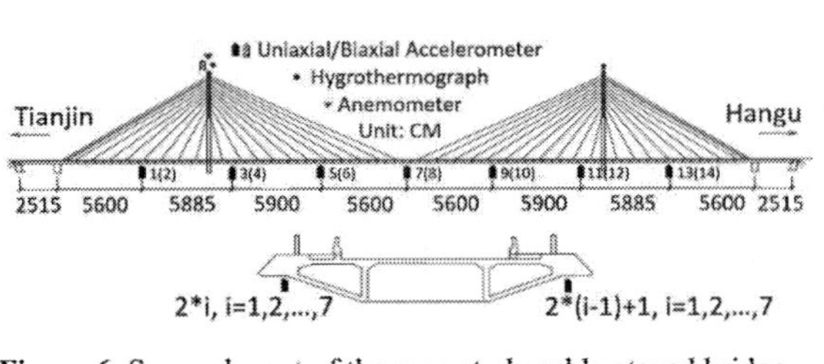

Figure 6. Sensor layout of the case-study cable-stayed bridge [15].

The analytical lowest natural frequency of the structure is 0.201Hz [15], which means the minimum segment length is 1.39hours. Applying the windowing strategy, Fig. 7 indicates the result of the optimal data segment. Since the highest frequency of interest for the flexible structure is 2.5Hz, 20Hz was chosen as the sampling frequency [15].

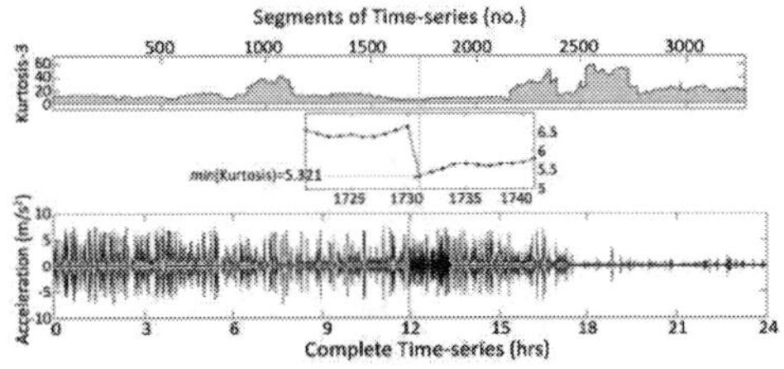

Figure 7. Selected data segment of lowest Kurtosis for "Damage status 1" data.

Applying the windowing strategy, the lowest modified Kurtosis is 5.3 while the maximum is around 60. Interestingly, we found the data from the mid-night have higher Kurtosis values than the segments of other times. This is contrary to the pre-knowledge that the structure in the night-time would more likely to have natural excitation. Besides, the bottom graph in Fig. 7 shows small impacts probably perhaps caused by off-peak heavy vehicles. Even though the magnitudes are small comparing to that in the daytime, these impacts may render low-Kurtosis data and are hence inappropriate for post-processing.SVD results from the worst, best, and complete 24-hour data have been compared together in Fig. 8.

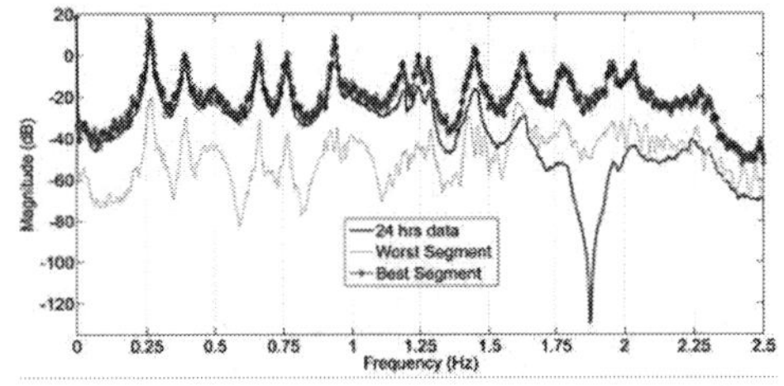

Figure 8. Spectrum of SVD values in FDD using different data segment of the case-study II structure (Data: Damage status 1)

Table 2. Identified natural frequencies by peak picking the spectrum in Figure 8 (Hz).

Mode	Identified natural frequencies (Hz)			Error
	Worst	Whole	Best	
1	0.2661	0.2637	0.2612	1.84%
2	0.3931	0.3906	0.3882	1.25%
3	0.6641	0.6641	0.6641	0
4	0.7666	0.7593	0.7666	0
5	0.9391	0.9424	0.9375	0.17%
6	*	1.194	1.189	n/a
7	*	1.245	1.245	n/a
8	1.289	1.284	1.282	0.54%
9	1.428	1.455	1.448	1.40%
10	1.458	1.633	1.626	11.52%

*Failed to identify as no dominant peaks observed

Again, obvious difference existed beyond 1.5Hz. Further, Table 2 shows the identified first 10 natural frequencies using the worst segment, the best segment, and the whole data session. In the 6th and 7th modes, noisy peaks are found for the dataset with weak randomness and no resonance could be determined. Worse, the frequency of the 10th mode is shifted by 11.52% comparing to the one estimated from the optimal dataset.

6 Conclusion

The assumption of Gaussian white noise for an OMA method is important and has yet been emphasised in literature. By analysing the raw modal test data of two case-study structures, this paper has addressed the impact of the randomness of the measurements in OMA-based identification methods, and has proposed a technique to filter the optimal data segment. One the one hand, the modal identification can be misled particularly at high frequencies if the data have high Kurtosis values. On the other hand, the proposed randomness check with the windowing strategy is able to filter out the data segment which is most likely of a Gaussian white noise distribution. The proposed method is straightforward and can be applied as a processing step prior to modal identification. The influence and the technique of data selection on other modal identification methods will be focused in future work.

Acknowledgement

The presented work is financially aided by the Special Supportive Grant for Returned Overseas Scholars (No. OTP-2014-003) from Beijing Academy of Science and Technology.

References

1. T. Chan, D. P. Thambiratnam, *Structural Health Monitoring in Australia* (2011)
2. A. A. Shama, J. B. Mander, S. S. Chen, A. J. Aref, Ambient vibration and seismic evaluation of a cantilever truss bridge. Eng. Struct. **23**, 10 1281–1292 (2001)
3. J. W. Brownjohn, F. Magalhaes, E. Caetano, A. Cunha, Ambient vibration re-testing and operational modal analysis of the Humber Bridge. Eng. Struct. **32,** 8, 2003–2018 (2010)
4. C. Z. Chrysostomou, T. Demetriou, A. Stassis, Health-monitoring and system-identification of an ancient aqueduct. Smart Struct. &Sys. **4,** 2 183–194 (2008)
5. S. El-Borgi, S. Choura, M. Neifar, H. Smaoui, M. S. Majdoub, D. Cherif, Seismic vulnerability assessment of a historical building in Tunisia. Smart Struct. & Sys. **4,** 2 209–220 (2008)
6. S. El-Borgi, M. Neifar, M. B. Jabeur, D. Cherif, H. Smaoui, Use of copper shape memory alloys in retrofitting historical monuments. Smart Struct. & Sys. **4,** 2 247–259 (2008)
7. A. El-Attar, A. Saleh, I. El-Habbal, A. H. Zaghw, A. Osman, The use of SMA wire dampers to enhance the seismic performance of two historical Islamic minarets, Smart Struct. &Sys. **4,** 2 221–232 (2008)
8. F. Magalhães, Á. Cunha, Explaining operational modal analysis with data from an arch bridge. Mech. Sys. & Sig. Proc. **25**, 5, 1431–1450 2011
9. R. Brincker, L. Zhang, L., P. Andersen, Modal identification from ambient responses using frequency domain decomposition, *Proc. 18th int'l modal anal. Conf.*, **1**, 625–630 San Antonio, TX. (2000)
10. H. Herlufsen, S. Gade, N. Møller, P. Andersen, Identification techniques for operational modal analysis – An overview and practical experiences. *IMAC-XXIV: Conf. & Expo. on Struct. Dyna.* St Louis, MO: Curran Associates, Inc. (2006)
11. P. D. Welch, The use of the fast fourier transform for the estimation of power spectra. IEEE Trans. **15**, 70–73 (1967)
12. D. N. Joanes, C. A. Gill, Comparing measures of sample skewness and kurtosis. J. Royal Statistical Society Series D: The Statistician, **47**, 1 183–189 (1998)
13. SVS (Structural Vibration Solutions A/S). How to set up the first measurement, *help manual of ARTeMIS Extractor.* Aalborg, East Denmark (2011).
14. L. Wang, *Innovative Damage Assessment of Steel Truss Bridges using Modal Strain Energy Correlation*, PhD Thesis, Queensland Univ. of Tech. (2012)
15. China Association for Engineering Construction Standardization, *Design standard for structural health monitoring system (CECS 333: 2012)*, China Architecture & Building Press, (2012)
16. H. Li, S. L. Li, J. P. Ou, H. W. Li, Modal identification of bridges under varying environmental conditions: Temperature and wind effects. Struct. Ctrl & Health Monit. **17**, 5, 495–512 (2010)

DOI: 10.1051/matecconf/20166814011

Study of Discharge Model in South-to-North Water Diversion Middle Route Project Based on Radial Basis Function Neural Network

Yusheng Cao[1], Jianxia Chang[1], Qiang Huang[1], Xiaonan Chen[2] and Yang Chen[2]

[1]*State Key Laboratory Base of Eco-hydraulic Engineering in Arid Area, Xi'an University of Technology, Xi'an 710048, P.R.China*
[2]*Administration of South-to-North Water Diversion Middle Route Project, Beijing 100038, P.R.China*

Abstract. The technology for water dispatch is very complex in South-to-North Water Diversion Middle Route Project, and it is necessary to take advantage of automation system for water delivery. The model for calculating flow rate is important to water dispatch, but traditional method often needs to rectify parameters manually. A model based on radial basis function neural network is established to describe the relationship between water level, gate opening and flux. The model uses the network to simulate the optimal function between water level, gate opening and flux coefficient, and calculates the flow rate by the coefficient. By taking the new method into South-to-North Water Diversion Middle Route Project and comparing the neural network model with traditional methods, the results show that the radial basis function neural network model has higher accuracy and efficiency

1 Introduction

South-to-North Water Diversion Middle Route Project (Middle Route Project) diverts water to Beijing and Tianjin cities, as well as Henan and Hebei provinces starting from Danjiangkou reservoir. The total length of the diversion canal is 1432km, and the main canal is 1277km long and Tianjin channel is 155km. Along the Middle Route Project there are 64 controlling gates and 97 water diversion gates. The joint dispatch for all the gates is in operation to achieve the smooth water supply in line with the water use discharge plan. On Dec.12th, 2014, Middle Route Project started its water supply. By the June 2015, around 700 million cubic meter water has been supplied to the cities and provinces along, effectively alleviated the water shortage in northern cities, especially the capital Beijing. The huge comprehensive benefits of Middle Route Project are being gradually revealed.

At present, domestic and international study of long-distance water dispatch technology have made many research achievements [1]-[8]. However, regulating and operating condition of Middle Route Project is complex and still in the early stages of running through the water, so learning from experience is hard. Many other study of water dispatch specifically for Middle Route Project is improving in practice continuously. Middle Route Project is a linear project which has no regulating reservoirs but many water diversion gates which is difficult to dispatch. For such a complex project, model is used as a basic tool in the dispatch process. Moreover, water level, gate opening and flow rate are the three most important factors of water dispatch. When water demands of the users change, water dispatch model is used to analyze and calculate conveyance flow and gate opening to adjust the flow of water from current state to a new equilibrium state by regulating the controlling gates according to water regime. The traditional method uses hydraulics empirical formula to calculate where the empirical data is adopted for hydraulic parameters. This method does not automatically analyze the data, but needs technical personnel to calibrate hydraulic parameters regularly based on measured hydrological data and the maintenance unit to modify in associated software source code, which is inconvenient and time- consuming.

To solve the problems, by combining Middle Route Project, the paper proposed radial basis function neural network (RBFNN) modeling technique to create and revise calculation mode automatically along with constantly updated measured data. Meanwhile, the measured flow data are generally concentrated in smaller range after water delivery becoming stable, if we use the radial basis function neural network to fit the relation of water level, gate opening and flow rate directly, it may lower the generalization ability of the model might. Taking the above into consideration, the paper uses RBFNN to establish relation of water level, gate opening and flux coefficient, and analyze flow rate through hydrodynamic formula by flux coefficient, which can automatically model and improve the generalization ability.

2 Methodology

2.1 Radial basis function neural network

Artificial Neural Network (ANN) is a mathematical model simulating biological neural network to process information and form a large-scale nonlinear dynamical system interconnecting a large number of neurons. With a strong self-learning ability, ANN has been widely used in forecasting, controlling, modeling and other fields [9]-[11]. The most common neural network models are BP neural network and RBFNN. Although BP neural network algorithm is widely used, it has drawbacks such as easily stacked into minimal value, long training time and slow convergence [12]-[13]. RBFNN is a nonlinear learning algorithm feedforward network of global convergence, with advantages such as high accuracy, fast learning and none local optimality problem [14]-[16].

RBFNN consists of input layer, hidden layer and output layer. Movement from input layer to hidden layer is nonlinear and movement from hidden layer to output layer is linear. Assuming the number of training sample sets is N, for any sample (X,Y), $X=(x_1,x_2,\cdots,x_n)$ is an n-dimensional input vector, $Y=(y_1,y_2,\cdots,y_m)$ is an m-dimensional output vector. The number of neurons of input layer is n, output layer is m, and hidden layer is h. The output function of hidden layer is RBF, generally choice of Gauss function:

$$\varphi(X)=exp\left(-\frac{\|X-C\|^2}{2\delta^2}\right) \tag{1}$$

where X is input vector, C is center of activation function, δ is width, $\|*\|$ is Euclid norm.
Result of output layer can be expressed as:

$$y_j=\sum_{i=1}^{h} w_{ij}\exp\left(-\frac{\|X-C\|}{2\delta^2}\right) \tag{2}$$

where y_j is the j th component, w_{ij} is the connecting weights from i th neuron in hidden layer to j th neuron in output layer, h represents the number of neurons in hidden layer.

It is using samples to train and calculate the coefficients of RBFNN including C, δ and w. Detailed procedures as follows:

(1) K-means clustering method to determine center

Step 1: Select h samples randomly to be the RBF center, where h is the number of neurons in hidden layer.

Step 2: Calculate Euclidean distance between all input samples and centers, and classify all the samples into h types in accordance with the principles of nearest distance to center.

Step 3: Calculate the average mean of each type, and set the sample mean as the new center.

Step 4: Repeat step 2 and step 3 until the center mean of each type doesn't change anymore.

(2) Determine the width

Step 1: Calculate the distance between centers to find the maximum distance;

Step 2: Calculate the width using the following formula:

$$\delta=\frac{C_{max}}{\sqrt{2h}} \tag{3}$$

where C_{max} is the maximum distance.

(3) Determine the connecting weight

When the centers and width of the hidden layer are determined, the connecting weight can be obtained by least square method.

Step 1: The output matrix in hidden layer is calculated through N samples $X_i, i=1,2\cdots,N$ is $H_{N\times h}$.

Step 2: Suppose the connecting weight from each neuron of hidden layer to jth of output layer is $W_j=(w_{1j},w_{2j},\cdots,w_{hj})^T$, the result of N groups of jth neuron in output layer $Y_j'=(y_{1j}',y_{2j}',\cdots,y_{Nj}')^T$ is:

$$Y_j'=H\cdot W_j \tag{4}$$

Step 3: Deviation from the ideal output $Y_j=(y_{1j},y_{2j},\cdots,y_{Nj})^T$ to Y':

$$\varepsilon=\left\|Y_j-Y_j'\right\|^2=\left\|Y_j-H\cdot W_j\right\|^2 \tag{5}$$

Step 4: Using the least square method to make ε minimum, and calculate W_j :

$$W_j=(H^TH)^{-1}H^T\cdot Y_j \tag{6}$$

2.2 Discharge model based on RBFNN

Based on hydraulics flow rate formula:

$$Q=\sigma_s\mu be\sqrt{2gH} \tag{7}$$

where Q is flow rate, m3/s; μ is flow coefficient; σ_s is submergence coefficient; b is gate bottom width, m; e is gate opening, m; H is front water depth, m. Merging μ and σ_s, we can get integrated flow coefficient m:

$$m=\sigma_s\cdot\mu \tag{8}$$

Flow rate formula can be simplified as :

$$Q=mbe\sqrt{2gH} \tag{9}$$

For controlling gates of Middle Route Project, the widths and bottom elevations are known, during the regulation, upstream water level, gate opening and flow rate can be obtained through automation system.

For specific controlling gate, the chamber width is fixed; flow coefficient m is function of gate opening and water level. Its upstream water level, gate opening and flow rate can be obtained during operation, and the integrated flux coefficient is calculated as follows:

$$m = \frac{Q}{be\sqrt{2gH}} \tag{10}$$

The RBFNN model can be established with gate opening, water level and flux coefficient, it consists of 3 layers. The number of neurons in input layer is 3, which corresponds to water level and gate opening respectively. The number of output neurons is 1 which is flux coefficient. The number of neurons in hidden layer can be obtained by trial calculation. Using known sample data, the network training can be achieved and a regression model can be established through RBFNN learning algorithms. Flux coefficient is calculated by RBFNN model, and then flow rate is calculated by function (7).

3 Case study

The 64 controlling gates of Middle Route Project deliver water to each water diversion gate smoothly through analysis of water balance and flow rate along during water dispatch. In this paper, the validity of the new model is verified by the example of Cihe controlling gate of Middle Route Project during a certain period. Cihe inverted siphon is first-grade project, of which main structure is designed as first-grade and affiliated project includes left and right bank diversion dikes of the channel. Cihe inverted siphon full-length is 579m, composed by inlet transition section, inlet overhaul gate chamber, pipe section, outlet controlling gate and overhaul gate chamber and outlet transition section. Cihe controlling gate is located in the inverted siphon outlet. The gate has 3 radial gates, each gate is 6m wide, and bottom elevation is 66.721m. The measured data of the gate of certain period is shown as follows:

Table 1. Data of water regime about Cihe gate

Serial Number	Upstream water depth (m)	Downstream water depth (m)	Gate opening (m)	Flow rate (m^3/s)	Integrated flux coefficient
1	6.662	5.934	1.05	55.92	0.259
2	6.647	5.929	1.05	56.84	0.263
3	6.687	5.919	1.00	54.93	0.267
4	6.707	5.889	0.85	47.07	0.268
5	6.697	5.849	0.65	47.56	0.355
6	6.797	5.779	0.65	37.39	0.277
7	6.807	5.689	0.50	29.80	0.287
8	6.777	5.669	0.50	31.35	0.302
9	6.766	5.664	0.45	25.57	0.274
10	6.727	5.649	0.50	29.84	0.289
11	6.757	5.659	0.50	29.86	0.288
12	6.757	5.639	0.55	32.93	0.289
13	6.762	5.644	0.55	31.29	0.275
14	6.747	5.649	0.55	33.72	0.296
15	6.722	5.644	0.55	30.61	0.269
16	6.715	5.642	0.55	32.48	0.286
17	6.717	5.639	0.55	31.94	0.281
18	6.714	5.631	0.55	31.33	0.276
19	6.709	5.615	0.55	32.44	0.286
20	6.697	5.609	0.55	32.90	0.290
21	6.682	5.614	0.55	32.56	0.287
22	6.677	5.607	0.55	31.76	0.280
23	6.689	5.599	0.55	32.34	0.285
24	6.669	5.594	0.55	31.70	0.280
25	6.660	5.587	0.55	32.06	0.283
26	6.662	5.589	0.55	32.56	0.288
27	6.657	5.589	0.55	32.14	0.284
28	6.657	5.584	0.55	33.04	0.292
29	6.662	5.579	0.55	32.02	0.283
30	6.652	5.579	0.55	32.39	0.287
31	6.647	5.574	0.55	33.41	0.296
32	6.647	5.579	0.55	33.37	0.295
33	6.652	5.554	0.50	29.97	0.292
34	6.652	5.559	0.50	29.56	0.288
35	6.647	5.584	0.50	28.32	0.276
36	6.652	5.594	0.50	28.77	0.280
37	6.647	5.604	0.50	27.72	0.270
38	6.637	5.609	0.50	29.89	0.291
39	6.627	5.604	0.50	27.89	0.272
40	6.607	5.604	0.50	28.23	0.276
41	6.617	5.609	0.50	28.13	0.274
42	6.617	5.599	0.45	25.76	0.279
43	6.612	5.589	0.42	23.21	0.270
44	6.587	5.569	0.42	24.15	0.281
45	6.562	5.559	0.42	23.42	0.273
46	6.547	5.529	0.39	22.59	0.284
47	6.527	5.499	0.39	21.75	0.274
48	6.522	5.484	0.39	23.35	0.294
49	6.522	5.474	0.39	23.40	0.295
50	6.507	5.449	0.39	22.94	0.289
51	6.492	5.434	0.39	22.26	0.281
52	6.479	5.419	0.39	22.02	0.278
53	6.477	5.394	0.36	21.29	0.292
54	6.459	5.379	0.36	20.37	0.279
55	6.455	5.377	0.32	19.13	0.295
56	6.458	5.329	0.28	17.52	0.309
57	6.439	5.349	0.23	17.52	0.377
58	6.437	5.329	0.19	10.94	0.285
59	6.438	5.304	0.19	11.61	0.302
60	6.436	5.302	0.21	12.72	0.300
61	6.445	5.304	0.21	12.48	0.294

In order to facilitate the calculation, the following formula can be used to normalize the input data. Since the value of integrated flux coefficient is already in the range of 0 and 1, there is no need to transform.

$$z' = \frac{z}{z_{max}} \quad (11)$$

where z' is the input data after normalization; z is the input data before normalization; z_{max} represents the maximum data which could be design water level, design water depth or design flow rate. For the controlling gate above, its design water level is 73.88m, design water depth is 7.16m, and design flow rate is 165 m^3/s.

The sample data for verification (not included in table 1) as follows:

Table 2. Data for verification

Serial Number	Upstream water depth (m)	Downstream water depth (m)	Gate opening (m)	Measured flow rate(m^3/s)
1	6.542	5.501	0.08	5.33
2	6.509	5.489	0.14	8.57
3	6.627	5.604	0.50	27.89
4	6.607	5.604	0.50	28.23
5	6.617	5.609	0.50	28.13
6	6.617	5.599	0.45	25.76
7	6.512	6.006	1.34	59.16
8	6.612	5.589	0.42	23.21
9	6.647	5.929	1.05	56.84
10	6.697	5.849	0.65	47.56

The paper compares the results of RBFNN model combining hydraulics (method 1) with the results by the neural network model based on water level, gate opening and measured flow rate(method 2) as well as the results by hydraulics methods. The results of analysis and comparison are shown below:

Table 3. Analysis and Comparison

Serial Number	Measured Flow Rate (m^3/s)	Hydraulics method		Method 1		Method 2	
		Calculated flow rate (m^3/s)	Error (%)	Calculated flow rate (m^3/s)	Error (%)	Calculated flow rate (m^3/s)	Error (%)
1	5.33	4.90	8	4.95	7	11.81	122
2	8.57	8.85	3	8.60	0	13.34	56
3	27.89	36.50	31	29.41	5	26.48	5
4	28.23	36.08	28	29.43	4	26.49	6
5	28.13	36.18	29	29.43	5	26.48	6
6	25.76	32.29	25	26.64	3	24.20	6
7	59.16	74.22	25	72.33	22	83.11	40
8	23.21	29.96	29	24.95	7	22.91	1
9	56.84	69.19	22	58.40	3	61.13	8
10	47.56	43.63	8	37.84	20	34.35	28

Analysis of data in Table 3 as follows:

1. Compare the calculated flow rate with measured flow rate in Table 3 of each method, the average error of method 1 which is the model established in this paper is 8%, while the hydraulics method is 21% and method 2 is 28%. The error of the established model was significantly lower than the other two methods while the accuracy is much higher. The average error is calculated by using the following formula:

$$E = \frac{1}{n}\sum_{i=1}^{n}\frac{|q - q'|}{q} \quad (12)$$

where E is average error, q is measured flow, q' is calculated flow, n is the number of samples for verification.

2. According to the calculation results of the 3rd, 4th, 5th, 6th, 8th row in test sample, average error of method 1 is 5%, method 2 is 5%, while hydraulics method is 28%, which shows the error of neural network calculation is significantly smaller than the hydraulic method. The output value of the samples is very close to the input value of training sample, indicating that the neural network has great data fitting ability. When the selected input data is within the range of training samples, we can get good effect of output reasoning.

3. According to the calculation results of the 1st and 2nd row in test sample, average error of method 1 is 4%, method 2 is 89%, while hydraulics method is 6%, of which the error of method 2 is the largest. The input range of the samples above is largely differentiated from that of the training sample, which indicates that the neural network has lower generalization ability. However, method 2 still achieved good reasoning effect, which indicates the method of fitting integrated flux coefficient by neural network can improve the generalization ability of the model.

4. Calculation error of the 7th row in test sample has the largest calculation error, average error value of hydraulics method is 25%, method 1 is 22%, and method 2 is 40%. The error is caused by the weak generalization ability that the gate opening of the training sample data are generally in the vicinity of 0.4m, while the opening of the data for the verification samples is 1.34m, the training ranges exist a large deviation. However, with the constant accumulation of sample data, ranges become wider, and the results will be more optimized.

5. Neural network model can automatically train and adjust model structure with updated sample data. Compared with the traditional hydraulics method, neural network can avoid modifying constant parameters in the source code, and it is adaptable, convenient, flexible, and saves human and financial resources for hydraulics parameter calibration.

4 Conclusions

With water diversion practice of Middle Route Project, this paper proposed discharge model based on RBFNN that fits the nonlinear function of water level, gate opening and flux coefficient through the neural network. The model uses hydraulics empirical formula to calculate water flow rate which is to play the advantages of high fitting accuracy and automatic modeling, and also improve generalization ability. The example showed that

RBFNN model is of high accuracy, convenience and adaptability, having strong practical value and generalizing significance.

References

1. H. Wan, Study on dispatch and control methods for long distance open-channel hydraulic systems. Hohai University, (2006)
2. A. J. Clemmens, E. Bautista, B. T. Wahlin, et al. Simulation of automatic canal control systems. Journal of Irrigation and Drainage, **131** 324-335, (2005)
3. H. Huang, Z. Liu, Jie Fan, W. Mao, Z. Wu, Study on initial control strategy of water conveyance dispatch of Middle Route Project of South-to-North Water Diversion. Yangtze River **43** 13-18, (2012)
4. D. Rogers, J. Goussard, Canal control algorithms currently in use. Journal of Irrigation and Drainage Engineering **124** 11–15, (1998)
5. T. Wang, K. Yang, New linear numerical hydraulic control model for water diversion projects. South-to-North Water Transfers and Water Science & Technology **3** 52-55, (2005)
6. M. A. Shahrokhnia, M. Javan, Dimensionless stage-discharge relationship in radial gates. Journal of Irrigation and Drainage Engineering **4** 180-184, (2006)
7. C. Zhang, X. Fu, G. Wang, One-dimensional numerical model for unsteady flows in long-route open channel with complex inner boundary conditions. South-to-North Water Transfers and Water Science & Technology **5** 16-20, (2007)
8. G. Corriga, A. Fanni, S. Sanna, et al., A constant volume control method for open channel operation, Journal of Modelling and Simulation **2** 108-112, (1982)
9. X. Zhang, Z. Deng, D. Li, et al., Simulation of hydrological response to land use/cover change in Hanjiang basin, Resources and Environment in the Yangtze Basin **23** 1451-1455, (2014)
10. T. Chen, Z. Dong, B. Jia, et al., Analysis of design flood of dam site of Xiaonanhai Hydropower Station considering flood control operation of upstream cascade reservoir group, Journal of Hohai University (Natural Sciences) **42** 476-480, (2014)
11. Q. Xu, K. Yang, J. Li, et al. Agent-based modeling of land use change and simulation of response on nonpoint source pollution for Erhai Lake watershed. Journal of Hydraulic Engineering **45** 1272-1283, (2014)
12. C. Yu, D. Lyu, Application of BP neural network in classified runoff simulation, South-to-North Water Transfers and Water Science & Technology **12** 109-112, (2014)
13. Y. Li, Q. Zhang, M. Li, et al. Using BP neural networks for water level simulation in Poyang Lake, Resources and Environment in the Yangtze Basin **24** 233-240, (2015)
14. X. Lan, D. Li. Application of radial basis function neural network in forecast network security, Computer Measurement & Control **22** 836-838, (2014)
15. G. Ji, Y. Yao, S. Jiang. Research on monthly rainfall forecast model based on RBF neural network, Computer Technology and Development **23** 186-189, (2013)
16. W. You, X. Ye, S. Tang. Research on forecast of the number of agricultural machinery based on RBF neural network, Journal of Chinese Agricultural Mechanization **34** 38-41, (2013)

Assessment on the Cohesion Pattern of Rail Transit: Integration of Rough Set and Grey Correlation Theory

Zhijie Yang[1], Xiaolong Chen[1] and Shiyun Lin[2]

[1]*Tongji University, School of Economic and Management, Shanghai, China*
[2]*Tongji University, School of Transportation and Engineering, Shanghai, China*

Abstract. Most existing project assessment relies on expert scoring, whose precision can be deteriorated by personal subjectivity. This paper presents an assessment method to compare the advantages and disadvantages of three cohesion patterns between suburban and urban rail transit network, which reduces the influence of subjective score. A modified rough set – grey correlation model is developed as a core of this assessment, based on the integration of rough set theory and grey correlational analysis, where an index system is set up for model calculation. A case study using the network in Ningbo is applied to demonstrate the effectiveness of the method, the results show that the method is more effective using discretely distributed data sensitive to sample size. The consistency of the results in comparison with marginal cost analysis can be a preliminary verification of the model.

1 Information

Scheme comparison is commonly used in rail transit planning; and the results are always sensitive to the assessment method. Currently, most of the assessment are based on expert scoring, in which the subjectivity can always be the problem that will cause uncertainty and inaccuracy of the results. Therefore, it is necessary to develop a data mining method to reduce the error in subjective judgement.

In the rail transit network planning, many cities has been faced with difficulty in selecting cohesion pattern connecting urban and suburban network, which can be categorized as (i) core-through; (ii) periphery-access; and (iii) core-terminate patterns. For instance, deeper extension of suburban line results in better accessibility; simultaneously increase the cost of project. To facilitate the rail transit builder to maximize the benefit, this paper developed an assessment method with a combined rough set-grey correlation model.

2 Literature review

Great efforts have been made to quantify the scheme evaluation. Currently, there are many popular methods that overcomes the drawbacks of the traditional qualitative method [1-3], such as marginal cost analysis, AHP (Analytic Hierarchy Process), fuzzy logic, grey system theory, clustering analysis and neural network. Some studies combine multiple methods to push forward a comprehensive assessment in order to make up the flaws of single method in complicated engineering system design, such as fuzzy AHP [4], fuzzy clustering analysis [5], neural AHP [6], grey fuzzy analysis [7] and so forth. Among these methods, there are still some problems: 1) the travel utility is calculated simply by the product of total travel time and unit value of time, which does not make sense in some cases; 2) the estimation of weight factors rely on the expert's score, which makes it difficult to eliminate the influence from personal subjectivity and uncertainty. To reduce the dependency on expert scoring, some scholars extended the rough set models in project scheme evaluation and decision making assistant, which incorporate features in processing data with small sample size, uncertainty and incompleteness.

The current research of grey rough can be divided into two categories: the first type is grafting of grey system and rough set theory [8-11], which can be regarded as combination of Grey Assessment Model from (i) the attribute reduction and attribute dependence of rough set, (ii) combination of grey clustering method and rough set, (iii) combination of grey relational analysis and rough set, etc., which is more interested in practical application. The second type contains four categories based on the model of grey information system and their attribute analysis and application, integration of grey system theory and rough set model [12-14], etc.

At present, the fusion of grey correlation analysis and rough set mainly focus on attribute-reduction, while scheme ranking is more concerned in practical application. Therefore, this study put forward a modified grey rough set model which can utilize flexible data input to calculate the significance index and assign weight (as an alternative of expert scoring). The model output contains not only the score but also the rank.

3 Index system

An index system is set up as an input for rough set calculation. Since rough set theory can only deal with discrete data, all the attributes must be discretized to fit the model. In addition, a decision table of non-decision attributes is set up for weight calculation, which contains only status variables instead of decision variables.

Considering the condition in China, where local government is in charge of rail construction and operation; and where the cost and benefit is sensitive to the decision for development of public transit system, we select the attributes in three aspects: the passenger utilization (mainly reflect the difference of travel time and the number of passengers), construction and operation (mainly reflect the difference of construction costs and operating costs), and the level of service (ultimately reflected in the number of passengers).

Table 1. Assessment index system.

Assessment **view**	assessment object	assessment index	code	Introuctions
Passenger utilization	Entire Network	The total in-vehicle time	C_1	Measure the difference of total travel time caused by different network turnover volume in different pattern
	Entire Network	The total transfer time	C_2	Measure the difference of total transfer time caused by different network transfer volume in different pattern
	Entire Network	The total waiting time	C_3	Measure the difference of total waiting time caused by different operation in different pattern
construction and operation	Entire Network	Construction cost of cohesion program	C_4	Measure the difference of construction cost by different network scale in different pattern
	Suburb rail network	Construction cost of core-outside line	C_5	Measure the difference of construction cost caused by different vehicle marshalling in different pattern
	Suburb rail network	Operation cost of suburb rail transit	C_6	Measure the difference of construction cost caused by different vehicle marshalling in different pattern

Definition of assessment index:

3.1 The total in-vehicle time

Unit: 10^4 min;
Calculation Method:

$$C_1 = T_{yc} + T_{qc} \tag{1}$$

where: T_{yc} represent the in-vehicle time of suburb rail network; T_{qc} represent the in-vehicle time of urban rail network;

Note: The in-vehicle time can be obtained according to forecast

3.2 The total transfer time

Unit: 10^4 min;
Calculation Method:

$$C_2 = (Q_{yh} + Q_{qh})T_h \tag{2}$$

where: Q_{yh} represent the transfer volume of suburb rail network;Q_{qh} represent the transfer volume of urban rail network;T_h represent the average transfer time;

3.3 The total waiting time

Unit: 10^4 min;
Calculation Method:

$$C_3 = Q_y B_y + Q_q B_q \tag{3}$$

where: Q_y represent the passenger volume of suburb rail network;B_y represent the average waiting time of suburb rail network;Q_q represent the passenger volume of urban rail network; B_q represent the average waiting time of urban rail network;

To Illustrate: if there is only one operation cross, we can reduce the waiting time by using short-marshalling train or improving the departure frequency in order to improve the service level; if the passenger flow section is non-uniform and the line is long, then multiple operation cross is needed, with the departure frequency of large cross restricted and increase of passengers' waiting time.

3.4 Construction cost of cohesion program

Unit: 10^4 yuan;
Calculation Method:

$$C_4 = D_{j1}C_{j1} + D_{j2}C_{j2} + D_{j3}C_{j3} \tag{4}$$

where: D_{j1} represent the length of underground line; C_{j1} represent the average construction cost of underground line;D_{j2} represent the length of over-ground line;C_{j2} represent the average construction cost of over-ground line;D_{j3} represent the length of elevated line;C_{j3} represent the average construction cost of elevated line;

3.5 Construction cost of core-outside line

Unit: 10^4 yuan;
Calculation Method:

$$C_5 = C_o + \Delta C \tag{5}$$

where: C_o represent construction cost of suburb line without cohesion scheme;ΔC represent construction cost increment bring by cohesion scheme;

$$\Delta C = N_e \Delta L C' + \Delta C_c \quad (6)$$

where: N_e represent the number of stations affected by different vehicles marshalling; ΔL represent the length increment of the station; C' represent unit mileage cost of station; ΔC_c represent vehicle purchase cost increment;

To illustrate: In different cohesion pattern, the high passenger flow sections of suburb rail are different, so vehicle types and vehicle marshalling are different. If the vehicle type selection is constant, the length of the station will be different, further to affect the construction cost of the station. At the same time, the total number of train may be different, further to affect the vehicle purchase costs.

3.6 Operation cost of suburb rail transit

Unit: 10^4 yuan;

Calculation Method:

$$C_6 = D_y C_y \quad (7)$$

where: D_y represent operation length of suburb rail, C_y represent vehicle kilometers operating cost;

To illustrate: Operation costs of rail transit are composed of running cost, management cost, financial expense and Non-operating expense. Because of the fixed scale and operation mode of urban network, the index can be calculated by the product of operating cost per vehicle kilometer and the operating mileage.

4 Model

4.1 Assumption

Model assumptions are as follows:

1) The urban rail network planning of the city is stable in the assessment year, where different cohesion pattern have no impact for the urban rail network structure and the operation;

2) The suburb rail network planning in the city's downtown area is the same for each cohesion patterns, the change brought about by different cohesion patterns change the urban section of the program, which affect the cost of the total construction and the operation;

3) The city has the economic strength and engineer condition to achieve any pattern of the three, while each cohesion pattern should be able to achieve the desired goal of traffic, and the capacity should be adapt to the traffic demand;

4) All cohesion programs can achieve the planning in the assessment year, the index value of the assessment system can be collected and accurate in the assessment system.

4.2 Weight definition based-on rough set

The concepts of rough set in the paper:

Definition 1: knowledge base

The pair *K= (U, R)* is called a knowledge base, where U is the universe of objects, R is a set of attributes. Different forms of U can be classified according to different R. In this way, the knowledge base expresses the basic classification methods of one or a group of intelligent organizations.

Definition 2: information table

The expression of knowledge is commonly in form of information table or for information system, which can be expressed as the quaternion orderly group $s = (U, a, V, f)$, where: U represent the nonempty finite set of a group of objects, called discourse domain; A represent the nonempty finite set of the attributes; V is the set of attribute range; f is an information function, $f = U \times A \to V$, for each $a \in A, x \in U$ there is $f(x,a) \in V_a$.

Definition 3: decision table

Set $A = (C \cup D)$, C is the set of conditional attribute, D is the set of decision attribute, and $C \cap D = \emptyset$, the knowledge expression system with different conditional attribute and different decision attribute is called decision table.

Definition 4: indiscernibility relation

Set an arbitrary subset $P \subseteq A$, if $\{(x,y)|f(x,a) = f(y,a), x \neq y, a \in P\}$ exist, called indiscernibility relation, denote by IND(P).

Definition 5: upper and lower approximation

Set X a nonempty subset of $U, P \subseteq A$ and $P \neq \emptyset$, upper approximation is $\overline{P}(X)$ and lower approximation is $\underline{P}(X)$, $\overline{P}(X) = \{x \in U | [x]_{IND(P)} \cap X \neq \emptyset\}$; $\underline{P}(X) = \{x \in U | [x]_{IND(P)} \subseteq X\}$ .Where, $\underline{P}(X)$ is called the positive region of X according to A, denote by $POS(X)$.

Definition 6: rough dependent degree

For the knowledge base $K = (U, P)$, P is a set of equivalence relations in U, and $R, Q \subseteq P$; when $IND(R) \subseteq IND(Q)$, called Knowledge Q depends on knowledge R, denote by $k = \gamma_R(Q) = \frac{|POS_R(Q)|}{|U|} = \frac{|\cup_{X \in U/Q} \underline{P}(X)|}{|U|}$, where $|U|$ is the number of set (i.e., the total number of set elements) base, called Knowledge Q K $(0 < K < 1)$ dependence on knowledge R.

Definition 7: discretization method

Because the rough set can only deal with the discrete data, in view of the decision table of this paper has no decision attribute (no-decision system is a set of attributes without decision attributes), use isometric discretization method:

Step 1: Calculate the Interval length of attribute

$$\overline{z} = (z_{imax} - z_{imin})/n \quad (8)$$

where z_{imax} is the maximum of attribute i, z_{imin} is the minimum of attribute i; n is the number of interval.

Step 2: Determine the interval range

$$[z_{imin}, z_{imin} + \overline{z}], [z_{imin} + \overline{z}, z_{imin} + 2\overline{z}], \ldots\ldots, [z_{imin} + (n-1)\overline{z}, z_{imax}] \quad (9)$$

Step 3: Calculate attribute value

Each attribute has a total of N intervals, for an attribute value Z, if it is located in the range i, its value is i.

Based on the theory above, the paper further study the determination method of weight: remove some of the attribute from the table, and then investigate the change of the classification out of the attribute removed. If the change of the corresponding classification is large, the intensity of the attribute is high. The importance of an index can be obtained by calculating the dependence degree.

$$k_C(C') = \gamma_{C'}(C) - \gamma_{C-C'}(C) = 1 - \gamma_{C-C'}(C) \quad (10)$$

Then the weight of each condition attribute index can be determined by the following formula

$$w_i = {k_i}/{\sum_1^n k_i} \quad (11)$$

4.3 Ranking based on grey correlation

The ranking process is applied using grey correlation theory. The correlation coefficients are the critical intermediate variables that clarify the system's structure and relationship based on the combination of intentions, opinions and demands.

The grey correlation degree is $\gamma(x_0, x_i)$, or γ_{0i}. correlation coefficient of point k is $\gamma(x_0(k), x_i(k))$, or $\gamma_{0i}(k)$. The grey correlation is calculated as follows:

Step 1: calculate difference sequence

$$\Delta i(k) = |x_0'(k) - x_i'(k)| \quad (12)$$

where: $\Delta i = (\Delta i(1), \Delta i(2), \cdots, \Delta i(k)), i = 0, 1, \cdots, m$

Step 2: calculate two stage maximum difference and minimum difference

$$M = max_i\, max_k\, \Delta i(k) \quad (13)$$

$$N = min_i\, min_k\, \Delta i(k) \quad (14)$$

Step 3: calculate correlation coefficient

$$\gamma_{0i}(K) = \frac{m + \xi M}{\Delta i(k) + \xi M}, \quad k = 1, 2, \cdots, n, i = 1, 2, \cdots, m \quad (15)$$

where: ξ is resolution ratio, $\xi \in (0, 1)$.

Step 4: calculate correlation degree

$$\gamma_i = \frac{1}{n}\sum_{k=1}^{n} \gamma_{0i}(k), i = 1, 2, \cdots, m \quad (16)$$

Step 5: rank correlation order

In order to evaluate the correlation degree of the subsequence to the sequence, the correlation degree is arranged according to the magnitude.

4.4 Solving Flow Chart

To solve the model, the solution approach can be divided into five steps.

Step 1: Establish the original index matrix

Assume m cohesion schemes, select n index for assessment, establish matrix as follow:

$$U = (U_{ik})_{m \times n} = \begin{bmatrix} u_{11} & \cdots & u_{1n} \\ \vdots & \ddots & \vdots \\ u_{m1} & \cdots & u_{mn} \end{bmatrix} \quad (17)$$

where: i means the NO of the scheme, i=1, 2, …, m; k means the NO of the index , k=1, 2, …, n; U_{ik} means the value of the index k in scheme I;

Step 2: obtain the decision table of assessment index system

Select the maximum value of each index in the column as the reference array:

$$U_0 = (u_{01}, u_{02}, \dots, u_{0k}), k = 1, 2, \dots, n \quad (18)$$

Compare the original matrix with the reference array, carry out non-dimensional treatment, obtain multi-scheme assessment matrix to form a decision table.

$$A = \frac{U}{U_0} = \begin{bmatrix} a_{11} & \cdots & a_{1n} \\ \vdots & \ddots & \vdots \\ a_{m1} & \cdots & a_{mn} \end{bmatrix} \quad (19)$$

Step 3: discrete the decision table and calculate the weights

Use the discretization method mentioned above to dispose the data of the decision table. Use the weight determination method according to the rough set, to calculate the weight of each index w_k.

Step 4: calculate the correlation of index system

Use grey correlation analysis method to calculate the initial image of the columns, difference sequence of decision table, two stage maximum difference sequence and two stage Minimum difference sequence. For the importance of each index is different, calculate the relevance value to take the weight multiplied by the correlation coefficient, the formula for the correlation value is as follow:

$$\gamma_i = \sum_{k=1}^{n} w_k \gamma_{0i}(k), i = 1, 2, \cdots, m \quad (20)$$

Step 5: sort the schemes

The γ in above formula is assessment result matrix, the greater of the value on behalf of the better scheme, the highest scoring scheme should be the recommended scheme ultimately.

5 Case study

5.1 Original data

The suburban and urban rail network planning of Ningbo, a port city in Zhejiang Province, China, is selected for our case study. We generate 4 sim-schemes based on three cohesion patterns, carry out passenger flow forecast for each one, the year of data in the paper is the forecast year when the scheme is assume to be achieved. We collected the forecast data, and to calculate the index data according to Eq. (1)~Eq. (7), then established the original data matrix.

The schemes is defined as set: $U = \{x_1, x_2, x_3, x_4\}$, the condition attribute set: $C = \{c_1, c_2, c_3, \dots, c_6\}$. The original data for calculation is listed in Table 2:

5.2 Weight

5.2.1 Reference array

$$A = [8300, 375, 900, 125, 13.1, 218]$$

5.2.2 Decision table

Table 2. Original data.

	C_1	C_2	C_3	C_4	C_5	C_6
x_1	8200	350	900	125	13.1	218
x_2	8300	375	750	58	0	24
x_3	8300	375	750	96	0	39
x_4	8225	350	900	108	7.3	173

Table 3. Decision table.

	C_1	C_2	C_3	C_4	C_5	C_6
x_1	0.99	0.93	1.00	1.00	1.00	1.00
x_2	1.00	1.00	0.83	0.47	0.00	0.11
x_3	1.00	1.00	0.83	0.77	0.00	0.18
x_4	0.99	0.93	1.00	0.87	0.56	0.80

5.2.3 Weight calculation

Using equal-distance method for discretization, the suggested discrete interval number is n = 2 according to the schemes amount. Since the smaller value of index means better in the decision table, the data of smaller value represent higher interval.

Table 4. Knowledge presentation system

	C_1	C_2	C_3	C_4	C_5	C_6
x_1	2	2	1	1	1	1
x_2	1	1	2	2	2	2
x_3	1	1	2	1	2	2
x_4	2	2	1	1	1	1

According to the classification rules:

$$U/C = \{(x_1), (x_2), (x_3), (x_4)\}$$

If remove the attribute c_1:

$$U/(C - c_1) = \{x_1, x_2, x_3 x_4\}$$

$$Pos_{C-c_1}(D) = \{(x_1, x_4), x_2, x_3\}$$

$$r_{C-c_1}(D) = |Pos_{C-c_1}(D)|/|U| = 0.5$$

Calculate the importance of the attribute c_1:

$$k_1 = 1 - 0.5 = 0.5$$

Similarly, calculate the importance of the other attributes:

$$k_2 = 0.5, k_3 = 0.5, k_4 = 1, k_5 = 0.5, k_6 = 0.5$$

calculate the weight of each attribute:

$$w_1 = \frac{k_1}{\sum_1^6 k_i} = \frac{0.5}{0.5 + 0.5 + 0.5 + 1 + 0.5 + 0.5} = 0.143$$

Similarly,

$$w_2 = 0.143, w_3 = 0.143, w_4 = 0.286, w_5 = 0.143, w_6 = 0.143$$

The index weight sequence:

$$W = (0.143, 0.143, 0.143, 0.286, 0.143, 0.143)$$

5.3 Correlation

Step 1: Select the optimal values of the columns as an assessment standard reference sequence

$$A_0 = (8200, 350, 750, 125, 13.1, 218)$$

Step 2: Calculate the difference sequence, two stage maximum difference sequence and two stage Minimum difference sequence

$$M = [100, 25, 150, 67, 13.1, 194]$$

$$N = [0, 0, 0, 0, 0, 0, 0]$$

calculate the correlation of index system, define $\xi = 0.5$

$$\gamma_{0i}(K) = \begin{bmatrix} 1.000 & 1.000 & 0.333 & 0.333 & 0.333 & 0.333 \\ 0.333 & 0.333 & 1.000 & 1.000 & 1.000 & 1.000 \\ 0.333 & 0.333 & 1.000 & 0.471 & 1.000 & 0.864 \\ 0.667 & 1.000 & 0.333 & 0.400 & 0.473 & 0.393 \end{bmatrix}$$

Step 3: Calculate the correlation of different schemes

$$\gamma_i = \sum_{k=1}^{6} w_k \gamma_{0i}(k)$$

Get the results

$$\gamma_i = [0.5238, 0.8095, 0.6389, 0.5237]^T$$

Sort the scheme according to the numerical value of the results above: x_2, x_3, x_1, x_4

5.4 Comparison with marginal cost analysis

To test the calculation result, the marginal cost analysis is deployed here. A non-cohesion scheme is assumed as a base scheme, a reference scheme for the 4 sim-schemes. The performance of each sim-scheme is measured by comparing the difference in cost and difference in value of time with the base scheme, which is formulated in Eq. (21). According to the original data, we assume the total passenger flow volume is 1 million, the total in-vehicle time is 82 million minutes, the total transfer time is 3.75 million minutes, the total waiting time is 7.5 million minutes, the vehicle marshalling is 2B, the peak departure frequency is 20 pair/h, the normal departure frequency is 12 pair/h, the time value is T.

The construction costs, operation costs and travel time value of the assumption scheme can be seen as fixed cost, cohesion scheme can be seen as the variable cost, take the passenger travel time saving as the benefits bring by the cohesion schemes [15].

Table 5. Decision table.

	ΔC_1	ΔC_2	ΔC_3	ΔC_4	ΔC_5	ΔC_6
x_1	100	25	0	125	13.1	218
x_2	0	0	150	58	0	24
x_3	0	0	150	96	0	39
x_4	75	25	0	108	7.3	173

$$M = \frac{\Delta C}{\Delta E} \tag{21}$$

where: M is the performance measure（Similar to the marginal cost), ΔC is the difference in cost, ΔE is the difference in value of time,Calculate with the original data:

$$M_1 = \frac{\Delta C}{\Delta E} = \frac{125 + 13.1 + 218}{(100 + 25) \times \mathrm{T}} = 2.84/T$$

$$M_2 = \frac{\Delta C}{\Delta E} = \frac{58 + 24}{150 \times \mathrm{T}} = 0.55/T$$

$$M_3 = \frac{\Delta C}{\Delta E} = \frac{96 + 39}{150 \times \mathrm{T}} = 0.90/T$$

$$M_4 = \frac{\Delta C}{\Delta E} = \frac{108 + 7.3 + 173}{(50 + 25) \times \mathrm{T}} = 2.89/T$$

According to the results above, it can be found: to save the time value of the same, scheme 2 will costs the least. This is consistent with conclusion calculate by the grey rough set model, the rationality and feasibility of assessment method is verified for a certain extent.

6 Conclusion

This paper presents an assessment method to compare the advantages and disadvantages of three cohesion patterns between suburban and urban rail transit network, which eliminates the dependency on expert scoring and output more comprehensive information, such as ranking and significance of selected attributes. A modified rough set – grey correlation model is developed as a core of this assessment. A index system is set up for rough set calculation, where all the variables needs to be discretized to be input to rough set model. A case study using the network in Ningbo is applied to demonstrate the effectiveness of the method. The consistency of the results in comparison with marginal cost analysis can be a preliminary verification of the model.

In addition, this method can also be applied in other types of scheme analysis. To enhance the capability of this method, there are some other issues that will be further studied in the future: 1) systemic methods to select attributes in the index system; 2) optimization of the variable's discretization method.

References

1. Z, P. and S. A, Rudiments of rough sets. Information Sciences, **1**, 3-27 (2007)
2. M, J.H. and C.S. K, Application of Fuzzy Decision Making Method to the Evaluation of Spent Fuel Storage Options. Progress in Nuclear Energy, **39**, (3-4 345-351 (2001)
3. Gu, W., et al., A new method of accelerated life testing based on the Grey System Theory for a model-based lithium-ion battery life evaluation system. Journal of Power Sources, **267**, 1 366-379 (2014)
4. Akkaya, G., B. Turanoğlu and S. Öztaş, An integrated fuzzy AHP and fuzzy MOORA approach to the problem of industrial engineering sector choosing. Expert Systems With Applications, **42**, 9565-9573 (2015)
5. Le Hoang Son, A novel kernel fuzzy clustering algorithm for Geo-Demographic Analysis. Information Sciences, **317**, 202–223 (2015)
6. Gomez-Ruiz, J.A., M. Karanik and J.I.P. A, Estimation of missing judgments in AHP pairwise matrices using a neural network-based model. Applied Mathematics and Computation, **216**, 10 2959–2975 (2010)
7. Dewangan, S., S. Gangopadhyay and C.K. Biswas, Multi-response optimization of surface integrity characteristics of EDM process using grey-fuzzy logic-based hybrid approach, **18**, 3 361–368 (2015)
8. H Y, W. and L. H Y, A hybrid approach to develop an analytical model for enhancing the service quality of e-learning. Computer & Education, **58**, 1318-1338 (2012)
9. M L, H., C. Y H and S. F F, Method of multi-attribute decision analysis based on rough sets dealing with grey information, in 2011 IEEE International Conference On Systems,Man and Cydernetics, 1457-1462 (2011)
10. H Q, H. and W. Y, S-grey rough set and its general characteristics. Journals of Information and Computational Science, **8**, 11 2077-2087 (2011)
11. S X, W., Study of grey rough set model based on tolerance relation. International Journal of Systems and Control, **2**, 1 74-83 (2007)
12. G, B.C. and S. J, Integrating sustainability into supplier selection with grey system and rough set methodologies. International Journal of Production Economics, **124**, 1 252-264 (2010)
13. Yamaguchi, A grey-based rough approximation model for interval data processing. Information Sciences, **177**, 21 4727-4744 (2007)
14. Yamaguchi, Rough set approximations in formal concept analysis. Lecture Notes in Computer Science, 226- 235 (2010)
15. R, K. and Krumwiede, Rewards and Realities of German Cost Accounting. Strategic Finance, (2005)

DOI: 10.1051/matecconf/20166815002

Modelling of the Marine Traffic Situation Complexity

Xiaoqiao Geng[1,2], Liang Huang[1,2] , Yuanqiao Wen[1,2] and Chunhui Zhou[1,2]

[1]*School of navigation, Wuhan University of Technology, Wuhan, 430063, China*
[2]*Hubei Key Laboratory of Inland Shipping Technology, Wuhan, 430063, China*

Abstract. With the growth of traffic density, traditional collision risk models cannot thoroughly describe the complex traffic any more. Therefore, in this paper, a marine traffic situation complexity model, which is based on the relationship between ships, is introduced to metric the complexity of vessel traffic. Firstly, traffic density factor and traffic conflict factors are considered to construct the traffic unit complexity model, reflecting the traffic complexity and traffic characteristics; Secondly, by extending the traffic unit complexity model into the own-ship complexity model, the influences of the positions of other ship on the complexity spatial distribution are analysed; Finally, two scenarios are imported to validate the availability of the complexity model. The verification results show that the complexity can respond to different route structures and traffic situations sensitively.

1 Introduction

The tendencies of the marine traffic density increasing has increased the difficulty of water traffic management. To make up for the limitations of traditional risk models that most of them only use DCPA and TCPA as the evaluation indicators [1], which only measure the spatial relationships and urgency degree of two ships, without considering their relative position, marine traffic complexity is introduced to describe the complicated situation of marine traffic. There have been some progresses in complexity researching in the fields of air transportation and road traffic: One the one hand, researchers use linear[2] and nonlinear[3] modelling methods to construct the dynamic density model of airspace complexity.[4] On the other hand, researchers think complexity can be considered as the subjective description of traffic situation using intrinsic properties[5]. Airspace complexity model was established based on the coupling relationship of traffic flow.[6,7] Therefore, it is needed to research into the traffic flow complexity in the marine traffic field. In this paper, intrinsic properties complexity is used to research the marine traffic situation, and two complexity factors of traffic density and traffic conflict are proposed to construct the complexity model, then verification is made.

2 Model of marine traffic situation complexity

In this paper, traffic situation complexity, which is related with the traffic flow characteristics is researched. Traffic flow is characterized by the vessel traffic density as well as the geometrical relationships between vessels. Therefore, researches on the traffic situation complexity can be carried from vessel traffic density and geometrical relationships between vessels. In the following research, the traffic relationship unit VC_{ij} (Vessel Couple) constructed by two ships of *i* and *j* is defined.

2.1 Influential factors of traffic density

Definition1: Traffic complexity directly caused by the traffic density is called as complicacy I C_{First_ij}. Traffic density described by the ships amount in unit water area can't fully represent the traffic complicacy, it is also related with the space among ships and the distribution of ships.

Definition2: The traffic density in the traffic unit is related with the relative distance of ships, and it equals the value of complicacy I C_{First_ij} . Then there is:

$$C_{First\_ij} = \begin{cases} \lambda \cdot e^{-\alpha \cdot \|\overrightarrow{D_{ij}}\| / R_{ij}} & , \|\overrightarrow{D_{ij}}\| \geq R_{ij} \\ \lambda \cdot e^{-\alpha} & , \|\overrightarrow{D_{ij}}\| < R_{ij} \end{cases} \quad (1)$$

In Eq.1, α , λ are correction parameters. R_{ij} is the safe distance within VC_{ij} . $\|\overrightarrow{D_{ij}}\|$ is the relative distance between ship *i* and *j*. C_{First_ij} is in nonlinear increase with the decrease of $\|\overrightarrow{D_{ij}}\|$.

2.2 Influential factor of traffic conflict

Definition 3: Traffic complexity caused by the traffic situation in the traffic unit is defined as the complexity factorII, which is called as conflict factor C_{Second_ij}. And $C_{Second\_ij} = \psi(\overrightarrow{D_{ij}}, \overrightarrow{v_{ij}}, \theta)$, $\overrightarrow{D_{ij}}$ is the relative distance, $\overrightarrow{v_{ij}}$ is the relative speed, θ is the track-crossing angle. In this paper, the process of traffic conflict is divided into four phases: judgement of converging tendency, estimation of converging consequence, measurement of urgency degree and difficulty of risk elimination.

2.2.1 Converging situations

When the relative distance of two ships is decreasing, then there is converging tendency:

$$d\left\|\overrightarrow{D_{ij}}\right\|/dt=\left\|\overrightarrow{v_{ij}}\right\|\bullet\cos(\overrightarrow{v_{ij}},\overrightarrow{D_{ij}})=\overrightarrow{v_{ij}}\bullet\overrightarrow{D_{ij}}/\left\|\overrightarrow{D_{ij}}\right\| \tag{2}$$

In Fig. 2, α is the angle between $\overrightarrow{v_{ij}}$ and $\overrightarrow{D_{ij}}$. When $\alpha\in(90,180]$, $\cos(\overrightarrow{v_{ij}},\overrightarrow{D_{ij}})<0$, then the two ships have converging tendency(shown in Fig. 1(A)), and there exists traffic conflict; when $\alpha\in[0,90]$, $\cos(\overrightarrow{v_{ij}},\overrightarrow{D_{ij}})\geq 0$, then the two ships have separating tendency(shown in fig.1(B)), and there is no traffic conflict. Defining function mix_{ij} to judge whether there exists converging tendency within VC_{ij}, then there is:

$$C_{mix\_ij}=\ell_{R^-}\left\{\overrightarrow{v_{ij}}\bullet\overrightarrow{D_{ij}}/\left\|\overrightarrow{D_{ij}}\right\|\right\}=\begin{cases}1, & \overrightarrow{v_{ij}}\bullet\overrightarrow{D_{ij}}/\left\|\overrightarrow{D_{ij}}\right\|<0\\ 0, & \overrightarrow{v_{ij}}\bullet\overrightarrow{D_{ij}}/\left\|\overrightarrow{D_{ij}}\right\|\geq 0\end{cases} \tag{3}$$

2.2.2 Converging consequence

DCPA (Distance to Closest Point of Approach) is used to measure the dangerous degree: When there is no converging tendency then there is no potential collision risk between these two ships; when there is converging tendency, then there exists traffic conflict. Using C_{Latent_ij} to represent the converging consequences of two ships within VC_{ij}, then the following equation can be established:

$$C_{Latent\_ij}=\varepsilon(D_{CPA},\left\|\overrightarrow{D_{ij}}\right\|)=\lambda\bullet(e^{-\alpha\bullet D_{CPA}/R_{ij}}-e^{-\alpha\bullet\left\|\overrightarrow{D_{ij}}\right\|/R_{ij}})$$
$$C_{mix\_ij}=1 \tag{4}$$

In Eq. 4, $DCPA=\left\|\overrightarrow{D_{ij}}\right\|\bullet\sin(\pi\text{-}\alpha)=\left\|\overrightarrow{D_{ij}}\right\|\bullet\sin(\alpha)$; α is the angle between $\overrightarrow{D_{ij}}$ and $\overrightarrow{D_{ij}}$; $\alpha=\arccos(\overrightarrow{v_{ij}}\bullet\overrightarrow{D_{ij}}/\left\|\overrightarrow{v_{ij}}\right\|\left\|\overrightarrow{D_{ij}}\right\|)$.

2.2.3 Urgency degree

In navigational practice, TCPA(Time to Closest Point of Approach)is used to represent the urgency degree of two converging ships. And $T_{CPA}=\left\|\overrightarrow{D_{ij}}\right\|\bullet\cos(\alpha)/\left\|\overrightarrow{v_{ij}}\right\|=\overrightarrow{D_{ij}}\bullet\overrightarrow{v_{ij}}/\left\|\overrightarrow{v_{ij}}\right\|^2$. T_{CR}(Time Collision Risk) is also introduced to measure the urgency degree of two ships, then there is:

$$T_{CR\_ij}=\tau(\overrightarrow{D_{ij}},\overrightarrow{v_{ij}})=\begin{cases}1, & T_{CPA}\leq t_1\\ [(t_2-T_{CPA})/(t_2-t_1)]^{3.03}, & t_1<T_{CPA}\leq t_2\\ 0, & T_{CPA}>t_2\ or\ T_{CPA}<0\end{cases} \tag{5}$$

In Eq.5, $t_2=\sqrt{12^2-D_{CPA}{}^2}/\left\|\overrightarrow{v_{ij}}\right\|$, $t_1=\sqrt{R_{ij}{}^2-D_{CPA}{}^2}/\left\|\overrightarrow{v_{ij}}\right\|$; R_{ij} is the radius of the security domain of ship i.

2.2.4 Difficulty degree of risk elimination

In navigational practice, the difficulty of collision risk elimination is not in linear relationship with the track-crossing angle. Using C_{angle_ij} to represent the difficulty degree of risk elimination in the condition of different track-crossing angle θ, their relationship is shown in Fig. 2.

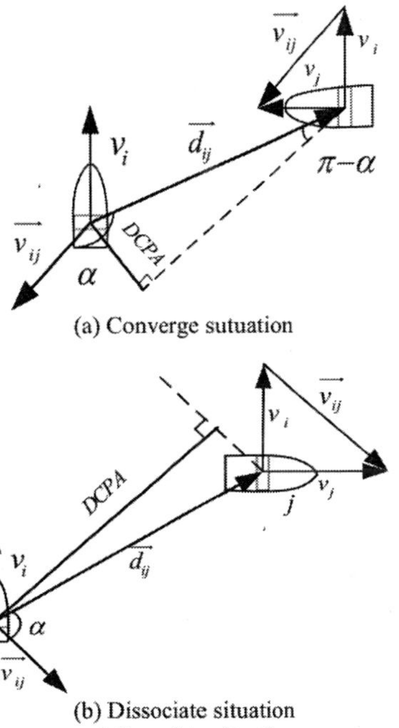

Figure 1. Converge and dissociate situation.

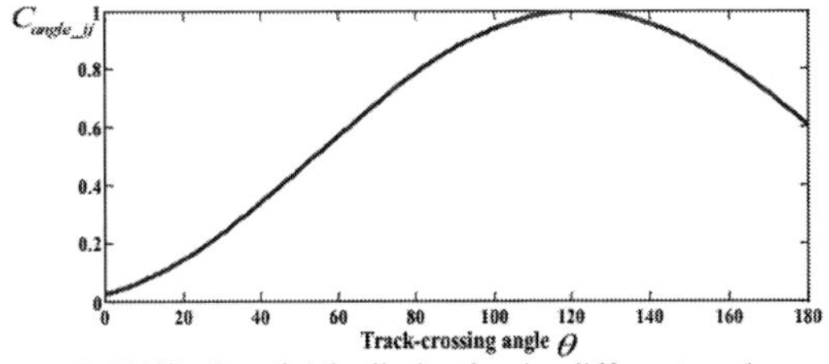

Figure 2. Difficulty of risk elimination in different angles.

And the mathematical expression is as follows:

$$C_{angle\_ij}=f(\theta)=\frac{1}{2}\bullet\left\{1-\cos\left[\left(\frac{180}{67.5}\right)\bullet\frac{\theta}{2}\bullet\frac{\pi}{180}+\frac{\pi}{10}\right]\right\} \tag{6}$$

From Fig. 2, we can see that on the condition that $\theta\in(0,120]$,difficulty degree of risk elimination increases and reaches its highest when θ=120 ; while on the condition that $\theta\in(120,180]$, it decreases. And there is $C_{angle\_ij}\in(0,1]$. In conclusion, the complexity increment of C_{Second_ij} caused by traffic conflicts can be measured as follows: C_{mix_ij} is used to judge whether traffic conflict exists within the traffic unit; C_{Latent_ij} is used to describe the complexity increment without any action taken. T_{CR_ij} is used to describe the urgency degree of the collision risk. C_{angle_ij} is used to describe the difficulty of collision risks elimination. Therefore, C_{Second_ij} can be expressed as Eq.7.

$$\begin{aligned} C_{Second\_ij} &= \psi(\overrightarrow{D_{ij}}, \overrightarrow{v_{ij}}, \theta) \\ &= C_{mix\_ij} \bullet C_{Latent\_ij} \bullet T_{CR\_ij} \bullet C_{angle\_ij} \\ &= l_{R^-} \left\{ \frac{\overrightarrow{v_{ij}} \bullet \overrightarrow{D_{ij}}}{\left\| \overrightarrow{D_{ij}} \right\|} \right\} \bullet \varepsilon(\overrightarrow{D_{ij}}, \overrightarrow{v_{ij}}) \bullet \tau(\overrightarrow{D_{ij}}, \overrightarrow{v_{ij}}) \bullet f(\theta) \end{aligned} \tag{7}$$

2.3 Complexity model

The traffic situation complexity of VC_{ij} consists of complicacy I and complicacy II. Therefore, it can be defined as C_{total_ij} in the adding format (Eq.8). And supposing that when the distance between two ships is less than 0.50nmile, the complexity reaches its maximum value, which is set as 80; when it decreases to 3nmile, corresponding actions should be taken and the complexity value is 0.25 times of the maximum one. Therefore, parameters can be set as: $\alpha = 2 \bullet \ln(2)/5$, $\lambda = 80 \times 2^{2/5}$.

$$C_{total\_ij}(\left\|\overrightarrow{D_{ij}}\right\|, \overrightarrow{v_{ij}}, \theta_{ij}) = C_{First\_ij}(\left\|\overrightarrow{D_{ij}}\right\|) + C_{Second\_ij}(\overrightarrow{D_{ij}}, \overrightarrow{v_{ij}}, \theta_{ij}) = \begin{cases} \lambda \bullet e^{-\alpha \frac{\left\|\overrightarrow{D_{ij}}\right\|}{R_{ij}}}, & C_{mix\_ij} = 0 \text{ or } T_{CR\_ij} = 0 \\ \lambda \bullet e^{-\alpha \frac{\left\|\overrightarrow{D_{ij}}\right\|}{R_{ij}}} + l_{R^-} \left\{ \frac{\overrightarrow{v_{ij}} \bullet \overrightarrow{D_{ij}}}{\left\|\overrightarrow{D_{ij}}\right\|} \right\} \bullet \varepsilon(\overrightarrow{D_{ij}}, \overrightarrow{v_{ij}}) \bullet \tau(\overrightarrow{D_{ij}}, \overrightarrow{v_{ij}}) \bullet f(\theta_{ij}), & C_{mix\_ij} \neq 0 \;\&\; T_{CR\_ij} \neq 0 \end{cases} \tag{8}$$

Complexity of the traffic unit can be calculated using Eq.8, then the distribution pattern of complexity, which can be applied to collision risk elimination, can be acquired using the interpolation formula. Supposing i as the own-ship, then in the perspective of ship i, recognize all the traffic units VC_{ij} concluding ship i, where, $j \in (1:n), (j \neq i)$. The complexity value of ship j can be calculated as $C_{total\_ij}(\overrightarrow{D_{ij}}, \overrightarrow{v_{ij}}, \theta_{ij})$. Then meshing is made to the area, and Eq.9 is used to calculate the complexity value of each grid. Finally, the spatial distribution pattern of this water area can be acquired.

$$C(a,b) = \sum_{j}^{n} \frac{C_{total\_ij}(\overrightarrow{D_{ij}}, \overrightarrow{v_{ij}}, \theta_{ij})}{d_j^3} \bullet \sum_{j}^{n} d_j^3 \tag{9}$$

In Eq.9, a,b represent the row and column number respectively after the meshing process; $C_{total\_ij}(\overrightarrow{D_{ij}}, \overrightarrow{v_{ij}}, \theta_{ij})$ represents the complexity value of ship i caused by ship j.

3 Model verification

3.1 Experimental design

In this paper, single ship type, of which the DWT is 70000, the ship length is 230m, the ship breadth is 32.2m, is selected [8]. And two experiment groups are designed: T1, which is set on the paralleled traffic, is to analyze the variation of complexity caused by relative distance; T2, which is set on the "cross" intersection, is to analyze the variation of complexity caused by traffic conflict. The number of ships in T1 and T2 are 9 and 8 respectively, and the ship speed is set as 8kn.

3.2 Experimental results

3.2.1 Experiment T1

In T1, the own-ship is in the location of (0, 0), ship course is 000, the distance to the other-ship is 0.9nmile, which is shown in fig.3(1). Fig.3(2) is the complexity distribution pattern in the initial moment. From fig.3(2) to (6), the time interval of the adjacent two pictures is 5min, and the solid line in red is the motion curve of the own-ship. From these figures, we can see that shorter the distance between other-ship and own-ship is, higher the complex degree of the own-ship caused by the other-ship will be, which well reflected the relationship between traffic density and complexity.

3.2.2 Experiment T2

In T2, the own-ship is in the location of (-4, 0), ship course is 045, distance between two adjacent ships is 0.85n mile. The course of the traffic flow in the left side of the own-ship is 045, and that in the right side is 315. And ships in both side are symmetry about the Y axis. Fig.4(1) is the complexity distribution pattern in the initial moment, and from fig.4(2) to (6), the time interval is 10min.

In fig.4(2) and (3), we can see that ships in the portside are close to the own-ship, therefore, the first ship in the portside has comparatively higher complexity, other ships are in the same course with the own-ship and there exists no converging tendency, so their complexity is lower. Ships in the starboard has encounter tendency with the own-ship, but their TCPA is large, so the urgency degree is low. Fig.4(4) to (6) show that as time goes on, ship density and complex degree in the portside change little, also has a little change. But in the starboard, the influence of the traffic conflict factors is obvious. In fig.4(4), when t=20min, as DCPA of the first ship in the starboard is small, its complex degree is higher than other ships, so the own-ship should stay alert. In fig.4(5), when t=30min, complexities between the own-ship and ships in both side have exceed 50, and actions should be taken. As time goes on, complexities of ships in the starboard keep increase, and when t=40min (shown in fig.4(6)), the first ship in the starboard collides with the own ship, and the complexity reaches its maximum value. Experiment T2 reflected the variation of complexity with the changing of traffic conflict, it is related with whether the converging tendency exists and the converging consequences.

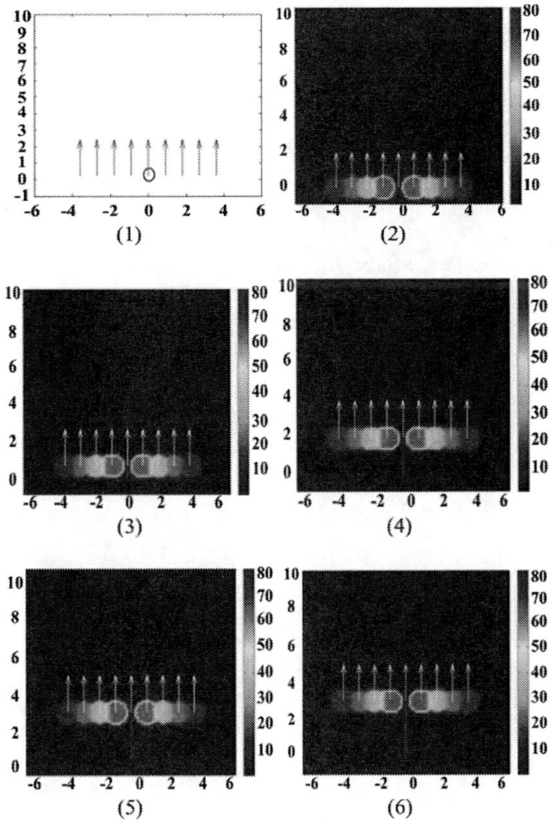

Figure 3. The spatial complexity distribution in parallel traffic.

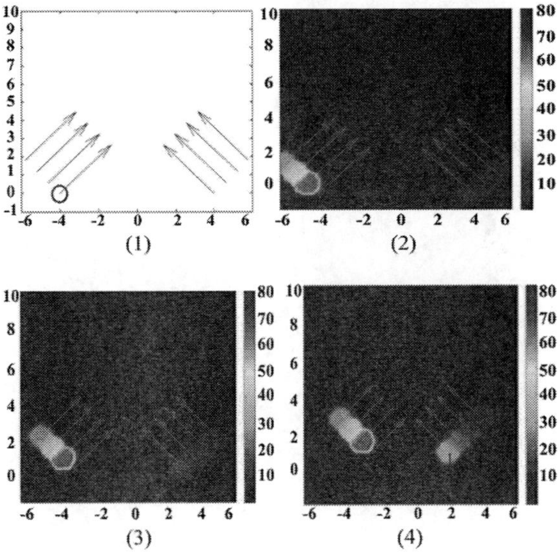

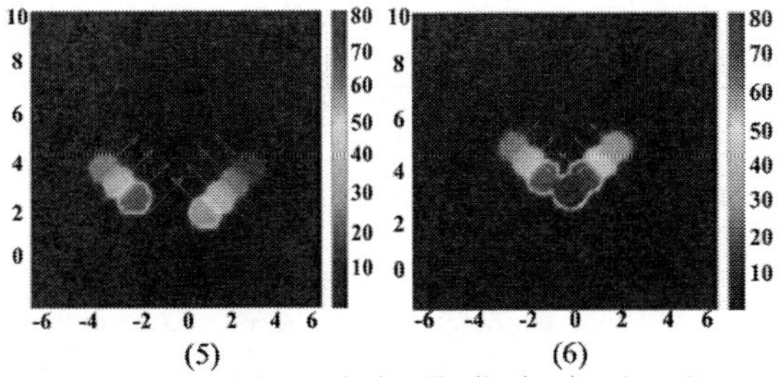

Figure 4. The spatial complexity distribution in "Cross" intersection traffic.

From T1 and T2, we can see that using marine traffic situation complexity model, the following two points can be realized: On the one aspect, traffic density factor can be used to recognize the dangerous degree of the surrounding traffic environment, and risk level of other-ship can be labeled. On the other aspect, traffic conflict factor can be used to recognize the surrounding ships which have potential conflicts with the own-ship or may cause operational constraints to the own-ship.

4 conclusions

In this paper, the marine traffic situation compleity, which is closely related with the traffic flow characteristics, is introduced to describe the complicated situation of marine traffic. Traffic flow is characterized by the vessel traffic density in the area as well as the geometrical relationships between vessels. Therefore, researches on the traffic situation complexity can be carried from two aspects, one is the vessel traffic density, the other is the vessel traffic conflict caused by the taffic situation. Then based on the analysis of marine traffic density and traffic conflict, the mathematical model of marine traffic situation complexity is established on the ideal condition. To verify the influences of different traffic conflict factors on the traffic situation complexity, senario simulations of different traffic structures are made. The experimental results demonstrated that: The model of marine traffic situation complexity proposed in this paper can well reflect the influences of the traffic density and traffic conflict on the marine traffic situation complexity. Not only the ships with collision risks can be identified, but ships with potential risks can also be recognized. Research in this paper can be a good basis for the future research of the security of the marine traffic system.

However, in this paper, influences of the mandatory rules (stipulated by Regulations for Preventing Collisions) and human factors on the traffic complexity are not considered, and influences of the ship types and channel conditions are not discussed as well. Therefore, how to describe the influences of these factors through mathematical expressions, and to establish the more universal model of marine traffic situation complexity need further research.

Acknowledgement

This work is supported by the National Science Foundation of China (NSFC) (No. 51579204), and is financially supported by self-determined and innovative research funds of WUT (No: 2016IVA064).

References

1. Y.Q. WEN, H. ZHANG, P. WAN, Risk degree analysis of ship collision risk in channel based on ordered probit model, J. China Safety Science Journal. **22**, 134-139 (2012).
2. LAUDEMAN I. V., SHELDEN S. G., BRANSTROMR, et al: Dynamic density: An air traffic management metric, R. NAS 1.15:112226, NASA/TM-1998-112226. USA Springfield: NTIS (1998).
3. CHATTERJI G. B., SRIDHAR B., Measures for air traffic controller workload prediction, C. // 1st AIAA Aircraft, Technology Integration, and Operations Forum. USA. Los Angeles: AIAA Press (2001).
4. F.Q. DAI, L.S. HONG, Study on Airspace Traffic Complexity Computation Methods, J. Journal of Civil Aviation University of China. **28**(3), 1-4 (2010). (In Chinese)
5. DELAHAYE D., PUECHMOREL S., Air traffic complexity based on dynamical systems, C. // 2010 49th IEEE Conference on Decision and Control (CDC), Atlanta: 0743-1546978-1-4244-7745-6.
6. C. ZHANG, J. ZHANG, M.H. HU, Air traffic complexity based on alliance effects, C. // The 28th Digital Avionics Systems Conference. USA Seattle: IEEE/AIAA Press (2009).
7. B.J. YE, M.H. HU, C. ZHANG, et-al, Traffic structure based air traffic complexity modeling, J. Journal of Transportation Systems Engineering and Information Technology. **12**(1), 166-172 (2012). (In Chinese)
8. A.M. DU, Study on port fairway capacity based on standard vessels, D. Dalian: Dalian University of Technology (2008).

MATEC Web of Conferences **68**, 16001 (2016) DOI: 10.1051/matecconf/20166816001
ICIEA 2016

The Linear Probability Density Function of Continuous Random Variables in the Real Number Field and Its Existence Proof

Yiyan Chen[1] and Ye Li[2]

[1]*College of Software, Tianjin University, Tianjin, China*
[2]*College of Mathematical Science, Capital Normal University, Beijing, China*

Abstract. For a continuous random variable in real number field, there must be a distribution and also a probability density function of this random variable. If there is a known function with this random variable as independent variable, its image is a smooth or piecewise smooth line, there must be at least one function that takes this random variable as its independent variable, these functions are bounded on the image of the first function. Any one of these functions conduct line integral operation to the line segment or arc length of the certain image of the first known function is the cumulative probability of this continuous random variable interval corresponding to the section of the image for line integral operation. A general designation for these functions are linear probability density function of continuous random variables. Conduct line integral operation to the linear probability density function and conduct integral operation to the probability density function have same results of the cumulative probability of continuous random variable. By the way, Line integration including curve integration. According to the uniqueness of the probability, the existence and the number of linear probability density function can be proved and calculated.

1 Introduction

For a continuous random variable as X, $X \in R$, which meet the condition $X \sim \Delta_X$, obviously, there is a probability density function $f(x)$, meet the condition that $\int_{-\infty}^{x} f(t)dt = F(x)$. $F(x)$ is the cumulative probability function of X. With reference to examples of curvilinear integral[1,2], If the integral domain of X, which equivalent to interval of the value of random variable X from real number axis, changes to the function image (or a part of it) of $g(x)$ which is a continuous function or countable piecewise function, there should be another function $h(x)$ in real number field, according to the uniqueness of probability, it meets $\int_{g(x)} h(u)du = F(x)$. The integration path of continuous random variable X is a moving path of the points $(x, g(x))$, composed of the value of X and its corresponding value of the value $g(x)$, on the $g(x)$ moving in the direction of x-axis. $h(x)$ are called the linear probability density function of the continuous random variable X in the real number field. To this end, its need to define the linear probability density function, and prove its existence and the number of the function in real number field.

2 Relevant definition

Here is a need to give definition and properties of linear integral, linear probability density function.

2.1 Linear integral

2.1.1 Definition of Linear integral

Set X, $X \in R$, X is a continuous random variable in the real number field, $X \sim \Delta_X$. Its probability density function is $f(x)$, cumulative probability function is $F(x)$, set $g(x)$ is a function of random variables with X as independent variable, its image L is smooth or piecewise smooth in plane xOy, there is another function $h(x)$ with independent variable the same as X, $h(x)$ is bounded on the image of $g(x)$, intercept an image L^* from L, its endpoints are $A(x_m, g(x_m))$ and $B(x_n, g(x_n))$, using unlimited division, insert a finite point range $M_1, M_2, \cdots, M_{n-1}$ in the L^*, divide L^* into n short sections, set the length of the ith section is Δ_{s_i}, also (ξ_i, η_i) is any point on the ith section, make product $h(\xi_i,\eta_i)\Delta_{S_i}$, $i=1,2,\cdots,n$, and also make product sums $\sum_{i=1}^{n} h(\xi_i,\eta_i)\Delta_{S_i}, i=1,2,\cdots,n$, in this case, if $\max\left[h(\xi_i,\eta_i)\Delta_{S_i}\right]=\lambda \to 0$, $\lim_{\lambda\to 0}\sum_{i=1}^{n} h(\xi_i,\eta_i)\Delta_{S_i}$ is always exist, then called this limit as a function $h(x,g(x))$ with doing linear integral to L^* on image of $g(x)$. The image L of $g(x)$ could be a straight line, could also be a curve, could be bounded but also could be unbounded, but must meeting the smooth or piecewise smooth.

2.1.2 Classification of solving linear integral

If the image L of $g(x)$ is bounded, set $L = L^*$, according to the methods just now, mark the linear integral as $\int_L h(x,g(x))ds$, there are:

$$\int_L h(x,g(x))ds = \lim_{\lambda\to 0}\sum_{i=1}^{n} h(\xi_i,\eta_i)\Delta_{S_i} \tag{1}$$

If L is unbounded, then $L^*\subset L$, two endpoints of L^* are $A(x_m,g(x_m))$ and $B(x_n,g(x_n))$, if it is need to required out linear integral of L^*, there are:

$$\begin{aligned}&\int_{L^*} h(x,g(x))ds\\ &= \int_{L(A\to B)} h(x,g(x))ds = \lim_{\lambda\to 0}\sum_{i=1}^{n} h(\xi_i,\eta_i)\Delta_{S_i}\end{aligned} \tag{2}$$

If L is unbounded, then $L^*\subset L$, if it is need to required out linear integral of L, insert an infinite point range $M_1,M_2,\cdots,M_{n-1},\cdots$ in L, divide L into r short sections, $r\to+\infty$, set the length of ith section is Δ_{S_i}, also (ξ_i,η_i) is any point on the ith section, make product $h(\xi_i,\eta_i)\Delta_{S_i}$, $i=1,2,\cdots,r$, in this case, also make $\sum_{i=1}^{r} h(\xi_i,\eta_i)\Delta_{S_i}$, if $\max\left[h(\xi_i,\eta_i)\Delta_{S_i}\right]=\lambda\to 0$, $\lim_{\substack{\lambda\to 0\\ r\to+\infty}}\sum_{i=1}^{r} h(\xi_i,\eta_i)\Delta_{S_i}$ is always exist, then called this limit as a function $h(x,g(x))$ with doing linear integral to L on image of $g(x)$. Still mark the linear integral of L as $\int_L h(x,g(x))ds$, there are:

$$\int_L h(x,g(x))ds = \lim_{\substack{\lambda\to 0\\ r\to+\infty}}\sum_{i=1}^{r} h(\xi_i,\eta_i)\Delta_{S_i} \tag{3}$$

In short, if it is a need to find out the entire or a part of image of $g(x)$ for operating linear integral, the method of infinite division could be used to realize to solve the linear integral calculation in traditional.

2.2 Linear probability density function

2.2.1 Definition of linear probability density function

According to the last section, reference to the definition and properties of probability density function and cumulative probability function[3,4,5], $h(x)$ is a function of X, $g(x)$ is also a function of X, set $F(x)$ as the cumulative probability function of X, $\int_{g(x)} h(u)du$ is operating linear integral by x-axis direction along the image of $g(x)$, if:

$$\int_{g(x)} h(u)du = F(x) \tag{4}$$

It is said $g(x)$ is path of linear integral of X, $h(x)$ is the linear density function of X. Another note that $h(x)=h(x,g(x))$, $g(x)$ is not a variable of two-dimensional random but a parameter of $h(x)$.

2.2.2 Properties of linear probability density function

It is obvious that linear integral of the linear probability density function is a change form of the curve integral, and the linear probability density function is also a change form of probability density function, so the linear probability density function has the characteristics of both of them.

Property 1. If $h(x)$ is the linear probability density function of a continuous random variable X, then:

$$h(x)\geq 0 \tag{5}$$

Property 2. If $h(x)$ is the linear probability density function of a continuous random variable X, $g(x)$ is the path of linear integral, $g(x|-\infty<x<+\infty)$ express the whole image of $g(x)$ on x-axis, then:

$$\int_{g(x|-\infty<x<+\infty)} h(u)du = 1 \tag{6}$$

Property 3. Set α,β as constants, $h(x)$ is the linear probability density function of a continuous random variable X, $g(x)$ is the path of linear integral, if $h(x)=\alpha h_1(x,g(x))+\beta h_2(x,g(x))$, then:

$$\begin{aligned}&\int_{g(x)}\left[\alpha h_1(u,g(u))+\beta h_2(u,g(u))\right]du\\ &=\alpha\int_{g(x)} h_1(u,g(u))du+\beta\int_{g(x)} h_2(u,g(u))du\end{aligned} \tag{7}$$

Property 4. $h(x)$ is the linear probability density function of a continuous random variable X, $g(x)$ is the path of linear integral, set the whole image of $g(x)$ on the x-axis is $g(x|-\infty<x<+\infty)$, in this case, make $g(x|-\infty<x<+\infty)=L$, if L could be divided into a finite number of short sections as L_i, $i=1,2,\ldots,n$,

$\sum_{i=1}^{n} L_i = L$, then:

$$\int_L h(u)du = \sum_{i=1}^{n} \int_{L_i} h(u)du \quad (8)$$

Property 5. $h(x)$ is the linear probability density function of a continuous random variable X, $g(x)$ is the path of linear integral, set the whole image of $g(x)$ on the x-axis is $g(x|-\infty < x < +\infty)$, in this case, make $g(x|-\infty < x < +\infty) = L$, if $h(x) \le t(x)$ on the L , then:

$$\int_L h(u)du \le \int_L t(u)du \quad (9)$$

Property 6. $h(x)$ is the linear probability density function of a continuous random variable X, $g(x)$ is the path of linear integral, $f(x)$ is the probability density function of X, $F(x)$ is cumulative probability of X, then:

$$\int_{g(x)} h(u)du = \int_{-\infty}^{x} f(t)dt = F(x) \quad (10)$$

Property 7 is the uniqueness of the probability, for a continuous random variable, it is bound to be the only existence of the cumulative probability in a given region, no matter what form of expression of probability. The following is a proof of it.

Proof: X is a continuous random variable, its probability density function is $f(x)$, its cumulative probability is $F(x)$, linear probability density function is $h(x)$, path of linear integral is $g(x)$, $g(x|a \le x \le b)$ express the image about $x \in [a,b]$ on the $g(x)$, if there is a specified integral region which like $a \le x \le b$, $\int_a^b f(x)dx \ne \int_{g(x|a \le x \le b)} h(x)dx$, set $\int_a^b f(x)dx = P_1$, set $\int_{g(x|a \le x \le b)} h(x)dx = P_2$, then $P(a \le x \le b)$ there are two values P_1 and P_2 , and meanwhile, the rest of the integral region of random variables satisfied $\int_{g(x)} h(u)du = \int_{-\infty}^{x} f(t)dt = F(x)$, set the cumulative probability of the rest of integral region of random variables as P_s, according to the regularity of $f(x)$, it is a need to meet $\int_{-\infty}^{+\infty} f(t)dt = 1$, but $P_1 \ne P_2$, obviously, $P_1 + P_s$ and $P_2 + P_s$ must have one not equal to 1, it is unsatisfied to the regularity of probability density function, therefore, for any x , $x \in X$, satisfied to $\int_{g(x)} h(u)du = \int_{-\infty}^{x} f(t)dt = F(x)$. That is to say, no matter the cumulative probability express by the integral of probability density function or linear integral of linear probability density function, for any value region of random variable X must be equal everywhere.

3 Proof of existence of linear probability density and its number

According to the uniqueness of the probability, it could prove the existence and number of linear probability density function. There are two kinds of situations, first one, $f(x)$ is known, $h(x)$ is unknown, $g(x) \ne h(x)$; Second one, $f(x)$ is known, $h(x)$ is unknown, $g(x) = h(x)$. The meaning of $f(x)$, $g(x)$ and $h(x)$ are the same as above section. If to prove the existence of linear probability density function, that is to prove the existence of $h(x)$, it could be based on the equation of the probability like $\int_{g(x)} h(u)du = \int_{-\infty}^{x} f(t)dt = F(x)$ to solve $h(x)$, if the equation has real roots, it is proved the existence of $h(x)$ in real number field, the number of real roots is the number of $h(x)$ in real number field.

3.1 First situation

Proof: $f(x)$ is known, $h(x)$ is unknown, $g(x) \ne h(x)$, The meaning of $f(x)$, $g(x)$ and $h(x)$ are the same as above section.To reference the content of calculation method of curve integral of arc length in mathematical analysis, it could provide that :

$$\int_{g(x)} h(u)du = \int_{-\infty}^{x} f(t)dt = F(x)$$

$$\Leftrightarrow \int_{-\infty}^{x} h(u)\sqrt{1+\left[g'(u)\right]^2}\,du = F(x)$$

Seek the first derivative on both sides, then:

$$\Leftrightarrow h(x)\sqrt{1+\left[g'(x)\right]^2} = f(x)$$

$$\Leftrightarrow h^2(x)\left\{1+\left[g'(x)\right]^2\right\} = f^2(x)$$

$$\Leftrightarrow h^2(x) = f^2(x)\Big/\left\{1+\left[g'(x)\right]^2\right\}$$

$$\Leftrightarrow h(x) = \pm\sqrt{f^2(x)\Big/\left\{1+\left[g'(x)\right]^2\right\}}$$

$f(x)$ and $g(x)$ are both real functions, obviously, $h(x)$ are also real functions, therefore, in real number field, the existence of $h(x)$ is proved. The number of $h(x)$ is 2.

3.2 Second situation

Proof: $f(x)$ is known, $h(x)$ is unknown, $g(x) = h(x)$, The meaning of $f(x)$, $g(x)$ and $h(x)$ are the same as above section.To reference the content of calculation method of curve integral of arc length in mathematical analysis, it could provide that :

$$\int_{g(x)} h(u)du = \int_{-\infty}^{x} f(t)dt = F(x)$$

$$\Leftrightarrow \int_{-\infty}^{x} g(u)\sqrt{1+\left[g'(u)\right]^2}du = F(x)$$

Seek the first derivative on both sides, then:

$$\Leftrightarrow g(x)\sqrt{1+\left[g'(x)\right]^2} = f(x)$$

$$\Leftrightarrow g^2(x)\left\{1+\left[g'(x)\right]^2\right\} = f^2(x)$$

$$\Leftrightarrow g^2(x)+g^2(x)\cdot\left[g'(x)\right]^2 = f^2(x)$$

$$\Leftrightarrow \left[g'(x)\right]^2 = \left[f^2(x)-g^2(x)\right]/g^2(x)$$

$$\Leftrightarrow \left[g'(x)\right]^2 = \left[1/g^2(x)\right]\cdot f^2(x)-1$$

Set $g(x)=1/k$, then $g'(x)=-1/k^2$, $k\neq 0$, substitute them to the equation, then :

$$\Leftrightarrow \left(-1/k^2\right)^2 = k^2 f^2(x)-1$$

$$\Leftrightarrow 1 = k^6 f^2(x)-k^4$$

Set $f^2(x)=C$, substitute it to the equation, then :

$$\Leftrightarrow \left(-1/k^2\right)^2 = k^2 C -1$$

$$\Leftrightarrow 1 = k^6 C - k^4$$

$$\Leftrightarrow k^6 C - k^4 - 1 = 0$$

$$\Leftrightarrow k x^6 - x^4 - 1 = 0$$

Solving higher order equation, as long as seeking out x , thus to seek out $h(x)$, with the aid of MATLAB, input code : syms x k;x=solve(k*x^6-x^4-1), it could get:

```
x =

1/6/k/(108*k^2+8+12*3^(1/2)*(27*k^2+4)^(1/2)*k)^(1/3)
*6^(1/2)*(k*(108*k^2+8+12*3^(1/2)*(27*k^2+4)^(1/2)*
k)^(1/3)*((108*k^2+8+12*3^(1/2)*(27*k^2+4)^(1/2)*k)
^(2/3)+4+2*(108*k^2+8+12*3^(1/2)*(27*k^2+4)^(1/2)*
k)^(1/3)))^(1/2)

-1/6/k/(108*k^2+8+12*3^(1/2)*(27*k^2+4)^(1/2)*k)^(1/
3)*6^(1/2)*(k*(108*k^2+8+12*3^(1/2)*(27*k^2+4)^(1/2)
*k)^(1/3)*((108*k^2+8+12*3^(1/2)*(27*k^2+4)^(1/2)*k)
^(2/3)+4+2*(108*k^2+8+12*3^(1/2)*(27*k^2+4)^(1/2)*
k)^(1/3)))^(1/2)

1/6/k/(108*k^2+8+12*3^(1/2)*(27*k^2+4)^(1/2)*k)^(1/3)
*3^(1/2)*(k*(108*k^2+8+12*3^(1/2)*(27*k^2+4)^(1/2)*
k)^(1/3)*(i*3^(1/2)*(108*k^2+8+12*3^(1/2)*(27*k^2+4)
^(1/2)*k)^(2/3)-(108*k^2+8+12*3^(1/2)*(27*k^2+4)^(1/
2)*k)^(2/3)-4+4*(108*k^2+8+12*3^(1/2)*(27*k^2+4)^
(1/2)*k)^(1/3)-4*i*3^(1/2)))^(1/2)

-1/6/k/(108*k^2+8+12*3^(1/2)*(27*k^2+4)^(1/2)*k)^(1/
3)*3^(1/2)*(k*(108*k^2+8+12*3^(1/2)*(27*k^2+4)^(1/2)
*k)^(1/3)*(i*3^(1/2)*(108*k^2+8+12*3^(1/2)*(27*k^2+
4)^(1/2)*k)^(2/3)-(108*k^2+8+12*3^(1/2)*(27*k^2+4)^
(1/2)*k)^(2/3)-4+4*(108*k^2+8+12*3^(1/2)*(27*k^2+4)
^(1/2)*k)^(1/3)-4*i*3^(1/2)))^(1/2)

1/6/k/(108*k^2+8+12*3^(1/2)*(27*k^2+4)^(1/2)*k)^(1/3)
*(-3*k*(108*k^2+8+12*3^(1/2)*(27*k^2+4)^(1/2)*k)^
(1/3)*(i*3^(1/2)*(108*k^2+8+12*3^(1/2)*(27*k^2+4)^
(1/2)*k)^(2/3)+(108*k^2+8+12*3^(1/2)*(27*k^2+4)^(1/
2)*k)^(2/3)+4-4*(108*k^2+8+12*3^(1/2)*(27*k^2+4)^
(1/2)*k)^(1/3)-4*i*3^(1/2)))^(1/2)

-1/6/k/(108*k^2+8+12*3^(1/2)*(27*k^2+4)^(1/2)*k)^(1/
3)*(-3*k*(108*k^2+8+12*3^(1/2)*(27*k^2+4)^(1/2)*k)
^(1/3)*(i*3^(1/2)*(108*k^2+8+12*3^(1/2)*(27*k^2+4)^
(1/2)*k)^(2/3)+(108*k^2+8+12*3^(1/2)*(27*k^2+4)^(1/
2)*k)^(2/3)+4-4*(108*k^2+8+12*3^(1/2)*(27*k^2+4)^
(1/2)*k)^(1/3)-4*i*3^(1/2)))^(1/2)
```

There are 2 real roots and 4 imaginary roots. Seek out $h(x)$ in real number field, seeking reciprocal ofreal roots of x by MATLAB, then:

```
x=
k*(108*k^2+8+12*3^(1/2)*(27*k^2+4)^(1/2)*k)^(1/3)*6
^(1/2)/(k*(108*k^2+8+12*3^(1/2)*(27*k^2+4)^(1/2)*k)
^(1/3)*((108*k^2+8+12*3^(1/2)*(27*k^2+4)^(1/2)*k)^
(2/3)+4+2*(108*k^2+8+12*3^(1/2)*(27*k^2+4)^(1/2)*k)
^(1/3)))^(1/2)

-k*(108*k^2+8+12*3^(1/2)*(27*k^2+4)^(1/2)*k)^(1/3)*
6^(1/2)/(k*(108*k^2+8+12*3^(1/2)*(27*k^2+4)^(1/2)*k)
^(1/3)*((108*k^2+8+12*3^(1/2)*(27*k^2+4)^(1/2)*k)^
(2/3)+4+2*(108*k^2+8+12*3^(1/2)*(27*k^2+4)^(1/2)*k)
^(1/3)))^(1/2)
```

Replace k with $f(x)$, then it could get :

```
h(x)= f(x)*(108* f(x)^2+8+12*3^(1/2)*(27* f(x)
^2+4)^(1/2)* f(x))^(1/3)*6^(1/2)/( f(x)*(108* f(x)^
2+8+12*3^(1/2)*(27* f(x)^2+4)^(1/2)* f(x))^(1/3)*
((108* f(x)^2+8+12*3^(1/2)*(27* f(x)^2+4)^(1/2)*
f(x))^(2/3)+4+2*(108* f(x)^2+8+12*3^(1/2)*(27*
f(x)^2+4)^(1/2)* f(x))^(1/3)))^(1/2)

h(x)=- f(x)*(108* f(x)^2+8+12*3^(1/2)*(27* f(x)
^2+4)^(1/2)* f(x))^(1/3)*6^(1/2)/( f(x)*(108* f(x)^
2+8+12*3^(1/2)*(27* f(x)^2+4)^(1/2)* f(x))^(1/3)*
((108* f(x)^2+8+12*3^(1/2)*(27*k^2+4)^(1/2)* f(x))
^(2/3)+4+2*(108* f(x)^2+8+12*3^(1/2)*(27* f(x)^2
+4)^(1/2)* f(x))^(1/3)))^(1/2)
```

$f(x)$ is real functions, obviously, $h(x)$ are also real functions, therefore, in real number field, the existence of $h(x)$ is proved. The number of $h(x)$ is 2.

In short, for two kinds of situation above, there are 4 $h(x)$ in real number field.

4 Brief summary

Linear probability density function is a kind of probability density function, operating linear integral for linear probability density function is another expression of seeking cumulative probability. It could according to the known specified path to obtain different linear probability density function.

References

1. Li Zhong, Fang Liping, *A course in mathematical analysis,* Higher Education Press, Beijing, **2** (2008): 70-75.
2. Department of Applied Mathematics of Tongji University, *Higher mathematics*, Higher Education Press, Beijing, **2** (2002): 126-130.
3. Mao Shisong, Pu Xiaolong, Cheng Yiming, *A concise on probability theory and mathmatical statistics*, Higher Education Press, Beijing, (2012): 42-44.
4. Mao Shisong, Lv Xiaoling, *Mathematical Statistics*(China Renmin University Press, Beijing, (2011), 35-45.
5. Wu Xizhi, *Statistics: from data to conclusions,* China Statistics Press, Beijing, (2011), 57-64

MATEC Web of Conferences **68**, 16002 (2016)
ICIEA 2016 DOI: 10.1051/matecconf/20166816002

Fatou type weighted pointwise convergence of nonlinear singular integral operators Depending on two parameters

Gumrah Uysal[1] and Sevilay Kirci Serenbay[2]

[1] *Karabuk University, Faculty of Science, Department of Mathematics, Karabuk, Turkey*
[2] *Baskent University, Faculty of Education, Department of Mathematic Education , Ankara, Turkey*

Abstract. In this paper we present some theorems concerning existence and Fatou type weighted pointwise convergence of nonlinear singular integral operators of the form:

$$(T_\lambda f)(x) = \int_{\mathrm{R}} K_\lambda\left(t-x; f\left(t\right)\right)dt, \ \ x \in \mathrm{R}, \ \lambda \in \Lambda$$

where $\Lambda \neq \varnothing$ is a set of non-negative indices, at a common generalized Lebesgue point of the functions $f \in L_{1,\phi}(\mathrm{R})$ and positive weight function ϕ. Here, $L_{1,\phi}(\mathrm{R})$ is the space of all measurable functions for which $\left|\frac{f}{\phi}\right|$ is integrable on R.

1 Introduction

In [1], Taberski analyzed the pointwise approximation to functions $f \in L_1\langle -\pi,\pi\rangle$ and their derivatives by a family of convolution type linear singular integral operators depending on two parameters of the form:

$$L_\lambda(f;x) = \int_{-\pi}^{\pi} f(t)K_\lambda(t-x)dt, \ \ x \in \langle -\pi,\pi\rangle, \ \lambda \in \Omega \subset \mathrm{R}_0^+$$

where the symbol $\langle -\pi,\pi\rangle$ stands for closed, semi-closed or open interval and $K_\lambda(t)$ is the kernel satisfying suitable assumptions. Subsequently, the pointwise convergence of the operators of above type was examined by Gadjiev [2] and Rydzewska [3] at generalized Lebesgue points and $\mu-$ generalized Lebesgue points of function $f \in L_1\langle -\pi,\pi\rangle$ respectively. Besides, Mamedov [4] and Taberski [5] obtained significant results on weighted pointwise convergence of linear singular integral operators. For some recent studies on singular integrals, authors refer to [6-8].

Later on, Musielak [9, 10] improved the notion of singularity to include the case of nonlinear integral operators of the form:

$$(T_w f)(y) = \int_G K_w(x-y; f(x))dx, \ y \in G, \ w \in \Lambda$$

where G be a locally compact Abelian group equipped with Haar measure and $\Lambda \neq \varnothing$ be an index set with any topology, via replacing the linearity property of the operators by an assumption of Lipschitz condition for K_w with respect to second variable. The studies, which were published until that time, showed that the singularity of the operators was related to their linearity [11]. Afterwards, Swiderski and Wachnicki [12] investigated the pointwise convergence of the operators of the preceding type at Lebesgue points of the functions $f \in L_p(-\pi,\pi)$.

Nowadays, approximation via nonlinear integral operators is extensively used in many branches of science such as medicine and engineering. Especially, effect of nonlinear integral operators in sampling theory must be emphasized here [11]. Further, signal and image processing are two major research fields around sampling theory. In view of this situation, one may choose to study the convergence of nonlinear integral operators rather than the convergence of linear integral operators. For further studies concerning Fatou type convergence of the singular integral operators, authors refer to [13-15].
The current manuscript deals with the Fatou type weighted pointwise convergence of nonlinear singular integral operators of the form:

$$(T_\lambda f)(x) = \int_{\mathrm{R}} K_\lambda\left(t-x; f\left(t\right)\right)dt, \ \ x \in \mathrm{R}, \ \lambda \in \Lambda$$

where $\Lambda \neq \varnothing$ be the set of indices, at generalized Lebesgue point of the functions $f \in L_{1,\phi}(\mathrm{R})$ and ϕ. Here, $L_{1,\varphi}(\mathrm{R})$ is the space of all measurable functions for which $\left|\frac{f}{\phi}\right|$ is integrable on R.

In Section 2, we introduce fundamental notions. In Section 3, we prove the existence of the operators. In

Section 4, we give a theorem concerning the pointwise convergence of $(T_\lambda f)(x)$.

2 Preliminaries

Following [12, 14], let G be the locally compact Abelian group equipped with Haar measure. In addition, $N(\vartheta)$ stands for the family of all neighborhoods of the identity element $\vartheta \in G$ and ϑ_0 is an accumulation point of G. Let $\Lambda \neq \varnothing$ be a set of indices with any topology and λ_0 be an accumulation point of this set in this topology.

Definition 1. (*Class* A_ϕ) Suppose that the weight function $\phi : \mathrm{R} \to \mathrm{R}^+$ is bounded on arbitrary bounded subsets of R and the inequality:

$$\phi(t+x) \le \phi(t)\phi(x), \quad t \in \mathrm{R},\ x \in \mathrm{R}$$

holds. We will say that the family of the functions $K_\lambda : G \times \mathrm{R} \to \mathrm{R}$, where $K_\lambda(\vartheta,u)$ is Haar integrable on G for all $u \in \mathrm{R}$, belongs to *Class* A_ϕ if the following conditions are satisfied:

a) $K_\lambda(\vartheta,0) = 0$ for every $\vartheta \in G$ and $\lambda \in \Lambda$.

b) There exists an integrable function $L_\lambda : G \to \mathrm{R}$ such that the following inequality:

$$|K_\lambda(t,u) - K_\lambda(t,v)| \le L_\lambda(t)|u-v|,$$

holds $\forall t \in G$, $\forall u, v \in \mathrm{R}$ and for each $\lambda \in \Lambda$.

c) $\forall u \in \mathrm{R}$, we have

$$\lim_{(x,\lambda)\to(x_0,\lambda_0)} \left| \int_G K_\lambda\left(t-x; \tfrac{u}{\phi(x_0)}\phi(t)\right)dt - u \right| = 0,$$

where $x_0 \in \mathrm{R}$ is a generalized Lebesgue point of ϕ.

d)

$$\lim_{\lambda\to\lambda_0} \sup_{t \in G\setminus\Omega} \left[\phi(t) L_\lambda(t)\right] = 0, \quad \forall \Omega \in N(\vartheta).$$

e)

$$\lim_{\lambda\to\lambda_0} \left[\int_{G\setminus\Omega} \phi(t) L_\lambda(t)\, dt \right] = 0, \quad \forall \Omega \in N(\vartheta).$$

f) For a given $\delta_1 > 0$, $L_\lambda(t)$ is non-decreasing function as a function of t on $\langle -\delta_1, 0]$ and non-increasing function as a function of t on $[0, \delta_1\rangle$.

g) $\|\phi L_\lambda\|_{L_1(G)} \le M < \infty, \quad \forall \lambda \in \Lambda.$

Throughout this article, we assume that $G = \mathrm{R}$ and K_λ belongs to class A_ϕ.

Definition 2. [16] A generalized Lebesgue point of a locally integrable function $g : \mathrm{R} \to \mathrm{R}$ is a point $x_0 \in \mathrm{R}$ satisfying

$$\lim_{h\to 0} \frac{1}{h^{\alpha+1}} \int_{x_0}^{x_0+h} |g(t) - g(x_0)|\, dt = 0, \quad 0 \le \alpha \le N.$$

Example 1. Let $f : \mathrm{R} \to \mathrm{R}$ be given by

$$f(t) = \begin{cases} te^{-t}, & t \in (0,1] \\ 0, & \text{otherwise.} \end{cases}$$

It is easy to see that $x_0 = 0$ is a generalized Lebesgue point of the given function for $\alpha = \frac{1}{4}$.

3 Existence of the Operators

Main result in this work is based on the following theorem.

Theorem 1. If $f \in L_{1,\phi}(\mathrm{R})$, then the operator $(T_\lambda f)(x) \in L_{1,\varphi}(\mathrm{R})$ *and*

$$\|(T_\lambda f)\|_{L_{1,\phi}(\mathrm{R})} \le \|\phi L_\lambda\|_{L_1(\mathrm{R})} \|f\|_{L_{1,\phi}(\mathrm{R})}$$

for all $\lambda \in \Lambda$.

Proof: Using Fubini's Theorem [17] we may write

$$\begin{aligned}
\|(T_\lambda f)(x)\|_{L_{1,\phi}(\mathrm{R})} &= \int_{-\infty}^{\infty} \frac{1}{\phi(x)} \left| \int_{-\infty}^{\infty} K_\lambda(t-x; f(t))dt \right| dx \\
&\le \int_{-\infty}^{\infty} \frac{1}{\phi(x)} \int_{-\infty}^{\infty} \left| K_\lambda(t-x; f(t)) - K_\lambda(t-x;\ 0) \right| dtdx \\
&\le \int_{-\infty}^{\infty} \frac{1}{\phi(x)} \int_{-\infty}^{\infty} \left| f(t+x) \frac{\phi(t+x)}{\phi(t+x)} L_\lambda(t) \right| dtdx \\
&\le \int_{-\infty}^{\infty} L_\lambda(t) \left[\int_{-\infty}^{\infty} \frac{\phi(t)\phi(x)}{\phi(x)} \left| \frac{f(t+x)}{\phi(t+x)} \right| dx \right] dt \\
&\le \|\phi L_\lambda\|_{L_1(\mathrm{R})} \|f\|_{L_{1,\phi}(\mathrm{R})},
\end{aligned}$$

where the norm of f is given by the following equality (see, e.g., [4]):

$$\|f\|_{L_{1,\phi}(R)} = \int_{-\infty}^{\infty} \left| \tfrac{f(x)}{\phi(x)} \right| dx.$$

4 Convergence at characteristic points

In this section we prove the Fatou type weighted pointwise convergence of the operators $(T_\lambda f)(x)$, i.e. the convergence will be restricted to the bounded planar subsets of $\mathrm{R} \times \Lambda$. For this purpose, we define the set:

$$Z=\{(x,\lambda)\in \mathrm{R}\times\Lambda\ :\ \Omega_\delta(x,\lambda)<C\},$$

where C is any positive constant and for δ values satisfying $0<\delta<\delta_1$, the function Ω_δ is defined by

$$\Omega_\delta(x,\lambda)=$$
$$\sup_{t\in(x_0-\delta,x_0+\delta)}\phi(t)\left(\begin{array}{l}\int\limits_{x_0-\delta}^{x_0+\delta} L_\lambda(t-x)\left|\{|t-x_0|^{\alpha+1}\}'_t\right|dt\\ +2L_\lambda(0)|x-x_0|^{\alpha+1}\end{array}\right),$$
$$0\le\alpha\le N.$$

Theorem 1. If $x_0\in\mathrm{R}$ is a common generalized Lebesgue point of the functions $f\in L_{1,\phi}(\mathrm{R})$ and ϕ then

$$\lim_{(x,\lambda)\to(x_0,\lambda_0)}|(T_\lambda f)(x)-f(x_0)|=0$$

as (x,λ) tends to (x_0,λ_0) whenever $(x,\lambda)\in Z$.

Proof: Let $0<x_0-x<\frac{\delta}{2}$ for δ which satisfies $\delta_1>\delta>0$. Set $I=|(T_\lambda f)(x)-f(x_0)|$. Now, using condition (c) of class A_ϕ, we may write

$$I=\left|\int\limits_{-\infty}^{\infty}K_\lambda(t-x;f(t))dt-f(x_0)\right|$$
$$=\left|\int\limits_{-\infty}^{\infty}K_\lambda(t-x;f(t))dt-\int\limits_{-\infty}^{\infty}K_\lambda\left(t-x;\frac{f(x_0)}{\phi(x_0)}\phi(t)\right)dt\right.$$
$$\left.+\int\limits_{-\infty}^{\infty}K_\lambda\left(t-x;\frac{f(x_0)}{\phi(x_0)}\phi(t)\right)dt-f(x_0)\right|.$$

Using condition (b) of class A_ϕ, the following inequality:

$$I\le\int\limits_{-\infty}^{\infty}\left|\frac{f(t)}{\phi(t)}-\frac{f(x_0)}{\phi(x_0)}\right|\phi(t)L_\lambda(t-x)dt$$
$$+\left|\int\limits_{-\infty}^{\infty}K_\lambda\left(t-x;\frac{f(x_0)}{\phi(x_0)}\phi(t)\right)dt-f(x_0)\right|$$
$$=I_1+I_2$$

is obtained. Using condition (c) of class A_ϕ, $I_2\to0$ as (x,λ) tends to (x_0,λ_0). By the hypothesis, it is easy to see that the following inequality for I_1,

$$I_1\le\varepsilon\Omega_\delta(x,\lambda)+\int\limits_{\mathrm{R}\setminus(x_0-\delta,x_0+\delta)}\left|\frac{f(t)}{\phi(t)}-\frac{f(x_0)}{\phi(x_0)}\right|\phi(t)L_\lambda(t-x)dt$$
$$\le\varepsilon\Omega_\delta(x,\lambda)+\phi(x)\|f\|_{L_{1,\phi}(\mathrm{R})}\sup_{|u|\ge\frac{\delta}{2}}[\phi(u)L_\lambda(u)]$$
$$+\phi(x)\left|\frac{f(x_0)}{\phi(x_0)}\right|\int\limits_{|u|\ge\frac{\delta}{2}}\phi(u)L_\lambda(u)du,$$

holds.

Therefore, if the points $(x,\lambda)\in Z$ are sufficiently near to (x_0,λ_0), then the first term on the right hand side of the above inequality tends to zero.

Also, remaining two terms tend to zero by conditions (d) and (e) of class A_ϕ, respectively. Using similar method, one may obtain the same result for the case $0<x-x_0<\frac{\delta}{2}$. Hence, the proof is completed.

5 Conclusion

In this paper, the weighted pointwise convergence for the family of nonlinear convolution type singular integral operators depending on two parameters is investigated. This study may be seen as a continuation and generalization of the previous studies, such as [1, 12]. Since the approximation results and the character of the kernel function are related, a special class of kernel functions, called class A_ϕ, has been defined. Therefore, the main result is presented as Theorem 1.

References

1. R. Taberski, Prace Mat. **7**, 173-179 (1962)
2. A. D. Gadjiev, *The order of convergence of singular integrals which depend on two parameters*, in: Special Problems of Functional Analysis and Their Applications to the Theory of Differential Equations and the Theory of Functions, Izdat. Akad. Nauk Azerbaidžan SSR., Baku, 40-44 (1968)
3. B. Rydzewska, Fasc. Math. **7**, 71-81 (1973)
4. R. G. Mamedov, Izv. Akad. Nauk. SSSR Ser. Mat. **27** (2), 287-304 (1963)
5. R. Taberski, Comment. Math. Prace Mat. 19 (1), 155-160 (1976)
6. G. Uysal, M. M. Yilmaz, E. Ibikli, J. Inequal. Appl. **2015**, 94 (2015)
7. G. Uysal, E. Ibikli, J. Pure Appl. Math. Adv. Appl. **14** (2), 151-166 (2015)
8. G. Uysal, E. Ibikli, New Trends Math. Sci. **4** (1), 104-114 (2016)
9. J. Musielak, *On some approximation problems in modular spaces*, in: Constructive Function Theory Publ. House Bulgarian Acad. Sci., Sofia , 455-461, (1983)
10. J. Musielak, Comment. Math. Prace Mat., **31**, 79-88 (1991)
11. C. Bardaro, J. Musielak, G. Vinti, *Nonlinear integral*

operators and applications, DeGruyter Series in Nonlinear Analysis and Applications, **9** (2003)
12. T. Swiderski, E. Wachnicki, Comment. Math. **40**, 181-189 (2000)
13. H. Karsli, E. Ibikli, Fasc. Math., **38,** 25–39 (2007)
14. M. Carlsson, Math. Scand. **102** (2), 231–252 (2008)
15. H. Karsli, Appl. Math. Comput. **246,** 221–228 (2014)
16. R. G. Mamedov, *A study of orders of convergence of one-dimensional and multidimensional singular integrals*, in: Studies in Theory of Differential Equations and Theory of Functions, Izdat. Akad. Nauk Azerbaĭdžan. SSR, Baku, 92-108 (1965)
17. P. L. Butzer, R. J. Nessel, *Fourier Analysis and Approximation*, vol. I. Academic Press, New York, London, (1971)

MATEC Web of Conferences **68**, 16003 (2016) DOI: 10.1051/matecconf/20166816003
ICIEA 2016

Parameter Derivatives of the Jacobi Polynomials with Three Variables on the Simplex

Rabia Aktaş

Ankara University, Faculty of Science, Department of Mathematics, 06100, Ankara, Turkey

Abstract. In this paper, an attempt has been made to derive parameter derivatives of Jacobi polynomials with three variables on the simplex. They are obtained via parameter derivatives of the classical Jacobi polynomials $P_n^{(\alpha,\beta)}(x)$ with respect to their parameters. The study is motivated by the expansions of parameter derivatives which are obtained for the classical Jacobi polynomials over the interval (-1, 1), Jacobi polynomials with two variables on the triangle and some other families of orthogonal polynomials in two variables.

1 Introduction

In literature, there are many studies on the parameter derivatives of various special functions which find applications in applied mathematics, several branches of mathematics, mathematical and theoretical physics. In [13, 14], author has studied the derivative of the Legendre function of the first kind, with respect to its degree ν, $[\partial P_\nu(z)/\partial \nu]_{\nu=n} (n\in\mathbb{N})$, and has obtained its some representations, which are encountered in some problems such as in the general theory of relativity and in solving some boundary value problems of potential theory, of electromagnetism and of heat conduction in solids. The derivatives of the associated Legendre function of the first kind with respect to its order and its degree (with applications to the construction of the associated Legendre function of the second kind of integer degree and order) have been obtained in [15, 16]. These derivatives of the associated Legendre function arise in solutions of some problems of heat conduction, theoretical acoustics and other branches of theoretical physics.

To obtain parameter derivative representations for orthogonal polynomials with one variable, a general method has been investigated in the following form

$$\frac{\partial P_n(\lambda;x)}{\partial\lambda}=\sum_{k=0}^{n} c_{n,k}(\lambda)P_k(\lambda;x) \qquad (1)$$

for orthogonal polynomials with variable x, with λ being a parameter, and such derivative representations have been obtained by many authors for orthogonal polynomials in one variable [6, 7, 8, 9, 11, 12, 17]. In particular, the explicit expansions of the form (1) have been investigated for dicrete Laguerre polynomials in [17], for Jacobi polynomials in [6], for Gegenbauer polynomials and the generalized Laguerre polynomials in [7]. The following representation given in [7] for the generalized Laguerre polynomials

$$\frac{\partial L_n^{(\alpha)}(x)}{\partial\alpha}=\sum_{k=0}^{n-1}\frac{1}{n-k}L_k^{(\alpha)}(x)$$

has applications in relativistic quantum mechanics. Also, the parameter derivative representations in the form of (1) for all the continuous and discrete orthogonal polynomials have been studied in [8]. Moreover, in [11], the following expansion form which is general than (1)

$$\frac{\partial^m P_n(\lambda;x)}{\partial\lambda^m}=\sum_{k=0}^{n} a_{n,k}(m,\lambda)P_k(\lambda;x) \quad (m\in\mathbb{N})$$

has been investigated and the recurrence relations for coefficients in the expansion has been obtained. In [9], a way to obtain parameter derivative representations of order m $(m\in\mathbb{N})$ for the families of almost all the classical orthogonal polynomials has been studied.

The representations of parameter derivatives for Jacobi polynomials $P_n^{(\alpha,\beta)}(x)$ obtained by Froehlich are as follows

$$\begin{aligned}\frac{\partial P_n^{(\alpha,\beta)}(x)}{\partial\alpha}&=[\psi(2n+\alpha+\beta+1)-\psi(n+\alpha+\beta+1)]P_n^{(\alpha,\beta)}(x)\\&+\sum_{k=0}^{n-1}\frac{(2k+\alpha+\beta+1)\Gamma(k+\alpha+\beta+1)}{(n-k)(n+k+\alpha+\beta+1)\Gamma(k+\beta+1)}\\&\times\frac{\Gamma(n+\beta+1)}{\Gamma(n+\alpha+\beta+1)}P_k^{(\alpha,\beta)}(x)\end{aligned} \qquad (2)$$

and

$$\frac{\partial P_n^{(\alpha,\beta)}(x)}{\partial\beta}$$
$$=[\psi(2n+\alpha+\beta+1)-\psi(n+\alpha+\beta+1)]P_n^{(\alpha,\beta)}(x)$$
$$+\sum_{k=0}^{n-1}\frac{(-1)^{n+k}(2k+\alpha+\beta+1)\Gamma(k+\alpha+\beta+1)}{(n-k)(n+k+\alpha+\beta+1)\Gamma(k+\alpha+1)}$$
$$\times\frac{\Gamma(n+\alpha+1)}{\Gamma(n+\alpha+\beta+1)}P_k^{(\alpha,\beta)}(x) \qquad (3)$$

for $\alpha,\beta>-1$ where $\psi(\xi)=\frac{1}{\Gamma(\xi)}\frac{d\Gamma(\xi)}{d\xi}$ is the digamma function, which satisfies

$$[\psi(\alpha+n)-\psi(\alpha)]=\sum_{k=0}^{n-1}\frac{1}{\alpha+k}$$

(see ([1], ? 6.3), [7]).

We remind that the classical Jacobi polynomials $P_n^{(\alpha,\beta)}(x)$ which are orthogonal polynomials with respect to the inner product

$$\langle f,g\rangle_{\alpha,\beta}:=\int_{-1}^{1}f(x)g(x)(1-x)^\alpha(1+x)^\beta dx$$

are defined through

$$P_n^{(\alpha,\beta)}(x)={}_2F_1[-n,\alpha+\beta+n+1;\alpha+1;\frac{1-x}{2}]$$
$$\times\frac{(\alpha+1)_n}{n!}$$

where ${}_2F_1$ denotes Gauss hypergeometric function and the Pochammer symbol is defined by

$$(\alpha)_0=1\ ,\quad (\alpha)_k=\frac{\Gamma(\alpha+k)}{\Gamma(\alpha)},\quad k=1,2,....$$

In the recent papers [2, 3, 4], author has studied similar expansion in the form of

$$\frac{\partial P_{n,k}(\lambda;x,y)}{\partial\lambda}=\sum_{m=0}^{n-1}\sum_{j=0}^{m}d_{n,j,m}P_{m,j}(\lambda;x,y)$$
$$+\sum_{j=0}^{k}e_{n,j,k}P_{n,j}(\lambda;x,y) \qquad (4)$$

for orthogonal polynomials with two variables, with λ being a parameter and $0\le k\le n; n\in\mathbb{N}_0$ and has given parameter derivative representations for Jacobi polynomials with two variables on the triangle and some families of orthogonal polynomials in two variables by using the expansions (2) and (3) given for Jacobi polynomials with one variable. Motivated and inspired by the works [2, 3, 4, 6, 7, 8, 9, 11, 12, 17], we derive representations of parameter derivatives for Jacobi polynomials with three variables on the simplex similar to the expansions in the forms of (1) and (4) by using the parametric derivatives of the classical Jacobi polynomials $P_n^{(\alpha,\beta)}(x)$.

2 The parameter derivatives of Jacobi polynomials with three variables on the simplex

The Jacobi polynomials with three variables on the simplex

$$T^3:=\{(x,y,z)\in\mathbb{R}^3\ :\ x,y,z\ge 0,1-x-y-z\ge 0\}$$

are orthogonal with respect to the weight function $W_{\alpha,\beta,\gamma,\delta}(x,y,z)=x^\alpha y^\beta z^\gamma(1-x-y-z)^\delta$ for $\alpha,\beta,\gamma,\delta>-1$ with respect to the inner product

$$\langle f,g\rangle_{\alpha,\beta,\gamma,\delta}:=\int_{T^3}f(x,y,z)g(x,y,z)W_{\alpha,\beta,\gamma,\delta}(x,y,z)dxdydz.$$

Let $V_n^3(W_{\alpha,\beta,\gamma,\delta})$ be the space of polynomials of degree n which are orthogonal with respect to this inner product. One of the bases for $V_n^3(W_{\alpha,\beta,\gamma,\delta})$ is as follows

$$P_{m,k}^n(x,y,z):=P_{n,m,k}^{(\alpha,\beta,\gamma,\delta)}(x,y,z)$$
$$=P_{n-m}^{(2m+\beta+\gamma+\delta+2,\alpha)}(2x-1)(1-x)^{m-k}P_{m-k}^{(2k+\gamma+\delta+1,\beta)}\left(\frac{2y}{1-x}-1\right)$$
$$\times(1-x-y)^kP_k^{(\delta,\gamma)}\left(\frac{2z}{1-x-y}-1\right)$$

for $0\le k\le m\le n$; $n\in\mathbb{N}_0$ where $P_n^{(\alpha,\beta)}(x)$ denotes the classical Jacobi polynomial of degree n (see [5]).

Since the simplex T^3 is symmetric under the permutation of $(x,y,z,1-x-y-z)$, by parametrizing the simplex differently, we can give the following orthogonal bases of $V_n^3(W_{\alpha,\beta,\gamma,\delta})$

$$T_{m,k}^n(x,y,z):=P_{n,m,k}^{(\beta,\gamma,\alpha,\delta)}(y,z,x),$$

$$S_{m,k}^n(x,y,z):=P_{n,m,k}^{(\gamma,\alpha,\beta,\delta)}(z,x,y)$$

and

$$Q_{m,k}^n(x,y,z):=P_{n,m,k}^{(\delta,\beta,\gamma,\alpha)}(1-x-y-z,y,z)$$

for $0\le k\le m\le n$; $n\in\mathbb{N}_0$. These bases of $V_n^3(W_{\alpha,\beta,\gamma,\delta})$ are given explicitly by

$$T_{m,k}^n(x,y,z)=P_{n-m}^{(2m+\alpha+\gamma+\delta+2,\beta)}(2y-1)(1-y)^{m-k}$$
$$\times P_{m-k}^{(2k+\alpha+\delta+1,\gamma)}\left(\frac{2z}{1-y}-1\right)(1-y-z)^k$$
$$\times P_k^{(\delta,\alpha)}\left(\frac{2x}{1-y-z}-1\right),$$

$$S^n_{m,k}(x,y,z)=P^{(2m+\alpha+\beta+\delta+2,\gamma)}_{n-m}(2z-1)(1-z)^{m-k}$$
$$\times P^{(2k+\beta+\delta+1,\alpha)}_{m-k}\left(\frac{2x}{1-z}-1\right)(1-x-z)^k$$
$$\times P^{(\delta,\beta)}_k\left(\frac{2y}{1-x-z}-1\right)$$

and

$$Q^n_{m,k}(x,y,z)=P^{(2m+\alpha+\beta+\gamma+2,\delta)}_{n-m}(1-2x-2y-2z)$$
$$\times(x+y+z)^{m-k}P^{(2k+\alpha+\gamma+1,\beta)}_{m-k}\left(\frac{2y}{x+y+z}-1\right)$$
$$\times(x+z)^kP^{(\alpha,\gamma)}_k\left(\frac{2z}{x+z}-1\right),$$

for $0\leq k\leq m\leq n$; $n\in\mathbb{N}_0$ (see [5]).

Now, by using the parameter derivative representation given by (3), let's obtain similar representations for Jacobi polynomials with three variables on the simplex.

Theorem 1. *For* $\alpha,\beta,\gamma,\delta>-1$, $0\leq k\leq m\leq n$; $n\in\mathbb{N}_0$, *Jacobi polynomials* $P^n_{m,k}(x,y,z)$ *satisfy the following representation for* $n\geq m+1$, $m\geq k\geq 0$

$$\frac{\partial P^n_{m,k}(x,y,z)}{\partial\alpha}$$
$$=[\psi(2n+\alpha+\beta+\gamma+\delta+3)-\psi(n+m+\alpha+\beta+\gamma+\delta+3)]\,P^n_{m,k}(x,y,z)$$
$$+\frac{\Gamma(n+m+\beta+\gamma+\delta+3)}{\Gamma(n+m+\alpha+\beta+\gamma+\delta+3)}$$
$$\times\sum_{s=0}^{n-m-1}\frac{(-1)^{s+1}\Gamma(n+m-s+\alpha+\beta+\gamma+\delta+2)}{\Gamma(n+m-s+\beta+\gamma+\delta+2)}$$
$$\times\frac{(2n-2s+\alpha+\beta+\gamma+\delta+1)}{(s+1)(2n-s+\alpha+\beta+\gamma+\delta+2)}P^{n-s-1}_{m,k}(x,y,z)$$

and for $n=m\geq k\geq 0$

$$\frac{\partial P^n_{n,k}(x,y,z)}{\partial\alpha}=0.$$

Proof. If we take derivative of the Jacobi polynomials with three variables $P^n_{m,k}(x,y,z)$ with respect to its parameter α and then take into consideration the relation (3), we find the desired result.
Similarly, one can easily obtain the following representations for the other bases $T^n_{m,k}(x,y,z)$, $S^n_{m,k}(x,y,z)$ and $Q^n_{m,k}(x,y,z)$.

Theorem 2. *For* $\alpha,\beta,\gamma,\delta>-1$, *we get the next results for* $n\geq m+1$, $m\geq k\geq 0$

$$\frac{\partial T^n_{m,k}(x,y,z)}{\partial\beta}$$
$$=[\psi(2n+\alpha+\beta+\gamma+\delta+3)-\psi(n+m+\alpha+\beta+\gamma+\delta+3)]T^n_{m,k}(x,y,z)$$
$$+\frac{\Gamma(n+m+\alpha+\gamma+\delta+3)}{\Gamma(n+m+\alpha+\beta+\gamma+\delta+3)}$$
$$\times\sum_{s=0}^{n-m-1}\frac{(-1)^{s+1}\Gamma(n+m-s+\alpha+\beta+\gamma+\delta+2)}{\Gamma(n+m-s+\alpha+\gamma+\delta+2)}$$
$$\times\frac{(2n-2s+\alpha+\beta+\gamma+\delta+1)}{(s+1)(2n-s+\alpha+\beta+\gamma+\delta+2)}T^{n-s-1}_{m,k}(x,y,z),$$

$$\frac{\partial S^n_{m,k}(x,y,z)}{\partial\gamma}$$
$$=[\psi(2n+\alpha+\beta+\gamma+\delta+3)-\psi(n+m+\alpha+\beta+\gamma+\delta+3)]S^n_{m,k}(x,y,z)$$
$$+\frac{\Gamma(n+m+\alpha+\beta+\delta+3)}{\Gamma(n+m+\alpha+\beta+\gamma+\delta+3)}$$
$$\times\sum_{s=0}^{n-m-1}\frac{(-1)^{s+1}\Gamma(n+m-s+\alpha+\beta+\gamma+\delta+2)}{\Gamma(n+m-s+\alpha+\beta+\delta+2)}$$
$$\times\frac{(2n-2s+\alpha+\beta+\gamma+\delta+1)}{(s+1)(2n-s+\alpha+\beta+\gamma+\delta+2)}S^{n-s-1}_{m,k}(x,y,z),$$

and

$$\frac{\partial Q^n_{m,k}(x,y,z)}{\partial\delta}$$
$$=[\psi(2n+\alpha+\beta+\gamma+\delta+3)-\psi(n+m+\alpha+\beta+\gamma+\delta+3)]Q^n_{m,k}(x,y,z)$$
$$+\frac{\Gamma(n+m+\alpha+\beta+\gamma+3)}{\Gamma(n+m+\alpha+\beta+\gamma+\delta+3)}$$
$$\times\sum_{s=0}^{n-m-1}\frac{(-1)^{s+1}\Gamma(n+m-s+\alpha+\beta+\gamma+\delta+2)}{\Gamma(n+m-s+\alpha+\beta+\gamma+2)}$$
$$\times\frac{(2n-2s+\alpha+\beta+\gamma+\delta+1)}{(s+1)(2n-s+\alpha+\beta+\gamma+\delta+2)}Q^{n-s-1}_{m,k}(x,y,z).$$

For $n=m\geq k\geq 0$,

$$\frac{\partial T^n_{n,k}(x,y,z)}{\partial\beta}=\frac{\partial S^n_{n,k}(x,y,z)}{\partial\gamma}=0$$

and

$$\frac{\partial Q^n_{n,k}(x,y,z)}{\partial\delta}=0.$$

3 Conclusion

This paper is motivated by the works on the representations of parameter derivatives for orthogonal polynomials with one variable and two variables. In this study, we give parameter expansions for Jacobi polynomials with three variables on simplex, which such parameter derivatives are derived via the expansion (3) obtained by means of the parameter derivatives of Gauss Hypergeometric function for the classical Jacobi polynomials $P_n^{(\alpha,\beta)}(x)$. Also, we investigate explicit parameter derivative representations for the other orthogonal bases on the simplex by parametrizing the simplex differently. Furthermore, the orthogonality properties of such parametric derivatives can be studied in the further research.

References

1. M. Abramowitz, I. A. Stegun, Handbook of Mathematical Functions. Dover Publ., New York, (1964)
2. R. Aktaş, A note on parameter derivatives of the Jacobi polynomials on the triangle, Appl. Math. and Comp., **247** (2014), 368-372.
3. R. Aktaş, On parameter derivatives of a family of polynomials in two variables, Appl. Math. and Comp., **256** (2015), 769-777.
4. R. Aktaş, Representations for parameter derivatives of some Koornwinder polynomials in two variables,Journal of the Egyptian Mathematical Society, In Press.
5. C. F. Dunkl, Y. Xu, Orthogonal Polynomials of Several Variables. Encyclopedia of Mathematics and its Applications 81. Cambridge: Cambridge University Press, (2001)
6. J. Froehlich, Parameter derivatives of the Jacoby polynomials and the gaussian hypergeometric function, Integral Transforms Spec. Funct., **2(4)** (1994), 253-266.
7. W. Koepf, Identities for families of orthogonal polynomials and special functions, Integral Transforms Spec. Funct., **5(1-2)** (1997), 69-102.
8. W. Koepf, D. Schmersau, Representations of orthogonal polynomials, J. Comput. Appl. Math., **90** (1998), 57-94.
9. S. Lewanowicz, Representations for the parameter derivatives of the classical orthogonal polynomials , Rend. Circ. Mat. Palermo, Ser. II, Suppl. **68** (2002), 599-613.
10. E.D. Rainville, Special Functions,The Macmillan Co., New York, (1960)
11. A. Ronveaux, A. Zarzo, I. Area, E. Godoy, Classical orthogonal polynomials: dependence on parameters,J.Comput. Appl. Math., 121 (2000), 95-112.
12. R. Szmytkowski, A note on parameter derivatives of classical orthogonal polynomials, arXiv: 0901.2639v3.
13. R. Szmytkowski, On the derivative of the Legendre function of the first kind with respect to its degree, J. Phys. A, **39** (2006), 15147-15172 [corrigendum: J. Phys. A, 40 (2007), 7819-7820].
14. R. Szmytkowski, Addendum to `On the derivative of the Legendre function of the first kind with respect to its degree', J. Phys. A, **40** (2007), 14887-14891.
15. Szmytkowski, On the derivative of the associated Legendre function of the first kind of integer degree with respect to its order (with applications to the construction of the associated Legendre function of the second kind of integer degree and order), J. Math. Chem., **46** (2009), 231-260.
16. Szmytkowski, On the derivative of the associated Legendre function of the first kind of integer order with respect to its degree (with applications to the construction of the associated Legendre function of the second kind of integer degree and order), J. Math. Chem., **49** (2011), 1436-1477.
17. M. Wulkow, Numerical treatment of countable systems of ordinary differential equations, Konrad-Zuse-Zentrum Berlin, Techn. Rep. TR 90-8, (1990)

MATEC Web of Conferences **68**, 16004 (2016) DOI: 10.1051/matecconf/20166816004
ICIEA 2016

KAOS: A Kinetic Theory Tool for Modeling Complex Social Systems

Dario Bruneo, Marina Dolfin, Francesco Longo, Marco Scarpa and Antonio Puliafito

Università degli Studi di Messina, Italy

Abstract. The kinetic theory approach is successfully used to model complex phenomena related to social systems, allowing to predict the dynamics and emergent behavior of large populations of agents. In particular, kinetic theory for active particles (KTAP) models are usually analyzed by numerically solving the underlying Boltzmann-type differential equations through ad-hoc implementations. In this paper, we present KAOS: a kinetic theory of active particles modeling and analysis software tool. To the best of our knowledge, KAOS represents the first attempt to design and implement a comprehensive tool that assists the user in all the steps of the modeling process in the framework of the kinetic theories, from the model definition to the representation of transient solutions. To show the KAOS features, we present a new model capturing the competition/cooperation dynamics of a socio-economic system with welfare dynamics, in different socio-political conditions

1 Introduction

Long since mathematical methods traditionally applied in physics have been applied in interdisciplinary fields such as biology, economy or social sciences. For a review of methods borrowed from statistical physics and applied to interdisciplinary fields, see [1], while an historical perspective can be found in [2]. The micro-scale approach to complex systems is that of agent-based models, where rule-based agents dynamically interact. The approach of Complex Adaptative Systems combines agent-based and complex network based models [3]. Widespread models using this approach are Cellular Automata models (for a review of applications see [4]). The approach at the macro-scale, basically borrows ideas from the principles of fluid dynamics (for examples in vehicular traffic, see [5, 6] and references therein). Frameworks as Sociodynamics [7] and Sociophysics [8] arise from a dynamical systems approach to the modeling of phenomena in social sciences. The kinetic theory mathematical framework has been applied to many different interdisciplinary contexts; for a review of the literature and many interesting examples, see [9]. The Kinetic Theory of Active Particles (KTAP) adopts the general approach of the classical kinetic theory, although introducing interactions modeled using methodologies borrowed from the evolutive theory of stochastic games. In many applications, the analytic solutions may be no longer feasible; in these cases, a software tool could help in testing in a simple and quick way the predictive ability of kinetic models with respect to the reference application. Examples can be the consensus vs. polarization in opinion formation [10] or the radicalization of the economic gap in a society [11]. In this paper, we present a new model of wealth redistribution testing the emergence of patterns in the wealth micro-state of individuals and we introduce the KAOS software tool, a kinetic theory of active particles modeling and analysis framework. We also show how KAOS can help in analyzing such complex models. In fact, it has been designed with the purpose of assisting domain experts in studying emergent behaviors in a simple way without knowing the mathematical details of the underlying kinetic theory. We show a set of numerical results obtained by changing the model parameters in order to represent different societies coupled to political actions.

2 Kinetic Theory for Active Particles

The founding idea of kinetic models for living systems is that a system composed of a sufficiently large number of living beings can be described using the laws of statistical mechanics, as it happens in a physical system composed of many interacting particles [9]. However, it is necessary to consider that the main actors are not particles but living beings and that their interactions can be described as stochastic evolutive games [12, 13] rather than as binary elastic collisions as in the classical Boltzmann equations. Hallmarks of KTAP, when applied to the modeling of social systems, are: 1. The overall system is a large population of entities, called active particles, whose micro-state is characterized is characterized by a variable called activity representing a social attribute. 2. The state of the system is defined by a suitable time dependent probability distribution over the micro-state space. 3. Interactions between active particles are modeled as stochastic evolutive games, where the state of the interacting particles and the output of the interactions are known only in probability. 4. The evolution of the probability distribution is obtained by a balance of particles within the elementary volume of the micro-state space, where the dynamics of inflow and outflow of particles is related to the interactions at the micro-scale. The general micro-state of the active particles can be represented as $u \in D_u \subset \mathbb{R}$, where u is the activity (which refers to a social state) and it is assumed that a measure of the distance between two element in D_u is defined. A distribution function on the micro-state space is introduced $f = f(t,u): [0,T] \times D_u \to [0,1]$, describing the overall state of the system, where T is a reference time. Three types of active particles are introduced: the candidate particle which interacts with a field particle and acquires, with a given probability, the state of a test particle; only binary interactions are considered. The rate of variation of the number of active particles is given by a balance of inlet and outlet fluxes in the micro-state space. In KTAP, interactions are considered as stochastic games whose payoff is the micro-state acquired by the interacting active particles which is given in probability and whose rules may depend on the distribution itself, thus evolving in time; in this sense one may speak of evolutive stochastic games. The specific 'game' played depends on the specific phenomenon under observation and it is phenomenologically determined. In the following we will briefly introduce two particular kind of interactions often used in the applications; for both of them a threshold triggers the dynamics.

Competition/Cooperation. The competition dynamics takes place when the activity distance related to the interacting particles, i.e. the difference between the activity variable values of the interacting particles, is less than a given threshold μ; it mimics the fact that when the particles share a common niche, they tend to behave competitively and the result of the interaction is that they may increase their activity distance. The cooperation dynamics takes place in the opposite situation, i.e. when the activity distance between the interacting particles is above the given threshold; in this case the active particles are induced to behave cooperatively, thus reducing their activity distance.

Consensus/Dissent. The consensus dynamics takes place when the activity distance between the interacting active particles is less than a given threshold and the dissent dynamics in the opposite situation. In the consensus dynamics the post-interaction activity distance is shorter than the pre-interaction one; basically, the active particles 'approach' each other due to the interaction. In the dissent dynamics, the post-interaction active distance is greater than the pre-interaction one; basically the active particles move away from one another, with respect to the activity variable. This dynamics generalizes the well-known Bounded Confidence Model of statistical physics [14].

2.1 Discrete activity models

If the variable $u \in D_u$ is discretized, activities are written as $\{u_i\}, i = 1, \dots, N$, and the probability distribution on the micro-state space is a vector $\boldsymbol{f}(t) = [f_1(t), \dots, f_N(t)]$, where $f_i(t) = f(t, u_i)$. Since we do assume that the number of particles does not change over time in the considered time period, the probability density property $\sum_{i=1}^{N} f_i(t) = 1,\ \forall\, t \in [0,T]$ holds. Interactions are stochastic games involving the candidate particle with activity u_h which, due to the interaction with the field particle with activity u_k attains the micro-state u_i of the field particle. The general Boltzmann-type equation takes the form of a system of ordinary differential equations

$$\frac{df_i}{dt} = \sum_{h,k \in D_u} \eta_{hk}\, \mathfrak{B}^i_{hk} f_h f_k - f_i \sum_{k \in D_u} \eta_{ik}\, f_k, \quad i=1,\dots,N \qquad \textbf{(2.1)}$$

Supplemented by the initial condition $f(0)$. In order to derive explicit expressions for Eq. (2.1), we need to specify the following quantities:

$\eta_{hk} = \eta(u_h, u_k)$ – rate of interactions between the candidate particle with micro-state u_h and the field particle with micro-state u_k.

$\mathfrak{B}^i_{hk} = \mathfrak{B}(u_h \to u_i, u_h, u_k, \mathbf{f})$ – transition probabilities (also called *table of games*) defining the probability for a candidate active particle characterized by u_h to shift to the micro-state u_i due to an interaction with a field active particle characterized by u_k,. It may depend on the distribution $\mathbf{f}$ through macroscopic quantities such as statistical moments, thus introducing nonlinearities in the model [11, 15]. We introduce here an explicit form regarding the transition probability matrix in the case of competition/cooperation dynamics of interactions, which is the case we consider as example is Sec.4. The interaction of the type cooperation applies when $|u_h - u_k| \geq \mu$, with $0 < \mu < N$ and, in the case of wealth dynamics it represents wealth redistribution (i.e. through taxation). Under the cooperation interaction, the candidate active particle has a given probability to shift to an activity value which is nearer to the activity value of the field active particle, then the output of the interaction is in the interval between the candidate active particle activity and the field particle one. To define the explicit form of $\mathfrak{B}$, let us define the following quantities: $d = \min(u_h, u_k)$ and $D = \max(u_h, u_k)$. The elements of the table of games related to the cooperation interaction are thus:

$$\mathfrak{B}^i_{hk} = \begin{cases} \alpha(u_i | u_h, u_k, \mathbf{f}) & if\ (|u_h - u_k| \geq \mu) \wedge (u_i \in [d, D]) \\ 0 & otherwise \end{cases}$$

where $\alpha(u_i | u_h, u_k, \mathbf{f})$ is a probability mass function defined on D_u. At the moment, in the KAOS tool we implemented two kind of distribution functions. The *first-neighbor* distribution allows the shift to the first neighbor activity value only. Thus, $\alpha(u_i | u_h, u_k, \mathbf{f}) = \mathrm{p} \neq 0$, for either $i = h+1$ or $i = h+1$ when $u_h < u_k$ or $u_h \geq u_k$ respectively, and $\alpha(u_h | u_h, u_k, \mathbf{f}) = 1 - \mathrm{p}$. The value of p

in KAOS could set as $p = \beta_0 + \beta\,\mathbb{M}$, where $\mathbb{M}$ is a macroscopic quantity, weighted by the nonlinearity factor β, and such that $0 \leq \mathbb{M} \leq 1$. As an example, $\mathbb{M}$ can be the first order moment normalized with respect to the number of clusters N, in opinion dynamics, modeling the stream induced by the mean opinion in the society (majority rule). The *linearity coefficient* β_0 instead models the intensity of the effect of the interaction. As a second possibility, the $\alpha()$ distribution could be set as a discrete uniform distribution on $[d, D]$. The *competition* interaction applied when the activity distance between the candidate and the field particle is less than the given threshold μ; in the case of wealth dynamics, it represents the economic competition that may start when two active particles shares the same economic niche. In this case, u_h and u_k are such that $|u_h - u_k| < \mu$. In the competition dynamics, the candidate active particle has a probability to shift to an activity value which is farther with respect to the activity value of the field active particle due to the interaction, thus approaching one of the two extremes of the activity state space. In this case, the table of games reads:

$$\mathfrak{B}_{hk}^{i} = \begin{cases} \alpha(u_i|u_h, u_k, \mathbf{f}) & if\ \ (|u_h - u_k| < \mu) \wedge (u_h \leq u_k) \wedge (u_i \in [0, u_h]) \\ \alpha(u_i|u_h, u_k, \mathbf{f}) & if\ \ (|u_h - u_k| < \mu) \wedge (u_h > u_k) \wedge (u_i \in [u_h, D_u]) \\ 0 & otherwise \end{cases}$$

Similarly to the cooperation interaction, $\alpha(u_i|u_h, u_k, \mathbf{f})$ can be set according to the first-neighbor or uniform distribution.

3 The KAOS software tool

We designed and implemented KAOS, a tool for the definition and the transient solution of KTAP models. KAOS is implemented as an open source tool and it is publicly available on the Web (https://github.com/MDSLab/KAOS). The modeler can graphically represent the KTAP model by enumerating the number of clusters (depicted as circles in the central panel) Model parameters can be set by filling the text fields in the *Model Parameters* box on the right. In particular, the number of clusters N, the encounter rate η, the interaction mode, and the transition probabilities (table of games) can be set. In the *Initial Condition* box, the modeler can set the initial condition in terms of the probability distribution on the clusters at $t = 0$. Finally, in the *Solution Parameters* box, the final time and the time step of the transient analysis can be set. At the time being, the tool supports competition/cooperation, consensus/dissent and user-defined interaction modes. If one of the two modes competition/cooperation or consensus/dissent is set, the modeler has to choose the value of the threshold μ. The tool allows to set a first neighbor approach or a uniform distribution. If the first neighbor approach is chosen, the modeler can choose between a linear solution (i.e. the transition probability is constant along the whole solution) or a non-linear one (i.e. the transition probability changes as a function of the probability distribution itself at the considered time instant through a non-linearity factor). In this case the table of games is automatically generated and updated by the tool at each step of the solution. All these choices are performed through user-friendly wizards. In the user-defined interaction mode, the modeler is free to set the table of games as he/she wants. KAOS helps the modeler in modeling the interactions by representing the candidate particle in gray and moving it to the central position. The field particle is represented in yellow and moved externally while the other particles are kept white and maintained in their original positions. Arrows represent transitions from the candidate particle to the test particles and the modeler is allowed to insert user-defined transition probabilities for each cluster transition. Once the model has been fully specified, the solution can be started. KAOS implements a transient analysis and it graphically provides the time trend of the expected value of the probability distribution on the clusters from $t = 0$ to $t = T_{max}$ with steps equal to $DeltaT$, as set by the modeler. Moreover, it provides the probability distribution on the micro-state space at $t = T_{max}$.

4 A KTAP welfare model with KAOS

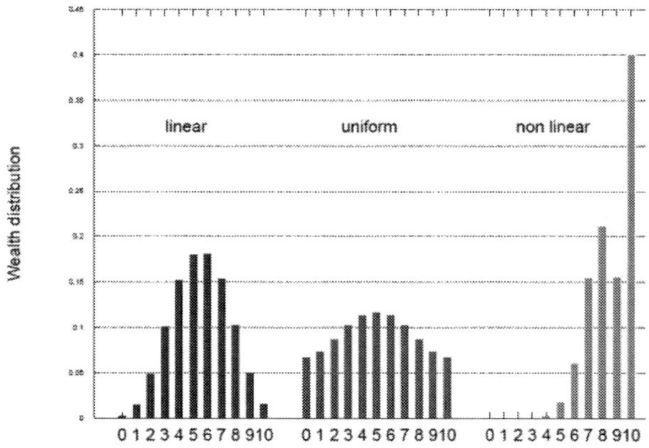

Figure 1. Steady-state wealth distribution (case study I, μ = 3).

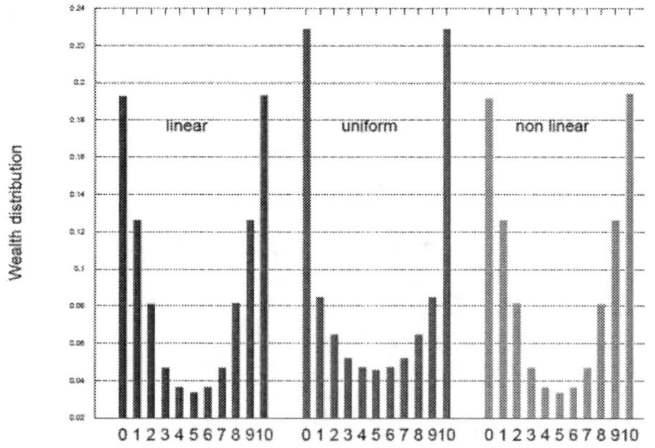

Figure 2. Steady-state wealth distribution (case study I, μ = 9).

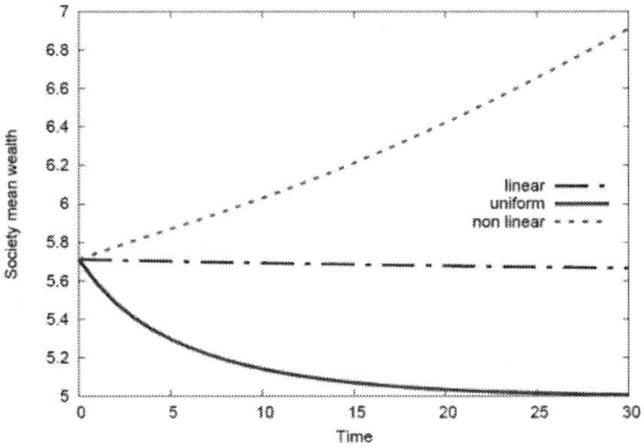

Figure 3. Time evolution of the society mean wealth (case study I, μ = 3).

We introduce a mathematical model for the redistribution of wealth in a society, numerically analyzing the resulting emergent behaviors. Active particles represent here individuals whose micro-state is characterized by individual's wealth. A wealth probability distribution characterizes the mesoscopic system and the related moments are the macroscopic variables representing the mean wealth within the society (first moment) or, in general, other macroscopic quantities (higher moments). We modela process based on a competition/cooperation dynamics. It mimics the fact that when the interacting active particles shae the same economic niche a competition between them may take place, whilst when they have wealth variables that are sufficiently distant a cooperation may start, as an effect of a welfare policy. The involved parameters are: the number of wealth classes *N* (*Number of clusters* in KAOS); the frequency of interactions between active particles η (*encounter rate* in KAOS); the degree of liberalism in the society μ (*Distance* in KAOS); the level of economic activity in competitive dynamics and the level of fiscal pressure in cooperative dynamics, both indicated as β_0 (*Neighbor Probability* in KAOS); the stream factor weighting the influence of macroscopic quantities on the dynamics β (*Non-linearity Factor* in KAOS). We used KAOS to numerically explore solutions of the model in the following two case studies.

Case study I. We model the effect of two different welfare policies on a society with an initial wealth distribution with prevalent presence of 'middle' wealth classes. The whole society is represented by $N = 11$ wealth classes. A low threshold represents a more enhanced welfare policy ($\mu = 3$) whilst an high threshold ($\mu = 9$) represents a policy enhancing economic competition. The initial distribution is chosen such that $f_i = 1/21$ for classes with $i = 0,1,2,3,4,9,10$ and $f_i = 1/6$ for classes with $i = 5,6,7,8$, while β_0=0.6 is the transition parameter. We compare the results obtained in the linear case, i.e. .for β_1=0, with those obtained in the nonlinear case, with β_1=0.4 (in both cases we used the *first-neighbor* table of games), with those obtained in the uniform distribution case. The obtained result are depicted in Figures 1, 2, 3 and 4. The value of μ can be considered, in this example of welfare policy, as the action of the goverrnment; smaller values model more cooperative actions among the wealth classes, whilst greater values introduce stronger competition. For a small value of the threshold, the society tends to evolve towards a middle class prevalence (Figure 1), instead in the case of an higher threshold the extreme competition creates higher inequalities in the society (Figure 2). In the non-linear case, we have chosen $\mathbb{M} = \sum_{i=6}^{11} f_i - \sum_{i=0}^{4} f_i$ quantifying a *social gap* [11], i.e. socio-economic inequalities in the society; in the case of low threshold the stream effect of the *social gap* of the society on the evolution of the wealth distribution has strong impact.

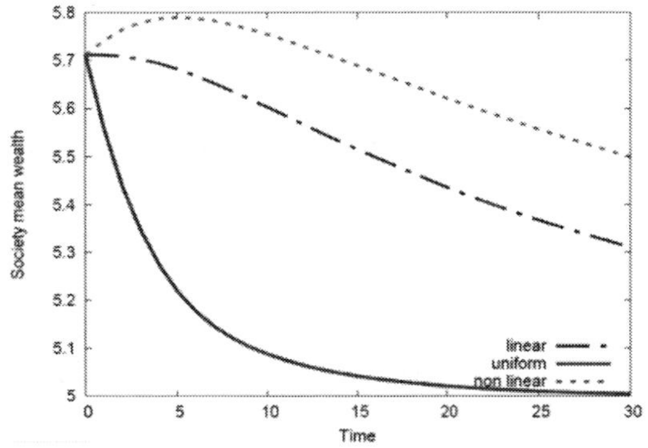

Figure 4. Time evolution of the society mean wealth (case study I, M = 9).

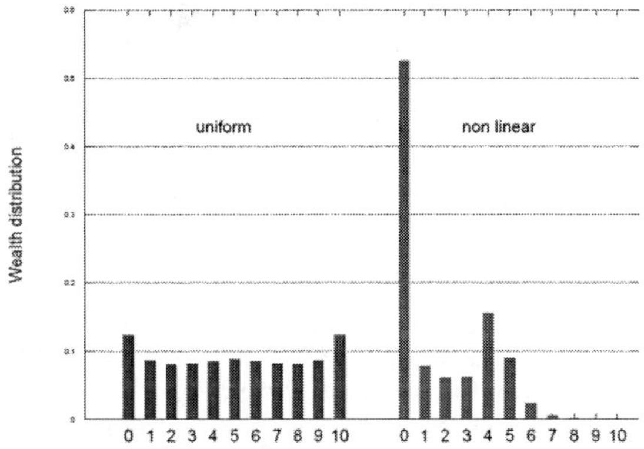

Figure 5. Steady-state wealth distribution for case study II with prevalent 'poor' classes initial distribution.

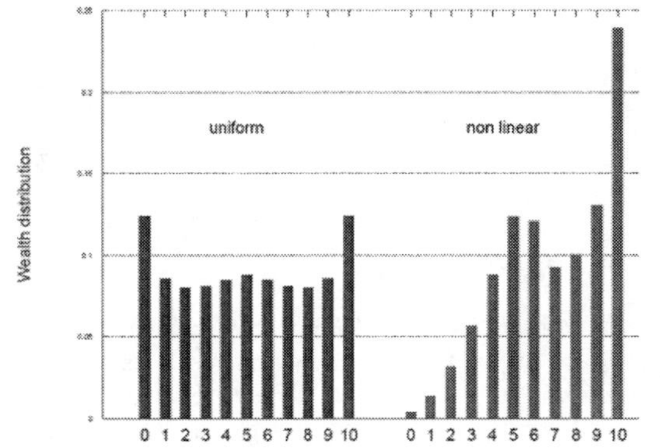

Figure 6. Steady-state wealth distribution for case study II with prevalent 'rich' classes initial distribution.

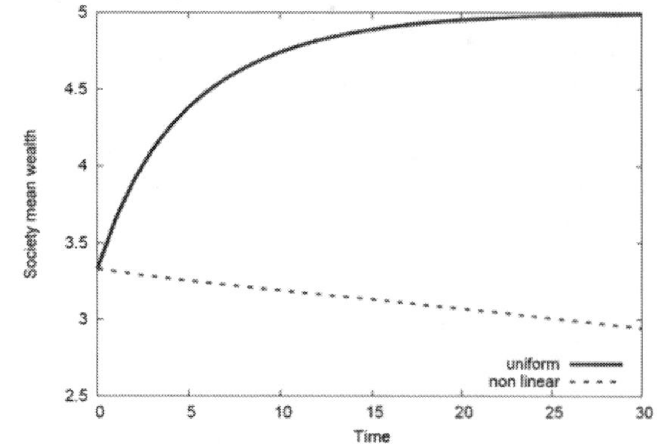

Figure 7. Time evolution of the society mean wealth for case study II with prevalent 'poor' classes initial distribution.

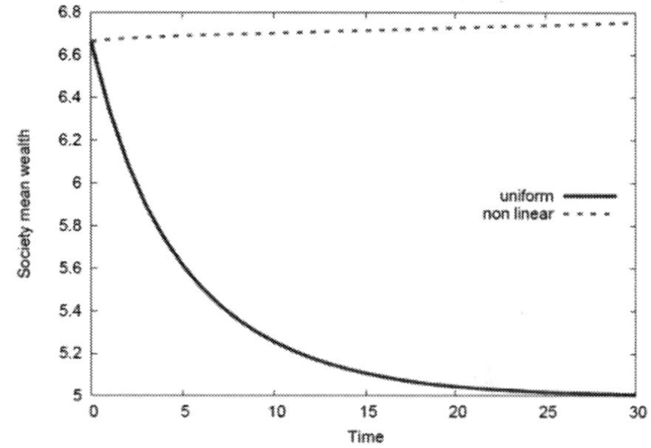

Figure 8. Time evolution of the society mean wealth for case study II with prevalent 'rich' classes initial distribution.

Case study II. We model the case of a welfare policy represented by a medium threshold ($\mu = 5$) on societies with different initial wealth distributions. In particular, we obtained the solutions for initial wealth distributions with prevalent 'poor' classes ($f_i = 1/6$ for classes with $i =$ 0,1,2,3 and $f_i = 1/21$ for classes with i = 4,5,6,7,8,9,10) and with prevalent 'rich' classes ($f_i = 1/21$ for classes with i =0,1,2,3,4,5,6 and $f_i = 1/6$ for classes with i = 7,8,9,10). In this case, for each solution we compare the results obtained in the case of non-linear first-neighbor table of games (β_1=0.4) with those obtained in the case of uniform distribution table of games. As in the previous case study, we used as nonlinearity function the *social gap*. The results obtained are depicted in Figures 5,6,7 and 8. The stream effect of the *social gap* of the society is evident for both nonlinear solutions.

5 Conclusions

KTAP theory is a powerful framework for the modeling of complex social systems. However, no comprehensive tools are available to help the modeler in describing and analyzing specific systems without the knowledge of the underlying mathematical details. In this paper, we introduce KAOS, a tool that fills this gap. We also proposed a model able to predict wealth dynamics in a complex society as a function of the society composition and the political governance, showing several numerical results. Even if in the proposed model the microscopic state of the agents is fully characterized by a single real variable and a single population, without the effect of external field agents, our final goal is to improve KAOS with the ability of managing many interacting populations on a network structure supporting the possibility to easily test many other already existing models.

Acknowledgements

We thank Giulio De Meo for the precious support in the implementation of the KAOS tool. This research was supported by University of Messina Research & Mobility 2015 Project (project code RES_AND_MOB_2015_DISTASO).

References

1. V. Castellano, S. Fortunato, Rev. Mod. Phys. **81**, 591 (2009)
2. P. Ball, Phys. A **314**, 1 (2002)
3. R. Wang, Q. Wang, Phys. Rev. E **84**,n 0361108 (2011)
4. B. Chopard, M. Droz, *Cellular Automata modeling of physical systems* (Cambridge University Press, Cambridge, 1988)
5. M. Dolfin, Applied Mathematics Letters **25**, 2163 (2012)
6. M. Dolfin, Applied Mathematics and Computation **234**, 260 (2014)
7. D. Helbing, *Quantitative Sociodynamics: Stochastic Methodsand Models of Social Interactions Processes* (Springer, 2nd edition, 2010)
8. S. Galam, *Sociophysics: A Physicist's Modeling of Psycho-political Phenomena* (Springer, 2nd edition, 2012)
9. L. Pareschi, G. Toscani, *Interacting Multiagent Systems – Kinetic equations and Monte Carlo methods* (Oxford Univ. Press, Oxford, 2014)
10. M. Dolfin, M. Lachowicz, Networks and Heterogeneous Media **20**, 877 (2015)
11. M. Dolfin, M. Lachowicz, Mathematical Models and Methods in Applied Sciences **24**, 2361 (2014)
12. G.A. Marsan, N. Bellomo, A.Tosin, *Complex Systems and Society: Modeling and Simulation* (Springer Briefs in Mathematics, Springer, New York City, 2013)
13. N. Bellomo, *Modeling Complex Living Systems. A kinetic theory approach and stochastic game approach* (Modeling and simulation in Science, Engineering and Technology, Birkhauser, 2008)
14. G. Weisbuch, G. Deffaunt, F. Amblard, J. Nadal, Complexity **7**, 55 (2002)

N. Bellomo, V. Coscia, AMS Series in Contemporary Mathematics **594**, 99 (2013)

On the Jost Solution and Its Properties of Quantum Difference Equations with Hyperbolic Eigenparameter

Yelda Aygar

Ankara University, Faculty of Science, Department of Mathematics, 06100, Ankara, Turkey

Abstract. This work is devoted to get Jost solution by the method of variation of parameters and properties of this Jost solution of a boundary value problem (BVP) depending on an hyperbolic eigenparameter. The results obtained are then used to find the Green function, resolvent and continuous spectrum of this BVP by using the methods of operator theory, methods of functional analysis and the general theory of difference equations. Moreover, using the Weyl compact perturbation theorem, we get that the operator L generated by the q -difference expression has continuous spectrum filling the segment $[-2\sqrt{q},2\sqrt{q}]$.

1 Introduction

Quantum calculus which known as the calculus without limits is a connection between mathematics and physics. In recent years, it has led to rapid development of the theory of q -difference equations. The study of q -difference equations has become an important area of research. Because such equations have an interest role due to their applications in several mathematical areas such as number theory, orthogonal polynomials, mathematical control theory, basic hyper-geometric functions and other disciplines including mechanic, theory of relativity, biology, economics. Since it has huge applications in several disciplines, spectral analysis of q -difference equations and operators have become a new research area for mathematicians. Spectral analysis of self-adjoint and nonselfadjoint q -difference equations has been investigated in [1, 2, 4, 7] both ordinary and matrix cases. Also BVPs consisting of q -difference equations and boundary condition depending on spectral parameter is studied in [3, 5]. Differently studies which are mentioned above, in this work the spectral analysis of a q -difference equation with boundary condition is investigated using hyperbolic eigenparameter. As a result of this, analytical region of Jost solution has changed from upper half-plane to left half-plane. Some different aspects on the appearance of Jost solution, resolvent and Green function are stand out. This new approach will provide a wide perspective on applications of these problems in physics, mathematics, economics and engineering. In this study, we let $q>1$ and use the notation

$$q^{\mathbb{N}_0} := \{q^n : n \in \mathbb{N}_0\},$$

where $\mathbb{N}_0$ denotes the set of nonnegative integers. $\ell_2(q^{\mathbb{N}})$ is a Hilbert space of complex-valued functions with the inner product

$$\langle f,g\rangle_q := \sum_{t\in q^{\mathbb{N}_0}} \mu(t)f(t)\overline{g(t)}, \quad f,g : q^{\mathbb{N}_0} \to \mathbb{C}$$

and the norm

$$\|f\|_q := \left(\sum_{t\in q^{\mathbb{N}_0}} \mu(t)|f(t)|^2\right)^{\frac{1}{2}}, \quad f : q^{\mathbb{N}_0} \to \mathbb{C}$$

where $\mu(t)=(q-1)t$ for all $t\in q^{\mathbb{N}_0}$. Let us consider the nonselfadjoint BVP consisting of the second-order q -difference equation

$$qa(t)y(qt)+b(t)y(t)+a\left(\frac{t}{q}\right)y\left(\frac{t}{q}\right)=\lambda y(t), \quad t\in q^{\mathbb{N}} \quad (1)$$

and the boundary conditions

$$\begin{gathered}(\gamma_0+\gamma_1\lambda)y(q)+(\beta_0+\beta_1\lambda)y(1)=0,\\ \gamma_0\beta_1-\gamma_1\beta_0\neq 0, \quad \gamma_1\neq \tfrac{\beta_0}{a(1)},\end{gathered} \quad (2)$$

where $\{a(t)\}_{t\in q^{\mathbb{N}_0}}$ and $\{b(t)\}_{t\in q^{\mathbb{N}}}$ are complex sequences, λ is an eigenparameter, $a(t)\neq 0$ for all $t\in q^{N_0}$, and $\gamma_i,\beta_i\in\mathbb{C}$, $i=0,1$. In this work, we discuss the Jost solution and Jost function of of the BVP (1)-(2) and after getting its Green function and Resolvent, we investigate the continuous spectrum of this BVP under the assumption

$$\sum_{t\in q^{\mathbb{N}}}(|1-a(t)|+|b(t)|)<\infty. \quad (3)$$

2 Main results

Under the condition , we get the solution of BVP (1)-(2) as

$$e(t,z)=\alpha(t)\frac{e^{\frac{\ln t}{\ln q}z}}{\sqrt{\mu(t)}}\left(1+\sum_{r\in q^{\mathbb{N}}}A(t,r)e^{\frac{\ln r}{\ln q}z}\right),\ t\in q^{\mathbb{N}_0} \quad (4)$$

for $\lambda=2\sqrt{q}\cosh z$ by using the similar method given in [5], where $z\in\overline{\mathbb{C}}_-:=\{z\in\mathbb{C}:z\le 0\}$, $\mu(t)=(q-1)t$ for all $t\in q^{\mathbb{N}_0}$ and $\alpha(t)$, $A(t,r)$ are expressed in terms of $\{a(t)\}$ and $\{b(t)\}$ as

$$\alpha(t)=\prod_{s\in[t,\infty)\cap q^{\mathbb{N}}}[a(s)]^{-1},$$

$$A(t,q)=-\frac{1}{\sqrt{q}}\sum_{s\in[qt,\infty)\cap q^{\mathbb{N}}}b(s),$$

$$A(t,q^2)=\sum_{s\in[qt,\infty)\cap q^{\mathbb{N}}}\left\{1-a^2(s)+\frac{b(s)}{q}\sum_{p\in[qs,\infty)\cap q^{\mathbb{N}}}b(s)\right\},$$

$$A(t,rq^2)=A(qt,r)$$

$$+\sum_{s\in[qt,\infty)\cap q^{\mathbb{N}}}\left\{(1-a^2(s))A(qs,r)-\frac{b(s)}{\sqrt{q}}A(s,qr)\right\},$$

for $t\in q^{\mathbb{N}_0}$ and $r\in q^{\mathbb{N}}$. Moreover it is given in [4] that $A(t,r)$ satisfies

$$|A(t,r)|\le C\sum_{s\in\left[tq^{\left\lfloor\frac{\ln r}{2\ln q}\right\rfloor},\infty\right)\cap q^{\mathbb{N}}}(|1-a(s)|+|b(s)|), \quad (5)$$

where $\left\lfloor\frac{\ln r}{2\ln q}\right\rfloor$ is the integer part of $\frac{\ln r}{2\ln q}$ and $C>0$ is a constant. Using (3) it is seen that $e(\cdot,z)$ is analytic with respect to z in $\mathbb{C}_-:=\{z\in\mathbb{C}:z<0\}$ and continuous in $\overline{\mathbb{C}}_-$. Using (4) = and the boundary condition , we define the function g by

$$g(z)=(\gamma_0+2\sqrt{q}\gamma_1\cosh z)e(q,z)$$
$$+(\beta_0+2\sqrt{q}\beta_1\cosh z)e(1,z).$$

The function g is analytic in $\mathbb{C}_-$, continuous in $\overline{\mathbb{C}}_-$, and $g(z)=g(z+2\pi)$. Analogously to the Sturm--Liouville differential equation, the solution $e(\cdot,z)$ and the function g are called the Jost solution and Jost function of (1)-(2), respectively [6]. Let $\phi(\lambda)=\{\phi(t,z)\}$ $t\in q^{N_0}$, be the solution of (1) satisfying the initial conditions

$$\phi(1,\lambda)=-(\gamma_0+\gamma_1\lambda),\quad \phi(q,\lambda)=(\beta_0+\beta_1\lambda).$$

If we define

$$\varphi(t,z)=\phi(2\sqrt{q}\cosh z)=\{\phi(t,2\sqrt{q}\cosh z)\}_{t\in q^{\mathbb{N}_0}},$$

then φ is an entire function and $\varphi(z)=\varphi(z+2\pi)$. Let us take the semi-strips $P_0=\{z\in\mathbb{C}_-:-\frac{\pi}{2}\le z\le\frac{3\pi}{2}\}$ and $P=P_0\cup\left[-\frac{i\pi}{2},\frac{3i\pi}{2}\right]$. The Wronskian of two solutions $y=\{y(t,\lambda)\}_{t\in q^{\mathbb{N}}}$ and $v=\{(t,\lambda)\}_{t\in q^{\mathbb{N}}}$ of (1) is defined by

$$W[y,v]=\mu(t)a(t)\{y(t,\lambda)v(qt,\lambda)-y(qt,\lambda)v(t,\lambda),\quad t\in q^{N_0}.$$

Hereafter, we find

$$W[\varphi(t,z),e(t,z)]=-\mu(1)a(1)g(z).$$

Since $\varphi(\cdot,z)$ and $e(\cdot,z)$ are independent solutions for all $z\in P$ with $g(z)\ne 0$, we find the Green function of the BVP (1)-(2)

$$G_{t,r}(z):=\begin{cases}-\frac{\varphi(r,z)e(t,z)}{qa(1)f(z)}, & r=tq^{-k},\ k\in\mathbb{N}_0\\ -\frac{e(r,z)\varphi(t,z)}{qa(1)f(z)}, & r=tq^{k},\ k\in\mathbb{N}.\end{cases} \quad (6)$$

It is clear that

$$R_\lambda h(t):=\sum_{r\in q^{\mathbb{N}}}G(t,r)h(r),\ h\in\ell_2(q^{\mathbb{N}}) \quad (7)$$

is the resolvent of the BVP (1)-(2). Related to the equation (1), we will recall the q -difference operator by L generated in $\ell_2(q^{\mathbb{N}})$ by the q -difference expression

$$(ly)(t):=a\left(\frac{t}{q}\right)y\left(\frac{t}{q}\right)+b(t)y(t)+qa(t)y(qt),\quad t\in q^N.$$

Note that (1) can be rewritten as $(ly)(t)=\lambda y(t)$. Let L_1 and L_2 denote the q -difference operators generated in $\ell_2(q^N)$ by

$$(l_1y)(t):=y\left(\frac{t}{q}\right)+qy(qt),\quad t\in q^N$$

and

$$(l_2y)(t):=\left(a\left(\frac{t}{q}\right)-1\right)y\left(\frac{t}{q}\right)+b(t)y(t)+q(a(t)-1)y(qt),\quad t\in q^N,$$

respectively. It is clear that L_1 is a self-adjoint operator and $L = L_1 + L_2$ [4]. It can be easily seen that $\frac{e^{\frac{\ln t}{\ln q}z}}{\sqrt{\mu(t)}}$ and $\frac{e^{-\frac{\ln t}{\ln q}z}}{\sqrt{\mu(t)}}$ are linearly independent solutions of $(l_1y)(t) = \lambda y(t)$ for $\lambda = 2\sqrt{q}\cosh z$. The Wronskian of these two solutions is

$$W\left[\frac{e^{\frac{\ln t}{\ln q}z}}{\sqrt{\mu(t)}}, \frac{e^{-\frac{\ln t}{\ln q}z}}{\sqrt{\mu(t)}}\right] = -\frac{2}{\sqrt{q}\,sinhz}, \quad z \in \mathbb{C}_-.$$

For all $z \in \mathbb{C}_- \setminus \{ik\pi : k \in \mathbb{Z}\}$, we define the green function and resolvent of L_1 by

$$S_{t,r}(z) := \frac{1}{2\sqrt{q}\sinh z}\begin{cases} \frac{e^{\frac{\ln r}{\ln q}z}}{\sqrt{\mu(r)}}\frac{e^{\frac{\ln t}{\ln q}z}}{\sqrt{\mu(t)}}, & r = tq^{-k}, \ k \in \mathbb{N} \\ \frac{e^{\frac{\ln r}{\ln q}z}}{\sqrt{\mu(r)}}\frac{e^{-\frac{\ln t}{\ln q}z}}{\sqrt{\mu(t)}}, & r = tq^{k}, \ k \in \mathbb{N}_0. \end{cases} \quad (8)$$

and

$$R_\lambda(L_1)\psi(t) := \sum_{r\in q^{\mathbb{N}}} S(t,r)\psi(r), \quad \psi \in \ell_2(q^{\mathbb{N}}) \quad (9)$$

respectively.

Lemma 1. *For every $\delta > 0$, there is a number such that C_δ such that*

$$\|R_\lambda(L_1)\|_q > \frac{C_\delta}{|sinhz|\sqrt{1-e^{2z}}}$$

for all $z \in \{z \in P_0 : |z| < \delta\}$.

Proof Let us define the function

$$f(r,z) := \begin{cases} \frac{e^{-\frac{\ln r}{\ln q}z}}{\sqrt{\mu(r)}}, & r = tq^{-k}, \quad k \in \mathbb{N} \\ 0, & r = tq^{k}, \quad k \in \mathbb{N}_0, \end{cases}$$

where $z = x + iy$. Thus, we obtain

$$\|f(\cdot,z\|_q^2 = \sum_{r\in(0,\frac{t}{q}]\cap q^{\mathbb{N}}} \left|e^{-\frac{\ln r}{\ln q}z}\right|^2 = \sum_{r\in(0,\frac{t}{q}]\cap q^{\mathbb{N}}} \left|e^{-2\frac{\ln r}{\ln q}z}\right| < \infty.$$

Since $\|f(\cdot,z)\|_q^2 < \infty$, $f(\cdot,z) \in \ell_2(q^{\mathbb{N}})$. Using the (8) and (9), we get

$$R_\lambda(L_1)f(\cdot,z) = \frac{e^{\frac{\ln t}{\ln q}z}}{2\sqrt{q}\sqrt{\mu(t)}\,sinhz}\|f(\cdot,z)\|_q^2.$$

Since $\left|e^{\frac{\ln t}{\ln q}z}\right| - \left|e^{\frac{\ln t}{\ln q}z}\right| > \frac{1}{2}e^{\frac{\ln t}{\ln q}z}$, we obtain that

$$\begin{aligned} \|R_\lambda(L_1)f(\cdot,z)\|_q^2 &= \frac{\|f(\cdot,z)\|_q^4}{4q|sinhz|^2}\sum_{t\in q^{\mathbb{N}}}\left|e^{\frac{\ln t}{\ln q}z}\right|^2 \\ &\geq \frac{\|f(\cdot,z)\|_q^4}{16q|sinhz|^2}\sum_{t\in q^{\mathbb{N}}}\left(e^{\frac{\ln t}{\ln q}z}\right)^2 \\ &= \frac{\|f(\cdot,z)\|_q^4}{16q|sinhz|^2}\sum_{t\in q^{\mathbb{N}}} e^{2\frac{\ln t}{\ln q}z} \\ &\geq \frac{\|f(\cdot,z)\|_q^4}{16q|sinhz|^2}\sum_{t\in(q,\infty)\cap q^{\mathbb{N}}} (e^{2z})^{\frac{\ln t}{\ln q}} \\ &= \frac{\|f(\cdot,z)\|_q^4 e^{2z}}{16q|sinhz|^2(1-e^{2z})}. \end{aligned}$$

By choosing $\frac{\|f(\cdot,z)\|_q^2 e^{2z}}{4\sqrt{q}}$, the proof is complete.

Theorem 1. $\sigma(L_1) = \sigma_c(L_1) = [-2\sqrt{q}, 2\sqrt{q}]$, *where $\sigma(L_1)$ and $\sigma_c(L_1)$ denote the spectrum and continuous spectrum of the operator L_1, respectively.*

Proof. Because of the operator L_1 has no eigenvalues, the spectrum of it consists of only its continuous spectrum. If $\lambda_0 = 2\sqrt{q}\cosh z_0 \in \sigma_c(L_1)$, then $\|R_{\lambda_0}(L_1)\| \to \infty$. Using (8) and (9), we have $z_0 \to 0$. If $z_0 \to 0$, then $\lambda_0 = 2\sqrt{q}\cosh z_0 \in [-2\sqrt{q}, 2\sqrt{q}]$. It follows from Lemma 1 that $\|R_\lambda(L_1)\|_q \to \infty$ for $\lambda = 2\sqrt{q}\cosh z \in [-2\sqrt{q}, 2\sqrt{q}]$. Now, if we show the range $\Re(L_1 - \lambda I)$ of values of the operator $L_1 - \lambda I$ is dense in $\ell_2(q^N)$, the proof is completed. The orthogonal complement of $\Re(L_1 - \lambda I)$ coincides with the space of solutions of the equation $(l_1^* y)(t) = \lambda y$ such that $y \in \ell_2(q^N)$. Since the operator L_1 is a selfadjoint operator and has no eigenvalues, the orthogonal complement of the set $\Re(L_1 - \lambda I)$ consists only of the zero element. It gives that the range $\Re(L_1 - \lambda I)$ is dense in $\ell_2(q^{\mathbb{N}})$.

Note that the different method of the proof of this theorem is given in [4].

Theorem 2. *Under the condition (3), we have* $\sigma_c(L) = [-2\sqrt{q}, 2\sqrt{q}]$.

Proof. Under the assumption (3), it is found that L_2 is a compact operator in $\ell_2(q^N)$ (see [8]). Then using Theorem 1 and the Weyl theorem [9] of a compact perturbation, it is seen that

$$\sigma_c(L) = \sigma(L_1) = \sigma_c(L_1).$$

3 Conclusion

In this study, first we give the Jost solution of the BVP (1)-(2), then using the asymptotic behavior and analytical properties of the Jost solution, we obtain the Green fuction and the resolvent of BVP (1)-(2). At the end, we show that the continuous spectrum of the operatör is filling the segment $[-2\sqrt{q}, 2\sqrt{q}]$ under the condition (3). The properties of eigenvalues and spectral singularities of this BVP can be invastigated in the further research. For example, a condition can be given which guarante that this BVP has finite eigenvalues and spectral singularities with finite multiplicities.

References

1. M. Adıvar, M. Bohner, Spectral analysis of q-difference equations with spectral singularities, Math. Comp. Model. **43** (2006), 695703.
2. Y. Aygar, Investigation of spectral analysis of matrix quantum difference equations with spectral singularities, Hacettepe Journal of Mathematics and Statistics (2015), DOI:10.15672/HJMS.20164513107
3. Y. Aygar, Principal vectors of second-order quantum difference equations with boundary conditions dependent on spectral parameter, Advances in Difference Equations **2015:249** (2015) DOI:10.1186/s13662-015-0587-3.
4. Y. Aygar, M. Bohner, A Polynomial-Type Jost Solution and Spectral Properties of a Self-Adjoint Quantum-Difference Operator, Complex Analysis and Operator Theory, (2015) DOI:10.1007/s11785-015-0463-x.
5. Y. Aygar, M. BohneR, On the spectrum of eigenparameter-dependent quantum difference equations, Appl. Math.& Inf. Scie. **9** (2015), 1725-1729.
6. M.A. Naimark, Linear differential operators. Part II: Linear differential operators in Hilbert space, Frederick Ungar Publishing Co., New York, 1968.
7. Y. Aygar, M. Bohner, Spectral Anlaysis of a matrix-valued quantum difference operator, Dynm. Equ. Appl. (2015), In press.
8. L.A. Lusternik, V.J. Sobolev, Elements of functional analysis, Hindustan Publishing Corp., Delhi, 1974.
9. I.M. Glazman, Direct methods of qualitative spectral analysis of singular differential operators, Israel Program for Scientific Translations, Jerusalem, (1965)

MATEC Web of Conferences **68**, 16006 (2016)
DOI: 10.1051/matecconf/20166816006
ICIEA 2016

Extension of Some Positive and Linear Operators to Domains with Curved Sides

Teodora Cătinaş

Babeş-Bolyai University, Faculty of Mathematics and Computer Science, Str. M. Kogălniceanu Nr. 1, RO-400084, Cluj-Napoca, Romania

Abstract. There are constructed some Cheney-Sharma type operators defined on a square with one curved side. They are extensions of the Cheney-Sharma type operators of second kind, given by E.W. Cheney and A. Sharma in [14], to the case of a curved sided domain. There are constructed the univariate Cheney-Sharma type operators, their product and boolean sum operators and there are studied their properties, their orders of accuracy and the remainders of the corresponding approximation formulas. Finally, there are given some illustrative examples.

1 Introduction

The aim of this paper is to construct some Cheney-Sharma type operators that have some interpolatory properties on a curved sided domain. There will be studied the interpolation properties, the orders of accuracy and the remainders of the corresponding approximation formulas.

Using the interpolation properties of such operators, blending function interpolants can be constructed, that exactly match function on some sides of a rectangular region. Important applications of these blending functions are in finite element method for differential equations problems with Dirichlet boundary conditions or for construction of surfaces which satisfy some given conditions.

2 Related reviews

Starting with the paper [3] of R.E. Barnhill, G. Birkhoff and W.J. Gordon, there have been constructed interpolation operators of Lagrange, Hermite and Birkhoff type, that interpolate the values of a given function or the values of the function and of certain of its derivatives on the boundary of a triangle with straight sides. These operators were applied in computer aided geometric design (see, e.g., [1]-[3], [5]]) and in finite element analysis (see, e.g., [1], [7], [8], [13], [17]-[19], [21], [22]).

In order to match all the boundary information on a curved domain (as in Dirichlet, Neumann or Robin boundary conditions for differential equation problems), there were considered interpolation operators on domains with curved sides (see, e.g., [4], [6], [9]-[12], [15], [16], [20], [21], [23]). Approximation operators on polygonal domains with some curved sides have also important applications especially in finite element method for differential equations with given boundary conditions and in the piecewise generation of surfaces in computer aided geometric design.

3 Methods

Let $m \in \mathbf{N}$ and β a nonnegative parameter. The Cheney-Sharma operators of second kind $Q_m : C([0,1]) \to C([0,1])$, introduced in [14], are given by

$$(Q_m f)(x) = \sum_{i=0}^{m} q_{m,i}(x) f\left(\frac{i}{m}\right), \tag{1}$$

$$q_{m,i}(x) = \binom{m}{i} \frac{x(x+i\beta)^{i-1}(1-x)[1-x+(m-i)\beta]^{m-i-1}}{(1+m\beta)^{m-1}}.$$

Remark. 1) Notice that for $\beta=0$, the operator Q_m becomes the Bernstein operator.

2) In [26], it has been proved that the Cheney-Sharma operator Q_m interpolates a given function at the endpoints of the interval.

3) In [14] and [26], it has been proved that the Cheney-Sharma operator Q_m reproduces the constant and the linear functions, so its degree of exactness is 1 (denoted $\mathrm{dex}(Q_m)=1$).

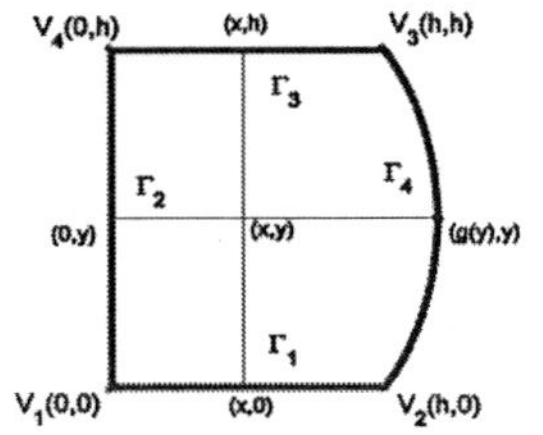

Figure 1. The square D_h.

For $h \in R_+$, let D_h be the square with one curved side having the vertices $V_1=(0,0)$, $V_2=(h,0)$, $V_3=(h,h)$ and $V_4=(0,h)$, three straight sides, Γ_1, Γ_2, along the coordinate axes and Γ_3 parallel to axis Ox, and the curved side Γ_4, that is defined by the function g, such that g(h)=g(0)=h (See Fig. 1).

4 Main results

4.1 Univariate operators

Let F be a real-valued function defined on D_h and (0,y), (g(y),y), respectively, (x,0), (x,h) be the points in which the parallel lines to the coordinate axes, passing through the point $(x,y) \in D_h$, intersect the sides Γ_2, Γ_4, respectively Γ_1 and Γ_3. We consider the uniform partitions of the intervals [0,g(y)] and [0,h], with $y \in [0,h]$:

$$\Delta_m^x = \{\frac{i}{m} g(y) \mid i=0,...,m\}$$

and

$$\Delta_n^y = \{\frac{j}{n} h \mid j=0,...,n\}.$$

For $m,n \in \mathbf{N}$, $b, \beta \in \mathbf{R}_+$, we consider the following extensions of the Cheney-Sharma operator given in (1):

$$(Q_m^x F)(x,y) = \sum_{i=0}^{m} q_{m,i}(x,y) F(\frac{i}{m} g(y), y), \tag{2}$$

$$(Q_n^y F)(x,y) = \sum_{j=0}^{m} q_{n,j}(x,y) F(x, \frac{j}{n} h),$$

with

$$q_{m,i}(x) = \binom{m}{i} \frac{\frac{x}{g(y)}(\frac{x}{g(y)} + i\beta)^{i-1}(1-\frac{x}{g(y)})}{(1+m\beta)^{m-1}} \cdot$$

$$\cdot [1 - \frac{x}{g(y)} + (m-i)\beta]^{m-i-1}$$

$$q_{n,j}(x) = \binom{n}{j} \frac{\frac{y}{h}(\frac{y}{h} + jb)^{j-1}(1-\frac{y}{h})[1-\frac{y}{h} + (n-j)b]^{n-j-1}}{(1+m\beta)^{m-1}}.$$

Remark. As Cheney-Sharma operator of second kind interpolates a given function at the endpoints of the interval, we may use the operators Q_m^x and Q_n^y as interpolation operators on D_h.

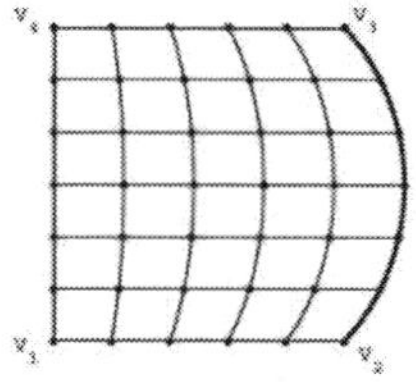

Figure 2.

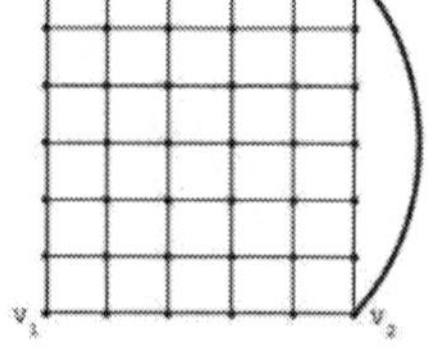

Figure 3.

In Figures 2 and 3 we plot the points $(\frac{i}{m} g(y), y)$, i=0,...,m, and respectively, $(x, \frac{j}{n} h)$, j=0,...,n, for x, $y \in [0,h]$, and m=5, n=6.

Theorem. If F is a real-valued function defined on D_h then

(i) Q_m^x F=F on $\Gamma_2 \cup \Gamma_4$,

(ii) Q_n^y F=F on $\Gamma_1 \cup \Gamma_3$.

Proof. (i) We may write

$$(Q_m^x F)(x,y) = \frac{1}{(1+m\beta)^{m-1}} \left\{ (1-\frac{x}{g(y)}) \left[1 - \frac{x}{g(y)} + m\beta \right]^{m-1} \right.$$

$$\cdot F(0,y) + \frac{x}{g(y)}(1-\frac{x}{g(y)}) \sum_{i=0}^{m} \binom{m}{i} (\frac{x}{g(y)} + i\beta)^{i-1} \tag{3}$$

$$\cdot \left[1 - \frac{x}{g(y)} + (m-i)\beta \right]^{m-i-1} F(\frac{i}{m} g(y), y)$$

$$\left. + \frac{x}{g(y)} (\frac{x}{g(y)} + m\beta)^{m-1} F(g(y), y) \right\}.$$

Considering (3) we may easily prove that

$$(Q_m^x F)(0,y) = F(0,y)$$

$$(Q_m^x F)(g(y),y) = F(g(y),y).$$

(ii) The proof follows the same steps as for (i).

Theorem. The operators Q_m^x and Q_n^y have the following orders of accuracy:

(i) $(Q_m^x e_{ij})(x,y) = x^i y^j, \quad i = 0,1; j \in N$;

(ii) $(Q_n^y e_{ij})(x,y) = x^i y^j, \quad i \in N; \; j = 0,1$, where $e_{ij}(x, y)=x^i y^j$, $i, j \in N$.

Proof. The proof follows by the property $dex(Q_m)=1$. We consider the approximation formula

$$F = Q_m^x F + R_m^x F,$$

where $R_m^x F$ denotes the approximation error.

Theorem. If $F(\cdot, y) \in C^2[0,g(y)]$ then

$$(R_m^x F)(x,y) = \frac{1}{2} F^{(2,0)}(\xi, y)[x^2 - (Q_m^x e_{20})(x,y)], \text{ for } \xi \in [0,g(y)].$$

Proof. Taking into account that $dex(Q_m^x) = 1$, and applying Peano's theorem (see, e.g., [25]), it follows

$$(R_m^x F)(x,y) = \int_0^{g(y)} K_{20}(x,y;s) F^{(2,0)}(s,y) ds,$$

where

$$K_{20}(x,y;s) = (x-s)_+ - \sum_{i=0}^{m} q_{m,i}(x,y)(\frac{i}{m} g(y) - s)_+ .$$

DOI: 10.1051/matecconf/20166816006

For a given $\nu \in \{1,...,m\}$, one denotes by $K_{20}^{\nu}(x,y;\cdot)$ the restriction of the kernel $K_{20}(x,y;\cdot)$ to the interval $\left[(\nu-1)\frac{g(y)}{m}, \nu\frac{g(y)}{m}\right]$, i.e.,

$$K_{20}^{\nu}(x,y;\nu) = (x-s)_+ - \sum_{i=\nu}^{m} q_{m,i}(x,y)(\frac{i}{m}g(y)-s),$$

whence,

$$K_{20}^{\nu}(x,y;s) = \begin{cases} x-s-\sum_{i=\nu}^{m} q_{m,i}(x,y)(\frac{i}{m}g(y)-s), & s<x \\ -\sum_{i=\nu}^{m} q_{m,i}(x,y)(\frac{i}{m}g(y)-s), & s \geq x. \end{cases}$$

It follows that $K_{20}^{\nu}(x,y;s) \leq 0$, for $s \geq x$.

For s<x we have

$$K_{20}^{\nu}(x,y;s) = x-s-\sum_{i=0}^{m} q_{m,i}(x,y)(\frac{i}{m}g(y)-s)$$
$$+\sum_{i=0}^{\nu-1} q_{m,i}(x,y)(\frac{i}{m}g(y)-s).$$

Applying the previous Theorem, we get

$$\sum_{i=0}^{m} q_{m,i}(x,y)(\frac{i}{m}g(y)-s) = (Q_m^x e_{10})(x,y) - s(Q_m^x e_{00})(x,y)$$
$$= x-s,$$

and it follows that

$$K_{20}^{\nu}(x,y;s) = \sum_{i=0}^{\nu-1} q_{m,i}(x,y)(\frac{i}{m}g(y)-s) \leq 0.$$

So, $K_{20}^{\nu}(x,y;\cdot) \leq 0$, for any $\nu \in \{1,...,m\}$, i.e., $K_{20}(x,y;s) \leq 0$, for $s \in [0,g(y)]$.

By Mean Value Theorem, one obtains $(R_m^x F)(x,y) = F^{(2,0)}(\xi,y)\int_0^{g(y)} K_{20}(x,y;s)ds$, $0 \leq \xi \leq g(y)$, with

$$\int_0^{g(y)} K_{20}(x,y;s)ds = \frac{1}{2}[x^2 - (Q_m^x e_{20})(x,y)].$$

Remark. Analogous results are obtained for the remainder $R_n^y F$ of the formula

$$F = Q_n^y F + R_n^y F.$$

4.2 Product operators

Let

$$(P_{mn}^1 F)(x,y) = \sum_{i=0}^{m}\sum_{j=0}^{n} q_{m,i}(x,y) q_{n,j}(i\frac{g(y)}{m}, y) F(i\frac{g(y)}{m}, j\frac{h}{n}),$$

$$(P_{nm}^2 F)(x,y) = \sum_{i=0}^{m}\sum_{j=0}^{n} q_{m,i}(x, j\frac{h}{n}) q_{n,j}(x,y) F(\frac{i}{m}g(j\frac{h}{n}), j\frac{h}{n}).$$

The nodes of $(P_{mn}^1 F)(x,y)$ and $(P_{nm}^2 F)(x,y)$ are given in Figures 4 and 5.

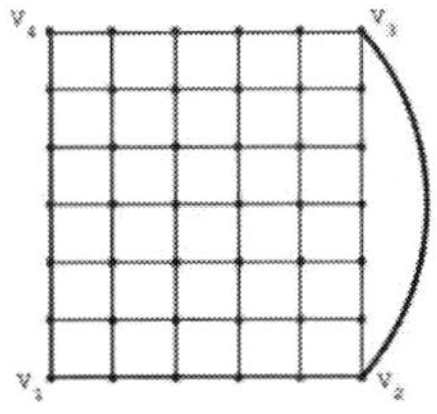

Figure 4. The nodes of $(P_{mn}^1 F)(x,y)$.

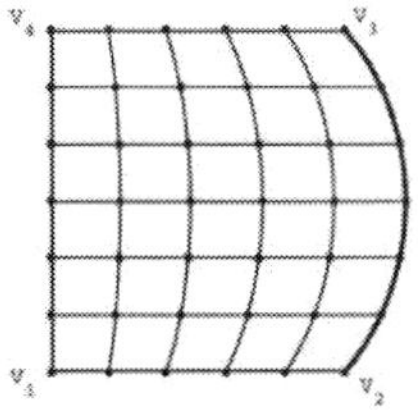

Figure 5. The nodes of $(P_{nm}^2 F)(x,y)$.

Theorem. If F is a real-valued function defined on D_h, then

(i) $(P_{mn}^1 F)(V_i) = F(V_i)$, $i = 1,...,4$

(ii) $(P_{nm}^2 F)(V_i) = F(V_i)$, $i = 1,...,4$.

Proof. The proof follows by a straightforward computation.

4.3 Boolean sum operators

We consider the Boolean sums of the operators Q_m^x and Q_n^y, i.e.,

$$S_{mn}^1 = Q_m^x \oplus Q_n^y = Q_m^x + Q_n^y - Q_m^x Q_n^y,$$

$$S_{nm}^2 = Q_n^y \oplus Q_m^x = Q_n^y + Q_m^x - Q_n^y Q_m^x.$$

Theorem. If F is a real-valued function defined on D_h, then

$$S_{mn}^1 F\big|_{\partial D_h} = F\big|_{\partial D_h}$$

and

$$S_{nm}^2 F\big|_{\partial D_h} = F\big|_{\partial D_h}.$$

Proof. The proof follows by a straightforward computation.

5 Numerical examples

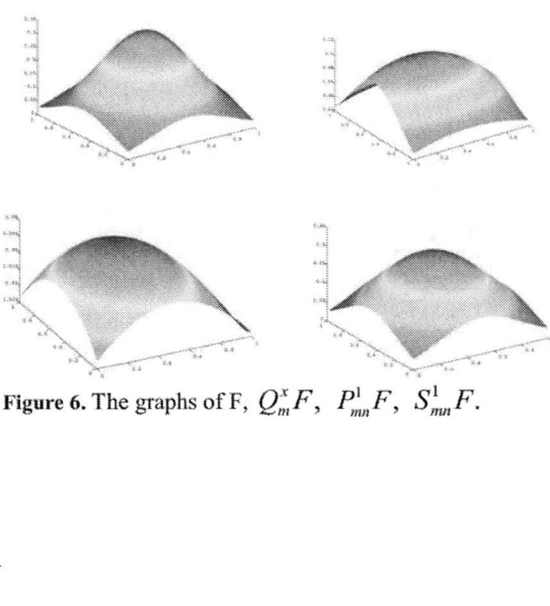

Figure 6. The graphs of F, $Q_m^x F$, $P_{mn}^1 F$, $S_{mn}^1 F$.

We consider the function:

Gentle: $F(x,y)=\frac{1}{3}\exp[-\frac{81}{16}((x-0.5)^2+(y-0.5)^2)]$, generally used in the literature, (see, e.g., [24]). In Figure 6 we plot the graphs of F, $Q_m^x F$, $P_{mn}^1 F$, $S_{mn}^1 F$, on D_h, considering $h=1$, $m=5$, $n=6$, $\beta=1$.

Table 1 contains the maximum approximation errors for approximation by $Q_m^x F$, $P_{mn}^1 F$, $S_{mn}^1 F$, considering $h=1$, $m=5$, $n=6$, $\beta=1$.

Table 1. The maximum approximation errors.

The approximant	The error
$Q_m^x F$	0.2205
$P_{mn}^1 F$	0.2858
$S_{mn}^1 F$	0.1277

6 Conclusions

The univariate operators obtained here, Q_m^x and Q_n^y, interpolate a function $F:D_h \to \mathbf{R}$ on two sides of the domain D_h, their products, P_{mn}^1 and P_{nm}^2, interpolate on the vertices of D_h, and their boolean sum operators interpolate on the entire frontier of D_h. There are generated the interpolation formulas and there are given the expressions of the corresponding remainders using Peano's theorem. The good approximation properties could be seen in the figures and in the table with errors presented in the previous section. Using the interpolation properties of these operators, blending function interpolants, that exactly match function on some sides of a curved domain region, can be constructed. Important applications of these blending functions are in finite element method for differential equations problems with Dirichlet boundary conditions or for construction of surfaces which satisfy some given conditions.

References

1. R. E. Barnhill, *Blending function interpolation: a survey and some new results*, Numerishe Methoden der Approximationstheorie, **30**, Birkhauser-Verlag, Basel, (1976), pp. 43-89.
2. R. E. Barnhill, *Representation and approximation of surfaces*, Mathematical Software III, (Ed. J.R. Rice, Academic Press, New-York, 1977), pp. 68-119.
3. R. E. Barnhill, G. Birkhoff, W. J. Gordon, *Smooth interpolation in triangles*, J. Approx. Theory, **8**, pp. 114-128 (1973).
4. R. E. Barnhill, J. A. Gregory, *Compatible smooth interpolation in triangles*, J. Approx. Theory, **15**, pp. 214-225 (1975).
5. R. E. Barnhill, J. A. Gregory, *Sard kernels theorems on triangular domains with applications to finite element error bounds*, Numer. Math., **25**, pp. 215-229 (1976).
6. C. Bernardi, *Optimal finite-element interpolation on curved domains*, SIAM J. Numer. Anal., **26**, no. 5, pp. 1212-1240 (1989).
7. M. Bernadou, *C1-curved finite elements with numerical integration for thin plate and thin shell problems, Part 1: construction and interpolation properties of curved C1 finite elements*, Comput. Methods in Applied Mechanics and Engineering, **102**, pp. 255-289 (1993).
8. M. Bernadou,, *C1-curved finite elements with numerical integration for thin plate and thin shell problems, Part 2 : approximation of thin plate and thin shell problems,* Comput. Methods in Appl. Mechanics and Engineering, **102**, pp. 389-421 (1993).
9. P. Blaga, T. Cătinaş, G. Coman, *Bernstein-type operators on tetrahedrons*, Studia UBB Math., **54**, no. 4, pp. 3-19 (2009).
10. P. Blaga, T. Cătinaş, Gh. Coman, *Bernstein-type operators on a square with one and two curved sides*, Studia UBB Math., **55**, no. 3, pp. 51-67 (2010).
11. P. Blaga, T. Cătinaş, G. Coman, *Bernstein-type operators on triangle with all curved sides*, Appl. Math. Comput., **218**, pp. 3072--3082 (2011).
12. P. Blaga, T. Cătinaş, G. Coman, *Bernstein-type operators on triangle with one curved side*, Mediterr. J. Math., **9**, No. 4, pp. 843-855 (2012).
13. J. C. Cavendish, W.J. Gordon, Ch.A. Hall, *Ritz-Galerkin approximations in blending function spaces*, Numer. Math., **26**, pp. 155-178 (1976).
14. E.W. Cheney, A. Sharma, *On a generalization of Bernstein polynomials*, Riv. Mat. Univ. Parma, **5**, pp. 77-84 (1964).
15. G. Coman, T. Cătinaş, *Interpolation operators on a tetrahedron with three curved sides*, Calcolo, **47**, no. 2, pp. 113-128 (2010).
16. G. Coman, T. Cătinaş, *Interpolation operators on a triangle with one curved side*, BIT Numer. Math., **50**, no. 2, pp. 243-267 (2010).
17. R. Dautray, J.L. Lions, *Analyse mathematique et calcul numerique*, Vol. 6, Methodes integrales et numeriques, Masson, Paris, pp. 798-834 (1988).
18. W. J. Gordon, Ch. Hall, *Transfinite element methods: blending-function interpolation over arbitrary curved element domains*, Numer. Math., **21**, pp. 109-129 (1973).
19. W. J. Gordon, J.A. Wixom, *Pseudo-harmonic interpolation on convex domains*, SIAM J. Numer. Anal., **11**, No 5, pp. 909-933 (1974).
20. J. A. Marshall, R. McLeod, *Curved elements in the finite element method*, Conference on Numer. Sol. Diff. Eq., Lectures Notes in Math., **363**, Springer Verlag, pp. 89-104 (1974)}.
21. J. A. Marshall, A. R. Mitchell, *An exact boundary technique for improved accuracy in the finite element method*, J. Inst. Maths. Applics., **12**, pp. 355-362 (1973).
22. J. A. Marshall, A. R. Mitchell, *Blending interpolants in the finite element method*, Inter. J. Numer. Meth. Engineering, **12**, pp. 77-83 (1978).

23. A. R. Mitchell, R. McLeod, *Curved elements in the finite element method,* Conference on Numer. Sol. Diff. Eq., Lectures Notes in Mathematics, **363**, pp. 89-104 (1974).
24. R. J. Renka, A. K. Cline, *A triangle-based C^1 interpolation method*, Rocky Mountain J. Math. **14**, pp. 223--237 (1984).
25. A. Sard, *Linear Approximation*, American Mathematical Society, Providence, Rhode Island, 1963.
26. D. D. Stancu, C. Cişmaşiu, *On an approximating linear positive operator of Cheney-Sharma*, Rev. Anal. Numer. Theor. Approx., **26**, pp. 221-227 (1997).

MATEC Web of Conferences 68, 16007 (2016) DOI: 10.1051/matecconf/20166816007
ICIEA 2016

The Research and Development of the External Magnetic Field Acting on Electro-Deposition Process

Menghua Wu[1], Weiping Jia[2]

[1] College of Mechanical Engineering, Dalian University, 116622, Dalian, China
[2] Key Lab of Non-traditional Machining& Functional Materials Preparing, Dalian University, 116622, Dalian, China

Abstract. The research and development status of the electro-deposition technology under the action of external magnetic field are introduced. The basic characteristics and applied manners of external magnetic field in electro-deposition process are summarized. The acting principle of external magnetic field, the effects of magnetic hydrodynamics (MHD) caused by the Lorentz force, and the acting of magnetic force on the metal ions and particles are described. The main actions of external magnetic field include MHD effect, magnetizing force, affecting the physical and chemical properties of the bath, affecting the disperse ability and coverage capacity of bath, affecting the mass transfer process of electro-deposition, affecting the chemical reaction process and current distribution of electrode surface. Some examples of electro-depositing single metal coatings, alloy coatings and composite coatings under action of magnetic field are explained. During the electro-depositing process, the external magnetic field has different degrees of impact on solution properties, mass transfer, charge transfer, content of composited nanoparticles, crystal growth and crystal orientation etc. The specific impact of magnetic field during the electro-depositing is also classified and summarized. The problems that existed in electro-deposition process while applying magnetic field and the next development trend were summarized.

1 Introduction

The electro-deposition preparation technology underaction of magnetic field is a integrated new technology which across the theory of magnetic field, electro-deposition processes and materials science areas, and it is gradually developing into an emerging interdisciplinary-Magnetic electrochemistry [1], [2]. As the magnetic fields posses many advantages, such as, high energy density and easy to be controlled, material is applied to the non-contact mode of action, so the thermal and kinetic energy may be transmitted to the material in the case of non-polluting, and so on. It has been concerned recently by the majority of researchers [3]. The magnetic field applied manners during electro-deposition process are mainly perpendicular to the current direction (B ⊥ J) and parallel (B // J), and it may generate the Lorentz force, magnetic force, the magnetic hydrodynamic forces, etc. These forces have a certain impact on the mass transfer process, ion sports, bath state, surface state of coating, and the metal crystallization process, etc. So, it can improve the quality and performance of coatings [4]-[6]. In this paper, some research results of external magnetic field applied to electro-deposition are speech summed, the working principle and the specific role forms of the external magnetic field in the electro-deposition process are described, the problems in the application stage and further research trend of applying external magnetic field in electro-deposition process are clarified.

2 Research overview of applying external magnetic field during the electro-deposition process

Since the pioneering research work about acting of magnetic field on electrolyte solution was completed by Faraday in 1930s, as a new form of energy, the magnetic field reflect many other advantages, such as, improving the quality of prepared materials, saving energy, improve the environment, etc. While external magnetic field is applied in electro-deposition process, the magnetic hydrodynamics (MHD) effects generated by magnetic field may affect the physical and chemical properties, dispersion capacity and coverage ability of the bath, and also have an effect on the mass transfer and the deposition rate of charged particles, the current distribution on the electrode surface, the discharge and grain growth process, and then improve the morphology and mechanical or physical properties of coatings.

Presently, the research about actions of the magnetic field in electro-deposition process mainly focused on two aspects: The acting directions and the intensity of applied magnetic field [3], [7]. The types of electro-deposition coatings which prepared under the action of magnetic field are as follow: Single metal coatings, alloy coatings and composite coatings [8], [9].

2.1 Electro-depositing single metal coatings under action of magnetic field

The research and application of magnetic field applied to prepare single metal coatings are more extensive and mature. So far, the research about electro-deposition process under the action of magnetic field for preparing single metal coatings, such as nickel, iron, cobalt, chromium, zinc, tin, lead, copper, bismuth etc., has been carried out and achieved certain results. For example:

Kristina Tschulik et al [3]. conducted electro-deposition of separated 3D metallic structures by pulse-reverse plating in magnetic gradient fields. Separate columns and stripes of Cu were generated by pulse reverse plating controlled by tailored magnetic field gradients. The experimental observations were explained by the action of the magnetic field gradient force on the paramagnetic Cu2+-ions during the deposition and the dissolution part of the cycle. Additional influences of the Lorentz force are also discussed.

Weiping Jia et al [10]. conducted research on surface morphology and texture of nickel crystal micro-casting by electro-forming under magnetic field. The results shown that the texture and surface morphology of the micro components are influenced by the current density and the magnetic field intensity. With the increase of current density, the size of crystal grains decrease firstly but increase later. And as the increase of the magnetic field intensity, the sizes of crystal grains become smaller and more uniform. Besides, the current density and the magnetic field intensity have a great impact on the texture of castings. As the increase of the current density, the texture orientation changes from (111) to (200). At the same time, along with the increase of the magnetic field intensity, the diffraction peak (200) is suppressed while the diffraction peak (111) is enhanced obviously.

2.2 Electrodepositing alloy coatings under action of magnetic field

How to change the composition ratio of alloy coating and improve its performance has become a hot field of electro-depositing alloy technique. If the external magnetic field is applied to electro-deposition process of alloy coatings, the characteristics of the external magnetic field could be taken full advantage for metal ion transmission and grain growth during the electro-depositing, so as to achieve the purpose of improving the coating properties. But it is not easy for electro-depositing alloy coatings to apply the external magnetic field due to the ferromagnetic, paramagnetic and diamagnetic of metal. So far, most relative research about electro-depositing alloy coatings under action of external magnetic field are focused on preparing nickel-iron or iron-nickel alloy coatings. For example:

Adriana Ispas et al [6]. conducted the study of external magnetic field effects on nickel–iron alloy electro-deposition based on linear and non-linear differential AC electrochemical response measurements. It was shown that by using linear electrochemical impedance spectrometry, it can be very difficult–if not almost impossible–to identify magnet hydrodynamic effects on the electro-deposition of Ni–Fe alloys. It was shown that by using a non-linear approach the real and imaginary parts of the total impedance of the system are sensitive to the presence of a magnetic field in the frequency range from 1 to 100 Hz.

Donggang Li et al [11]. conducted the evolution of morphology in electro-deposited nanocrystalline Co–Ni films by in-situ high magnetic field application. The effect of high magnetic fields may up to 12 T. Their research results shown that along with the increase of magnetic flux density, the grain size and the surface roughness of the films increased to reach a maximum value at a field of 9 T. Meanwhile, higher magnetic flux density could improve cobalt atomic percentage in the film due to the impacts of magnet hydrodynamic effect. However, at a high field of 12 T, the paramagnetic force played a predominant role in a decrease of mass transport, resulting in minimum grain size and roughness. The average values of surface roughness and lateral feature size, which can be used to characterize the grain size, were calculated using standard Nova software of AFM.

Marek Zieliński [12] presented a study of the effect of constant magnetic field (CMF) on the basic processes of cobalt-tungsten alloys electro-deposition. The exposure to CMF caused also an increase of the volume fraction (by about 9% by volume) of the dominant phase (Co3W and Co7W6) in the alloy. The reason for these changes was the fact that the Lorentz force, generated in CMF, caused the magnetic hydrodynamic (MHD) effect. This induced movement of the electrolyte. The Nernst diffusion layer was depleted, whereas a new Naviere Stokes hydrodynamic layer appeared, which determined the velocity of electro active molecules flow to the working electrode under CMF conditions.

P. Żabiński et al [7]. conducted the research on electro-catalytically active Co-W and Co-W-C alloys electro-deposited in a magnetic field. The influence of sodium tungstate concentration, arginine concentration and orientation of applied magnetic field (MF) on composition, morphology, structure, cathodic current efficiency of deposition process and electro-catalytic activity of deposited alloys were examined. The positive effect of carbon addition on catalytic activity of hydrogen evolution was observed. The influence of electro-deposition parameters on size of micro cracks formed during deposition processes was also investigated.

2.3 Electro-depositing composite coatings under action of external magnetic field

The content of nanoparticles in composite coating is a key factor to determine the composite coating performance, and how to improve the content of nanoparticles in the composite coating has become a recently hot research. When external magnetic field is introduced into electro-depositing process, it may not only produce a beneficial stirring to the bath, let nanoparticles uniformly and stably suspended in the bath, but also accelerate the transmission process of particles, so as to provide effective way of solving agglomeration of the nanoparticles. Also, the content of nanoparticles in composite coatings may be increased to a certain extent, thereby the performance of the composite coatings may be greatly improved. For example:

Pengwei Zhou et al [2], [13]. conducted the research on behavior of Fe/nano-Si particles composite electro-deposition with a vertical electrode system in a static parallel magnetic field. Compared to the rather smooth surface morphology of Fe-nano Si composite coating without magnetic field, it became wavy and striped obviously with imposing magnetic field. The nano silicon particles dispersed uniformly on the surfaceof coating when no magnetic field was applied, whereas the nano silicon particles content of coatings presented a trend of increasing with increasing the magnetic flux density. The influence of an external parallel strong parallel magnetic field (respect to current) on the electro-deposition of nano-silicon particles into an iron matrix had been studied in their another paper. Test results show that magnetic field has a great influence on the distribution of silicon, as well as the surface morphology and the thickness of the composite coatings. The thicker and rougher composite deposits appearin the edge region (Lor R region), and the thinner and smoother ones appear in the middle region (M). The distribution curve of silicon content looks like a "pan" along the centre line of coatings.

Denny Thiemig et al [14]. studied the effect of an external magnetic field on the electro-deposition of composites consisting of either Co or magnetite nanoparticles in a Ni matrix. The particle incorporation showed a distinct dependency on the orientation of an externally applied magnetic field. While the particle incorporation increased in a perpendicular field (perpendicular with regard to the electrode surface), it decreased in a parallel orientation. This result was explained with the dominating action of the magnet phoretic force. The structure and the properties of the Ni layers were significantly affected by the particles co-deposition. A refinement of the Ni grains was found with increasing plating current density and as a result of the nanoparticle incorporation. The magnetic hardness and the Vickers micro hardness of the films increased significantly due to the incorporation of the nanoparticles.

Chao Wang et al [15]. applied an external high parallel magnetic field into the co-deposition process of Ni/nano-Al2O3 composites. It was found that the distribution of nanoparticles showed the network shape when super imposing a magnetic field, and the average size of the single network increased with increasing the current density. High concentration of nanoparticles could be obtained at low current density with a magnetic field, while a high current density was needed to realize that when without a magnetic field. The super imposed magnetic filed would improve the current efficiency.

3 The working principle of external magnetic field

The magnetic field is produced by moving charges or changing electric field. The basic characteristics of magnetic field is that it is able to apply force to the electric charges which moving in magnetic field, i.e., the energized conductor is affected by the magnetic force in magnetic field [16]. When magnetic field was applied in the material preparation process, the electromagnetic force acting on the prepared material may be expressed as follow equation [17], [18]:

$$F = j \times B + \rho_e E + (\mu_0 M \cdot \nabla)H + (P \cdot \nabla)E \tag{1}$$

In the right side of above equation, the first term is the Lorentz force, the second term is the Coulomb force, the third term is the magnetizing force, and the fourth term is the polarization force. The main forces which ions or materials subjected to in magnetic field are Lorentz force and magnetic force, and the two main forces produced by magnetic field are normally used to improve the preparing process of materials. When the magnetic field strength reaches a certain value (a few T to dozens T), it can even have an effect of Lorentz force and magnetic force on a very weak current and non-magnetic materials, thus largely expand its scope of application. Influence of magnetic field on material preparation process may performance in many ways: distribution of particles in composite coatings may be affected by the Lorentz force of the magnetic field, mass transfer process and crystal orientation may be affected by magnetizing force, in addition, chemistry reaction also may be affected to some extent by the magnetic field [19], [20]. Under acting of 10T-20T, even 100T super magnetic field, there are very significant influences on the chemical reaction system, and thus indirectly affect the performances of prepared materials.

With the development of super conducting technology, it will be more easy to obtain strong magnetic field. Super conducting magnets with 10T magnetic field have been commercially available in the international, so the non-magnetic material can easily be magnetized, and so as to the possibility of using the method of materials preparation to access to new theories and discoveries have also been increased [21], [22].

4 The action of external magnetic field on electro-deposition process

Normally, the electro-deposition process can be roughly divided into the following steps: Mass transfer, electrochemical and electrical crystallization [23], [24]. If the magnetic fields are applied during the electro-deposition process, it would have a corresponding effect on the steps mentioned above, and ultimately affect the morphology and properties of the coatings. These effects embodied in the following aspects [7], [12]:

The Magnetic Hydro Dynamic (MHD) effect caused by Lorentz force [9], [17]. The Lorentz force caused by Interaction of magnetic fields and current may generate the macroscopic MHD effects and micro-MHD effects, these effects not only enhance the transmission of mass, but also play the role of stirring the solution and make the solution more uniform, thereby improving the quality of the coatings. In the electro-deposition process, the macroscopic MHD effects perform the action of Lorentz force when plasma flow cutting magnetic lines. The result is that the bath is stirred, and this effect can be observed when applying relatively weak magnetic field. When applying strong magnetic field, the MHD effect will act on the trajectory of each ion, and the Lorentz force will also act on the micro current between the oxidation particles and reduction particles. This phenomenon is called micro-MHD effect [12]. When applying the vertical magnetic field, the electric force and the Lorentz force act at the same time on the charged particles in solution, and the charged particles will perform a spiral movement, then disturbance the solution to form a MHD effect. Finally, it will result changing of solution performance and variation of grain size in depositing layer.

Action of magnetizing force. Atomic electrons carried spin motion on the one hand, and kept movement around the nucleus on the other hand. Electrons can be equivalent to this feature as a small ring current, called molecular currents [8]. In the absence of external magnetic field, the orientation of these molecular current is random and does not exhibit any macroscopic effects. Once the magnetic field was applied, these molecular current appear regular orientation and form macroscopic magnetization current. This phenomenon is called the magnetic substance. When the applied magnetic field is parallel to the current manner during the electro-deposition process, with no Lorentz force, the magnetization force will become very apparent. Thus, the metal ions and particles will be subject to the ferromagnetic matrix metal and be easy to deposit with lowest energy state on the surface of the base body during depositing process. So that the coating is more compact with fewer defects [9]. The intrinsic reasons that magnetizing force may impact on the electro-deposition process are as follow: Coating crystal exists magnetic anisotropy, there are differences in the magnetic energy of the various ion in bath.

Affecting the physical and chemical properties of the bath. The magnetic field may cause the physical properties of water change, such as, reducing the surface tension, improving wettability, altering the process of dissolution, crystallization and condensation, etc [25], [26]. Since the majority of the bath is aqueous system, so applying magnetic field may accordingly impact the bath solubility, viscosity, wettability and liquidity, and furthermore impact the electro-deposition process.

Affecting the disperse ability and coverage capacity of bath. The MHD effect generated by the magnetic field can accelerate the mass transfer process, make the viscosity coefficient of bath be smaller, enhance conductivity and cathodic polarization, thereby improving the disperse ability of the bath. The electric line distribution at concave surface of the coating tends to be uniform under the action of magnetic field. Ion trajectory will be changed while the ions in bath are affected both by electric and magnetic forces, and the shielding effect of protrusions surface will be weakened, so that coverage capacity of bath is improved [6], [10].

Affecting the mass transfer process of electro-deposition. The electrode process of electro-deposition involves liquid mass transfer process and the electrode reaction process. The liquid mass transfer process include three transfer forms: Convection, diffusion and electro-migration. These three mass transfer processes always occur simultaneously. However, under certain circumstances, there are normally one or two transfer forms play a major role. After applying a magnetic field, the magnetic induced convection effect which generated by Lorentz force will act on the bath ions and cause convection diffusion, thereby thinning the thickness of the diffusion layer and accelerating the mass transfer process [6], [20]. Special attention is required because the magnetic field not only affect the mass transfer rate but also affect the charge transfer [8].

Affecting the chemical reaction process. Magnetic field can not only control the macro physical and chemical reactions of material, but also influence and regulate the inside microscopic state of substances [9]. Magnetic field may affect chemical reaction heat, pH value, direction of chemical reaction, reaction rate, activation energy, entropy and many other factors, thereby affecting the electro-deposition process and coating performance.

Affecting the current distribution of electrode surface. Under the action of magnetic field, the polarization will increase where the current density of cathode surface are larger [11]. The increasing polarization, in turn, contribute to the current distribution of cathode surface tends to uniform, thereby improving the coating uniformity on electrode surface. Up to now, some research achievements of applying magnetic field in the field of electro-deposition have been gained. Scholars from various countries have experimentally studied the electro-deposition process under the action of different magnetic field strength and different acting directions, and the acting mechanism of magnetic field on electro-deposition process have been carried out more in-depth discussion and analysis [4], [27], [28]. Learned that the MHD effect and action of magnetizing force generated by the magnetic fluid effect were the major factors which can affect the process of electro-deposition, and the affect mechanism of each factor on electro-deposition process were specifically analyzed. The MHD effect is macroeconomic performance of Lorentz force generated by perpendicular magnetic field, it may accelerate the mass transfer process and improve the disperse ability of bath by stirring the bath, and then improve the performances of coatings [25, 29]. The functions of parallel magnetic field are mainly performance the effect of magnetizing force, this effect may have a significant impact on transporting of particles, the chemical reaction rate, charge transfer, and the crystal growth process, thereby affecting the morphology, tissue and properties of coating [27], [28].

5 The problems of applying external magnetic field during the electro-deposition process

Although the research work about applying magnetic field on electro-deposition process has been made some progress, and to some extent, has been applied. However, according to the situation of the current research, the research work still exist some problems and shortcomings:

Firstly, the affecting mechanism of the magnetic field on electro-deposition process is now more clearly that the MHD effect and the magnetic force effect generated by the magnetic field on the electro-deposition process. However, it is not very clear about the variation laws of physical and chemical properties of the bath and electrochemical reaction rate.

Secondly, the current experiments are mostly simple magnetic field effect on the electro-deposition process, but the composite experimental studies of magnetic field and other physical fields (such as ultrasonic wave, flow field generated by mechanical stirring , etc.) have not been fully conducted, and it is unclear about the composite affecting mechanism of multi-physical fields. Therefore, the research depth of which should be extended.

Thirdly, there are relatively more research about affecting of weak magnetic field on the electro-deposition process, and relevant theories are more mature, but there are less research or just get starting about affecting of strong magnetic fields (10T or more) on the electro-deposition process. Strong magnetic field will not only have an magnetization effect on ferromagnetic material, but also have an magnetization effect on the diamagnetic and paramagnetic materials, so the electro-deposition process under acting of strong magnetic field will become more complex. Therefore, it is urgent to prove the acting mechanism of strong magnetic field on electro-deposition process and performance of coatings.

Fourthly, there are relatively more experimental study of preparing magnetic materials and anisotropic materials by applying magnetic field. However, it is unclear about the affecting laws of strong magnetic fields on the mass transfer process of ferromagnetic ions and particles in bath, and on the nucleation and growth pattern of grains. So the experimental study in this regard should be strengthened.

Fifthly, since the metal itself has a ferromagnetic, paramagnetic and diamagnetic, the electro-depositing alloy coating process under action of magnetic field will become complicated. Thus the experiments of electro-depositing different types of alloy coating under action of magnetic field will be limited and constrained.

In short, there is lot of research space for applying the magnetic field in the field of electro-deposition technology. Therefore, to carry out further research and exploration is very necessary.

6 Conclusions

Combining the current research status and problems about applying magnetic field in the electro-deposition process, the next research trends in this field be summarized as follows:

a) In order to more comprehensive and in-depth grasp the influence of magnetic field on the electro-deposition process, the research about acting mechanism of magnetic field on the performance of bath, the electrochemical reaction rate, the crystal growth process, the plasma discharge process and other aspects should be strengthened.

b) The compound electro-deposition experiments and research, which combined the action of magnetic field and other fields (such as ultrasonic wave, flow field electric field, etc.), should be conducted for proving the complex acting mechanism of multi-physics fields applied in electro-deposition process.

c) As strong magnetic field may have magnetization to the non-magnetic materials, the experimental study of electro-deposition under the action of strong magnetic field should be conducted, so as to master and apply the electro-deposition process under the action of strong magnetic field.

d) The further study of electro-depositing alloy process under the action of the magnetic field should be conducted, so as to investigate the effect of magnetic field on the composition and properties of alloy coating.

e) As the substrates for electro-deposition are diamagnetic material or ferromagnetic material, so it is necessary to compare the action of magnetic field on the different substrate bath.

Acknowledgements

This work was supported by the Dalian Science and Technology Project (2015A11GX003) and the Science &Technology Research Project of Education Department of Liaoning province (L2015029).

References

1. K. Tschulik, J. A. Koza, M. Uhlemann, Electrochemistry Communications. J. **11**(2009), p.2241-2244
2. Pengwei Zhou, Yunbo Zhong, Huai Wang, Lijun Fan, Electrochimica Acta. J. **111** (2013), p. 126-135
3. Kristina Tschulik, Ralph Sueptitz, Margitta Uhlemann, Electrochimica Acta. J. **14** (2011), p. 5174-5177
4. Yanling Wen, Yunbo Zhong, Zhongming Ren, The Chinese Journal of Nonferrous Metals. **4** (2006), p. 715-720
5. Yundan Yu, Lixia Sun, Junwei Lou, Hongliang Ge, Electroplating & Pollution Control. J. **4** (2012), p. 1-3 "In Chinese"
6. Adriana Ispas, Hisayoshi Matsushima, Andreas Bund, Electro-analytical Chemistry. J. **2** (2011), p.197-203
7. P. Żabiński, K. Mech, R. Kowalik, Electrochimica Acta. J. **1** (2013), p. 542-548
8. Chi-Cheng Hung, Wen-Hsi Lee, Japanese Journal of Applied Physics. **48** (2009), p.076506-1
9. Mehdi Ebadi, W. J. Basirun, Yatimah Alias, Journal of Chemical Sciences. **2** (2010), p.279-285
10. Weiping Jia, Fan Yang, Menghua Wu, Advanced Materials Research. J. **535-537** (2012), p.450-454
11. Donggang Li, Alexandra Levesque, Agnieszka Franczak, Talanta. J. **110** (2013), p. 66-70

12. Marek Zieliński, Materials Chemistry and Physics. J. **1** (2013), p.370-377
13. Pengwei Zhou, Yunbo Zhong, Huai Wang, Qiong Long, Applied Surface Science. J. **282**(2013), p. 624– 631
14. Denny Thiemig, Clemens Kubeil, Christian P. Gräf, Andreas Bund, Thin Solid Films. J. **517** (2009), p.1636–1644
15. Chao Wang, Yunbo Zhong, Weili Ren, Zuosheng Lei, Applied Surface Science. J. **254** (2008), p.5649–5654
16. W.Szmaja, W. Kozowski, K. Polanski, J. Balcerski, Mater. Chem. Phys. J. **132** (2012) , p.1060–1066
17. Nana Liu, Menghua Wu, Zhi Li, Yuangang Wang, Journal of Rare Metal Materials & Engineering. **3** (2013), p. 649-655 "In Chinese"
18. S.A. Lajevardi, T.Shahrabi, J.A. Szpunar, Applied Surface Science. J. **279**(2013), p.180-188
19. O. Ergeneman, K.M. Sivaraman, S. Pane, E. Pellicer, Electrochimica Acta. J. **56** (2011), p.1399-1408
20. Shuhong Wang, Huanping Zhang, Hui Du, Shaogang Hou, Yong Zheng, Anlin Li, Applied Chemical Industry. J. **44** (2015), p.732-735 "In Chinese"
21. P. Żabiński, K. Mech, R. Kowalik, Archives of Metallurgy and Materials. J .**57** (2012), p. 127-133
22. Xiao-qin Ouyang, Linyan Zhou, Bin Yu, Ying Wan, Bin-bin Zhang, Shuilian Hu, Changjie Feng, Plating and Finishing, J. **36** (2014), p.20-29 "In Chinese"
23. Jiao Feng, Nanjing University of Aeronautics and Astronautics, D, (2009). "In Chinese"
24. F. C. Walsh*, and C. Ponce de Leon, Transactions of the IMF. **2** (2014), p. 84-98
25. C. Wang, Y. B. Zhong, J. Wang, Journal of Electroanalytical Chemistry. **1-2** (2009), p.42-48
26. Lorena M.A. Monzon, J.M.D. Coey, Electrochemistry Communications. J. **42** (2014), p. 38-41
27. Peng Gao, Zhongdong Yang, Xiaolian Xu, Transaction of Materials & Heat treatment. J. **4** (2012), p.136-140 "In Chinese"
28. Jianming Lu, David Dreisinger, Thomas Glück, Hydrometallurgy. J. **141** (2014) , p.105-116
29. Yundan Yu, Lixia Sun, Junwei Lou, Hongliang Ge, Electroplating & Pollution Control. J. **4** (2012), p. 1-3 "In Chinese"

Design and Validation of Affective Warning Pictorial on Cigarette Labels

Pat-Arin Chanduen and Arisara Jiamsanguanwong

Department of Industrial Engineering, Faculty of Engineering, Chulalongkorn University, Bangkok, Thailand

Abstract. The purpose of present study were to design and validate affective warning pictorials for cigarette label in Thailand. Brainstorming and survey techniques were used to collect the idea of possible warning pictorials. All ideas were grouped for finding candidated pictorials. Then, primary sixty warning pictorials were collected and equally classified into three affective warning pictorial groups as positive, neutral, and negative. Sixty Thai male engineering students participated in affective validation of warning pictorials using SAM rating. The International Affective Picture System (IAPS) was used to manipulate the affective state of participants to neutral affective state before the experiments. The results revealed that all affective warning pictorials were successfully evoked target affective states on participants. After refining, thirty affective warning pictorials were provided as positive, neutral, and negative affective warning pictorials for using on cigarette labels. Implications on the affective warning pictorials design and validation.

1 Introduction

Warning pictorials are widely used on packaging and labelling of tobacco products or cigarette for anti-smoking campaigns in various countries such as Canada, Brazil, Singapore, and Thailand [1]-[3]. In order to satisfy the criteria of the World Health Organization Framework Convention on Tobacco Control (WHO-FCTC), the warnings should be presented colorful graphics about health effect, hazards or emissions, and its size should be 50% or more of display areas, large, clear, visible, and legible [4]. While the goal of warning is to inform people in order to appreciate potential hazards [5], [6], to help reduce or eliminate risks [6]-[8], and attempt to promote safe behavior while reduce those unsafe behavior [6]. Thus, the design of warning pictorial is critical since it relates to the effectiveness of safety communication.

Several studies informed that current warning pictorials used on cigarette labels generated negative emotional responses. Mostly Canadian reported fear and disgust after viewing warning pictorial on cigarette label [1]. In Brazil, people described their emotional responses from warning pictorials of cigarette label on rating scales as unpleasant and moderately aroused [2]. While smoking people in Thailand also reported negative emotional responses, but still smoking [3]. Those people also reported that they attempted to avoid seeing warning pictorials on cigarette label by using their personal packaging cover, painting over warning pictorial, or importing cigarette from other countries [1]-[3].

There were some suggestions that warning pictorials used on cigarette label should not only focus on presenting the negative effect of smoking, but should provide positive information including benefits of quit smoking [9]. Azagba and Sharaf also suggested that the effectiveness of warning pictorials used on cigarette warning labels was mainly based on emotional responses and label presentation form [10]. Also, there was a study tried to improve the effectiveness of warning pictorials by introducing the idea of affective design of warning pictorials [11]. Affective design is defined as the design which capable to evoke appropriate affects in users [12]. In this paper, the term affect includes emotions, moods, and feelings.

Jiamsanguanwong and Umemuro examined the effect of positive, neutral, and negative affective states of people on their comprehension and hazard perception of standard industrial warning pictorials [13]. They manipulated affective states of participants using International Affective Pictures System (IAPS) [14] with an intention to introduce the concept of affective warning pictorial that pictorial itself could evoke target affective states on users. Result showed that people who were in positive affect condition perceived greater hazard from warning pictorials than those in neutral, and negative affect condition. The concept of affective warning pictorial should be validated in the context of cigarette label in order to confirm that some specific affective design of warning pictorial could improve the effectiveness of warning pictorial used on cigarette label. However, to the best of our knowledge, no one has design and validate the affective warning pictorial for using on cigarette label.

Thus, the purpose of present study were to design and validate affective warning pictorials for cigarette label.

The results would provide three groups of affective warning pictorials as positive, neutral, and negative warning pictorials for using on cigarette label.

2 Methods

2.1 Initial step

The goal of this step is to find the possible warning pictorials used on cigarette label. First of all, ideas about the characteristics of warning pictorial used on cigarette label for anti-smoking campaign were collected by survey and brainstorming methods.

Fourteen Thai undergraduate and graduate engineering students participated in this brainstorming session. Five are males. They ranged in age from 21 to 26 years (M = 23.1, SD = 1.9). Participants were asked to write down the ideas of possible warning pictorials used on cigarette label for anti-smoking campaign.

Thirty Thai smoking and non-smoking people participated in survey session. 56.7% are males. They ranged in age from 13 to 57 years (M = 29.5, SD = 12). Participants were asked to draw or write down at least three ideas of possible warning pictorials used on cigarette label for anti-smoking campaign.

Total 198 ideas were collected and grouped into 5 groups; family, health, socials, finance, and others. Then, sixty warning pictorials were collected from free sources such as Thai health [15], and Tobacco Labelling Regulations [16], following the grouping ideas from brainstorming and survey session with cultural concerned. After that, all pictorials were primary equally classified into each positive, neutral, and negative warning pictorial groups for affective validation step.

2.2 Affective validation

This step is to validate affective warning pictorials from initial step. There were limited to a maximum of five participants per experimental session.

2.3 Participants

Sixty Thai undergraduate and graduate male engineering students participated in this study. They ranged in age from 20 to 28 years (M = 21.5, SD = 1.69).

2.4 Stimulus material

2.4.1 Affective picture stimuli

Ten International Affective Picture System (IAPS) pictures were selected based on valence rating as neutral for males provided by manual of IAPS [14]. Their valence scores ranged from 4.05 to 5.19 (M = 4.56, SD = 1.54) and arousal scores ranged from 4.10 to 5.00 (M = 4.56, SD = 2.02). The valence rating ranged from 1: unpleasantness, to 9: pleasant, and arousal rating ranged from 1: calm, to 9: excited. For ethical reasons, pictures of mutilated and erotic imagery were excluded.

2.4.2 Warning pictorials

Sixty warning pictorials from the initial step were primary equally classified into positive, neutral, and negative pictorial groups. The majority of positive warning pictorials were from family group idea, neutral warning pictorials were from social and finance group idea, negative warning pictorials were from health group idea which are currently used on cigarette labels [16].

2.5 Apparatus

Affective picture stimulus and warning pictorials were presented by microsoft powerpoint and 4K ultra HD, 65 inch television. The experiments were conducted in 6 x 4 m private laboratory room.

2.6 Measurement

The Self-Assessment Manikin (SAM) [17] was used to assess valence and arousal affect of participants (Figure 1). Affective valence ranges from 1: unpleasant to 9: pleasant with 5 as neutral. Affective arousal ranges from 1: calm to 9: excited. Participants were asked to rate their affective states after affect manipulation and after viewing each of sixty warning pictorials.

2.7 Procedure

Prior to beginning, participants were asked to read and sign a consent form for their participation. Then, participants were given a brief instructions, the purpose of the research, and were informed the purpose of viewing the set of images was to manipulate their emotions.

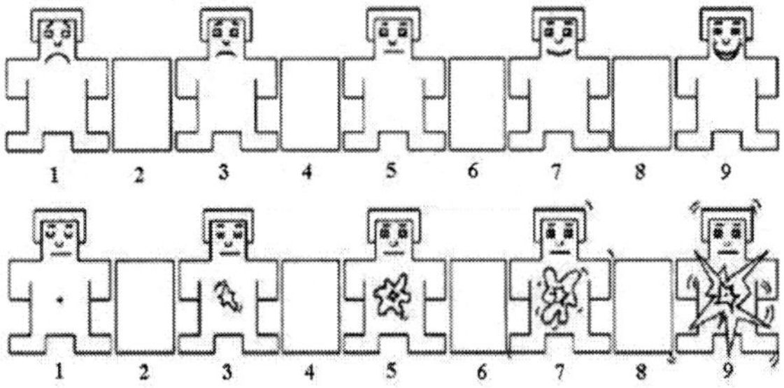

Figure 1. The Self-Assessment Manikin (SAM) [17]: valence (top panel), and arousal (bottom panel)

Ten neutral affective pictures were each presented for 6s, with a 2-s interval between pictures. After that, participants were asked to complete the Self-Assessment Manakin (SAM) questionnaire [14] to rate their present emotional valence and arousal (SAM I).

Then participants were given brief instructions and shown an example pictorial for practice. All sixty pictorials were randomly presented for 20 s, with 10-s interval between pictures. Participants were asked to immediately rate their emotional valence and arousal responses (SAM II) to each warning pictorials.

3 Results

3.1 Affect manipulation

For the SAM I, sixty participants were achieved target neutral affective states. Their valence scores ranged from 3 to 6 (M = 4.50, SD = 0.89) and arousal score ranged from 1 to 7 (M = 3.57, SD = 1.70). This result indicated that all participants were successfully manipulated their emotion to neutral affective state.

3.2 Affective validation

Two participants were identified as outliers and excluded from the following analysis. Mean and standard deviation of valence and arousal scores for each group of affective warning pictorials were shown in table 1.

One-way analysis of variance (ANOVA) were conducted with primary classified group of affective warning pictorial (positive, neutral, and negative) on the SAM II scores (valence, and arousal). The main effect of warning pictorial conditions were significant on valence score (F (2,3477) = 3332.23, p<.001) and on arousal score (F (2,3477) = 351.61, p<.001). Post hoc analyses revealed that participants perceived warning pictorials from positive group more positive as they rated higher valence rating scores than those warning pictorials from neutral group (p<.001), or warning pictorials from negative group (p <.001). Participants also rated higher

Table 1. Range of valence and arousal score of each warning pictorial groups. Before and after refining pictorials.

Before refining pictorials							
Warning Pictorial Groups	N	Valence score			Arousal score		
		Max	Min	Mean (SD)	Max	Min	Mean (SD)
Positive	20	7.21	6.12	6.69 (0.33)	4.22	3.14	3.72 (0.32)
Neutral	20	5.34	4.03	4.61 (0.37)	4.38	3.17	3.73 (0.28)
Negative	20	3.64	1.97	2.74 (0.54)	6.66	4.10	5.54 (0.78)
After refining pictorials							
Warning Pictorial Groups	N	Valence score			Arousal score		
		Max	Min	Mean (SD)	Max	Min	Mean (SD)
Positive	10	7.21	6.76	6.97 (0.15)	4.22	3.21	3.84 (0.30)
Neutral	10	4.94	4.45	4.56 (0.11)	4.09	3.28	3.67 (0.28)
Negative	10	2.72	1.97	2.28 (0.25)	6.66	5.57	6.20 (0.42)

Valence score of warning pictorials from neutral group than those warning pictorials from negative group (p<.001). While, warning pictorials from negative group had significantly higher arousal scores than those warning pictorials from positive group (p<.001), or warning pictorials from neutral group (p<.001). These results indicated that these affective warning pictorials successfully evoked target affective states on participants.

3.3 Refining pictorials

Ten out of twenties warning pictorials in each group were selected as final affective warning pictorial sets in order to reduce the variations of affective responses (SAM II) in each group (Figure 2). Mean and standard deviation of valence and arousal scores each group after refining the number of pictorials were shown in Table 1. The criteria used for selecting the pictorials in each group based on their target valence score. For example, ten positive warning pictorials (Figure 2a) were selected by their top ten mean valence scores among other pictorials in positive group. Ten negative warning pictorials (Figure 2c) were selected by their least mean valence scores among other pictorials in negative group. Ten neutral warning pictorials (Figure 2b) were selected by the least mean absolute error from valence score of five as represented neutral rating score on SAM.

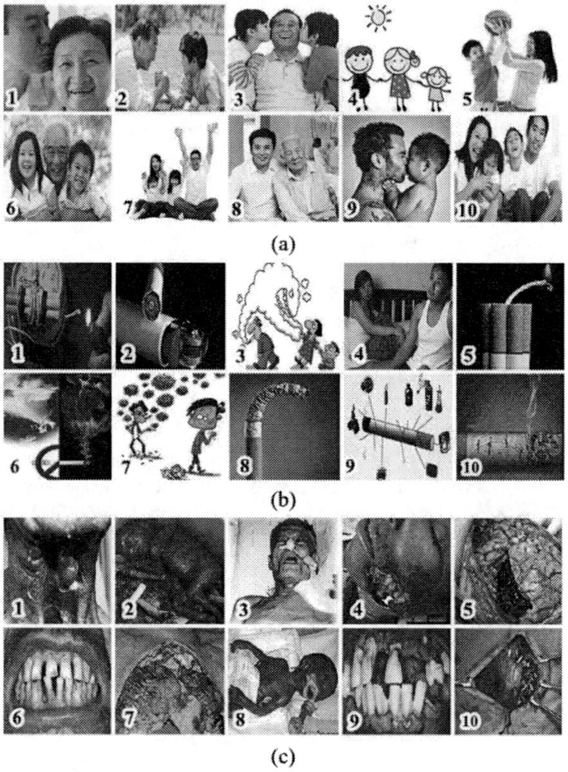

Figure 2. Final affective warning pictorials (a) positive warning pictorials (b) neutral warning pictorials (c) negative warning pictorials.

4 Discussion

Final affective warning pictorials were shown in Figure 2. The results confirmed its signiticantly evoked target affective states on participants to specific affective states such as positive, neutral, and negative affect. Ten neutral affective warning pictorials mainly presented the effect of smoking in terms of health, finance, and social (Figure 2b). The pictorials used in this group had both real photos and cartoon pictorials. Positive warning

pictorials used in present study were majority pictorials of family reflecting the effect of smoking on their beloved people (Figure 2a). The pictorials of son kissing mother (Figure 2a-1) received highest rating valence scores among other pictorials (M = 7.21, SD =

0.97). People may feel and reflect from the pictorial that he could be a good son with no smoking and takes care his mother's well-being. The positive rating responses from participants revealed that participants could be evoked positive affective states by this warning pictorials. While the least valence (*M* = 6.76, *SD* = 1.08) and arousal (*M* = 3.21, *SD* = 1.68) rating responses from participants among ten final set of pictorials was an only cartoon pictorial of family in this group (Figure 2a-4).

Negative warning pictorials in ten final pictorials were depicting health effect from smoking. These pictorials were actually the pictures used for reduce smoking campainge [15] such as chronic sore and cancer. The most negative pictorial which received the lowest valence rating scores (*M* = 1.97, *SD* = 0.99) from participants was a pictorial of mouth cancer (Figure 2c-4). Moreover, participants rated their arousal scales (*M* = 6.66, *SD* = 1.77) as highest for the pictorials (Figure 2c-2) in this negative group. Participants may feel frightened from the pictorials as they represented real human body of dreadful health effect from smoking.

After considered all the affective pictorials used in present study, cartoon-like pictorials evoked moderate affective responses on participants comparing to real images of people or cigarette. Although, there are one cartoon-like pictorials in final set of positive warning pictorial group, that pictorials were the least valence rating scores comparing other pictorials in same group. While there were no cartoon-like pictorial in negative warning pictorial group. The characteristics of pictorial (cartoon-like and real people pictorials) and its differences in people's affective responses should be investigated in future study. This may give some guideline for designing future affective pictorials.

5 Conclusions

This study aimed to design and validate affective warning pictorials used on cigarette label. Thirty affective warning pictorials (positive, neutral, and negative) were successfully validated as able to evoke specific target affective states on people.

There were limitations regarding to the scope of this study. The participants were limited to only male adults from engineering school. It is expected to have some different degree of affective responses regarding to the gender, culture, and generations in the results. Thus, it would be worth investigating with broader range of people from different generations, people with different background, and also female participants in future study. Moreover, future study should validate these affective warning pictorials in terms of warning effectiveness such as comprehension and hazard perception before real implementation.

Acknowledgment

This work was supported by Grants for Development of New Faculty Staff, Ratchadaphiseksomphot Endoment Fund, Chulalongkorn Univeristy, Bangkok, Thailand.

References

1. D. Hammond, G.T. Fong, P.W. McDonald, K.S. Brown, R. Cameron, AM J. Public Health, **94,** 1442-1445 (2004)
2. B.E.M. Nascimento, L. Oliveira, A.S. Vieira, M. Joffily, S. Gleiser, M.G. Pereira, T. Cavalcante, E. Volchan, Tob Control, **17,** 405-409 (2008)
3. S. Osiri, S. Kongkasawad, HSRI Journal, **1,** 280-288 (2007)
4. WHO, Guidelines for implementation of the WHO Framework Convention on Tobacco Control, Article 11 (2005)
5. C.C. Braun, P.B. Mine, N.C.Silver, Int. J. Ind. Ergonom., **15,** 179-187, (1995)
6. M.S. Wogalter, Ergodesigne USIHC, *Rio de Janeiro: FAPERJ*, 25-36 (2004)
7. M.F. Lesch, J. Safety Res., **34,** 495-505 (2003)
8. M.F. Lesch, Accident Anal. Prev., **40,** 1005-1012 (2008)
9. D. Sujirarat, P. Silpasuwan, C. Viwatwongkasem, N. Sirichothiratana, *Southeast Asian J. Trop. Med. Public Health*, **42,** 134-141 (2011)
10. S. Azagba, M.F. Sharaf, Nicotine Tob Res., **15,** 708-717 (2013)
11. N. Schwarz, H. Bless, G. Bohner, *Adv. Exp. Soc. Psychol.*, **24,** 161-199 (1991)
12. H. Umemuro, *Human-Computer Interaction*, **5612,** 683-692 (2009)
13. A. Jiamsanguanwong, H. Umemuro, *Appl. Ergon.*, **45,** 1362-1366 (2014)
14. P.J. Lang, M.M. Bradley, B.N. Cuthbert, International Affective Picture System (IAPS) NIMH Center for the Study of Emotion and Attention, 39-58 (1997)
15. P. Vartesatokkit, Luckpim Publishing, *Warning Pictorial on Cigarette*, (2009)
16. Tobacco Labelling Resource Centre. *Thailand.* (2013) Retrieved September 20, 2015, from http://www.tobaccolabels.ca/countries/thailand
17. M.M. Bradley, P.J. Lang, *J. Behav. Ther. Exp. Psy.*, **25,** 49-59 (1994)

 DOI: 10.1051/matecconf/20166817002

Automatic Traffic Sign Detection and Recognition Using Colour Segmentation and Shape Identification

Karel Horak[a], Pavel Cip and Daniel Davidek

Brno University of Technology, Department of Control and Instrumentation, 61600 Brno, Czech Republic

Abstract. The paper describes a colour-based segmentation method of European traffic signs for detection in an image and a feature-based recognition method for categorizing them into given classes. At first, we have performed analysis of several well-known colour spaces as the RGB, HSV and YCbCr often used for segmentation purposes. The HSV colour space has been chosen as the most convenient for segmentation step and colour-based models of traffic signs representatives were created. Next, the fast radial symmetry (FRS) detection method and the Harris corner detector were used to recognize circles, triangles and squares as main geometrical shapes of the traffic signs. For these purposes a new gallery of real-life images containing traffic signs has been created and analysed. Overall efficiency of our recognition method is approx. 93 % on our gallery and is usable for real-time implementations.

1 Introduction

Machine vision in a traffic is common technology at these days. A plenty of various system for traffic signs recognition were developed and some of them has ambition to be a part of superior car system of online navigation and for autonomous driving [1, 2]. Such systems are generally referenced as the driving assistance system. Some automotive manufacturers like Daimler-Benz, Opel or BMW already uses their own technologies for assisting driver or preceding accidents [3]. Despite a lots of designed and implemented systems there is no one generally efficient all over the world. Relatively simple system introduced in this paper takes into account a region of Europe and its scheme of traffic signs.

1.1 Traffic signs differences around the world

In different countries are very different opinions on how to give instructions or provide information about traffic conditions to drivers and other road users. The basic concept of the traffic signs used in the European Union is based on a usage of simple graphical pictogram inside a plain geometrical shape of one principal colour. Similar idea determines traffic signs in the United States, nevertheless more additional text is used on a plenty of signs and thus subconscious perception of the traffic signs is slightly more complicated and is not as straightforward as in the EU. On the contrary, almost arbitrary traffic meaning (intent) can be visualised by this more flexible way established in the US. As an example of such typical difference the informative traffic sign *"One-way street"* in the last row of the Table 1 can be pointed out.

For quick and clear overview of differences in the EU and US region, the Table 1 shows two example signs in each row related to one of the five basic traffic sign classes: warning, regulatory, obligatory, priority and informative.

Table 1. Different signs in EU and US region.

	EU	US
Warning		
Regulatory		
Obligatory		
Priority		
Informative		

An exact meaning of the traffic signs in the Table 1 is following:

[a] Corresponding author: horak@feec.vutbr.cz

Warning:	
Curve to the right	Traffic lights ahead
Regulatory:	
No entry for vehicles	Speed limit 80 km/h
Obligatory:	
Turn right ahead	Cycle left, pedestrian right
Priority:	
Yield (give way)	Stop
Informative:	
One-way street	Dead end (no through road)

The overall difference of traffic signs within the EU are conceptually reduced by a multilateral treaty "Vienna Convention on Road Signs and Signals" from 1968. Such treaties effectively limit differences in traffic signs and thus facilitate deploying of autonomous recognition systems despite of high number of all traffic signs defined. For example more than 200 traffic signs, each assigned to one of the five main classes, are defined in traffic regulations of the Czech Republic in the EU. Algorithms and methods designed and described in this paper take into account the only EU traffic signs concept.

1.2 Traffic sign recognition approaches

There are two principally different approaches for traffic sing detection and recognition task in a theory of the computer vision. First one, the template-based approach computes similarity between unknown signal (part of an input image with some object) and a set of given templates of traffic signs. A template of traffic sign of the highest similarity with queried sub-image is then chosen as the correct one.

The other approach, a so-called feature-based, computes a set of features on the input image with unclassified objects and classifying mechanism searches for similar structure and values of these features in an internal database. Corners, radial symmetry, SIFT, SURF, BRISK, MSER and plenty of other descriptors can be used for detection and description of the so-called points of interest.

1.3 Our design

Briefly, we have selected the feature-based approach in this paper. At first, the input image is transformed from a native camera picture format to the HSV 8-bit colour coding. Further, a colour-based segmentation is used for a detection of red, blue and yellow objects in the input image. These objects are then filtered by means of given geometrical and colour conditions valid for traffic signs e.g. only circles and rectangles are interesting in blue colour etc. A shape identification and recognition step is carried out as last step before final sign classification. Well-known features as Harris corners and Fast Radial Symmetry (FRS) are used at this stage. Next chapters introduce mentioned approach in more details.

2 Colour-based segmentation

A colour information in an image is one of several fundamental features of each object together with e.g. edges, corners, brightness level etc. Because of exact definition of appearance of traffic signs in various national regulations of each well urbanized state, it is convenient to use entire or only part of these definitions. Traffic signs generally have well defined geometrical shape and very limited number of permissible shapes (often only circle, triangle, rectangle and octagon). The other well defined feature of traffic sign is its colour. Only the one specific red, blue and yellow colours are used for definition of the most crucial signs.

Plenty of various colour spaces and coding schemes are introduced in a theory and only some of them are used in practical implementations [4]. We have analysed almost 350 own images of traffic scene with at least one traffic sign present on each image. The analysis has been performed in the RGB, HSV and YCbCr colour schemes. Results of the RGB and HSV colour-based segmentation are brought out in the following subsections.

2.1 Segmentation in RGB colour space

The RGB (Red-Green-Blue) colour space is the most frequent output format of almost all image acquisition devices as webcams, industrial cameras, handy-cams, etc. This general popularity of the RGB is given by a construction of CCD and CMOS sensors equipped with the well-known Bayer colour mask R-G-G-B. Nevertheless, this format is not usually convenient for image processing and more generally for computer vision tasks because of lack of an explicit brightness channel and thus its implicit presence directly in all the R, G and B channels values.

As can be seen in the Fig. 1 on the left, the RGB colour space is in form of cube. Colour-based segmentation in such space is simply given by limits of min and max values along all three axis related to R, G and B channel respectively. These limits are given by a model of an object looked for. The limits can form either a simple orthogonal block in the RGB colour space or some more sophisticated shape as a sphere, ellipsoid, octahedron etc.

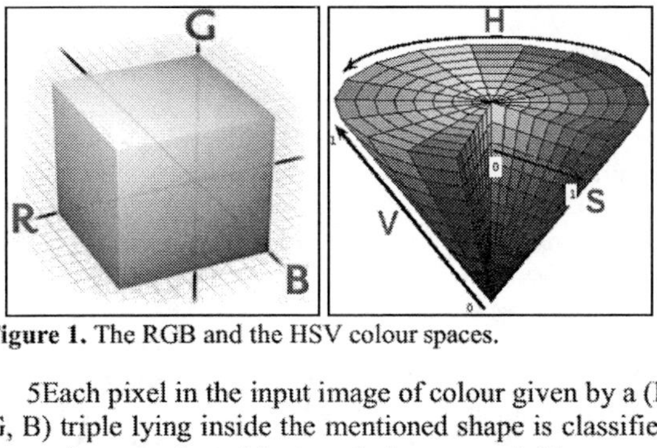

Figure 1. The RGB and the HSV colour spaces.

5Each pixel in the input image of colour given by a (R, G, B) triple lying inside the mentioned shape is classified as an element of an objects and vice versa. A binary map of pixels memberships' to particular colour is a result of such colour-based segmentation.

We have used a so-called comparative method for our experiments described later. This method applies limits mentioned above to a ratio of two particular colours R, G and B as it is introduced by the equation (1) for the red component.

$$map\_R(x,y)=\begin{cases}1 & \frac{R(x,y)}{G(x,y)}\geq k_{RG}\wedge\frac{R(x,y)}{B(x,y)}\geq k_{RB}\wedge\frac{B(x,y)}{G(x,y)}\geq k_{BG}\\ 0 & otherwise\end{cases}\qquad(1)$$

Binary maps for the other two colours G and B are given by analogical equations. All three maps can be finally visualized in only one combined RGB image.

2.2 Segmentation in HSV colour space

The HSV space (Hue-Saturation-Value) is closer to intuitive perception of colours by human eye and is cone shaped. As can be seen in the previous Fig. 1 on the right, the Hue component is angle from 0° to 360° and represent a basic shade of colour. The other two components are numbers from 0 to 1 and represent saturation (purity) of colour and value (brightness) respectively.

Segmentation in the HSV colour space is very similar to the previous one with only one difference that shape determining an area of object colours is distorted in accordance with a shape of the HSV space.

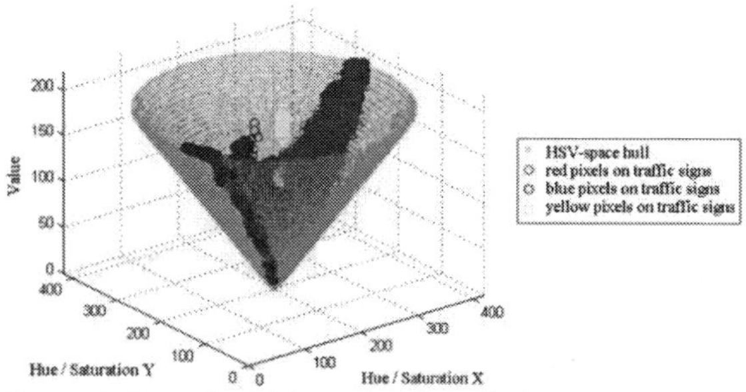

Figure 2. Sets of red, blue and yellow pixel representatives in images with traffic signs.

In the Fig. 2, the three clusters of pixels related to red, blue and yellow parts of the traffic signs can be seen. Geometrical representatives (hulls of clusters) serve as segmentation rule at the same way as in the case of the RGB colour space segmentation.

3 Shape recognition

One or more regions of interest (ROIs) are an output of the previous colour-based segmentation step. Now classification of its shape has to be carried out. The task of traffic sign recognition is very limited by itself: only circles, triangles and rectangles are considered as permissible classes of shapes.

A so-called fast radial symmetry detection method (FRS) introduced in [5] has been employed for circles recognition in [6] and was employed also in our experiments. This algorithm outputs a grayscale map with a peak at that position, where any circle is present in the input image. The input ROI and a response of the FRS algorithm are shown in the Fig. 3.

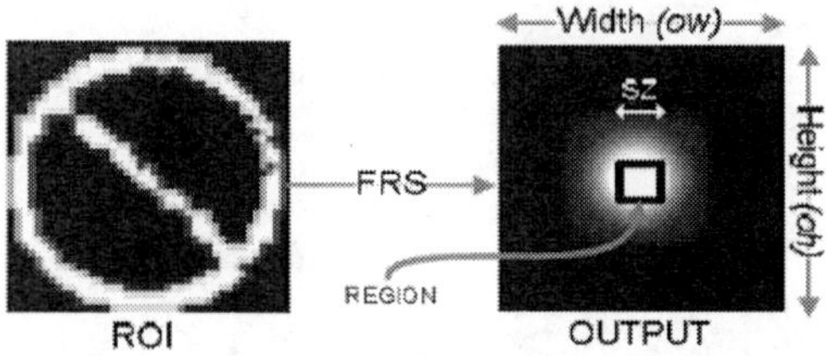

Figure 3. Fast radial symmetry detection on circle object [6].

More circle-shaped the input object is, higher response the FRS method gives at its output. For our case of traffic signs, two examples of responses of the FRS detection method can be seen in the Fig. 4.

Figure 4. Difference of results of the fast radial symmetry algorithm on circular and triangular traffic sign.

Note that almost ideal circle-shaped sign *No entry* (on the left) results in a clear and high peak at output image. On the contrary, the triangle-shaped sign *Warning* does not create any significant peak and the response is blurred. Recognition of valid circle-shape object in the input image is then done based on the peak level and its compactness.

The other two basic shapes (triangle and rectangle) are distinguished by means of the Harris corner detector [7]. We use the knowledge from the previous steps that object of traffic sign (or other similar object) is normalized in the ROI and is straight in it. Under these conditions, the Harris corner detector gives a strong response at the four principal corners of the ROI in case of rectangular object in the input image and the three corners in case of triangular object. Based on mutual position of these corners the proper class of input object can be specified [6].

4 Experimental results

We have carried out a series of experiments in both laboratory and real life. Whole procedure is described in following chapters in detail.

4.1 Image gallery

First of all an image gallery was created on our own. We have acquired many hundred images and almost 350 of them were selected as suitable for next processing during our experiments. An example of two gallery images is depicted in the following Fig. 5.

Figure 5. Weak and good contrast input images with signs.

The gallery has been employed for creating of model representatives of the red, blue and yellow signs classes at first stage. Hundreds of pixels were semi-manually picked out to create a robust and reliable colour models (red, blue and yellow). See the Table 2 containing numbers of images and related numbers of pixels selected for individual colour classes of traffic signs.

Table 2. Image gallery for basic classes of traffic signs.

Traffic sign	Images in gallery	Pixels picked out
Red circular	180	65000
Red triangular	47	14000
Blue circular	16	10500
Blue rectangular	82	40250
Yellow squared	18	8500
Total	343	138250

4.2 Colour segmentation

In accordance with the theory described in the previous chapters, a colour-based segmentation in the RGB, HSV and additionally for comparison in the YCbCr colour spaces have been performed on the whole gallery during the tests.

Even though the HSV segmentation needs a transformation step RGB→HSV, its results are more suitable for traffic sign's shape recognition. The most important is a lower over-segmentation factor and a higher compactness of segmented objects in comparison with the RGB model. It is caused by pure colour specification in the HSV colour model with strong ability to handle over-exposed and under-exposed images.

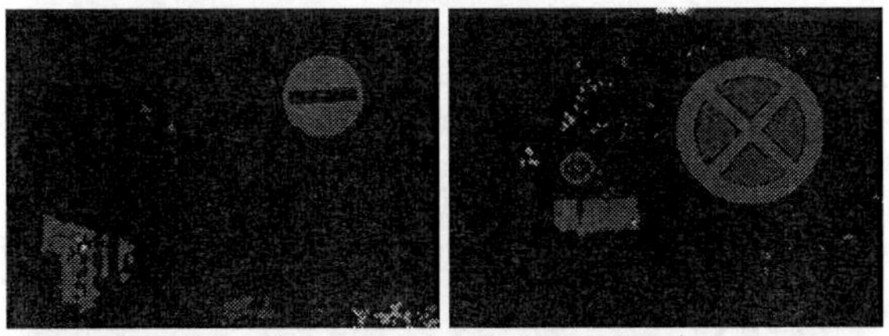

Figure 6. Results of segmentation in the RGB colour space.

Figure 7. Results of segmentation in the HSV colour space.

In the previous two figures you can see results of segmentation step in the RGB and HSV colour space applied on the two images from the Fig. 5. All the three binary maps of the red, blue and yellow colours are showed in one image together by the red, blue and green colour respectively. Overall efficiency of segmentation algorithm is recorded in the following Table 3 in association with an average time required by the algorithm for processing of each one image from the gallery. A long duration of segmentation in the HSV colour-space has been anticipated due to RGB→HSV.

Table 3. Comparison of colour spaces for segmentation.

Colour space	Duration [ms]	Efficiency
RGB	200	93 %
HSV	420	97 %
YCbCr	351	82 %

Images of reduced dimensions 640 by 480 pixels have been taken into account during this measuring. Efficiency was defined as successful/unsuccessful segmentation of complete area of traffic sign without any connection with its neighbourhood. The shape recognition algorithm can be efficiently applied on such distinctive regions in the next step of image processing.

4.3 Region normalization

Traffic signs in the both input images in the Fig. 5 are almost ideally perpendicular to an optical axis of a camera used to capture images. However, absolute majority of traffic signs captured do not fulfil this special case and it is the reason why a normalization of object of presumable traffic sign is performed.

An example of the normalization procedure is shown in the Fig. 8. An original image of the traffic sign *No left turn* is depicted on the left in grayscale. Its binary map containing corresponding ellipse with highlighted major and minor axis is shown on the right.

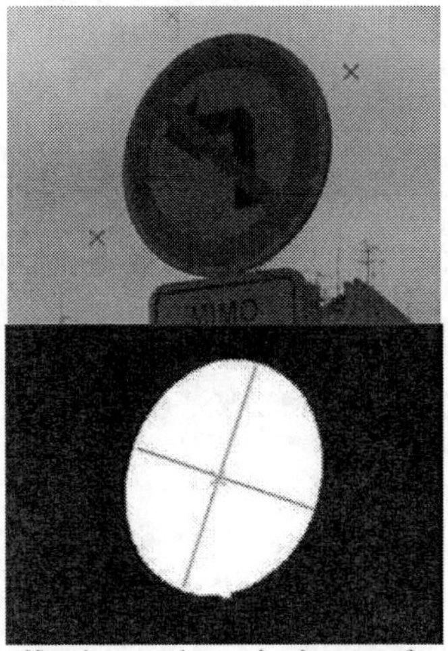

Figure 8. Traffic sign on the projective transformation in the input image and its binary map.

Each ellipse-shaped object always corresponds to some circular traffic signs (or less probably some other circle object). The ellipse is then transformed to a circle-shaped object with minor and major axis of the same length and with the major axis to be exactly vertical. Result of this transformation is depicted in the Fig. 9 on the left. In the middle and on the right pictures are illustrated rotational correction of the transformation and a trimmed image as final output of segmentation and normalization steps.

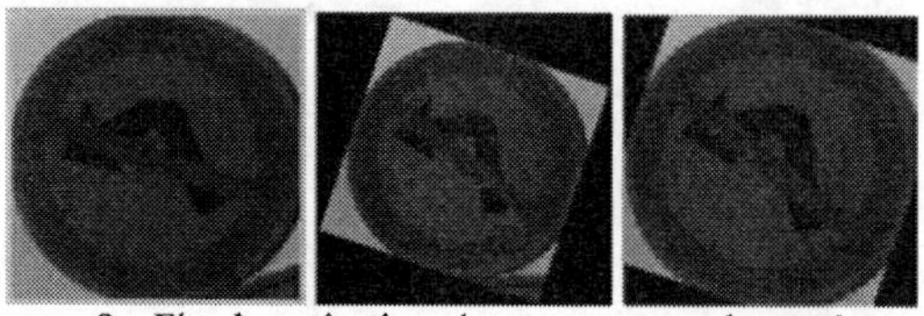

Figure 9. Fixed projection image, corrected rotation and trimmed image of the traffic sign.

Reliable region normalization is crucial stage for later recognition of type of a traffic sign. An example set of normalized traffic sign objects resampled to dimensions of 100 by 100 pixels is introduced in the Fig. 10.

Figure 10. Normalized traffic sign objects as input for recognition algorithm.

4.4 Real-scene recognition experiments

By implementation the HSV colour-based segmentation algorithm, the shape recognition algorithm and a pictogram inside a traffic sign detector, we obtain interesting results. At first, responses of the FSR method to the two input images introduced in the Fig. 5 can be seen in the following Fig. 11.

Figure 11. FRS responses to the input images from Fig. 5.

It is quite clear that a recognition of circular traffic signs is performed by evaluation of height of peaks in these maps. The presence of triangular or rectangular traffic sign was determined from a distribution of the strongest corners within a region of an object.

In the next Fig. 12 two semi-synthetic testing images are depicted in association with imprinted position and types of detected traffic signs.

Figure 12. Verification of traffic sign recognition on semi-synthetic images.

Figure 13. Traffic sign recognition system results.

Finally, we have applied our design directly in urban traffic as the driver assistance system aforementioned in the introductory chapter. Several selected pictures of this system are introduced in the previous Fig. 13. Such images serve as system's output for a driver on the LED display inside a car.

5 Conclusions and future work

During our experiments we have implemented a combination of the colour-based and shape-based methods to achieve better results in recognition of traffic signs. Results obtained were really highly promising, nevertheless a deep verification both on a broad gallery and in the traffic should be carried out.

Future work lies in an improvement of basic colour segmentation in cases, when a traffic sign is connected with some other near object due to its similar colour. Next, a better algorithm for pictogram recognition has to be developed instead of correlation technique used here.

Acknowledgements

This paper was supported by the grant The research of new control methods, measurement procedures and intelligent instruments in automation (FEKT-S-14-2429) of the Internal Grant Agency at the Brno University of Technology and by the project Centre for Applied Cybernetics 3 (TE01020197) under the Technology Agency of the Czech Republic.

References

1. N. Barnes, A. Zelinsky, L. S. Fletcher, Real-Time Speed Sign Detection Using the Radial Symmetry Detector, Intelligent Transportation Systems, IEEE Transactions on , **9(2)**, pp. 322-332, June (2008)

2. A. Soetedjo, K. Yamada. *An Efficient Algorithm for Traffic Sign Detection*. JACIII, **10**, 3, pp. 409-418, September (2006)
3. D. M. Gavrila. *Traffic Sign Recognition Revisited*. 21. DAGM-Symposium Bonn, Springer Berlin Heidelberg, pp. 86-93, September (1999)
4. A. Escalera, L. Moreno, M. Salichs, J. Armingol, *Road traffic sign detection and classification,* IEEE transactions on industrial electronics, **44(6)**, Dec. (1997)
5. G. Loy, and A. Zelinsky, *Fast Radial Symmetry for* Detecting *Points of Interest*, IEEE Transactions on Pattern Analysis and Machine Intelligence, **25(8)**, pp 959-973, August (2003)
6. C. F. Paulo, P. L. Correia, Automatic Detection and Classification of Traffic Signs, in Image Analysis for Multimedia Interactive Services, WIAMIS '07. Eighth International Workshop on, pp.11-11, 6-8 June (2007)
7. C. Harris and M. J. Stephens. *A Combined Corner and Edge Detector*, Alvey Vision Conference, pp. 147-152, (1988)

MATEC Web of Conferences **68**, 17003 (2016) DOI: 10.1051/matecconf/20166817003
ICIEA 2016

The Application of Marker Based Segmentation for Surface Texture Characterization

Nuraini binti Che Pin[1,a], Mohd Fauzi bin Ismail[1] and Siti Noraini binti Sulaiman[2]

[1]*Faculty of Mechanical Engineering, University Technology Malaysia*
[2]*Faculty of Electrical Engineering, University Technology Mara, Malaysia*

Abstract. Structured surfaces have been increasingly used in industry for a variety of applications, including improving the tribological properties of the surfaces. Surface metrology plays an important role in this discipline since with the help of surface metrology technology, surface texture can be measured, visualize and quantified. Traditional surface texture parameters, such as roughness and waviness, cannot be related to the function for structured surfaces due to the less statistical description and little information. Therefore, a new approaches based on characterizing the structured surface is introduces where this paper focus on type of edges grain surface. To identify features, it is a must to detect the location of the edges and segmented the features based on the detected edges. Hence characterization of surface texture segmentation based on the edges detection is developing using Marker Based segmentation and it is prove that this method is possible to be used in order to characterize the structured surface.

1 Introduction

All surfaces whether at the nano or macro scale have made up of collection of fundamental features parameter which can be visualize in surface topography [1]. There are many type of surface of the product design but the new generation of surface nowadays is so called as structured surfaces. This type of surface has dominant deterministic pattern design to produce a specific function [2]. Besides, this kind of surface also has been increasingly used in industry for a variety of applications, including improving the tribological properties of the surfaces. High potential application of structures surfaces related product also found in the diverse automotive, biotech, medical and consumer application and are one of the current drivers of future growth.

Surface metrology plays an important role in this discipline since with the help of surface metrology technology, surface texture can be measured, visualize and quantified of their feature parameter on the surface. Over the years, the characterization and evaluation of engineering surface texture has constitutes a challenging metrological problem that has remained open so far, especially when high precision or structured surface with functional performance requirements exist. Besides that, it has been found that the standard areal parameters are often insufficient for characterizing structured surface, primarily because the areal surface texture parameters capture properties that are related to the entire measured topography whereas structured surface often exert their function through the action of individuals or multiple, localized surface features [3].

In this paper, a new approaches based on characterizing the structured surface is introduces. Initially it is useful to classify structured surface into two classes each based on their generic geometric form. This paper will focus on structured steps include surfaces that have grain, edges and facets [4]. Example of surface that has been tested is shown in Figure 1. This edges grain surface is characterize to extract the feature parameter such as the area, perimeter of the specific feature that have been selected on the surface. Consequently, this characterization will indirectly help the manufacturer to quantify the feature on the surface since in manufacturing process, if the measurement parameter is constant from any work piece to another work piece then the process is said to be under control.

Figure 1. Sample of diamond grinding surface.

[a] Corresponding author: nuraini_chepin@yahoo.com

DOI: 10.1051/matecconf/20166817003

To identify features on the structured surface, an important method required is to segment the surface into regions on interest [5]. Since this paper is focusing surface with edges features, therefore it is a must to detect the location of the edges and segmented the features based on the detected edges. Hence, this paper will develop the characterization of surface texture segmentation based on the edges detection method.

2 Measurement method

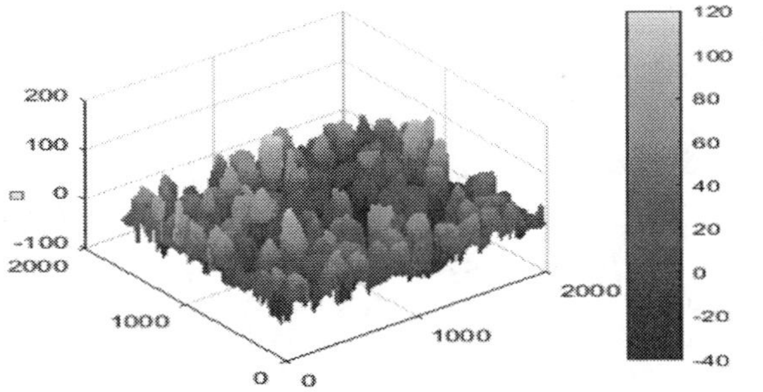

Figure 2. Topography surface of diamond grinding.

The confocal laser scanning is used to measure the surface of diamond grinding in the present study which shown in Figure 2. Surfaces are acquired point-by-point and reconstructed with a computer, allowing three-dimensional reconstructions of topologically complex objects were obtained [6]. The surface topography profile gives true colour information that can be measured in 3D data. This data structure consist of a matrix in rectangular shape with regular x, y and z grid where rows and columns referring to x and y while height referring to z.

3 Characterization method

In order to characterize surface topography of the diamond grinding, the surface need to go preprocessing steps which are leveling and filtering. Function to do leveling to the surface is to provide a consistent reference plane so that it will become easier to characterize the whole surface later on. While filtering is done so that the outliers and noise of the measured data surface can be remove. This is due to the fact that measured data always have some irrelevant information that is not interested in.

Morphological filter is used in this research for the filtration process. This type of filtration is choosing with the fact that it can help to preserve the edges features on the surface [7]. Specific morphological filtration is used which are opening and closing by reconstruction. Reconstruction by opening and closing are more effective than standard opening and closing at removing small outlier without affecting the overall shapes of the surface. Disk is the structuring element that has been used with appropriate radius. Experimental testing is done to determine the best radius of disk to be used.

4 Marker based segmentation

4.1 Detection of significant edges

Since this research study focus on edges segmentation, therefore a critical step is to detect the significant edges on the surface. Sobel Operator is the method of edges detection approaches that have been used to detect the significant edges. The size of this kernel of operator is set to be fixed with 3 x 3 size of kernel.

4.2 Creating marker controlled based

In this research, simple method is used to create the foreground and background marker which is by using morphological reconstruction. In morphological reconstruction, structuring element used was playing an important role since it gives big impact to the number of segmented features later on. Two steps involved in this steps which are erosion based reconstruction and followed by dilation based reconstruction. Regional maxima also being calculated in order to get smooth edge foreground objects. Later on, these two steps and the edges detection surface are been superimposed on the surface that has been filtered [8].

4.3 Segmentation using watershed

Application of watershed segmentation will gives final segmentation results of the surface. Therefore, full algorithm of marker based segmentation is shown on the flowchart in Figure 3.

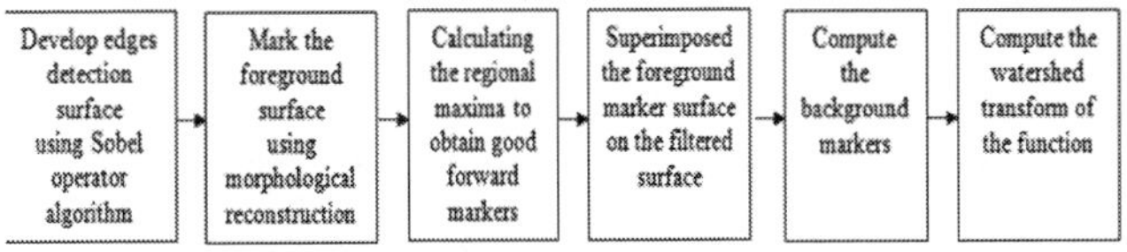

Figure 3. Algorithm for maker based segmentation.

5 Discussion

In order to get better segmentation results, radius of disk in morphological filtration need to choose correctly. Different radius of disk is tested to see the effect of the segmentation results. Radius of the disk is tested starting from 1μm, 3μm, 5μm, 7μm and finally 9μm. The graph in Figure 4 shows the results from the testing.

Based on the graph in Figure 4, it shows that the number of segmentation is decrease accordingly with the increase of radius of the structuring element until one point that the number of segmentation becomes more stabilized. Besides, it can be compared that the count of number of segmentation features at 1μm radius of disk, the number of segmentation has sharply reducing at 3μm radius. Then the count becomes linear after 5μm of radius of the disk.

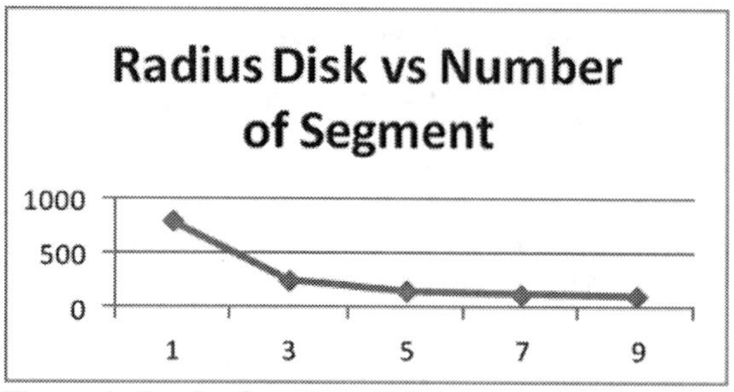

Figure 4. Graph of number of features segmentation versus radius of structuring element.

To make it more clear, Figure 5 shows the view from the top of the segmentation results after undergo Marker Based Segmentation. When radius of the structuring element used was too small, over segmentation is happen to the surface. While after radius of the structuring element is increase, number of feature segmentation become more stable and linear.

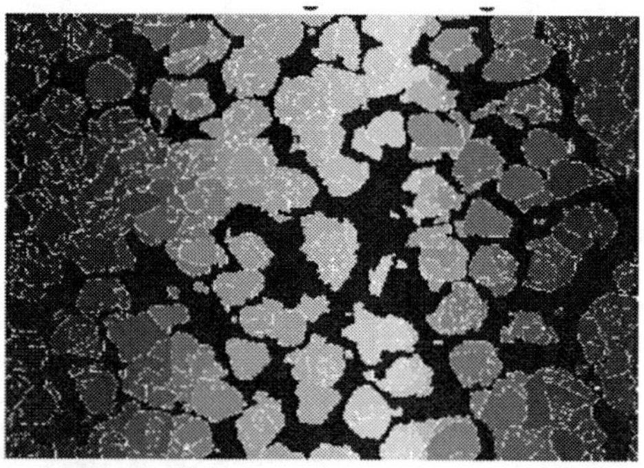

800 number of features

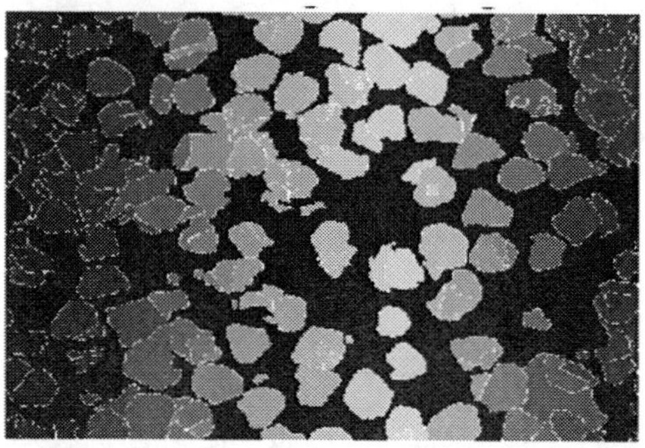

254 number of features

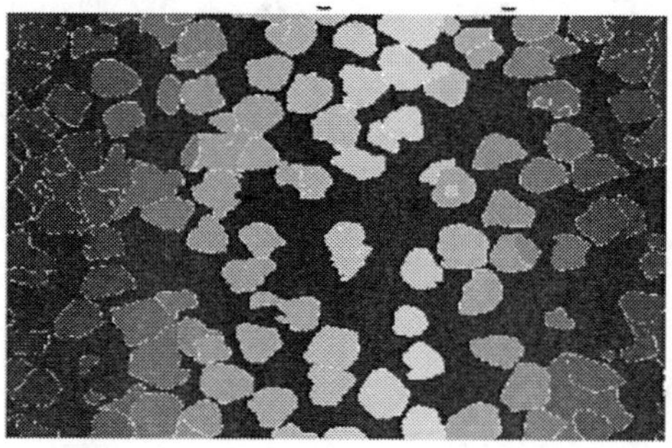

160 number of features

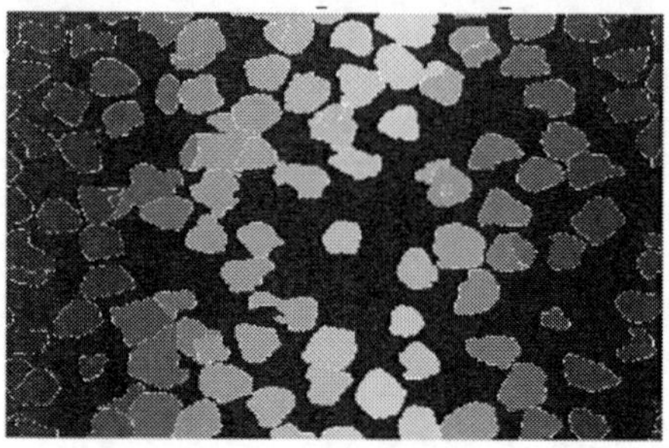

115 number of features

Figure 5. Different number of features segmentation based on different radius of disk.

6 Conclusion

As the conclusion, Marker Based Segmentation is possible to be used to detect the edges on the surface data topography and finally segment the surfaces following the shaped of the features. Morphological Filtration, Sobel Operator, Regional Maxima and also Watershed Segmentation are the best method to be paired in group for the overall algorithm of the Marker Based Segmentation. For further work, all the related feature parameter such as area, perimeter, volume and others will be specifically identified.

Acknowledgement

The authors gratefully acknowledge University Technology MARA and Ministry of Higher Education Malaysia for the financial support under Fundamental Research Scheme (Grant no: FRGS/1/2014/TK01/UITM/02/6).

References

1. L. Blunt and S. Xiao, "The use of surface segmentation methods to characterise laser zone surface structure on hard disc drives," Wear, **271**, 3–4 604–609 (2011)
2. G. D. MacAulay, N. Senin, C. L. Giusca, R. K. Leach, and A. Ivanov, "Review of feature boundary identification techniques for the characterization of tessellated surfaces," Surf. Topogr. Metrol. Prop., **3**, 1 013002 (2015)
3. G. D. Macaulay, N. Senin, C. L. Giusca, and R. K. Leach, "Comparison of Segmentation Techniques to Determine the Geometric Parameters of Structured Surfaces," Surf. Topogr. Metrol. Prop., (2014)
4. X. Jiang, P. Scott, and D. Whitehouse, "Freeform Surface Characterisation - A Fresh Strategy," CIRP Ann. - Manuf. Technol., **56**, 1 553–556 (2007)
5. A. G. Demir, P. Maressa, and B. Previtali, "Fibre laser texturing for surface functionalization," Phys. Procedia, **41**, 759–768 (2013)
6. J. Peters, J. B. Bryan, W. T. Estler, C. Evans, H. Kunzmann, D. a. Lucca, S. Sartori, H. Sato, E. G. Thwaite, P. Vanherck, R. J. Hocken, J. Peklenik, T.

DOI: 10.1051/matecconf/20166817003

Pfeifer, H. Trumpold, and T. V. Vorburger, "Contribution of CIRP to the Development of Metrology and Surface Quality Evaluation during the last fifty years," CIRP Ann. - Manuf. Technol., **50**, 2 471–488 (2001)

7. P. Dobrzaski and P. Pawlus, "A study of filtering techniques for areal surface topography assessment," Proc. Inst. Mech. Eng. Part B J. Eng. Manuf., **225**, 11 2096–2107 (2011)
8. R. Maini and J. S. Sohal, "Performance Evaluation of Prewitt Edge Detector for Noisy Images," Int. J. Graph. Vis. Image Process., **6**, 3 39–46 (2006)

MATEC Web of Conferences **68**, 17004 (2016) DOI: 10.1051/matecconf/20166817004
ICIEA 2016

Research on Image Reconstruction Algorithms for Tuber Electrical Resistance Tomography System

Zili Jiang , Xiang Deng , Shuyi Liu and Yu Tang

School of Mechanical, Electronic and Control Engineering, Beijing Jiaotong University, Beijing.100044, P.R.China

Abstract. The application of electrical resistance tomography (ERT) technology has been expanded to the field of agriculture, and the concept of TERT (Tuber Electrical Resistance Tomography) is proposed. On the basis of the research on the forward and the inverse problems of the TERT system, a hybrid algorithm based on genetic algorithm is proposed, which can be used in TERT system to monitor the growth status of the plant tubers. The image reconstruction of TERT system is different from the conventional ERT system for two phase-flow measurement. Imaging of TERT needs more precision measurement and the conventional ERT cares more about the image reconstruction speed. A variety of algorithms are analyzed and optimized for the purpose of making them suitable for TERT system. For example: linear back projection, modified Newton-Raphson and genetic algorithm. Experimental results showed that the novel hybrid algorithm is superior to other algorithm and it can effectively improve the image reconstruction quality.

1 Introduction

Precision agriculture is a new trend of modern agricultural development in the world, and it is the system driven by information technology and a full set of modern agricultural management and production means with the locating, timely, quantifiably and the spatial variation characteristics [1]. At present, the research of electrical resistance tomography (ERT) technology is mainly focused on the industrial multiphase-flow field. In the paper, the application of ERT technology in the agriculture is described and the concept of TERT (Tuber Electrical Resistance Tomography) is proposed. TERT is a new research field. By the TERT system, the growth morphology of plant tubers and soil environment of the plant growth can be better monitored timely.

Image reconstruction of TERT system is a key problem and the research result can provide powerful tool for the visualization monitoring of the growth process characteristics of plant tubers such as precious and Chinese herbal medicines ginseng, etc. Therefore, it is very meaningful to study the image reconstruction algorithms of plant tubers tomography.

In order to meet the accuracy and real-time requirements of the application in ERT system, a variety of image reconstruction algorithms[2][3][4] had been proposed in the past. In this paper, on the research of TERT Image reconstruction algorithms, the algorithms such as linear back projection algorithm[5](LBP), Newton-Raphson algorithm[6] (MNR) and genetic algorithm[7] (GA), applied to the TERT system, are systematically studied and optimized under different conditions.

The LBP algorithm is the simplest and most rough image reconstruction algorithm. Although it has the advantages of simple and fast, the experiments show that it has better reconstruction effect only when the actual resistivity distribution is homogeneous. MNR algorithm is a kind of iterative reconstruction algorithm with a perfect and precise characteristics and better practical application, but it is not ideal in most cases because of the ill conditioned Hessian matrix. In some related literature, the genetic algorithm was used to study the imaging reconstruction of ERT [8][9]. But it has defects of premature and the small processing dimension, easy to reach the local optimal solution, and is difficult to achieve effective convergence with hundreds of split units. So it is difficult to completely meet the system requirements using one algorithm alone.

Based on the research and analysis above, a hybrid type imaging algorithm is proposed and discussed.

2 The basic working principle of TERT

Figure.1 shows the simple working principle of TERT system. Under a certain current excitation, the change of the electrical conductivity in area Ω will cause the change of the measured boundary voltage V_{ij}. Thus by measuring the boundary voltages, the object (tuber) in the area Ω could be detected

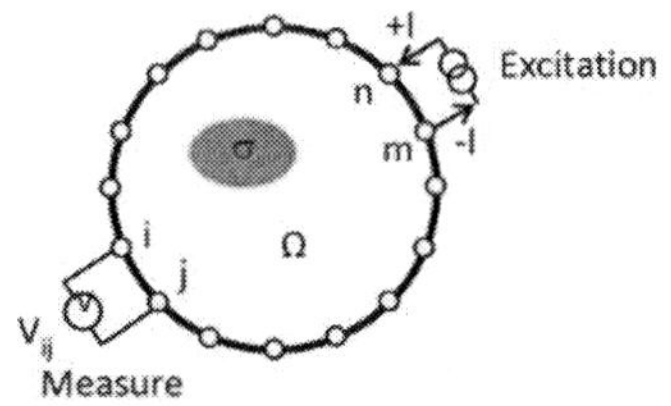

Figure 1. diagram of the TERT work.

The mathematic model of the forward problem of the TERT system is described as:

$$\begin{cases} \nabla \cdot (\sigma \cdot \nabla \varphi) = 0 \\ \sigma \dfrac{\partial \varphi}{\partial n}\Big|_{\partial \Omega} = +I \\ \sigma \dfrac{\partial \varphi}{\partial n}\Big|_{\partial \Omega} = -I \end{cases} \tag{1}$$

where I is injected current, σ is the potential, $\sigma \frac{\partial \varphi}{\partial n}\Big|_{\partial n}$ is the derivative in the outward normal direction along the boundary, Ω is the whole target area.

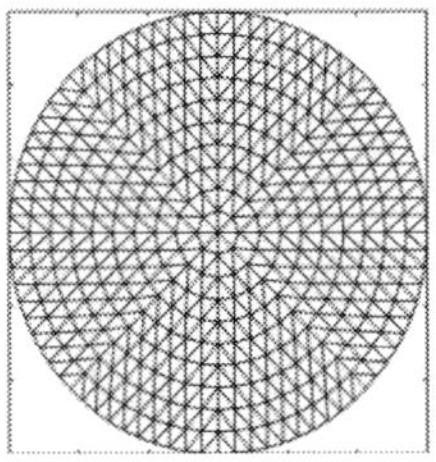

Figure 2. The pictorial diagram of the finite element model.

The finite element method is used to solve the forward problem in TERT system. The basic idea is to discrete the continuous solution domain into an assembly of finite elements. Such an assembly can better simulate or approximate the solution domain. If the mesh unit meets the convergence requirements, the approximate solution will converge to the exact FEM. The detailed steps including: 1) Determine of domain section parameters; 2) Selecting of field variable interpolation function in a unit; 3) Forming element characteristic matrix and element nodal equation; 4) Computing system matrix and integral equation; 5) Solving linear equations to get the field variable value.

Different topological types of finite element model will cause symmetry error and structural error [10]. Triangular element finite element method was used in the paper as shown in Fig. 2.

In Fig. 2, electrode number is 16, the simulation electrode is point electrode mode, the total number of nodes is 441, the number of boundary nodes is 80, the number of finite elements is 800, field radius is 0.15m, the excitation current intensity is 30mA, the data acquisition using adjacent excitation mode.

In the TERT system, plant tuber is grown in soil, so the field can be approximated as two different kinds of material. By measuring, the set average resistivity of a tuber is about $6353\,\Omega \cdot \mathrm{m}$ and the set average resistivity of the soil environment is about $1869\,\Omega \cdot \mathrm{m}$

3 Simulation research on image reconstruction algorithms of TERT

The analysis result shows that the resistivity data obtained by MNR has obvious ladder and partition, and the number of units in the area is about 20. So the number of chromosomes in the GA algorithm is 30, the maximum iteration number is 200, the crossover and mutation probability are 0.7 and 0.1 respectively.

Fitness function is:

$$f = \frac{2}{\left\| V^k - V_0 \right\|^2} \tag{2}$$

In addition, in order to further measure the quality of the reconstructed image, the spatial image error SIE and space occupied area error E are introduced.

The spatial image error can represent the error of the position and the distribution of the reconstructed image:

$$SIE = \frac{\sum_{i=1}^{M} \left| G_S^*(i) - G_r^*(i) \right|}{\sum_{i=1}^{M} G_S^*(i)} \tag{3}$$

G_s^*, G_r^* represent the set model and the reconstructed image respectively.

The space occupied area error can represent the relative value of the difference between the actual object area and the reconstructed image area is indicated.

$$E = \frac{[S_S - S_R]}{S_S} \tag{4}$$

TERT system is built to study the growth of plant tubers in the soil which are roughly round, so the simulation calculation is completed based on circular and ellipse. The simulation resistivity distribution is as follows:

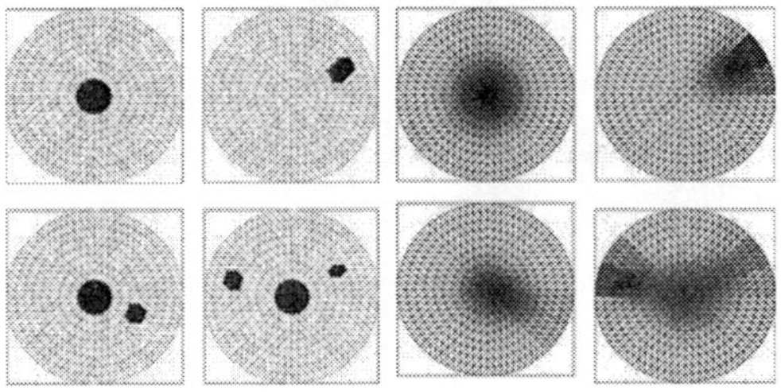

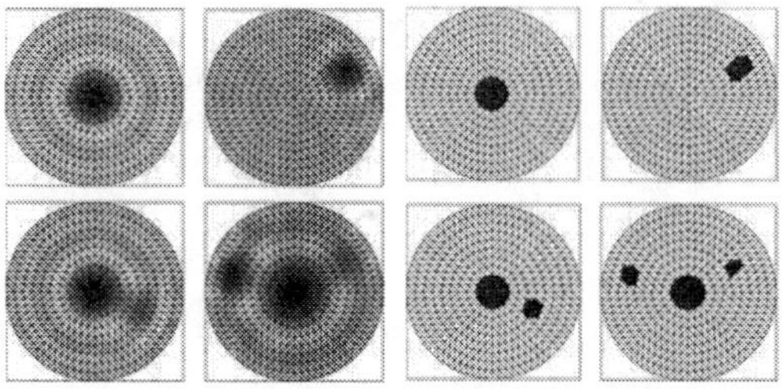

Figure 3. Image reconstruction results by using different algorithms.

Fig. 3 shows that the proposed algorithm LMG has obvious improvement. This algorithm can reconstruct clear field distribution patterns which are difficult with other two algorithms. Compared with the SIE parameters in Table.1, the hybrid algorithm is the most effective, which can improve the quality of the reconstructed image.

Table 1. Compare date results of different algorithms.

Field	Par	L	LM	LMG
A	*SIE*	0.2483	0.1835	0
	E	0.12	0.05	0
B	*SIE*	0.3648	0.2372	0
	E	0.105	0.035	0
C	*SIE*	0.3852	0.3024	0
	E	0.07875	0.06625	0
D	*SIE*	0.04254	0.02879	0
	E	0.1225	0.10375	1.25e-4

4 Initial experimental results and analysis

In order to verify the practical application effect of the combination algorithm, the developed TERT system was used to obtain the actual experimental data. As shown in Figure 4, a composite array of electrodes is used in the system. The field diameter is about 300 mm, the excitation frequency can be adjusted from 1 kHz to 5 kHz, and the excitation current size can be adjusted from 0 to 30mA.

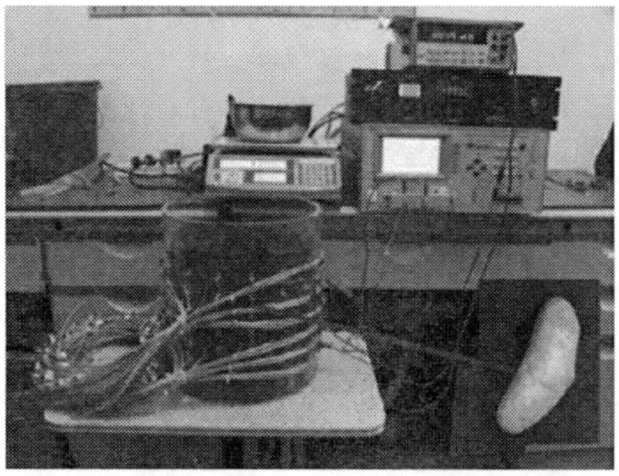

Figure 4. The developed TERT system with array electrodes.

In the actual experiment, the soil was collected in the farmland and adds water to saturation, the plant tuber (potato) was buried as shown in Fig. 5.

TERT system equidistant set up 6 layers of electrode array from top to bottom, spacing is 5cm. The tuber can only across the 2, 3, 4 layer electrode imaging area, therefore, only these three layers were imaged as follow:

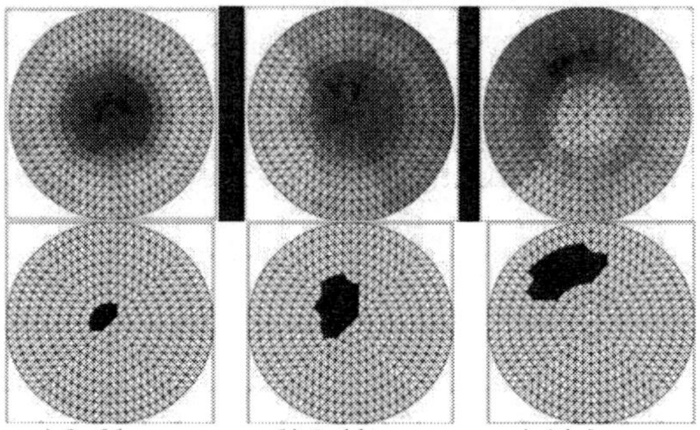

a) 2nd layer b) 3rd layer c) 4th layer

Figure 5. Image reconstruction results by using LM and LMG

In the comparison of the actual imaging parameters, the MNR evaluation parameter was used to represent the similarity between the calculated field and the actual field.

Figure 5 is image reconstruction results by using LM and LMG. Table 2 shows, in the actual experiment, due to the error of finite element model and the noise interference, the effect of hybrid algorithm is slightly worse, but still significantly outperforms the LBP and MNR algorithms, and can clearly reconstruct the potato profile.

Table 2. Actual imaging parameters.

Field	Parameters	LM	LMG
A	F	2.1325	0.0352
B	F	2.5864	0.00543
C	F	3.7412	0.0624

5 Conclusion

In this paper, linear back projection algorithm, modified Newton-Raphson algorithm and interval splitting were introduced to genetic algorithm. And a hybrid TERT image reconstruction algorithm was proposed which makes up for the deficiency of the single algorithm. It has effectively eliminated the transition region in the reconstructed image and achieved the transformation of imaging simulation from fuzzy to clear.

Experiments show that the hybrid algorithm has the following advantages: 1) this algorithm not only has a good effect on the field reconstruction of a single tuber, but also can reconstruct clear field distribution patterns which are difficult with other two algorithms. 2) it has effectively overcome the problem of poor local search ability of genetic algorithm, and improve the quality of the reconstructed image. 3) it has strong robustness.

 DOI: 10.1051/matecconf/20166817004

Acknowledgement

The authors would like to thank the National Natural Science Foundation of China (No.: 61271307).

References

1. Jian Ning, Yitian Sun, Qing Liu, Tianshi Ma, Hui Liu, Zhiming Wei. Development trend of intelligent precision agricultural equipment [J]. Mechanical and electrical products development and innovation. Volume **24**(2), 2011.
2. Yanjun Zhang, Yu Chen, Deyun Chen. Sensitivity analysis and image reconstruction of electrical resistance tomography system [J]. Computer Science, 2010, **37** (8): 257-261.
3. Deyun Chen, Yanjun Zhang. An image reconstruction algorithm for electrical resistance tomography based on algebraic neural network [J]. Computer Engineering and application, 2009, **45** (32):19-21.
4. Liqing Xiao, Xiaogen Shao. Analysis and design of the finite element simulation model for electrical resistance tomography [J]. Journal of Chinese and instrument, 2008, **29** (2):354.360.
5. Peng, X. M.; Tan, W. X.; Wang, Y. P. Convolution back-projection imaging algorithm for downward-looking sparse linear array three dimensional synthetic aperture radar[J]. PROGRESS IN ELECTROMAGNETICS RESEARCH-PIER. 2012, **129**:287-313.
6. Liqing Xiao, Huaxiang Wang, Xiaoju Xu etc. Improved Newton - Raphson electrical resistance tomography image reconstruction algorithm [J]. The Chinese Electrical Engineering Journal, 2012, **32** (1): 198-200. (8): 91-97.
7. Xia Bian, Liang Mi. Research Progress on genetic algorithm theory and Application [J]. Computer Application Research, 2010, **27** (7): 2425-2429.
8. Sanaz Asgarifar, Javad Frounchi. A novel two-stage genetic algorithm for image reconstruction of electrical resistance tomography [J]. All Publications, 2012, **01**(4).
9. Furman, A; Ferre, TPA; Warrick, AW. Optimization of ERT surveys for monitoring transient hydrological events using perturbation sensitivity and genetic algorithms [J].VADOSE ZONEJOURNAL, 2004, **3**(2):1230-1239.
10. Liqing Xiao, Xiaogen Shao, Linlin Wang, et. Analysis and design of the finite element simulation model for electrical resistance tomography [J]. Journal of Chinese and instrument, 2008, **29** (2):354.360.

Evaluating the Improvements of *StarCraft* Gameplay in the ABL Agent EISBot by Implementing Dynamic Specificities through an External Component

Jared Okun

Byram Hills High School, USA

Abstract. I studied how artificial intelligence (AI) adapts to failure. I worked with AI framework ABL, which uses a behavior tree. The tree begins by creating behaviors to accomplish an AI goal. Behaviors are chosen based on priorities. However, these priorities are not effective in all situations, and cannot change in ABL, which prevents the AI to learn from failure. For my study, I gave ABL the ability to change its priorities. This essentially allows an ABL AI to adapt from its mistakes. This ability was then tested for effectiveness using the ABL AI EISBot (an AI that plays videogames) and the game of *StarCraft*. The new ability caused EISBot to improve significantly showing that the ability to change priorities improves an AI's performance.

1 Introduction

A main goal of Artificial Intelligence (AI), and intelligent agents more specifically, is to mimic high level decision making such as that of a human in computer systems [1]. Games are useful test beds for assessing our AI approaches because the actions of AI players (agent), and their results are observable in real-time. My study used Real-Time Strategy (RTS) games, where the player acts as an off-field general for an army, usually with the goal of destroying the opponent. The genre of RTS games is of particular interest because it requires an agent to exhibit multi- scale reasoning and concurrent planning, which requires learning and recognizing an opponent's plan from their observed actions, along with spatial and temporal reasoning [1]. Expressive Intelligence Studio Bot (EISBot) is an example of such an agent [1]. EISBot plays the RTS game *StarCraft: Brood War* and attempts to emulate human-level reasoning with its unique decision-making process [1].

EISBot's decision-making process is implemented in A Behavioral Language (ABL) using an Active Behavior Tree (ABT). The ABT contains all the goals that EISBot is pursuing. An example of an ABT can be seen in Figure 3 in Section 4.2. Examples of these goals could be to collect a specific amount of resources or train a certain number of units. To attain these goals, open leaf nodes are used, which are behaviors available from a hand-authored behavior library within EISBot that are selected to be enacted by EISBot based on a priority that is determined by the behavior's specificity to complete these goals. These specificities exist within EISBot as integer values. The one with the greatest specificity has the highest priority of being chosen. If there is a tie, the behavior is chosen at random from the behaviors with the highest specificity. These behaviors are comprised of a list of steps that could be mental actions (e.g., mathematical operations), physical actions (e.g., moving units), or actions that create goals which produce more open leaf nodes to accomplish the created goals. Ultimately, the chosen leaf node is expanded into a set of steps that will be added to the ABT in the form of *children* to that leaf node. In other words, the steps of the behavior are children created by the *parent leaf node*. If one of the children steps fails, the leaf node it was expanded from is aborted and the leaf node with the next highest specificity is chosen.

However, there is a problem with this method. The priorities are not changed after a failure has occurred; hence, the node with the next highest priority may contain similar actions that caused the failure of the previously selected node, thus causing the leaf node to likely fail again. Obviously, this is detrimental to winning the game. While EISBot spends time by continuing to try to accomplish the desired goal, the opponent may use this time to set up and strengthen their army. This was demonstrated in the 2011 game of EISBot versus a well-known professional player that was documented in a video recording on YouTube [2]. EISBot attacked the human player with dragoons, which are offensive units in *StarCraft*, and in response, the human player produced hydralisks and zerglings, which are units that counter dragoons. After again attacking the human player with dragoons, EISBot's army was destroyed by the human player's hydralisks. However, EISBot still continued to primarily produce dragoons that attacked the human player and failed, giving the human player the time to produce even more hydralisks and zerglings. Ultimately,

EISBot lost the game to the human player's army o f hydralisks and zerglings. In other words, EISBot lost that game because it did not respond to the human player creating units that countered EISBot's army. Importantly, if EISBot had the ability to "learn" from its failure, then it might have chosen to cease building dragoons and instead produce units to counter the human player's army. Repeating useless behavior is counterproductive to EISBot's success. The cause of this problem is that the priorities of EISBot's nodes remain stagnant throughout the game. These priorities do not change even after a failure has occurred. Unlike EISBot, humans will normally learn from mistakes and try not to repeat what caused the mistake. To address this problem, I proposed a solution: I modified EISBot's source code to automatically reduce the priority of leaf nodes based on their similarity to the one that failed. These revisions will be based on the steps that the nodes and failed leaf nodes both contain. These changing priorities are known as *dynamic specificities* or *dynamic priorities*.

In my study, EISBot was edited to allow the use of dynamic specificities. This new agent is called *EISBot Dynamic Specificities* (EISBotDS). I tested EISBotDS against EISBot in the game *StarCraft: Brood War* by playing them both against the pre-existing StarCraft agent in a pre-made scenario and comparing their number of wins over time. This is the first study on adding dynamic specificities to ABL, as ABL does not support them. The implementation of dynamic specificities into EISBot will future agents created with ABL to use dynamic specificities, which should increase their performance.

2 Hypotheses

H_0: EISBot and EISBotDS will record the same total wins in the scenario

H_1: EISBotDS will record more total wins in the scenario than EISBot

3 Objectives

1. Implement dynamic specificities into ABL so EISBot can use dynamic specificities.
2. Create a controlled scenario that both agents will play to determine whether implementing specificities has a positive effect on EISBotDS's ability to play *StarCraft*.
3. Test EISBotDS against EISBot with the scenario and compare total wins.

4 Methods

I created this study under the tutelage of my mentor, Dr. David W. Aha of the Naval Research Laboratory's Navy Center for Applied Research in Artificial Intelligence in Washington, D.C. He introduced me to Dr. Ben Weber's agent, EISBot, which plays *StarCraft: Brood War*. After reading Weber's dissertation on the subject and watching it play for a period of a few months, I began to focus on its ABL implementation. I decided I wanted to begin a project where I would improve upon EISBot's method for responding to failing behaviors. For this study, I created an implementation of dynamic specificities and the test scenario with the help of Michael Leece and Dr. Mark (Mak) Roberts, who are collaborators with Dr. Aha. I also ran the experiment and performed the statistical tests reported below.

4.1 StarCraft

StarCraft and its expansion, *StarCraft: Brood War*, are futuristic RTS games developed by Blizzard Entertainment in 1998. Unlike many video games, *StarCraft: Brood War* is played at a competitive level due to its fast, deep, and unforgiving gameplay [1]. This game is used in AI research because gameplay requires many aspects that are needed in human-level AI, and actions in game are observable and can be repeated.

In a standard game of *StarCraft*, the goal is to destroy all of your opponents' structures. Each player has the choice of three races at the start of the game. These races are the *Terran, Zerg,* and *Protoss*. Each race has its own unique units, structures, and mechanics. For example, the *Zerg* and *Protoss* have limited areas where they can construct structures while the *Terran* can make structures on any level ground as long as there is room. Mastering any race in this game requires a lot of practice for a human player. The races are considered to be balanced as all races are viable competitively.

In a game of *StarCraft*, a player performs multiple tasks. These include producing an army to counter an opponent, micromanaging each unit in the army to make sure they are doing their optimal work, expanding their race's tech tree to make the best units and upgrades, and managing their resources to be able to produce units and upgrades. Since this game is also multiplayer, the player needs to keep track of his opponent. If the player does not know what his opponent is doing, he will be unable to prepare and effectively counter the opponent's strategy. Figure 1 shows a screenshot of StarCraft's user interface.

StarCraft and *StarCraft: Brood War* have a built-in AI that players can choose to play against. The disadvantage of this AI is that it is inherently random. That is, in the same repeated scenario, it could choose different units to create or attack in different ways.

4.2 ABL

A Behavior Language (ABL) is a behavioral language created by Michael Mateas and Andrew Stern in 2004 based on the language HAP used in the Oz Project in order to make life-like agents [3]. One of ABL's most notable advantages is that it allows the construction of multi-scale AI because the agents it creates can pursue multiple goals simultaneously and each pursuit of a goal can communicate with any other pursuit.

Bots made with ABL can be given a set of goals to complete. These goals are then pursued by selecting a behavior from a hand-authored library. In other words, each goal can be achieved by selecting a sequence of actions that was created by the agent's developer.

Each behavior is itself a list of actions that the agent performs. These actions can be (1) physical actions that cause the agent to interact with the medium in which it is acting upon, (2) mental actions that cause the agent to perform mathematical operations or perform external functions, (3) actions that create goals and subgoals, or (4) actions that change elements in the working memory. Each behavior can have specificity, which are priorities that determine what order the behaviors will be run in if they fail. In other words, if a behavior fails, the agent will check the current specificity of the behavior that failed and then choose the next behavior that has specificity less than the current one. If the specificities of multiple behaviors are the same, the behavior that is run among them is randomly selected. Behaviors may also be given preconditions that have to be met in order for the behavior to run in the first place. Behaviors can also have parameters that serve a similar function to the parameters of methods in the programming language, Java. This means that the behaviors can be given variables to be used during execution. Behaviors can also be sequential or parallel. Sequential behaviors run their actions one after another while parallel behaviors run their actions simultaneously. A behavior that was used in this study is displayed in Figure 2.

Figure 1. A screenshot taken in *StarCraft: Brood War*. In the top right corner is the player's resources. The bottom left corner is the minimap. The center screen shows the blue *Protoss* player under attack by a teal *Protoss* player.

```
sequential behavior spendExcess(int minerals, int gas) {
    precondition {
        (minerals>=600 && gas>=300)
        (PlayerWME supplyUsed::used supplyTotal::total)
        (used < total)
        (CyberneticsCoreWME)
        (GatewayWME ID::unitID active==true trainTimer==0 buildTimer==0)
        (Logger.print("ABL", "Excess spending on Dragoons"))
    }

    specificity 3;

    act train(unitID, Protoss_Dragoon);
}
```

Figure 2. An example of the spendExcess behavior in EISBot's behavior library. This behavior is sequential and contains a precondition, a specificity, and a physical action. [4].

All active behaviors are stored in the Active Behavior Tree (ABT). During each execution cycle, goals that are not being pursued are added to a set. For each goal, a behavior is chosen to be executed based on the specificity and preconditions of that behavior. This behavior then adds its actions to the tree as children of the behavior. An example of an ABT is shown in Figure 3.

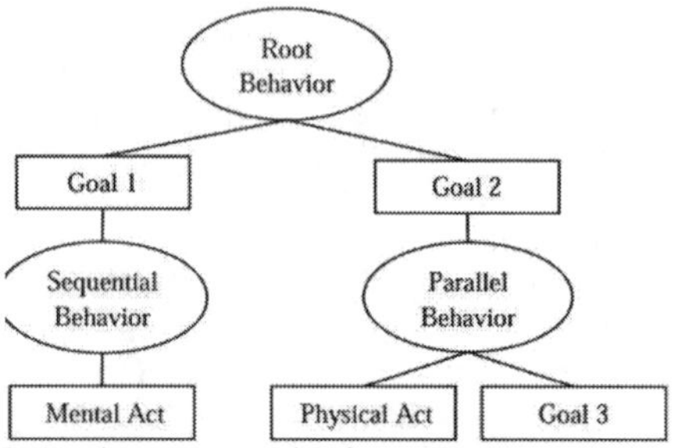

Figure 3. An abstract example of an ABT. The behavior 'Root Behavior' was a parallel behavior with two goals. The behavior 'Sequential Behavior' that was chosen to pursue 'Goal 1' executes a mental action. The behavior 'Parallel Behavior' that was chosen to pursue 'Goal 2' executes a physical action and creates another goal.

One of ABL's advantages for use in game AI is the scheduling and plan monitoring of the AI can be handled by the language itself without the need for external software [1]. This is most relevant for physical actions as some require multiple frames in the game to run. These physical actions will be set to executing, and any actions after that in a sequential behavior will be unable to run ABL has its own mechanism for handling failures while the code is running. If a behavior fails for any reason, the behavior and its actions are then removed from the tree. The goal that the failed behavior was pursuing will once again be available to be pursued. The behavior chosen to achieve this goal cannot be one that had been previously chosen and failed. However, the next behavior chosen can be similar to the one that failed.

4.3 EISBot

EISBot (Expressive Intelligence Studio Bot) is a AI software system that was developed by Dr. Ben Weber while he was with the Expressive Intelligence Studio at the University of California, Santa Cruz, and it plays the game, *StarCraft: Brood War*. EISBot was developed using a combination of Java and ABL. EISBot interfaces with *StarCraft* through the use of JNIBWAPI, Control Process, and ProxyBot [1].

Table 1. This lists the purpose of some of the managers in EISBot. I used behaviors from the Strategy manager in this study.

Manager	Purpose
Strategy	Responsible for choosing strategy and the timing to execute actions
Tactics	Responsible for combat and micromanagement of units
Worker	Responsible for resource income, workers, and expansion
Construction	Manages requests for construction
Supply	Manages the supply resource. This resource is required to build more units in *StarCraft: Brood War*
Build Order	Manages the order in which structures are built, units are trained, and upgrades are researched

EISBot is composed of multiple managers that are associated with different aspects of play in *StarCraft* [1]. Its managers include: Strategy manager, Tactics manager, Worker manager, Construction manager, Supply manager, and Build Order manager [4-5]. The purpose of each relevant manager is listed in Table 1. The managers are all interfaced with the various components and sensors that are related to the work they perform. Each of these interfaces also determines how working memory elements are edited by each manager.

Working memory elements are the primary source of communication of the various managers. Working memory elements can also act as a blackboard for the managers and external components to post information about the current game such as the opponent's predicted strategy [1]. The external planner of EISBot also uses working memory. The planner adds to working memory, which causes actions and behaviors to activate in the ABT.

4.4 Implementation

In order to add dynamic specificities to EISBot, I had to bypass ABL's normal specificities because they are static. I did this by creating a text file called "buildSpecificities" that contained the specificities and a Java class called "SpecificityEditor" that had methods to interact with the ABL code. This study specifically worked with the spendExcess behaviors from the Strategy manager in EISBot.

SpecificityEditor contains local variables that corresponded with the specificities of the relevant spendExcess behaviors. The file buildSpecificities also contains values that corresponded with the specificities of the spendExcess behaviors. Upon a new instance of SpecificityEditor being created, the variables in SpecificityEditor would be set to the values in buildSpecificities. The current behavior that EISBotDS was executed is assigned to a variable, currentBuild, in SpecificityEditor. I created a method that rewrites the buildSpecificities file with new specificities was also added to SpecificityEditor. I also added a static SpecificityEditor object that refers to the SpecificityEditor used during each run of EISBotDS.

Implementing SpecificityEditor with ABL required making the specificities of the spendExcess behaviors equal. This allowed EISBotDS to choose between these behaviors "randomly." To prevent the behaviors from being chosen randomly, each behavior was given an extra precondition. This precondition used the getters from SpecificityEditor to emulate the function of specificities. In other words, this precondition requires the behavior that is chosen to have the highest specificities among the other specificities that can be chosen. A mental action that sets the value of currentBuild in SpecificityEditor to correspond with the behavior being executed was also added to each of these behaviors.

The specificities are currently set to change at the end of each game that EISBotDS plays. EISBot already has a function that returns the victor of the game. Using the value that this returns, SpecificityEditor rewrites buildSpecificities with new values for the specificities. If the game was won, the last behavior chosen would have its specificity increase by one. If the game was lost, the last behavior chosen would have its specificity decrease by one. I chose to change the last behavior as EISBotDS will use only one behavior in each scenario if that behavior has the highest specificity. If there is a tie, EISBotDS will be choosing its behaviors randomly and therefore the specificity it chooses to lower will also be random. EISBotDS was not able to change its specificities in the middle of the scenario.

5 Experimental Setup and Results

For this study, EISBot was integrated with the ability to use dynamic specificity for its spendExcess behaviors. In other words, the only action that changes between EISBot and EISBotDS is the units that they are building. In order to perceive if this change affects EISBot, I tested EISBot and EISBotDS by playing them on a scenario 100 times. This scenario required the agents to defend the base against a wave of enemy units controlled by *StarCraft*'s built-in AI. Since EISBot plays as the *Protoss* and by extension EISBotDS plays as the *Protoss*, both of the agents began with two structures used in the spendExcess behaviors to build units, the various other structures required to advance the tech tree to build all the available units and upgrade them, and sufficient resources. The agents were given two minutes to build and upgrade the units before the enemy wave arrived at their base. The enemy wave would always consist of the same units. The way that the enemy wave attacks is partially random because the built-in *StarCraft* AI is controlling it. The enemy wave could spread out their units, which would cause the unit that EISBot uses to be less effective. On the other hand, the enemy wave could also focus on attacking structures instead of units, which would generally help the agent it is playing against. If the current agent can destroy all enemy units in a wave, it wins. However, like the game, if the current agent loses all its structures, they lose. Because the scenario does not start like a normal match would begin, the Construction manager, the Worker manager, the Supply manager, and the Build Order manager were disabled in both agents.

Table 2. The total wins that EISBot and EISBotDS had at the end of the 100 scenarios

	EISBot	EISBotDS
Total Wins	41	88

For each run of the scenario, the total wins up to that game were collected. The data can be seen in Table 2. EISBotDS had a statistically greater proportion ($p<0.0001$) of wins than EISBot using a two proportion z-test. The win counts of both agents were normalized to the current game number. The results of normalization can be seen in Figures 4 and 5.

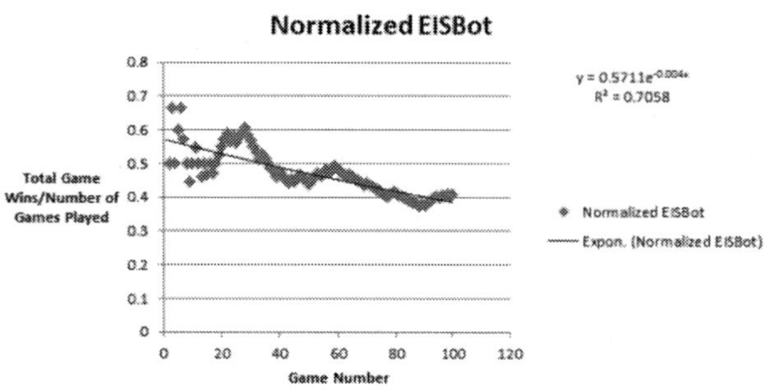

Figure 4. A graph showing the normalized wins of EISBot. The first point of the graph at (1,0) was removed which produced an exponential regression with an R^2 above 0.7.

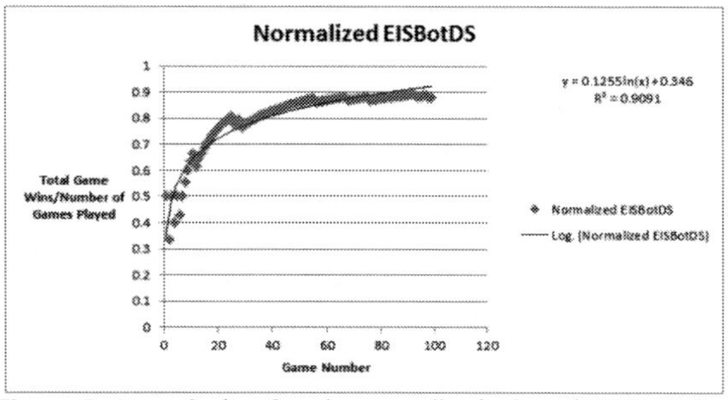

Figure 5. A graph showing the normalized wins of EISBotDS. The first point of the graph at (1,1) was removed which produced a logarithmic regression with an R^2 above 0.9.

6 Discussion

I rejected H_0. The changes made to EISBot to allow it to use dynamic specificities have shown, for the scenario used in Section 5, a significant (p<0.0001) improvement in its gameplay in terms of wins. The results support H_1. Based on both agents' proportion of wins to the 100 games played, it is reasonable to state that EISBotDS is better than EISBot at *StarCraft*, versus the built-in game AI, in scenarios that are similar to the given scenario. From this experiment, there is reason to believe that dynamic specificities will help EISBot succeed at other scenarios with different behaviors chosen. As can be seen in Figure 5, EISBotDS was able to improve over time as its wins to games played ratio approached one over 100 games; in contrast, EISBot's performance decreased, as shown in Figure 4. Because adding dynamic specificities to EISBotDS caused it to perform better at *StarCraft* than EISBot, other creations using the ABL framework may also benefit from the addition of dynamic specificities. This shift to using dynamic specificities could possibly result in a change of the ABL language or the creation of a new language that uses a similar framework to ABL but also has the ability to use dynamic specificities.

This study has implications for ABL agents. All behaviors that were created for these agents had to use static specificities. However, with the use of external components like SpecificityEditor, ABL agents can now use dynamic specificities. As can be seen by the results of the study, giving an ABL agent the ability to use dynamic specificities could cause it to be more effective in the scenarios it is being run in. For example, an external component like SpecificityEditor could be used for an ABL agent that controls a drone in a flight simulation. If the agent is running a behavior with static specificities that resulted in a crash, the agent will continue to crash in subsequent runs. If the same agent was implemented with dynamic specificities, it would run a different behavior to avoid the crash.

This study was limited in that it tested only one set of behaviors with one scenario. Another limitation is that the game AI of *StarCraft* is inherently random and does not act like a human player. This limitation can be fixed by using another external agent like EISBot as the player that controls the units.

7 Conclusion

All of the objectives of this study were met. *This study is the first of its kind to develop an external component to ABL to allow it to use dynamic specificities.* EISBotDS was significantly improved for the given scenario and therefore I accepted H_1. This experiment shows that adding dynamic specificities to ABL can give it the ability to perform better at complex tasks.

In future work, different behaviors will be tested in more scenarios. An example of another behavior that can be tested is the one that chooses how units attack, which would be found in the Tactics manager. The SpecificityEditor class was also specialized for this experiment. The implementation of SpecificityEditor needs to be simpler with more behaviors instead of having to add more methods and variables. The option might also present itself when dynamic specificities can be directly integrated into the ABL framework so an external component is not needed. Another area of future research is modifying EISBotDS to update its specificities in the middle of the game. This would improve EISBotDS's ability to play longer games where similar situations are occurring multiple times. Currently, EISBotDS can only update specificities at the end of the game based on wins. It would be preferable to change specificities in the middle of game so it could change its strategy in the middle of the game. This would require a new metric to determine how to change the specificities instead of wins and where to implement this change in the code.

This research validated the use of dynamic specificities in ABL. These dynamic specificities have applications in other AI that use the ABL framework. The SpecificityEditor class that was developed in this study can also be used in future ABL agents. ABL can be used for more than just games. AI is used for language identification, autocorrect, drones, robots, self-driving cars, searching, and various other applications. ABL with dynamic specificities can be the language that is used to create AI for all of these and should be considered to be used in future AI research.

Acknowledgements

This project would not have been possible without the overarching guidance of my mentor, Dr. David Aha. I would like to thank Michael Leece and Dr. Mark (Mak) Roberts, who helped me when I was working on the scenario and SpecificityEditor. I would also like to thank the science research teachers Mrs. Stephanie Greenwald and Mr. David Keith for pointing me down the right path for my project. I would also like to thank my family and friends for supporting me through this difficult endeavor.

References

1. B. Weber. Integrating learning in a multi-scale agent. *Doctoral dissertation: Department of Computer Science, University of California, Santa Cruz, CA.* (2012)
2. B. Weber [UCSCbweber]. EISBot vs Dennis Fong (Thresh). Retrieved from https://www.youtube.com/watch?v=txGPuN7PXAY &list=UUB0UyZ9THzSjcXaCkGCQLMg&index=14 (2011)
3. M. Mateas & A. Stern. A Beh. Lang. Joi. Act. and Beh. Id. *Like-like Characters: Tools, Affective Functions, and Applications* 135 (2004)
4. B. Weber. EISBot [computer software]. Available from https://github.com/bgweber/eisbot (2012)
5. B. Weber, M. Mateas, A. Jhala. Building Human-Level AI for Real-Time Strategy Games. *Proceedings of the AAAI Fall Symposium on Advances in Cognitive Systems* (2011)

Pattern Recognition of mtDNA with Associative Models

María Elena Acevedo, Marco Antonio Acevedo, Federico Felipe and David Aquino

Instituto Politécnico Nacional, Escuela Superior de Ingeniería Mecánica y Eléctrica, Mexico City, Mexico

Abstract. In this paper we applied an associative memory for the pattern recognition of mtDNA that can be useful to identify bodies and human remains. In particular, we used both morphological hetroassociative memories: *max* and *min*. We process the problem of pattern recognition as a classification task. Our proposal showed a correct recall, we obtained the 100% of recalling of all the learned patterns. We simulated a corrupted sample of mtDNA by adding noise of two types: additive and subtractive. The memory showed a correct recall when we applied less or equal than 55% of both types of noise.

1 Introduction

The scientific and technologic advances in the tools used in genetic have allowed new applications in other disciplines. More frequent, there is a collaboration between forensic groups and other medical specialities. Nowadays, the applications of computational algorithms in several areas have become essential. In this work, we applied artificial intelligent tools to create a software that is able to recognize patterns of mtDNA as a forensic application.

Mitochondria [1] are structures within cells that convert the energy from food into a form that cells can use. Although most DNA is packaged in chromosomes within the nucleus, mitochondria also have a small amount of their own DNA. This genetic material is known as mitochondrial DNA or mtDNA. This type of DNA resists adverse conditions without being degraded.

Some approaches are applied to analyse and recognize DNA. Craven and Shavlik [2] applied artificial neural networks to recognize promoters in DNA. A genetic algorithm was applied to find variable length motifs (transcription factor binding sites) [3]. A Self-Organizing Neural Network [4] was used to identify motifs. A related work with motifs applied the RISO software and box-link structures [5]. In addition, voting and matching pattern algorithms [6] were used to identify motifs. Wells and Sperling [7] used the software PAUP 4.0 to analyse and identify mtDNA of the blow fly subfamily of Chrysomyinae to help a forensic-entomological analysis. In other paper, a neural network based multi-classifier system [8] for the identification of Escheri-chia coli promoter sequences in strings of DNA is presented. Pushpalatha and Mukunthan [9] applied a Neural-fuzzy Mapping to identify DNA Finger Printing. A genetic algorithm [10] was used as a preprocessing tool to Identifying significant genes from DNA microarray. Inbamalar and Sivakumar [11] applied the DWT Coiflet 5 to detect the protein coding regions in DNA sequences.

In this work, we applied the Morphological Associative Memories to identify people from the mtDNA. We use the two types of heteroassociative memories: *max* and *min*. The results from these memories were mapping to binary numbers and then, we applied the AND operation between the two binary results.

2 Basic concepts

2.1 Associative memory

Associative Memories (AM) [12] associate patterns **x** with **y**, which can represent any concept: faces, fingerprints, DNA sequences, animals, books, preferences, diseases, etc. We can extract particular features of these concepts to form patterns **x** and **y**.

There are two phases for designing an associative memory: Training and Recalling.

In the Training Phase, the process of associate patterns **x** with patterns **y** is performed. Now, we say that the memory is built.

The input and output patterns are represented by vectors. The task of association of these vectors is called Training Phase and the Recognizing Phase allows recovering patterns. The stimuli are the input patterns represented by the set $\mathbf{x} = \{x^1, x^2, x^3, \ldots, x^p\}$ where p is the number of associated patterns. The responses are the output patterns and are represented by $\mathbf{y} = \{y^1, y^2, y^3, \ldots, y^p\}$. Representation of vectors x^μ is $x^\mu = \{x_1^\mu, x_2^\mu, \ldots, x_n^\mu\}$ where n is the cardinality of x^μ. The cardinality of vectors y^μ is m, then $y^\mu = \{y_1^\mu, y_2^\mu, \ldots, y_m^\mu\}$. The set of associations of input and output patterns is called the fundamental set or training set and is represented as follows: $\{(\mathbf{x}^\mu, \mathbf{y}^\mu) \mid \mu = 1, 2, \ldots, p\}$

2.2 Morphological memories

The basic computations occurring in the proposed morphological network [13] are based on the algebraic lattice structure (**R**, ∨, ∧, +), where the symbols ∨ and ∧ denote the binary operations of maximum and minimum, respectively. Using the lattice structure (**R**, ∨, ∧, +), for an *m* x *n* matrix *A* and a *p* x *n* matrix *B* with entries from **R**, the matrix product $C = A \nabla B$, also called the *max* product of *A* and *B*, is defined by equation (1).

$$c_{ij} = \bigvee_{k=1}^{p} a_{ik} + b_{kj} = (a_{i1} + b_{1j}) \vee \ldots \vee (a_{ip} + b_{pj}) \quad (1)$$

The *min product* of A and B induced by the lattice structure is defined in a similar fashion. Specifically, the *i,j*th entry of $C = A \Delta B$ is given by equation (2).

$$c_{ij} = \bigwedge_{k=1}^{p} a_{ik} + b_{kj} = (a_{i1} + b_{1j}) \wedge \ldots \wedge (a_{ip} + b_{pj}) \quad (2)$$

Henceforth, let $(\mathbf{x}^1, \mathbf{y}^1), (\mathbf{x}^2, \mathbf{y}^2), \ldots, (\mathbf{x}^p, \mathbf{y}^p)$ be *p* vector pairs with $x^k = (x_1^k, x_2^k, \ldots, x_n^k)^t \in \mathbf{R}^n$ and $y^k = (y_1^k, y_2^k, \ldots, y_m^k)^t \in \mathbf{R}^m$ for $k = 1, 2, \ldots, p$. For a given set of pattern associations $\{(\mathbf{x}^k, \mathbf{y}^k) \mid k = 1, 2, \ldots, p\}$ we define a pair of associated pattern matrices (**X**, **Y**), where $X = (\mathbf{x}^1, \mathbf{x}^2, \ldots, \mathbf{x}^p)$ and $Y = (\mathbf{y}^1, \mathbf{y}^2, \ldots, \mathbf{y}^p)$. Thus, **X** is of dimension *n* x *p* with *i,j*th entry x_i^j and *Y* is of dimension *m* x *p* with *i,j*th entry y_i^j. Since $\mathbf{y}^k \nabla (-\mathbf{x}^k)^t = \mathbf{y}^k \Delta (-\mathbf{x}^k)^t$, the notational burden is reduced by denoting these identical morphological outer vector products by $\mathbf{y}^k \times (-\mathbf{x}^k)^t$.

With these definitions, we present the algorithms for the training and recalling phase.

Training Phase

1. For each *p* association $(\mathbf{x}^\mu, \mathbf{y}^\mu)$, the minimum product is used to build the matrix $\mathbf{y}^\mu \Delta (-\mathbf{x}^\mu)^t$ of dimensions *m* x *n*, where the input transposed negative pattern $\mathbf{x}^\mu$ is defined as $(-\mathbf{x}^\mu)^t = (-x_1^\mu, -x_2^\mu, \ldots, -x_n^\mu)$.
2. The maximum and minimum operators (∨ and ∧) are applied to the *p* matrices to obtain *M* and *W* memories as equations (3) and (4) show.

$$M = \bigvee_{k=1}^{p} \left(y^k \otimes (-x^k)^t\right) \quad (3)$$

$$W = \bigwedge_{k=1}^{p} \left(y^k \otimes (-x^k)^t\right) \quad (4)$$

Recalling phase

In this phase, the minimum and maximum product, Δ and ∇, are applied between memories **M** or **W** and input pattern $\mathbf{x}^\omega$, where $\omega \in \{1, 2, \ldots, p\}$, to obtain the column vector **y** of dimension *m* as equations (5) and (6) shows:

$$\mathbf{y} = M \oplus \mathbf{x}^\omega \quad (5)$$

$$\mathbf{y} = W \oplus \mathbf{x}^\omega \quad (6)$$

3 mtDNA recognition

In this section, we will describe the dataset that we used in this work, after, we will describe the algorithm to store and recognize the mtDNA.

3.1. mtDNA dataset

The mtDNA dataset [14] was obtained from National Center for Biotechnology Information (NCBI) that is part of the National Library of Medicine of United States. Besides, the NCBI offers some bioinformatics tools for the analisys of DNA sequences, RNA and proteins, and they are free and online.

The dataset has 276 patterns; each pattern has 60 characters such as adenine (a), cytosine (c), guanine (g) y thymine (t), which represent the four nitrogenous bases that are contained in DNA.

3.2 Algorithm

The data was stored in an Excel file. The data (characters) are converted to integers by the use of the ASCII code. Due to lower case characters start at the decimal number 97, we subtracted 97 to all the dataset, therefore, the numbers representing the characters are: 0-a, 2-c, 6-g and 19-t. For example, one of the sequences is,

gatcacaggt ctatcaccct attaaccact cacgggagct ctccatgcat ttggtatttt

then, the corresponding decimal numbers represented as a vector,
[6,0,19,2,0,2,0,6,6,19,2,19,0,19,2,0,2,2,2,19,0,19,19,0,0,2,2,0,2,19,2,0,2,6,6,0,6,2,19,2,19,2,2,0,19,6,2,0,19,19,19,6,6,19,0,19,19,19,19]

This vector corresponds to an input pattern x. Then, we have 276 input patterns with dimension of 97 and, we have 276 output patterns, y, with dimension of 276. When we use the max memory, the vectors y are built as follows,

$\mathbf{y}^1 = [500\ 0\ 0\ 0 \ldots 0_{(100)}\ 0_{(101)} \ldots 0_{(275)}\ 0_{(276)}]$
$\mathbf{y}^2 = [0\ 500\ 0\ 0 \ldots 0_{(100)}\ 0_{(101)} \ldots 0_{(275)}\ 0_{(276)}]$
$\mathbf{y}^3 = [0\ 0\ 500\ 0 \ldots 0_{(100)}\ 0_{(101)} \ldots 0_{(275)}\ 0_{(276)}]$
...
$\mathbf{y}^{100} = [0\ 0\ 0\ 0 \ldots 500_{(100)}\ 0_{(101)} \ldots 0_{(275)}\ 0_{(276)}]$
$\mathbf{y}^{101} = [0\ 0\ 0\ 0 \ldots 0_{(100)}\ 500_{(101)} \ldots 0_{(275)}\ 0_{(276)}]$
...
$\mathbf{y}^{275} = [0\ 0\ 0\ 0 \ldots 0_{(100)}\ 0_{(101)} \ldots 500_{(275)}\ 0_{(276)}]$
$\mathbf{y}^{276} = [0\ 0\ 0\ 0 \ldots 0_{(100)}\ 0_{(101)} \ldots 0_{(275)}\ 500_{(276)}]$

The vectors to build a min memory are similar but we changed the value of 500 for -500. We used this value because we have observed that the value in the diagonal must be greater than the maximum value of the elements of the input vectors [15]. We want to highlight that we see the recognition problem as a classification task. Therefore, every pattern of mtDNA is seen as a class, then, we have 276 different classes. That is the reason for which the output vectors have a number 500 or -500 in the index that indicates the number of the class.

Now, we applied the training phase to build both memories: *max* and *min* by the use of equations (3) and (4). Then, we present an input pattern to the memories to recall its corresponding output vector. The following example illustrates these operations. We used vectors with similar values but with lower dimension, and we just used *max* memory

$$\mathbf{x}^1 = [6\ 0\ 19\ 2\ 0] \rightarrow \mathbf{y}^1 = [500\ 0\ 0\ 0]$$
$$\mathbf{x}^2 = [6\ 6\ 19\ 2\ 19] \rightarrow \mathbf{y}^2 = [0\ 500\ 0\ 0]$$
$$\mathbf{x}^3 = [6\ 6\ 6\ 0\ 6] \rightarrow \mathbf{y}^3 = [0\ 0\ 500\ 0]$$
$$\mathbf{x}^4 = [0\ 0\ 2\ 2\ 0] \rightarrow \mathbf{y}^4 = [0\ 0\ 0\ 500]$$

Now, we calculate the first association as follows.

$$\mathbf{y}^1 \otimes (-\mathbf{x}^1)' = \begin{bmatrix} 500 \\ 0 \\ 0 \\ 0 \end{bmatrix} \otimes -\begin{bmatrix} 6 & 0 & 19 & 2 & 0 \end{bmatrix}$$

$$\mathbf{y}^1 \otimes (-\mathbf{x}^1)' = \begin{bmatrix} 500\text{-}6 & 500\text{-}0 & 500\text{-}19 & 500-2 & 500-0 \\ 0-6 & 0-0 & 0-19 & 0-2 & 0-0 \\ 0-6 & 0-0 & 0-19 & 0-2 & 0-0 \\ 0-6 & 0-0 & 0-19 & 0-2 & 0-0 \end{bmatrix}$$

$$\mathbf{y}^1 \otimes (-\mathbf{x}^1)' = \begin{bmatrix} 594 & 500 & 481 & 498 & 500 \\ -6 & 0 & -19 & -2 & 0 \\ -6 & 0 & -19 & -2 & 0 \\ -6 & 0 & -19 & -2 & 0 \end{bmatrix}$$

We calculate the remain asociations and the results are,

$$\mathbf{y}^2 \otimes (-\mathbf{x}^2)' = \begin{bmatrix} -6 & -6 & -19 & -2 & -19 \\ 494 & 494 & 481 & 498 & 481 \\ -6 & -6 & -19 & -2 & -19 \\ -6 & -6 & -19 & -2 & -19 \end{bmatrix}$$

$$\mathbf{y}^3 \otimes (-\mathbf{x}^3)' = \begin{bmatrix} -6 & -6 & -6 & 0 & -6 \\ -6 & -6 & -6 & 0 & -6 \\ 494 & 494 & 494 & 500 & 494 \\ -6 & -6 & -6 & 0 & -6 \end{bmatrix}$$

$$\mathbf{y}^4 \otimes (-\mathbf{x}^4)' = \begin{bmatrix} 0 & 0 & -2 & -2 & 0 \\ 0 & 0 & -2 & -2 & 0 \\ 0 & 0 & -2 & -2 & 0 \\ 500 & 500 & 498 & 498 & 500 \end{bmatrix}$$

Now, we obtain the greatest number of each element of each association and we have as a result the max memory.

$$\mathbf{M} = \begin{bmatrix} 494 & 500 & 481 & 498 & 500 \\ 494 & 494 & 481 & 498 & 481 \\ 494 & 494 & 494 & 500 & 494 \\ 500 & 500 & 498 & 498 & 500 \end{bmatrix}$$

In the recalling phase (equations (5) and (6)), we present an input pattern to **M** to recall its corresponding output pattern.

$$\mathbf{M} \oplus \mathbf{x}^1 = \begin{bmatrix} 494 & 500 & 481 & 498 & 500 \\ 494 & 494 & 481 & 498 & 481 \\ 494 & 494 & 494 & 500 & 494 \\ 500 & 500 & 498 & 498 & 500 \end{bmatrix} \oplus \begin{bmatrix} 6 \\ 0 \\ 19 \\ 2 \\ 0 \end{bmatrix}$$

$$\mathbf{M} \oplus \mathbf{x}^1 = \begin{bmatrix} 494+6 \wedge 500+0 \wedge 481+19 \wedge 498+2 \wedge 500+0 \\ 494+6 \wedge 494+0 \wedge 481+19 \wedge 498+2 \wedge 481+0 \\ 494+6 \wedge 494+0 \wedge 494+19 \wedge 500+2 \wedge 494+0 \\ 500+6 \wedge 500+0 \wedge 498+19 \wedge 498+2 \wedge 500+0 \end{bmatrix}$$

$$\mathbf{M} \oplus \mathbf{x}^1 = \begin{bmatrix} 500 \wedge 500 \wedge 500 \wedge 500 \wedge 500 \\ 500 \wedge 494 \wedge 500 \wedge 500 \wedge 481 \\ 500 \wedge 494 \wedge 513 \wedge 502 \wedge 494 \\ 506 \wedge 500 \wedge 517 \wedge 500 \wedge 500 \end{bmatrix} = \begin{bmatrix} 500 \\ 481 \\ 494 \\ 500 \end{bmatrix}$$

All the recalled vectors are,

$$\mathbf{yrec}^1 = [500\ 481\ 494\ 500]$$
$$\mathbf{yrec}^2 = [500\ 500\ 500\ 500]$$
$$\mathbf{yrec}^3 = [487\ 487\ 500\ 498]$$
$$\mathbf{yrec}^4 = [483\ 481\ 494\ 500]$$

From these results, we can observe that only the two last patterns are recalled or well classified. We conclude that because the number 500 appeared in the correct index of the vector.

The recalled patterns when we use the *min* memory **W** are,

$$\mathbf{yprec}^1 = [-500\ {-500}\ {-487}\ {-483}]$$
$$\mathbf{yprec}^2 = [-481\ {-500}\ {-487}\ {-481}]$$
$$\mathbf{yprec}^3 = [-494\ {-500}\ {-500}\ {-494}]$$
$$\mathbf{yprec}^4 = [-500\ {-500}\ {-498}\ {-500}]$$

Then, we transform the recalled vectors by placing a number 1 where the value of the element is 500 or -500, and we write a 0 in the rest of the elements. The new vectors are the following.

$$\mathbf{yt}^1 = [1\ 0\ 0\ 1],\ \mathbf{ypt}^1 = [1\ 1\ 0\ 0]$$
$$\mathbf{yt}^2 = [1\ 1\ 1\ 1],\ \mathbf{ypt}^2 = [0\ 1\ 0\ 0]$$
$$\mathbf{yt}^3 = [0\ 0\ 1\ 0],\ \mathbf{ypt}^3 = [0\ 1\ 1\ 0]$$
$$\mathbf{yt}^4 = [0\ 0\ 0\ 1],\ \mathbf{ypt}^4 = [1\ 1\ 0\ 1]$$

The last step is to apply an and operation between **yt** and **ypt** vectors.

$$\mathbf{yt}^1 \mathit{and}\ \mathbf{ypt}^1 = \begin{bmatrix} 1 \\ 0 \\ 0 \\ 1 \end{bmatrix} \mathit{and} \begin{bmatrix} 1 \\ 1 \\ 0 \\ 0 \end{bmatrix} = \begin{bmatrix} 1 \\ 0 \\ 0 \\ 0 \end{bmatrix}$$

The last result indicates that the first pattern was well classified, *i.e*, the pattern corresponds to the class 1, because the number 1 is located at the first position. Now, we do the same with the remain vectors.

$$\mathbf{yt}^2 \mathit{and}\ \mathbf{ypt}^2 = \begin{bmatrix}1\\1\\1\\1\end{bmatrix} \mathit{and} \begin{bmatrix}0\\1\\0\\0\end{bmatrix} = \begin{bmatrix}0\\1\\0\\0\end{bmatrix}$$

$$\mathbf{yt}^3 \mathit{and}\ \mathbf{ypt}^{31} = \begin{bmatrix}0\\0\\1\\0\end{bmatrix} \mathit{and} \begin{bmatrix}0\\1\\1\\0\end{bmatrix} = \begin{bmatrix}0\\0\\1\\0\end{bmatrix}$$

$$\mathbf{yt}^4 \mathit{and}\ \mathbf{ypt}^4 = \begin{bmatrix}0\\0\\0\\1\end{bmatrix} \mathit{and} \begin{bmatrix}1\\1\\0\\1\end{bmatrix} = \begin{bmatrix}0\\0\\0\\1\end{bmatrix}$$

We can observe that in all the cases the patterns were classified in the correct class.

4 Results

We developed a software in Matlab 2013, implemented in a laptop Dell® with an Intel® Core i7 processor.

We built two morphological associative memories: max and min with the 276 patterns of the dataset.

In the recalling phase we recovered all the patterns. This means that our memory does not have forgetting factor: everything is learned everything is recalled. Then we can say that our memory has a correct recall.

We simulated the problem when the sample of DNA is corrupted with noise (or incomplete).

We applied additive and subtractive noise to the patterns. In Table 1, we show the results when we apply a percentage of additive noise.

Table 1. Recalling of our memory with additive noise

% additive noise	Correct recall?
10	yes
20	yes
30	yes
40	yes
50	yes
51	yes
52	yes
53	yes
54	yes
55	yes
56	no
57	no
58	no
59	no
60	no

From Table 1, it can be observed that we have the correct recall when the percentage of additive noise is less or equal than 55.

We observed the same behaviour when we applied subtractive noise to the patterns: from the 56% of subtractive noise, there is not correct recall.

5 Conclusions

We applied an artificial intelligent tool to recognize patterns of mtDNA. We chose this kind of DNA because it is resistant to degradation.

The Morphological Associative Memories were a suitable algorithm for this application. This model has a low complexity due to their main operations: addition, subtraction and max and min operators.

We deal the pattern recognition problem as a classification task. This is the reason that we built the output vectors with value of 500 or -500 in the index that indicates the number of its corresponding class.

Our proposal recalled all the patterns that were trained, therefore, it has a correct recall and it showed a forgetting factor of zero.

If a sample of DNA is corrupted, there is the possibility that the sequence cannot be complete or some elements of the sequence can be change, even when this type of DNA is resistant. This problem were simulated by adding noise to the patterns. We applied additive and subtracting noise. We found that when the patterns contained more than the 55% of noise, the memory was not able to recognize the corresponding class.

Acknowledgments

The authors would like to thank the Instituto Politécnico Nacional (COFAA and SIP), and SNI for their economical support to develop this work.

References

1. Genetics Home Reference, A service of the U.S. National Library of Medicine, http://ghr.nlm.nih.gov/mitochondrial-dna
2. M. Craven and J. Shavlik. Proc. of the Tenth Int. Conf. on Machine Learning (1993).
3. S. Vijayvargiya and P. Shukla. Int. J. of Adv. In Tech. **2**(1), (2011).
4. Derong Liu, Xiaoxu Xiong and Bhaskar DasGupta. Networking, Sensing and Control. IEEE (2005).
5. Alexandra M. Carvalho, Ana T. Freitas, Arlindo L. Oliveira, and Marie-France Sago. IEEE/ACM Trans. Comp. Biol. Bioinfo. **3**, 2 (2006).
6. M. Abbass and H. Bahig. Math. Comput. Sci. **7** (2013)
7. J.D. Wells and Felix Sperling. Forensic Sci. Int. **120** (2001).
8. R. Ranawana and V. Palade. Neural Comput. & Applic. **14** (2005).
9. A. Pushpalatha and B. Mukunthan. Int. J. Comp. App. **13**(3), (2011).

10. K. Dheenathayalan and J. RamSingh. Intern. Conf. Intellig. Comput. App. (2014).
11. T. M. Inbamalar and R. Sivakumar. The Sci. World J. Article ID 786497 (2015).
12. C. Yáñez-Márquez, Associative Memories Based on Order Relations and Binary Operators (In Spanish). PhD Thesis. Centro de Investigación en Computación, Mexico (2002).
13. G. X. Ritter, P. Sussner and J. L. Diaz de León. IEEE Trans. on Neur. Net. **9** (1998).
14. http://www.ncbi.nlm.nih.gov/nuccore/AY195748
15. R. Navarro, E. Acevedo, M. Acevedo and F. Martínez. MCPR 2012, LNCS 7329, (2012).

Demand of Innovative Services on a Computed Tomography Scan

Chamaiporn Sudasna Na Ayudhya[1], Achara Chandrachai[2] and Kriskrai Sitthiseripratip[3] and Nattee Niparnan[4]

[1]*Chulalongkorn University, Technopreneurship and Innovation Management Program, Thailand*
[2]*Chulalongkorn University, Faculty of Commerce and Accountancy, Chulalongkorn Business School, Thailand*
[3]*National Metal and Materials Technology Canter, Head of Biomedical Engineering Research Unit, Thailand Science Park, Thailand*
[4]*Chulalongkorn University, Faculty of Engineering, Department of Computer Engineering, Thailand*

Abstract. This paper seeks to investigate the needs of service, innovative Mobile Computed Tomography (CT) Scan service, from the leading level of Neurosurgeon in Bangkok, Thailand. The research is proposed investigating Neurosurgeon's expectations for Mobile CT Scan and current problems of CT scan operating. This is a survey research approach, performing in-depth interview which collected data from 21 Neurosurgeons, 12 hospitals in Bangkok and surrounding areas. Due to the number of samples limited, the descriptive statistics were used to analyze the data. The research address two research questions: According to the first research question, the authors' findings that most of Neurosurgeon need to use Mobile CT Scan at Operation Room (OR), Intensive Care Unit (ICU), and Emergency Room (ER), respectively. Regarding the second research question, the authors' analysis reveals purposes and found the current problems that when Neurosurgeon use CT Scan they faced low resolution and slice, moving unstable patients, wriggle patients, and Claustrophobia patients. Therefore, what are the findings of the survey are very useful in the design of new services for Mobile CT Scan to work in the next step.

1 Introduction

Presently, medical science has made great progress on surgery to correct a defect in the skull, face and jaws as well as in Surgery, Orthopaedic and Dental. They use rapid prototyping technology which consist of Computed Tomography (CT) and Medical imaging to make 3D medical model for diagnosis and treatment plan prior to surgery along with design and build implant materials by using basic medical imaging. Compared to the old tricks, Neurosurgeon use only basic medical imaging for surgery plan and implant forged by hand in the operation room after wound opened, its shown that 3D rapid prototyping technology increase the quality and safety of the surgery treatment substantially. Moreover, it also reduce surgery time and risk that occurred in patients which improves quality of Thai patients' life and bring the life as usual to them.

Computed Tomography is mainly to detect abnormalities of organs in the body by X-ray beam thro ugh the organ in vertical transversal that needs examination then create medical images. CT scan is divided into 4 systems, brain, abdomen and chest, bone muscles joints and spine, and vascular system. As a result, with a special display of vertical transversal image from CT scan combined with the present (2015) Thailand has only 2 mobile CT scan and hospitals do not allow others to use it, they have less utilization. Accordingly, National Metals and Materials Technology Center (MTEC) is interesting to study and develop a smaller and movable CT scan which using cone beam technology instead of fan beam technology that most manufacturers widely used today. Since, the cone beam is one more compact, cheaper and lower dose, although the quality of the images coming out is still not equivalent fan beam, but it is likely to develop the area and can be used as a tool to help diagnose correctly. Thus, MTEC has built up this machine, which is cheaper by almost half, to support and disseminate CT and medical imaging technology available in every hospital in the country.

This research focus on finding the need has not been met (unmet needs) by the user then the author has led to the design of new services in response to the new format or develop a difference services to be comfortable and able to meet the unmet needs. In addition, the research also specific considered in the diagnostic vascular system, since the increasing rate of patients starting from 2004 to 2008 that reported by Bureau of Policy and Strategy office of the Permanent Secretary, Ministry of Public Health, Thailand, From figure 1, compared number of vascular system patients with other diseases, it can be seen that the percentage of 4-year average growth of vascular system, with an average of more than 20%, which is greater than the other three diseases. However, in addition to the author is interested in finding the unmet needs of the user, the author also believe that period of waiting for a CT scan is a barrier to diagnosis to treatment. Because it large and immobile so whether it is diagnosed brain, spine, abdominal, muscles, vascular, bone or small joint in the body, it must have been waiting for this big machine. Therefore, to find unmet needs and proven authors' belief of objective above, it is necessary to continue to enhance the diagnosis and treatment of patients in Thailand by providing new service of mobile CT scan.

2 Background literature and research questions

2.1 Background literature

This research is a study of demand of innovative services On a Computed Tomography scan, so the literature review were necessary to study on current problems and theory that associated with the use CT scan today.

Problems were expected to arise from the use of CT scan was currently handling the treatment time in queues. In 2009, L. Siciliani et al. were studied in time to make the lowest overall cost which collected data from 137 hospitals in English National Health Service from 1998 to 2002 and found the level of anticipation that makes the lowest overall cost of hospital must not exceed 10 days. If not, the cost will increase as the process management increased. [2] In the same story, but change the view to the government policy of the Netherlands that deal with the problem of waiting time surgery. F. T. Schut and M. Varkevisser point that the government has started to use static control policy budget coupled with the ability to maintain the treatment in 1990, as a result, the waiting time of the patients dropped too much lower than other countries in the Organization for Economic Co-operation and Development (OECD). In 2001, afterwards, the government has changed the hospital financial from static control policy budget to Activity-based funding and during the year 2000-2006, with a competition between hospitals and more severe, the period of waiting for surgery even drop further. Finally, in 2011 in the Netherlands, it can shorten the waiting time for surgery shortened to no more than 5 weeks, so today government does not have to care about controlling the period of waiting longer, but to turn to the subject of price competition in the provision of hospital services in line with the state health insurance for future development plans. [3]

When looking deeper into the matter of reducing the cost of the hospital. There was also research that studies the management of patient care from primary or first aid in England by a cross-sectional study in patients 5 million from the 8,000 actual of 10 chronic disease through patient records, data treatment, and logs that can reduce the cost of hospital or not. The results of the study found, first aid that was made in the ambulance or in the emergency medical service would be able to contribute to the cost of hospital reduced, because this could reduce the number of emergency patients and patients with clinical symptoms of acute disease were down by more than the treatment of patients with lower prices or patient selection for admission to the hospital. [4]

Review of the literature mentioned, researchers saw a gap in the research of queue management of patients that will contribute to reducing the cost of hospital down. The literature review related to queuing theory in the healthcare system of more than 40 articles, will have overall management system that queuing was necessary to consider the three parts together. The first part of the design of the health care that the ambulance unit in the hospital and pharmaceutical division, the second part was about the management system that deals with the allocation of resources and priorities patient, the last was part three as a matter of analyzing the health care system by reducing cost, the ability to queue, and manage time efficiently queues. [5, 6]

In a matter of prioritizing patients, 1952, N.T.J. Bailey has proposed that variable to measure the performance of patients in the grading system was a period of time to get appointments and counseling, rather than finding a balance between long queues of patients with spare time consultant. By ratio time lost all patients have equal time to the counsel of advisors which refers to the period of consultation to coincide with the patient, the second to arrive. [7] Afterward, 1991, M. Brahimi and D.J. Worthington have designed a patients grading system that could reduce the number of patients in the queue at any time while reducing the waiting period of patients without increasing the physician's time. [8] In 2005, on the part of the Department of Radiation Oncology at close to the research being conducted there, S.S. Vasanawala and T.S. Desser in the Radiotherapy Department, typically it would be analyzed radiology daily schedule already but every day of work, it would have an emergency that requires the table to a new analysis by the importance and urgency of the case. Therefore, the use of queuing theory to calculate the probability of the table daily analysis and the results show that there will be free time was about 1%-5% of the daily schedule analysis in order to receive emergency to be analyzed and not make the same analysis table was corrupted or must defer to the analysis [9].

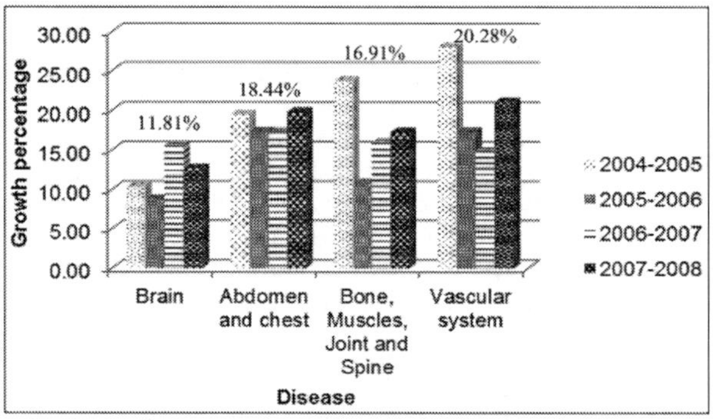

Figure 1. To compared number of vascular system patients with other diseases starting from 2004 to 2008 that reported by Bureau of Policy and Strategy office of the Permanent Secretary, Ministry of Public Health, Thailand [1].

Review of the literature on the theory of queuing in the healthcare system in terms of grading the patients, it was seen that most articles have shown a relationship in the long queues that would be able to contribute to the cost of the hospital, but did not find the article about how long the queues that will affect the effectiveness of patient treatment with the focus on bringing people into the engine or engines were adopting the approach. Therefore, it was one of the research gaps and see what the point in further studies.

2.2 Research questions

1. What are the needs of the services from CT scan to stroke patients in the hospital network in Bangkok?
2. What are the expectations in Mobile CT scan when new service implemented in the hospital network in Bangkok?

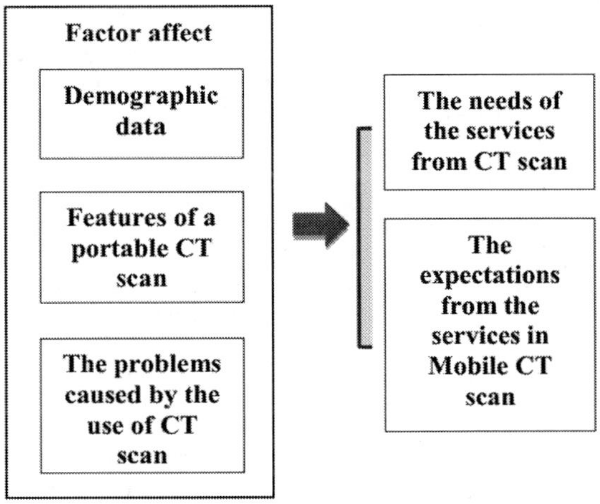

Figure 2. Conceptual framework for the study.

3 Methodology

The research "Demand of innovative services on a Computed Tomography scan" was a survey research by means of collecting data using the in-depth interview with the method as conceptual framework is presented in figure 2.

3.1 Population and sample

The population in this research was a Neurosurgeon at participating centres of excellence in computer applications for Biomedical Engineering (Phase1: The National Centre of Excellence), the project was finished in the year 2007, which all came from hospitals in the area of Bangkok. The sampling method as follows:

The selection of the sample. The Non-Probability Sampling along with Purposive Sampling were used in this study, since the researchers studied the CT scan machine with a Neurosurgical physician, Physicians Nervous system, Medical diagnostic and Radiology residency. However, researchers interested in specific Neurosurgical physician because the doctors group that was active most CT scan machine.

Due to the study use the Non-Probability Sampling along with Purposive Sampling and the researchers want to study a particular group that the population in this research was a Neurosurgeon which the Report on Public Health Resource in the year 2014 that reported by Bureau of Policy and Strategy, Permanent Secretary Offices, Ministry of Public Health shown that Thailand has a total of 343 Neurosurgical Medical Group physicians. Because of this study, the researchers chose to study in Bangkok area which was reported that 110 doctors were in Bangkok. [10] Moreover, with access to information and the person may hinder the collection of information, the researchers have to use the connection with a research project in centres of excellence in computer applications for Biomedical Engineering with a population of 39 doctors in the project. Therefore, the study population was selected to implement on 39 doctors as mentioned above. However, the major drawback of this research was that the researchers have tried to contact and approach the doctor difficulty before they accept the interview arrangement. Finally, the researchers got an appointment for the in-depth interview with 21 people, response rate at 53.85%, and about the determine how to collect data in this study, the researchers collected primary data from in-depth interview by themselves.

3.2 The variables used in research

3.2.1 Independent variable

This including demographic characteristics (gender, age, education, job, and years of service), this was the first organization to work, under the hospital, and the current problems caused by the use of CT scanning of stroke patients.

3.2.2 Dependent variable

- The service acceptance of a small and movable CT scan of physicians in the hospitals. (Ordinal scale)
- The expectations from the services of a small and movable CT scan to a network hospital in Bangkok. (Interval scale)

3.3 The instrument used in research

The instrument used for data collection was attachment interview, which the researchers developed the attachment based on the research in the past along with the accepted theory of consumer-related services in the hospital that can be divided into five parts. [11-14]

3.3.1 Demographic data

It was referred to overview of the respondents with 3 models was questioned, Close-ended response question, Multiple choice question, and Dichotomous question.

3.3.2 Factors affecting the service acceptance of a small and movable CT scan of physicians in the hospital network in Bangkok

The factors included the acceptance of a consumer product or service four parts together, products and services, pricing, distribution, and promotion. The researchers designed the attachment interview filled out the sequence by the respondents (Ranking question) and the respondents chose multiple answers (Checklist question).

3.3.3 The expectations from the services of a small and movable CT scan to a network hospital in Bangkok

The researcher has proposed that all 15 deals with a Likert scale.

3.3.4 Interest in the service

The researcher has designed the attachment interview that there were many answers to choose from (Multiple choice question) and the respondents to select multiple answers (Checklist question) mixed with Open-ended question.

3.3.5 The problems caused by the use of CT scan of patients with coronary artery disease from the hospital network in Bangkok

The researcher has designed the attachment interview style Open-ended question.

4 Finding and conclusion

4.1 Finding

From research "Demand of innovative services on a Computed Tomography scan" that used the attachment interview for in-depth interview to collect data directly from 21 Neurosurgeon in Bangkok area which there were issues that can be discussed below.

According to the data from five parts, the researchers chose to extract the needs of the services from CT scan and the expectations from the services in mobile CT scan which are presented in Table 1 and Table 2 in order to answer the research question as the two mentioned above.

Table 1. The result of the needs of the services from CT scan according to the in-depth interview.

Needs	Frequency	Percentage
Movable and easy to use.	17	80.95
Solve the problem of moving the patient unstable.	16	76.19
Have CT scan in OR, ICU and ER.	15	68.18
Have a lease CT scan.	11	52.38
Have CT scan in ambulance.	10	47.62
Have CT scan in country hospitals and public hospitals.	5	22.73

From Table 1, the demand trends of the needs of the services from CT scan were about movable and easy to use up to 81% and due to this core, other related requirements were followed such as if CT scan can movable then it helps the hospitals to solve the problem of moving the patient unstable, it can be moved between Operating Room (OR), Intensive Care Unit (ICU) and Emergency Room (ER), moreover, it was also desirable to put them in the ambulance. About the lease CT scan, the respondents have shared the idea that the choice of two promotions, one-time purchase and leasing, were associated with the nature of each hospital, if the hospital has many surgical cases it also want to use CT scan in one-time purchase and installation was based at the hospital. And on the other hand, if the case surgery and patients who require CT scan was not much the hospital might to use the format for a time. Furthermore, depending on whether the unit to a large hospital in the city must attract executives or to the public or country hospitals must be priced no higher than at the hospital to pay as well.

Table 2. The result of the expectations were particularly from the services in Mobile CT scan according to the in-depth interview.

Expectations	Frequency	Percentage
In urgent situations, the doctor or nurse can make a decision to move mobile CT out of the examination room.	18	85.71
Scanned image does not exceed 5 minutes.	17	80.95
The effectiveness of diagnosis and treatment by a physician must be faster.	15	75.00
Nurse or medical technician can use the machine as it was under the control of the doctor.	14	66.67
Doctors can diagnose and symptoms faster.	16	76.19
No matter where can access to CT small portable.	9	52.94

As of the third part which have 15 optional to answer, the researchers have been extracted that information with conditions, the expectations were particularly from the services in Mobile CT scan according to the in-depth interview, are shown in Table 2. The expectations from the services of a small and movable CT scan to a network hospital in Bangkok, all through the analysis, it concluded that doctors have expectations the urgent situation a doctor or nurse can be moved CT scan outside the examination room for diagnosis, up to 86%, and the effectiveness of treatment by a physician must be faster, around 75%. With regards to the ability to read the results from CT scan has to be at least 5 minutes after the patient has photographed which was up to 81%. However, all three were expected to follow the researchers hypothesized. This was also what the researchers do not think the doctors know what to expect, but from the results of this study shown the nurse or medical technician can operate the machine itself was under the control of the doctor, which has 67% approximately. In addition to the foregoing, it was evident that the issue of access at anywhere any time was also subject to the respondents to focus on more than 53%, since it reduced the rate of readmission of patients.

Table 3. The result of interest in the services from Mobile CT scan according to the in-depth interview.

Interest	Frequency	Percentage
Interested	18	85.71
Uncertain	3	14.29
Not interested	0	0

Table 3 shows data from the fourth part, interest in the services, it was found that although mobile CT scan was meet the doctors expected and medical attention, but since every doctor does not have the power to decide to order a medical device itself, so, if the research aims to provide the new service and innovation to physicians and patients that require the management of the hospital, the work requires planning and good marketing to offer and convince the management that benefit from the introduction of this system to improve the diagnosis and treatment. These data were the results of a study that supports physicians surveyed those were interested in this service up to 86%. Another case of Tele medicine services via mobile phones, it's not that it can help doctors decide on urgent cases faster. For that reason, some doctors do not make sure to use this machine, mobile CT scan, and the key finding of this research was not based on the assumption of the set before this study. Therefore, the result has to be studied further to get the answer correct and most suitable for future research.

Here are the problems caused by the use of CT scan of patients with coronary artery disease from the hospital network in Bangkok that got from the respondents' conversation, the analysis then showed, the research hypothesize that problems in the long queues of patients using CT scan was not a real problem after obtaining information from Neurosurgeon medical group. But the real problem, it was about to remove the patient from OR or ICU or ER to CT scan room in the hospital. It can be seen that, this conclusion was consistent with the needs in Table 1. Consequently, this is another point that researchers need to bring this issue to research information that is currently available to fix this or not, by any means and there are points that can be synthesized innovative services or innovative product or process innovation in the future.

4.2 Discussion and conclusion

As paper seeks to find unmet needs of the users, the leading level of Neurosurgeon in Bangkok, which collect data from 21 Neurosurgeons, 12 hospitals in Bangkok and surrounding areas. The methodology of data collection was in-depth interview and analysed by descriptive statistics. The research address two research questions, the first was what are the needs of the service from CT scan to stroke patients, the authors' finding that most of Neurosurgeon need to use mobile CT scan at the OR, ICU and ER, respectively. According to the second research question, what are the expectations in mobile CT scan when new service implemented in the hospital?, the authors reveals and found the Neurosurgeons expect it can play a role in solving the current problems, low resolution and slice of current CT scan, hard to use it with moving unstable patients, wriggle patients and Claustrophobia patients, and the most important was unable to move it to elsewhere in hospital that all affect their treatment decision.

For that reason, the researchers want to solve the problem of low utilization of mobile CT scan in Thailand using mobile CT scan that MTEC has developed. In order to solve this problem, apart from increasing the number of mobile CT scan to meet the requirements of the job, in the first phase of the solution also requires management to allocate transportation schedule with the resources are limited, too. The researchers conducted this research in parallel with finding unmet need and get the results as described above then bring all data together and extracting to models and algorithms for scheduling which determine the timetable and the use of mobile CT scan in the hospital to meet the needs of medical users. This is likely to result in effective treatment and access to public services with mobile CT scan increasing as well.

References

1. Bureau of Policy and Strategy office of the Permanent Secretary. (2014, August 15). Retrieved from http://bps2.moph.go.th/
2. L. Siciliani, A. Stanciolec, R. Jacobs, J. Health Econ. **28**, 771–780 (2009)
3. F. T. Schut, M. Varkevisser, Health Policy **113(1-2)**, 127-133 (2013)
4. M. Dusheiko, H. Gravelle, S. Martin, N. Rice, P. C. Smith, J. Health Econ. **30**, 919– 932 (2011)
5. N.T.J. Bailey, Queueing for medical care, J R STAT SOC C-APPL. **3**, 137–145 (1954)
6. P. DeLaurentis, R. Kopach, R. Rardin, M. Lawley, K. Muthuraman, H. Wan, L. Ozsen, P. Intrevado, *IIE Annual Conference and Exposition*, 2006
7. N.T.J. Bailey, J R Stat Soc Series B. **14 (2)**, 185–199 (1952)
8. M. Brahimi, D.J. Worthington, J Oper Res Soc. **42**, 733–746 (1991)
9. S.S. Vasanawala, T.S. Desser, J. Radiol. **235**, 244–249 (2005)
10. Bureau of Policy and Strategy office of the Permanent Secretary, *The Report on Public Health Resource in the year 2014* (Permanent Secretary Offices, Ministry of Public Health, 2014)
11. P. Kotler, *Marketing Management: Analysis, Planning, Implement, and Control* (9th ed., Prentice-Hall, Inc., 1997)
12. A. Chandrachai, *Business Excellence: Strategic planning and balance score card preparation guide* (13th ed.,Chulalongkorn University Press, 2012)
13. K. Vanichbuncha, *Statistic for research* (6th ed., Chulalongkorn University Press, 2012)
14. K. Vanichbuncha, *SPSS for Windows: Data analysis* (18th ed., Chulalongkorn University Press, 2011)

Relationships between Locus of Control, Self-Efficacy, Efforts and Academic Achievement among Engineering Students

Maizam Alias[1] , Zainal Abidin Akasah[2] and Mohd Jahaya Kesot[3]

[1]*Faculty of Technical and Vocational Education, Universiti Tun Hussein Onn Malaysia, 86400 Batu Pahat, Johor DT, Malaysia*
[2]*Faculty of Civil and Environmental Engineering, Universiti Tun Hussein Onn Malaysia, 86400 Batu Pahat, Johor DT, Malaysia*
[3]*Centre for Diploma Studies, Universiti Tun Hussein Onn Malaysia, 86400 Batu Pahat, Johor DT, Malaysia*

Abstract. The aim of this study is to investigate the relationships between the affective learning needs namely, self-efficacy and locus of control, learning efforts and academic achievement among engineering students. For this purpose, a survey was conducted on first year engineering students from two technical universities in Malaysia. Self-efficacy and locus of control were assessed using existing instruments while learning efforts were assessed using a specifically designed instrument based on Carbonaro's model of learning effort. Academic achievement data were based on cumulative grade point average (CGPA) obtained from self-report by participants. The findings indicate that females engineering students tend to have higher self-efficacy compared to males while both groups have similar locus of control and invest in similar learning efforts. Only locus of control is found to be related to academic achievement while self-efficacy is found to be related to efforts. In conclusion, locus of control seems to be an important factor in predicting academic achievement among engineering students.

1 Introduction

The need to produce engineers who have high technical skills has resulted in engineering educators giving high emphasis on cognitive learning needs in their teaching and learning practices. However, achievement of cognitive learning goals is not solely dependent on meeting learners' cognitive learning needs but also meeting their affective learning needs as learning involves affects such as feeling and emotion [1]. Furthermore, with additional demand on engineers to have more people skills, affective learning outcomes are recently identified as necessary learning outcomes of engineering education. Thus, understanding the affective domain within engineering education teaching and learning is beginning to be emphasized in engineering education research.

Affective learning needs include having the self-perception of being able to succeed on task (self-efficacy) and feeling of being in control of event outcomes (locus of control). Self-efficacy and locus of control are two concepts that were introduced within the Social learning theory framework by Bandura [2] and Rotter [3] respectively. Self-efficacy is the self-confidence of a person that is related to a specific task, challenge or endeavor which means that self-efficacy may differ according to situations. Locus of control on the other hand is a person feeling of attribution towards a factor on the outcome of an event. A person can thus have an intrinsic or extrinsic locus of control. A person with an extrinsic locus of control believe that luck, chance or fate plays an important role in their life while a person with an intrinsic locus of control believe they have control over their life. Since the introduction of Rotter's idea on the general locus of control, specific locus of control has been suggested. One of it is the academic locus of control by Trice [4] and [5]. Thus some studies use the general locus of control while others focus on the specific locus of control in studying academic achievement.

Locus of control has been shown to be an important factor in predicting academic achievement [6]. Thus, it is expected that with greater belief in one's ability to influence success (high self-efficacy and having an internal locus of control), greater efforts will be invested that will lead to academic success. This hypothesis is supported by previous studies such as [7], [8], [9], [10] and [11]. Higher self-efficacy was found to be associated with higher mathematic achievement score as well as higher Grade point Average (GPA) among electronic engineering students [7] and Grade point Average (GPA) [9]. Self-efficacy is also found to be positively associated with efforts [11] and locus of control [10].

Findings on associations between learning efforts and academic achievement are inconsistent; while [10] found positive association between efforts and academic achievement, none was found in [12]. However, consistency in effort is found to be related to academic achievement instead [12] which support previous suggestion that management and control of learning efforts on classroom academic tasks is an important element in explaining academic success [13].

Based on the literature reviewed, it can be safely concluded that the findings on relationships between self-efficacy, efforts and academic achievement among engineering students is inconclusive and more studies are necessary to have a better understanding of how these parameters influence academic achievement. The purpose of this study was to investigate the relationships between self-efficacy and locus of control (two affective learning needs), efforts and academic achievement among engineering students. Specifically, the objectives were:

i. To assess the self-efficacy, locus of control and efforts among engineering students.
ii. To determine the relationships between self-efficacy, locus of control, efforts and academic achievement.

2 Methods

The participants were 410 first year engineering students from two public technical universities in Malaysia. Similar entry requirements were required for all engineering programmes. Thus, participants were expected to be similar in academic ability across programmes. However, each programme differs in terms of its general course content. For example, the electrical engineering programme in these two universities has greater mathematics content compared to the civil and mechanical electrical engineering programme which may create varying learning efforts from students.

Three instruments were used for data collection on self-efficacy, locus of control and learning efforts. Self-efficacy was assessed using an instrument developed by Klobas, Renzi, and Nigrelli [14]; locus of control was assessed using the Trice Academic Locus of Control (ALoC) Scale [4] and learning efforts were assessed using a specifically designed instrument based on the efforts model from Carbonaro [15].

The Self-efficacy scale has 30 items. Participants were asked to rate a statement on a scale of 1 to 10 on their agreements to given statements. The maximum score is 270 and the minimum score is 0. The average score is set at 135. The Self-efficacy scale has a reliability of á=.70 indicating moderate and adequate reliability for the purpose of this study.

The Trice ALoC scale which consists of 28 items is similar to the Rotter's Locus of Control Scale but with a difference in focus. While Rotter's measure general locus of control, Trice ALoC measures locus of control that is specific to academic setting. The maximum probable score is 28 while the minimum score is zero and the average score is 13. In this study, a person who scores above 13 is considered to have an extrinsic locus of control. The Cronbach Alpha reliability of Trice ALoC Scale is á =.71, indicating moderate reliability and acceptable quality for the purpose of the study. A high stake assessment would demand a higher reliability than currently achieved.

The Learning efforts instrument was developed based on Carbonaro's model on learning efforts [15]. According to Carbonaro, efforts can be divided into three components, rule oriented efforts, procedural effort and intellectual effort. The quantity of each type of effort can be measured according to their frequency and duration. A maximum efforts score possible is 150; a minimum score is 30, and the middle score is 90. Thus, the learning efforts instrument assesses three categories of effort, rule oriented efforts, procedural efforts and intellectual efforts; measured according to their frequency and duration. The reliability of the learning effort instrument is á=.88 indicating a high reliability.

The instruments were distributed to participants with the help of lecturers of the respective classes of participants. Permissions were sought through the various university authorities prior to the data collection activity. At the onset of the data collection activity, the aim of the study was explained to the participants and confidentiality was assured to ensure that participants give honest responses. Each participant was given a small token for his/her assistance in the study. Completed instruments were gathered on the same day that they were distributed.

3 Results and Discussions

Instruments were distributed to 410 participants, and 360 (87%) completed questionnaires were returned. Distributions of respondents are shown in Table 1.

Table 1. Response rate according to programme and gender.

	Female	Male	Total
Civil engineering (CE)	86	51	137
Electrical engineering (EE)	44	35	79
Mechanical engineering (ME)	45	99	144
Total	**175**	**185**	**360**

The civil engineering participants have the highest mean on CGPA (M = 3.11, SD=0.50) followed by the electrical engineering (M=3.05, SD=0.45) and mechanical engineering participants (M=2.62, SD=0.69).

3.1 Self-Efficacy, Locus of Control and Learning Efforts among Engineering Students

Overall mean for locus of control is slightly above 14 (Mean=14.56, Standard Deviation=4.22) indicating slightly extrinsic locus of control. Mean (M) and Standard Deviation (SD) for locus of control, learning efforts and self efficacy according to gender are given in Table 2. Similar means are observed among male and female participants (Table 2). Overall mean on learning efforts is above average (M=111.73 which is greater than 90) with females scoring approximately 5 points higher than males. Above average self-efficacy is reported with higher self-efficacy among females.

The independent t-test results on means between genders indicate that only the difference on learning efforts is statistically significant (t=3.108, df=358, p=.002). This result in combination with Table 2 indicates that females are making greater learning efforts compared to males.

Table 2. Locus of control, Learning efforts and self-efficacy of engineering students according to gender.

	Female (n=175)		Male (n=185)	
	M	SD	M	SD
Locus of control	14.37	4.22	14.75	4.21
Learning efforts	114.04	13.28	109.55	14.11
Self-173.57 efficacy		33.95	175.67	33.92

The current finding indicate that engineering students were slightly more extrinsic in their locus of control which means that they perceive external factors as more influential in determining the outcomes of their efforts. The findings contradict previous finding [17]. However, in [17] the Rotter's locus of control scale which measures the general locus of control was used instead. Students may feel more in control on general matter but may feel less in control where academic matter is concerned. So the current finding is not necessarily indicative of contradiction to [17].

Similar locus of control trends are also observed among male and female participants indicating that both males and females similarly perceive the external factors as being influential in determining the outcomes of events. Similarity in locus of control between males and females engineering students support the findings from a previous study by [19] and [20]. The current finding is however, in contrast to that of [21] where Jordanian males engineering were found to be more intrinsic in their locus of control compared to females meaning that males feel that they are more in control of the outcome of events compared to females. The difference could be due culture where males are generally expected to take up the leadership roles. Females engineering students in the current study seem to be making more efforts than male engineering students are. This could be due to the biased nature of engineering programmes that necessitates females to try harder to succeed.

Statistics according to programmes which are shown in Table 3 indicate that differences exist between group means on all three variables under study.

Table 3. Locus of control, learning efforts and self-efficacy according to programmes of study.

	CE	EE	ME
	M (SD)	M (SD)	M (SD)
Locus of control	13.2 (3.9)	14.5 (3.3)	15.9 (4.6)
Learning Effort	112.3 (12.4)	114.9 (12.9)	109.5 (15.4)
Self-Efficacy	179.1 (34.6)	170.8 (31.8)	172.6 (34.1)

When differences in means were tested using Analysis of Variance (ANOVA) method, only differences between groups on locus of control (F=15.471, df=2, p=.000)) and learning efforts (F=4.143, df=2 and p=.017) were statistically significant. The statistically significant difference indicates that there is a statistically significant difference between at least two groups. A pos-hoc analysis indicates that civil engineering students are more intrinsic in their locus of control compared to the mechanical engineering group.

The difference observed between electrical and mechanical engineering participants on learning efforts indicates that electrical engineering participants make greater learning efforts compared to mechanical engineering participants. Difference in self-efficacy was not statistically significant between programmes (p>.05) indicating that all groups have similar self-efficacy level.

Self-efficacy of engineering participants were found to be above average, similar to previous findings by [17] and [18] who found that engineering students have average or high self-efficacy. Perception of self-efficacy is also found to be similar between males and females which support findings from a previous study on Indian engineering students [19]. Current finding is different from [18] where American female engineering students were found to be more self-efficacious compare to males. Cultural influence could be a factor here as the studies in [19] and [20] are within the Eastern culture while the study in [18] is on American students (Western culture). Further study is needed to verify this hypothesis.

When the three parameters were compared among the different engineering disciplines, interesting findings emerged. Participants from different discipline differ in their locus of control and learning efforts. Civil engineering participants were the most intrinsic in their locus of control and has the highest on the self-efficacy measure. This finding is in accordance with earlier expectation where high belief in ability to succeed and ability to control an outcome is associated with better academic success. In this case, the civil engineering participants who have the highest perception on the ability to succeed and the ability to control outcome have actually gotten the highest mean on CGPA as previously stated.

Electrical engineering participants on the other hand reported making the most efforts but have the lowest self-efficacy. This could mean that having weak belief in their natural ability to succeed, the electrical engineering participants thus invest in more learning efforts to ensure that they can actually succeed in their programme. Another possibility is that the greater efforts required is due to the high mathematical content of electrical engineering programmes that may demand greater learning efforts from the electrical engineering students. Mechanical engineering is the most extrinsic with the lowest effort mean score. This could mean that believing the external factors such as luck and teachers to be in control of outcomes, mechanical engineering students may then decide not to invest in much effort which reflects their lowest mean score. This interpretation is in line with earlier expectations and supported by previous studies.

However, the greater investment in learning efforts observed within the electrical engineering group could simply be due to a manifestation of the gender effect as the electrical engineering group has more females than males and females in the study have been shown to invest in more learning efforts irrespective of

DOI: 10.1051/matecconf/20166818004

programme of study.

Variations in efforts for different programmes of study have been observed in previous study. In [21] software engineering students work harder compared to Bachelor in Information Technology and Bachelor in Computer Science students who seem to invest less effort to achieve higher grades. This indicates that different programmes may demand less or more efforts from students depending on the nature of programmes.

3.2 Relationship between Locus of Control, Learning Efforts, Self-Efficacy and Academic Achievement

Statistically significant correlations are only found between CGPA and locus of control r=-.223, p<.01) and between self-efficacy and learning efforts (r=.294, p<.01). Statistically significant correlation is not found between locus of control and learning efforts (p>.05). Furthermore, no statistically significant correlations are found between learning efforts or self-efficacy and academic achievement (p>.05).

The results indicate that those who score low on locus of control tend to score high on CGPA meaning intrinsic locus of control is associated with higher CGPA. The current finding supports findings from previous studies in [23] and [21]. Data from the current study also indicate positive association between self-efficacy and learning efforts which is similar to [10] but contradicts [20]. Similar to [10] and [12] the current study fails to establish correlation between effort and CGPA. Consistency in effort instead of total effort was associated with academic achievement [12]. In contrast to [9], [10] and [23] the current study fails to establish associational relationship between self-efficacy and CGPA. No correlation between self-efficacy and locus of control was found in this study which is similar to [20] and [23].

However, when the data were analyzed separately for extrinsic and intrinsic locus of control group, the results were different (Table 4); there is a statistically significant correlation between locus of control and learning efforts while the other relationships remain similar in directions and strengths.

Table 4. Correlations between locus of control, learning efforts, self-efficacy and CGPA for group with extrinsic locus of control (N=206).

	CGPA	Locus of Control	Self-efficacy	Learning efforts
CGPA	1	-.219**	-.007	.057
Locus		1	-.047	.195**
Self-efficacy			1	.383**
Efforts				1

The schematic representation of the relationships based on statistics in Table 4 is given in Figure 1. The arrows indicate direction of influence. Thus it can be seen that locus of control influences CGPA and learning efforts, while self-efficacy only influences learning efforts.

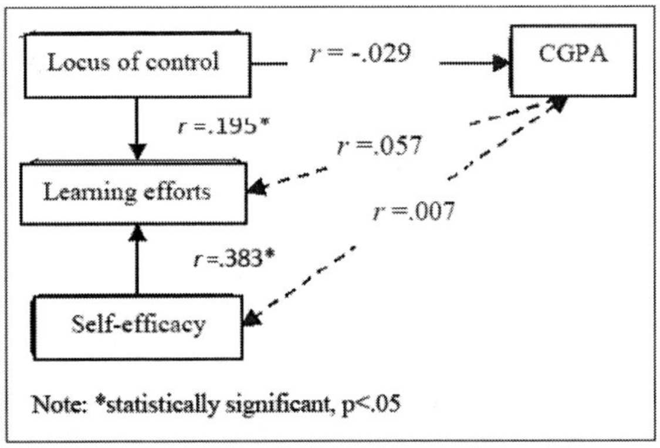

Figure 1. Relationships between locus of control, learning efforts, self-efficacy and CGPA.

4 Conclusions

The study was set out to determine the level of locus of control, self-efficacy and learning efforts among engineering students in Malaysia and ultimately to identify the connections between these parameters and academic achievement. The data support the conclusion that Malaysian engineering students tend to have high self-efficacy with females having higher self-efficacy compared to males engineering students.

In general, engineering students tend to have extrinsic locus of control with the exception the engineering students in the civil engineering students who tend to be intrinsic in their locus of control. This finding can arise out of two possibilities i.e., a specific programme of study can predispose a learner to be intrinsic or extrinsic in locus of control or having a specific type of locus of control can predispose a person to choose a certain engineering programme. More research is needed to have a definitive answer.

The data also support the conclusion that self-efficacy and effort are indirectly related to academic achievement for students with extrinsic locus of control but not for those with intrinsic locus of control. This is an interesting finding that demands further research. The inconclusive findings on some hypotheses tested indicate that the relationship between affective learning needs namely self-efficacy and locus of control, effort and academic achievement among engineering students are still in need of further research to be fully understood.

Acknowledgement

The authors would like to express their gratitude to The Ministry of Education that provided the financial support in the form of FRGS grant vot 1449. Our thanks also go to Mashirnani Mohd Tahir, our Master in Education students for her assistance in the data collection process.

References

1. J. Dirk, *New Direction for Adult and Continuing Education*, A Publishing Unit of John Wiley & Sons Inc . (2001)

2. A. Bandura, The Evolution of Social Cognitive Theory. In K.G. Smith & M.A. Hitt (Eds.) *Great Minds in Management.* (p.1) (2005). Oxford: Oxford University Press.
3. J. B. Rotter, Internal Versus External Control of Reinforcement: A Case History of a Variable, *Am. Psychologist*, April 1990, 490-493, (1990).
4. A. Trice, An academic Locus of Control scale for college students. *Perceptual and Motor Skills*, **61**, 1043-1046 21, (1985).
5. A.Trice, et al. Concurrent validity of the Academic Locus of Control Scale. Educational and Psychological Measurement, **47**, 483-486(1987).
6. A. Anderson, J. Hatie, R. Hamilton, Locus of Control, Self ‑ Efficacy, and Motivation in Different Schools: Is moderation the key to success? *Educ. Psy.* **25**, 5, 517-535 (2005)
7. C. W. Loo, J. L. F. Choy, Source of Self-efficacy influencing academic performance of engineering students. *Am. J. of Educ. Res.* **1**,3, 86-92, (2013).
8. A. Nasrollahi, Barjasteh, H. Iranian Students' Self Efficacy and Their Language Achievements. *Theory and Pract. in Lang. Stud.*, **3**, 10, 1837-1843 (2013).
9. R. Shkullaku, The Relationship Between Self – Efficacy and Academic Performance in the Context of Gender Among Albanian Students. *Euro. Acad. Res.* **1**, 4, 467-478 (2013). www.euacademic.org
10. K. A. Li , A Study of the Attitude, Self-Efficacy, Effort and Academic Achievement Among CityU Students Towards Research Methods and Statistics. *SS Student E-Journal.* **1**, 154-183 (2012).
11. A.Valle, J. C. Núñez, R. G. Cabanach, J. A. González-Pienda, S. Rodríguez, P. Rosário, et al. Academic goals and learning quality in higher education students. *The Spanish J. of Psy.*, 12(1), 96-105(2009).
12. H. Patron, S. Lopez, Student Effort, Consistency and Online Performance. *The J. of Educ. Online*, **8** (2), 1-12 (2011)
13. P. R. Pintrich, E. V, d Groot, Motivational and self-regulated learning components of classroom academic performance. *J. of Educ. Psy.* **82** (1), 33-40. (1990).
14. J. E. Klobas, S. Renzi, M. L. Nigrelli, Dondena Working Paper No. 2: Measuring self-efficacy for learning. Carlo F. Dondena Centre for Research on Social Dynamics. (2007) URL: www.dondena.unibocconi.it/wp2
15. W. Carbonara, Tracking, Students' Effort, and Academic Achievement. *Socio. of Educ.*, **78**, 27-49 (2005).
16. M. Alias, Z. A Akasah, M. J. Kesot, Self-efficacy, Locus of Control and Attitude among Engineering Students: Appreciating the Role of Affects in Learning Efforts. *Procedia-Social and Behavioral Sciences,* **56**, 8, 183–190 (2012).
17. C. J. Burger, J. A. Raelin, R. M. Reisberg, M. B. Bailey, D. Whitman, The Effect Of Gender On Support and Self-Efficacy and Self-Efficacy in Undergraduate Engineering Programs. *2010 ASEE South-eastern Section Annual Conference, Blacksburg, VA.* Accessed from http://se.asee.org/proceedings/ASEE2010/Papers/PR2010Bur184.PDF
18. J. W. Elliott, Self-Efficacy, Motivation, and Locus of Control, Among Male and Female Construction Management Students. Paper ID #9531. *121st ASEE Annual Conference and Exposition*, Inidianaolis IN June 15-18, 2014
19. J. Parameswari, K. Shamala, Academic Motivation and Locus of Control Among Engineering Students *J. of Psy. Res.* **7** (1), 159 -167 (2012)
20. R. M. Mazjub, The Relationship between Locus of Control and Academic Achievement and Gender in Selected higher Education Institution In Jordan. *Proceedings of the 8th WESEAS International Conference on Education and Educational* Technology , 215-220. (2009).
21. B. R. von, Konsky, J. Ivins, M. Robey, Using PSP to Evaluate Student Effort in Achieving Learning Outcomes in a Software Engineering Assignment. *Proceedings of the 7th Australasian conference on Computing education*, **42,** 193-201 (2005).
22. S. S. Hassan, and Khalid Ruhi, Academic Locus of Control of High and Low Achieving Student. *J. of Res. and Ref. of Educ.* **8**,1, 22-33 (2014).
23. A. O. Ogunmakin, Academic Self Efficacy, Locus of Control and Academic Performance of Secondary School Student in Onde State Nigeria. (2013). Accessed from http://web.stanford.edu/dept/SUSE/projects/ireport/articles/self-regulation/self-regulated%20learning-motivation.pdf

Research and Systematic Design on Autonomous Mission Management Technology of Agile Satellite

Zhu Jian-bing, Zhao Wei, Wang Lu-yuan and LI Wen-juan

CAST, Beijing Institute of Spacecraft System Engineering, 100094, Youyi Road HaiDian District No.104, China

Abstract. Compared to traditional non-agile imaging satellites, the imaging efficiency of agile satellites are far higher and the imaging requirements are correspondingly more sophisticated, which in turn calls for better design and implementation of the onboard autonomous mission management system. According to the characteristics of the autonomous mission management system of agile satellites, this paper presents and illustrates the functional modules and key technology required by the autonomous mission management system. This paper proposes a layered architecture design, which employs a progressive and gradually detailing way to facilitate the systematic modular design and to effectively reduce the complexity of the implementation of the design. Combined with mission plan, emergency replan, image parameter real-time calculation technique, autonomous monitoring technique of plan implementation and other techniques, this design makes the agile imaging mission more scientifically designed, properly scheduling and safely implemented.

1 Introduction

With the advancement of space technology, the operating mode of remote sensing satellites has changed from traditional passive scanning of single degree of freedom to active scanning which change directions freely in roll, pitch and yaw axis. This considerably enhances the imaging efficiency and imaging quantity. This new type of imaging satellites is internationally called agile satellite, the typical examples of which are the American WorldView, the French Pleiades, the Britain TopSat, etc. The traditional satellite control and management mode of "ground planning and onboard implementation" can't fulfill the need of multi-mode and multi-mission imaging per revolution. Therefore, all the space powers are carrying out research on autonomous mission management technology. Based on EO-1satellites, NASA has developed two mission plan system of satellites, ASPEN [1] and CASPER [2]. ASPEN can generate daily command of EO-1 automatically by users' mission. CASPER, by using a real-time replanning mechanism, achieves the goal of recovering the real-time condition of the system and new objections on a fixed time step or event step and realizes necessary replans based on the this real-time information. PROBA [3], supported by ESA, studied and validated certain space-borne autonomous planning techniques. Trials on autonomous mission plan system of satellites [4] are also done by FSA. Many domestic scholars and engineers as well focuses on the research on remote sensing satellite mission plan technology. HE Ren-jie and LI Ju-fang from National University of Defense Technology give model and algorithm of multiple satellites mission plan on ground and develops an integrated mission plan system on ground which is applied to the daily control and management of remote sensing satellites [5]. Yet currently this system is mainly employed on the ground and applied to non-agile satellites. TIAN Zhi-xin from Beijing institute of Spacecraft System Engineering put forward the algorithm of generating satellite mission based on directed graph model [6], which initiated the disintegration of instructions from mission level to execution level so that the complexity of satellite control was effectively reduced and improved the data injection efficiency of satellites. This method, however, is mainly aimed at non-agile satellites without taking the sophisticated imaging mission like active scan and corridor scan of agile satellites into consideration.

This paper analyzes the features of the autonomous mission management system of agile satellites first, and then demonstrates the four capabilities essential to the autonomous mission management system: mission planning, real-time high frequency calculation of the image parameter (the integral time), mission replanning and autonomous state monitoring throughout the procedure of imaging. At the same time, this paper studies the critical techniques which includes mission plan technique, emergency replan technique, image parameter real-time calculation technique, autonomous monitoring technique. The autonomous mission management system has been used on certain agile remote sensing satellite platform, which meets the requirements of multi-mode and multi-mission per revolution imaging for agile satellites.

 DOI: 10.1051/matecconf/20166818005

2 The features of onboard autonomous mission management system

Compared with the traditional satellites mission management system, the agile satellites autonomous mission management system has the following features:

1) **More missions per revolution should be supported, requiring a stronger ability of onboard mission planning.**
An agile satellite has a higher imaging efficiency and shorter mission switching, the number of the mission per revolution has a large increase compared with traditional satellite. The number of telecommand to upload will increase in exponential level if still use the traditional way of satellite control, can't meet the requirement of uploading all telecommand in a limited observe and control time window, so the onboard autonomous mission management system should have the ability of converting the telecommand from high mission level to low control level, to provide technology support for more imaging target.

2) **The imaging mode becomes more complicated, requiring the satellite has the ability of real-time high frequency calculation of the camera parameter (integral time).**
In addition to the imaging mode of passive scanning, an agile satellite also supports many complicated imaging mode like active scanning, scanning not along the track[7], etc.(see Figure 1),the attitude changes flexible during the imaging procedure, the satellite can move on three axes and the angle velocity of the three axes can be nonzero. When the satellite images with nonzero angle rate, the severe change of slant distance not only brings the scale change in image, but also the change of image motion's angle velocity, resulting in the severe changes of integral time parameters during the imaging procedure. So the onboard autonomous mission management system should have the ability of real-time high frequency calculation of the integral time, so as to guarantee the imaging quality in agile mode.

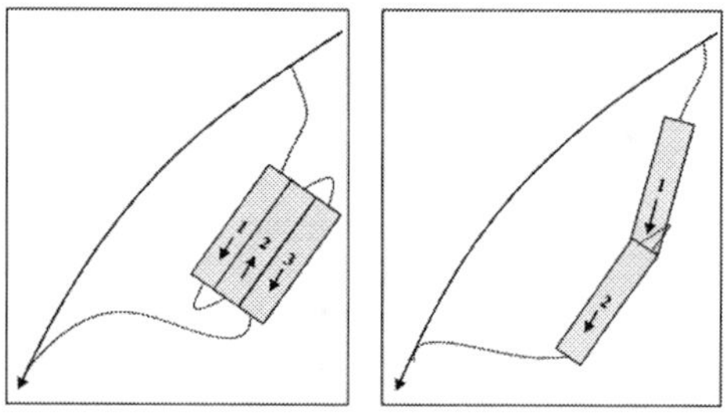

Figure 1. Complicated imaging mode of agile satellite.

3) **The response speed of the emergency mission becomes faster, requiring the satellite has the ability of mission replanning.**
The agile satellite has a stronger ability of attitude control, more flexible imaging, faster task switching, so the onboard autonomous mission management system should have the ability of mission replaning, providing rapid response services for users' emergency mission, to enhance the emergency support capability of satellite effectively.

4) **The processing logic of the satellite becomes more complicated, requiring the satellite has a stronger ability of autonomous monitoring throughout the mission execution.**
As agile satellites is required to implement more missions, to have more sophisticated imaging modes and to respond to emergencies more rapidly, there is greater probability of satellite abnormity throughout the mission execution, which in turn makes the capability of autonomous monitoring of the mission execution necessary for the satellites, so as to ensure the operation of satellite safe and reliable.

3 The research of key technology

As can be seen from the discussion about the features of autonomous mission management system, an agile satellite has a variety image mode, complicated data flow control , one of the key technology of the system is how to design a good system architecture. Secondly, the system should breakthrough three key technology including mission planning and replanning, real-time calculation of the integral time and autonomous monitoring throughout the mission execution, enabled the system has the four kind of ability discussed above.

3.1 Layered architecture design of autonomous mission management system

Since the Autonomous Mission Management System of agile satellites is characterized by diverse missions and difficult control of information flow, a series of proper system structures are demanded to facilitate the modular design of the system and effectively diminish the complexity of the execution of the system. As is displayed in Figure 2, this paper employs the idea of layered architecture to divide the system into application layer, planning layer, service layer and execution layer. On-board computer fulfills the planning, scheduling and execution of agile imaging in a layered progressive and gradually detailing way.

(1) Application Layer

This layer is composed of two modules, meta task receive module and meta task pre-process module. The meta task recerve module receives the demands from the ground, conducts batch management and submit the demands to meta task pre-process module on the basis of orbit time period. The module of pre-process sequences the to-be-planned meta task of current batch in time order, do preliminary check (e.g. the intervals of meta tasks, accumulated time length of imaging, etc.) and form the logical sequence of meta tasks to be used by mission plan module.

(2) Planning Layer

This layer contains three modules, mission plan module, mission replan module and instruction generating module. Mission plan module receives meta tasks from the pre-process module of application layer, invokes the mission planning algorithm and outputs mission plan. The instruction generating module transform the mission plan into instructions that can be directly executed. The mission replan module deals with emergency mission from the ground, generate renewed plan and invokes the instruction generating module to produce instructions.

(3) Service Layer

This layer includes two modules, module of autonomic computing of parameter and module of mission execution monitoring. Module of autonomic computing of parameter provides parameters as integral time for mission plan and mission replan. Module of mission execution monitoring the running state of the satellites and rejects the instructions that cannot be executed.

(4) Execution Layer

This layer is formed of two modules, instruction verifying module and instruction execution module. Instruction verification makes judgment on the safety of the instructions sending safe instructions to the instruction execution module and rejecting unsafe ones. Instruction execution module activates the hardware of the satellite directly.

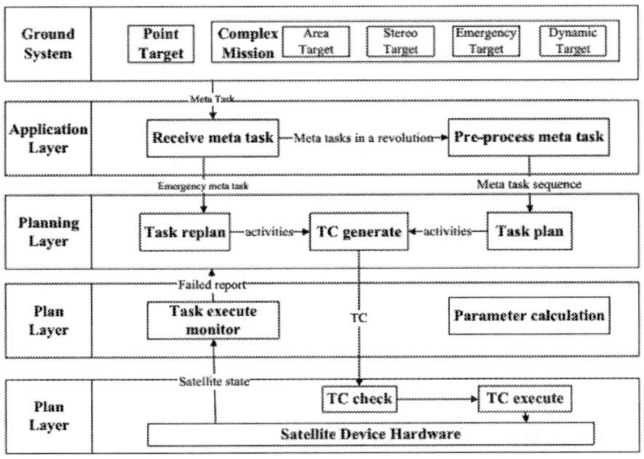

Figure 2. Architecture of autonomous mission management system.

3.2 Techniques of mission plan and emergency replan

To meet the imaging mission requirement, the agile satellite performs multiple different mode per revolution, the system needs the ability of rapidly mission planning. In order to reduce programming complexity and facilitate the implementation onboard, this paper adopts the strategy of planning missions next track before the track starts and form a preliminary plan. The planning process builds the constraint model of attitude control, power, storage first, planning algorithm calls constraint model to process the constraint checking of the missions contain. Instruction generation module rapidly translated the planning action sequence into the instruction can be directly executed according to the instruction template. If the system receives emergency meta task, it needs to insert the rapid emergency meta task into planned meta task queue in time order and analysis relations of conflict with the existing meta task time constraint. Deleting the existing tasks until meet the time constrains if there are conflicts and forming a feasible sequence of meta tasks. Then performing the mission planning process again (including constrains checking and action planning) and forming a task action sequence .The instruction generation module rapidly translated the planning action sequence into the instruction can be directly executed according to the instruction template at last. The whole procedure is shown in Figure 3.

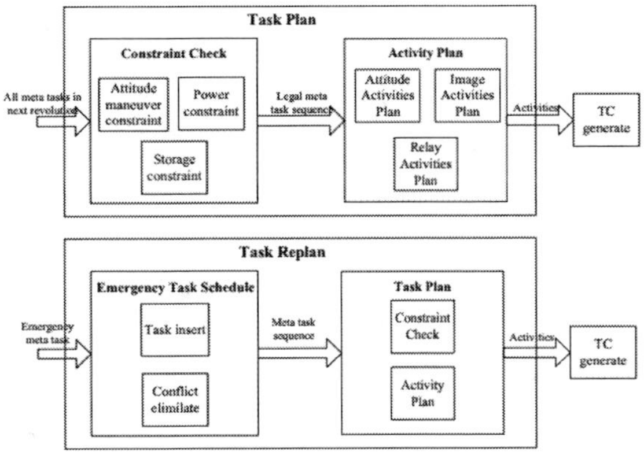

Figure 3. The procedure of mission planning and replanning.

3.3 Techniques of onboard real-time calculation of the integral time

Integral time is an import parameter for TDICCD camera in agile satellite. The accuracy of its value will affect image quality directly. The attitude of agile satellite changes quickly which lead the value of integral time wave largely, so traditional calculate mode cannot meet the requirement of agile satellite. We modify the velocity and height mode on paper [8], and use the new mode to calculate the integral time in agile satellite.

3.3.1 The modification of shoot point velocity mode

The attitude angle rate was nonzero when image in agile mode, so we should consider attitude angel rate when calculate vector of satellite coordinate angel rate. The calculate formula is:

$$w_b = w_n + w_s \tag{1}$$

w_b is the vector of satellite coordinate angle rate, w_n is the vector of orbit angle rate, w_s is the vector of satellite attitude angle rate

So the shoot point velocity mode of agile satellite has difference compared to the traditional calculate mode, the calculate formula is:

$$v = w_e \times R - (w_n \times r + w_b \times H + v_r) \tag{2}$$

w_e is the vector of earth rotation angle rate, R is the vector from earth center point to shoot point , r is the

 DOI: 10.1051/matecconf/20166818005

vector from earth center point to satellite, H is the vector from satellite to shoot point, v_r is the vector of satellite speed along to the earth center point.

3.3.2 The modification of height mode

Height calculate mode is a simple and low-accuracy mode, it doesn't consider the ground height, we should use the digital ground height diagram to modify the height between satellite and shoot point for more accurate integral time value. Figure 4 shows the new calculate mode.

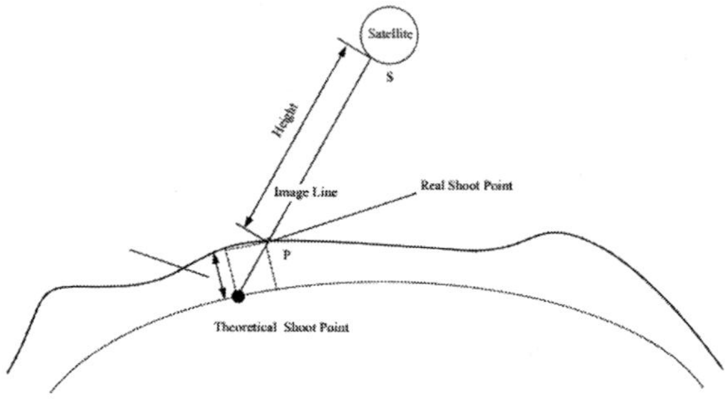

Figure 4. The new height calculate mode.

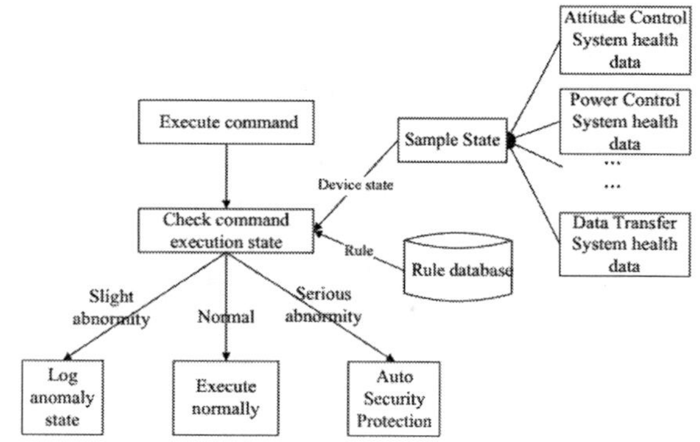

Figure 5. The procedure of autonomous monitoring throughout the mission execution.

3.4 Techniques of autonomous monitoring of mission execution

As agile satellites is required to implement more missions, to have more sophisticated imaging modes and to respond to emergencies more rapidly, there is greater probability of satellite abnormity throughout the mission execution, which in turn makes the capability of autonomous monitoring of the mission execution necessary for the satellites. The fundamental part of the mission execution monitoring is to guarantee safe execution of each instruction of each subsystem during the mission, thus real-time checking of the results of each instruction is necessary. State sampling module collects real-time state of each subsystem, with the integration of the judgment rule database in the computer memory of the satellites, to make judgments on the normality of the instruction execution. Slight abnormity is logged and regularly sent back to the ground by event reports so that the designer can analyze the cause of the abnormity. If there is serious abnormity, autonomous protection mode will be switched on and the satellite loading will be safely shut down, in which way the satellite will be running safely at the minimum function level. If the execution is normal, next instruction will be executed. The whole procedure is illustrated in Figure 5.

4 Applications

Key technology based design of autonomous mission management system has been applied in an agile platform of remote sensing satellite and its data flow diagram is showed in Figure 6.

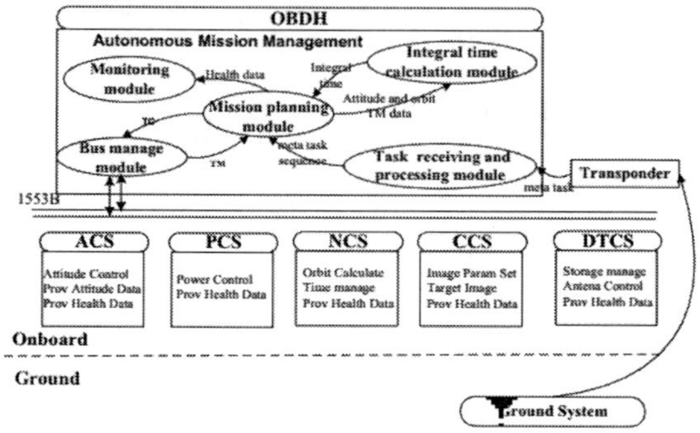

Figure 6. Data flow diagram of autonomous mission management system in an agile platform.

Onboard transponder receives meta task from ground and sends them to OBDH subsystem. OBDH subsystem using these meta task data and each subsystem TM data including attitude data, orbital data and subsystem health data etc., from 1553B bus to complete mission management functions such as the mission planning, integral time calculation, mission execution monitoring. Current system operation condition conforms to expectation design and the system can meet the multi-mode multi-mission per revolution imaging needs of an agile satellite. Compared with traditional remote sensing satellite, the advantages of using onboard autonomous mission management functions are shown in Table 1.

Table 1. Comparison of applied and note applied the onboard autonomous mission management.

	Applied	**Not-applied**
The number of missions per revolution	High	Low
Task constrained check ability	Yes	No
Single Mission injection data volume/Byte	*40*	220
Rapid response ability to Emergency Mission	*Yes*	No
Integral time real-time frequent calculation ability	*Yes*	No
Mission autonomous monitoring ability	*Yes*	No

5 Conclusions

High autonomy is a significant field in spacecraft development. Comparing with TM&TC based traditional mission operation and management mode, onboard autonomous mission management technology can sharply improve the satellite whole application efficiency and reduce the cost of operation and management on the ground. This paper proposes an autonomous mission management system onboard based on the key technologies researched, the system has been used in certain agile platform and will meet the requirement of new earth observing satellites in future.

References

1. S.Chie, G.Rabideau. ASPEN-auto-mated planning and scheduling for space mission operations[C]//6th International Symposium on Space Missions Operations and Ground Data System. France, Toulouse, 2000.
2. Cichy B, Chien S. Validating the EO-1 autonomous science agent[C]//International Workshop on Planning and Scheduling for Space. Germany, Darmstadt, 2004.
3. Bamsley MJ, Settle JJ, Cuter MA, te al.The PROBA/ CHRIS mission: a low-Cost smallsat for hyperspectral multiangle observations of the earth surface and atomosphere[J]. IEEE Transactions on Geoscience and Remote Sensing, 2004, 42(7): 1512-1520.
4. Biachessi N.Planning and Scheduling problems for earth observing satellites: models and algorithms[D]. Milano: Universita degli Studi di Milano,2006.
5. HE Ren-jie, GAO Bao-cun, LI Ju-fang. Models, algorithms and applications to the mission planning system of imaging satellite[J] Systems engineering – Theroy&Practice, 2011,31(3): 412-413(in Chinese)
6. TIAN Zhixin, Cui Xiaoting. Remote Sensing Satellite Autonomous Command Sequences Generation Based on Digraph-model[J]. Spacecraft Engineering, 2014, 23(6):55-57(in Chinese)
7. ZHANG XinWei, DAI Jun. Research on Working Mode of Remote Sensing Satellite with Agile Attitude Contro l[J]. Spacecraft Engineering, 2011, 20(4) :34-37(in Chinese).
8. CHEN Shaolong Computation Model of Veer Angle and Velocity – Height Ratio for Space Camera in Roll Attitude [J]. Spacecraft Engineering, 2010, 19(1):38-39 (in Chinese)

MATEC Web of Conferences **68**, 18006 (2016) DOI: 10.1051/matecconf/20166818006
ICIEA 2016

An Efficient Virtual Trachea Deformation Model

Cui Tong[a] and Zhao Haifeng

Nanjing Research Institute of Electronics Engineering, 210007, Nanjing, China
Science and Technology on Information Systems Engineering Laboratory, 210007, Nanjing, China

Abstract. In this paper, we present a virtual tactile model with the physically based skeleton to simulate force and deformation between a rigid tool and the soft organ. When the virtual trachea is handled, a skeleton model suitable for interactive environments is established, which consists of ligament layers, cartilage rings and muscular bars. In this skeleton, the contact force goes through the ligament layer, and produces the load effects of the joints , which are connecting the ligament layer and cartilage rings. Due to the nonlinear shape deformation inside the local neighbourhood of a contact region, the RBF method is applied to modify the result of linear global shape deformation by adding the nonlinear effect inside. Users are able to handle the virtual trachea, and the results from the examples with the mechanical properties of the human trachea are given to demonstrate the effectiveness of the approach.

1 Introduction

Trachea surgery is still one of the most demanding fields in surgery to cure tumours, that is, one of the leading causes of death worldwide. Teaching and learning trachea surgery and internal trachea anatomy have been restricted to merely watching trachea operations and videos, studying conventional diagnostic images [1].

The most challenging task in virtual trachea surgery is to build an efficient and real-time force/deformation model in the surgery simulation. However, high-quality feedback demands a high level of physical realism as well as a high update rate in rendering, which are often conflicting requirements. Over the past two decades, many models have been developed to simulate deformable objects for a broad range of computer graphics applications. An alternative approach to physically based deformation modelling are spring-based methods, which model a deformable object as a collection of mesh nodes connected by springs in a mesh skeleton. In general the computations involved in spring models are relatively simple and easy to implement. But the human trachea is non-isotropic and made up of anisotropic materials, so that it is difficult to render physically believable behaviours. In addition, the large number of springs and mass points composing the model makes it fastidious to define parameters and control the elongation, flexion and torsion at a macroscopic level [2].

The proposed approach is a physically based model more apt for modelling force and deformation under simple contact or multiple contact conditions. The paper presents a skeleton model to simulate the trachea deformation in virtual surgery and we discuss implementation and experimental results including the single finger contact.

[a] Corresponding author: cuitong_seu@sina.cn

2 Trachea skeleton model

Due to the complex shape and non-homogeneous material properties, we give some assumptions about human trachea model and restrict our study in this paper to the trachea model scenarios. And a novel skeleton modelling approach is introduced to simulate the trachea deformation under the external force.

2.1 Assumptions about trachea skeleton

In this paper, we assume human trachea is elastic model and mainly made of two homogeneous materials — cartilage rings and muscular tissues. And each material has a constant elastic coefficient , in which the external pressure distributes evenly.

In the real trachea, two or more of the cartilages often unite, partially or completely, and they are sometimes bifurcated at their extremities. And the outer surfaces of the cartilage rings are flattened in a vertical direction, but the internal are convex, the cartilages being thicker in the middle than at the margins [3] as shown Fig.1. However, to simplify the case, the cartilage rings, which have the same geometry shape and material properties, are placed horizontally above each other and separated symmetrically by narrow intervals.

As the cartilage material is stiffer than the muscular tissue to resist the external forces, we only consider the effect of the cartilage rings under the press and ignore the elasticity of the muscular tissue between the cartilage rings. In addition, the gravity force of the trachea itself as a thin shell tube is much smaller than the react forces, and therefore the gravity force is neglected in the experiments. And the larynx connecting with the trachea is stiff solid,

DOI: 10.1051/matecconf/20166818006

when it is pulled by the stretch from the trachea, its shape keeps invariable but its displacement is changed. Similarly, the displacement of the primary bronchi with the trachea fork end is proportional to the stretch from the trachea end.

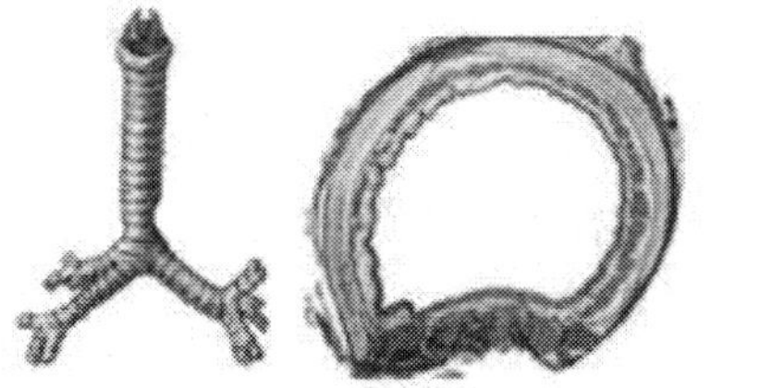

Figure 1. Trachea and its crossing-section with cartilage ring

Duffing Equation [4] is used to characterize the nonlinear force response F and the magnitude D of the displacement vector of contact point at a quasi-static state, which is the qualitative analysis of the deformation behaviour manually during the operation.

$$||F|| = m_0|\omega_0^2 D + 0.75\beta_0^2 \varepsilon D^3| \quad (1)$$

where m_0 is the mass of the virtual rigid tool, and $\omega_0, \beta_0^2\varepsilon$ are the parameters of linear and nonlinear restoring terms in the Duffing Equation respectively.

2.2 Trachea skeleton model

The skeletal trachea is a bony tube that commences at the larynx and bifurcates into the primary bronchi, and is an important part of the vertebrate respiratory system. It is flexible and forms the vacuum trunk of an upside-down tree. The trachea is mainly made up of the elastic smooth ligament tissues with about 15-20 incomplete C-shaped cartilaginous rings, which keeps the trachea open so the head and neck may twist and bend during the process of breathing. The each cartilage of the trachea occupies the anterior two-thirds, being deficient behind, where the tube is completed by unstrapped muscular fibbers. The trachea has an inner diameter of 20-25 mm and a length of 10-16 cm in adults [5]. Seen in Fig. 2.

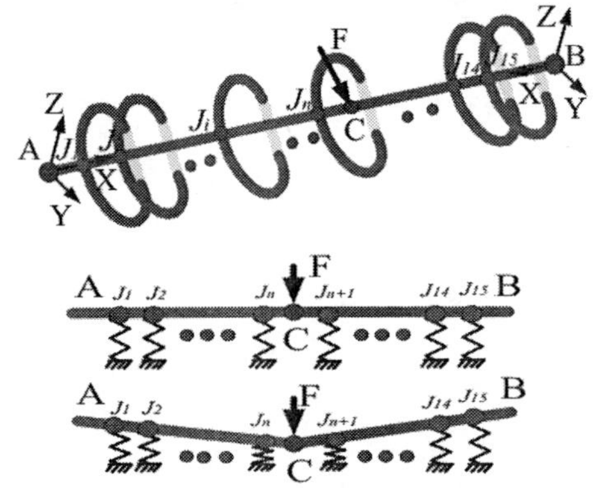

Figure 2. The trachea skeleton

What is meant by skeleton decisions and skeleton design is the process of arriving at a suitable system to support a form or shape and to prevent it from collapsing [6]. In order to compute the real-time global deformation of the trachea efficiently, we present a unique skeleton model for the trachea to simulate force and deformation when the contact occurs. The trachea skeleton involves a serial of highly elastic cartilage rings (they may become calcified in advanced life), the ligament layer between the cartilage rings, and the smooth muscle connecting the two ends off the incomplete cartilaginous ring open in the back.

And the ligaments tend to be more flexible and be able to accommodate movement, while cartilage rings tend to be rigid and unable to accommodate movement to avoid some damage to the skeleton. So the skeleton can guarantee the stability by being placed in such a way as to procure balance when contacted.

As the ligament tissue is a longitudinal and transverse layer, once a contact point C is formed, between the trachea fork end A and the larynx end B, two linear continuous ligament beams AC and BC are established at the same time, which are parallel with the central line of the tube and joint the cartilage rings through the ligament layer. And the bottom elastic muscular layer resists the pressure and supports the cartilage rings, which can be viewed as a spring mat to sustain a serial of two-hinge arches. To simplify this case, we design a continuous ligament beam with spring supports to simulate the skeleton deformation as shown in Fig.2.

In our application, the force *F* applied to the contact point can be regarded as applied to the common end C of two separate beams. Establish two coordinate systems A (B)-*xyz* for the two beams respectively, where the origin is set at one fixed end A (B) of the beams, and x-axis is along the central line of each beam before it is bent and pointing to the contact point C. The y-axis follows the bending force direction, and the *z*- axis is orthogonal to both *x* and *y* axis following the right-hand rule.

The contact force F as the sum of the force FAC and FBC applied to the beam AB and BC, so the deformation of each beam can then be computed separately. In this case, the two forces can be both decomposed into the bending case along *x*-axis direction and compression/expansion case along *y*-axis direction. According to the Bernoulli-Euler bending beam theory, the relation between the distributed forces of each beam can be expressed as:

$$F_{AC}^x : F_{BC}^x = \frac{D_x S_B}{L_{AC}} : -\frac{D_x S_B}{L_{BC}} = -\frac{L_{BC}}{L_{AC}} \quad (2)$$

$$F_{AC}^y : F_{BC}^y = \frac{3\,D_y E_B I_B}{L_{AC}^3} : \frac{3\,D_y E_B I_B}{L_{BC}^3} = \frac{L_{BC}^3}{L_{AC}^3} \quad (3)$$

where D_x, D_y are the displacements of the contact point C along *x*-axis and *y*-axis; LAC, LBC are the lengths of two beams, EB and IB are the Young module and the moment of inertial of the ligament beam, and SB is the cross-section area, whose diameter is the thickness of the ligament layer. Note that plus in the equation (2) show one beam is contracted and the other is extended.

The two ligament beams connect the cartilage ring *i* (i=1, 2, …, 15) with the joint J_i. So the pressure P_i from the ligament beams can not only make the joint J_i

displacement but also change the shapes and positions of its cartilage ring. For each cartilage ring open in the back, its two ends are connected by a muscular bar viewed as a spring tie, which allows the cartilage ring to stretch and contract in breathing or swallowing. Seen in Fig. 3. We can compute general contact/deformation effects resulted from the beams or arches based on Structure Analysis Theory, as detailed in the following sections.

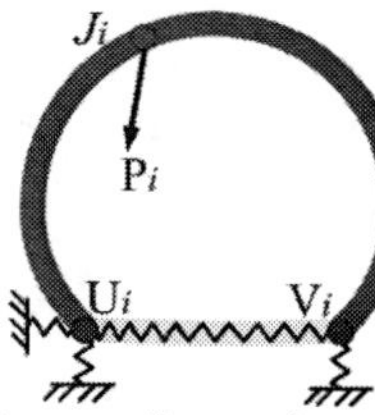

Figure 3. The two- hinge cartilage arch model

2.3 Skeleton shape deformation

We model the shape deformation of the elastic trachea under the contact force from the virtual rigid object. As the trachea skeleton is a statically indeterminate structure with redundant, Force Method as a general approach is used to analyze the skeleton with the determination of the degree of static indeterminacy and identification of the redundant [7].

In this section, we consider how to use Force Method to compute the deformation of the joint supports under the continuous ligament beam, which is commonly subjected to transverse loads. We shall ensure the degree n of statically indeterminate beam, so it is necessary to find n stress resultants in the beam from conditions of compatibility of displacements. That is, the beam is statically determinate when n external reaction supports have to be released. In this case, each redundant i in the continuous ligament beam is assumed to be replaced by the joint force P_i, so a stable and statically determinate skeletons becomes a simply supported beam with the fixed end as shown in Fig. 4.

The n support forces will be introduced at released supports from J_i to J_n along the continuous beam; thereby the beam comprises n spans of length l and one span of length a near the contact point. And at any joint J_i $(1 \leq i \leq n)$ of the continuous beam there is a joint force P_i, which produces the joint deflection in the beam. The joint bending moment m_i from the support i and the load bending moment $m_{F_{AC}^y}$ can be calculated by the compatibility equations [8] based on Diagrammatic Multiplication Method [7]. Note the contact force component F_{AC}^x have no moment effect on the continuous beam.

$$m_i = ilp_i$$

$$m_{F_{AC}^y} = (nl + a)F_{AC}^y$$

And each flexibility influence coefficient means a displacement made by a unit support force, which is along the direction of the support force on one joint.

$$\delta_{ij} = \delta_{ji} = \int \frac{m_i m_j}{E_B I_B} dx = \frac{(3j-i)i^2 l^3}{E_B I_B} \quad (1 \leq i \leq j \leq n)$$

where the first subscript of this influence coefficient means the index of the joint site and its displacement vector is along the direction of the elastic force of the joint spring support, and the second subscript denotes the index of the support force. The continuous beam has a constant Young's Module EB, moment of inertia IB.

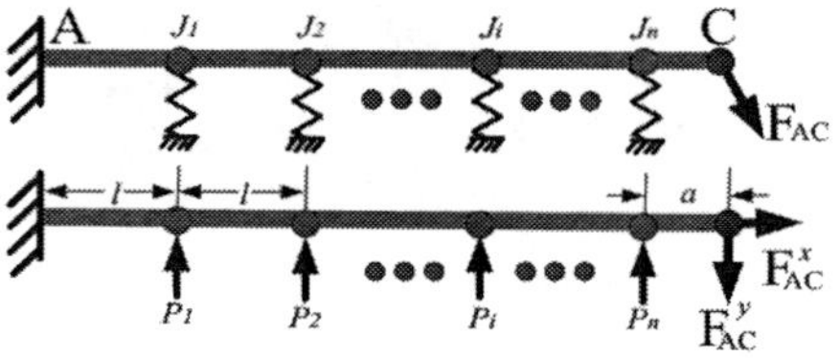

Figure 4. Continuous beam model and redundant force

And then we set up the flexibility matrix δ of the released beam to express n simultaneous equations as following.

$$\delta = \begin{bmatrix} \delta_{11} + \frac{1}{k} & \delta_{12} & \cdots & \delta_{1n} \\ \delta_{21} & \delta_{22} + \frac{1}{k} & \cdots & \delta_{2n} \\ \delta_{n1} & \delta_{n2} & \cdots & \delta_{nn} + \frac{1}{k} \end{bmatrix}$$

Similarly, the deflection $\Delta_{iF_{AC}^y}$ at the release i caused by the applied contact force F_{AC}^y are calculated as

$$\Delta_{iF_{AC}^y} = \int \frac{m_i m_{F_{AC}^y}}{E_B I_B} dx = \frac{3(n+a) - i}{6(n+a)E_B I_B} i^3 l^3 F_{AC}^y$$

where $m_{F_{AC}^y}$ is the bending moment distribution in the released skeleton due to the applied loading.

The compatibility function under the deformation condition at the releases is the following equation, and the joint force P_i at each resultant i is what we want to find.

$$\sum_{i=1}^{n} \delta_{ij} P_i + \Delta_{iF_{AC}^y} = 0$$

That is,

$$\begin{bmatrix} \delta_{11} + \frac{1}{k} & \delta_{12} & \cdots & \delta_{1n} \\ \delta_{21} & \delta_{22} + \frac{1}{k} & \cdots & \delta_{2n} \\ \delta_{n1} & \delta_{n2} & \cdots & \delta_{nn} + \frac{1}{k} \end{bmatrix} \begin{bmatrix} P_1 \\ P_2 \\ \cdots \\ P_n \end{bmatrix} + \begin{bmatrix} \Delta_{1F_{AC}^y} \\ \Delta_{2F_{AC}^y} \\ \cdots \\ \Delta_{nF_{AC}^y} \end{bmatrix} = \begin{bmatrix} 0 \\ 0 \\ \cdots \\ 0 \end{bmatrix}$$

So we can obtain the displacement d_i^y of the each joint J_i, which is always along the direction of the joint load component F_{AC}^y.

$$d_i^y = \frac{P_i}{k}$$

where the internal stiffness k is achieved by the combination of the cartilage materials and the bottom muscular layer. To simplify the case, we let k equal to the elastic coefficient k_M of the muscular layer approximately because the cartilage material is more rigid than the muscular tissue.

However, the external force FAC applied to a ligament beam is not necessarily normal to the beam central line.

2.4 Global deformation rendering

The trachea model has a general surface and let S be the set of its surface. Such a surface S can be partitioned into many single-surface patches S_k, which represents the region between two neighbourhood cartilage rings. As shown in Fig.5, the contact point (if it exists in this patch), two joints (J_i, J_i +1)and four ring ends (V_i, V_i +1, U_i, U_i +1) are viewed as the key points in this patch S_k. So we can divide the patch S_k into three regions G_i (U_i, U_i +1, J_i, J_i +1) , G_i +1 (J_i +1, J_i, V_i, V_i +1) and G_i +2 (V_i, V_i +1, U_i +1, U_i) according to the distribution of the ligament beam and the muscular layer.

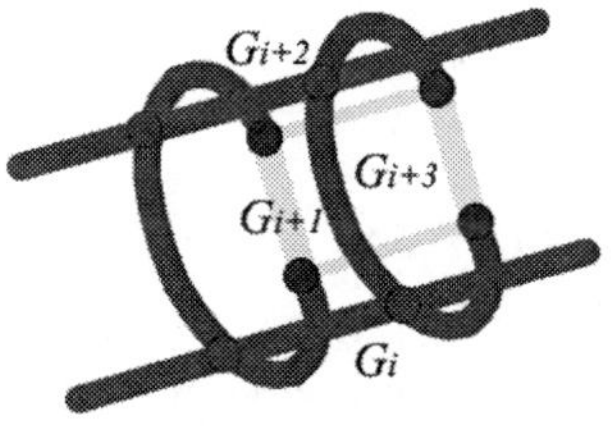

Figure 5. The Region distributions in the S_k patch

Phong shading method is an interpolation method, which is used for linear interpolation of vectors at vertices bounding a polygon across internal points of the polygon [9]. In each region G of the surface patch S_k, let θ_i represents the displacement angle of key point i, which means the angle between the direction of displacement and the outward normal direction of the key point i. So the directions of displacement vectors across the region G are obtained by linearly interpolating the displacement angles θ_i in the same region, and the magnitudes of displacement vectors across the region G are obtained by linearly interpolating the magnitudes of displacement vectors of the key points in this region. We achieve the deformed shape by performing the above shading on all surface patches. The global shape deformation of the trachea model obtained from interpolating these displacements is smooth and satisfies that the closer a surface point to the contact point, the greater the deformation change is at this point [4].

3 Implementation and test results

In this section, we show a virtual trachea surgery method to simulate the actual surgical cases. Our implementation is done in a virtual environment on a PC with dual Intel Xeon 3.0 GHz Processors and 2 GB system RAM. The ultimate goal of this study is to obtain a real-time virtually simulated result of the trachea deformation. In order to accomplish this, we apply the virtual fingers and tools controlled by keys on the computer keyboard or via a haptic device for the real-time interaction to test the simulation algorithm.

The operator controls the virtual rigid finger move and touch the model surface. There, we assume the finger can press or draw the trachea and move with it together. And the Fig. 6 shows the contact force/deformation rendering between the rigid finger and the trachea. As for the compliant motion, we will research in the future work.

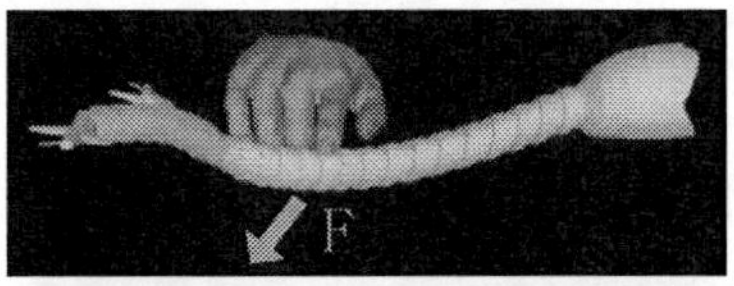

(a) The haptic rendering under the normal force

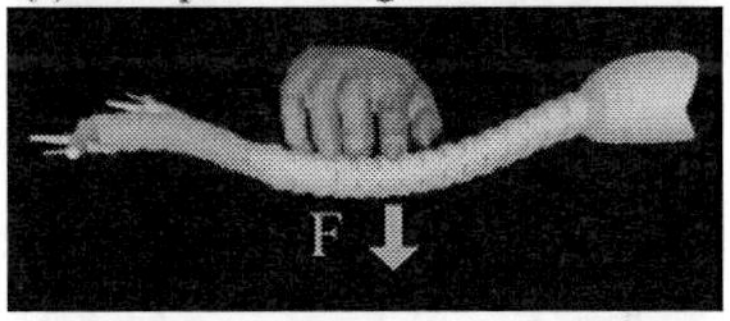

(b) The haptic rendering under the force pointing to left-down direction

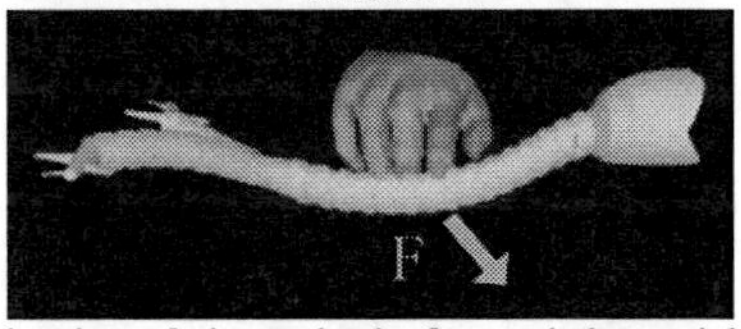

(c) The haptic rendering under the force pointing to right-down direction

Figure 6. Trachea haptic rendering in the simple contact

Fig. 7 describes the above experiment and represents the relations between the force and deformation in the experiments. When the contact occurs, each joint load P_i (from the 1th to the 15th) has two components: P_i^x and P_i^y. Here during the pressing, the value of each load P_i is changed from zero to the maximum. The nearer to the contact point, the larger the magnitude of the force component P_i^y is. Here, the 6th joint achieves the largest force P_6^y (in the Fig. 7a). And then the other force component P_i^x has the approximately same value at each joint. Similar to the case of the joint load Pi, the deformation D_i (D_i^x , D_i^y) of each joint i can also be shown in Fig. 7b.

Here the deformation component d_i^x of each joint is proportional to the distance from the contact point. Due to the normal force, the increasing of the deformation d_i^y is faster than that of the d_i^x component. And for the ends of the cartilage ring, the greater force causes the larger deflection when pressing, and we take the 6th cartilage ring for the example as shown in Fig. 7c.

DOI: 10.1051/matecconf/20166818006

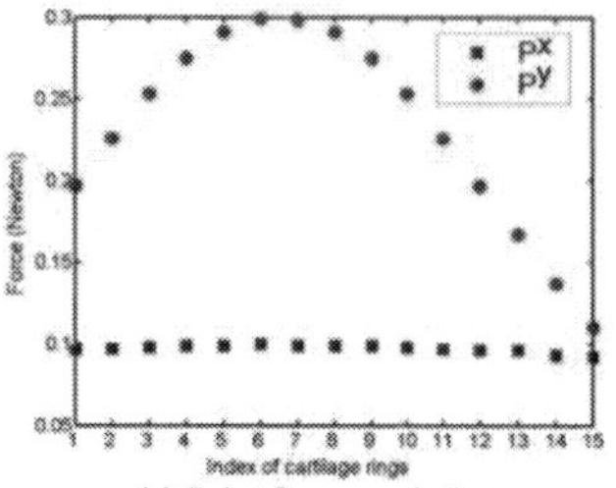

(a) Joint force rendering

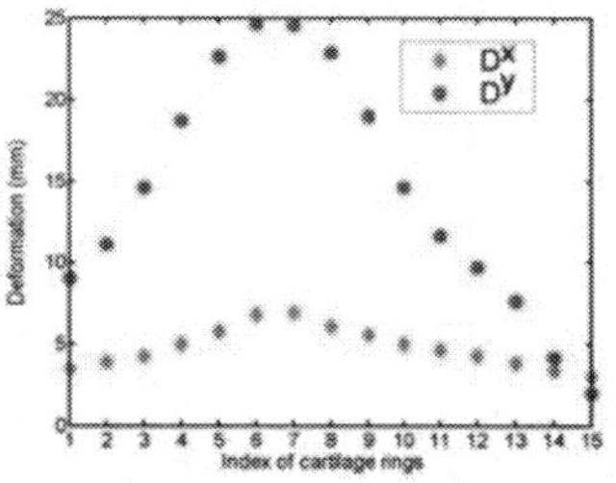

(b) Joint deformation rendering

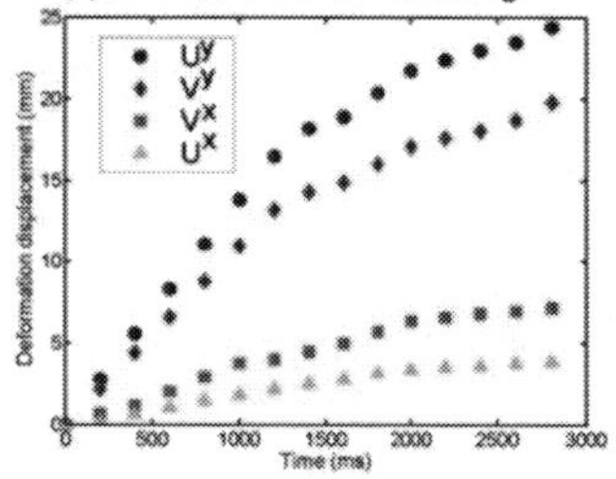

(c) Comparison of the deformation in two ends of the sixth cartilage ring

Figure 7. Force/deformation rendering

Table 1. Our experimental parameters

Our experimental parameters		
Symbols	**Parameters**	**Related value**
M_0	Trachea mass	30g
L_0	Trachea length	160 mm
D_0	Inner diameter of the trachea	20 mm
h_C	Thickness of the cartilage rings	2.6 mm
h_M	Thickness of muscular layer	2 mm
A_C	Area of the cartilage rings	4 mm^2
A_L	Area of the cartilage ligament	6 mm^2
E_C	Young Module of cartilage rings	5.8 MPa
E_L	Young Module of ligament	0.65 MPa
β	Friction parameter	0.66

In all experiments, we use the real-time GJK method [10] to realize the contact detection and provide the distance information between the two objects, and then compute the penetration depth by Duffing Equations. The computation about skeleton model has an update rate of approximately 10 kHz. However, it is negligible compared with the time needed for real -time contact detection and shape updating for the trachea and rigid tool models in the order of kHz, where the number of triangles in the trachea mesh model is 6856. Table 1 lists the parameters used in our experiments.

4 Conclusions

Trachea is a non-homogenous tube form larynx to the bronchial tubes, which consists of a supporting layer of connective and ligament tissue in which elastic cartilage rings are embedded and encircle the front of the tube. In this paper, we presented a novel skeleton model for a real-time virtual trachea surgical simulation. Our focus is to establish the trachea model, and use the skeleton approach to simulate the trachea deformation. The approach is able to model one contact applied by the virtual rigid finger, which can be handled by a user via a haptic device, and achieve the global or local shape deformation of the trachea model due to contacts. In the virtual surgery system, we implemented the approach and demonstrated its effectiveness, it is possible to perform surgical maneuvers at a speed approaching that of the actual procedure, while experiencing a sense of touch.

Future research will focus on implementing an implicit integration scheme for the simulation of visco-elastic materials. And considering the incision to keep the surgery integrity.

Acknowledgment

The research work was supported by National Natural Science Foundation of China under Grant No. 61402426 and partially supported by Collaborative Innovation Center of Novel Software Technology and Industrialization.

References

1. J. Wu, C. Dick, R. Westermann, *Efficient collision detection for composite finite element simulation of cuts in deformable bodies*, Visual Computer, **29**(6), pp. 739-749 (2013)
2. L.P. Nedel and D. Thalmann, *Real time muscle deformations using mass-spring systems*, Computer Graphics International, pp. 156–166 (1998)
3. J. Russo, R. Robinson and M. J. Oldham, *Effects of cartilage rings on airflow and particle deposition in the trachea and main bronchi*, Medical Engineering and Physics, **30**(5), pp. 581-589 (2008)
4. Q. Luo and J. Xiao, *Contact and Deformation Modeling for Interactive Environments*, IEEE Transactions on Robotics, **23**, 3, pp.416-430 (2007)
5. Z.R. Liu, Z.Z. Teng, Y.Q. Wang and W.C. Tang, *The circumferential residual strain and its nonuniform distribution in rat trachea*, Chinese Journal of Biomedical Engineering, **22**, 2, pp. 149-152 (2003)
6. Y.C. Fung, *Biomechanics*, pp.400-406 (1981)
7. L.C. Teague, Jr., C.W. Pidgeon, *Structured analysis methods for computer information systems* , 1985

8. A. E. Jr. Anuta, *Compatibility of the modal equations, Computers and Structures*, **33**, 5, pp. 1229-1232 (1989)
9. A.Watt, *3D Computer Graphic*, 3rd ed. Reading, MA: Addison Wesley, 2000
10. S. Cameron, *Enhancing GJK: Computing minimum and penetration distances between convex polyhedra*, Proc. IEEE International Conference on Robotics and Automation, pp. 3112–3117 (1997)

The Study of Land Use Classification Based on SPOT6 High Resolution Data

Wu Song[1], Jiang Qigang[2]

[1]College of Earth Sciences, Jilin University, Changchun, China
[2]College of Geo-Exploration Science and Technology, Jilin University, Changchun, China

Abstract. A method is carried out to quick classification extract of the type of land use in agricultural areas, which is based on the spot6 high resolution remote sensing classification data and used of the good nonlinear classification ability of support vector machine. The results show that the spot6 high resolution remote sensing classification data can realize land classification efficiently, the overall classification accuracy reached 88.79% and Kappa factor is 0.8632 which means that the classification result of support vector machine is ideal and better than other traditional image classification method. So, the method which is used high-resolution satellite provide a rapid and feasible way for classification of land use types.

1 Introduction

Research of land use classification provides the basic work for technical support, such as land planning and management, land change mechanism analysis and environmental protection. Remote sensing technology has become the most effective means for the acquisition of land use information as the technology has many characteristics , such as macroscopic, dynamic and rapid .At the same time , using satellite remote sensing data for automatic classification of land use and thematic information extraction has been the forefront direction of remote sensing technology application [1], [2]. So many scholars at home and abroad carried out the research about this, and the support vector machine (SVM) technology has been widely used in the automatic classification of land use with its characteristics of small sample training, support high dimensional feature space and fast convergence.

With the rapid development of high resolution remote sensing technology, the use of high resolution remote sensing data at home and abroad for the research of automatic classification of land is increasing, and has already obtained rich success [3], [6]. SPOT satellite is wide band, high spatial resolution remote sensing satellite and has been applied in many industries due to its good application performance, but its application is less applied for the automatic classification and information extraction of land use. In this paper, the high resolution remote sensing image and image automatic classification technology has combined organically [7], [8]. Analysis of the spectral characteristics and high resolution of SPOT6 satellite based on support vector machine (SVM) classification principle which realize land information rapid extract classification, that provide evidence for monitoring land use situation, formulate comprehensive control measures and use policy [9].

2 Introduction in the study area and SPOT6 data preprocess

2.1 The geographical situation in the study area and the selection of the test data

The study area is located in the middle of Morocco, this experiment selected SPOT6 satellite images in May 23, 2013 as the remote sensing data. The image range from 30 ° 29 '34 "N ~ 30 ° 33' 56" N to 527 ° 08 '59 "W ~ 09 W ° 05' 07" and we choose a typical area in this area. The total of land use category in the area is 5,its mainly has plough, forest land, construction land, water and other classes, so it can test the classification method effectively as shown in Fig. 1.

Figure 1. SPOT RS images of study area.

2.2 SPOT6 high resolution data preprocess

Spot 6 earth observation satellite was produced by the European space technology company in September 9, 2012, and it was lunched by the PSLV carrier rocket in India successfully. In 22th September, SPOT6 entered the orbit that in the same orbit plane with Pleiades 1A satellite which was 695 Km high, after January 2013, the Satellite was into the formal business operation. The satellite can obtain image data of the spatial resolution of panchromatic 1.5M and multi-spectrum 6M, and it can receive 600 square kilometers image a day. The service objects mainly distributed in ecological environmental, geology and mineral resources, agriculture, forestry, environmental protection and disaster monitoring, telecom network planning, surveying and mapping, city planning and national defense [10], [11].

Table 1. Satellite parameters of SPOT6.

Bands	Spectral range (μm)	Resolution (m)	Imaging Swath (km)	Angle of incidence (°)
1	0.45~0.52	6	60	±30
2	0.53~0.59	6		
3	0.62~0.69	6		
4	0.76~0.89	6		
5	0.45~0.75	1.5		

2.3 Remote sensing data preprocess

2.3.1 Band synthetic

When using remote sensing technology to extract information, it is necessary to study the spectral characteristics which will be beneficial for the combination of band that identify the target. In this test, first of all, in the experimental zone, the spectral characteristics of each band of multi-spectral data will be analyzed ,then calculate the mean, standard deviation and information entropy of the image grey value respectively, add up and compare the information reflected by the band, as it is shown in Table 2.

From the statistical analysis of the results ,it can be seen that the statistical indicators of the fourth band is greater than the other bands, which means that the information concluded in the fourth band is the greatest, and the fourth band (near infrared) has a significant role in the vegetation type classification, in the same time it has minimum correlation with other bands, so through combining the bands together based on the forth band which is fixed on the green channel, we get experimental comparison results, then integrate with experimental zone concluded with the class situation, select 431 band as the band combination (figure 1a), the feature difference of this combination is big and it contains rich amount of information which is beneficial for visual discrimination and the study of classification automatically.

Table 2. Basic statistical information of multispectral images

Basictats	Min	Max	Mean	Stdev	Eigenvalue
Band 1	0	788	339.878473	97.053276	126.748302
Band 2	0	1359	603.766984	198.886201	219.758493
Band 3	0	1243	538.598726	189.697247	63.63284
Band 4	0	1357	573.856963	211.784032	45.268089

2.3.2 Image fusion processing

Image fusion technology can make use of the different characteristics of the data in a maximize way in order to improve the visual effect of image and the ability of image feature recognition and make the image has a higher spectral and spatial resolution in the same time[12]. Make the panchromatic and multi-spectral data fusion processing can let the SPOT6 data play a role in a maximum way, there are many algorithms for remote sensing data fusion. Principal component analysis (pca), IHS transform method and wavelet transform method were selected for fusion of resolution in this text. And the information after the process of fusion was calculated (table 3). From table 3 we can see that 3 kinds of fusion method have their advantages and disadvantages respectively, but wavelet transform method is superior to other fusion methods in spectral information keeping and peak signal to noise ratio. Therefore, the method of wavelet transform was adopted to process image fusion in this text based on the aim of image automatic classification.

Table 3. Satellite parameters of SPOT6.

	Band	Mean	Standard deviation	Snr	Entropy	meangradient
PCA	R	228.159	37.455	9.435	4.518	4.235
	G	454.573	85.401	12.163	5.034	5.435
	B	239.548	48.629	10.006	4.229	5.654
IHS	R	231.617	99.102	10.801	4.309	4.345
	G	500.855	196.096	12.436	5.805	5.257
	B	253.535	99.644	9.925	4.336	4.286
WAVELET	R	333.918	89.633	12.354	4.011	4.387
	G	698.465	187.232	13.198	5.903	5.376
	B	355.011	109.918	8.765	4.678	5.587

3 Image classification principle based on support vector machine

Support vector machine (SVM) classification algorithm is a kind of machine learning algorithms based on statistical learning theory, which is used structural risk minimization principle of SRM through solving quadratic programming problem under the inequality constraints [13]-[15].

Structural classification hyperplane in the training set T= { (,) ,……(,)} , , {-1, 1} , i=1, 2, ...n. Assuming that the classification of equation is <X• >+b=0, the equation should be satisfied with type (1):

$$Y_i\ (<X \bullet \omega>+b)\ -1 \geq 0 \qquad (1)$$

The class interval according to analytic geometry is $D=2/\|\omega\|$.

The problem can be converted into that introduce Lagrange function which is used for solving this optimization problem in order to make the function $\phi(\omega)=\|\omega\|^2/2$ minimum[16~17].

$$L = \|\omega\|^2/2 - \sum_{i=1}^{n} \alpha_i Y_i (< X_i \bullet \omega > +b) + \sum_{i=1}^{n} \alpha_i \qquad (2)$$

the α_i>0 is the Lagrange multiplier, the solution of the problem must be satisfied type (3) according to the KKT conditions:

$$\alpha_i \{[< X_i \bullet \omega > +b] Y_i - 1\} = 0 \qquad (3)$$

Therefore, the resulting discriminant function (4):

$$f(X) - \mathrm{sgn}\{\sum_{i=1}^{n} Y_i \alpha < X_i \bullet X > +b\} \qquad (4)$$

In general case, most of α_i are 0, the others are not 0, the samples which the α_i corresponded is SV, b calculated by whichever SV. Considering some samples that could not be classified by the hyperplane correctly, convert the optimization problem to constrain condition by introducing slack variable:

$$Y_i(< X_i \bullet \omega > +b) \geq 1 - \xi_i\ ,\ \ \xi_i \geq 0 \qquad (5)$$

In the constrain condition above, there is $\min \frac{1}{2}\|\omega\|^2 + C\sum_{i=1}^{n}\xi_i$ (C>0) .

In the result, the most α_i are 0, referrer the sample that α_i is not 0 as support vector. The function which the support vector defined is SVM. Usually, we called the training sample which has a small amount samples as support vector that means SVM has the advantages of sparseness, so the classification speed of SVM is better than others[18-21].

The essence of SVM: first of all, to transform the input space into a higher dimensional space through the nonlinear transformation which is defined by the appropriate product function, then to achieve the optional linear classification surface by linear regression in the higher dimensional space.

4 The process of SVM image classification

4.1 Remote sensing data preprocess

Support vector machine (SVM) classification algorithm is a kind of machine learning algorithms based on statistical learning theory, which is used structural risk minimization principle of SRM through solving quadratic programming problem under the inequality constraints[13-15].

First of all, feature extraction for high resolution data, then selecting the features as extraction algorithm, such as the mean、 the standard deviation and k-l transform.

Selected the interested area of the image which the training and choice of the size of texture window is the result of test for many times. If the size is too small that will make not only the training speed reduced strongly, but also the result has no significant change; oppositely, if the size is so big that the precision will be low [29].

As it is shown in t Fig. 2.

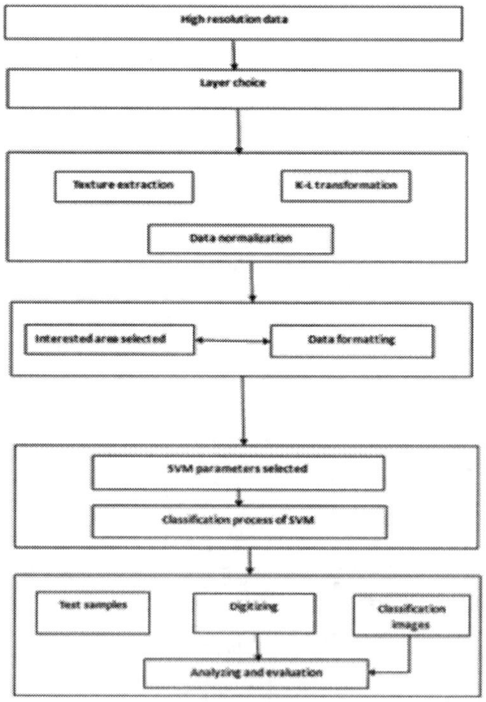

Figure 2. The flow chart of feature extraction.

4.2 Classification results and validation of precision

4.2.1 Image fusion processing

Table 4. The confusion matrix of maximum likelihood classification.

Category	Water	Plough	Forestland	Construtions	Others	Amounts
water	604	0	4	0	25	633
Plough	0	651	183	6	0	840
Forestland	0	16	418	4	0	438
Construtions	0	7	9	69	0	85
Others	0	170	102	0	107	379
Amounts	604	844	716	44	132	2375

Table5. Confusion matrix of Markov distance method of classification.

Category	Water	Plough	Forestland	Construtions	Others	Amounts
water	469	0	166	3	0	638
Plough	0	914	0	0	0	914
Forestland	0	0	504	10	0	514
Construtions	145	0	0	60	0	205
Others	171	221	118	6	132	648
Amounts	604	1135	788	79	132	2919

Table 6. Confusion matrix processed by SVM classification used by spectral characteristics.

Category	Plough	Forestland	Water	Construtions	Others	Amounts
Plough	299	32	0	1	0	332
Forestland	50	259	0	8	0	317
Water	0	0	296	0	0	296
Contructions	0	0	0	61	0	61
Others	0	0	0	19	72	91
Amounts	349	291	296	89	72	1097

Table 7. Comparing with classification precision.

Methods	Overall accuracy (%)	Kappa Coefficient
Maximum likelihood method	77.85	0.7013
Mahalanobis distance method	71.22	0.6218
SVM classification method	89.79	0.8632

In order to verify the applicability of the SVM which is used for high resolution image classification, using the markov distance method and the maximum likelihood for classification , Calculating the belonging category and making the land use classification figure of test area to the study area, as it is shown in Fig. 3.

(a)SVM (b)Maximum likelihood classification
(c)Mahalanobis distance classification

4.2.2 Classification accuracy evaluation

This experiment adopts widespread confusion matrix method to analysis classification results, selected the test sample randomly corresponding to various land use types In remote sensing image, then calculated its classification confusion matrix and its related precision index respectively based on the different results of classification above. The results as shown in table 4,5,6.

The comparison table shows(Table 7): the classification accuracy of application of SVM method is superior to maximum likelihood classification and Markov distance classification method, which verified the superiority of support vector machine (SVM) on the nonlinear classification problem of small sample. The overall classification accuracy Kappa coefficient reached 88.79% and 0.8632.

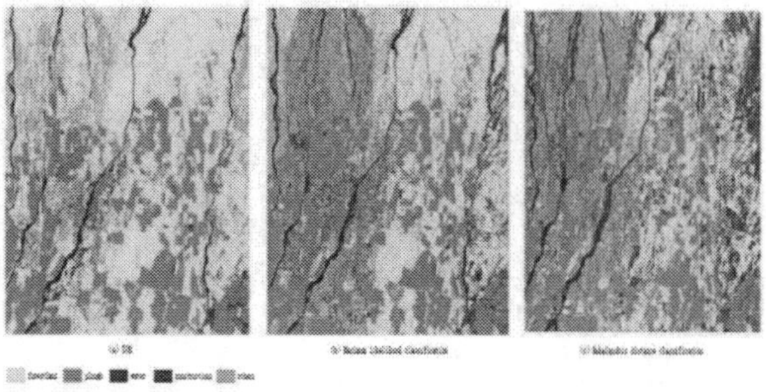

Figure 3. Comparison of classification result.

5 Conclusion

The method which realizing the rapid division of land use type based on SPOT6 high-resolution satellite data and image automatic classification technology has improved the recognition efficiency of agricultural land types.

Spot6 data was processed by training sample and predicting classification through their spectral information and the SVM classification method. The classification results shows that not only the algorithm precision of support vector machine is superior to the traditional classification algorithms, characterized by strong adaptability. The phenomenon of fault classification and miss classification is less, but also it has a high degree of stability. Therefore, in remote sensing image classification, the selection of support vector machine method for land use to classification research based on high resolution images can improve the classification precision and has a great advantage.

References

1. Luo Jiancheng, Zhou Chenghu, Yang Yan. Land-cover and land-use classification based on remote sensing intelligent Geo-interpreting model[J].Journal of Natural Resources, 2001, **16(2)**:179~183.(in Chinese)
2. Li Xiubin. A review of the international researches on land use/land cover change [J].ACTA Geographica Sinica, 1996, **51(67)**:553~557.(in Chinese)
3. Mai Kaile, Zhang Wenhui. Object-oriented classification approach for remote sensing imagery information extraction in loess hilly-gully region[J].Transactions of the Chinese Society for Agricultural Machinery2011, **42(4)**:153-158. (in Chinese)
4. Zhao Chunhui, Qiao Lei. Classification of hyperspectral remote sensing image using improved LS-SVM[J]. Applied Science and Technology, 2008, **35(1)**:44~52.(in Chinese)
5. Fu Wenjie, Hong Jinyi, Lin Mingsen. A method of land use classification from remote sensing image based on support vector machines and spectral similarity scale[J]. Remote Sensing Technology and application, 2006, **21(1)**:25~30.(in Chinese)
6. Du Peijun,Liu Sicong, Zheng Hui. Land cover change detection over mining areas based on support vector machine [J]. 2012, **41(2)**:262~267. (in Chinese)
7. Sun Danfeng,Yang Yihong, Liu Shunxi. Application of high-spatial IKONS remote sensing images in land use classification and change monitoring[J]. Transactions of the CSAE, 2002, **18(2)**:160~164. (in Chinese)
8. Chen Qihao, Liu Zhimin, Liu Xiuguo, et al. Element-oriented land-use classification of mining area by high spatial resolution remote sensing image[J]. Earth Science — Journal of China University of Geosciences, 2010, **35(3)**:453~458. (in Chinese)
9. Zhou Pei, Zhou Shenli. Effect of land use on ecological benefit of farm belt in suburbs[J]. Journal of Ecology and Rural Environment, 2007, **23(4)**: 6~10. (in Chinese)
10. Zhou Yi, Wu Juan, Li Qi, et al. Test and analysis for detecting land use change by using CBERS-02C satellite image[J].Mineral Exploration.2012,**3(5)**: 688~694. (in Chinese)
11. Ma Ligang, Zhang Leping, Zheng Jinsong, et al. Land use classification using ZY1- "02C" remote sensing images [J]. Journal of Zhejiang University: Engineering Science ,2013, **47(8)**:1508~1516. (in Chinese)
12. Wang Haihun, Peng Jiaxiong, Wu Wei, et al. A study of evaluation methods on performance of the multi-source remote sensing image fusion[J].Computer Engineering and Applications,2003,(25):33~37. (in Chinese)
13. hang Xuegong. Introduction to statistical learning theory and support vector machines[J]. Acta Automatica Sinica, 2000, **26(1)**:32~42. (in Chinese)
14. Vapnik V N. The Nature of Statistical Learning Theory[M].Berlin: Springer-Verlag Berlin Heidelberg,1998.
15. Hsu C, Lin C.A comparison of methods for multiclass support vector machines[J]. IEEE Transactions on Neural Networks, 2002,**13 (2)**:415~425.
16. Zhang Quanming, Liu Huijin. Application of LS-SVM in classification of power quality disturbances[J]. Proceedings of the CSEE,2008,**28(1)**:106~110. (in Chinese)
17. Yang Jiajia，Jiang Qigang，Chen Yongliang，et al. Lithology division for large-scale region segmentation based on LS-SVM and high resolution remote sensing images[J].Journal of China University of Petroleum(Edition of Natural Science)，2012,**36(1)**:60~67. (in Chinese)
18. Wang Kai,Hou Zhurong,Wang Congli. Intrusion detection based on cross-validation SVM[J]. Journal of Test and Measurement Technology，2010, **24(5)**:419~423. (in Chinese)
19. Zhang Jinshui, He Chunyang, Pan Yao, et al. The high spatial resolution RS image classification based

on SVM method with the multi-source data, 2006,**10(1)**:49~57. (in Chinese)

20. Fauvel M, Chanussot J, Benediktsson J A. Kernel principal component analysis for the classification of hyperspectral remote sensing data over urban areas[J]. EURASIP Journal on Advances in Signal Processing, 2009:1~14
21. Guo Hui, Wang Ling, Liu Heping. Integrating kernel principal component analysis with least squares support vector machines for time series forecasting problems[J]. Journal of University of Science and Technology Beijing, 2006, **28(3)**:303~306. (in Chinese)

A Fast Time-Delay Calculation Method in Through-Wall-Radar Detection Scenario

Zhang Qi [1], Lei Wentai[1], Wang Qianzhe [2]

[1] *School of Information Science and Engineering, Central South University, No. 932, Lushan South Road, Changsha, Hunan, P. R. China*
[2] *Department of Scientific Research, Air Force Engineering University, No.1, Changle East Road, Xi'an, Shanxi, P. R. China*

Abstract. In TWR (Through Wall Radar) signal processing procedure, time delay estimation is one of the key steps in target localization and high resolution imaging. In time domain imaging procedure such as back projection imaging algorithm, round trip propagation time delay at the path of "transmitter-target-receiver" needs to be calculated for each pixel in imaging region. In typical TWR scenario, transmitter and receiver are at one side and targets at the other side of a wall. Based on two-dimensional searching algorithm or solving two variables equation of four times, traditional time delay calculation algorithms are complex and time consuming, and cannot be used to real-time imaging procedure. In this paper, a new fast time-delay (FTD) algorithm is presented. Because of that incident angle at one side equals to refracting angle at the other side, an equation of lateral distance through the wall can be established. By solving this equation, the lateral distance can be obtained and total propagation time delay can be calculated subsequently. Through processing simulation data, the result shows that new algorithm can be applied effectively to real-time time-delay calculation in TWR signal processing.

1 Introduction

TWR is now widely used in various application scenarios, while it ensures that researchers obtain quantitative and qualitative information in the media [1]. The quantitative information is obtained by full wave transposition, including electromagnetic property. Qualitative information can be extracted from TWR echo to estimate properties such as size, shape, permittivity ect. of subsurface target by using inverse scattering imaging algorithms. TWR becomes a hot research area and a lot of achievements have obtained [2,3]. Time delay calculation is very important in time domain imaging. E.M.Johansson and J.E.Mast use linear approximation method to get the electromagnetic waves propagate time-delay [4]. Rappaport.C.M describes that the waves propagate with a wavefront shape that resembles a hyperbola when enter the medium, and a formula is provided to reduce the error between the approximate hyperbola and the exact shape [5]. Lei et al. present a new fast time-delay calculate method based on one-dimensional searching and solving quartic equation [6]. Chen et al. give an approximate round-trip time delay algorithm to reduce the wall clutter [7]. Protiva et al. present a wall parameter estimation method that utilizes time-delay-only measurements of echoes backscattered from a wall [8]. However these methods are complex, time-consuming, and not applicable at real time imaging in TWR, so this paper proposes an improvement method to solve this problem.

In this paper, a fast time-delay calculation (FTD) is presented. By analyzing electromagnetic wave propagation path, we can get the lateral distance in the wall and total path, and then time delay can be calculated. Through processing simulation data, the results show that the algorithm can be applied effectively to time-delay calculation.

2 Theoretical analysis

A physical model of electromagnetic waves propagation in TWR application is shown in Fig. 1. Transmitter and receiver are located at point A and target is located at point B. An impulse signal is transmitted from transmitting antenna located at point A and reached point B after two times reflection at two sides of a wall respectively.

Electromagnetic waves propagate from the starting point A on the one side of the wall, and the incident point C is at the upper surface of the wall, then waves refract out of the wall on the lower side at point D after transmitting a distance in the wall, finally reach the receiving point B. ε_1 is the relative permittivity of the environment at both two sides of the wall, ε_2 is the relative permittivity of the wall, d is the thickness of the wall. We define the top surface of the wall as $z=0$, z is the vertical coordinate, the starting point's coordinate position is (x_a, z_a), and the receiving point's coordinate position is (x_b, z_b).

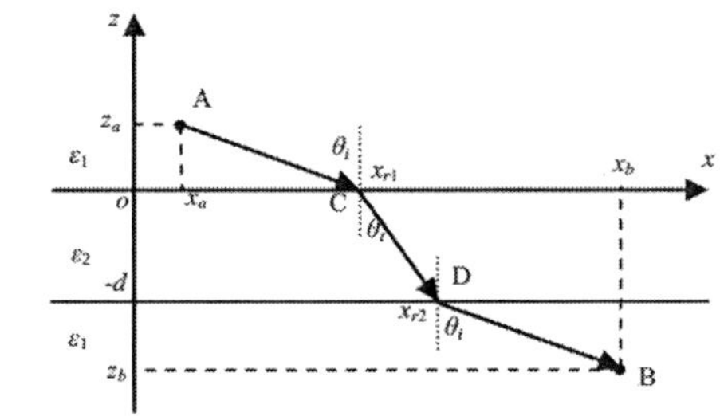

Figure 1. Electromagnetic waves propagation model.

2.1 Traditional calculation method

Traditional method for calculating electromagnetic wave propagation time-delay through wall is finding the minimum value of one-way propagation time-delay by two dimensional searching, shown below:

$$\begin{cases} \min \tau = \frac{1}{c}\Big[\sqrt{\varepsilon_1}\sqrt{z_a^2+(x_a-x_{r1})^2}+\sqrt{\varepsilon_2}\sqrt{d^2+(x_{r2}-x_{r1})^2} \\ \quad +\sqrt{\varepsilon_1}\sqrt{(z_b+d_1)^2+(x_b-x_{r2})^2}\Big] \\ x_{r1}\in[x_a,x_b] \\ x_{r2}\in[x_a,x_b] \end{cases} \tag{1}$$

where x_{r1} and x_{r2} are points' positions of axis X on upper and lower surfaces of the wall.

Or according to Snell's Law, a bivariate quartic equation can be expressed as Eq.2:

$$\begin{cases} \varepsilon_1(x_{r1}-x_a)^2\left[d^2+(x_{r2}-x_{r1})^2\right] \\ \quad =\varepsilon_2(x_{r2}-x_{r1})^2\left[z_a^2+(x_a-x_{r1})^2\right] \\ \varepsilon_2(x_{r2}-x_{r1})^2\left[(z_b+d)^2+(x_b-x_{r2})^2\right] \\ \quad =\varepsilon_1(x_b-x_{r2})^2\left[d^2+(x_{r2}-x_{r1})^2\right] \end{cases} \tag{2}$$

where x_{r1} and x_{r2} are the two refract points' positions of axis X on the wall. Then we get the one-way time-delay by Eq.3:

$$\tau=\frac{1}{c}\Big[\sqrt{\varepsilon_1}\cdot\sqrt{z_a^2+(x_a-x_{r1})^2}+\sqrt{\varepsilon_2}\cdot\sqrt{d^2+(x_{r2}-x_{r1})^2} \\ +\sqrt{\varepsilon_3}\cdot\sqrt{(z_b+d)^2+(x_b-x_{r2})^2}\Big] \tag{3}$$

But the two dimensional searching method is relatively slow and solving the bivariate quartic equation is time-consuming.

2.2 Fast time-delay calculation method (FTD)

A new time delay calculation method (FTD) is presented. Total propagation path can be divided into three parts. The first part is the path in the wall; the other two parts are two paths at both sides. Because of the same environment outside the wall, we can confirm that incident angle at point B equals to refracting angle at point C. Defining the top surface of the wall as $z=0$, the starting point's coordinate is (x_a, z_a),and the receiving point's coordinate is (x_b, z_b), d is the thickness of the wall. The lateral distance between point A and point B is $l_{total}=|x_b-x_a|$, and can also presented as $l_{total}=l_{inner}+l_{outer}$,where l_{total} is the lateral distance in the wall, l_{outer} is the lateral distance outside the wall.

The incident angle at point C is equal to the angle refract at point D, so the ratio of the lateral distance and the vertical distance outside the wall is the same.

Then l_{outer} can be divided into two parts, the distance up the wall and the distance down the wall:

$$l_{outer}=l_{up}+l_{down} \tag{4}$$

where $l_{up}=\frac{|z_a|}{|z_a|+|z_b+d|}l_{outer}$, $l_{down}=\frac{|z_b+d|}{|z_a|+|z_b+d|}l_{outer}$.

According to Snell's Law, we can get Eq.5:

$$\frac{l_{up}^2}{l_{up}^2+z_a^2}\cdot\frac{l_{inner}^2+d^2}{l_{inner}^2}=\frac{\varepsilon_2}{\varepsilon_1} \tag{5}$$

where $\frac{\sin\theta_i}{\sin\theta_t}=\frac{\sqrt{\varepsilon_2}}{\sqrt{\varepsilon_1}}$, $\sin\theta_i=\frac{l_{up}}{\sqrt{l_{up}^2+z_a^2}}$,

$\sin\theta_t=\frac{l_{inner}}{\sqrt{l_{inner}^2+d^2}}$.

In addition, another equation according to the geometric similarity relationship between the incident angle and the refracting angle can be expressed as Eq.6:

$$l_{up}=\frac{|z_a|}{|z_a|+|z_b+d|}\left(|x_b-x_a|-l_{inner}\right) \tag{6}$$

A quartic equation about l_{inner} can be deduced by two simultaneous equations Eq.5 and Eq.6:

$$s_4l_{inner}^4+s_3l_{inner}^3+s_2l_{inner}^2+s_1l_{inner}+s_0=0 \tag{7}$$

where $s_4=c_a^2(\varepsilon_2-\varepsilon_1)$, $s_3=2c_ac_b(\varepsilon_2-\varepsilon_1)$, $s_2=\left[\varepsilon_2(c_b^2+z_a^2)-\varepsilon_1(c_a^2d^2+c_b^2)\right]$, $s_1=-2\varepsilon_1c_ac_bd^2$, $s_0=-\varepsilon_1c_b^2d^2$, $c_a=-\frac{|z_a|}{|z_a|+|z_b+d|}$, $c_b=\frac{|z_a|\cdot|x_b-x_a|}{|z_a|+|z_b+d|}$.

The lateral distance in the wall l_m can be calculated by solving the simple quartic equation, and the time delay can be shown as Eq.8:

$$\tau=\frac{l_p}{c} \tag{8}$$

where

$l_p=\sqrt{\varepsilon_1}\sqrt{l_{up}^2+z_a^2}+\sqrt{\varepsilon_2}\sqrt{l_m^2+d^2}+\sqrt{\varepsilon_1}\sqrt{l_{down}^2+(z_b+d)^2}$,

$l_{up}=\frac{|z_a|}{|z_a|+|z_b+d|}\left(|x_b-x_a|-l_m\right)$, $l_{down}=\frac{|z_b+d|}{|z_a|}l_{up}$.

3 Simulation experiment

In this example, both sides are air, the relative permittivity of air and wall are $\varepsilon_1 = 1$, $\varepsilon_2 = 8$ respectively, the thickness of the wall is 0.5m, the coordinates position at starting point (x_a, z_a) position is (1,0.2) , the imaging region is $x_o \in [-1,3]$, $z_o \in [-3,-1.5]$.

Firstly, the accurate time-delay of the imaging region points calculated by traditional method can be shown as Fig.2 (a).The computation time is 7.3291s on the computer which has 2G RAM and Intel-Core CPU with 2.17GHz frequency. Next applying the fast time-delay calculation method (FTD), the computation time is 0.2726s, which only about 3.72 percent of the tradition method. The relatively error between the traditional method and fast calculation method is about 10^{-5} , and the relatively error is zero in some imaging pixels as shown in Fig.2 (b).So, the FTD method is faster than traditional calculation, but the relatively error is very small.

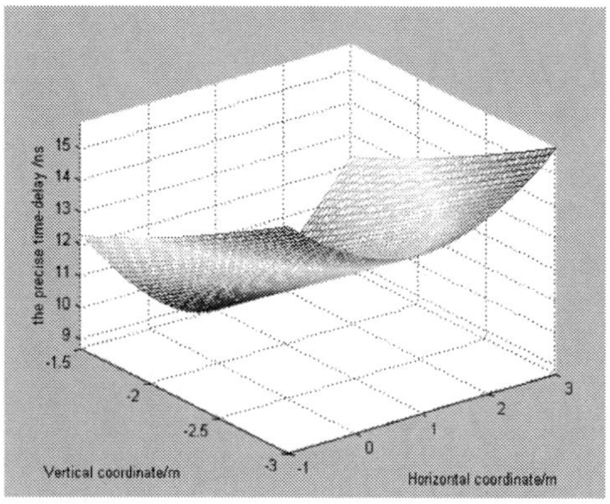

(a)time-delay calculated by traditional method

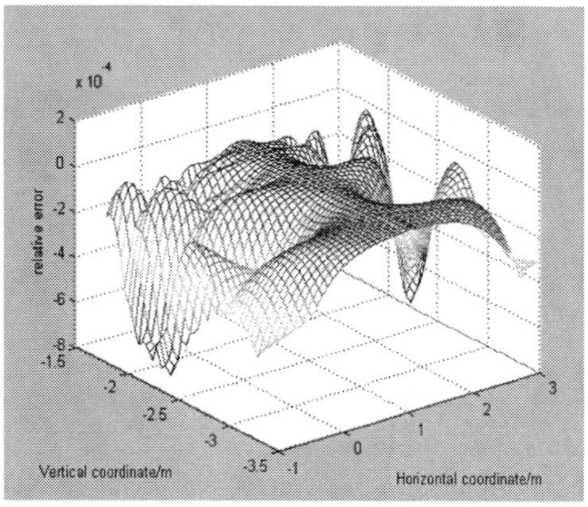

(b)relatively error between method and FTD method

Figure 2. The simulation experiment of FTD method

4 Conclusion

In this paper, a new fast time-delay calculation method is presented. By analyzing the electromagnetic waves propagate path, the lateral distance and total path; the time-delay can be got quickly. Simulation experiment shows the new algorithm is highly efficient and the error rate is low, and suitable for real-time positioning and image processing in TWR.

Acknowledgement

This work was financially supported by Science and Technology Project of Hunan Province (Grant No. 2015JC3043) and Research Fund of Shanxi Key Laboratory of Electronic Information System Integration (Grant No.2015AA018).

References

1. Jol H M.Ground Penetrating Radar Theory and Applications[J]. Ground Penetrating Radar Theory & Applications, (2009):509–524.
2. Dogaru T, Le C. SAR Images of Rooms and Buildings Based on FDTD Computer Models.[J]. IEEE Transactions on Geoscience & Remote Sensing, (2009), **47**(5):1388 - 1401.
3. Ahmad F, Amin M G, Kassam S A. Synthetic aperture beamformer for imaging through a dielectric wall[J]. IEEE Transactions on Aerospace Electronic Systems,(2005), **41**(1):271-283.
4. Johansson E M, Mast J E. Three-dimensional ground-penetrating radar imaging using synthetic aperture time-domain focusing[C]//*SPIE's 1994 International Symposium on Optics, Imaging, and Instrumentation. International Society for Optics and Photonics*, (1994): 205-214.
5. Rappaport C M. Accurate Determination of Underground GPR Wavefront and B-Scan Shape From Above-Ground Point Sources[J]. IEEE Transactions on Geoscience & Remote Sensing, (2007), **45**(8):2429-2434.
6. Wen T L, Yu J S. An Improved Iteration Back Projection Imaging Technique for Surface Penetrating Radar Applications. Applied Mechanics & Materials, (2013), 1509-1513.
7. Chen X, Xue S, Feng D, et al. A new wall clutter reduction strategy for through wall radar imaging[C]//Signal Processing (ICSP), *2014 12th International Conference on. IEEE*,(2014): 2102-2105.
8. Protiva P, Mrkvica J, Machac J. Estimation of Wall Parameters From Time-Delay-Only Through-Wall Radar Measurements[J]. IEEE Transactions on Antennas & Propagation, (2011), **59**(11):4268-4278.

Forex Market Prediction Using NARX Neural Network with Bagging

Nima Shahbazi[1], Masoud Memarzadeh[2] and Jarek Gryz[1]

[1]*Department of Electrical Engineering and Computer Science, York University, Canada*
[2]*Department of Mathematics, Tehran University, Iran*

Abstract. We propose a new methodfor predicting movements in Forex market based on NARX neural network withtime shifting bagging techniqueand financial indicators, such as relative strength index and stochastic indicators. Neural networks have prominent learning ability but they often exhibit bad and unpredictable performance for noisy data. When compared with the static neural networks, our method significantly reducesthe error rate of the responseandimproves the performance of the prediction. We tested three different types ofarchitecture for predicting the response and determined the best network approach. We applied our method to prediction the hourly foreign exchange rates and found remarkable predictability in comprehensive experiments with 2 different foreign exchange rates (GBPUSD and EURUSD).

1 Introduction

The Forex Market is the largest and most liquid market around the word with an estimated $2 trillion traded every day.Modelling the system is difficult as currency market is characterized by significant non-linearity and low signal to noise ratio. Also, foreign exchange rates are affected not only by economic news but alsopolitical and psychological factors. Not surprisingly, predicting the movement of foreign exchange rates or finding a trend in the market is very difficult.

There are several neural network approaches [1] and [2] which ledto successful predictions. However, these approaches use moving averages for the input of their networks and do not use any bagging which,as we will showlater, improves the response prediction. It has been shown in [3] that the relationship between neural networks and conventional statistical approaches for time series forecasting are complementary. It has also been indicated in [4] that traditional statistical techniques for forecasting have reached theirlimit inapplications with nonlinearities in the data set (such as stock indices).Recently, Dynamic Neural Networks such as NARX have been used in many businessapplications especially when the problem domain involves predictions [5] but again, there is no bagging or indicator selection.

This paper proposes a new method for predictingforex market movement by using NARX neural network with bagging technique. Instead of using the next close price or moving averages, we uses RSI (Relative Strength Index) indicator for the output response. We feed data aboutForex news to the networkasthere is strong correlation between the news piece and the stock price behavior[6].

Our contributions can be described as three phases:

Phase 1: We find the most relevant financial indicators to be fed into the network by sensitivity analysis regarding the response, the input/output indicators, and the architecture.Then we train the network with news data with static neural network (Section 0).

Phase 2: Using the same inputs and output,we use the NARX neural Network architecture (Section 0). By usingdynamic network both the error rates as well as the directional success improve(Section 0).

Phase 3: In the last phase, time shifting bagging method (Section 0)is added to the system and shown to improve the test error and directional success rate on the test data.

After presenting the experimental results, a discussion on future research concludes the paper.

2 Architecture and learning algorithm

2.1 NARX neural network as a forecasting tool for Forex

Neural Networks (NN) based prediction is a suitable tool for forecasting in non-linear time seriesapplications. With neural network approach to time series it is not necessary to know any information regarding the cause that generates the signal.

The architectural approach proposed here to deal with Forex market is based on *Nonlinear Autoregressive models with eXogenous input* (NARX model)[7]. All other dynamic networks have either been focused networks, with the dynamics only at the input layer, or feed-forward networks. NARX is a recurrent dynamic network, with feedback connections enclosing several layers of the network. It is based on the linear ARX model, which is commonly used in time-series modeling.This is a powerful class of models which has been demonstrated to be well suited for modeling nonlinear systems and especially time series. It has been shown that in NARX networks learning is more effective than in other neural networks and that these networks converge much faster and generalize better than the other networks [8].

The defining equation for the NARX model is:

$$y(t) = F(y(t-1), \dots y(t-2), y(t-n_y), u(t-1), u(t-2), \dots, u(t-n_u)) \quad (1)$$

where the next value of the dependent output signal $y(t)$ is regressed on previous values of the output signal and previous values of an independent $u(t)$(exogenous) input signal. One can implement the NARX model by using a feed-forward neural network to approximate the function F.

There is another important configuration for NARX model (useful in training) that we use in our approach. Let the output of the NARX network be an estimate of the output of some nonlinear dynamic system that we are trying to model. The output is fed back to the input of the feed-forward neural network as part of the standard NARX architecture. Since the true output is available during the training of the network one could create a series-parallel architecture [9] in which the true output is used instead of feeding back the estimated output as shown in Figure 1. This has two advantages. First, the input to the feed-forward network is more accurate. Second, the resulting network has purely feed-forward architecture and static back propagation can be used for training as our dynamic network is now a feed-forward neural network without any feedback.

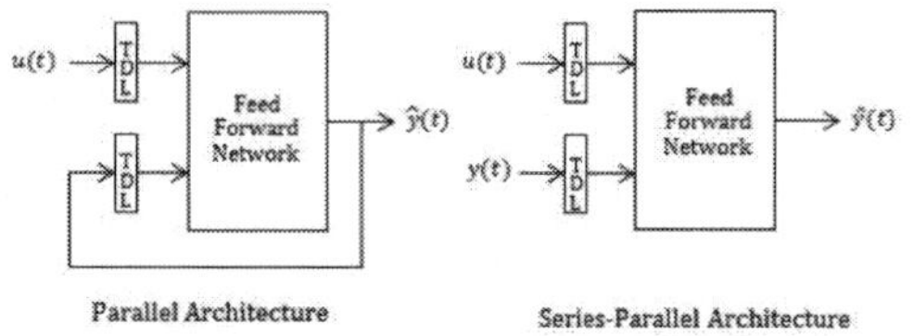

Figure 1. Serial/parallel architecture for NARX model.

2.2 Learning algorithm

For learning, a dynamic back-propagation algorithm is needed to evaluate the gradients. This is more computationally than static back-propagation. In addition, the error can be larger for dynamic than for static networks. Training is more likely to fall into local minima [10]. But asshownin Figure 1we can use instead a series-parallel architecture.Thus, it is possible to use true output instead of the estimated output to train the network with the feedback connections.

The process of training a neural network involves iteratively presenting it with the input data so that it is calibrated and can be used later as a forecasting tool. The objective of the training is to minimize a defined error function, which implies that the neural network fit the input data, given the expected result as outputs. In this approach we use RMSE (Root Mean Square Error) aserror function:

$$RMSE = \sqrt{\frac{1}{N}\sum_{i=1}^{N}(e_i)^2} = \sqrt{\frac{1}{N}\sum_{i=1}^{N}(y_i - t_i)^2} \quad (2)$$

where t_i is the target and y_i is the predicted value. In this paper, the training function that updates the weights and bias values is the Levenberg-Marquardt function. In general, in function approximation problems, for networks that contain up to a few hundred weights, the Levenberg-Marquardt algorithm will have the fastest convergence. This advantage is especially noticeable if very accurate training is required. Also,it was shown that the neural network training can be made more efficient if certain preprocessing steps on the network inputs and targets are performed. The *normalization* of the input and target values simplifies the problem of the outliers in the network. The normalized inputs and targets that are returned will all fall in the interval[-1, 1].Another factor in the neural network isthe transfer function. The best transfer function is chosen by the best performance seen on the test result. In this problem Tangent-Sigmoid makes the learning phase faster and increases the neural network accuracy.

3 Model construction and evaluation

3.1 Dataset and input/output predictors

The prediction is done hourly. The dataset used here is pulled from Alpari-Forex.com[11]. For the news release data, two sources on relevant currencies are used: (1) Fxstreet.com[12], economic calendar from 2001; and (2) Bloomberg.com [13], economic calendar from 2001.

We define six hour window, three hours before the related news release and three hours after the news release, and gather the data (Open, Low, High and Close) every minute in the six hour window for training network. We divided the data into (70%, 15% and 15%) with 70% use for training, 15% for validation,and the rest for testing.

Once the raw data have been chosen by six hour window, a set of indicators based on those values might be developed. In this work we used five indicators for input/output of the neural network shown in Table 1. For all formulas below, the subscript $t\pm 1$ indicates time t plus or minus one hour.

Differential RSI (Relative Strength Index) as Input/Output

1. A technical momentum indicator that compares the magnitude of recent gains to recent losses in an attempt to determine overbought and oversold conditions of an asset. RSI ranges from 0 to 100. We use differential RSI as:

$$DRSI(t)_{input} = RSI_t - RSI_{t-1} \quad (3)$$

The output of the network will be:

$$DRSI(t+1)_{ouput} = RSI_{t+1} - RSI_t \quad (4)$$

2. Differential EMA (Exponential Moving Average) as Input

Table 1. Selected features and their formulae: C, closing price; H, high price; L, low price; LL_n, lowest low price in the last n samples; HH_n, highest high in the last n samples; Up, upward price change; Dw, downward price change.

Name of features	Formula
STOCH %D	$\sum_{i=0}^{n-1} \frac{C_t - LL_{t-5}}{HH_{t-5} - LL_{t-5}} \times \frac{100}{n}$
PR(William %R)	$\frac{H_n - C_t}{H_n - L_n} \times 100$
RSI	$100 - 100 / (1 + (\sum_{i=0}^{n-1} Up_{t-i} / n) / (1 + (\sum_{i=0}^{n-1} D$
EMA	$EMA_t = EMA_{t-1} + \alpha \times (Price_t - EMA_{t-1})$
MACD	$EMA_{fast} - EMA_{slow}$

Exponential moving average (EMA), also known as an exponentially weighted moving average (EWMA), is a type of infinite impulse response filter that applies weighting factors that decrease exponentially. We use Differential EMA as:

$$DEMA(t)_{input} = EMA_t - EMA_{t-1} \quad (5)$$

3. Differential STOCH (Stochastic Oscillator %D) as Input

In technical analysis of securities trading, the stochastic oscillator is a momentum indicator that uses support and resistance levels. The term stochastic refers to the location of the current price in relation to its price range over a period of time. This method attempts to predict price turning points by comparing the closing price of a security to its price range. STOCH ranges from 0 to 100. We use Differential STOCH as:

$$DSTOCH(t)_{input} = STOCH_t - STOCH_{t-1} \quad (6)$$

4. MACD (Moving Average Convergence/Divergence) as Input

MACD makes use of moving averages of different time frames to indicate momentum changes and swings in the mood of the crowd, to give buying and selling signals that catches the big moves. The MACD indicator measures the difference between two moving averages (EMA):

$$MACD(t)_{input} = MACD_t - MACD_{t-1} \quad (7)$$

5. Differential PR (William %R) as Input

Williams's %R, or just %R, is a technical analysis oscillator showing the current closing price in relation to the high and low of the past N values (for a given N). Its purpose is to tell whether a stock or commodity market is trading near the high or the low, or somewhere in between, of its recent trading range. The oscillator is on a negative scale, from -100 (lowest) up to 0 (highest). We use Differential PR as:

$$DPR(t)_{input} = DPR_t - DPR_{t-1} \quad (8)$$

In this paper, the hourly period sample for each of the indicator is given in Table 2.

Table 2. Period (in hours) samples used in this work.

Name of features	Period
STOCH %D	5:sampling 3:smoothing
PR(William %R)	14
RSI	14
EMA	14
MACD	Fast:12 Slow=26

3.2 Construction phases and trading strategy results

In order to track the improvement of the method we divided the problem into three phases.All networks were trained between January 2001 and May 2014. For each experiment, a range of different training parameters and numbers of hidden nodes were tried, and the results are listed here. For each experiment, error in predicting training and test set outputs are reported, in terms of RMSE, along with the directional success.

While the accuracy of a model may be measured using RMSE (the error in directional success) what is ultimately needed is a measure of the effectiveness of the model in relation to its use in driving decisions to buy/sell shares. As the output of the model is $DRSI(t+1) = RSI_{t+1} - RSI_t$, the trading strategy will be:

$$\begin{cases} if\ DRSI(t+1) > \alpha\ Buy\ Signal \\ if\ DRSI(t+1) < \alpha\ Sell\ Signal \end{cases} \quad (9)$$

Here α is non-negative tuning parameter. The value of the tuning parameter is traded off between the number of trades and directional success that can be evaluated from the back testing result on historical data (we ended up using 1.22 for α). In all phases, a range of different parameter settings and configurations of hidden nodes were evaluated and the most successful ones are documented. The results given in tables are for both

GBPUSD/EURUSD rates. The training parameters lay within the following ranges:

1. Learning rate, (0.001 to 0.3)
2. Momentum, (0.002 to 0.4)

3.2.1 Phase 1

In this phase we use static feed forward network for training. The prediction errors on train/test sets are shown in Table 3.

Table 3. Phase 1 train/test errors.

Set	RMSE	
	GBPUSD	EURUSD
Training Set	3.92	3.66
Test Set	4.72	4.42

3.2.2 Phase 2

In the second phase a series-parallel architecture as shown in Figure 1is used, so that now NARX is used as the network architecture.The improvements in RMSE from using this network are shown inTable 4.

Table 4. Phase 2 train/test errors.

Set	RMSE	
	GBPUSD	EURUSD
Training Set	3.76	3.36
Test Set	4.32	4.02

3.2.3 Phase 3

In the last phase of the work we used the time shifting bagging method. Time shifting bagging method works as follows: we gather the data in a minute and produce the hourly rates based on M1 (minutes) data. We can sample the data in 60 different ways for hourly sampling. Thus, we can start sampling in M1 data at time t and sample repeatedly at times t-60, t-120, t-180, …. Therefore, the sampling has 60 steps starting at time t. Or we can start the sampling in M1 data at time t-1, and sample repeatedly at times t-61, t-121, t-181, … . Clearly, we can do this kind of sampling in 60 different ways to prepare the data for the input of the network. In this work we sample in 12 different ways starting at t, t-5, t-10... and t-55 and compute each of the indicators based on these datasets. Finally, we train the network and average the result of the output for 12 ways. Prediction errors are shown in Table 5.

Table 5. Phase 3 train/test errors GBPUSD.

Prediction Error	RMSE	
	GBP	USD
Training Set	3.53	3.23
Test Set	4.17	3.87

The results in Table 5 shows that the error rates for train/test improve remarkably when the time shifting average method is used.There is a significant change in test error rate when bagging method is used in combination with NARX neural network. InTable 6the directional success for all the phases is listed.

Table 6. Directional success on all the phases.

Phase No.	Directional success rate on test date	
	GBPUSD	EURUSD
1	54%	57%
2	60%	63%
3	68%	72%

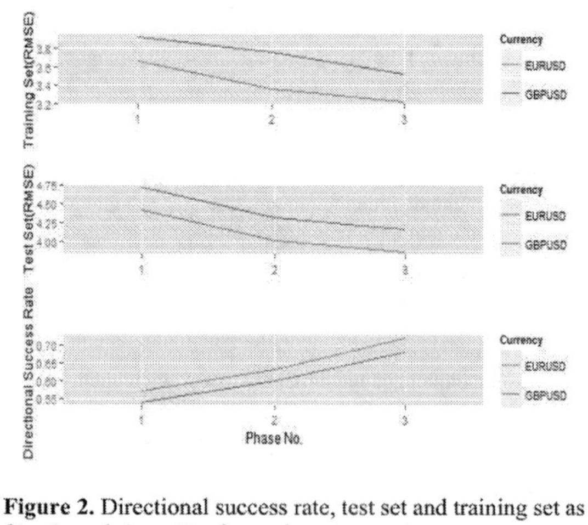

Figure 2. Directional success rate, test set and training set as a function of phase No. for each currency.

3.3 Discussion

The purpose of Phase 1 was to determine the performance of a neural network trained solely on the hourly time series data. This showed thedirectional success of 54%(GBPUSD) and 57%(EURUSD). The results indicate that this is useful; yet the success rate is not good enough to attempt to use this method for profit in the real market.

In Phase 2, the architecture was changed to NARX neural network and the directional success rate became 60%(GBPUSD) and 63%(EURUSD). By changing the method to time shifting bagging, as we did in Phase 3, we improved the rate to 68%(GBPUSD) and 72%(EURUSD) which is a significant rate of success.

4 Conclusions

There is an enormous range of indicators that are not accounted for in this paper. The analysis presented here shows only the benefit of considering NARX neural network and time shifting bagging method with some external indicators when predicting Forex market movement. The use of other indicators or even different bagging methods should be explored further. Also, this study was on two pairs of currencies only and should be repeated on other currency combinations.

References

1. J. Yao and C. Tan, "A case study on using neural networks to perform Technical Forecasting of Forex," Neurocomputing, **34**, 79-98 (2000)
2. F. Ferna´ndez-Rodrı´guez, C. Gonza´lez-Martel and S. Sosviall-Rivero, "On the profitability of technical trading rules based on artificial neural networks: Evidence from the Madrid Stock Market," Economics Letters, **69**, 89-94 (2000)
3. H. White, "Learning in artificial neural networks: A statistical perspective," Neural Computing, **1**, 425-464 (1989)
4. A. Refenes and G. F. A. Zapranis, "Stock performance modeling using neural networks: A comparative study with regression models," Neural Network, **5**, 961-970 (1994)
5. L. Haizhon and K. Robert, "A Dynamic Neural Network Method for Time Series Prediction Using the KIII Model," in Proceeding of the 2003 International Joint Conference on Neural Networks, (2003)
6. G. Fung, "The Predicting Power of Textual Information on Financial Markets," IEEE Intelligent Informatics Bulletin, **5**, 1-10 (2005)
7. S. Haykin, Neural Networks and Learning Machines (3rd Edition), Pearson Education, (2008)
8. T. Lin, C. L. Giles, B. G. Horne and S. Kung, "A Delay Damage Model Selection Algorithm for NARX Neural Networks," IEEE Transactions on Signal Processing, "Special Issue on Neural Networks", **45**, 2719-2730 (1997)
9. K. S. Narendra and K. Parthasarathy, ""Learning Automata Approach to Hierarchical Multiobjective Analysis,"," IEEE Transactions on Systems, Man and Cybernetics, **20**, 263-272 (1991)
10. K. S. Narendra and K. Parthasarathy, ""Learning Automata Approach to Hierarchical Multiobjective Analysis,"," IEEE Transactions on Systems, Man and Cybernetics, **20**, 263-272 (1991)
11. "Alpari," 1 09 2013. [Online]. Available: www.alpari-forex.com.
12. "FXstreet," 1 09 2013. [Online]. Available: www.fxstreet.com.
13. "Bloomberg," 01 09 2013. [Online]. Available: www.bloomberg.com.

Modeling Philippine Stock Exchange Composite Index Using Weighted Geometric Brownian Motion Forecasts

Willy Gayo[1] and Guido David[2]

[1]*College of Science, Polytechnic University of the Philippines, Manila, Philippines*
[2]*Institute of Mathematics, University of the Philippines, Quezon City, Philippines*

Abstract. Philippine Stock Exchange Composite Index (PSEi) is the main stock index of the Philippine Stock Exchange (PSE). PSEi is computed using a weighted mean of the top 30 publicly traded companies in the Philippines, called component stocks. It provides a single value by which the performance of the Philippine stock market is measured. Unfortunately, these weights, which may vary for every trading day, are not disclosed by the PSE. In this paper, we propose a model of forecasting the PSEi by estimating the weights based on historical data and forecasting each component stock using Monte Carlo simulation based on a Geometric Brownian Motion (GBM) assumption. The model performance is evaluated and its forecast compared is with the results using a direct GBM forecast of PSEi over different forecast periods. Results showed that the forecasts using WGBM will yield smaller error compared to direct GBM forecast of PSEi.

1 Introduction

The Philippine Stock Exchange Composite Index (PSEi), formerly known as Phisix, provides investors and other market participants a point of reference that can measure the performance of Philippine stock market. It is composed of 30 common stocks of listed companies, which are carefully selected by the Philippine Stock Exchange (PSE). To be selected, a company must rank among the top 25% in terms of median daily value in nine out of the twelve-month period in review, and have at least 12% free float level. The Philippine Stock Exchange (PSE) chooses the top 30 among qualified companies based on the full market capitalization. PSEi serves as an indicator in the price level changes of the entire Philippine stock market. It also provides a useful benchmark against which to measure an investor's portfolio. Because of these, there is a need to model and forecast the value of PSEi.

There have been different methods in modeling stock composite indices. Cointegration and Granger causality tests were used by Albu [1] in choosing what variables or indicators estimated the composite index of economic activity in Romania. Hussain and Li [2] used generalized extreme value (GEV), generalized logistic (GL) and generalized Pareto distributions to model the distribution of the extreme daily returns of the Shanghai Stock Exchange Composite Index from 1991 to 2013. They also used power-weighted method to estimate their model parameters. Leigh, Hightower and Modani [3] used a nonlinear neural network model that featured information on volume increase, interest rate change and previous price behavior in modeling the New York Stock Exchange (NYSE) Composite Index.

There were also other papers that studied PSEi. Sudirman and Darmayanti [4] used Box Jenkins methodology to estimate the best model for the stock exchange composite indices of Indonesia, Malaysia, Singapore, Philippines and Thailand. Dolores [5] applied Monte Carlo simulations based on a regime-switching lognormal model to project returns of PSEi. To estimate the parameters of the model, he used maximum likelihood estimation. He also compared the results obtained to the results using a simple lognormal model.

Stock prices and stock indices are stochastic in nature, which means they are not constant over time. Here, we model PSEi using Geometric Brownian Motion (GBM). Moreover, PSEi values are calculated based on a weighted mean of the component stocks. PSE is the governing agency that assigns the weight of each component stock. These weights are periodically subject to review and revision. Unfortunately, these weights are not publicly disclosed.

In this paper, we model and forecast PSEi using Monte Carlo simulations based on weighted GBM forecasts. The weights of the component stocks are estimated using constrained multilinear regression. We refer to this method as WGBM. The forecasts of WGBM are then compared with those obtained by modeling PSEi as GBM, referred to as GBM.

2 Methods

Let $t = 0,1,2,\ldots,T$ where T is the end horizon, W_t be a Brownian motion and $\mathcal{F}_t, t \geq 0$, be an associated filtration. Define a geometric Brownian process by the stochastic differential equation

$$dS_t = \alpha\, S_t dt + \sigma\, S_t dW_t. \quad (1)$$

By Itô's formula, assuming constant α, σ, (1) can be solved as

$$S_t = S_0 \exp\left(\left(\alpha - \frac{1}{2}\sigma^2\right)t + \sigma W_t\right). \quad (2)$$

The parameters α and σ are the drift rate and volatility of the GBM, respectively.

Let $Y_t = \log(S_t/S_{t-1})$ for $t = 1,2,\ldots,T$. Then the volatility σ is

$$\sigma = std(Y_t) \times \sqrt{N} \quad (3)$$

where $std(Y_t)$ is the standard deviation of Y_t for $t = 1,2,\ldots,T$ and N is the number of trading days in a year. The drift of S_t is calculated using linear regression on the log price process, i.e. we seek the parameters a and b such that minimizes

$$\varepsilon = \sum_{t=1}^{T}(\log S_t - b - at)^2. \quad (4)$$

The drift rate α is then obtained by

$$\alpha = Na + \sigma^2/2. \quad (5)$$

Let S_t be the actual value, $\hat{S}_t$ be the forecast value and $e_t = S_t - \hat{S}_t$ be the forecast error or residual for each time $t = 1,2,\ldots,T$. To evaluate the performance of the model, we use the root mean square error (RMSE) and mean absolute percentage error (MAPE). RMSE measures the square root of the average squared deviation of the forecasted and actual values, given by:

$$RMSE = \sqrt{\frac{1}{T}\sum_{t=1}^{T} e_t^2}. \quad (6)$$

MAPE measures the percentage of average absolute deviation of the forecasted and actual values, given by:

$$MAPE = \frac{1}{T}\sum_{t=1}^{T}\frac{|e_t|}{S_t}. \quad (7)$$

Let $i = 1,2,\ldots,n$ be the index of the component stock. Given P_t the PSEi value at time t, S_t^i and V_t^i the closing price and outstanding volume of stock i at time t, respectively, and w^i the weight of each stock i, then the weights of the component stocks are calculated using constrained multilinear regression, i.e. we solve

$$\mathbf{P} = \mathbf{A}\mathbf{w} \quad (8)$$

for **w**, subject to $w^i \geq 0$ for $i = 1,2,\ldots,n$. Here the entries of the design matrix **A** are given by $A_t^i = S_t^i \cdot V_t^i$ for each t and i.

3 Results

We used historical data of the closing prices of the component stocks and PSEi for every trading day from 2012 to 2014. The first method (WGBM) modeled PSEi using 100,000 Monte Carlo simulations of the 30 component stocks, each assumed to follow GBM process, repeated 30 times. In the second method (GBM), PSEi was modeled using Monte Carlo simulations of the PSEi index, assumed to follow GBM process, repeated 30 times. The Monte Carlo simulations were implemented using Matlab®. We varied the range of historical data used and the forecast period to evaluate model performance.

Each component stock of PSEi was modeled and forecasted using WGBM where the parameters α and σ were estimated using (5) and (3), respectively. The forecasted closing price of component stock i for $i = 1,2,\ldots,n$ was denoted by $\hat{S}_t^i$ for each time t. We then used (8) to determine the weight of each component stock. The forecasted value of PSEi, denoted by $\hat{P}_t$, was computed by

$$\hat{P}_t = \sum_{i=1}^{n}(\hat{S}_t^i \cdot V_t^i)\, w^i. \quad (9)$$

For the first forecast period, we used PSEi data from April 2012 to December 2013 to model and forecast PSEi for January to March 2014. There were 411 trading days during this time frame. Table 1 shows the computed weight for each component stock where each weight was scaled by 10^{-8}.

Table 1. Weights of the 30 component stocks ($\times 10^{-8}$) for the period from April 2012 to December 2013.

i	Symbol	w^i	i	Symbol	w_i
1	AC	0.1126	16	JFC	0.1569
2	AEV	0.0507	17	JGS	0.0000
3	AGI	0.0805	18	LTG	0.0129
4	ALI	0.2328	19	MBT	0.1991
5	AP	0.2159	20	MEG	0.0000
6	BDO	0.1531	21	MER	0.0360
7	BLOOM	0.0924	22	MPI	0.2224
8	BPI	0.0519	23	PCOR	0.0005
9	DMC	0.0961	24	RLC	0.2327
10	EDC	0.2128	25	SCC	0.0000
11	EMP	0.0000	26	SM	0.0075
12	FGEN	0.2174	27	SMC	0.1482
13	GLO	0.1304	28	SMPH	0.0000
14	GTCAP	0.2082	29	TEL	0.0000
15	ICT	0.3857	30	URC	0.1002

The component stocks were individually modeled using the GBM process given in (2). Table 2 shows the drift rate (α^i) and volatility (σ^i) of component stock i, for $i = 1,2,\ldots,30$. We then used these parameters to forecast each component stock for the period from January to March 2014.

Table 2. Drift rate and volatility of the component stocks.

i	α^i	σ^i	i	α^i	σ^i
1	0.2474	0.3112	16	0.3870	0.3693
2	0.0174	0.3345	17	0.1758	0.3119
3	0.6014	0.3571	18	0.9135	0.5612
4	0.2313	0.3569	19	-0.0063	0.3165
5	-0.0116	0.2549	20	0.3877	0.3621
6	0.1782	0.3267	21	0.1270	0.3066
7	0.0147	0.4736	22	0.1361	0.3247
8	0.2031	0.2816	23	0.2346	0.2492
9	-0.0941	0.3001	24	0.1635	0.3318
10	-0.0781	0.3376	25	0.1380	0.2907
11	1.0194	0.8852	26	0.1112	0.3655
12	-0.0668	0.2934	27	-0.2991	0.2952
13	0.3180	0.3191	28	0.1570	0.3906
14	0.3684	0.3071	29	0.0821	0.2147
15	0.2779	0.3276	30	0.5471	0.3393

Note that Emperador Inc. (EMP) had the highest drift rate and highest volatility, while San Miguel Corporation

(SMC) had the lowest drift rate and was among six stocks that had negative drift rates, indicating a general downward trend over the period of study. Philippine Long Distance Telephone Company (TEL) had the lowest volatility. Using the estimated drift rate and volatility, the forecasted values of each component stock were multiplied by the corresponding outstanding volume (i.e. number of outstanding shares) and the weight computed using constrained multilinear regression in Table 1 to determine the predicted value of PSEi. Averaging over 100,000 runs and repeated 30 times, the forecasted values of PSEi using weighted Geometric Brownian Motion (WGBM) are shown in Figure 1. The performance of the forecasts was evaluated using RMSE and MAPE given by (6) and (7). The computed mean RMSE and MAPE are 132.1974 and 1.6717%, respectively.

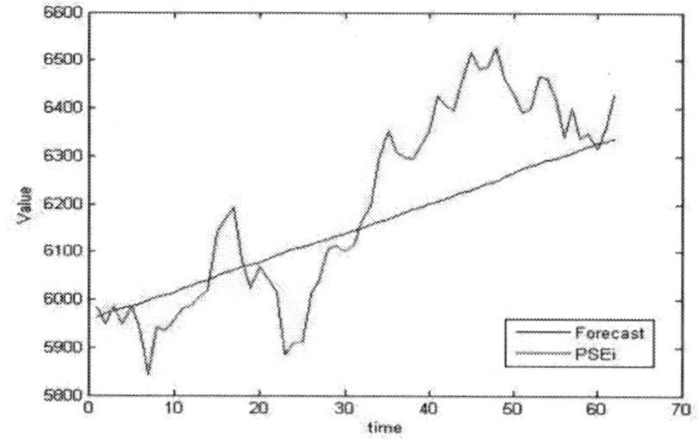

Figure 1. Graph of PSEi with WGBM forecasts from January to March 2014.

We extended the period of study to 2.25 years, 2.5 years, 2.75 years and 3 years and used WGBM to model the value of PSEi by recalculating the weight of each component stock using constrained multilinear regression, subject to nonnegative parameters. Then, we forecasted the values of PSEi for the succeeding quarter following the period of study. We conducted 100,000 runs each simulation, repeated 30 times and recorded the RMSE for each simulation. Afterwards, we computed the mean and standard deviation of the errors compared with the actual values of PSEi in the forecast period.

Table 3 shows the weight of each component stock obtained over the study period from April 2012 to March 2014. Figure 2 shows the forecast for April to June 2014 with mean RMSE and MAPE of 203.6851 and 2.5987%, respectively.

Table 3. Weights of the 30 **component** stocks ($\times 10^{-8}$) for the period from April 2012 to March 2014.

i	Symbol	w^i	i	Symbol	w_i
1	AC	0.1189	16	JFC	0.2099
2	AEV	0.1918	17	JGS	0.0301
3	AGI	0.0016	18	LTG	0.0001
4	ALI	0.1721	19	MBT	0.3115
5	AP	0.0000	20	MEG	0.0001
6	BDO	0.0000	21	MER	0.0343
7	BLOOM	0.1015	22	MPI	0.2600
8	BPI	0.1703	23	PCOR	0.0005
9	DMC	0.0900	24	RLC	0.0005
10	EDC	0.0798	25	SCC	0.0002
11	EMP	0.0000	26	SM	0.0148
12	FGEN	0.2704	27	SMC	0.1570
13	GLO	0.1962	28	SMPH	0.0001
14	GTCAP	0.8554	29	TEL	0.0000
15	ICT	0.3858	30	URC	0.0005

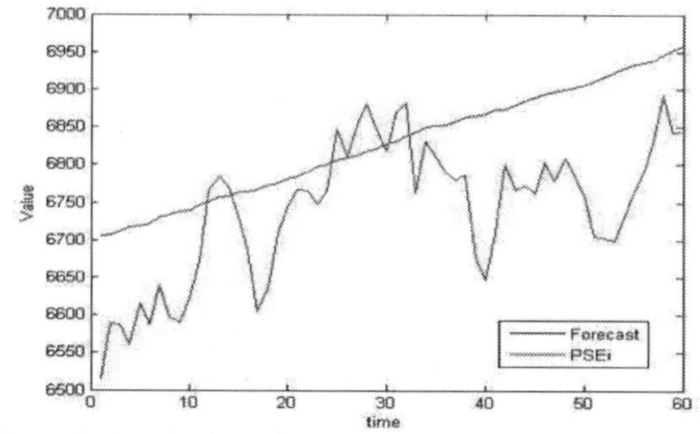

Figure 2. Graph of PSEi with WGBM forecasts from April to June 2014.

For the study period from April 2012 to June 2014, Table 4 shows the weight of each component stock and Figure 3 shows the forecast for July to September 2014. The mean RMSE and MAPE of the forecast are 191.0458 and 2.5645%, respectively.

Table 4. Weights of the 30 component stocks ($\times 10^{-8}$) for the period from April 2012 to June 2014.

i	Symbol	w^i	i	Symbol	w_i
1	AC	0.0958	16	JFC	0.2473
2	AEV	0.2172	17	JGS	0.0141
3	AGI	0.0000	18	LTG	0.0000
4	ALI	0.1532	19	MBT	0.3626
5	AP	0.0000	20	MEG	0.0000
6	BDO	0.0000	21	MER	0.0369
7	BLOOM	0.0765	22	MPI	0.2317
8	BPI	0.1986	23	PCOR	0.0000
9	DMC	0.1138	24	RLC	0.0007
10	EDC	0.0476	25	SCC	0.0000
11	EMP	0.0000	26	SM	0.0214
12	FGEN	0.3488	27	SMC	0.1131
13	GLO	0.1943	28	SMPH	0.0000
14	GTCAP	0.7567	29	TEL	0.0000
15	ICT	0.3870	30	URC	0.0000

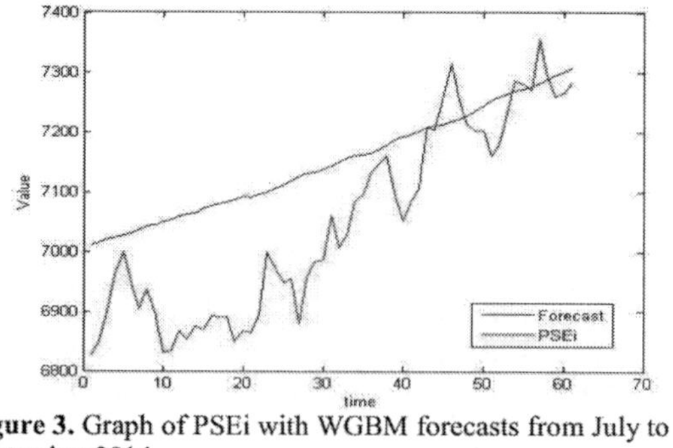

Figure 3. Graph of PSEi with WGBM forecasts from July to September 2014.

For the study period from April 2012 to September 2014, Table 5 shows the weight of each component stock and Figure 4 shows the forecast for October to December 2014. The mean RMSE and MAPE of the forecast are 97.4783 and 1.1812%, respectively.

Table 5. Weights of the 30 component stocks ($\times 10^{-8}$) for the period from April 2012 to September 2014.

i	Symbol	w^i	i	Symbol	w_i
1	AC	0.0262	16	JFC	0.3167
2	AEV	0.2769	17	JGS	0.0021
3	AGI	0.0000	18	LTG	0.0001
4	ALI	0.2003	19	MBT	0.1230
5	AP	0.0000	20	MEG	0.0000
6	BDO	0.0000	21	MER	0.0950
7	BLOOM	0.0588	22	MPI	0.4486
8	BPI	0.1711	23	PCOR	0.0000
9	DMC	0.0067	24	RLC	0.0002
10	EDC	0.2616	25	SCC	0.0000
11	EMP	0.0000	26	SM	0.0451
12	FGEN	0.1944	27	SMC	0.1519
13	GLO	0.1564	28	SMPH	0.0000
14	GTCAP	0.7407	29	TEL	0.0000
15	ICT	0.2849	30	URC	0.0001

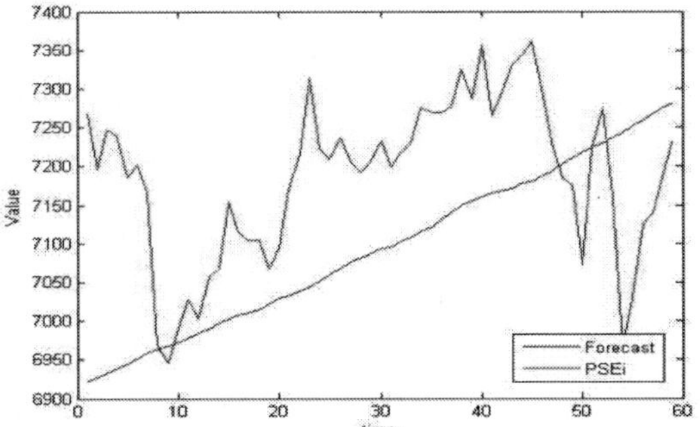

Figure 4. Graph of PSEi with WGBM forecasts from October to December 2014.

Lastly for the study period from April 2012 to December 2014, Table 6 shows the weight of each component stock and figure 5 shows the forecast for July to September 2014. The mean RMSE and MAPE of the forecast are 273.5275 and 3.3441%, respectively.

Table 6. Weights of the 30 component stocks ($\times 10^{-8}$) for the period from April 2012 to December 2014.

i	Symbol	w^i	i	Symbol	w_i
1	AC	0.0628	16	JFC	0.2698
2	AEV	0.2938	17	JGS	0.0000
3	AGI	0.0000	18	LTG	0.0000
4	ALI	0.2036	19	MBT	0.1393
5	AP	0.0000	20	MEG	0.0000
6	BDO	0.0000	21	MER	0.0881
7	BLOOM	0.0424	22	MPI	0.4690
8	BPI	0.2165	23	PCOR	0.0000
9	DMC	0.0000	24	RLC	0.0000
10	EDC	0.1213	25	SCC	0.0000
11	EMP	0.0000	26	SM	0.0370
12	FGEN	0.1977	27	SMC	0.1529
13	GLO	0.1726	28	SMPH	0.0000
14	GTCAP	0.4910	29	TEL	0.0000
15	ICT	0.3232	30	URC	0.0000

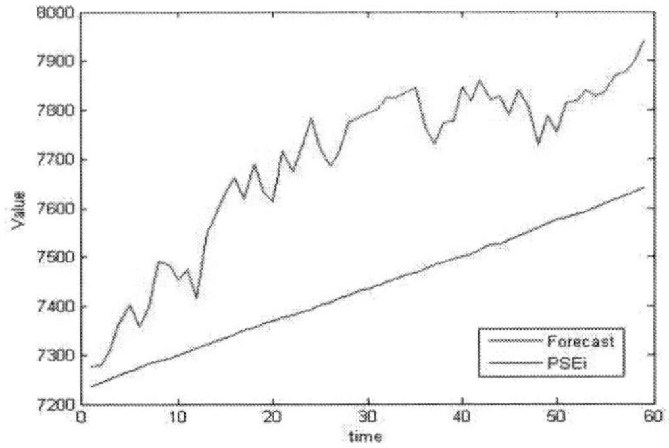

Figure 5. Graph of PSEi with WGBM forecasts from January to March 2015.

As comparison for WGBM, we then modeled PSEi directly using Monte Carlo simulations based on GBM. The drift and volatility of PSEi were computed similarly using (5) and (3). Table 7 shows the drift rate and volatility of the value of PSEi for the different periods of study. Figures 6 to 10 show the forecast for different periods of the study.

Table 7. Drift rate and volatility of PSEi modeled using GBM.

Forecast Period	Drift Rate	Volatility
2 years	0.1693	0.1972
2.25 years	0.1254	0.1902
2.5 years	0.1200	0.1828
2.75 years	0.1206	0.1757
3 years	0.1189	0.1727

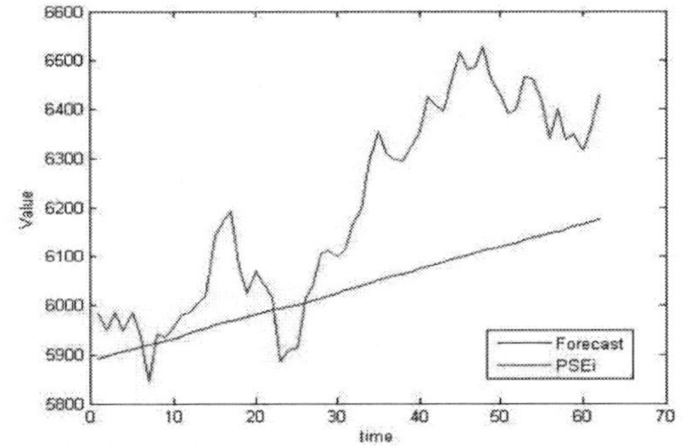

Figure 6. Graph of PSEi and forecast for January to March 2014 using GBM.

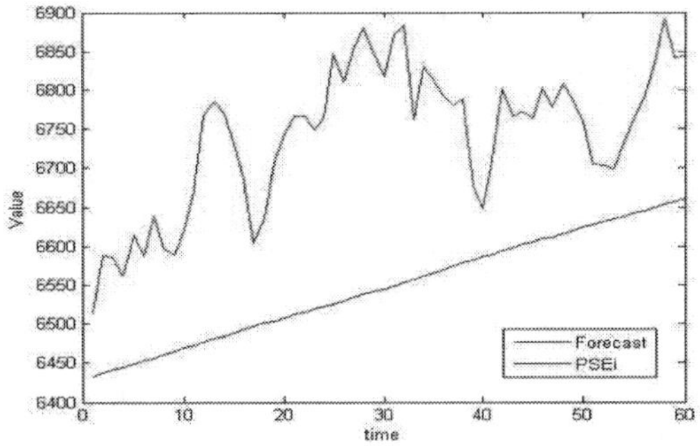

Figure 7. Graph of PSEi and forecast for April to June 2014 using GBM.

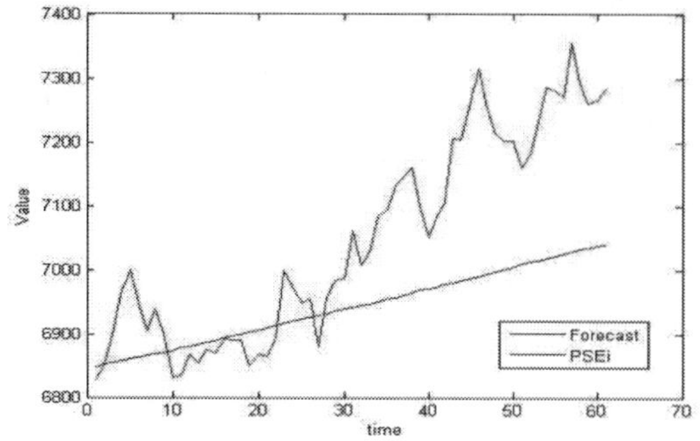

Figure 8. Graph of PSEi and forecast for July to September 2014 using GBM.

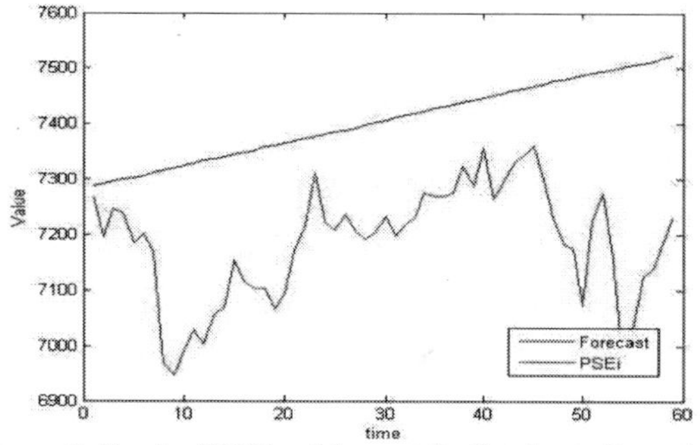

Figure 9. Graph of PSEi and forecast for October to December 2014 using GBM.

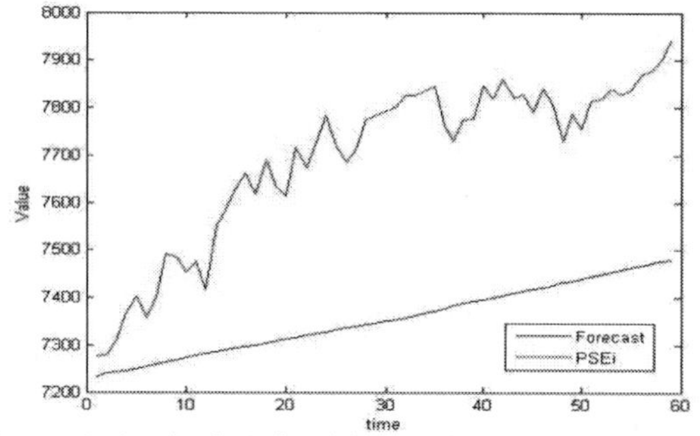

Figure 10. Graph of PSEi and forecast for January to March 2015 using GBM.

Table 8 shows the mean and standard deviation RMSE of the model performance over different forecast periods. WGBM has smaller RMSE compared to GBM for all forecast periods. Based on the results of this study, WGBM obtained better forecasts compared to directly using GBM on PSEi, with difference in errors as high as 100%.

Table 8. Evaluation for different forecast periods using GBM.

Forecast Period	RMSE	MAPE
2 years	211.3857	2.7490%
2.25 years	207.2595	2.8616%
2.5 years	147.9104	2.6473%
2.75 years	247.8183	3.1248%
3 years	356.4535	4.3538%

4 Conclusion

In this paper, the Philippine Stock Exchange index was modeled using Monte Carlo simulations based on weighted Geometric Brownian Motion forecasts and a direct application of Geometric Brownian Motion. Simulations were done for 100,000 iterations and repeated 30 times. The performance of each model was evaluated using root mean square error. Numerical results showed that the first method (WGBM) yielded a significantly smaller RMSE compared to direct GBM. Hence, WGBM was a better method in modeling PSEi, in this study. WGBM method should be applicable to any stock index data, and even other stochastic data such as interest rates, exchange rates, mutual funds and others. It may also be used in weather data such as precipitation, rainfall and others. This study may be extended further using other stochastic models for forecasting. The estimation of the weights of the component stocks may also be improved.

References

1. L.L. Albu, RJEF **9** Issue 4,111-124 (2008)
2. S.F. Hussain, S. Li, Int. Fin. Markets, Inst. and Money **34**, 263–276 (2015)
3. W. Leigh, R. Hightower, N. Modani, Expert Systems with Applications **28**, 1–8 (2005)
4. M.S.N. Sudirman, P.A. Darmayanti, Conference Paper in Forum Manajemen Indonesia ke-5 (2014)
5. P.A. Dolores, Technical Paper (University of the Philippines)
6. S.E. Shreve, Stochastic Calculus for Finance II: Continuous-Time Model, Springer Finance, United States of America, (2004)
7. S.M. Ross, An Elementary Introduction to Mathematical Finance 3rd Ed., Cambridge University Press, (2011)

MATEC Web of Conferences 68, 20001 (2016) DOI: 10.1051/matecconf/20166820001
ICIEA 2016

The Advance Experience of Foreign Top-level Think Tanks and its Enlightment to the Construction of Chinese University Think-Tank

Yong Tao[1], Qiushi Li[1], Gang Zhao[2] and Wen Zhao[1]

[1]*National Laboratory on Aerospace/The Chinese Institute of Aeronautical Engineering Development Strategies, Beihang Univerisy, Beijing, China*
[2]*School of Mechanical Engineering and Automation, Beihang University, Beijing, China*

Abstract. This paper expounds the development course and current situation of foreign top-level Think Tank, and analyzes the important effects of foreign colleges and universities Think Tank in supporting national scientific decision-making and rapid development. It analyzes the development course and advanced experience of top-level Think Tank in America and other countries. The paper summaries the advance experiences, combines with the background and features of Think Tank construction of colleges and universities in China. It puts forward that Chinese colleges and universities Think Tank should aim at the national strategic needs, play the advantages of talents gathering and main innovation force, insist on opening, and set constructing professional colleges and universities Think Tank as the breakthrough point. The think tanks produce strategic consulting research results with high quality. The paper proposes the reference effect of foreign Think Tank on characteristic Think Tank construction of colleges and universities in China and the development strategy recommendations.

1 Introduction

Think Tank is also called as "Idea Tank". The internationally accepted definition on Think Tank is the public research institution that is composed by experts, combines multi disciplines, gives advice and suggestions for decision maker to deal with social, economic, scientific, military, diplomatic and other aspects issues, then makes investigation, analysis and research on political, economic, social, cultural, military and other policies, so as to provide the best theories, strategies, methods, ideas, etc. it is an important force to influence government's decision-making and promote social development [1].

In order to respond all the challenges faced with by industries, respond the difficult issues during development, we not only need enough courage, but also need enough wisdom, we should agglomerate the wisdom of the whole industry, even all humans, we need the Think Tank with irreplaceable role in making decision, as the compass of industry, since Think Tank has strong ability of research and participating in politic, thus Think Tank is the intelligence gathering, communication and application platform, it is the intelligent output platform of experts in various industries, and is also the intelligence platform of intelligent and capital.

The development level of Think Tank has become the important component of country's "soft power", since the Third Plenary Session of the Eighteenth Central Committee of Party, nation and government appeal to construct the new think tank with Chinese characteristics, so as to improve national governance system and governance capacity building, then announce Chinese voice in international stage. Under this background, new colleges and universities Think Tank ceremoniously emerges with unprecedented posture, thus the colleges and universities Think Tank construction obtains more and more attentions.

Colleges and universities Think Tank is the organization belongs to university that is engaged in policy research and decision-making consultation, this kind of Think Tank is constructed by university independently or with the support of other organizations and groups. Most of the researchers are the scholars in various disciplines of university or the researchers employed from other universities or research institutes, the service objects and research topics are also very wide. The advance of colleges and universities Think Tank is that, gathering high level talents, leading the disciplines with strong comprehensive nature, rich data and unobstructed information, relaxed academic research atmosphere, etc. these advances can often make the colleges and universities Think Tank generate innovative ideas, methods, results and policy recommendations.

2 Development course and current situation of colleges and universities think tank abroad

Think Tank in the modern sense is originated in America during the Second World War, later, it refers to research institutions that regard policy research as own duty, and affecting public decision-making and public opinion as purpose in general. World famous Think Tanks include America's Rand Corporation, America's Stanford International Consultative Institution, Hoover Institution, Rockefeller Foundation, International Institute for Applied Systems Analysis, British Royal Society, Japan's Nomura Institute of Research and Germany Industrial Equipment Business Corporation, etc. According to the definition of Frank Koelb Mo who is the founder of international famous Think Tank-Rand Corporation, Think Tank is a 'strategic thinking center' that dares to transcend all existing wisdom, dares to challenge and defy existing authority, it is a Think Tank specializing in study of soft science.

It has nearly hundred years of history since the foundation of first American Think Tank. It is not only a symbol of America's soft power, and also provides strong support for countries to make scientific decisions. For a long time, university research institutions in America not only pay attention to talent cultivation and academic research, and also bear 'Think Tank' national heavy responsibility-directly serving for the needs of international strategy through academic research [1].

As the development of America's economy, science and technology and so on various aspects, the development of colleges and universities Think Tank is more and more rapidly, all the industries of society pay more attention to it. Before the 1970s, universities had been playing the function of country's main Think Tank, then as the specialization of country's decision-making consulting services, professional Think Tank generated and selectively replaced part of work of colleges and universities Think Tank, this made its Think Tank dominant position gradually weaken. But it is still an important part of Think Tank, and constitutes 'three-in-hand' of American Think Tank with official Think Tank and independent Think Tank [2]. For example, Hoover Institution in Stanford University, Fairbank Center for East Asian Studies in Harvard University, East Asian Institute in Columbia University and Media Lab in Massachusetts Institute of Technology, etc. are current universities Think Tanks with important international influence. In 2006, a famous Ivy League university Princeton University in America published a national security report entitled Princeton Project, the report attracted great attention of international community, and played an important role in Obama Administration's foreign policy making afterwards. This undoubtedly expresses colleges and universities Think Tank's position in American development. Under America's highly value and developing, among current big three Think Tanks in America, they account for more than 50% of dependency degree on colleges and universities Think Tank in all aspects. At the same time, American government invests more and more in colleges and universities Think Tank.

The development of Think Tank in America comes from support from government, the people and universities three aspects. According to the Global Think Tank Report 2014 report [3] continuously published by America's University of Pennsylvania in 'Think Tank and Civil Social Programs' on January 22, 2015, nowadays, there are 6681 Think Tanks worldwide, including 1830 in America and 429 in China, they are two countries with the largest number of Think Tanks around the world. America's Brookings Institution, British Royal Institute for International Affairs, America's Carnegie Endowment for International Peace, America's Center for Strategic and International Studies, Belgium's Think Tank Bruegel, Sweden's Stockholm International Peace Research Institute, Rand Corporation, America's Council on Foreign Relations, British International Institute for Strategic Studies and America's Woodrow Wilson International Center respectively rank the top ten in world's top-level Think Tanks. In China, a total of seven Think Tanks enter into 'top 150 of world's top-level Think Tanks', respectively are Chinese Academy of Social Sciences, China Institute of International Studies, China Institutes of Contemporary International Relations, Development Research Center of the State Council, Institute of International and Strategic Studies of Peking University, Shanghai Institutes for International Studies and Chongyang Institute for Financial Studies, Renmin University of China. Compared with last year, Chongyang Institute for Financial Studies, Renmin University of China is a newly increased one. Among them, there are still only two universities Think Tanks [4].

For any Idea Tank, 'quality, independence and influence' are core values that it must conform to, the reasons why Idea Tanks in American universities can have strong influence on public policy and public opinion, are mainly because:

Firstly, the think Tanks in colleges and universities have strong specialty. Think Tank in every university will have its own adept field of study, this is the so-called 'specializing in', and it has set up a certain brand and has important influence on areas of specialization compared with well-known Think Tanks.

Secondly, the perfect and scientific management operation mechanism of Think Tank in colleges and universities ensures its innovation ability and independence; based on independence of scientific research, scientific nature of management, diversification and flexibility of talent cultivation and management model, Think Tanks abroad make them become more neutral in terms of research results and policy suggestions, good reputation and strong strength also make their research achievement transformation be more smooth so as to exert its influence.

Thirdly, the strong social demand and all-around thought spread mechanism ensure the maximization of its influence. Think Tank researchers in American universities are more diversified, most are scholars with professional background and doctoral degree, part of them are former government officials, and there are still senior media practitioners, diversified staff composition ensures close combination of Think Tank research results and social practice, and also ensures the possibility of transformation between thought study and political practice [5].

3 Foreign think tank's important role

3.1 Inspire thought for a variety of policy schemes and suggestions

The most fundamental task of public decision-making Think Tank is to provide professional advisory opinions and solutions according to decision-making demands, general products include strategic decision report, prospective studies, monographic study report, survey report and information collection analysis, etc. The essence of these services is to provide real information and ideas with reference value so as to make decision-makers more accurately, systematically and deeply understand objective fact and then more scientifically and reasonably make decisions.

Compared with other research institutions, in the case of mutual combination of talent cultivation and disciplinary research, research institutions in universities are more easily to inspire thinking and create knowledge so as to produce creative thought achievements and policies, and then put forward new ideas and views for reference of policy makers [6].

3.2 Cultivate talents, affect government public decision-making through talent flow

The connection between Think Tank of foreign universities and government has various channels. In addition to researchers' regular contact with decision-making departments or inviting of government officials to give speeches in Think Tank, more is flow between staff. 'Revolving Door' mechanism is the most distinctive phenomenon in American Think Tank, namely realizing mutual flow between scholars and officials. 'Revolving Door' in American Think Tank makes Think Tank effectively combine with politics, this not only makes American politics maintain vitality and effectiveness, and also makes Think Tank become harbor which cultivates and reserves talents for government [7]. But foreign top-level Think Tanks also stipulate that it is disallowed that current officials participate in research in principle, and emphasize that 'it must be avoided to get involved in political debates which have nothing to do with study itself', and maintain the purpose of 'not serving for any individual or power groups'. Think Tank experience abroad shows that maintaining objectivity and independence of research work, and maintaining close ties with government are complementary.

3.3 Educate the masses, guide government and society to reach a consensus so as to promote public policy-making

You are free to use colour illustrations for the online version of the proceedings but any print version will be printed in black and white unless special arrangements have been made with the conference organiser. Please check with the conference organiser whether or not this is the case. If the print version will be black and white only, you should check your figure captions carefully and remove any reference to colour in the illustration and text. In addition, some colour figures will degrade or suffer loss of information when converted to black and white, and this should be taken into account when preparing them. Think Tank has numerous influencing channels for public opinion, generally including the several following ways: one is to participate in government policy consultation, Think Tank members' statement tends to have important reference effect by virtue of in-depth study of professional field. The second is to release policy reports, publish related policy research reports, works and journals, etc. as reference publications of government officials and researchers. The third is to give an interview and make comments in mainstream media, it is a kind of two-way dependence relationship between Think Tank and media, and this makes ideas of Think Tank be promoted. The fourth is to organize various BBS, lectures and training classes, these projects are usually faced with all industries of government and society, they not only have enlightening effect and also are helpful for Think Tank to strengthen the establishment of contacts with government and all the industries.

3.4 Track implementation effect of evaluation policy and feedback public decision-making

As evaluator of policy, Think Tank tracks policy implementation, analyzes problems, predicts consequences and proposes solutions, this has positive significance in government's understanding of policy implementation and effect. First of all, Think Tank's evaluation on policy effect can go beyond limited understanding of a few policy makers, it independently and objectively evaluates policy, integrates attitudes and tendency of various aspects of society toward policy, timely adjusts and supplements policies, and effectively improves policy goals. Secondly, Think Tank's participation of policy evaluation and perfection is conducive to resolving contradiction between interested parties, reflecting fairness and impartiality of policy, and promoting scientification and democratization of government decision-making.

4 Foreign think tanks' reference role in China's construction of colleges and universities think tanks

As an important content of promoting the modernization of national governance system and governance ability, Think Tank construction needs to use other countries' mature experience for reference, and also needs to use Chinese wisdom of ruling state and dealing with politics in the history of thousands of years, 'take history as a mirror, make the past serve the present', innovatively develop new colleges and universities Think Tank with Chinese characteristics and international influence. As a knowledge-intensive organization, core competitiveness of colleges and universities Think Tank is intellectual capital. Only to practically improve colleges and universities Think Tank's core competitive advantage, institutional guarantee, talent resources and communication capacity, can China's colleges and universities Think Tank truly become one of the sources of power of national rising and truly push China move towards Think Tank power from Think Tank giant.

4.1 Colleges and universities Think Tank is talent gathering highland, and main force of unique and innovative ideas

Think Tank development cannot be separated from talent gathering, while colleges and universities have always been places that talents gather, at the same time, colleges and universities are places for cultivating talents, they not only open various majors and gather all kinds of talents, directions and specialties of subject majors in universities specialize in different things, high-level talents have obvious advantages. With the development of informatization, talent requirements are more urgent, the gathering of Think Tank talents in colleges and universities can't be ignored.

As an 'Idea Tank', Think Tank's characteristic is thought gathering. Colleges and universities are just a necessary path of imparting ideological knowledge, faced with good environmental atmosphere and rich foundation of teachers, development of thought must be boundless. In terms of special social status of colleges and universities, their ideas have certain differences with society. Ideological differences often bring new breakthrough, so colleges and universities Think Tank plays an important role in economic and social development [8, 9].

4.2 Practice national requirements, actively construct characteristic colleges and universities Think Tank, and pay attention to the combination of government and universities.

As huge impact brought by foreign Think Tank, China constantly strengthens efforts in construction of Think Tank. Since policy summary in the Sixth Plenary Session of the Seventeenth Central Committee of Party, Chinese government continuously strengthens efforts in Think Tank construction. Under this policy environment, the construction of colleges and universities Think Tank is also placed on the agenda, faced with numerous policy requirements, government must thoroughly practice national policy requirements and implement various national indicators. At the same time, it must actively construct colleges and universities Think Tank with project characteristics, and offer policy support and encouragement for construction of colleges and universities Think Tank. It should also offer moderate aid for the development and construction of colleges and universities Think Tank, at the same time, it can take measures of the combination of government and universities to seek for mutual development in mutual cooperation [10]. At the same time, it should pay attention to achievement construction of colleges and universities Think Tank, government interior and its development must be in harmony and hand in hand.

4.3 Take specialized Think Tank as breakthrough to cultivate a batch of new colleges and universities Think Tanks

New Think Tank with Chinese characteristics should seriously grasp product construction, product of Think Tank is thought, really good thought is necessary. World famous Think Tank development experience shows that selecting strategic and forward-looking major issues for preceding study tends to get immeasurable achievement effect. To enhance China Think Tank' core competitiveness, new universities Think Tank with characteristics must clearly find its own positioning, deeply makes prospective study faced with significant practical issues, long-term issues and strategic issues, and systematically puts forward constructive and operational policy suggestions. Colleges and universities Think Tank must start from intellectual products and focus on the improvement of research quality, including whether issues are accurate, whether proofs are rigorous and whether arguments are powerful, hereby creating a batch of high-end international leading Think Tanks with strong competitive power and high professional level [11].

4.4 Provide high quality thought products, comprehensively upgrade Think Tank's influence

Constructing colleges and universities Think Tank with Chinese characteristics, the first is to strengthen Think Tank cultural construction. Based on neutrality, scientific nature and constructive nature, we should stick to Chinese cultural characteristics, adhere to 'seeking truth from fact, basing on the tradition and making innovations', regard interests of the state and people as own duty, and have the courage to pioneer and innovate. Secondly, Think Tank itself should also expand academic, policy, social and international big four influences. In order to expand academic influence, the Chinese Think Tank in university should publish high level academic papers in the famous academic publications. In order to expand policy influence, Think Tank should strengthen study of forward-looking, reserved and strategic issues, obtain more thoughtful policy thinking and innovative research results, and build an information bridge between academic circle and decision-making circle so as to expand Think Tank's influence on high-level decision-making [12, 13]. In order to expand social influence, Think Tank should actively cooperate with mainstream media and fully propaganda valuable research results so as to influence public opinion and lead social ideological trend. In order to expand international influence, Think Tank should represent Chinese academic frontiers and academic schools to take the initiative to participate in global cooperation and competition of Think Tank, and make efforts to expand China's international discourse right and influence.

4.5 Insist on open operation mode of Think Tank and comprehensively upgrade global influence

Take advantage of global and international vision to plan the development of Think Tank. Today's world is undergoing profound and complicated changes, global cooperation expands to be multilevel and all-dimensional, as first-class Think Tank, if it wants to seize this opportunity and challenge, and directly respond to new situations, new problems and solutions triggered by integration of world economy, it must strengthen global consciousness of Think Tank, focus on studies of major issues and improve China's ability in dealing with international affairs [14, 15]. At the same time, it should strengthen international exchanges between Think Tanks, energetically develop a variety of long-term or short-term international topic cooperation, and timely exchange understandings and viewpoints. It should strengthen international platform construction of Think Tank, further play the role of Global Think Tank Summit established by domestic Think Tank, create it as an international Think Tank exchange platform with important influence, and make it become famous brand of showing China's soft power.

5 Conclusion

Think Tank is the important sign of country's "soft power", Chinese colleges and universities Think Tank has the necessary condition to develop into the big country idea tank with international influence power. This paper analyzes the development course and advanced experience of top-level Think Tank in America and other countries, based on this, it puts forward that Chinese colleges and universities Think Tank should aim at the national strategic needs, play the advantages of talents gathering and main innovation force, insist on opening, and set constructing professional colleges and universities Think Tank as the breakthrough point, thus produce strategic consulting research results with high quality, then construct and train cultivate a series of new colleges and universities Think Tank.

Acknowledgements

Thank you for the support of strategic consulting research project of Chinese Academy of Engineering.

References

1. Stella Ladi. Think Tanks and Policy Advice in the US: Academics, Advisors, and Advocates.Public Administration. **87**,1 (2009)
2. LiLi W.The development course and challenges facing of American Think Tanks[J].Red Flag Manuscript. **4**:33-36 (2009)
3. James G. McGann. 2014 Global Go To Think Tank Index Report. The University of Pennsylvania. **1** (2015)
4. Think Tank Research Center of SASS. An Empirical Research on the Influence of Chinese Think Tanks and Policy Advices[J].Journal of Social Sciences. **4**:4-21 (2014)
5. Wang L. International Experience and Its Prospect of Think-tanks in University[J]. Chongqing Social Sciences. **8**:106-109 (2012)
6. Sun Z, Zhang Z. On the Development of Think Tank: Based on the Content Analysis of its Academic Definitions[J]. Journal of Intelligence. **1**,1 (2011)
7. Research Team on Strengthening Studies on the Building of New-Type Think Tanks with Chinese Characteristics, CCIEE . Building of New-Type Think Tanks with Chinese Characteristics: Present Situation,Problems, and Countermeasures[J]. Globalization.**2**:107-119(2015)
8. AnGang H. Building a new think tank with Chinese characteristics [J]. Tsinghua Journal of Education. **5**: 1-4 (2013)
9. Angang H. Building New Think Tanks with Chinese Features: Practice and Summary [J]. The Journal of Shanghai Administration Institute.**25**:4-11 (2014)
10. YaMeng Z.The construction method exploration of University Think-tanks influence[J]. Speed Reading.**8**:292 (2015)
11. ZhenQiang G. Attribute Characteristics of Local University Think-tank and Its Development Strategies[J]. Higher Education Development and Evaluation. **3**:37-43 (2014)
12. Li W.Think about University Think Tanks construction[J]. Science & Technology Information. **27**:200-201 (2013)
13. XiaoHu Xu, Qi Chen. Development Process and Prospect of Think Tank [J]. Forum on Science and Technology in China. **7**:63-68 (2012)
14. Wei Z, Chen G. The Current Situation of the Construction of Chinese New Think Tanks and the Approaches of Improvement[J]. Education Science. **3**:66-70 (2014)
15. Yang J, Chen Y. Reasearch on the construction of University think-tank under theory of Cooperative Innovation[J]. Science & Technology Progress & Policy. **7**:7-11 (2015)

MATEC Web of Conferences **68**, 20002 (2016) DOI: 10.1051/matecconf/20166820002
ICIEA 2016

Virtual Reality As an Effective Simulation Tool for OSH Education on Robotized Workplace

Miroslav Janak, Tomas Cmorej, Tomas Vysocky, Marek Kocisko and Monika Teliskova

Technical University in Kosice, Faculty of Manufacturing Technologies with a Seat in Presov, Sturova 31 08001 Presov, Slovakia

Abstract. In last decade, the virtual reality became a huge trend in the field of visualization not only for simple elements, but also for complex devices, their actions and processes, rooms and entire areas. This contribution focuses on the possibilities of utilization of the elements of virtual reality for educational purposes regarding the potential employees and their OSH (Occupational Safety and Health) training at the specialized workplaces. It points to the possibility of using the applications and simulations of the workplaces and the processes for better understanding of potential danger and eventual prevention options. Such an application simulates the real workplace including the technological processes and provides the user with a motion in 3D environment of virtual scene observing and interacting in actual work zones

1 Introduction

The creation of the application that includes the elements of the virtual reality it is necessary to build a bridge between the real world, which presents the basics of the data transfer on the input, and the program section for control of virtual elements. This will practically enable the combination of real and virtual space. Currently there is a lot of ways how to manage the 3D data using the sensors and commercial software or even some Open Source products. After considerations about several software solutions, the application that presents an output described by this paper was designed and created in two well-known software products: Blender and Unity 3D.

2 Blender

It is an open source software which despite its price serves as an effective tool for displaying of 3D computer graphics with integration of animations, rendering, motion systems and physics. As a typical Open Source product, it is created, released and supported by the community. It can be freely downloaded in full versions providing all the tools and functions. The functions integrated in Blender can be extended and upgraded by the scripts programmed in Python language. Lot of scripts is already available in the program, but many others can be implemented or programmed directly according the requirements of the application creator. Use of this software allows the integration of the formulas, equations and calculations necessary for the successful and effective run of the application. Entire programming and also operational environment can be adjusted for the needs of the application that is created. All main programming operations are handled with use of visual scripting, which is based on connecting of the programs blocks and creation of the loops that finally builds the program sections generating the logical structure of the application.

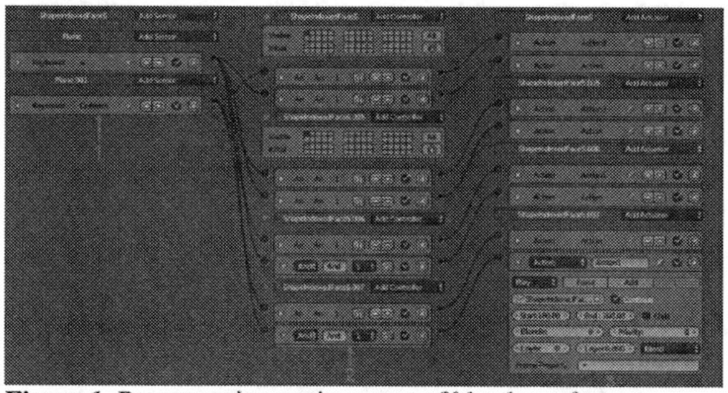

Figure 1. Programming environment of blender software

Script on Fig. 1 shows the program section that controls the settings for arms of robotic device. This script determines the actual positions and behavior (translation and rotation) of these arms which realize the continuous motion between two working units. The initiation element for the operation is selected keyboard key, which starts the complex manipulation process of the robot. In case of using the external initiation, the script can be changed Accordingly to provide the manipulation with any control device. The motions of the robot can be controlled even by the human, using the camera or Kinect device to capture and transcript live gestures.

On the figure, the first column describes the input device – initiation keyboard key. The second column builds the connection between input and output unit.

Third one represents the motions and actions that should be realized (translation and rotation motions).

3 Virtual environments

For the generation of virtual environment, it is first necessary to create the 3D models of all included objects, machines, tools and fixtures that relate the manufacturing process. Such models can be modeled directly in Blender software, or (usually) imported from other graphic or CAD system. The compatibility rule must be followed when using the models from external source. Next step is to locate the machines and manipulators according to the requirements of the client and to create the connections between the input devices and the objects integrated in created environment. After creation of the compact program unit, the manipulator can realize the operations and motions assigned to it in the environment of C# programming language. Visual of particular devices responds to the quality and level of details for each model. The output is then presented as a whole unit which includes all the objects and their necessary motions and actions with integrated spatial dispositions and export option for further eventual use of the graphics in other software.

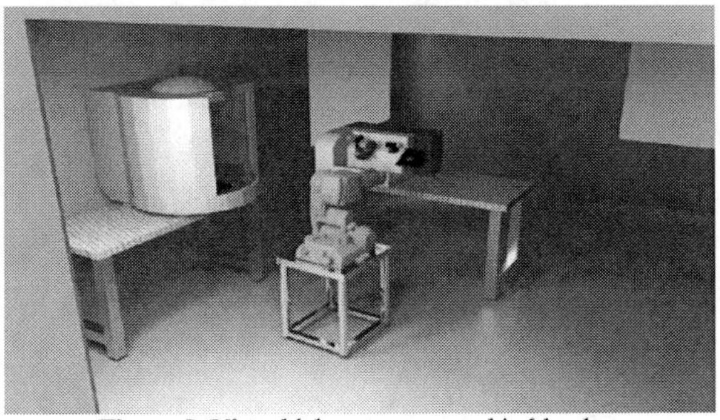

Figure 2. Virtual laboratory created in blender.

4 Unity 3D

This software belongs to the group of Open Source solutions for creation of virtual environment and applications. It is closely related to the Blender. It can work with programming languages C#, Python and Javascript. It also includes the elements for modeling, so in case of need the models of new or additional objects can be created right here. It main feature though lies in scripting and creation of complex applications with all necessary parts. After importing the model from the Blender environment to the Unity all the scripts, graphics and relations are preserved and valid. So the motions and connections of particular models do not need any adjustments or settings, functions and intentions created in any previous phase of creation of the application (Blender or other software) work just fine.

The material settings and texture properties are also preserved, basically all except the lights and room attributes, which need to be created or simply selected among the existing ones. Unity provides very powerful rendering tools, which assure an excellent reproductions of colors and textures. In combination with great light calculations, this gives the user serious feeling of reality while watching the virtual animation.

Unity also handles the visual output of entire application. There are limitless options for settings of the lights, different effects and hidden interactive control elements etc.

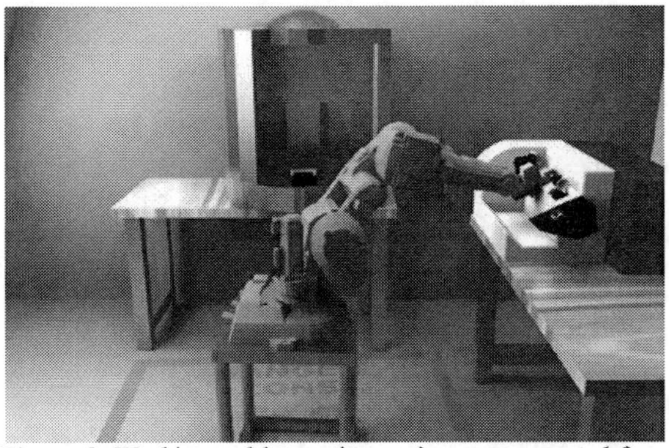

Figure 3. Working and interactive environment exported from unity software.

Output generated by the Unity software is presented as an independent application (executive file for operation on PC) or as an installation package (apk format) for devices operating on Android. While exporting for mobile devices (cellular phones) it is necessary to know the displaying parameters of the device, as it has the influence on type of glasses needed for virtual displaying. For glasses with single lens the application can run in fullscreen mode. For glasses with two lenses the application must be duplicated on the displaying surface of used device, so the user will get the proper final view.

It is hard to decide which kind of glasses for virtual reality is better. In case of the glasses with two lenses it is necessary to assure the right setting for different users regarding the eye distance and focusing. With glasses with single lens it is enough to assure possibility of the adjustment motion of the lens in the look direction, what will provide the focusing.

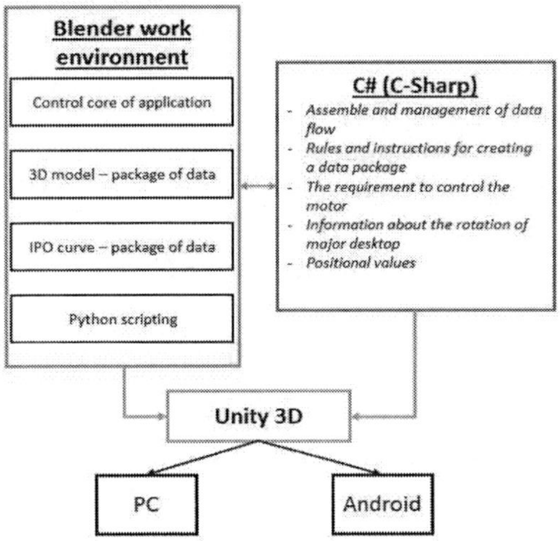

Figure 4. Logical structure of application creation process.

5 Occupational safety and health

Main document for determination of applicable rules from the area of OSH in Slovakia is # 124/2006 law – Safety and Health Protection at Work. Part of this law is forcing the duty of an employer to inform their employees about the situation, facts and actions in subject field and to provide them with relevant education and training. The employer (usually through the person specialized in realization of training processes) guarantees the fulfillment of the law.

Person (trainer) selected for realization of the training and other educational activities should have relative certification obtained from National Labor Inspectorate. Experiences and praxis (or at least relative education) are often needed to be declared. The law also asks to create the sample training project of specified content and extent. This project should include the forms and methods that are planned for educational purposes. Created 3D application could be integrated into the educational process as a new form, individual demonstrational method. Individual approach would be realized using the PC, tablet or other displaying device. Each employee would have the access to the displaying device with pre-installed application and visualization of entire potential workplace. Training supervisor would determine the training modules or specific parts and workplaces from the simulation that should be taken by the employee.

6 Application

Proposed virtual workplace presents the real sources and equipment of the laboratories at the Faculty of Manufacturing Technologies with Seat in Presov, Slovakia. Machines are located in the working area of robotic device IRB 140 from ABB company, so the robot could manipulate with the parts and thus to simulate real working conditions. Application can be run on the computer, tablet or cellphone using either Windows or Android platform.

6.1 Spatial motions

Motion around the created space is provided after pressing the keyboard arrow buttons. In case of using the tablet or phone we can also employ remote controller for motion control using the Bluetooth technology. The user is situated in the environment with standard graphical and animation tools and also with features considering the physics and physical behavior of objects. Of course it laso has a definition of collision boundaries for created virtual scene. The motion around the entire scene is unlimited; limitations are presented only by the models of real manufacturing devices executing their standard tasks.

6.2 Safety actions

Safety actions are provided using the safety zone created in order to point out an existing danger for the user who may eventually enter the working zone with a certain risk level. If the safety zone is entered, the application gives the error message which signals that some rule has been broken. Beside this, the working cycle is stopped automatically. Safety zone is displayed as a green square shown on Fig. 5 This square located on the floor serves also as the sensor reference.

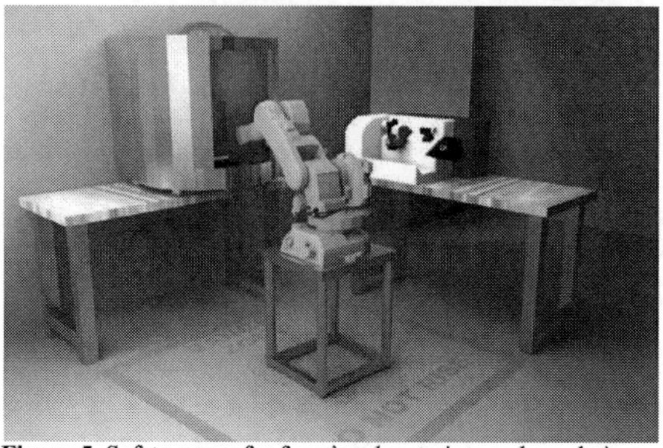

Figure 5. Safety zones for functional restraints and regulations.

The user will learn to move around the working zone without affecting the real working process by interruptions or creating any dangerous situations. Safety zones can be created in the working zones of manufacturing machines, manipulation devices or any other mechanisms, elements or systems. The zones can be also used for the definition of proper working area and for understanding of dispositional relations between all the objects, even among more employees. In case of complex mechanism and if the situation requires so, the safety zone can be created for particular section of the system which has some elements of the danger, and then to change the color of specific object after entering its safety zone. This method can be used not only for the virtual workplace but also for the simulation and visualization of entire company.

Virtual applications and applications of virtual reality provides many types of visualization features (animations, visual graphic cycles, information elements, additional audio instructing and guidance) for realization of which some constraints and logical loops needs to be created. Using them allows the user for flexible reactions to the events and situations while the application is running. After debugging, the logical core of the application realizes the generation of action of the visual element in real time without any mistakes.

Table 1 represents basic differences between the training of the employees with and without the utilization of the application for virtual reality. The users in general find this virtual way to be more attractive, as they can see the workplace they are supposed to work at yet in the educational phase in a comfortable and attractive way. Efficiency of the education is equal from the viewpoint of the output, but considering other actualities and things that can be prevented using the VR (none risk of injury for the employee and none damage caused for the company), the application exploitation is more effective. Overall time span necessary for training itself will be decreased and there is no need to get the employees to the operation workplace at all times, what leads to the decreasing of the potential danger in educational period. Economically, both methods are quite demanding for the employer. In case he decides for the virtual application, there are some initial investments necessary to buy the displaying devices and creation of virtual workplace. Estimated return of the costs can be expressed in the form of lower level of occupational accidents in specific time period.

Table 1. Educational methods and their strong sides.

	Training method	
	without VR	***with VR***
Atractivity	✗	✓
Efficiency	✓	✓
Time	✗	✓
Economy	✓	✓

6.3 Data collection

During the educational process using the application the trainer can record entire activity of a trainee, including the mistakes done during specific training phase. These data can be stored in external application which could evaluate the knowledge level of the trainee regarding the safety at the workplace. On the base of data collected from more training cycles, the application should calculate the percentage of tested and verified knowledge in order to confirm the readiness of the employee for the integration in the real working process. Collected data can also be used as an input for eventual application adjustments. Therefore such recording, storing, processing and evaluation of obtained data presents high additional value to the application itself.

7 Conclusion

Virtual reality presents the area that finds its utilization as en effective working tool in many spheres. Currently the elements of the virtual reality are more oriented to the areas of gaming industry and advertisement, but VR can be as well used in manufacturing area. Proposed method of education of new employees removes large portion of certain actualities related to standard training process and makes the user able to learn even at home. Training personnel is able to provide, observe and supervise the training process without the need to travel directly to the subject workplace. The application possibilities of such a program are really wide, although this area needs to be better acknowledged, understood and developed.

The paper deals with the possibility to use the application of virtual reality as an effective tool for education and training of employees in the area of OSH. It describes the innovative way of teaching the employees and informing them about the potential risks and actions necessary to avoid them.

Acknowledgment

Ministry of Education, Science, Research and Sport of SR supported this work, contract VEGA No. 1/0032/12, KEGA No. 002TUKE-4/2012 and ITMS project 26220220125.

Agentúra
Ministerstva školstva, vedy, výskumu a športu SR
pre štrukturálne fondy EÚ

We support research activities in Slovakia/
Project is cofinanced from fonds of EU.
ITMS project code: 26220220125
Európska únia

References

1. J.N. Marcinčin, J. Barna, Ľ.N. Marcinčinová, V. Fečová, Principles of product assembly visualization by augmented reality usage, In: Annals of DAAAM for 2011 and Proceeding of the 22nd International DAAAM Symposium, Viena, 2011, 1049-1050
2. J.N. Marcinčin, J. Barna, Proceedings in Manufacturing Systems, **6**, 2, 101-104 (2011)
3. J.N. Marcinčin, M. Janák, J. Barna, J. Török, Ľ.N. Marcinčínová, V. Fečová, Verification of a program for the control of a robotic workcell with the use of AR [et al.] - 2012, International Journal of Advanced Robotic Systems, **9**, 1-7 (2012)
4. J. Vallino, K.N. Kutulakos, Augmenting reality using affine object representations. Fundamentals of weareable computers and augmented reality, W. Barfield and T. Caudell (Ed.), Mahwah: Lawrence Erlbaum Assoc. Publ., 157-182
5. W.R. Sherman, A.B. Craig, Understanding of virtual reality, Interface, Application, and Design, USA, Elsevier Science, 2003

 DOI: 10.1051/matecconf/20166820003

Personalized Education Approaches for Chemical Engineering and Relevant Majors

Feng-qing Zhao

Chemical Department of Hebei University of Science & Technology, Shijiazhuang 050018, PR China

Abstract. Personalized education has drawn increasing attention in universities these years. With the purpose of improving the students' comprehensive ability and developing teaching strategies to ensure students' education is tailored to their needs, we proposed Three-Stage Approach (TSA) to enhance personalized education for chemical engineering and relevant majors: professional tutorial system--equipping with professional guidance teachers for freshman students to guide their learning activities and provide professional guidance; open experimental project--setting up open experimental projects for sophomore and junior students to choose freely; individualized education module--setting up 10 different individualized education modules for senior students to select. After years of practice, the personalized education model is improved day by day and proved effective and fruitful.

1 Introduction

Personalized education is a teaching and learning approach which is centred on the needs, aptitudes, and interests of individual students. It has got increased popularity in the last 15 years or so in a number of countries including United Kingdom, United States of America, Australia, New Zealand and China. It has been promoted as a key learning approach to prepare young people for the demands of the twenty-first century and the expectations placed on them by society [1].

Personalized education emphasizes the main position and the motivation of their subjective initiative. The educated acquire knowledge and enhance the ability through their active participation [2]. Personalized education, showing sufficient respect for the individual differences and personal willingness, pays attention to the main role of the students' personality in education, which is conducive to the development of the intellectual potential of the students and the improvement of their creative ability. Personalized education is to achieve the goal of self-development and self-exceeding of individuals on the basis of general education. With this concept of education, the individual is no longer a passive recipient who takes understanding, accepting and adapting the socio-culture as the sole purpose of education, but a person with sound personality, creativity and pioneering talent who is bound to meet the needs of the society through education, turning "want me to do" into "I want to do". It should be stressed that the personalized education should meet the social demands for talents, but also respect and develop the students' personality, so that each student gives full play to their creativity by meeting the requirements of the society. The personalized education and general education are like the "required action" and "freestyle" in gymnastics, both of which are interrelated and mutually reinforce, deserving to be balanced [3].

The Outline of China's National Plan for Medium and Long-term Education Reform and Development (2010-2020) says universities at all levels should 'innovate and reform talents training modes' and 'explore a variety of training modes'. Responding to this requirement, we came up with personalized education as an important way to put forward the innovation and reform of talents training modes [4-5]. Launched in 2005, the reform practice of personalized education has been actively and innovatively promoted until a system was finally formed with a clear goal, of distinct characteristics and rich in both contents and forms [6].

In order to improve students' comprehensive quality, we are going further to develop a Three-Stage Plan with the characteristics of specialties on the basis of a large amount of research and years of practice. The approaches in TSA can effectively improve students' innovation ability and practice ability and thus, help the cultivation of the professional quality and engineering practice ability of chemical and pharmaceutical engineering students.

2 Personalized education strategies and approaches

Developing the characters of college students and stimulating their creativity is the trend of higher education. To make students adapt to the society, the local universities are required to implement personalized education, develop innovative education, cultivate high-quality and innovative talents. The essence of personalized education is "student-centred" education. When carrying out personalized practical strategies, priorities should be set to reforming the traditional teaching mode, establishing the system of personalized training system and constructing the teachers' team as well as creating a relaxed and open teaching environment. Through years practice, we set up a TSA plan in implanting the personalized education for chemical and pharmaceutical engineering Majors.

2.1 Assign professional tutor for students as the entrance to the university

Tutorial system was an innovative measure for the cultivation of the undergraduate proposed by the university in 2006, the core value of which is "personal growth" system is embodied in the "tutorial system", which is currently the most complete system for the personalized education. Under this tutorial system, the students receive individual instruction in major; guidance ideologically and in life, psychological counselling and all-round development, thus the tutorial system has become a "guiding man" to guarantee the healthy growth of the students. The tutors' duties include assisting students in their courses selection, helping students make their personal development plan in university even the life plan, conducting regular discussions and counselling.

Four documents are designed for this program, the main contents is listed in Table 1.

Table 1. Documents in personal tutor project.

Document	Content
Student information registration form	The name, gender, nationality, class, home address, telephone number, expertise, hobbies, family members and resume.
Students' individual growth training program	Personal occupation ideal, Research field of interest, Environment and the difficulties faced, Training plan design
Tutor work records	Name of students, guiding place, guidance form, records, summary of the work and the supervisor's signature
Assessment form	The assessment item, the evaluation of students by tutor and score, including: Ideological and moral, Academic and learning, Self-planning in occupation, Individualized training, Participation in extracurricular activities and practice.

2.2 Open experimental projects for sophomore and junior students

Open experimental projects are usually designed for sophomore and junior students to choose freely. Open experimental projects generally refers to that the experiment project is not provided by the teacher in advance, but by the students from their professional characteristics, hobbies and learning situation. The students independently carry out corresponding experimental activities under the teacher's appropriate guidance. At present, the open experimental projects in our college include following three types:

- Experimental projects provided in innovative open laboratories, which are most often national or provincial key labs. Some specially set up laboratory which managed by the students are used for this purpose.
- Research projects come from the national, provincial and university, or from community activities, personal hobbies. The students organize research groups to complete independently research project using the experimental facilities in the laboratories.
- Students, as members of the project group, are directly involved in teachers' research projects, mostly undertake one sub-projects. Of course, these were implemented in the tutors' research labs.

Specially, Hebei University of Science and Technology gives "Science and Technology Innovation Fund Project for College Students" project every year. In 2014, 510 projects were declared in 2014, and nearly 90% of them were approved and can get scientific and technical funds financial support. In order to guarantee the smooth implementation and achieved substantial results for the innovation fund projects, some measures were outlined. The Fund for the project is divided into stages, after the start of the project and the end of the project. 50% of the fund was allocated each stage. Those projects completing with patent application (including software copyright rights registration) and paper publishing, the first author must be the student, not the tutor. All the project must be examined through expert review (expert review by the university academic committee). If needed, project evulation and review meeting is held. The students make a report about the implementing process and results obtained, the expert group consisting of teachers and students evaluate the project. The Pass rate is about 90%, while the other 10% were rejected and continue to further improved.

2.3. Module setup of personalized education for chemical engineering and relevant majors.

In the process of personalized education practice, we kept modifying and adjusting the education approach as well as expanding its dimensions in order to suit the major requirements and students' and personal interests. We set up 10 personalized education modules with 15 affiliated personalized education groups. Students can choose which group to join in on the basis of their personal interests. In each group, the tutors are responsible for the implementation of the experiments and the assessment of students' performances. Table 2 gives the current modules for personalized practice for chemical engineering and relevant specialties in Hebei University of Science and Technology.

It should be noted that the internship module might continue until graduation. In this period, the students can be put to work in a plant, taking parts in the actual production activities. First, they are given a three-level (plant, workshop and group level) safety education and training which teaches safety knowledge, production process, work content and precautions After that, the workshop and the human resources department make a joint decision after discussion on which posts these students should be assigned to. Once taking their posts, these students get into a "one-to-one" pattern (A skilled worker is assigned to a student as his/her tutor). Over the whole practice period, a student is usually arranged to work in 2-4 different production sections where he/she must master the technological processes of each section, get familiar with the corresponding operations, and write the internship summary.

Table 2. Current modules for personalized practice.

Module	Content	Major
Professional development	Join in the research work of the tutors, finishing literature search, experiment design and planning.	Chemical engineering; Pharmaceutical Engineering; Pharmacy; Pharmaceutical preparation
Professional skills	Study in pharmaceutical preparations training base where real pharmaceutical process is simulated.	Pharmacy; Pharmaceutical preparation
Software applications	Focusing on study of Aspen Plus, a process simulation tool, and Gaussian, a molecular simulation software.	Chemical engineering; Pharmacy;
Design experiment for Unit operations	Teachers propose goals of experiments; students write project plans, make preparations, install devices, make equipment, sort data, and finally write lab reports.	Chemical engineering; Pharmaceutical Engineering; Pharmacy; Pharmaceutical preparation
Comprehensive design	Students conduct operations to form a chemical system, such as installation, pressure testing, leak testing, draw equipment and piping layout of device.	Chemical engineering Pharmaceutical Engineering
Project feasibility	Market demand forecast, production scale, technology & equipment & site selection, project implementation, environment protection and safety, economic evaluation.	Chemical engineering Pharmaceutical Engineering
Internship in factories	Students go to work in a plant, taking parts in the actual production activities.	Chemical engineering; Pharmaceutical Engineering; Pharmaceutical preparation
Safety evaluation	Basics of safety, safety assessment criteria and how to compile engineering design documents in the actual process of safety evaluation.	Chemical engineering; Pharmaceutical Engineering; Pharmacy; Pharmaceutical preparation
Instrumental analysis	Learn to use gas chromatography, liquid chromatography, infrared analysis, UV analysis, DTA, atomic absorption, XRD and XEM, etc.	Chemical engineering; Pharmaceutical Engineering; Pharmacy; Pharmaceutical preparation
Basic knowledge strengthening	Deepening the knowledge of Unit operations by selected case studies.	Chemical engineering; Pharmaceutical Engineering

3 Results and discussion

In the period of implementing personalized education, the students' interest in learning was remarkably aroused. Significant results were obtained.

The students completed, in the past three years, more than 120 extracurricular scientific and technological projects and won 18 state-level awards and 30 province level achievements, among which, include 1 first prize and 1 second prize in "Challenge Cup" National Science and Technology College of Extra-curricular Academic Works Competition, which is one of the top level academic contest for Chinese university students. Besides, the achievement in taking part in scientific contests, such as, "National University Energy Saving and Pollution Reduction" and "Mitsui Cup" National University Chemical Design Competition is outstanding.

The students are satisfied with personalized education approach. Their satisfaction was measured via open-ended questions. In June 2014, a sample survey was performed on the effect of personalized education approach in 580 participating students. Of them, 521 students recognized the value of this approach. Only 14 students expressed no opinion on, or were not dissatisfied with this approach. But most of them suggested that this approach is good for them in seeking jobs. Almost all the participating students gave positive evaluative comments. As a student put it, "It was great to get some experience with personalized education approach, I can do what I please; It was really fun." [7]

Above 98 percent of students took part in various scientific activities, and finished 56 China national patent applications, including 26 patents of invention and 30 patents of utility model in 2012-2014. Some of these patents have found practical applications and made remarkable economic and social benefit. Besides, more than 200 research papers, the students as first author, were published in academic journals, in which over 30 % were cited in Science Citation Index Expanded or EI Compendex web.

With the implementation of personalized education, the students' overall ability was obviously improved. The employment rate is an index closely related with education method. According to statistics, the employment rate (including those pursue further study) of the graduates of our department in the three consecutive years (2011-2013) was more than 95%, ranking among the tops of similar universities in China.

After years of practice, although positive results were obtained, there are still much left to do in the future for TSA approach.

Firstly, lack of matched incentive policies, the teachers' enthusiasm as professional tutor for students as the entrance to the university is not so high.

Secondly, there exhibited uneven participation situation among the 10 education modules for students, some modules, for example, basic knowledge strengthening module, draws a major of the students annually, while the professional development module seems unwelcomed by the students, which draws only 5% student to participate every year. The quantity and types of instruments are still limited in instrumental analysis module, which cannot satisfy the need of different majors.

Thirdly, the independent innovation laboratories which equipped with various devices and accessories are not sufficient. Most students applied for this module have to carry out the project in the research labs of the tutors.

Last but not the least, the organization and management of TSA approach is still need to be improved. We have set up a management system, led by the dean of the college and joined by every department director. In the future, we will to let more off-campus specialist, outstanding alumni and the students in college take part the organization and management. If possible, we will try to corporate with production enterprises, design institute and research institutions in implementation personalized education.

4 Concluding remarks

China is now vigorously promoting the Excellent Engineer Training Program and Engineering Education Accreditation, the former is initiated by the Ministry of Education with the aim of producing millions of graduates with top engineering talent within a decade. The latter is moved by China Engineering Education Accreditation Association, which is authorized by the Ministry of Education as the only legitimate organization to conduct engineering educational accreditation in China, aiming to promote international accreditation of Chinese engineering education and to establish sound foundation for international competition for engineering talents. In the implementation of these works, we studied the contents and development trend of the professional accreditation standards in international engineering education, and developed a professional accreditation standards for Chemical engineering and technology major late in 2015, aiming at improving the quality of engineering education and realizing the essence of equivalent. China's engineering education professional accreditation standards includes two standards: general standard and professional supplemented standard. The former is the requirement that all the engineering majors should comply with, and the latter is the specific requirements for different professional characteristics on the basis of the former. In recent years, we have been exploring the personalized education approaches in chemical engineering and relevant majors according to the requirements of professional accreditation and Excellent Engineer Training Program. The combination of personalized education program, Excellent Engineer Training and professional certification, greatly improved the quality of education for chemical engineering and relevant majors.

As an important idea of higher education and teaching reform, personalized education has found its way to the whole teaching process of the faculty of our college. With years of exploration and practice, a personalized education mode and operating mechanism which benefit our courses has taken shape initially, with a positive atmosphere gradually built for the implementation of its educational ideas, of which student-centeredness has also won our faculty's recognition. But we have already clearly recognized that personalized education activities are carried out according to students' different personalities, and the students' differences in personalities and needs make personalized education a complex work of teaching, the form and content of which need to be adjusted and updated at any moment if necessary. Therefore, we should be careful to stay student-oriented, recognizing their individual differences, respecting their individual needs, focusing on their personality development, exploring their personal potential, constantly improve their practical ability and professional quality, and fully develop their scientific quality and innovation, so as to promote the development of personalized education which fosters students' creative and innovative talents.

In this paper, we proposed Three-Stage Approach to enhance personalized education for chemical engineering and relevant majors: professional tutorial system--equipping with professional guidance teachers for freshman students to guide their learning activities and provide professional guidance; open experimental project--setting up open experimental projects for sophomore and junior students to choose freely; individualized education module--setting up 10 different individualized education modules for senior students to select. A lot of positive results were obtained in the implementation of this approach. It is proved good for the students. Just as I stated above, there are still much left to be improved, nevertheless, the future is bright for this approach.

Acknowledgements

In this paper, I would like to show the best grateful to Mr. Yifeng Yu , Xinyu Rong, Zhao Ma, Ran Liu, Jie Yao, Su-gai Cao,Zhiguo Zhanmg and Shaojie Liu as they had given a lot of valuable ideas and support to the experiments, which made a great contribution to the paper.

References

1. Bentley, T. " *Personalised learning in the UK: from vision to strategy?* " (Personalised learning: high expectations symposium, Department of Education and Training, Sydney.2005)
2. X. Hao, W. Shen, W. Cong, The implementation of personalized education to enhance the quality of personnel training, China Adult Education, **2**,11 (2013)
3. F. Zhao, Z. Ma, *Promoting the Innovative Personnel Cultivation System of Local Universities.* (Eastern Academic Forum, Australia. 2015)
4. W. Ji, Education mode and validation of local college personalized construction, Journal of Hebei University of Science and technology (Social Sciences), **13**,1(2013)
5. W. Ji, Practice verification of the value of individualized education and deepening approach, Journal of Hebei University of Science and technology (Social Sciences), **11**,82(2011)
6. X. Liu. Exploring the personalized education system of undergraduate students, Research on Higher Engineering Education,**6**,105(2012)
7. F. Zhao, Y. Yu, S. Reng, S. Liu, X. Rong, Improving Engineering Practice of Chemical and Pharmaceutical Engineering Majors In Chinese Universities, J. Chem. Educ., **91**,2(2014)

MATEC Web of Conferences 68, 20004 (2016) DOI: 10.1051/matecconf/20166820004
ICIEA 2016

Course Selection with AHP & PROMETHEE Methods for Post Graduate Students: An Application in Kirikkale University Graduate School of Natural and Applied Sciences

Neşet Bedir[1], Emir Hüseyin Özder[2,a] and Tamer Eren[1]

[1]Kirikkale University, Industrial Engineering Department, 71450 Kirikkale, Turkey
[2]Baskent University, Management Information Systems Department, 06810 Ankara, Turkey

Abstract. Post Graduate Study has a very important role in people's career planning. Besides, it helps for gaining expertising on their fields. Specializing must be in the right area to reach their targets in line with people's interests. The aim of the study is to propose a model for students that determines which courses will be chosen on master course selection and this model gives an application example. In this study, course selection problem is discussed for post graduate students in Industrial Engineering Department of Kirikkale University. All criteria that are effected for selecting a course are identified by the help of survey then, significance levels are determined with using Analytical Hierarchy Process (AHP) which is a Multi-Criteria Decision Making Method. According to the weights which are determined before in AHP results, six classes weighted by PROMETHEE method in Industrial Engineering Department.

1 Introduction

Students always complain about not to acquaint with some information on selection of the elective courses throughout time in their post graduate education life. Necessary information cannot be given on time and in detail. Bottlenecks occur in the selection of elective courses. For this reason, some students cannot choose the classes they want [1]. Elective courses should respond to the demands of students in the way of diversity. Some arrangement would be made about courses that students want to take. Students must decide when choosing the course of considering multiple factors. For doing it right, there are various methods in the literature. Multi-Criteria Decision Making (MCDM) method is one of the technique located in literature as intuitive. Lots of MCDM methods are frequently encountered in the literature like; AHP, PROMETHEE, ELECTRE, DEMATEL, VIKOR and TOPSIS.

The aim of this study is to analyze the problems faced by students in course selection and to suggest a model for the integrated course selection. Post graduate courses selection problems are discussed of Kirikkale University Institute of Science and Technology Department of Industrial Engineering students. AHP and PROMETHEE methods were used as a solution method. Five factors, that influence the course selection, are weighted with AHP technique. Six courses are ordered with an alternative method PROMETHEE.

The outline of the paper are as follows: In the second section, importance of course selection problems has explained. In the third section, AHP and PROMETHEE methods are described which multi-criteria decision-making methods used in the solution of the problem. In the fourth chapter, the application made in Kirikkale University for course selection are described. The fifth and the last part, the results are given of the practices and some suggestions are made for future work.

2 Course selection

In post graduate education, students take elective compulsory courses on issues the requirements of the program and that take courses that related to their professional interests and personal skills for developing themselves. Students should know the features and benefits of the courses and they should select the appropriate course that suit for their skills.

3 Multi-criteria decision making methods

The optimization of several criteria at the same time makes complicate the decision-making process. In this case, the problem can be solved by applying MCDM. Multi-criteria decision-making methods are used in many fields. In this study, AHP and PROMETHEE multi-criteria decision-making methods are used for the problem of the choosing of elective courses.

[a] Corresponding author: ehozder@baskent.edu.tr

3.1. Analytical Hierarchy Process (AHP) Method

AHP method was developed by Thomas L. Saaty in 1980. Analytic Hierarchy Process; because the visualization of hierarchical basis to complex, multi-person and multi-criteria problems, to make this technique can be understood by managers. Selection process that uses quantitative and qualitative criteria, so the decision makers determine the relative importance of each criterion is based on the transmitter and then decide to choose between alternatives by each criterion [2]. The steps of the Analytic Hierarchy Process Many applications which form is listed below [3];

Step 1: Identification of the Problem: The identification of decision points and the decision-making problem occurs in the definition of the factors affecting them. The decision point is extremely important in terms of the consistency of the results to be defined clearly. The hierarchical structure of the identified problems are created at this stage. Located problems discussed at the beginning of the hierarchy. By affecting the intended destination of the problem it is written below. It is located in the lower part of the criteria comparison option to do.

Step 2: Creating a Pairwise Comparison Matrix:
At this stage, factor from comparison matrix is a square matrix of size. Matrix components on the diagonal of this matrix takes the value 1. If the number of criteria is indicated as n, pairwise comparison square matrix will be m*n dimension.

$$A=\begin{bmatrix} a_{11} & a_{12} & \cdots & a_{1n} \\ a_{11} & a_{22} & \cdots & a_{2n} \\ \vdots & \vdots & \cdots & \vdots \\ a_{n1} & a_{n2} & \cdots & a_{nm} \end{bmatrix} \tag{1}$$

Step 3: Determine the weight of the criteria
Decision matrix by weight is made to obtain normalized. Normalization process, each element the column vector b is calculated by dividing the total column of each element in the matrix of pairwise comparisons. Equality is calculated by the Equation 2.

$$b_{ij} = \frac{a_{ij}}{\sum_{i=1}^{n} a_{ij}} \tag{2}$$

The resulting normalized matrix is formed as in Equation 3.

$$C=\begin{bmatrix} c_{11} & c_{12} & \cdot & c_{1n} \\ c_{21} & c_{22} & \cdot & c_{2n} \\ \vdots & \vdots & \vdots & \vdots \\ c_{n1} & c_{n2} & \cdot & c_{nn} \end{bmatrix} \tag{3}$$

Taking advantage of the C matrix by showing the distribution of the obtained percent value relative importance value to each other. As shown in Equation 4 using the arithmetic mean for each row to the priority vector C matrix in Equation 5, and W column vector is obtained.

$$w_i = \frac{\sum_{i=1}^{n} c_{ij}}{n} \tag{4}$$

$$W = \begin{bmatrix} w_1 \\ w_2 \\ \cdot \\ \cdot \\ w_n \end{bmatrix} \tag{5}$$

Step 4: Calculation of Consistency Rate: Consistency Ratio (CR) calculation is based on the comparison of the the basic value and the number of factors (λ). Λ calculating the obtained first column vector D. A vector is obtained as shown in Equation 6. and W multiplied by the weight vector.

$$D=\begin{bmatrix} a_{11} & a_{12} & \cdot & a_{1n} \\ a_{21} & a_{22} & \cdot & a_{2n} \\ \vdots & \vdots & \vdots & \vdots \\ a_{n1} & a_{n2} & \cdot & a_{nn} \end{bmatrix} x \begin{bmatrix} w_1 \\ w_2 \\ \cdot \\ \cdot \\ w_n \end{bmatrix} \tag{6}$$

Column D and W column vector corresponding elements of each part as a result of Equation 7 for each vector is obtained by using the basic E values.

$$E_i = \frac{d_i}{w_i} \qquad n\ i,...,2,1 \tag{7}$$

By dividing this number by the total value of λ in Equation 8 is obtained.

$$\lambda = \frac{\sum_{i=1}^{n} E_i}{n} \tag{8}$$

After calculating λ Consistency Index (CI) is calculated with Equation 9.

$$CI=\frac{\lambda-n}{n-1} \tag{9}$$

In the last step; CI Random Index (RI) and named as indicated in Table 1 by dividing the standard correction value C is obtained in Equation 10.

$$CR = \frac{CI}{RI} = \frac{(\lambda_{max}-n/(n-1)\,)}{RI} \tag{10}$$

Table 1. Randomness Index

n	**1**	**2**	**3**	**4**	**5**	**6**	**7**
R.I.	0	0	0,58	0,9	1,12	1,24	1,32
n	**7**	**9**	**10**	**11**	**12**	**13**	**14**
R.I.	1,14	1,45	1,49	1,51	1,48	1,56	1,57

3.2 PROMETHEE Method

PROMETHEE is developed by Brans [4] as a multi-decision procedure. Decision point determines the main stage of order with PROMETHEE I (partial order) and PROMETHEE II (full sequence) Methods. Promethean method is based on binary comparison of the decision point by the evaluation factors. But the main difference from other multiple decision-making process, In addition to the weight of the evaluation factors indicating the level of importance of the relationship between them is that each evaluation factor to consider its own internal affair. PROMETHEE method, compared to other multi-criteria decision-making methods in terms of scope of application and can be adapted to a number of criteria that can be expressed with the actual values is a simple method. PROMETHEE method consists of 7 steps [5]:

Step 1: Creating the Data Matrix: w = (w1, w2, ..., wk) k by c with weight = (f1, f2, ..., fk) to the alternative being considered by = (a, b, c, ...) for data matrixIt is created as shown in Table 2.

Table 2. Data Matrix

		Evaluation Factors				
		f_1	f_2	f_3	...	f_k
Decision Points	A	f_1(A)	f_2(A)	f_3(A)	...	f_k(A)
	B	f_1(B)	f_2(B)	f_3(B)	...	f_k(B)
	C	f_1(C)	f_2(C)	f_3(C)	...	f_k(C)
	...	...	...	...	...	...
Weights	w_i	w_1	w_2	w_3	...	w_k

Step 2: Identification of preferred function Criteria: Six different preference function is used for implementation.

Table 3. Preference Functions

Type	Parameters	Function	Graph, $p(x)$
Type-I (usual)	-	$p(x)=\begin{cases}0, & \forall x \le 0\\ 1, & \forall x \ge 0\end{cases}$	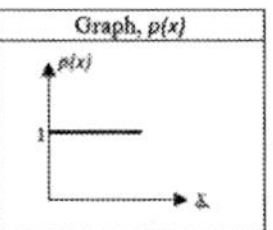
Type-II (U-type)	l	$p(x)=\begin{cases}0, & x \le l\\ 1, & x \ge l\end{cases}$	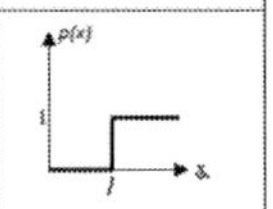
Type-III (V-type)	m	$p(x)=\begin{cases}x/m, & x \le m\\ 1, & x \ge m\end{cases}$	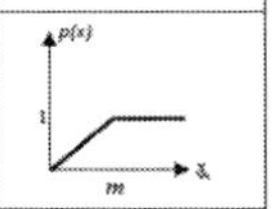
Type-IV (level)	q,p	$p(x)=\begin{cases}0, & x \le q\\ \frac{1}{2}, & q < x \le q+p\\ 1, & x > s+r\end{cases}$	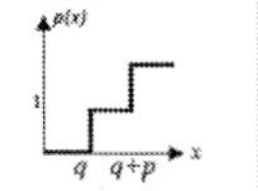
Type-V (Linear)	s,r	$p(x)=\begin{cases}0, & x \le s\\ (x-s)/r, & s < x \le s+r\\ 1, & x > s+r\end{cases}$	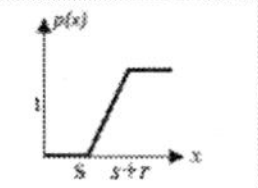

Step 3: Determination of Preference Function: Preferred alternative identified for common functions are shown in Figure 1 a and b common preference for alternative function equation (11) is created.

$$P(a,b) = \begin{cases} 0 & ,f(a) \le f(b) \\ p[f(a) - f(b)] & ,f(a) > f(b) \end{cases} \quad (11)$$

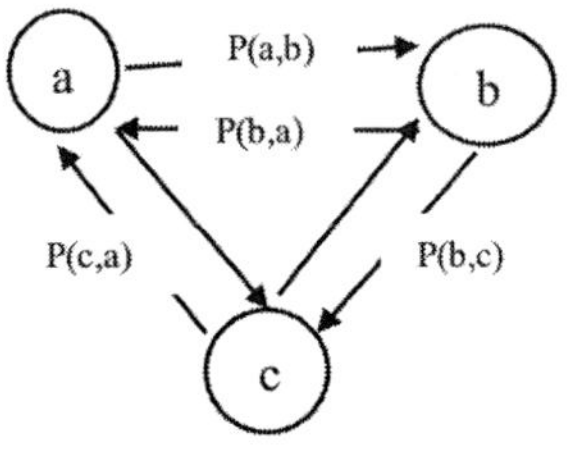

Figure 1. Schematic representation of the common preference function

Step 4: Determine the preferred index: the choice of the Common functions can be determined preference index for each pair of alternatives.

W (i = 1,2, ... k) evaluated by weight by having a k and b the alternative preferred index are calculated by Equation 12.

$$\pi(a,b) = \frac{\sum_{i=1}^{k} w_i * P_i(a,b)}{\sum_{i=1}^{k} w_i} \quad (12)$$

Step 5: Positive (Φ^+) and Negative (Φ^-) Superiority Alternative rule for determining: Alternatives are calculated with (13) and (14).

$$\Phi^+(a) = \frac{1}{n-1}\sum \pi(a,b) \quad (13)$$

$$\Phi^-(a) = \frac{1}{n-1}\sum \pi(b,a) \quad (14)$$

Step 6: Determination of PROMETHEE I Partial Priority for Alternatives: In some cases are involved in the determination of the two alternatives A and B for partial priority.

Situation I: Equation 15, 16, 17 If either of the conditions, a preferable alternative to the alternative b.

$$\Phi^+(a) > \Phi^+(b) \text{ and } \Phi^-(a) < \Phi^-(b) \quad (15)$$

$$\Phi^+(a) > \Phi^+(b) \text{ and } \Phi^-(a) = \Phi^-(b) \quad (16)$$

$$\Phi^+(a) = \Phi^+(b) \text{ and } \Phi^-(a) < \Phi^-(b) \quad (17)$$

Situation II: If the condition does not allow the following, A and B alternative is identical.

$$\Phi^+(a) = \Phi^+(b) \text{ and } \Phi^-(a) = \Phi^-(b) \quad (18)$$

Situation III: If either of the following conditions A alternative, comparable to the B alternative.

$$\Phi^+(a) > \Phi^+(b) \text{ and } \Phi^-(a) > \Phi^-(b) \quad (19)$$

$$\Phi^+(a) < \Phi^+(b) \text{ and } \Phi^-(a) < \Phi^-(b) \quad (20)$$

Step 7: PROMETHEE II complete with identification of priorities for alternatives: The following equation is calculated with the help of exactly the priorities for each alternative. Calculated values of all alternatives with full priority ranking is determined by assessing precisely the same plane.

$$\Phi(a) = \Phi^+(a) - \Phi^-(a) \quad (21)$$

Depending on the exact priority value calculated for A and B are two alternative decisions are given below.
If $\Phi(a) > \Phi(b)$, A Alternative is superior.
If $\Phi(a) = \Phi(b)$, A and B are identical alternatives.

4 Literature review

In this section, studies in the literature regarding course selection were examined.
Woolnough [6] measured the factors of selecting the elective courses of freshmen who are studying biology by applying survey in his study. Results show that women are less interested in science than men.
Bewick and Southern, [7] were intended to determine the cause of gender trends arising from the difference between the students who choose mathematics by using survey on 198 students. Among the reasons for choosing math students indicated that they have significant differences due to gender.
Isobel [8] examined the factors that influence students' choice of university. The author used the questionnaire for identifying the factors of decision process of the students.
Hodgkinson and Innes [9] studied the effect of environmental factors of course selection on freshmen who are studying in different departments at the same university.
Dündar [10] investigated the problem selection of elective courses on business students. Some of criteria are verified like Skills of the instructor, course name and content, information gathered by students and those criteria's priority is specified by AHP Method.
Tezcan ve Gümüş [11] tried to find factors of choosing an elective course from the perspective of students. And also they tried to find the selection order of priority.
Salomon et al. [12] have studied in Brazil and used the AHP for the selection of instructor for a faculty.
Aydin et al. [13] aims to select the best location for a new hospital which have planned to open in Ankara., it has been determined the most appropriate site selection by taking the opinion of experts to implement the project using AHP technique.
Kutlu et al. [14] studied course selection problem in Kirikkale University Industrial Engineering for 3. and 4. grade students. They used AHP and TOPSIS Method as Multi-Criteria Decision Making Methods. The weights of the factors affecting the selection of courses are calculated with AHP, the courses are sequencing with TOPSIS Method.
Bansal and Kumar [15], Bedir and Eren [16], Bedir et al. [17] and Kazan et al. [18] are used AHP and PROMETHEE methods in their studies together for doing selection.

5 Case study

Our problem is choosing the most appropriate elective course for graduate students within certain criteria. Steps are shown in Figure 2.

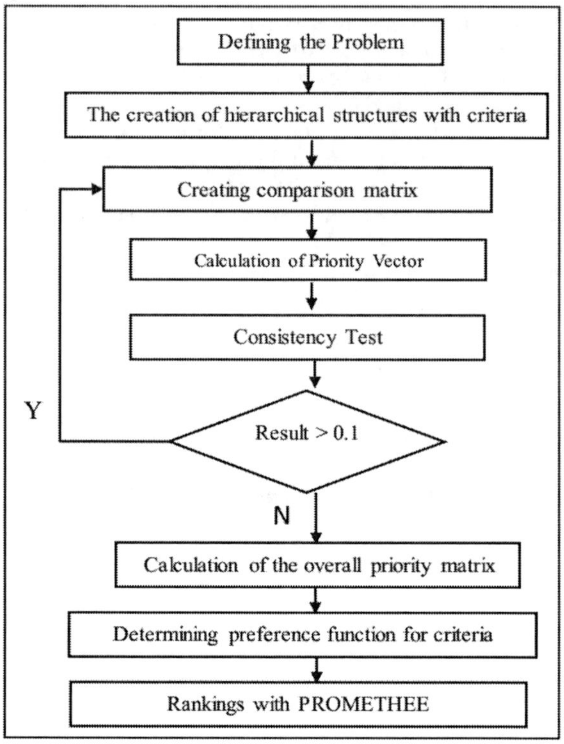

Figure 2. Application Steps

5.1. Defining the Problem

Our problem is choosing the most appropriate course in accordance with certain criteria of graduate students in Kirikkale University Industrial Engineering Department.

Elective Courses for Graduate Students:

- Integer Modelling and Optimization
- Sequencing and Scheduling
- Simulation Languages
- Artificial Intelligence and Expert Systems
- Intelligent Business Systems
- Stochastic Processes

Determination of criteria that influence the choice of the course:
In determining the effective factors in the selection of courses in industrial engineering department results of a survey administered to students who are studying it is determined as follows.

- Content of the Course
- Day of Course
- Interests of Students
- Instructor's Attitude
- Academic Expectations

5.2. Solution of the Problem

Determination of Criteria Weight with AHP

Step 1: Creating a hierarchical structure: In the light of gathering data results, hierarchical structure is getting shaped and shown in Figure 3.

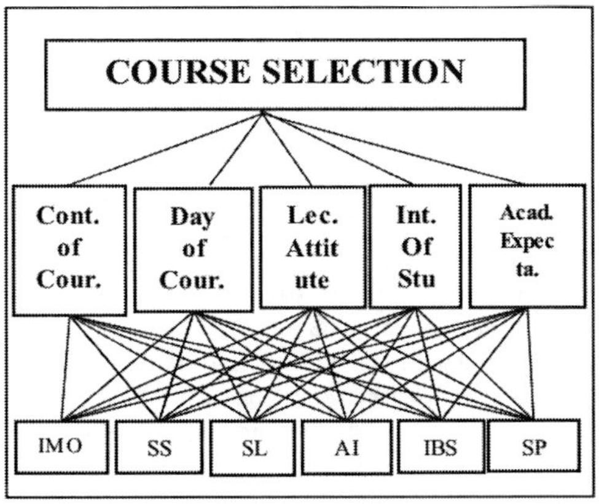

Figure 3. Hierarchical Structure

Step 2: Determination of Priority: The hierarchical structure created after the scale is determined by the criteria for comparison. Unlike other multi-criteria decision-making methods, in the AHP, Satty [19] is used 1-9 scale by the gain on the literature which is shown in Table 4.

Table 4. Saaty's Scala of 1 - 9

Intensity of Importance	Definition	Explanation
1	Equal Importance	Two activities contribute equally to the objective
3	Moderate importance	Experience and judgement slightly favour one activity over another
5	Strong importance	Experience and judgement strongly favour one activity over another
7	Very strong or demonstrated importance	An activity is favoured very strongly over another; its dominance demonstrated in practice
9	Extreme importance	The evidence favouring one activity over another is of the highest possible order of affirmation
2,4,6,8	Average (compromise) values	Reconciliation between two consecutive values to be used in court if necessary

Step 3: Bilateral Comparison Matrix and Solution: The geometric mean of the answers given by the students to create the pairwise comparison matrix on Table x below were created according to the factors like Content of the Course, Day of Course, Interests of Students and Instructor's Attitude.

Table 5. Pairwise Comparison Matrix

Criteria	CC	DL	IS	LA	AE
CC	1,000	3,000	1,000	4,000	1,000
DL	0,333	1,000	0,333	0,500	0,500
IS	1,000	3,000	1,000	2,000	1,000
LA	0,250	2,000	0,500	1,000	2,000
AE	1,000	3,000	1,000	3,000	1,000

Step 4: Normalization and Relative Importance Weights: Pairwise comparison matrix forming in the previous step will be normalized. The relative importance weights by averaging each row in the resulting matrix is obtained Table 6.

Table 6. Relative Importance Weights with Criteria

Criteria	Weights
CC	0,295
DL	0,078
IS	0,245
LA	0,173
AE	0,208

Step 5: Calculation of Ratio Consistency: Consistency of actions taken in this step is tested. Consistency is calculated by this formulation:

$$CR = \frac{CI}{RI} = \frac{(\lambda_{max} - n/(n-1))}{RI} \quad (22)$$

Randomness index are shown in Table 7.

Table 7. Randomness Index

n	1	2	3	4	5	6	7
R.I.	0	0	0,58	0,9	1,12	1,24	1,32
n	7	9	10	11	12	13	14
R.I.	1,14	1,45	1,49	1,51	1,48	1,56	1,57

$$CI = \frac{5,396-5}{5-1} = 0,099 \quad (23)$$

$$RI = 1,12\ (n=5) \quad (24)$$

$$CR = \frac{0,099}{1,12} = 0,088 < 0,10 \quad (25)$$

Our result is consistent.
Step 6: Final Determination of the sequence: In this step PROMETHEE method using the weights obtained by the AHP was applied in ranking the alternatives.
Sorting the Courses with PROMETHEE
The second survey is applied to the Graduate students and obtained data are shown in Table 8.

Table 8. Data Matrix

	Candidate No	Evaluation Factors				
		CC	DL	LA	IS	AE
Decision Points	D1	3	2	5	3	4
	D2	4	3	4	4	5
	D3	4	3	5	5	5
	D4	3	4	4	5	4
	D5	5	3	4	5	4
	D6	4	3	5	4	3

Weights	w_i	0,295	0,078	0,245	0,173	0,208

After the data matrix is done, preference functions is verified by Brans [20] who created specifically for PROMETHEE Method. This functions are shown in Figure 3. For Quantitative criteria; third or fifth type (linear) preference functions are used and for qualitative data first (ordinary) type or fourth (level) preferred type functions are used [21]. Visual PROMETHEE program is used to sort with the help of PROMETHEE Method. A data input program used is shown in Figure 4.

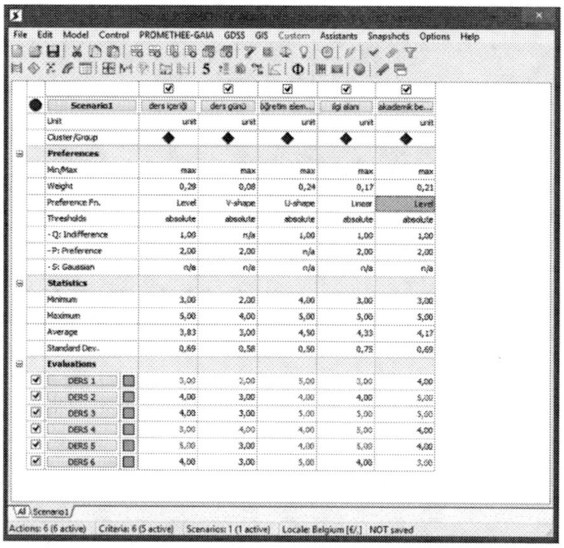

Figure 4. Program Data Input

After the obtained data are verified as input, the program output is shown in Figure 5.

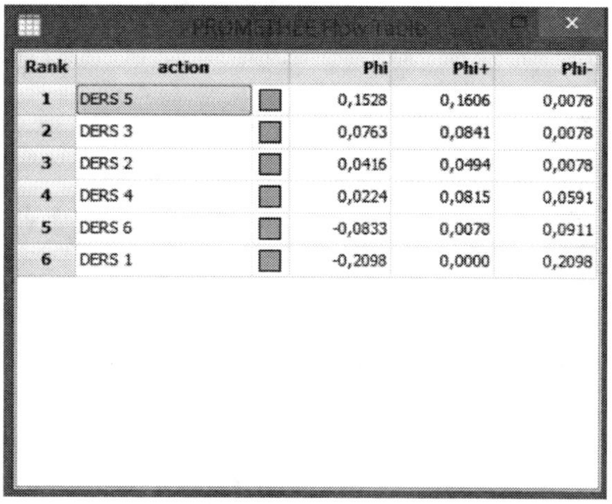

Rank	action	Phi	Phi+	Phi-
1	DERS 5	0,1528	0,1606	0,0078
2	DERS 3	0,0763	0,0841	0,0078
3	DERS 2	0,0416	0,0494	0,0078
4	DERS 4	0,0224	0,0815	0,0591
5	DERS 6	-0,0833	0,0078	0,0911
6	DERS 1	-0,2098	0,0000	0,2098

Figure 5. Program Result

6 Results

Today, the management processes become more complex and lots of factors that influence the decision-making processes. People are forced to make a choice from using different technologies, systems, policies and strategies to find solutions for the problems. The excess of the criteria to be considered in deciding complicates the decision problem. At this point, use of multi-criteria decision-making methods are extremely important in terms of consistency of the results. In this study, course selection is discussed problems of graduate students with using a multi-criteria decision making method as AHP and PROMETHEE methods in Kirikkale University, Institute of Science and Technology Department of Industrial Engineering. Criteria which are affecting students in course selection are determined with a survey applied to the students. The resulting criteria are weighted with AHP. Given the weight obtained by the AHP was sequenced with courses PROMETHEE method. In future studies, new criteria may be added and consulted other multi-criteria decision-making methods can be used in.

References

1. A. Demir, *Üniversitedeki Seçmeli Ders Uygulamasının Öğrenciler ve Öğretim Üyelerince Değerlendirilmesi*, PDRD, **2** 7 (1996).
2. Z. Anık, *Nesne Yönelimli Yazılım Dillerinin Analitik Hiyerarşi ve Analitik Network Prosesi ile Karşılaştırılması ve Değerlendirilmesi*, GÜ, (2007).
3. T. L. Saaty, *Decision making with the analytic hierarchy process*. Int. JSS, **1** 1 (2008).
4. J. P. Brans, P. H. Vincke, *A preference ranking organization method: The PROMETHEE method*, MS, **31** 647–656 (1982).
5. M. Dağdeviren, & E. Eraslan, *Promethee Sıralama Yöntemi ile Tedarikçi Seçimi*, GÜMMF, **23** 69-75. (2008).
6. B. E. Woolnough, *Why Students Choose Physics, or Reject It. Physics Education*, **29** 368-374 (1994).
7. V. Bewick, & J. Southern, *Factors Influencing Students' Choice of Mathematics at A-Level.* TMIA **16** 74- 78 (1997).
8. F. R. Isobel, *Influences on Choice of Course Made by University. Year 1 Bioscience Students a Case Study*. IJSE, **22**, 1201-1218. (2000).
9. S. P. Hodgkinson, & J. M. Innes, *The Attitudinal Influence of Career Orientation in 1 st-Year University Students: Environmental Attitudes as a Function ofDegree Choice.* JEE, **32**, 37-40. (2001).
10. S. Dündar, *Ders Seçiminde Analitik Hiyerarşi Proses Uygulaması.* SDUJFE & AS, **13** (2008).
11. H. Tezcan, & Y. Gümüş, *Üniversite Öğrencilerinin Seçmeli Ders Tercihlerine Etki Eden Faktörlerin Araştırılması.* GEFD, **28** 1-17 (2008).
12. V. A. Salomon, F. S. Duarte, J. L. Junior, & N. Paula, *Faculty selection for a Brazilian private higher education institution.* In International Symposium on the Analytic Hierarchy Process, **1** (2009).
13. Ö. Aydin, S. Öznehir, E. Akcali, *Ankara İçin Optimal Hastane Yeri Seçiminin Analitik Hiyerarşi Süreci İle Modellenmesi.* SDÜİ ve İBFD, **14** 69-86 (2009).

14. B. S. Kutlu, Y. A. Abalı, T. Eren, *Çok Ölçütlü Karar Verme Yöntemleri İle Seçmeli Ders Seçimi*. KÜSBD, **2** 5-25 (2012).
15. A. Bansal, P. Kumar, *3PL selection using hybrid model of AHP-PROMETHEE*, IJS & OM, **14** 373-397 (2013).
16. N. Bedir, T. Eren, *AHP-PROMETHEE Yöntemleri Entegrasyonu İle Personel Seçim Problemi: Perakende Sektöründe Bir Uygulama*, 16. Uluslararası Ekonometri, Yöneylem Araştırması Ve İstatistik Sempozyumu, **16** (2015).
17. N. Bedir, E. H. Özder, T. Eren, *The Third Party Logistics Firm Selection Using AHP-PROMETHEE Methods*. 13th International Logistics and Supply Chain Congress: **13** (2015).
18. H. Kazan, S. Özçelik, & E. H. Hobikoğlu, *Election of Deputy Candidates for Nomination with AHP-Promethee Methods*. P-SBS, **195** 603-613 (2015).
19. T. L. Saaty, *How to Make a Decision : The Analytic Hierarchy Process*, Interfaces, **6** 19 – 43 (1994),
20. J. P. Brans, P. H. Vincke, B. Mareschall, *How to select and how to rank projects:The PROMETHEE method*, EJOR, **14** 228-238 (1986).
21. R. S. Bhatti, P. Kumar, & D. Kumar, *A Fuzzy AHP model for 3PL selection in Lead Logistics Provider scenarios*. EIS & IIICI, **1** 261-277 (2010).

MATEC Web of Conferences 68, 20005 (2016) DOI: 10.1051/matecconf/20166820005
ICIEA 2016

Construction of TCM English Intelligent learning system in the Era of Internet +

Yanxia Wu[1] and Weicai Song[2]
[1] Institute of humanities, Jiangxi University of Traditional Chinese Medicine, Nan chang,330004, China
[2] Institute of Computer Science, Jiangxi University of Traditional Chinese Medicine, Nan chang,330004, China

Abstract. At present, the teaching materials of TCM (Traditional Chinese Medicine) English which learners can choose are limited and which some important updated information about TCM are not provided timely in. A platform of resources and communication is urgently needed to provide high quality service for TCM English learners. The advent of the era of "Internet +" brings opportunities and challenges to the development of TCM English education. Driven by the trend of "Internet +", we try to establish TCM English Intelligent learning system which will conform to the trend of "Internet +" in ecological environment, resource supply, system design to accelerate the pace of TCM English education modernization.

1 Introduction

On March 5, 2015, Chinese prime minister, Li Keqiang, in the *Report on the Work of the Government* put forward the concept of "Internet +"for the first time ([1,2]). The potential meaning of "Internet +" can be interpreted by means of a formula," Internet + ?=?". In this formula, nothing more than one factor is clear, that is the "Internet", and other factors can be freely imaged. When internet + traditional market, *Taobao*, the largest and most popular Internet site, appeared; When internet + traditional Banks, Alipay, a Web payment processing service founded by Alibaba Group, appeared; When internet + mobile phone, WeChat appeared……([3, 4, 5]) Driven by the trend of "Internet +", the TCM English education will establish "Internet + TCM English" running mechanism and construct TCM English intelligent learning system. With the deep fusion of the internet

Innovation and the TCM English teaching, the teaching activities will typify the internet characteristics of "personalized, mobile, social, digital". There by at the present and the future, our TCM English teaching team will multiply our efforts to make the following preparations for the TCM English Intelligent learning system:

2 The ecological environment

Although teachers were trained in information technology in the past few decades time and time again, but the teaching style in the classroom merely took on a small range change, in which we diminutively employed the PPT in the place of the traditional blackboard writing. It seems the impact of internet development on education is not enough to fundamentally change the ecological environment of the education. Therefore actions might as well be taken. Firstly we are supposed to establish the ecological environment of TCM English learning with an advanced education thoughts as the instruction; Secondly we should build the pleasant learning environment of free exchange, discussion, collaboration rather than the teaching process; Finally we will promote the sharing and evolution of the knowledge and develop students' ability.

3 Resource supply

3.1. Co-construction of the teachers and students

Against the network background knowledge is ceaselessly fused, updated and expanded. The complexity of the knowledge is intensified and the available resources is luxuriant, but there is a knowledge mingling of good and evil. In such cases, the filtering and sorting of knowledge is particularly important. The construction of TCM English intelligent learning system adheres to the following principles: Co-construction of the teachers and students. Teachers design system framework, students design creative work. In other words, teachers cultivate creative students and student produce creative works. Give full play to the collective ability and form into the collective wisdom.

3.2 Balance the form and the content

To avoid the problem of paying more attention to the form and less attention to content, the pattern of "Internet + TCM English" innovate not only the forms, but also the teaching content that are conveyed by the forms. The students personally arrange and iterate the personalized learning contents based on the learning condition, interests, and methods, and teachers carry on the teaching design and targeted training based on students' learning records on the system so that a pointed practice can be provided.

3.3 Match the practice with the theory

TCM English learning in Intelligent learning system is synchronized with the solution of the problem. We are prepared to distribute the assignment, completing creative works, to students from the beginning, which will drive students to learn by doing. Whereby we will transform the obsolete method of the acquiring the knowledge into the superior methods of the applying the knowledge. TCM English online learning system provides the theoretical knowledge and practical field, render students to connect the knowledge with the real world, and upgrade learning from knowledge continuity to knowledge creation.

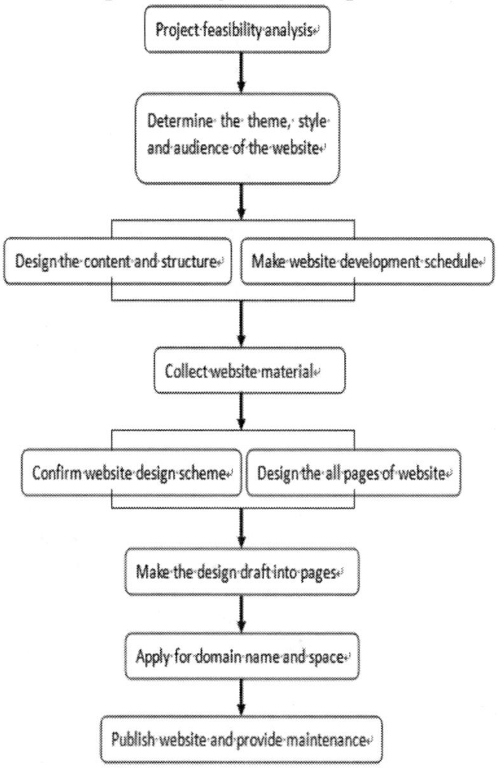

Figure1. Development process of the system

4 The design of the system

4.1 Design orientation of TCM English Intelligent learning system

We have to determine the system design orientation before the design of TCM English Intelligent learning system. Put simply, it is to determine the theme and the style of the system according to audience, subject characteristics and teaching law. The determination of the theme and the style is precondition of the design and production.Theme of the system is the details that the viewers browse from the system. The style of the system is this specific content to show the details. The former is the core of the system production, the direction of system development, and the latter is the manifestation of this specific content. Establishing superior system depends on a fine theme. Setting up your style, mainly decides two aspects: the visual effect and the functional content of the system.

4.2 Constructive process of TCM English Intelligent learning system

The system design the static website by virtue of Photoshop + Dreamweaver + Flash and achieve dynamic development of interactive features by means of IIS + SQL server + Asp.net. The development process is shown in Fig. 1.

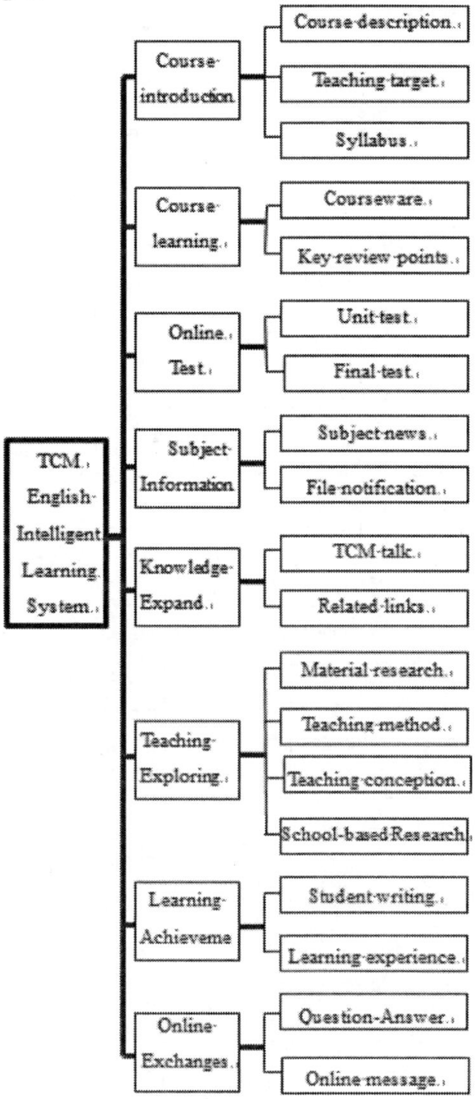

Figure 2. Function modules of the system

4.3 Function modules of TCM English learning system

As for an excellent learning system, its knowledge structure should be complete and comprehensive, and its presentation should be switchable and flexible. Content focus should be put on the knowledge navigation, providing learners with the knowledge structure of the entire website at any time so that students always know which part of the knowledge structure what they are currently studying is located on and know the contextual content ([6], [7]).Thus the learners will not get lost in the vast ocean of knowledge. Based on design orientation of TCM English learning system, the system is divided into such eight functional modules as course descriptions, course learning, online testing and others in terms of the function module setting, shown in Fig. 2 and Table 1:

Table 1. Function module pages

System modules	System module pages
Course Homepage	Index.aspx
Course Introduction	Introduction.aspx
Course Learning	Learning.aspx
Online Testing	Test.aspx
Subject Information	Information.aspx
Knowledge Expand	Knowledge.aspx
Teaching Exploring	Explore.aspx
Learning Achievement	Achievement.aspx
Online Exchanges	Exchange.aspx

SQL server2008 database is used here. Based on the design of the database concept and the analysis of the system function module, two main databases are designed: the database of Users' information and the database of Administrators' information ([8]). The data sheets are designed as following Table 2 and Table 3.

Table 2. Users' information

Column Names	Data Types	Field Length	Null/ Not Null
ID	char	8	Not null
User Name	varchar	10	Not null
Password	char	9	Not null
Sex	char	2	Not null
Birthday	Date time	8	Not null
School and Specialty	varchar	40	Not null
E-mailIn Campus	varchar	20	Not null

Table 3. Administrators' information

Column Names	Data Types	Field Length	Null/ Not Null
ID	char	8	Not null
AdminClass	varchar	10	Not null
AdminUserID	varchar	8	Not null

5 The application of system

When completed, system will bring forth a lot of teaching innovation.

5.1 Innovation of the teaching application

Based on the TCM English Intelligent learning system we will set up new teaching. Pay special attention to the deep integration of digital resources and teaching and give full play to the advantage of the system's convenience and quick in order to promote the efficiency of TCM English teaching. Relying on the HD recorded classroom, we will start course-lecturing and course-competing activity to build the high-quality class. Actively carry out sapiential classroom teaching pilot and flip classroom teaching research, explore effective ways of students' autonomous learning against the network environment, promote the teaching reform, study the teaching mode based on large data analysis and strengthen the pertinence and effectiveness of teaching. Give full play to the role of TCM intelligent learning system, share education and teacher resources, actively carry out intercollegiate exchange and research and obtain complementary advantages and win-win cooperation.

5.2 Innovation of the learning ways

Based on the TCM English Intelligent learning system we will set up students' learning mode. Through QQ or WeChat group we will push learning content and surrounding the topic we will carry out learning exchange activities once a month, guiding students to learn spontaneously and lightheartedly.

By means of reading together on the system, we start such activities as "reading classic and current affairs", "establishing reading database on the system", "timely delivering articles and bibliography" to expand the space-time of the students' learning. Relying on online communication, we proceed learning communication, task arrangement, the online FAQ, inquiry learning and the like to promote interaction between teachers and students, and between students and students to ensure that students can preview before class and review after class. Guide students to use mobile terminals to carry out autonomous learning, cooperative learning, network learning and boost net learning become a habit.

5.3 Innovation of the student assessment

Based on the TCM English Intelligent learning system we will set up student evaluation mechanism. In virtue of the evaluation system on digital system which will assess students' comprehensive quality. We will construct the scientific student evaluation mechanism to meet the following intentions: digital evaluation means, diversified evaluation subject and scientific evaluation content. We will negotiate actively to expand the digital evaluation function and open the evaluation way of the mobile version in order to make the evaluation process more convenient, quick, and timely, assisting students to make scientific evaluation and judgment about their development and providing scientific basis for the students' long-term development p.

6 Conclusion

The advent of the era of "Internet + education" brings opportunities and challenges to the development of TCM English. Thereby we should seize opportunities and think calmly. Primarily we should adhere to the "Education is the core, and the Internet is the tool". No matter how technology develops, we should always bear in mind that education is the core target, and the Internet is the tool that is merely used to improve and change education technology and that is supposed to used reasonably. Secondly TCM English education should dare to shift from the knowledge education to the thinking education. In the face of the impact of the "Internet +", the TCM English education take more responsibility to "teach them the fishing", giving priority to the concise rather than fussy know ledge and transforming the knowledge-oriental education to thinking-oriental one.

References

1. 2015 the Report on the Work of the Government[EB/OL].http://www.people.com.cn/n/
2. Guiding opinions of the State Council on promoting "Internet+" action. [EB/OL].
3. http://news.xinhuanet.com/politics/2015-07/04/c_1115815944.htm.,2015-7-4.
4. W. Cai, "Education Reform in the era of "Internet +". China Education Newspaper, **4** (2015)
5. A. G. Tang, G. Yang, Study on the Construction of University Online Teaching System against "Internet +". Computer Knowledge and Technology, **11**(2015)
6. X. H. Zhu, Promote the Modernization of Education via "Internet +". Xuezhoukan, **5** (2016)
7. X. K. Wang, L. L. Liu, ASP.NET Full Record of Atypical Module Development. Beijing: Tsinghua University Press, (2010)
8. Y. Yang, Y. X. Zhang, Chinese Version of Dreamweaver + Flash + Photoshop Web Production: from Entry to the Master. Beijing: Tsinghua University Press, (2010)
9. L. Wei, SQL Server 2008 Database application and development course. Beijing: Tsinghua University Press, (2011)

DOI: 10.1051/matecconf/20166820006

Digital Trainer for the Development of the Fine Motor Ability in Children with Cerebral Palsy

Jonnathan Berrezueta-Guzmán, Luis Serpa-Andrade, Vladimir Robles-Bykbaev and Eduardo Pinos-Velez

GIIATa, Universidad Politécnica Salesiana, Calle Vieja 12-30 y Elia Liut, Cuenca, Ecuador

Abstract. The leading cause of disability in Ecuadorian children is cerebral palsy, this disorder in most cases produces a deficiency of the ability to move fingers, hands and wrists at various levels, this happens too with the intellect of the patient. Many of the treatments and therapies are seeking that the patient can develop all of your motor ability and intellectual skills, using activities that involve the part Intellectual and practicality of their extremities. Today technology gives us the opportunity to manage devices of aid and assistance that not only complement the daily activities that are performed during the therapies in the help centers, they need to give results that show leaps and bounds in the progress that you want to get. The purpose of this project is to make a device that helps a patient to develop their fine motor ability to the patient can use their hands, fingers and wrists movements in various ways in coordination with their vision in conjunction with occipital lobe causing that brain activity in the patient, present alterations of amplitude in the beta waves in the hemispheres of the brain that allow move muscles, with only maneuver a fully digital device and low cost. These brain and muscles signals will be analyzed in this project, to test the efficiency of this project.

1 Introduction

These days, the cerebral palsy is associated as synonym for the spastic diplegia, in the past this disability was known as "Little disease" in honor to the doctor Wiiliam Little who wrote in 1844 about the influence of the abnormal childbirth in a clinical picture. It is so spoken of cerebral palsy from the years 1889 and has detailed as the result of encephalitis due to congenital factors and prenatal complications. [1].

INEC (Ecuadorian Institute of statistics and censuses) in the publication of the final results of the censuses of population and housing in 2010, shows that in Ecuador the 2.77% of the population has some kind of disability (in numbers means 401.538 people), this value is observed in Figure 1, 48% (in numbers means value of 191.384 people) has some sort of physical or motor disability and 22% (90.259 people) disability intellectual, these percentages it is estimated that one-third is associated with people who suffer from cerebral palsy and that his medical chart shows these disabilities are most present in children, so has come to regard the cerebral palsy as the main cause of physic and intellectual disability in children in Ecuador, data are worrisome concerned to the population of an underdeveloped country, since in many cases there is not possible to establish what was the real cause of cerebral palsy in a child. [2]

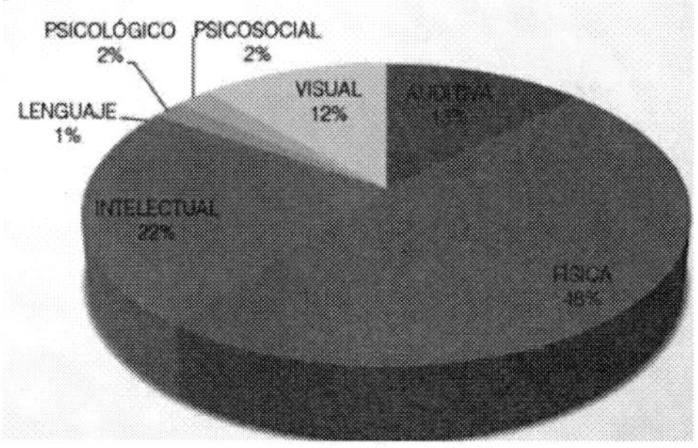

Figure 1. Statistics of classification of disabilities in Ecuador. Data obtained from 2010 in a population census and this data is thinks that is increasing with the population increasing and underinvestment in special medical care, compared to other countries [3].

The Institute of Cerebral Palsy (IPCA), is located in the province of Azuay, this is an institution non-profit which is home to almost a hundred of people with cerebral palsy like children and adults from different social classes. Here each patient presents a different medical chart, each child or teen is a one world seeing from a medical point, and these patients have from a low to a high level of disability.

The implemented project, aims to cover treatment for the improvement of the fine-visual moving ability, giving the patient a physical interaction with the device through hand, fingers and wrists movements doing nervous stimuli in the hemispheres of the brain with the occipital lobe to work in conjunction with the muscular forearms (flexor and extensor) to stimulate to the patient to resolve the problems raised by the device.

2 Demonstration and impairment of fine motor action in patients with CP

Cerebral palsy is manifested in different ways, whether it is affecting the physical part as the intellectual part of the patient, or both simultaneously, depending on the features of disabilities, to cerebral palsy can classify it from different points of view, in various bibliographic materials are classifications that include topography, degree, tone and even dysfunctions associated with cerebral palsy , but the most general is the classification according to the type; of this classification exists:

• Spastic type, is which interfere with the realization of voluntary movements due to muscle spasms that the patient presents at the time of performing some action. [4]
• Type athetosis, which has involuntary movements in the extremities that generate the bad ability to maneuver with their hands, fingers and wrist. [5]
• Type ataxia which is characterized by the lack of coordination in the March and the motor ability both fine as thick, due to alterations in posture of the body. [6]
• Finally a patient may show a picture that presents the combination of these three types, which is called mixed type by this combination. [7]

As denoted in the classification, motor action is the most involved, this is due to damage in the afferent and efferent pathways of the cortex, where the first receive sensory information and the latter make the motor activity, within these lines stands a pyramid system and an extrapyramidal and is the first of these which is in charge of voluntary fine movements, then here the communication of the cortex of the brain with the marrow is affected since the Synapse between the medullary and cortical neuron only performs all the way and there is a dysfunction in the efferent pathways the pyramidal system that collapses, and does not perform the communication necessary to transmit stimulation action members who make up the fine motor skills as well as postural balance mechanisms creating even abnormal reflexes. [8]

3 Pedagogical approach to the treatment of the CP

The physiotherapy field considers the educational part as a key pillar in the rehabilitation medical, guiding their treatments to passive and active muscles stimulating because that has been shown in previous studies that the various applications that are carried out in the physical therapy schools encourage the development of motor and intellectual part of the patient. However although not always the therapists work with multidisciplinary psycho-pedagogical insight that develops a necessary educational technology, then these are not applied and therefore is not always fulfilled the objectives mentioned above. [9]

Develops a process in which is set a note of voluntary movements not reflexes, this serves to be able to estimate the capacity of development of motor coordination with the intellect and thus establish what the activity neuro-developmental of the child to be a success this observation technique of motor behaviors with the intellectual dexterity Unable to provide any technical and technological aid that each case requires. [10, 11]

4 The proposed system

User machine is an embedded system that resembles a digital whiteboard, which offers the user a very attractive and eye-catching interface for interaction, controls are friendly so that the therapist can provide first configurations so that the patient can use. Hardware is resistant and comes recorded instructions, so it can be used properly.

The purpose of this device is to improve fine motor ability of the patient with cerebral palsy, promoting exercise and coordination of the fingers of the hands, but above all the rotation of the wrists doing that the muscles of the anterior compartment (flexor) and the muscles of the posterior compartment (extensor) of the forearm work of more intensely in coordination with vision.

The trainer is a combination of an Android device with a driver that allows the user to contact indirect with the Tablet so that controls programmed in the PIC may serve as machine user interface to enable the patient to perform the activities planned in the application.

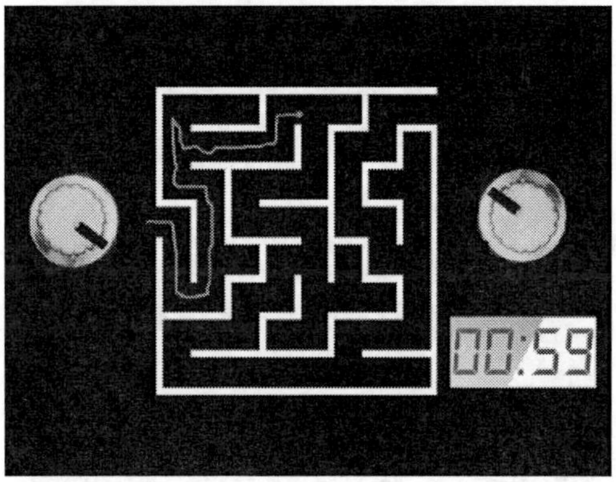

Figure 2. Screenshot of the device interface in configuration mode use of knobs, where tells you the movement to be carried out and realize the exercise and the time spent in this.

The device has knobs, sliders and led´s buttons that act depending on settings preprogrammed by the therapist, each control has a purpose in specific, either be required fill a picture with colors or solve a maze, and these controls are intended to put to the test the movements of fingers, hands and wrists in coordination with the vision.

The device is intended to train the patient, so that her fine motor ability see developed, every case is a different section but is observed that in the majority of cases these children with the disabilities present a pattern, is a signal that although distorted (as usually the brain signals) has a repeating pattern can be identified, this tells us the lack of stimulation in brain activity then this repeating pattern must be broken, and to know if it is fulfilling this goal, is analyzed brain signals issued by the child when is using the device, this happens same with the muscles signals in fingers, hands and wrists.

5 Preliminary results

The device was tested in the Cerebral Palsy Institute of Azuay (IPCA), a Center non-profit whose primary function is to provide specialized care to people who have this disability, tests were considered successful since it obtained an excellent acceptance of the patients, in addition with the encephalographic analysis was seen that it breaks completely preliminary settings of brain activity obtained prior to the use of the device as shown in Figure 3. Same it happens in the muscle analysis as shown in Figure 4.

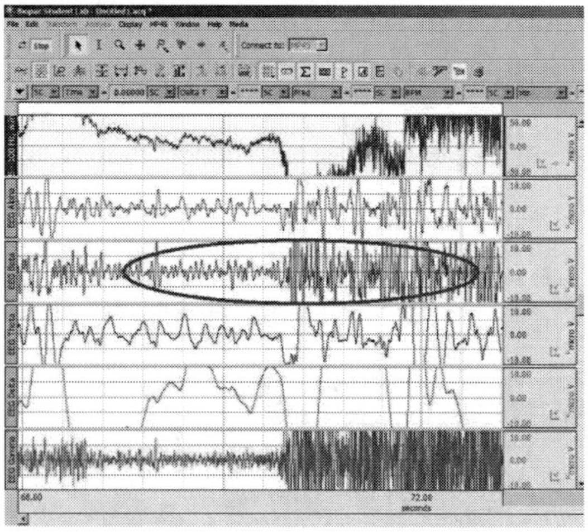

Figure 3. Results of the cerebral signal in beta waves that is obtained by the device "Student Lab MP45 BIPAC"; is observed that the amplitude and frequency changes drastically pattern that was established, when it starts to solve a problem in the trainer device.

Beta waves was seen be altered in its amplitude and frequency, which indicates that the patient concentrates her brain activity in solving the programmed problem in the device and tries to coordinate their hands, fingers and wrists for the resolution of what he sees.

The most appreciated in visual form results, we can corroborate that the patient shows a promising picture on their fine motor mobility at the moment which is using device, so it is said that the device is very effective making that patient can be greater activity in their left and right hemispheres so they coordinate their right hand and left respectively as well as the back of the brain at the time of visually guided in the development of games of speed and visual coordinating.

If we go to spectral analysis of muscle activity that runs on the hands and wrists when using the device fingers shows that the myoelectric signals received respond with a significant excitement about when these limbs are at rest, so muscle activity with brain activity represent a continuous improvement to the fine motor skills of the child.

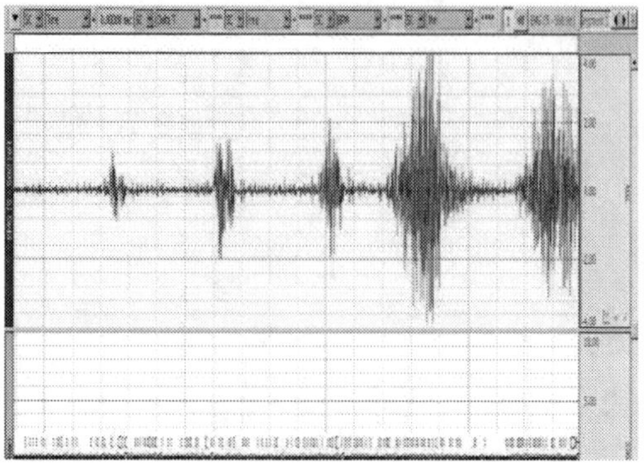

Figure 4. The electrical activity in the muscles that are running movements of the fingers, hands and wrists are affected very significantly as the device operation progresses, this electrical signal is coordinated with the brain signal.

After some weeks of tests, we set a track record that shows patients who used the device and the results obtained in all of them. Table 1 shows how many and ages of the children who were to implement the corresponding tests with the device.

Table 1. Ages and number of children with CP that participated on the experiment with authorization of the institute and their parents respectively.

Age (years)	Boys	Girls
4-6	9	6
7-9	18	18
10-12	14	9

Figure 5 shows the results of feasibility and acceptance of the device, this graphic is very promising and expected results the device had a 92% acceptance in boys and girls of the Institute, also in this sample was obtained 100% of alteration in brain waves that breaks repetitive parameters of these signals and amplifies the signals Beta (signals issued by the brain when it is in a moment of concentration or solving a problem) as shown in Figure 2, the same happens to the electrical impulses emitted by muscle activity, these signals are coordinated with the cerebral activity, such as shown in Figure 4.

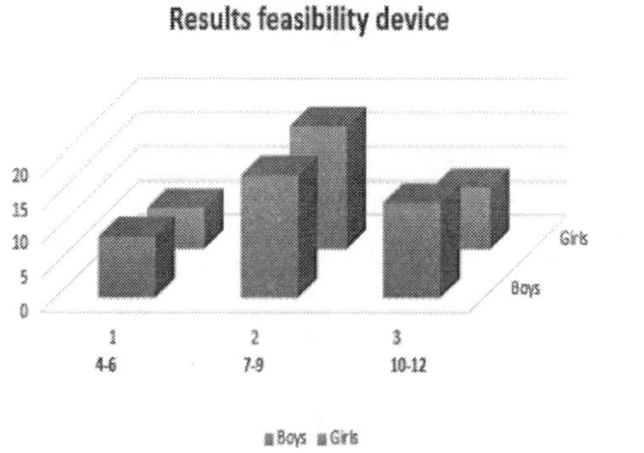

Figure 5. Preliminary results of the pilot experiment (in acceptance terms and good results).

6 Conclusions

This paper shows how we designed a device that combines a Smart technology with a control system that helps a patient with cerebral palsy improve their fine motor skill in coordination with visual activity, it also checks that intellectual activity also is affected, run different exercises established problems.

The benefits that shows the device is showing greater effectiveness the rest of the toys and teaching material that was used so far, however this device no longer only a support to the therapies carried out the teacher or therapist from the Center, but is a joint work with excellent results.

The results of the analysis of brain activity and muscle shows the coordination were expected from the use of this device, testing continues and results are increasingly promising, the therapist and the patient can see the changes that are generated gradually in their ability to move fingers, hands and wrists in a coordinated manner, not only by using this device but also to perform other coordination activities like holding objects and move them.

Cerebral palsy is a field of study that gives us many issues, but the engineer can respond with different ways of development of elements and equipment for remedy, therefore, for a disease that has no cure, the only thing we look for is help for improve the quality of life of these patients (mostly children) putting into practice our knowledge.

Acknowledgements

We would like to thank Dr. Hernán Tenorio and all the collaborating people from Instituto de Parálisis Cerebral del Azuay (IPCA).

References

1. Cruickshank, W. M., Hallahan, D. y Bice, H. V. Personality and Behavioral Characteristics. En W.M. Cruickshank (ed.): Cerebral Pal.vy. A developmental disabiliry. New Yark: Syracuse University Press, (1976)
2. INSTITUTO NACIONAL DE ESTADÍSTICA Y CENSOS. Boletas Censales período 1950-2010. Quito,<http://www.ecuadorencifras.gob.ec/documentos/webinec/Publicaciones/Evolucion_variables_1950_2010_24_04_2014.pdf>
3. CONSEJO NACIONAL DE IGUALDAD DE DISCAPACIDADES, Ecuador sacado de: < http://www.consejodiscapacidades.gob.ec/wpcontent/uploads/downloads/2015/09/estadistica_conadis.pdf >
4. Illescas Vásquez, Margarita Raquel, Desarrollo de un software educativo para el Instituto Fiscal Especial "Stephen Hawking" en el área de Lenguaje y Comunicación para niños con parálisis cerebral nivel 1; directora Paola Cristina Ingavélez Guerra. Cuenca [Ecuador], (2010)
5. Gesell, A. y Amatruda, C. S. Developmental diagnosis: Normal and abnormal child development. Hagerstown, Md: Harper and Row (trad. cast.: Diagnóstico del Desarrollo Nomnl y Anormal del nifio. 1979. Buenos Aires: Paidós.), (1941)
6. Crickmay, M. (1966). Speech Theropy and the Bobaih Apprmch ro Ci;ebral Pnlsy. Springíield. Illinois: Thomas (trad. cast.: Logopedia y enfoque Bobath en Parálisis Cerebral. Buenos Aires: Panamericana), (1974)
7. Bobath, K. y Kong, E. Zerebrale Bewqungsst8rungen Beim Kind. Müchen: S. Karger-Basel (trad. cast.: Transtornos Cerebromotores en el niiio. 1986. Buenos Aires: Panamericana), (1976)
8. Hernández Gómez, R. DejFciencias Cerebrales Infantiles. Madrid: Pablo del Rio, (1977)
9. Delmas, A. Vias y Centros Ndrviosos. Barcelona: Toray-Massoy, (1973)
10. Phelps, W. M. The role of physical therapy in cerebral plasy and bracing in the cerebral palsied. En orrhopaedie appliances Arlas, **1**, 251-522 1952
11. Milani-Comparatti, A. y Gidoni, E. Pattern analysis of motor development and its disorders. Developmenral Medicine and Child Neurology, **9**, 625-630 (1967)

AUTHOR INDEX

AUTHOR INDEX

AUTHOR INDEX